2nd edition

anatomy

The National Medical Series for Independent Study

2nd edition
anatomy

Ernest W. April, Ph.D.

*Associate Professor of Anatomy
and Cell Biology
College of Physicians and Surgeons
Columbia University
New York, New York*

Illustrated by

Anne Erickson

Salvatore Montano

A WILEY MEDICAL PUBLICATION
JOHN WILEY & SONS
New York • Chichester • Brisbane • Toronto • Singapore

Harwal Publishing Company, Media, Pennsylvania

Editor: Jane Edwards
Editorial assistant: Michael N. Samsot
Production coordinators: Keith LaSala, Judy Johnson
Layout artists: Adriana Kulczycky, Steffany Verna Trueman
Compositors: June Sangiorgio Mash, Richard Doyle

Library of Congress Cataloging-in-Publication Data

April, Ernest W.
   Anatomy/Ernest W. April; illustrated by Anne
Erickson and Salvatore Montano.—2nd ed.
   p.   cm.—(The National medical series for independent study) (A Wiley medical publication)
   ISBN 0-471-61666-4
   1. Human anatomy—Outlines, syllabi, etc.
2. Human anatomy—Examinations, questions, etc.
I. Title. II. Series. III. Series: A Wiley medical
publication.
   [DNLM: 1. Anatomy—examination questions.
2. Anatomy—outlines.
QS 18 A654a]
QM31.A67   1990
611′.002′02—dc20
DNLM/DLC
for Library of Congress   89-26907
                                                      CIP

10 9 8 7 6 5 4 3 2 1

Contents

Preface

I. PURPOSE

A. Overall objectives

1. **To organize** and **list** items and topics for easier comprehension, study, and review.

2. **To maximize** use of the limited time available to the student physician or dentist.

3. **To direct** the student toward the most fruitful aspects of anatomic study by illustrating and emphasizing structural and functional detail pertinent to the skillful practice of up-to-date medicine.

B. Intermediate objectives

1. **To provide** the student physician or dentist with **a framework** upon which he or she may construct a working knowledge of human anatomy by presenting **basic anatomy** with illustrations, diagrams, and tables; **functional anatomic concepts**; and **clinical notes**.

   a. **To serve as an outline** and companion for human anatomy courses that are part of the health sciences curricula.

   b. **To assist** the student physician or dentist in determining the relative importance of anatomic structures.

      (1) The subject material does not include detailed discussions of every topic in anatomy nor does it belabor the major points of anatomy, which are usually straight-forward and easily assimilated.

      (2) For the clinician in training, some anatomic structures and concepts are more important than others and these are frequently emphasized with a clinical note. Effort has been made to distinguish between the included *minutia* (clinically important detail) and the excluded *trivia* (inconsequential detail of interest only to anatomists and specialists).

2. **To provide** a **review** of human anatomy for:

   a. **In-course examinations** in the professional health sciences curricula.

   b. **Subsequent courses** and clinical clerkships in the medical curriculum.

   c. **Licensure examinations:**

      (1) **NBME** (National Board Examinations), **Part I**

      (2) **FLEX I** (Federated Licensure Examination)

      (3) **FMGEMS** (Foreign Medical Graduate Examination in the Medical Sciences)

      (4) **Specialty board examinations**

II. SUGGESTED METHOD OF USE

A. The **subject presentation is regional**, which is the method used in most medical and dental programs.

1. The **section order** is only one of a number of sequences that have proved workable, but one for which I have a preference.

2. **Extensive cross-referencing** of topics enables the user of this outline and study guide to proceed in any order.

B. **To benefit maximally** from this outline and study guide:

1. Use it in conjunction with any textbook or atlas of human anatomy.

2. Use the questions at the end of each section and the challenge exam to determine whether a basic knowledge of anatomy and a working knowledge of human anatomy has been attained.

III. SPECIFIC TO THE SECOND EDITION

A. Reorganization. In response to student input and feedback, numerous changes were made.

1. **All major outline points** are now presented with headings. To facilitate use, these headings were made as parallel as possible both between and within chapters.

2. **All outline points** were evaluated and only the most significant were retained.

3. **Major revisions** were made in Part II (Upper Extremity), Part V (Back), and Part VII (Lower Extremity), as well as in Chapter 17 (Gastrointestinal Tract).

4. **Minor revisions** were made in Chapter 11 (Thoracic Cage), Chapter 14 (Superior Mediastinum and Posterior Mediastinum), and Chapter 21 (Perineum). In addition, the new chapter sequence in Part IX promotes a more fluid flow of material.

5. **Eleven new tables** appear throughout the text.

6. **Updated nomenclature** has been incorporated.

B. **Additional questions with explanations** have been included in response to student requests.

1. The **number of study questions** was doubled in each chapter. In keeping with NBME format, more A-type questions (one best answer) and B-type questions (matching) were added. However, because this book is also used for course examinations, K-type questions (one or more correct answers) were retained.

2. A **new Challenge Exam,** similar in format to NBME examinations, consists of A-type, B-type, and C-type questions and contains nearly triple the number of questions found in the former Post-test.

Ernest W. April

Acknowledgments

Because the function of a text or review book is to present concisely the basic information that forms an accepted body of knowledge, only the organization and presentation of those facts and concepts may be original. Therefore, I humbly acknowledge the numerous anatomic reference texts against which the material presented herein has been checked for accuracy. Even more humbly, I acknowledge those uncounted and anonymous deceased individuals as well as my numerous academic and clinical colleagues who have contributed over the years to my fund of knowledge and, finally, the student physicians against whom this anatomic knowledge has been honed. In addition, I am indebted to Dr. Timothy Chuter, a surgical colleague, for his thorough reading of the first edition manuscript, his lively discussions, his contributions to the section on the extremities, and his assistance in writing much of the section on the head and neck in the first edition.

The anatomic illustrations of Anne Erickson and Salvatore Montano confirm the importance of the visual aspects of anatomy. Much of their work has been based upon their own dissections.

To the Reader

Since 1984, the *National Medical Series for Independent Study* has been helping medical students meet the challenge of education and clinical training. In this climate of burgeoning knowledge and complex clinical issues, a medical career is more demanding than ever. Increasingly, medical training must prepare physicians to seek and synthesize necessary information and to apply that information successfully.

The *National Medical Series* is designed to provide a logical framework for organizing, learning, reviewing, and applying the conceptual and factual information covered in basic and clinical studies. Each book includes a concise but comprehensive outline of the essential content of a discipline, with up to 500 study questions. The combination of distilled, outlined text and tools for self-evaluation allows easy retrieval and enhanced comprehension of salient information. Each question is accompanied by the correct answer, a paragraph-length explanation, and specific reference to the text where the topic is discussed. Study questions that follow each chapter use current National Board formats to reinforce the chapter content. Study questions appearing at the end of the text in the Challenge Exam vary in format depending on the book; the unifying goal of this exam, however, is to challenge the student to synthesize and expand on information presented throughout the book. Wherever possible, Challenge Exam questions are presented in the context of a clinical case or scenario intended to simulate real-life application of medical knowledge.

Each book in the *National Medical Series* is constantly being updated and revised to remain current with the discipline and with subtle changes in educational philosophy. The authors and editors devote considerable time and effort to ensuring that the information required by all medical school curricula is included and presented in the most logical, comprehensible manner. Strict editorial attention to accuracy, organization, and consistency also is maintained. Further shaping of the series occurs in response to biannual discussions held with a panel of medical student advisors drawn from schools throughout the United States. At these meetings, the editorial staff considers the complicated needs of medical students to learn how the *National Medical Series* can better serve them. In this regard, the staff at Harwal Publishing Company welcomes all comments and suggestions. Let us hear from you.

Part I
Introductory Overviews

1
Introduction

I. HUMAN ANATOMY

A. Origins. The oldest medical science, human anatomy traces its origins to early Greek civilizations.

B. Derivation. The Greek term for anatomy (*anatome*) means taking apart, as does the Latin term for dissection (*dissecare*).

C. Offshoots. From anatomy came the daughter medical sciences of pathology (morbid anatomy), physiology, neuroanatomy, histology and cytology (microscopic anatomy), embryology (developmental anatomy), and physical anthropology. Several of these, in turn, gave rise to other disciplines, such as biochemistry, microbiology, cell biology, and molecular biology.

II. GROSS (MACROSCOPIC) ANATOMY

A. Systematic (systemic) anatomy is organized according to the following systems:

1. Integument (Chapter 2)

2. Musculoskeletal system (Chapter 3)

3. Nervous system (Chapter 4)

4. Circulatory system (Chapter 5)

5. Viscera

6. Endocrine glands

B. Regional anatomy is concerned with the systems found within a discrete portion of the body.

1. Back and extremities (Parts II, V, and VII)

2. Thorax (Part III)

3. Abdomen (Part IV)

4. Pelvis (Part VI)

5. Head and neck (Parts VIII and IX)

C. Functional anatomy concerns the correlations between structure and function. For example, parasympathetic innervation to the male genital organs mediates erection, and sympathetic innervation mediates ejaculation.

D. Clinical anatomy emphasizes structure and function as it relates to the practice of medicine and other health professions. For example, a displaced fracture of the humerus may injure the adjacent radial nerve, resulting in paralysis of the extensor muscles of the wrist and loss of sensation over most of the dorsum of the hand.

E. Objectives. This book uses many of these approaches to cover the following:

1. Normal macroscopic structure of organs and systems

2. Topographic relations between structures

3. Developmental aspects as they apply to adult body structure

4. Histologic structure as it elucidates function

5. Neurologic considerations as they apply to normal function and loss of function

6. Functional considerations as they apply to understanding structure

7. Clinical considerations as they apply to functional disorders of structure

III. ANATOMIC TERMINOLOGY

A. Anatomic position

1. All structures are described and frequently named with reference to the anatomic position.

2. In the anatomic position, the person is erect (or lying supine as if erect) with the arms by the sides, palms facing forward, the legs together, and the feet directed forward (Fig. 1-1).

B. Anatomic planes (see Fig. 1-1)

1. The midsagittal median plane is vertical between the anterior midline and the posterior midline, dividing the body into left and right halves.

2. Parasagittal (paramedian) planes are parallel to the midsagittal plane.

3. Coronal planes are vertical and perpendicular to the midsagittal plane. The **midcoronal (frontal) plane** divides the body into anterior and posterior halves.

4. Transverse (horizontal) planes are mutually perpendicular to the midsagittal and coronal planes, dividing the body by cross sections.

5. An axis is defined by the intersection of any two mutually perpendicular planes.
   a. The vertical axis is defined by the intersection of midsagittal and midcoronal planes.

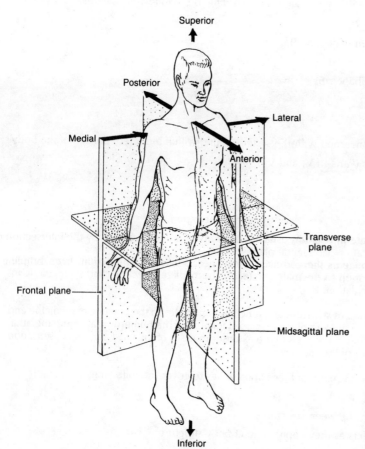

Figure 1-1. *Anatomic planes.* The body is in the anatomic position. The principal anatomic planes and some anatomic adjectives are indicated.

      b. Anteroposterior axes are defined by the intersection of transverse and sagittal planes.

      c. Bilateral axes are defined by the intersection of coronal and transverse planes.

C. Anatomic adjectives are arranged as pairs of opposites (see Fig. 1-1).

  1. Anterior/posterior

      a. Anterior (ventral) is toward the front aspect of the body.

      b. Posterior (dorsal) is toward the back aspect of the body.

      c. Palmar is the ventral side of the hand.

      d. Plantar is the sole of the foot.

  2. Proximal/distal

      a. Proximal is close to the median or near the origin of a structure.

      b. Distal is away from the origin of a structure.

  3. External/internal

      a. External (superficial) is close to the surface of the body.

      b. Internal (deep) is close to the center of the body.

  4. Superior/inferior

      a. Superior (cephalad, craniad, cephalic, rostral) is toward the head.

      b. Inferior (caudad, caudal) is toward the tail or feet.

  5. Medial/lateral

      a. Median is in the midsagittal plane.

      b. Medial is toward the median.

      c. Lateral is away from the median.

  6. Central/peripheral

      a. Central is toward the center of mass of the body.

      b. Peripheral is away from the center of mass of the body.

  7. Prone/supine

      a. Prone is ventral surface down.

      b. Supine is ventral surface up.

D. Anatomic movements are usually described as pairs of opposites (Fig. 1-2).

  1. Flexion/extension usually occurs in the midsagittal or a parasagittal plane.

      a. Flexion brings primitively ventral surfaces together (e.g., bending the arm at the elbow).

        (1) Plantar flexion is downward flexion (flexion) of the foot at the ankle joint.

        (2) Dorsiflexion is upward flexion (extension) of the foot at the ankle joint.

        (3) Radial deviation (flexion) is abduction of the hand at the wrist joint.

        (4) Ulnar deviation (flexion) is adduction of the hand at the wrist joint.

      b. Extension is movement away from the ventral surface (e.g., straightening the leg at the knee joint).

  2. Abduction/adduction usually occurs in the midcoronal plane.

      a. Abduction (lateral flexion) is movement away from the median, away from the middle finger, or away from the second toe (e.g., directing the eye laterally).

      b. Adduction is movement toward the median, toward the middle finger, or toward the second toe (e.g., bringing one leg adjacent to the other).

  3. Medial rotation/lateral rotation usually occurs about a line described by the intersection of coronal and sagittal or parasagittal planes.

      a. Medial rotation is movement of a ventral surface toward the median (e.g., bringing a flexed arm across the chest).

      b. Lateral rotation is movement of a ventral surface away from the median (e.g., directing the head toward one side).

  4. Elevation/depression

      a. Elevation raises or moves a structure cephalad (e.g., shoulder shrug).

      b. Depression lowers or moves a structure caudally (e.g., directing the eye downward).

  5. Protraction/retraction

      a. Protraction moves a structure anteriorly (e.g., jutting out the jaw).

      b. Retraction moves a structure toward the median (e.g., withdrawing a protracted tongue into the oral cavity).

  6. Pronation/supination refers to rotations of specific regions.

      a. Pronation of the arm, for example, is a medial rotation so that the palm faces posteriorly.

      b. Supination of the arm is a lateral rotation so that the palm faces anteriorly.

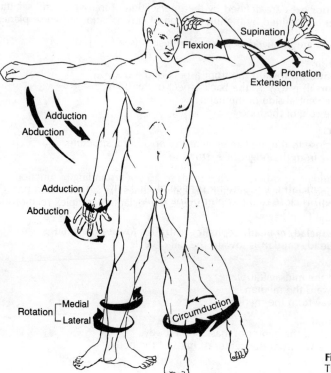

Figure 1-2. *Anatomic movements.* The principal anatomic movements are depicted.

7. **Inversion/eversion** refers to rotations of specific regions.
   a. **Inversion** of the foot, for example, rotates the plantar surface inward.
   b. **Eversion** of the foot rotates the plantar surface laterally.

8. Other terms pertain to movement of particular structures.
   a. **Intorsion/extorsion** of the eye refers to rotation about an axis through the pupil.
   b. **Opposition/reposition** of the thumb is a uniquely human characteristic, which refers to rotation about a complex axis.
   c. **Circumduction** is a combined movement, involving flexion/extension with abduction/adduction.

E. **Anatomic vocabulary—the language of medicine.** Although the vocabulary of anatomy contains approximately 5000 terms, usually there is a rationale to the term that makes it easy to remember.

1. **Guidelines for naming anatomic structures** have been established by the International Anatomical Nomenclature Committee to simplify a potentially complex terminology.
   a. There shall be only one name for each structure with no alternatives, and eponyms shall be discarded.
   b. Names shall be in Latin whenever practical.
   c. Terms shall be as short, simple, and informative as possible.
   d. Spatially related structures shall have similar names whenever possible (e.g., femoral artery, femoral nerve, femoral vein, femoral ring, and femoral canal).
   e. Differentiating adjectives shall be arranged as opposites (e.g., major, minor; medial, lateral).

2. **Most structures are named in the following ways:**
   a. In ancient or obsolete languages (e.g., esophagus, Greek; ileum, Latin; liver, Anglo-Saxon)
      (1) Usually the name translates into a meaningful description [e.g., duodenum, L. 12 (finger breadths long)].
      (2) Many of the original names have been transliterated into modern English (e.g., arteria profunda brachii becomes deep brachial artery).
   b. By descriptive terms (vermiform appendix)
   c. According to their relative position in the body (external intercostal muscle)
   d. According to function (levator scapulae muscle)

      e. By eponymic names associated with mythology (Achilles tendon), the first person to describe the structure (circle of Willis), or the first person to associate the structure with a malformation or disease state (Hunter's canal)

         (1) Most anatomy books, in agreement with the International Anatomical Nomenclature Committee, do not acknowledge eponyms.

         (2) Because the clinical use of many eponyms persists and because eponyms reflect the rich history of anatomy and medicine, the more common eponyms are acknowledged in this book.

3. Competence with the vocabulary of the body is only the first step in the mastery of anatomy. Nomenclature aside, the study of anatomy is best approached with logic rather than rote memory.

    a. Comprehension of anatomy to the point of being able to use it as a medical tool requires integration of structure as it relates to function.

      (1) An understanding of structure often leads to an understanding of function.

      (2) An understanding of function often logically justifies the structure of a part.

    b. Understanding the development of a structure often clarifies complex relations (e.g., the innervation of the diaphragm).

4. An anatomic principle is frequently the basis for the diagnosis or choice of treatment for a clinical problem (see study questions at the end of Part I). Anatomy establishes the foundation for clinical observation, physical examination, and the interpretation of clinical signs and findings as well as the basis for treatment.

2
Integument and
Mammary Glands

I. INTEGUMENT (SKIN)

A. Surface area. The integument may be considered the largest organ of the body with a surface area somewhat less than 2 m².

B. Multiple functions of the skin include the following:

1. Protection by:
   a. Preventing fluid loss (the greatest problem in burn patients)
   b. Reducing abrasive trauma

2. Sensation mediated by general sensory afferent nerve endings (pain, touch, and temperature)

3. Secretion by:
   a. Sweat glands for temperature regulation mediated by general visceral efferent (sympathetic) nerves
   b. Mammary glands, which are modified sweat glands, under the primary control of endocrine hormones

C. Divisions. The integument consists of two principal layers:

1. Epidermis
   a. The epidermis is a superficial cellular layer of stratified epithelium.
   b. It is between 20 and 1400 μm thick, depending on location.

2. Dermis
   a. The dermis is an underlying layer of loose, irregularly arranged connective tissue.
   b. It is between 400 and 2500 μm thick, depending on location.
   c. It contains accessory structures, such as hair follicles, sweat glands, mammary glands, blood vessels, lymphatics, nerves, and special nerve endings.

D. Cleavage lines (of Langer) are important surgical considerations (Fig. 2-1).

1. The meshwork of collagen fibers (elastic fibers to some extent) provides overall mobility of the skin. The elasticity of the connective tissue fibers creates a slight tension. Although the connective tissue fibers in the dermis appear randomly oriented, there is a prevailing direction in each area of the body, which is denoted by the crease lines, or **Langer's lines.** When the connective tissue fibers are severed, they retract so that wounds gape.
   a. Incisions made *across* the prevailing direction of the connective tissue fibers will retract considerably, producing gaping wounds and resulting in prominent scars upon healing.
   b. Incisions made *parallel* to Langer's lines will sever fewer connective tissue fibers, thereby decreasing the tendency to retract and resulting in less unsightly scars.

2. The direction of Langer's lines on the thorax, the base of the neck, and the abdominal wall are characteristic (see Fig. 2-1). On the female breast, Langer's lines tend to be circumferential to the nipple, an important consideration, considering the frequency of tumor biopsy and cosmetic mammaplasty. The primary surgical consideration is always adequate exposure; cosmetic considerations are, of course, secondary.

E. Innervation of the integument

1. In general, the skin is innervated in a segmental pattern. Most higher invertebrates and all chordates exhibit somatic segmentation (*soma*, Gr. body) or metamerism. As a first approximation, the body may be thought of as developing from a series of 42–46 identical **somites**

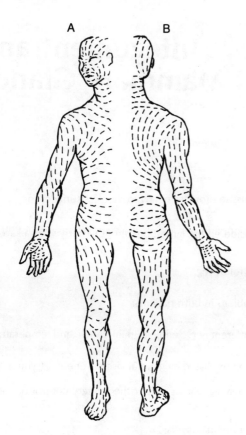

Figure 2-1. *Cleavage lines of the skin.* A, Anterior surface; *B*, Posterior surface.

(segments). Subsequently, there is differential growth and development so that segments may lose individual identity, such as in portions of the head, or develop extensively, such as the cervical and lumbosacral contributions to the extremities.

2. The adult body retains a segmented vertebral column. Between each two vertebrae, a left and right mixed (afferent plus efferent) spinal nerve arises to supply a **dermatome** and a **myotome**. These are discrete portions of the skin and muscle of the body wall or limb that originated from that segment. The superficial branches of each spinal nerve terminate in nerve endings of corresponding dermatomes.

3. Adjacent dermatomes usually overlap somewhat. Understanding the dermatomal arrangement is essential to the conduct and interpretation of the physical examination of a patient. Reference landmarks include the nipple and umbilicus, which are in the fourth thoracic and tenth thoracic dermatomes, respectively (see Fig. 4-3).

F. Significance of the integument

1. General medicine: manifestations of systemic disease, such as vasoconstriction (cold and clammy skin), vasodilation (flushed skin), eruptions or rash, petechiae, ecchymoses, and edema

2. General and plastic surgery: cosmetic incisions, skin grafts, and loss of body fluids in severe burns

3. Dermatology: skin disease

4. Neurology: manifestations of neurologic dysfunction and disease

II. FASCIAL LAYERS

A. Superficial fascia (subcutaneous tissue) underlies the integument. It is composed of loose, irregularly arranged connective tissue between the dermis and the deep (investing) fascia and contains varying amounts of fat, depending on genotype, phenotype, and location. This fascial layer serves as a loose packing material and matrix through which course medium-sized (distributing) vessels and nerves. It is divisible into:

 1. **Superficial layer of the superficial fascia**, which is fatty
     a. On the abdominal wall, this layer is known as Camper's fascia.
     b. In the perineum, this fascia is referred to as Cruveilhier's fascia.
 2. **Deep layer of the superficial fascia**, which is membranous
     a. On the abdominal wall, this layer is referred to as Scarpa's fascia. Only Scarpa's fascia will support sutures.
     b. In the perineum, this fascia is known as Colles' fascia.

 B. Deep (investing) fascia, a membranous layer that usually overlies muscle as epimysium, defines fascial planes between muscles.

 1. With the aid of extracellular fluid, this tissue provides nearly frictionless surfaces for the motion of one muscle over another.
 2. Unfortunately, these layers form potential pathways for infection or extravasation of fluids.

III. MAMMARY GLAND OR BREAST

 A. Structural considerations (Fig. 2-2)

 1. **The breast** consists of 15–20 pyramidal **lobes** of glandular tissue contained entirely within and considered part of the superficial fascia. These lobes are only slightly developed in the male.
     a. Lactiferous ducts, each of which opens onto the mammary papilla (nipple), drain the lobes.
     b. Suspensory ligaments (of Cooper), which are connective tissue septa, define and interconnect the lobes.
         (1) The suspensory ligaments attach the breast to the skin and deep layer of the superficial fascia and determine the posture of each breast.
         (2) Tumors (benign or malignant) will often displace a suspensory ligament, causing retraction of the breast surface, which is usually a diagnostic sign.
     c. Glandular tissue of each lobe is surrounded by varying amounts of fat that determine the shape of each breast. Both are under hormonal control.
         (1) The interlobular fat is particularly sensitive to estrogen titers.
         (2) The proliferation of glandular tissue during pregnancy is due to estrogen and progesterone; secretory activation of glandular tissue is due to prolactin.
 2. **The nipple**, or **mammary papilla**, receives 15–20 lactiferous ducts and is variable in size. (A large nipple may appear on a chest x-ray and be mistaken for a small pulmonary mass.)

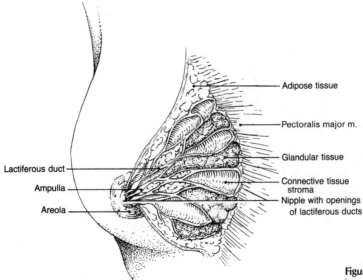

Adipose tissue

Pectoralis major m.

Glandular tissue

Connective tissue stroma

Nipple with openings of lactiferous ducts

Lactiferous duct

Ampulla

Areola

Figure 2-2. *The breast.* The right breast is partially dissected, showing the secretory portion.

      3. The areola, a continuation of pigmented skin beyond the nipple, contains smooth muscle fibers. (The pigmentation is irreversibly darkened after a first conception in some women.)

      4. The axillary tail (of Spence) is a normal extension of breast tissue toward or into the axilla.

      5. Due to the shape of the thoracic wall and the teardrop shape of the breast with the axillary tail, about 75% of the mammary tissue lies laterally to the nipple.

B. Blood supply and lymphatic drainage of the mammary gland are of especial importance because of the high incidence of malignant tumors of this organ. Lymphatic drainage of the mammary gland generally follows the blood supply.

      1. The lateral and inferior portions, comprising approximately 75% of the breast, drain along the **thoracoacromial** and **lateral thoracic vessels** toward the **axillary nodes**.

      2. The medial portions of the breast tend to drain along the **anterior intercostal vessels** toward the **internal thoracic (parasternal) nodes**, which lie in a chain along the **internal thoracic artery** deep to the costal cartilages and parallel to the sternum.

      3. A small superior portion tends to drain toward the **supraclavicular nodes**.

      4. The superficial lymphatics may drain across the midline to the contralateral breast or along the anterior abdominal wall.

C. Anomalies

      1. Beyond an extensive normal range, the breasts frequently are overdeveloped or underdeveloped, both of which conditions are amenable to mammaplasty.

      2. Accessory breasts and supernumerary nipples may develop along the ''milk line'' from the axilla to the groin.

      3. Abnormal bilateral hypertrophy or hyperplasia of the male breasts (gynecomastia) is usually due to endocrine disorders, impaired liver function, or drugs.

3

Musculoskeletal System

I. SKELETON

A. Bone, a calcified connective tissue, forms most of the adult skeleton, which consists of approximately 206 bones. The **axial skeleton** consists of the bones of the head, the bones of the vertebral column, the ribs, and the sternum. The **appendicular skeleton** consists of the bones of the extremities.

1. Basic functions of the skeleton
   a. Bone supports and protects certain internal organs.
   b. Bones act as biomechanical **levers** on which muscles act to produce motion.
   c. As a tissue, bone is in dynamic equilibrium with its bathing medium and serves as a reservoir of ions (Ca^{++}, PO_4^-, and $CO_3^=$) in mineral homeostasis.
   d. The bone marrow in the adult is the source of red blood cells, granular white blood cells, and platelets.

2. Composition of bone. Bone is composed of living cells and an organic intercellular matrix with an inorganic component.
   a. Connective tissue cells become **osteocytes** as the collagen meshwork that they secrete undergoes calcification and ossification.
   b. The collagenous matrix provides tensile strength. If the mineral content of bone is removed by acid, the remaining collagenous meshwork is flexible but not particularly extensible.
   c. The mineral content, a crystalline hydroxyapatite complex of calcium, provides shear strength and compressive strength. If the collagenous matrix is removed by incineration, the remaining inorganic matrix is very brittle.

3. Types of bone
   a. Long bones, such as those of the extremities (Fig. 3-1):
      (1) Develop by replacement of hyaline cartilage
      (2) Usually provide the levers for movement
      (3) Receive their blood supply through one or more nutrient foramina
      (4) Have structurally distinct regions
         (a) The diaphysis (shaft) of long bones is composed of a thick collar of dense compact bone (cortical bone) beneath which is a thin layer of spongy trabecular bone adjacent to the marrow cavity.
            (i) The microstructure of cortical bone is arranged for maximal strength.
            (ii) Growth in thickness occurs by circumferential apposition of bone.
         (b) The metaphyses (ends) of long bones are composed of a trabecular bony meshwork surrounded by a thinner collar of compact bone. The spicules are arranged along the lines of stress.
         (c) The epiphyses, toward the ends of long bones, are separated from the metaphyses in a young person by cartilaginous growth plates, the **epiphyseal disks**.
            (i) Longitudinal growth occurs both proximally and distally from the epiphyseal plate by cartilage proliferation, calcification, and remodeling, until epiphyseal fusion occurs at about the time of puberty.
            (ii) The epiphyseal disks become ossified as longitudinal growth ceases.
            (iii) Unfused epiphyses may look like fractures radiographically.
   b. Flat (squamous) bones, such as ribs and the bones of the cranium:
      (1) Develop by replacement of connective tissue
      (2) Generally serve protective or reinforcement functions
      (3) Consist of two plates of compact bone separated by spongy bone (**diploë**) that bridges the marrow cavity

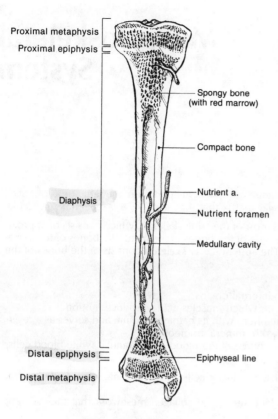

Proximal metaphysis
Proximal epiphysis

Spongy bone
(with red marrow)

Compact bone

Nutrient a.

Nutrient foramen

Diaphysis

Medullary cavity

Distal epiphysis

Epiphyseal line

Distal metaphysis

Figure 3-1. *Structure of a long bone.* The tibia is shown partially cut open to reveal the internal organization of the bone. The epiphyseal lines represent fusion of the growth plates. The trabeculae of the spongy bone of the metaphyses align along the lines of force and transmit these forces to the compact bone of the shaft. The principal blood supply to the marrow cavity enters via a nutrient foramen.

 c. Sesamoid bones, such as the patella and the pisiform bone:
   (1) Develop within tendons
   (2) Serve to reduce attrition on the tendon
   (3) Increase the lever arm of the muscle by moving the tendon away from the fulcrum

 4. Functional and clinical considerations. As a living tissue, bone is in dynamic equilibrium with the body fluids and reacts to external forces. These two factors interrelate in the process of bone remodeling and healing.
  a. Bone remodeling
   (1) The trabecular meshwork that comprises the internal structure of the metaphyses as well as the collagenous substructure of compact bone of the diaphysis develop along the lines of force (compression and stress) established by the mass of the body in response to gravitational force and voluntary exertion.
   (2) As these forces change over time (e.g., the result of a change in body weight, pregnancy, or exertion), bone undergoes a subtle remodeling to maximize intrinsic strength.
   (3) The constant pull of a muscle at its sites of attachment produces ridges, crests, tubercles, and trochanters by the remodeling reaction.
   (4) If bone is not stressed, due to illness, injury, or the weightlessness of outer space, calcium is rapidly resorbed.
  b. Bone fractures. Fractures can be classified according to the degree of displacement (**nondisplaced** or **displaced**), whether there is compression of the bone fragments (**comminution**), and whether the skin is torn by displaced fragments (**compound fracture**).
   (1) Fractures are usually visible radiographically. However, small fractures with no compression or displacement may not be visible radiographically until some bone resorption occurs, usually in about a week.
   (2) Fracture of the cartilaginous epiphyseal plate in a young person may be difficult to detect unless there is compression or displacement. Healing of epiphyseal fractures in growing young persons can interfere with subsequent growth.
   (3) Fragments of fractured long bones may jeopardize adjacent soft tissues, especially neurovascular bundles.
   (4) In the head, many nerves and vessels pass through bony canals and foramina in flat bones. Fractures of these may compress or lacerate the nerve or vessels as well as compress or lacerate the brain.

        (5) If there is a single major nutrient foramen involved, a fracture may isolate a significant portion of a long bone from its blood supply with resultant avascular necrosis.

  c. Bone healing

        (1) Fracture of long bones may result in loss of integrity of the lever arm; fracture of flat bones results in loss of protective function.

        (2) Fibroblasts peripheral to bone in the region of the fracture proliferate and secrete a collar of collagen about the fracture, the **callus**. The callus calcifies, providing an internal splint for the fracture.

        (3) There is some bone resorption on either side of the fracture followed by unorganized proliferation of connective tissue across the fracture. The unorganized connective tissue in the fracture calcifies within 6 weeks, thereby healing the fracture.

        (4) Over the next several months, remodeling occurs in the region of the fracture as well as in adjacent regions; at least in the young person, little evidence remains of the fracture.

B. Cartilage, a dense irregular connective tissue, forms a small portion of the skeleton. Cartilage is formed by living cells and the intercellular matrix that they secrete. It is essentially avascular. The composition of the intercellular matrix determines the type of cartilage.

  1. Hyaline cartilage:

    a. Has an intercellular matrix especially rich in hyaluronic acid and mucopolysaccharides, which are natural lubricants

    b. Forms the anterior portion of most ribs to complete the rib cage and provide the resiliency necessary for ventilation

    c. Forms the articular cartilage in most joints

    d. Provides the anlage for long bone development

  2. Fibrocartilage:

    a. Has an intercellular matrix rich in mucopolysaccharides and bundles of collagenous fibers

    b. Has a very high osmotic pressure and a high water content as a result of a high concentration of mucopolysaccharides

    c. Is an especially resilient and durable form of cartilage

    d. Forms most symphyses, such as the pubic symphysis and the intervertebral joints, as well as certain joint disks, such as that of the temporomandibular joint

  3. Elastic cartilage:

    a. Has an intercellular matrix rich in mucopolysaccharides and bundles of elastic fibers, which provide a strong, yet flexible, support.

    b. Forms the skeletal structure of the external ear and the tip of the nose.

C. Articulations (Fig. 3-2). Most bones articulate with each other.

  1. Three types of articulations or joints

    a. Synarthroses, or fibrous joints, are barely movable or nonmovable, as in a **suture** (sagittal suture, Fig. 3-2*A*), **syndesmosis** (tibiofibular joint, Fig. 3-2*B*), and the **gomphosis** (tooth joint).

    b. Amphiarthroses, or cartilaginous joints, such as the pubic symphysis (Fig. 3-2*C*) and intervertebral disks, allow limited motion.

    c. Diarthroses, or synovial joints, permit relatively free motion about at least one axis of rotation, as in the shoulder, elbow, and hand (Fig. 3-2*D–H*).

      (1) Common characteristics of diarthrodial joints include the following:

        (a) Joint (synovial) capsule

        (b) Articular cartilage, which covers the articular surfaces of the bones. Articular cartilage is usually hyaline cartilage but may occasionally be fibrous cartilage.

          (i) Because articular cartilage is somewhat fluid, it can change shape so that mechanical forces are distributed over the largest possible surface area within the joint.

          (ii) The degeneration of articular cartilage, **degenerative arthritis**, results in loss of the smooth gliding surface and pain accompanying joint motion.

        (c) Reciprocally concave and convex shapes of the two articular surfaces

        (d) Synovial membrane, which encloses the diarthrodial joints

          (i) The synovial cavity between the synovial membrane and the bone or cartilage is filled with **synovial fluid**.

          (ii) Rich in hyaluronic acid, synovial fluid functions as a lubricant for the articular surfaces.

      (2) Joints act as fulcrums for bony levers so that motions about the fulcrums can be produced.

      (3) Motions at all diarthroses can be reduced to rotations in one or more mutually perpendicular planes. All rotary movement takes place about an axis of rotation.

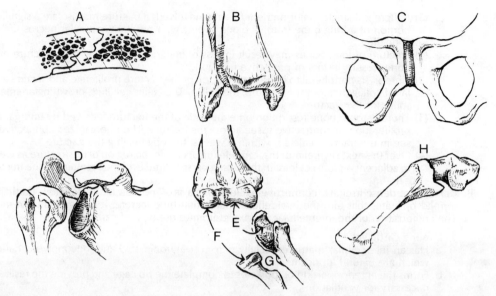

Figure 3-2. *Articulations. A*, Cranial suture; *B*, tibiofibular syndesmosis; *C*, pubic symphysis; *D*, glenohumeral joint (ball-and-socket, three degrees of freedom); *E*, humeroulnar joint (hinge, one degree of freedom); *F*, the humeroradial joint (ball-and-pivot, two degrees of freedom); *G*, radioulnar joint (gliding pivot, one degree of freedom); and *H*, first carpometacarpal joint (saddle, two degrees of freedom).

   (a) Relative movements of bones at the common articulation are given pairs of descriptive names that describe the motion that occurs about a single axis of rotation:
   - **(i) Flexion/extension**
   - **(ii) Abduction/adduction** (or sometimes, **elevation/depression**)
   - **(iii) Internal rotation/external rotation**
   - **(iv) Circumduction**, which is produced by a combination of flexion/extension and abduction/adduction
   (b) Axes of rotation. Joints can have one, two, or three axes of rotation with one degree of freedom for each axis. Some diarthroses permit only one pair of motions, others permit two, and still others permit three pairs of motions.

2. **Ligaments** are bands of dense, regularly arranged connective tissue that cross joints and frequently form an **articular capsule** about the joint.
   a. Ligaments are embedded in bone on either side of a joint by Sharpey's fibers in order to:
      (1) Connect the bones
      (2) Reinforce the articulations
      (3) Contribute to joint stability
   b. Torn ligaments, sometimes referred to as sprains, present difficult clinical problems.
      (1) Because ligaments are relatively avascular, unlike bone, tears heal slowly.
      (2) When a ligament is detached from bone, its fibers do not grow back into the bone as extensively as before the injury; thus, a healed ligament is usually weaker, predisposing to subsequent injury.
      (3) Torn ligaments destablilize the joint and predispose to dislocation of the joint.
         (a) Loss of joint stability is a sign of a torn ligament.
         (b) Although ligaments usually cannot be visualized radiographically, displacement of bones at an articulation is a sign of torn ligaments.

II. MUSCLE

A. **Composition and actions of muscle.** Approximately 48% of the body is muscle mass.

   1. **Composition of muscle.** Muscles are composed of bundles of muscle cells (muscle fibers), which are contractile. **Skeletal muscle** runs between two **points of attachment** on related bones or even on bones separated by a considerable distance.
      a. **Tendons (bundles) and aponeuroses (sheets)** are formed by dense, regularly arranged connective tissue into which each end of a muscle inserts, and which, in turn, attach to the outer layer of the periosteum; a few tendons attach directly to bone through Sharpey's fibers.
         (1) The more proximal attachment site of a muscle is often referred to as the **origin**; the more distal attachment site of a muscle, the **insertion**.

(2) Tendons may continue into the muscle as **septa**.

(3) Where a tendon is subjected to intense friction, a sesamoid bone may form in the tendon.

(4) Muscle pulling at the sites of tendinous attachment to bones produces a remodeling re-action within the bones; thus, attachment sites are indicated by ridges, crests, tuber-cles, and trochanters.

b. The line of action of the muscle is the line that best describes the mean direction of the muscle between the centers of any two such attachments (see Fig. 3-3).

c. The lever arm of a muscle is a line drawn perpendicular to the line of action of that muscle through the axis of rotation (fulcrum) of the joint (see Fig. 3-3).

d. Arrangement of fascicles (groups of muscle fibers) within the muscle falls into several pat-terns.

(1) Fusiform. Fascicles of muscle fibers lie parallel to the line of action along the long axis of the muscle (e.g., the sartorius muscle).

(2) Pennate. Fascicles lie at an angle to the long axis of the muscle. Vectorial analysis dem-onstrates that the same amount of contraction produces slower movement but more force in a pennate muscle than in a fusiform muscle.

(a) Unipennate. Fascicles lie at the same angle on one side of the tendon (e.g., the flexor pollicis longus muscle).

(b) Bipennate. Fascicles lie at an angle on either side of a tendinous septum (e.g., the soleus muscle).

(c) Multipennate. Fascicles reach the tendinous septa from many directions (e.g., the deltoid muscle).

2. Muscle action

a. Movement. By shortening, muscles act on the bony levers to produce motion in one or both of the bones to which they attach.

(1) All muscles exert *equal* and *opposite* tension at both attachments.

(a) Any muscle whose line of action crosses an unconstrained axis of rotation at a joint must produce movement at that joint.

(b) Movement at a joint is determined by the sum of the activity of all the muscles whose lines of action cross the axis of rotation.

(c) The bone that is least stabilized will move.

(2) The strength of the muscle is a function of its cross-sectional area (4 kg/cm²) and the length of its lever arm (mechanical advantage).

b. Group actions. Regardless of the specific innervation, some muscles may act together as **synergists** to produce a specific motion; other muscles act together as **antagonists** to op-pose this motion.

(1) Synergistic muscles cross the same side of the axis of rotation; antagonistic muscles pass over opposite sides.

(a) In most instances, motion at a joint is initiated by one set of synergistic muscles and brought to a close by the antagonists. For example, controlled flexion of the fore-arm at the elbow joint is initiated by flexor muscles and brought to a close at any desired position by extensor muscles.

(b) Simultaneous contraction of both synergists and antagonists produces maximal joint stability with little or no movement.

(2) Because muscles can only shorten actively, lengthening of a muscle requires contrac-tion of an antagonist on the opposite side of the axis of rotation.

B. Force-generating capacity of muscle is a function of muscle stretch rather than overall muscle length.

1. Contraction of muscles

a. Isometric contraction. Muscles exert force without producing a rotation at the joint (e.g., the elbow flexors trying to lift a weight that is too heavy to move).

b. Isotonic contraction. Muscles shorten to produce motion (e.g., the elbow flexors lifting a manageable weight).

2. Measurement of force at a series of lengths

a. Maximum isometric force is produced when the muscle is stretched to its rest length. (For a flexor, this occurs in full extension; for an extensor, this occurs in full flexion.) When a muscle reaches half its resting length, it is no longer able to generate significant contractile force.

b. This force–length relation has its morphologic basis in the ultrastructure of the muscle fibers.

(1) Maximum force of 4 kg/cm² occurs when there is just maximal overlap between myosin rods and actin filaments.

        (a) At just maximum overlap, there can be a maximum number of force-generating cross-bridge interactions between these structural proteins.

        (b) This occurs in maximally stretched muscle in situ.

    (2) Shortening results in double overlap of filaments in the center of the contractile unit.

        (a) Double overlap of filaments interferes with cross-bridge interaction, resulting in loss of force-generating capacity.

        (b) This occurs in maximally shortening muscle in situ, which is approximately one-half of the stretched length.

    (3) Because maximal force is 4 kg/cm², the greater the cross-sectional area of a muscle, the greater the number of contraction filaments acting in parallel, the greater the force-generating capacity.

        (a) Muscle use stimulates synthesis of contraction fibers, resulting in muscular hypertrophy.

        (b) Disuse or paralysis has the opposite effect, resulting in muscular atrophy.

C. Mechanical advantage of a muscle is a function of the length of the lever arm.

  1. Effect of lever arm. In addition to the intrinsic force–length effect, as the joint is flexed the length of the lever arm of a muscle acting across that joint changes; thus, the mechanical advantage of the muscle either increases or decreases. This is shown in Figure 3-3.

    a. For example, on full extension of the elbow, the lever arm of the flexor is relatively short because the joint angle is nearly 180° and the flexor muscle lies close to the joint. The mechanical advantage of the short lever arm is minimal.

    b. As the elbow is flexed to 90°, the lever arm of the flexor is longer because the line of action of the flexor moves away from the joint. Here the mechanical advantage of the muscle is maximal.

    c. With additional flexion of the elbow, the lever arm again becomes shorter and the mechanical advantage decreases.

  2. Integrative summary

    a. Although the intrinsic force-generating capacity of the flexors is maximal in the fully extended arm, the lever arm of the flexor is minimal; hence, the muscle is strongest at the point at which its mechanical advantage is least (see Fig. 3-3A).

    b. As the arm flexes, the force-generating capacity diminishes because of the force–length relation, but the mechanical advantage increases as the length of the lever arm increases; hence, as the muscle weakens, its mechanical advantage increases. The net effect is maintenance of fairly constant strength.

    c. Beyond 90° of flexion, the force-generating capacity of the flexor muscle and length of the lever arm both diminish; hence, the muscle becomes weaker and loses its mechanical advantage; the effect is a rapid loss of strength (see Fig. 3-3B).

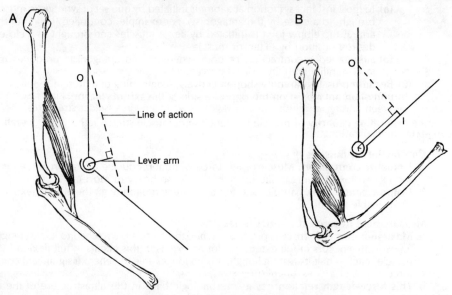

Figure 3-3. *Musculoskeletal action. A,* The stretched brachialis muscle, while intrinsically most powerful in the stretched position, has a short lever arm that provides little mechanical advantage. *B,* Upon isotonic contraction, while intrinsically less powerful in the shortened position, this muscle has a long lever arm that provides greater mechanical advantage.

  d. Finally, the speed of contraction is a function of both the weight of the load to be moved and the pennation.

 D. Nervous control of musculoskeletal movement

  1. Physiologic recording demonstrates that electrical excitation passes along nerves to the muscles. This is the basis for diagnostic **nerve conduction studies**.

  2. Nerve electrical activity causes release of a neurotransmitter at the neuromuscular junctions, initiating electrical excitation along the muscle fiber and inducing contraction. This is the basis for diagnostic **electromyography**.

  3. When gravity, friction, and inertia are overcome, the nerve becomes relatively silent, and motion continues because of inertia.

  4. To halt motion, the nerve to the antagonistic muscle becomes active, and the antagonist contracts sufficiently to cease movement.

  5. Voluntary control of muscle is from high centers of the brain. Reflex control and muscle tone are accomplished by neurons within the spinal cord.

III. SOMATIC FASCIA

 A. Fascia is loose, irregularly arranged connective tissue composed of fibroblasts, collagen bundles, and some elastic fibers, which forms planes. Fibroblasts secrete tropocollagen, which aggregates into liquid-crystalline collagen bundles. The elasticity of the collagen comprising the fascial planes permits adjacent muscles to shorten and lengthen independently.

 B. Fascial subdivisions

  1. Superficial fascia is relatively mobile in most regions of the body; notable exceptions include the palms and soles. Superficial fascia consists of two layers:
   a. Superficial layer of superficial fascia (of Camper):
    (1) Is predominantly fatty—panniculus adiposus
    (2) Is of variable thickness and serves as insulation and padding
    (3) Contains the superficial arteries, veins, lymphatics, and nerves
    (4) Is particularly sensitive to estrogenic hormones
   b. Deep layer of superficial fascia (of Scarpa):
    (1) Is membranous and relatively thin
    (2) Holds sutures
    (3) Fuses with the deep fascia

  2. Deep (investing) fascia cannot be stripped completely from the structures that it invests (i.e., it becomes continuous with periosteum, perimysium, perineurium, and other adventitial layers. Deep fascia consists of three layers:
   a. Outer investing fascia overlies the musculature beneath the superficial fascia.
   b. Inner investing fascia underlies the musculature of the body wall and supports the transversalis fascia and endopelvic fascia.
   c. Intermediate investing fasciae are septa arising from the outer investing fasciae that run between and around individual muscles as well as neurovascular structures.

 C. Fascial specializations

  1. Retinacula are strong fascial bands in the regions of joints that prevent tendons from "bow-stringing" away from the joint.

  2. Bursae are fluid-filled openings between or within fascial planes that reduce friction between tendons, muscles, ligaments, and bones.

  3. Synovial tendon sheaths are fluid-filled tunnels about muscle tendons that permit a considerable degree of movement and reduce the friction.

 D. Clinical considerations

  1. Fascial planes are easily opened by surgical blunt dissection and by extravasation of fluid, such as blood, urine, and pus.

  2. Spread of infection across fascial planes is limited.

  3. Infection may track along fascial planes; a classic example is the spread of tuberculosis of the lumbar vertebrae beneath the psoas fascia to present as an infection in the femoral triangle.

4
Nervous System

I. INTRODUCTION

A. The nervous system is a complex organ system. It provides a mechanism by which the organism can monitor the ever-changing external and internal environments.

1. **Sensory.** Signals originating in the sensory receptors are monitored by, processed in, and transmitted through the nervous system.
   a. These inputs may reach the conscious sphere or may be used at subconscious and reflex levels.
   b. Sensation is mediated through the somatic sensory system and the visceral sensory system.

2. **Somatic motor.** The nervous system controls and integrates various parts of the body (Fig. 4-1).
   a. The neural messages modulating and regulating motor activity are processed in and conveyed through the nervous system to muscles and glands.
   b. These outputs may be voluntary, involuntary, or the result of reflex mechanisms.
   c. These actions are mediated through the somatic motor system.

3. **Visceral motor.** The nervous system maintains the internal environment within narrow limits mediated through the autonomic motor system.

B. Neurons (nerve cells) comprise the principal units of the nervous system. Neurons typically have two types of processes.

1. **Dendrites** are afferent processes that typically receive synaptic contact from receptor cells or the axons of other neurons.

2. **The axon** is the single efferent process through which each neuron communicates with other neurons and effectors (i.e., muscles and glands). Axons may be wrapped in multiple layers of membranes (myelin), which insulate the axon, thereby promoting conduction.

C. Subdivisions. The bilaterally symmetric nervous system is subdivided anatomically into the **central nervous system** and the **peripheral nervous system** and subdivided functionally into the **somatic nervous system** and the **autonomic nervous system** (see Fig. 4-1).

1. **Central nervous system (CNS)**
   a. The CNS is composed of the brain and spinal cord, which are encapsulated within the skull and vertebral column, respectively.
   b. It is the center of perception and the site of integration of sensory information as well as initiation and coordination of motor activity.

2. **Peripheral nervous system (PNS)**
   a. The PNS includes the cranial nerves and spinal nerves that arise from the brain and spinal cord, respectively.
   b. It conveys neural impulses:
      (1) To the CNS as input from the sense organs and sensory receptors of the body
      (2) From the CNS as output to the muscles and glands of the body

3. **Somatic nervous system**
   a. **The afferent portion** includes the neural structures of the CNS and PNS that are involved in conveying and processing conscious and unconscious sensory information.
   b. **The efferent portion** includes the neural structures of the CNS and PNS that are involved in motor control of voluntary muscle.

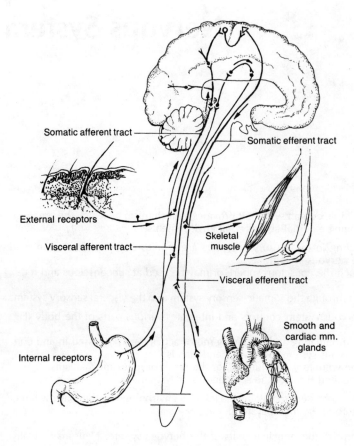

Somatic afferent tract

Somatic efferent tract

External receptors

Visceral afferent tract

Skeletal muscle

Visceral efferent tract

Smooth and cardiac mm. glands

Internal receptors

Figure 4-1. *Basic organization of the nervous system.* Somatic and visceral sensory input is schematized on one side while the somatic and autonomic motor output is schematized on the opposite side.

   4. **Visceral nervous system**
      a. **The afferent portion** includes neural structures that convey sensory information from the splanchnic (visceral) portion of the body.
      b. **The autonomic nervous system (ANS)** is motor only.
         (1) It is composed of the neural structures of the CNS and PNS that are involved in motor activities influencing the involuntary (smooth) and cardiac musculature and glands of the viscera and skin.
         (2) It has two divisions: the **sympathetic system** and the **parasympathetic system**.

II. CENTRAL NERVOUS SYSTEM

   A. **Brain**

      1. **Morphology.** The brain is the enlarged, convoluted, and highly developed rostral portion of the CNS. The average adult human brain weighs about 1400 g, approximately 2% of the total body weight. The brain can be divided into the **cerebrum, cerebellum**, and **brain stem**. The **gray matter** of the brain contains neuron cell bodies; the **white matter** consists of pathways (tracts) containing the axons.

      2. **Meninges.** The gelatinous brain is protected by an outer rigid capsule, the **bony skull**, and is invested by a succession of three connective tissue membranes, the meninges.
         a. **Pia mater**
            (1) The pia mater is intimately attached to the brain.
            (2) It contains the small blood vessels that supply the brain and brain stem.
         b. **Arachnoid**
            (1) The arachnoid is a thin membranous layer external to the pia mater and connected to it by web-like trabeculations—hence its name.
            (2) It supports the large distributing cortical arteries.
            (3) It delimits the **subarachnoid space**.
               (a) The subarachnoid space is located between the arachnoid and pia mater.
               (b) It surrounds the brain but does not dip into the sulci.

(c) This space is filled with **cerebrospinal fluid (CSF)** and may contain blood on hemorrhage of a cerebral artery. The brain floats in CSF, which supports it and acts as a shock absorber when the head moves or is jarred.

c. Dura mater

(1) The dura mater consists of two adherent fibrous membranes (hence its name) external to the arachnoid.

(a) The outer dural layer is the periosteum of the cranial vault.

(b) The inner dural layer is the true dura.

(c) In certain locations, venous sinuses (draining blood from the brain) course between the two layers of dura mater.

(2) The dura mater delimits the **subdural space**.

(a) The subdural space is a potential space located between the arachnoid and dura.

(b) It does not contain CSF.

(c) This space is the location of a subdural hematoma, which is usually a low-pressure venous hemorrhage.

(3) The dura mater defines the **epidural space**.

(a) The epidural space is a potential space that lies between the inner and outer layers of the dura mater.

(b) It contains the meningeal arteries and dural venous sinuses.

(c) This space is the location of an epidural hematoma, which is usually a high-pressure arterial hemorrhage.

3. Functions of brain. The brain functions in perception of sensory stimuli, in integration and association of stimuli with memory, and in neural activity, resulting in coordinated motor response to stimuli.

a. Input to the brain is from the spinal cord as well as from cranial nerves.

b. Output from the brain is through the brain stem and spinal cord as well as through cranial nerves.

B. Spinal cord

1. Morphology. The cylindric spinal cord is located in the upper two-thirds of the **vertebral canal** of the bony vertebral column.

a. A continuation of the brain stem, the spinal cord extends from the foramen magnum at the base of the skull to its termination as the **conus medullaris**, usually located at the caudal level of the first lumbar vertebra in the adult.

b. The non-neural **filum terminale** continues caudally as a filament from the conus medullaris to the coccyx.

c. The spinal cord is enlarged in those segments that innervate the extremities.

(1) The **cervical (brachial) enlargement** extends from spinal levels C5 to T1, the segments that innervate the upper extremities.

(2) The **lumbosacral enlargement** extends from spinal levels L3 to S2, the segments that innervate the lower extremities.

d. Because the vertebral column continues to grow after birth and the spinal cord grows very little in length, the adult spinal cord (ending at the level of the L1 vertebra) is much shorter than the bony vertebral column.

(1) The spinal nerves emerge from the vertebral column lower than the spinal cord segments from which the corresponding rootlets originate.

(2) The lumbar and sacral nerves develop long roots that extend from the spinal cord as the **cauda equina** (horse's tail) within the **lumbar cistern**.

2. Spinal meninges are continuous with those of the brain.

a. Pia mater

(1) The pia mater is intimately attached to the spinal cord and its roots.

(2) It continues beyond the termination of the spinal cord as the filum terminale, which attaches to the coccyx.

(3) It contains the blood supply to the spinal cord.

b. Arachnoid

(1) The arachnoid is a thin membranous layer external to the pia mater and connected to it by web-like trabeculations—hence its name.

(2) It supports the large distributing vessels.

(3) It delimits the **subarachnoid space**.

(a) The subarachnoid space is located between the arachnoid and pia mater.

(b) It surrounds the spinal cord and its roots.

(c) It is filled with CSF.

(d) This space extends caudally to the level of the second sacral vertebra and is wider between vertebral levels L1 and S1—the **lumbar cistern**, which is normally devoid of spinal cord.

   (i) A **lumbar tap** to sample CSF is usually accomplished by inserting a needle in the midline between vertebrae L3 and L4 or L4 and L5.

   (ii) CSF contains blood after hemorrhage of a cerebral artery.

   (iii) CSF contains many polymorphonuclear leukocytes if bacterial meningitis is present.

   (iv) **Spinal anesthesia** is accomplished by infusing the anesthetic about the nerve roots in the lumbar cistern through a midline needle between vertebrae L3 and L4 or L4 and L5.

c. Dura mater

   (1) The dura mater is a tough fibrous membrane (hence its name) external to the arachnoid. It does not fuse to the vertebral periosteum.

   (2) It delimits the **subdural space**.

   (a) The subdural space is a potential space located between the arachnoid and dura.

   (b) It does not contain CSF.

   (3) It defines the **epidural space**.

   (a) The epidural space lies between the dura mater and the periosteum of the vertebral column.

   (b) It contains profuse venous plexuses and fat.

   (c) **Epidural anesthesia** is accomplished by perfusing the anesthetic agent about the spinal nerves in the epidural space.

III. PERIPHERAL NERVOUS SYSTEM

A. Definition

   1. A nerve bundle consists of axons (fibers) in the PNS. Depending on its location, such a bundle may be called a rootlet, root, trunk, division, cord, ramus, or branch.

   2. A plexus is a network or interjoining of nerves.

   3. A ganglion is an aggregation of nerve cells (cell bodies and processes).

B. Spinal roots. Nerve fibers emerge from the spinal cord in a paired, uninterrupted series of dorsal (input, sensory, and afferent) and ventral (output, motor, and efferent) rootlets, which join to form 31 pairs of dorsal and ventral roots (Fig. 4-2).

   1. Dorsal (sensory) roots

   a. The afferent fibers that comprise the dorsal root convey input from the sensory receptors in the body via the spinal nerves to the spinal cord.

   b. The cell bodies of the neurons lie in the **dorsal root ganglion**, which is located within the intervertebral foramen.

   c. Some fibers of the dorsal root of each spinal nerve supply the sensory innervation to a skin segment known as a **dermatome**, whereas other fibers provide nerve endings to deep structures (Fig. 4-3).

   d. There is usually no C1 or Co1 dermatome.

   e. Adjacent dermatomes overlap, and the loss of one dorsal root results in diminished sensation (not a complete loss) in that dermatome.

   f. Anesthesia or paresthesia involving an entire dermatome is indicative of spinal cord injury or root damage. Partial dermatome involvement is indicative of peripheral nerve damage.

   2. Ventral (motor) roots

   a. The motor nerves, which comprise the ventral root, convey output from the spinal cord. cord.

   b. The cell bodies lie in the **ventral horn** of the spinal cord gray matter and project axons which may:

   (1) Innervate voluntary striated muscles (general somatic efferent)

   (2) Synapse with neurons in peripheral ganglia, which, in turn, innervate involuntary smooth muscles and glands (general visceral efferent)

   (3) Innervate the striated muscle derived from the branchiomeric (gill) structures (special visceral efferent)

   c. The fibers of the ventral root of each spinal nerve supply the motor innervation to specific groups of voluntary muscles known as **myotomes**.

   d. Usually several myotomes act over any joint (see Fig. 4-3).

   e. There is growing evidence that some visceral afferent fibers enter the spinal cord through the ventral root.

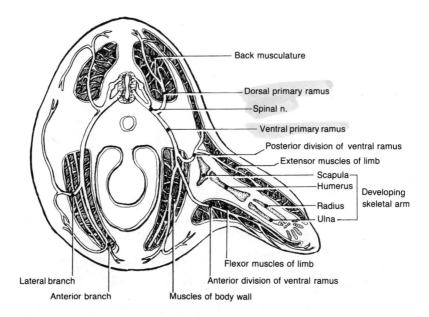

Figure 4-2. *Organization of the spinal nerve.* One side depicts the basic pattern of the spinal nerve with division into dorsal and ventral primary rami and the lateral branch of the ventral ramus. The other side depicts the formation of the brachial and lumbosacral plexuses with the lateral branch of the ventral ramus forming the posterior division while the continuation of the ventral ramus forms the anterior division.

    f. A central lesion will involve to some extent all of the muscles of a myotome while a peripheral nerve injury will involve only a portion of a myotome.

C. Spinal nerves. In the vicinity of an intervertebral foramen, a dorsal root and a ventral root meet to form a spinal nerve, which supplies the innervation of a segment of the body.

    1. Organization. Spinal nerves are numbered in association with vertebral levels.
        a. The thoracic, lumbar, and sacral nerves are numbered by the vertebra just rostral to the intervertebral foramen through which they pass (e.g., nerve T4 emerges below vertebra T4).
        b. The cervical nerves are numbered by the vertebrae just caudal to them (e.g., nerve C7 is rostral to vertebra C7) except that nerve C8 exits caudally to vertebra C7 and rostrally to vertebra T1.
        c. In all, there are 31 pairs of spinal nerves:
            (1) Cervical: C1–C8
            (2) Thoracic: T1–T12
            (3) Lumbar: L1–L5
            (4) Sacral: S1–S5
            (5) Coccygeal: Co1

    2. Composition. Each spinal nerve contains both motor and sensory axons although C1 and Co1 have only ventral roots (i.e., whereas there are C1 and Co1 myotomes, there are no C1 and Co1 dermatomes).

    3. Subdivisions. The short mixed spinal nerve divides almost immediately into two primary rami and two secondary rami (see Fig. 4-2).
        a. The dorsal primary ramus of each spinal nerve arches dorsally to innervate the skin (the posterior portion of the dermatome) and muscles of the back.
        b. The ventral primary ramus continues anteriorly to innervate the muscle and skin of the lateral and ventral aspects of the body.
            (1) In the thoracic region, the ventral primary rami are termed **intercostal nerves**.
                (a) The 12th intercostal nerve is termed the subcostal nerve.
                (b) In the lumbar region, the anterior primary rami form other named nerves (e.g., iliohypogastric and ilioinguinal nerves).
            (2) When the ventral primary ramus (e.g., intercostal nerve) reaches a point in line with the axilla, it gives off a **lateral cutaneous branch**, which penetrates the muscle to innervate the overlying skin.

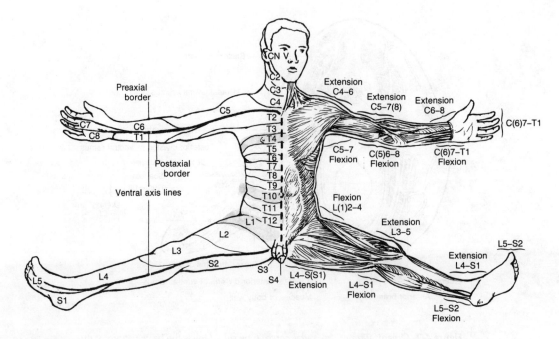

Figure 4-3. *Cutaneous and muscular nerve distributions in the primitive position.* The body is positioned in such a way that the primitive ventral surfaces are anterior. A typical arrangement of the dermatomes on the right side of the body depicts the dermatomal migration into the extremities during development. The innervating root numbers to the muscles that act across the joints of the upper and lower left extremities also show evidence of myotomal migration.

     (a) The homologues of these branches form the **posterior divisions** of the brachial plexus and lumbosacral plexus.
     (b) The lateral branch of the ventral primary ramus then divides into anterior and posterior perforating branches, which innervate muscle and the lateral portion of the dermatome.
   (3) The ventral primary ramus continues anteriorly toward the midline after giving off the lateral cutaneous branch.
     (a) The homologues of these branches form the **anterior divisions** of the brachial plexus and lumbosacral plexus.
     (b) Just lateral to the midline the ventral ramus terminates as the **anterior cutaneous branch**, which innervates the anterior portion of the dermatome.
   (4) The cutaneous branches of the spinal nerves overlap in their distribution so that every region of the body is ensured innervation. The dermatomes also overlap in the anterior and posterior midline.
  c. The meningeal ramus is actually the first branch of the spinal nerve.
  d. Rami communicantes
   (1) White ramus communicans:
     (a) Connects the spinal cord and the sympathetic chain between T1 and L2, inclusive
     (b) Conveys sympathetic preganglionic myelinated fibers to the paravertebral and prevertebral ganglia
     (c) Conveys visceral afferent fibers to the spinal nerve
   (2) Gray ramus communicans:
     (a) Connects the sympathetic chain and the spinal nerves at every level
     (b) Conveys sympathetic postsynaptic unmyelinated fibers from the paravertebral ganglia to the spinal nerve to reach the skin

IV. SOMATIC NERVOUS SYSTEM

  A. Afferent nerves (Fig. 4-4). The afferent nerves convey sensory input from free nerve endings and special receptors to the CNS.

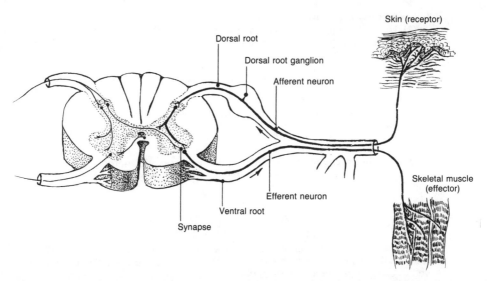

Figure 4-4. *Somatic reflex arc.* Somatic afferent neurons with cell bodies in the dorsal root ganglion convey sensory information from the periphery to the spinal cord. The information is passed through small internuncial neurons to the motor neuron. The efferent neurons with cell bodies located in the ventral horn of the spinal cord pass along the spinal nerve to contact effector organs.

1. Classification of afferent innervation
   a. General somatic afferent nerves convey pain, temperature, touch, and proprioception from the head, body wall, and extremities to the CNS.
   b. Special somatic afferent nerves convey vision, hearing, and balance to the CNS.
   c. General visceral afferent nerves (frequently included in this classification for functional reasons) convey sensory information from the visceral organs to the CNS.
   d. Special visceral afferent nerves (frequently included in this classification for functional reasons) convey smell and taste sensations to the CNS.

2. Afferent pathway. One peripheral neuron and at least one central neuron are involved.
   a. Peripheral neuron
      (1) The cell bodies of these pseudounipolar sensory neurons lie in the **dorsal root ganglia** of the spinal nerves or the cranial nerve equivalents.
      (2) The distal end of the axon either is a free nerve ending or makes contact with a specialized receptor; the proximal end of the axon enters the spinal cord through the dorsal rootlets.
      (3) The axons may synapse with a central neuron in the gray matter of the spinal cord or run cranially in the white matter to a region of gray matter in the brain stem at which synaptic contact is made.
   b. Central neurons lie in the gray matter of the spinal cord or brain stem and send axons to higher centers of the brain at which synaptic contact is made.

B. Somatic efferent nerves (see Fig. 4-4). The basic role of the **somatic motor nerves** is to regulate the coordinated muscular activities associated with voluntary motion and the maintenance of posture.

1. Classification of somatic efferent innervation
   a. General somatic efferent nerves supply the muscles of the head, body wall, and extremities, which arise from myotomes of the embryonic somites.
   b. Special visceral efferent nerves supply the muscles of the head and neck, which arise from branchiomeric (gill) structures.

2. Somatic efferent pathway. Two sets of neurons are involved in producing voluntary somatic motor activity—one central and one peripheral.
   a. Upper motor neurons
      (1) The cell bodies of the upper motor neurons lie in the gray matter of the motor areas of the cortex and various nuclei of the brain stem.
      (2) The upper motor neuron passes through the cerebrum, brain stem, and the white matter of the spinal cord to contact a lower motor neuron.

b. Lower motor neurons

    (1) The cell bodies of the lower motor neurons lie in the gray matter of the brain stem for cranial nerves or in the ventral horns of the spinal cord gray matter for the spinal nerves.

    (2) Each somatic lower motor neuron has an axon that courses through a cranial nerve or spinal nerve to make synaptic connections with voluntary muscle fibers at myoneural junctions.

3. Reflex arcs. Simple reflex arcs are composed of one sensory neuron and one motor neuron. Complex reflex arcs may have one or more interneurons (located in the gray matter) intercalated between the sensory and motor neurons.

C. Visceral afferent nerves are frequently included with somatic afferents, but they are more conveniently discussed in association with the visceral efferent innervation because the pathways are similar.

V. AUTONOMIC NERVOUS SYSTEM

A. Composition. The ANS is composed of the neural structures of the CNS and PNS that are involved with the motor activities that influence the involuntary (smooth) and cardiac musculature as well as the glands of the viscera and skin.

B. Classification

    1. The ANS is classified as **general visceral efferent** and supplies smooth muscles, cardiac muscle, and glands.

    2. It is often called the visceral or vegetative motor system because the effectors are associated with systems (e.g., cardiovascular, digestive, respiratory, perspiratory, and vasodilatory) over which only minimal direct conscious control can be exerted.

C. Visceral efferent pathway. Three sets of neurons are involved in producing motor activity—one central and two peripheral.

    1. Upper motor neuron

      a. The cell bodies of the upper motor neurons are located in the gray matter of the autonomic areas of the brain.

      b. The axons pass in the descending tracts at the brain stem and spinal cord.

    2. Preganglionic neuron

      a. The preganglionic neuron, which may be seen as the equivalent of a lower motor neuron, originates in the gray matter of the brain stem or the lateral horn of the spinal cord gray matter.

      b. Its myelinated axon courses through a cranial nerve or spinal nerve.

      c. It terminates by synapsing with a postganglionic neuron located in a ganglion outside the central nervous system.

    3. Postganglionic neuron

      a. The postganglionic neuron is unique to the ANS.

      b. The cell body of the postganglionic neuron is located in an autonomic ganglion.

      c. The postganglionic neuron has an unmyelinated axon that extends peripherally to terminate in endings associated with smooth muscles, cardiac muscle, or glands.

D. Divisions

    1. Sympathetic division (Fig. 4-5). As a very rough rule of thumb, the sympathetic system stimulates activities that are mobilized by the organism during emergency and stress situations—the so-called *fight, fright, and flight* responses. The sympathetic system is also called the **thoracolumbar** or **adrenergic system.**

      a. Sympathetic preganglionic fibers

        (1) Sympathetic preganglionic fibers originate from cell bodies located in spinal levels T1 through L2, hence thoracolumbar, and pass successively through the:

          (a) Ventral roots, referred to as the thoracolumbar outflow

          (b) Anterior primary rami of the spinal nerves

          (c) Myelinated **white rami communicantes.** Only spinal levels T1 through L2 have white rami communicantes.

          (d) Sympathetic chain (composed of **paravertebral ganglia** and interconnecting fiber bundles)

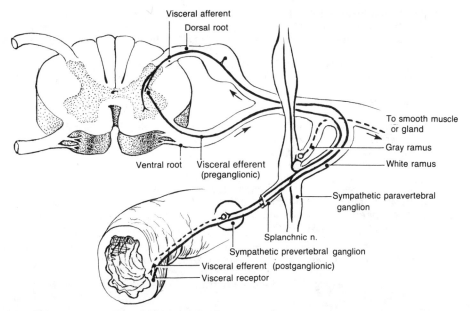

Figure 4-5. *Autonomic and visceral afferent neurons.* Presynaptic sympathetic neurons with cell bodies in the lateral horns of the spinal cord pass along the spinal cord and along a white ramus communicans to reach the sympathetic chain where they either synapse or pass along a splanchnic nerve to synapse in a prevertebral ganglion. The postganglionic neurons contact the effector organs.

(i) In the sympathetic chain, the preganglionic fibers terminate by synapsing with a postganglionic neuron in the paravertebral ganglion associated with the spinal nerve level.

(ii) They course up or down the sympathetic chain to synapse with a postganglionic neuron located in a ganglion several levels away from that of entry.

(iii) They pass through the sympathetic chain without synapse to form a splanchnic nerve that leads to a **prevertebral ganglion** from the cervical to coccygeal regions.

(2) The preganglionic sympathetic neurons are **cholinergic** because the neurosecretory transmitter released is **acetylcholine**.

b. Sympathetic postganglionic fibers

(1) Sympathetic postganglionic fibers from cells in the paravertebral ganglia pass either:

(a) Back to a spinal nerve along an unmyelinated **gray ramus communicans** from a paravertebral ganglion at *every spinal level* to terminate in the dermal sweat glands or the smooth muscles of blood vessels and hair (arrector pili) muscles of the body wall and extremities. Every spinal nerve receives a gray ramus from the sympathetic chain.

(b) To splanchnic nerves and perivascular plexuses, leading to the visceral structures of the head, neck, thorax, abdomen, and pelvis.

(2) The sympathetic postganglionic fibers from cells in the prevertebral ganglia course in the perivascular plexuses to innervate the abdominal and pelvic viscera.

(3) The sympathetic postganglionic neuron is **adrenergic** because the neurotransmitter is usually **norepinephrine**.

(4) The cells of the **medulla** of the **adrenal gland** are specialized postganglionic sympathetic neurons.

(a) Preganglionic cholinergic fibers stimulate the adrenal chromaffin cells to release both norepinephrine and epinephrine into the circulatory system, which distributes these neurosecretions throughout the body with a time delay.

(b) In conjunction with the immediate norepinephrine release mediated by the sympathetic postganglionic fibers, the adrenal neurosecretions prolong the actions on the receptive tissues.

c. Sympathetic ganglia. Because the paravertebral and prevertebral ganglia are located some distance from the organ innervated, the sympathetic preganglionic fibers are shorter, whereas the postganglionic fibers are longer as compared to the parasympathetic division.

2. Parasympathetic division. Following the same rough rule of thumb, the parasympathetic system stimulates those activities associated with conservation and restoration of body resources. The parasympathetic system is also called the **craniosacral system** or **cholinergic system**.
   a. The cranial preganglionic fibers emerge with cranial nerves III, VII, IX, and X.
      (1) These supply the parasympathetic innervation to the head, thorax, and most of the abdominal viscera.
      (2) Parasympathetic function decreases heart rate, increases gastrointestinal activity, and increases secretory activity.
   b. The sacral parasympathetic outflow is through sacral spinal levels S2 through S4 as the **nervi erigentes** or **pelvic splanchnic nerves**.
      (1) The sacral spinal cord supplies innervation to lower abdominal and pelvic viscera.
      (2) It is involved with urination, defecation, and sexual function.
   c. Both the preganglionic and postganglionic neurons are **cholinergic** because the neurotransmitter secreted at the terminals is **acetylcholine**.
   d. Parasympathetic ganglia. Because the postganglionic cell bodies are located close to the organ innervated, the preganglionic fibers have relatively long axons, whereas the postganglionic fibers have very short axons as compared to the sympathetic division.

3. The enteric nervous system is comprised of the neural networks and plexuses of the gastrointestinal canal, which are considered a distinct division of the autonomic nervous system.

E. Visceral afferents (see Fig. 4-5). Afferent innervation to the viscera usually is not considered part of the ANS.

1. Afferent fibers from the viscera pass retrogradely along the autonomic pathways.
   a. In the cervical region, visceral afferents travel along cervical splanchnic nerves, such as the cardiac accelerator nerves, to reach the sympathetic chain, thence down the chain to the white rami communicantes of the upper thoracic levels to gain access to the spinal nerves and the upper thoracic levels of the spinal cord.
   b. In the thorax and abdomen, they pass along splanchnic nerves to the sympathetic chain.
      (1) On reaching the sympathetic chain, the afferents pass through the white rami communicantes to gain access to a spinal nerve.
      (2) If there is no white ramus communicans (above T1, below L2), the afferents course down or up the sympathetic chain until a white ramus is reached so that the spinal cord may be accessed.
   c. In the pelvic region, there are two distinct afferent pathways.
      (1) From upper pelvic viscera, afferent neurons travel along sympathetic pathways to the lumbar splanchnic nerves, thence along white rami communicantes to the lumbar spinal nerves that bring the sensory information to the upper lumbar levels of the spinal cord.
      (2) From the lower pelvic viscera, afferent neurons travel along the parasympathetic nervi erigentes (pelvic splanchnic nerves) to reach midsacral (S2–S4) levels of the spinal cord.

2. The visceral afferent pathways provide the anatomic basic for **referred pain**, whereby sensation from a visceral structure appears as if it originates from the somatic dermatome associated with the spinal level at which the visceral afferents enter the spinal cord.

5

Circulatory System

I. INTRODUCTION

A. The circulatory system includes the heart and vessels (arteries, capillaries, and veins) that conduct the blood through the body (see Fig. 5-1).

  1. The heart is a muscular pump that propels blood through the circulatory system. There are two circulatory loops.

    a. Pulmonary circulation. Blood is pumped from the right side of the heart through the lungs and returned to the left side of the heart.

    b. Systemic circulation. Blood is pumped from the left side of the heart through the tissues of the body and returned to the right side of the heart.

  2. Arteries conduct blood from the heart to the capillary beds.

  3. Capillaries provide for gaseous diffusion and exchange of nutrients and waste products.

  4. Veins collect blood after its passage through capillary beds and return it to the heart.

B. The lymphatic system, a of part of the circulatory system, includes the **lymphatic vessels**, a set of channels that begins in the tissue spaces and returns excess tissue fluid to the bloodstream.

II. HEART

A. Cardiac pump. The heart is a set of two adjacent folded tubes of cardiac muscle (see Fig. 5-1).

  1. Systemic and pulmonary circulation. The right side of the heart pumps blood from the venous system through the lung capillary beds and into the left side of the heart, which pumps blood through the systemic circulation.

  2. Cardiac dynamics. Contraction of the muscular walls propels blood out of the chambers.

    a. Valves guarding the lumina maintain the blood flow in a fixed direction.

    b. The work done by the heart muscle is the product of the pressure and the volume of blood pumped.

  3. Cardiac control. The pumping rate of the heart is regulated by the autonomic nervous system, which controls an internal pacemaker (the sinoatrial node).

    a. An internal-impulse conducting system of modified muscle cells (atrioventricular node and bundle) coordinates the contraction in the ventricular chambers of the heart.

    b. Pathology in this impulse initiation and conduction system can lead to heart block or fibrillation (uncoordinated contractions).

B. Coronary circulation supplies the **myocardium** with blood.

  1. Right and **left coronary arteries** begin in the sinuses behind the right and left semilunar cusps of the aortic valve. They distribute blood in large part to their own half of the heart.

  2. Blood flow in the coronary arteries is maximal during diastole and minimal in systole.

C. Fetal circulation is different from circulation in the adult. In the fetus, oxygenation of the blood occurs in the placenta, and circulation through the lungs occurs to a much lesser degree than in postnatal life.

  1. The right heart pumps very little blood to the collapsed fetal lungs.

  2. The left heart pumps blood through the systemic circulation and the placenta.

3. A system of shunts operates before birth to bypass partially the lungs and liver. These shunts close postnatally, and the adult circulatory pattern is established. Failure of these shunts to close at birth represents a series of possible defects that may be surgically correctable.
  a. The foramen ovale shunts blood from the right atrium to the left atrium, bypassing the pulmonary circulation.
  b. The ductus arteriosus shunts blood from the left pulmonary artery to the aorta, bypassing the pulmonary circulation.
  c. The ductus venosus shunts blood from the umbilical vein to the inferior vena cava, bypassing the liver.

III. ARTERIAL SYSTEM

A. Conducting arteries are the largest vessels of the body (Fig. 5-1). The arterial system begins with the **aorta**, a single artery with a large diameter (3 cm).

  1. Wall structure is related to pressure.
    a. The elastic tissue in the conducting arteries permits both stretch of the wall during the ejection phase of cardiac *systole* and a propulsive elastic recoil during ventricular *diastole*.
    b. The aorta gives off distributing arteries with small diameters.
      (1) The higher the pressure, the greater the amount of elastic tissue relative to muscular tissue.
      (2) As the arteries get smaller, intermittent pulsations (from cardiac systole) become a steady, continuous flow, and no pulsation is observed in the capillary bed.

  2. Arterial hemodynamics
    a. The elastic tissue in these vessels enables the pulse to be palpated.
    b. Blood pressure is measured by auscultation of the bruit produced by the intermittent flow of blood through an artery.
      (1) When sufficient pressure is applied by a pneumatic tourniquet to compress the artery completely, no bruit is heard.
      (2) As pressure is released, an intermittent bruit is heard when systolic pressure just exceeds that applied by the tourniquet.

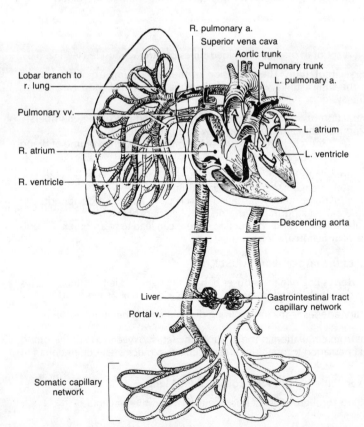

Figure 5-1. *Schematic representation of the circulatory system.* The pulmonary circulation is associated with the right side of the heart, and the systemic circulation is associated with the left side of the heart.

(3) As pressure is further released, the diastolic pressure equals that applied by the tourniquet, flow becomes continuous, and the bruit disappears.

B. **Distributing arteries** constitute the remainder of the named arteries of the body (see Fig. 5-1). The distributing arteries progressively divide into arteries with smaller and smaller diameters (although the total cross-sectional area increases).

1. **Main arterial pathways** have the following characteristics:
   a. They take the shortest possible course—for example, in the limbs they:
      (1) Run on flexor surfaces
      (2) Usually do not pass directly through muscles, avoiding compression
      (3) Are somewhat extensible
   b. They branch in particular patterns such that:
      (1) The angle of branching is related to hemodynamic factors and minimizes pull between parent stem branches.
      (2) Branches of equal size (e.g., the common iliac arteries) make equal angles with the parent stem; the ideal angle is approximately 75°.
      (3) With smaller branches, whose diameters barely affect the size of the parent stem (e.g., the intercostals), the angle can range from 70°–90°.

2. **Arterial anastomoses** permit equalization of pressures and alternate channels of supply.
   a. Abundant anastomoses occur in the regions of joints in which movement might temporarily occlude the main channel.
   b. In the brain, the circle of Willis serves to equalize the blood supply to brain.
   c. In the gastrointestinal tract, abundant anastomoses occur between some regions, whereas anastomoses are sparse between others.
   d. The surgeon must differentiate between *potential* and *functional* anastomoses and must be aware of variability, which might result in sparse or nonfunctional anastomoses where profuse functional anastomoses are expected.

3. **End-arteries** supply discrete regions of tissue that have no collateral supply. End-arteries are found in the heart, kidneys, liver, brain, and organs of the gastrointestinal tract.
   a. There are no direct anastomoses between end-arteries.
   b. A thrombosis or embolus lodged in an end-artery produces ischemia and necrosis (infarct) of the tissue supplied exclusively by that vessel.
   c. Potential anastomoses may not be functional at a crisis unless collateral channels develop with a slow-onset pathology.

C. **Arterioles**, the terminations of the smallest arteries, are nearly as small as capillaries and are not merely conducting channels, but regulate the distribution of blood. The presence of muscles in the walls of the arterioles provides the basis for nervous regulation of the size of the vessel lumen. This muscular tissue runs in spiral patterns.

IV. CAPILLARY BEDS

A. **Capillaries** are the exchange sites of the circulatory system (Fig. 5-2).

1. **Total cross-sectional area of the capillary bed** is approximately 800 times that of the aorta.
   a. **The diameter of a capillary** is about 5 μm, just large enough to enable a red blood cell to squeeze through.
   b. **The velocity of circulation** changes from 0.5 m/sec in the aorta to 0.5 mm/sec in the capillaries.

2. **Diffusion barrier.** Capillary walls are formed by a single layer of endothelial cells. This endothelium and a thin layer of connective tissue fibers (the basal lamina) constitute the diffusion barrier.
   a. **Exchange of gas.** Gaseous diffusion occurs according to partial pressure gradients.
      (1) In the peripheral tissues, oxygen released from arterial (oxygenated) blood diffuses across the endothelium into the tissue spaces. Carbon dioxide diffuses from the tissue spaces into the blood.
      (2) In the lungs, carbon dioxide diffuses from the blood into the alveolar air, and oxygen diffuses from the alveolar air into the venous (unoxygenated) blood.
   b. **Exchange of fluid.** Fluid movement occurs according to differentials between blood pressure and osmotic pressure.
      (1) At the arterial end of the capillary bed, blood pressure exceeds the tissue osmotic pressure.
      (a) An ultrafiltrate of nutrient-rich plasma exudes from the arterial end of the capillaries into the tissue spaces.

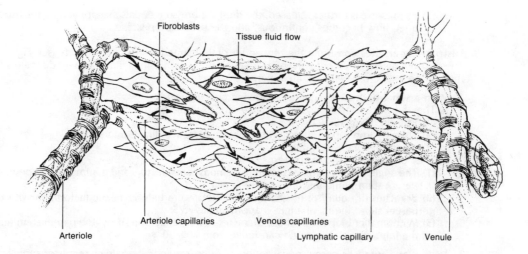

Fibroblasts

Tissue fluid flow

Arteriole capillaries

Venous capillaries

Arteriole

Lymphatic capillary

Venule

Figure 5-2. *Capillary bed.* The continuum from arteriole through the capillary bed to venule is depicted. *Arrows* indicate the flow of fluid. Blind lymphatic capillaries drain excess fluid from the tissue spaces.

        (b) This raises the osmotic pressure of the blood plasma remaining in the capillary and lowers the tissue osmotic pressure.

      (2) At the venous end of the capillary bed, the blood pressure is less and the plasma osmotic pressure is greater than the tissue osmotic pressure.

        (a) There is osmotic uptake of tissue fluid that is high in metabolic waste products into the venous capillaries.

        (b) This restores the osmotic pressure of the tissue spaces and blood plasma in a constant, finely balanced process.

    3. Density of a capillary bed in a tissue correlates with the metabolic need (e.g., endocrine glands are very vascular, and cartilage is almost avascular).

B. Sinusoids substitute for capillaries in some organs, such as the liver, spleen, and red bone marrow, where circulation is slow. Sinusoids often contain phagocytic cells.

C. Clinical considerations

    1. Edema is the collection of excess fluid in the tissue spaces.

      a. When associated with trauma or inflammation, edematous exudate is caused by increased permeability of the walls of the capillary bed whereby red blood cells are contained but plasma proteins leak out and raise the tissue osmotic pressure, thus interfering with the hydrostatic–osmotic pressure balance that normally results in fluid uptake at the venous end of the capillary bed.

      b. When associated with elevated venous pressure, edematous transudate is caused by imbalance of the hydrostatic–osmotic pressures that normally result in fluid uptake at the venous end of the capillary bed.

    2. Hematoma (black-and-blue mark associated with edema) results from a major loss of integrity of capillary walls or the walls of other blood vessels; red blood cells, in addition to plasma, leak into the tissue and cause discoloration.

V. VENOUS SYSTEM

A. Capillary beds drain into **venules**, which come together to form veins (see Fig. 5-2).

B. Veins return blood to the heart (see Fig. 5-1).

    1. Venous blood flow. The veins must carry the same volume of blood as the arteries but at a lower pressure.

      a. Compared to arteries, veins have large and somewhat more irregular lumina and thin walls. Thus, veins are relatively compressible by external forces, a fact that aids in the flow of the venous blood.

    b. Muscles in venous walls tend to be arranged in a loop near the point of drainage of the tributary to act as a "sluice gate" (e.g., the penile veins and the point at which the hepatic veins enter the vena cava). The walls of large veins have some elastic tissue to resist the pressure of right atrial systole.

    c. Many veins contain valves that permit proximal flow only.

       (1) Valves occur primarily in veins of limbs and movable viscera but not in the cerebral veins.

       (2) Valves are especially prevalent at junctions between tributaries and large veins.

  2. Parallel pattern of arteries and veins

    a. Venous patterns are far more variable than arterial patterns.

       (1) Large veins are usually single.

       (2) Medium-sized veins are frequently doubled (venae comitantes).

       (3) In several regions, arterial patterns and venous patterns are quite separate and distinct (e.g., in the brain, liver, lungs, and penis).

    b. The angulation of entry of tributaries is less related to hemodynamic factors than it is in arteries because venous pressure is lower.

    c. The frequency of venous cross-anastomoses reflects retention of an embryonic condition in the adult.

    d. The "counter-current" pattern of parallel arteries and veins is often related to transfer of water or heat, as in the kidney and the extremities, respectively.

  3. Venous hemodynamics. Pressure gradients between the periphery and the right side of the heart control venous flow. These pressure gradients are established by:

    a. An arterial pressure of approximately 10 mm Hg transmitted through the capillary bed to the venous side

    b. The sucking bulb-syringe action of the heart during right ventricular diastole

    c. The negative pressure, relative to atmospheric pressure, produced by the thoracic cage during inspiration

    d. The flow in the superficial veins, which can be occluded by application of mild pressure

       (1) A snug tourniquet occludes superficial veins and causes distal pooling of blood.

       (2) The superficial veins provide convenient sites for obtaining blood samples.

    e. The contractile activity of the muscles of the extremities, which "milks" the venous blood toward the heart

       (1) This action is aided in the limbs by the disposition of the deep and superficial veins.

         (a) The muscles of the extremities, between which the deep veins lie, are surrounded by relatively inelastic fascial septa.

         (b) Upon shortening, muscles become wider, thereby increasing the pressure on the deep veins and moving the blood in a direction guided by the internal valves.

       (2) The movements of the segments of the limbs help move blood in the superficial veins.

C. Portal systems begin in capillaries and subsequently form a large vein before breaking up into sinusoids. The hepatic portal system, for example, drains most of the venous blood from the capillary beds of the intestinal tract to the sinusoids of the liver.

D. Arteriovenous (AV) anastomoses permit direct transfer of blood from arterial to venous channels, bypassing the capillary bed.

  1. Distribution. AV anastomoses are widely distributed.

    a. These vessels rise as a side branch of a terminal arteriole and join a venule.

    b. Each AV anastomosis has a thick muscular wall, abundantly supplied with vasomotor nerves (general visceral efferent, sympathetic), thereby forming a sphincter.

  2. Functions. AV anastomoses usually occur in organs whose functions are intermittent.

    a. In the gut, AV anastomoses are open except during periods of digestion.

    b. In the skin, these connections are especially numerous in apical parts (e.g., fingers, nose, lips, and ears), where they serve in temperature regulation.

VI. VASCULAR PATTERNS

A. Development. Although embryonic vessels develop before the initiation of circulation and are, therefore, preadaptive, additional development is functionally regulated.

  1. Remodeling of the embryonic vascular pattern (e.g., aortic arches and cardinal veins) produces the adult patterns with great variation.

    a. An increase in vascular pressure is followed by an increase in endothelial budding that forms new vessels to relieve the load.

b. An increase in vascular pressure, which exerts a circumferential stress on the walls, results in an increase in the diameter of the vessel and, because the amount of muscular and elastic tissue is a response to stress, an increase in wall thickness.

2. Vascular malformations result from incomplete remodeling.

B. Characteristic patterns. The vascular supply to some organs, such as the kidney, is altered on migration within the body, whereas other organs, such as the gonads and the gastrointestinal tract, carry their original vascular supply with them.

VII. LYMPHATIC SYSTEM

A. Composition. The lymphatic system is composed of an extensive network of extremely variable **lymphatic vessels** and **lymph nodes**, which serve as filters and a source of lymphocytes and plasma cells.

B. Function. The system has evolved as a specialized mechanism to return to the bloodstream fluids that were not taken up by the blood capillaries.

1. The lymphatic vessels contain **lymph**, which is clear and colorless. When lymph contains fat droplets, it is called **chyle**.
   a. These fatty substances are absorbed by the lymphatics from the gastrointestinal tract.
   b. Chyle is channeled to the blood for distribution to appropriate organs.

2. Proteins are constantly entering the lymph stream from the tissue spaces.
   a. These proteins filter through lymph nodes on their way to the blood vascular system.
   b. This association is of particular importance if the protein is an antigen (i.e., "foreign").

C. Structure (see Fig. 5-2)

1. Lymph capillaries begin as cul-de-sacs that drain the tissue spaces.
   a. The lymph capillaries are wider than blood capillaries and are irregular in diameter (from a few microns to 1 mm).
   b. The lymphatic capillaries are numerous in the mucous membranes, serous surfaces, and dermis of the skin. The brain, spinal cord, and eyeball lack lymphatic capillaries; however, these structures have other channels for draining tissue fluids. Lymphatic capillaries are also absent in bone marrow, parenchyma of the spleen, and superficial fascia.

2. Lymphatic vessels are formed by the convergence of lymph capillaries.
   a. Lymphatics have valves that give them a beaded appearance.
   b. They are more plentiful than veins, which they tend to accompany.
   c. The walls of the larger lymphatics become somewhat thicker as they acquire small amounts of smooth muscle and elastic tissue.
   d. Intercalated along the lymphatic vessels are the lymph nodes.

3. Lymph ducts are formed by the convergence of lymphatic vessels. The lymphatics from the lower extremities converge on lymph nodes located anteriorly and superficially in the uppermost part of the thigh.
   a. From the inguinal lymph nodes, the main duct from each lower extremity enters the abdomen.
   b. After passing through the lumbar lymph nodes, the main duct becomes known as the **lumbar duct**, which enters the cisterna chyli.

4. The cisterna chyli is a dilated sac located between the diaphragmatic crura, opposite the first lumbar vertebra and behind the right side of the aorta.
   a. The cisterna chyli can vary in size and location.
   b. It contains some smooth muscle and is somewhat pulsatile.
   c. In addition to the lumbar lymphatics, the cisterna chyli also receives the large **common duct**, which conveys lymph from most of the intestinal tract.

5. The thoracic duct originates from the upper end of the cisterna chyli.
   a. The thoracic duct ascends upward on the anterior aspect of the vertebral column.
   b. It inclines slightly to the left and ends at the base of the neck, usually by entering the left brachiocephalic vein.
      (1) The thoracic duct thus receives the lymph from both legs, the pelvis, the abdomen, and the left side of the thorax.
      (2) It may drain lymph from the left arm and from the left side of the head and neck, because the left jugular lymph duct and the left subclavian lymph duct may join the thoracic duct in the base of the neck.

(a) The **left jugular lymph duct** drains lymph from the left side of the head and neck, accompanying the left internal jugular vein in the neck.

(b) The **left subclavian lymph duct** drains the left axillary and subclavicular nodes of the left upper extremity and accompanies the left subclavian vein.

(c) The **left bronchomediastinal lymph duct**, which frequently joins the thoracic duct, drains the thoracic viscera on the left.

   c. In normal situations, the large volume of lymph flowing through the thoracic duct (1–2 ml/kg/hr) is derived mainly from the liver and alimentary tract.

6. **The right lymph duct** receives lymph from the right side of the head and neck, the right upper extremity, and the right side of the thorax and returns lymph to the great veins at the base of the neck on the right side. The channels are the:
   a. **Right jugular lymph duct** from the right side of the head and neck
   b. **Right subclavian lymph duct** from the right axillary and subclavicular nodes
   c. **Right bronchomediastinal lymph duct** from the right thoracic viscera

D. **Lymphaticovenous communications** function when the lymphatic pressure is increased by blockage of the usual channels. Direct lymphaticovenous connections have been demonstrated between the thoracic duct and the hemiazygous vein and the abdominal lymphatic ducts and the inferior vena cava. It is postulated that these connections are the result of enlargements of preexisting channels that normally convey little or no lymph.

E. **Lymph flow.** The flow of lymph from a region is generally unidirectional toward the large veins at the base of the neck. The progression and directionality result from:

1. **Hydrostatic pressure** from the volume of tissue fluids taken up by the lymphatic capillaries

2. **Mechanical forces**
   a. The pressure resulting from the voluntary muscular activity on the valved lymph vessels (e.g., the flow of lymph in an immobile limb is almost negligible but becomes significant during muscular activity)
   b. Respiratory movements: The alternation of pressures within the thorax during ventilation propels lymph in the valved lymphatics of the mediastinum.
   c. The contractions of the muscles of the abdominal wall on expiration, coughing, and straining: These produce a positive abdominal pressure on the cisterna chyli that propels lymph.
   d. The pulsations of adjacent blood vessels: These have a massaging effect on lymph vessels to aid the flow of lymph.
   e. In some parts of the body, rhythmic flow may be aided by the contraction of the smooth muscles in the walls of the lymphatic vessels and the cisterna chyli.

3. **Valves** in the thoracic duct and right lymph duct, which prevent backflow

F. **Clinical considerations**

1. **Lymphatics** are the major route by which carcinoma metastasizes.
   a. Malignant cells may be trapped in lymph nodes where they proliferate.
   b. Malignant cells may be delivered to the venous system.
   c. Surgical removal of malignant tumors involves dissection of the major lymphatic vessels and nodes draining the involved region.
   d. Wound healing requires the regeneration of the lymphatics as well as the growth of blood capillaries.

2. **Thoracic duct**
   a. Injury to the thoracic duct may result in accumulation of fluid in thoracic or pleural cavities (*chylothorax*). The diagnosis is made by the high lymphocyte count in a pleural aspirate.
   b. Thoracic duct obstruction with backup of intestinal lymph to the kidney lymphatic capillaries may produce *chyluria*.

3. **Lymph nodes**
   a. Lymph vessels drain into nodes of lymphatic tissues, which are usually grouped in specific regions and named accordingly.
   b. Lymph vessels drain the nodes, often terminating in another group of nodes. Lymph may filter through several nodes before reaching a major trunk.
   c. Bacteria or antigens filtered by the phagocytes of the lymph nodes may induce inflammation or a cell-mediated immune reaction, either of which can produce swelling of the node (swollen glands).
   d. When nodes are swollen by inflammation or blocked by metastatic cells, edema or drainage along alternate lymphatic pathways may occur.

4. Lymphatic drainage pattern

a. Lymphatic absorption of fluid in the peritoneal cavity is almost entirely through the peritoneal surface of the diaphragm. While the omentum has lymphatics, the absorption is minor compared to that of the diaphragmatic peritoneum. The rate of absorption from the peritoneal cavity is:

(1) Exceedingly rapid (labeled protein injected intraperitoneally is detected in the thoracic duct in 10–15 minutes)

(2) Normally about 1 L per day

b. Lymph from liver contributes a considerable part of lymph flow of the thoracic duct.

(1) Hepatic lymph drains along the hepatic pedicle toward the hepatic nodes and thence to cisterna chyli.

(2) Ascites is the accumulation of extracellular fluid in the peritoneal cavity.

(a) Ascitic fluid usually is a transudate from dilated subcapsular and hepatic hilar lymphatics.

(b) It is related to increased capillary permeability, a hydrostatic–osmotic pressure imbalance, or altered lymphatic flow.

(c) It is commonly, but not exclusively, associated with conditions that raise intrahepatic blood pressure (e.g., right-sided heart failure or cirrhosis).

c. Lymphatic drainage of the lungs is via the bronchomediastinal duct.

(1) Pulmonary edema is caused by either increased permeability or a hydrostatic–osmotic pressure imbalance in the pulmonary vascular bed with fluid accumulation in the tissue spaces and transudation of fluid into the alveoli.

(2) Pulmonary effusion (hydrothorax) can be caused by similar processes but with transudation into the pleural cavity.

d. Lymph return from the trunk and extremities is to axillary and inguinal nodes.

(1) Lymph follows two sets of channels.

(a) The superficial lymphatics accompany the superficial veins to the axillary and inguinal nodes. Infection may spread along superficial lymphatics, causing fine striae.

(b) The deep lymphatics accompany the deep veins to the axillary and inguinal nodes.

(2) Lymphedema is the accumulation of tissue fluid as a result of lymphatic obstruction (e.g., *elephantiasis* caused by *Filaria bancrofti*).

Part I Introductory Overviews

STUDY QUESTIONS

Directions: Each question below contains five suggested answers. Choose the **one best** response to each question.

1. Circumduction is a movement that involves

(A) abduction/adduction and flexion/extension
(B) flexion/extension and medial rotation/lateral rotation
(C) medial rotation/lateral rotation and abduction/adduction
(D) one axis of rotation
(E) ball-and-socket joints only

2. The principal lymphatic drainage of the right breast is toward the

(A) right axillary nodes
(B) right internal mammary nodes
(C) right internal thoracic nodes
(D) right supraclavicular nodes
(E) thoracic duct

3. Which of the following statements about the lobes of the breast is true?

(A) They are defined by connective tissue septa, which determine the posture of the breast
(B) They are located between the deep layer of the superficial fascia and the deep fascia
(C) They contain fatty tissue, which is the source of breast size variation over the menstrual cycle
(D) They empty at the nipple via a common ampulla
(E) They number between 40 and 60

4. Longitudinal growth in a long bone may be interrupted prematurely by a fracture through the

(A) anatomic neck
(B) diaphysis
(C) epiphyseal disk
(D) epiphyseal line
(E) metaphysis

5. The muscles of the back receive motor innervation from

(A) dorsal primary rami
(B) dorsal roots
(C) posterior branches of lateral perforating nerves
(D) ventral primary rami
(E) none of the above

6. Pain of visceral origin is *not* referred to dermatomes L3 through S1 because there is an *absence* of

(A) paravertebral sympathetic ganglia below L2
(B) sympathetic efferent (motor) supply to spinal nerves L3–S1
(C) visceral afferents in the lumbar splanchnic nerves
(D) visceral afferents in the sacral section of the spinal cord
(E) white rami communicantes to spinal nerves L3–S1

Directions: Each question below contains four suggested answers of which **one or more** is correct. Choose the answer

A   if **1, 2, and 3** are correct
B   if **1 and 3** are correct
C   if **2 and 4** are correct
D   if **4** is correct
E   if **1, 2, 3, and 4** are correct

7. An incision made through the skin across the prevailing direction of the connective tissue fibers will

(1) tend to gape
(2) always provide optimum surgical exposure
(3) heal with a prominent scar
(4) parallel the lines of cleavage (of Langer)

8. The healing of a bone fracture involves

(1) formation of a cartilaginous callus about the fracture
(2) calcification of the callus
(3) bone resorption at the fracture with deposition of connective tissue across the fracture
(4) bone remodeling across the fracture

SUMMARY OF DIRECTIONS

A	B	C	D	E
1, 2, 3 only	1, 3 only	2, 4 only	4 only	All are correct

9. Correct statements concerning the epiphyseal disk in a child include which of the following?

(1) It may be mistaken for a fracture of the bone on x-ray
(2) It consists of fibrocartilage
(3) It can cause asymmetrical growth if fractured
(4) It is responsible for appositional (transverse) bone growth

10. Correct statements about torn ligaments include which of the following?

(1) They interrupt the connection between bone and muscle
(2) They can usually be visualized radiographically
(3) They, like bone, repair rapidly and strongly
(4) They destabilize the joint

11. The strength of a muscle is a function of the

(1) length of the lever arm
(2) cross-sectional area
(3) degree to which the muscle is stretched in situ
(4) length of the muscle in its fully stretched state in situ

12. The subarachnoid space between the arachnoid and pia meningeal layers contains

(1) blood upon hemorrhage of the middle meningeal artery
(2) blood upon hemorrhage of a cerebral artery
(3) blood upon hemorrhage of a cerebral vein
(4) cerebrospinal fluid

13. A white ramus communicans, which extends between an anterior primary ramus of a spinal nerve and a sympathetic trunk ganglion, contains

(1) axons of a preganglionic ganglion
(2) fibers that conduct general visceral afferent impulses
(3) myelinated fibers
(4) fibers that directly activate smooth muscle in the skin

14. Sympathetic postganglionic neurons, which join each spinal nerve via a gray ramus communicans, innervate

(1) arrector pili muscles
(2) smooth muscle of the peripheral vascular bed
(3) sweat glands and modified sweat glands
(4) skeletal muscle

15. A blood clot that develops in a peripheral vein will lodge in the vascular bed of the

(1) brain
(2) heart
(3) extremities
(4) lungs

16. Correct statements about end-arteries include which of the following?

(1) They form abundant anastomotic networks
(2) They are found around most highly movable joints
(3) They provide a high margin of safety in the event of vascular blockage
(4) They are found in vital organs

17. Valves occur in which of the following vessels?

(1) Veins of the viscera
(2) Lymphatics
(3) Veins of the extremities
(4) Cerebral veins

18. In addition to the body below the rib cage, which of the following regions usually drain via the thoracic duct?

(1) Left arm and thoracic cage
(2) Right arm and thoracic cage
(3) Left side of the face and neck
(4) Right side of the face and neck

Directions: The group of questions below consists of lettered choices followed by several numbered items. For each numbered item select the **one** lettered choice with which it is **most closely** associated. Each lettered choice may be used once, more than once, or not at all.

- **A** if the item is associated with **(A) only**
- **B** if the item is associated with **(B) only**
- **C** if the item is associated with **both (A) and (B)**
- **D** if the item is associated with **neither (A) nor (B)**

Questions 19–21

For each question, select the fascial layer from the list below that is most aptly described by it.

(A) Superficial layer of superficial fascia
(B) Deep layer of superficial fascia
(C) Both
(D) Neither

19. Membranous fascia that anchors sutures well

20. Connective tissue that overlies muscle and neurovascular structures, defining planes between them

21. Fatty layer that contains numerous blood vessels and nerves

Directions: The groups of questions below consist of lettered choices followed by several numbered items. For each numbered item select the **one** lettered choice with which it is **most** closely associated. Each lettered choice may be used once, more than once, or not at all.

Questions 22–25

For each question involving bone articulations, select the joint classification from the list below that is most appropriately associated with it.

(A) Amphiarthrosis
(B) Diarthrosis
(C) Synarthrosis
(D) Syndesmosis
(E) None of the above

22. The nonmovable cranial sutures

23. The slightly movable intervertebral disk

24. The highly mobile glenohumeral joint

25. A joint with a synovial capsule and synovial fluid

Questions 26–30

For each question describing the location of a cell body, select the most appropriate neuron from the accompanying list.

(A) Upper motor neuron
(B) Lower motor neuron
(C) Peripheral sensory neuron
(D) Central sensory neuron
(E) None of the above

26. Dorsal root ganglion

27. Dorsal horn of the spinal cord or brain stem

28. Gray matter of the cerebral cortex

29. Ventral horn of the spinal cord

30. White matter of the spinal cord or brain

ANSWERS AND EXPLANATIONS

1. The answer is A. [Chapter 1 III D 8 c] Circumduction is a combined movement involving both abduction/adduction and flexion/extension. It must, therefore, occur at joints with these two degrees of freedom.

2. The answer is A. [Chapter 2 III B 1; Chapter 5 VII C 5 b (2) (b)] Approximately 75% of the drainage of the breast is through the axillary nodes with a relatively small proportion draining through the internal thoracic (parasternal and internal mammary) nodes. Only the left axillary nodes drain into the thoracic duct.

3. The answer is A. [Chapter 2 III A 1 b] The 15–20 lobes of the breast are separated by connective tissue septa (Cooper's liagments), which run between the epidermis and the deep layer of the superficial fascia. The suspensory ligaments determine the posture of the breast.

4. The answer is C. [Chapter 3 I A 3 a (4), 4 b (2)] Fracture through the unfused cartilaginous growth plate (epiphyseal disk) in a young person may impede or terminate longitudinal growth in a long bone. The cartilaginous epiphyseal disk is generally located toward one end of a long bone between the diaphysis of compact bone and the metaphysis of trabecular bone. Fusion of the growth plate upon cessation of growth leaves the epiphyseal line.

5. The answer is A. [Chapter 4 III C 3 a, b] The muscles of the back are innervated by dorsal primary rami of the spinal nerves. The ventral primary rami, giving off lateral and ventral perforating branches, innervate the lateral and ventral regions of the body wall. The dorsal roots are sensory only.

6. The answer is E. [Chapter 4 V E 1 b, c] The chain of paravertebral sympathetic ganglia extends along the length of the vertebral column. Every spinal nerve receives a sympathetic component from an associated gray ramus communicans; white rami communicantes occur only between T1 and L2. Visceral pain fibers traveling along sympathetic pathways gain access to the spinal nerves via white rami communicantes. In those regions of the body where cervical or lumbar splanchnic nerves bring afferent fibers to paravertebral ganglia that have no associated white ramus communicans, the sensory fibers follow the sympathetic chain to reach the first white ramus and its associated spinal nerve and the corresponding level of the spinal cord. The pain is perceived as originating from (referred to) the dermatome associated with that level of the spinal cord. Thus, visceral pain of abdominal origin is not referred to levels that lack a white ramus communicans. In the pelvic region, the visceral afferents may be carried by the nervi erigentes (pelvic splanchnics) to nerves S2–S4 and subsequently to the midsacral levels of the spinal cord. Visceral pain of pelvic origin is referred to dermatomes associated with those sacral levels of the spinal cord.

7. The answer is B (1, 3). [Chapter 2 I D] Langer's cleavage lines parallel the prevailing direction of the collagenous connective tissue fibers of the skin. An incision made parallel to these crease lines will sever fewer connective tissue fibers and, therefore, will gape less and will heal with a more cosmetic scar. However, other factors must be considered, such as whether or not an incision along Langer's lines will provide adequate surgical exposure.

8. The answer is E (all). [Chapter 3 I A 4 c] The initial step in fracture healing is the formation of a cartilaginous callus about the fracture, which subsequently calcifies to provide a splint. After a few days, bone is resorbed from the fracture site, and an ingrowth of vessels and connective tissue spans the fracture. The unorganized connective tissue also calcifies, but the union is not particularly strong. Remodeling occurs along the lines of force and compression so that the new bone becomes as strong as the uninjured bone, often without a trace of the fracture.

9. The answer is B (1, 3). [Chapter 3 I A 3 a (4) (c), 4 b (2)] The epiphyseal plate in a child, composed of hyaline cartilage, is radiolucent on x-ray and, thus, appears as a fracture. The disk is the site of longitudinal bone growth, and a fracture of the disk, causing premature closure, results in asymmetrical growth.

10. The answer is D (4). [Chapter 3 I C 2 b] Ligaments connect bones and stabilize joints. Since ligaments are radiolucent, radiographic evidence of a ligamentous tear or sprain occurs only if a misalignment of bony structure accompanies such a tear. While ligaments heal by scar formation, they do not undergo the extensive remodeling seen in bone, and thus a healed ligament usually is not as strong.

11. The answer is A (1, 2, 3). [Chapter 3 II A 2 a (2), B 2, C] The strength of a muscle is a function of the cross-sectional area (the number of contractile units in parallel), the length of the lever arm (the mechanical advantage), and the degree to which the muscle is stretched in situ (the amount of interaction between contractile filaments). The relative length of a muscle in its fully stretched state (the number of contractile units in series) has no bearing on the strength.

12. The answer is C (2, 4). [*Chapter 4 II A 2 b (3)*] The subarachnoid space contains cerebrospinal fluid (CSF), which bathes and protects the brain. The large distributing cerebral arteries lie in the subarachnoid space so that hemorrhage of one of these vessels will be evident by blood in the CSF. A middle meningeal hemorrhage produces an epidural hematoma, and a cerebral vein usually produces a subdural hematoma, neither of which has access to the subarachnoid space.

13. The answer is A (1, 2, 3). [*Chapter 4 V D 1 a (1) (b), (c), E 1 b (1)*] Each white ramus communicans of the sympathetic chain is composed of myelinated presynaptic sympathetic neurons, which leave the spinal nerve and travel to a paravertebral or prevertebral ganglion. These fibers directly innervate neither the viscera nor sweat glands and smooth muscle associated with hair follicles and arterioles of the skin. Visceral afferent fibers reach the spinal nerve through the white rami.

14. The answer is A (1, 2, 3). [*Chapter 4 V D 1 a (1), b (1) (a)*] Sympathetic preganglionic neurons leave the spinal cord in segments T1 through L2, gain access to the sympathetic chain via white rami communicantes, course up and down the chain, and synapse with a postsynaptic neuron. Those postsynaptic neurons that join a spinal nerve via gray rami communicantes innervate the arrector pili muscles, sweat glands, and the smooth muscle of the peripheral vascular bed.

15. The answer is D (4). [*Chapter 5 I A 1; II A 1; Figure 5–1*] A thromboembolus originating in the peripheral venous system will lodge in the vascular bed of the lung, producing a pulmonary embolus. Emboli that arise in the pulmonary veins or in the left side of the heart may lodge in the systemic vascular beds (i.e., the brain, heart, kidneys, and extremities).

16. The answer is D (4). [*Chapter 5 III B 3*] End-arteries supply distinct regions of tissue with no collateral blood supply. Thus, there is a low margin of safety so that blockage of an end-artery results in ischemia and necrosis. End-arteries tend to be found in vital organs, such as the brain, heart, kidneys, and organs of the gastrointestinal tract.

17. The answer is A (1, 2, 3). [*Chapter 5 V B 1 c*] As a rule of thumb, the veins of the dependent portions of the body, including the extremities and viscera, have valves; the cerebral veins are valveless. Large lymphatic vessels, lymph trunks, and the thoracic duct all have valves.

18. The answer is B (1, 3). [*Chapter 5 VII C 5*] The thoracic duct, originating as the cisterna chyli in the abdomen, receives lymph from the entire body below the diaphragm as well as from the thoracic viscera, the left side of the thoracic cage, the left arm, and the left side of the face and neck. It enters the systemic venous system at the junction where the left subclavian and internal jugular veins join to form the left brachiocephalic vein. The right arm, right side of the thoracic cage, and the right side of the head and neck usually drain into the right lymph duct, which empties into the right brachiocephalic vein.

19–21. The answers are: 19-B, 20-D, 21-A. [*Chapter 2 II A, B*] The superficial fascia is divided into two layers. The superficial layer of superficial fascia (Camper's fascia) is fatty and contains abundant blood vessels, lymphatic vessels, and nerves. The deep layer of superficial fascia (Scarpa's fascia) is membranous and, while very thin, holds sutures well. The deep (investing) fascia, not superficial fascia, surrounds neurovascular bundles and muscles, providing fascial planes between these structures.

22–25. The answers are: 22-C, 23-A, 24-B, 25-B. [*Chapter 3 I C 1 a–c*] Cranial sutures are synarthroses in which little or no movement is possible and in which there is no articular cartilage. The fibrocartilaginous intervertebral disks are amphiarthroses in which the attachment of the cartilaginous connective tissue to the adjacent bones severely limits the amount of movement. The shoulder joint, like all joints with a capsule, synovial fluid, and articular cartilage capping the bones, is a diarthrosis, which allows extensive movement. A syndesmosis (a type of synarthrosis) is a broad fibrous joint, which attaches two bones with little movement permitted, such as occurs between the tibia and fibula.

26–30. The answers are: 26-C, 27-D, 28-A, 29-B, 30-E. [*Chapter 4 III B 1 b, 2 b; IV A 2, B 2*] The somatic nervous system is divided into sensory and motor divisions, each of which has central and peripheral portions. The peripheral sensory neuron, which is stimulated by receptors, has its cell body in a dorsal root ganglion or a homologous cranial nerve ganglion. The initial central sensory neuron, which receives input from the peripheral neuron, is located in the dorsal horn gray matter of the spinal cord or brain stem. On the motor side, the upper motor neuron is located in the gray matter of the cerebral cortex. It synapses with a lower motor neuron located in the ventral horn gray matter of the spinal cord or brain stem. The lower motor neuron travels in the peripheral nerves to a voluntary muscle. All central neuron cell bodies are located in gray matter; their myelinated processes form white matter.

Part II
Upper Extremity

6

Shoulder Region

I. INTRODUCTION TO THE UPPER EXTREMITY

A. Basic principles

1. **Primitive position.** The basic pattern of the upper extremity can be understood by simulating the primitive position; that is, by abducting the arm until it is horizontal with the palm facing forward so that the upper extremity approximates the pectoral fin of a fish (see Fig. 4-3).
   a. **Primitive surfaces.** The **primitively ventral surface** generally corresponds to the anterior or flexor side; the **primitively dorsal surface** generally corresponds to the posterior or extensor side.
   b. **Axial borders.** An imaginary plane drawn through the long axis of the extremity will differentiate between a cephalad **preaxial border** and a caudal **postaxial border**. While there generally is overlap between sequential dermatomes, there is *no* overlap across axial lines (see Fig. 4-3).

2. **Early terrestrial adaptation.** Flexing the elbow 90° and extending the wrist 90° from the finlike position of the extremity approximates the amphibian/reptilian upper extremity.

3. **Late terrestrial adaptation.** In mammals, humans in particular, the amphibian/reptilian brachium has rotated 90° caudally, bringing the upper extremity alongside the lateral body wall so that in the erect posture the primitively ventral surface faces anteriorly. The human upper and lower extremities are homologous, and the structural differences between them reflect different functions.
   a. The upper extremity of the human, freed from its original role of support and propulsion, has developed high intrinsic mobility.
   b. The function of the human upper extremity is to place its effector end, the hand, in a position to manipulate the environment. The hand can also be positioned so that it touches every part of the body.
   c. The hand is developed for prehension, and only the human hand has a sophisticated two-point grasp between the thumb and index finger.
   d. The hand is also a very sophisticated sensory organ.

B. Regions of the upper extremity

1. **The shoulder** consists of two bones, the **clavicle** and the **scapula**.

2. **The arm** or **brachium** contains one bone, the **humerus**.

3. **The forearm** or **antebrachium** contains two bones, the **radius** and the **ulna**.

4. **The wrist** or **carpus** contains the following eight bones: a proximal row of three carpal bones (**scaphoid**, **lunate**, and **triquetrum**) and one sesamoid bone (**pisiform**); a distal row of four carpal bones (**trapezium**, **trapezoid**, **capitate**, and **hamate**).

5. **The hand** contains the following bones: the five **metacarpals** and fourteen **phalanges** (two in the thumb and three in each finger).

C. Basic organization. The free portion of the upper extremity (arm, forearm, wrist, and hand) is suspended from the axial skeleton by the embedded portion (the pectoral girdle).

1. **Bony support.** The only bony articulation between the pectoral girdle and the axial skeleton is through the clavicle at the sternoclavicular joint.

2. **Muscular support.** The attachment of the upper extremity to the thorax is primarily muscular.
   a. The muscles of the pectoral girdle provide support and stability and produce movement.
   b. The pectoral girdle is very mobile at the expense of stability.

II. PECTORAL REGION

A. Composition. The **pectoral (shoulder) girdle** consists of the clavicles and the scapulae.

B. Bony landmarks

1. **The clavicle**, an **S**-shaped strut between the scapula and sternum, is an easily observed and palpable subcutaneous bone.

2. **The scapula**, including its spine, acromion process, coracoid process, vertebral border, and inferior angle (apex).

3. **The humerus**, including its head, greater tuberosity, intertubercular (bicipital) groove, and lesser tuberosity.

III. BONES OF THE PECTORAL GIRDLE

A. Clavicle (Fig. 6-1)

1. **Location.** The clavicle connects the scapula to the sternum. It articulates with the sternum at the **sternoclavicular joint** and with the scapula at the **acromioclavicular joint**.

2. **Ossification.** It is the first bone to begin ossification, starting in the fifth week of fetal development, but it is the last to complete ossification at about the twenty-first year. It is the only long bone formed by intermembranous ossification.

3. **Clavicular fracture.** The clavicle is one of the most commonly fractured bones in individuals below middle age.
   a. During a fall onto an outstretched arm, forces transmitted through the clavicle from the arm to the sternum produce shear near the middle of the S-shaped clavicle.
   b. Fracture of the clavicle in the middle third (the most common fracture site) results in upward displacement of the proximal fragment, due to the pull of the sternomastoid muscle, and downward displacement of the distal fragment, due to the pull of the deltoid muscle and the effect of gravity on the arm.
   c. Because major neurovascular structures lie between the clavicle and the first rib, clavicular fracture can produce nerve damage as well as severe, even fatal, internal bleeding.

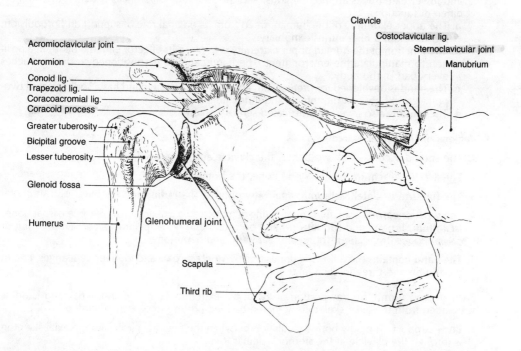

Figure 6-1. *Right pectoral girdle.* The right shoulder is shown in anterior view. The sternoclavicular and acromioclavicular joints are strongly reinforced by ligaments, while the glenohumeral joint is stabilized primarily by muscles.

B. Scapula (see Fig. 6-1)

   1. Characteristics. The scapula lies against the posterior aspect of the rib cage. It is a flat bone that is shaped like an inverted triangle with superior, vertebral, and lateral borders and an inferior apex.

   a. The spine of the scapula (subcutaneous and horizontal) is a posterior reinforcing ridge. It divides the dorsal surface into a **supraspinous fossa** and an **infraspinous fossa**.

   b. The acromion process is the lateral expansion of the spine. It articulates with the clavicle and provides an attachment for the trapezius and deltoid muscles.

   c. The coracoid process projects anteriorly. A remnant of the third bone of the pectoral girdle, it is palpable anteriorly, just medial to the head of the humerus. It provides attachment for several muscles and ligaments, one of which, the **coracoacromial ligament**, transmits tensile force from the coracoid process to the acromion process and spine of the scapula.

   d. The glenoid fossa is the site of articulation with the humerus. It faces laterally but slightly anteriorly and slightly superiorly.

     (1) Even though this shallow fossa is deepened slightly by a lip of fibrocartilage, the **glenoid labrum**, the glenohumeral joint is predisposed to dislocation.

     (2) The supraglenoid and infraglenoid tubercles provide attachments for the long heads of the biceps and triceps muscles, respectively.

   2. Articulations. The scapula articulates with the sternum, body wall, and the humerus (see IV).

C. Humerus. The proximal end of the long bone of the arm (humerus) consists of a head and two tuberosities separated by an intertubercular groove (see Fig. 6-1).

   1. The head of the humerus is covered by hyaline cartilage and articulates with the scapula at the glenohumeral (scapulohumeral) joint.

   2. The anatomic neck separates the head from the metaphysis and marks the location of an epiphyseal plate that normally fuses between the nineteenth and twenty-first years.

   3. The lesser tuberosity lies on the anterior surface of the humerus, just distal to the anatomic neck, and provides an attachment for the subscapularis muscle.

   4. The greater tuberosity lies on the lateral surface of the humerus, just lateral to the anatomic neck, and provides attachments for the supraspinatus, infraspinatus, and teres minor muscles.

   5. The intertubercular groove lies between the greater and lesser tuberosities and contains the tendon of the long head of the biceps brachii muscle. It is bridged by the **transverse humeral ligament**. The lateral and medial lips provide muscular attachments for the pectoralis major and teres major muscles, respectively, while the latissimus dorsi muscle inserts into the floor.

IV. ARTICULATIONS OF THE SHOULDER GIRDLE

A. Sternoclavicular joint (see Fig. 6-1)

   1. Structure. The clavicle articulates with the **manubrium** of the sternum, proximally. The joint has a fibrocartilaginous articular disk that divides the joint into two synovial capsules.

   2. Movement. The sternoclavicular joint has two degrees of freedom.

   a. Elevation/depression of the shoulder

   b. Protraction/retraction of the shoulder

   c. Circumduction of the shoulder (the combined result of the two pairs of movements)

   3. Support. The fibrous joint capsule is reinforced by strong ligaments so that dislocation of the sternoclavicular joint is uncommon.

   a. The anterior and posterior sternoclavicular ligaments run between the clavicle and the manubrium.

   b. The costoclavicular ligament runs between the clavicle and the first rib (see Fig. 6-1).

B. Acromioclavicular joint (see Fig. 6-1)

   1. Structure. The clavicle articulates with the acromion process of the scapula.

   2. Movement. The acromioclavicular joint is a sliding one with two degrees of freedom. Rotation of the scapula occurs primarily at this joint. Other movements at this joint are mainly accommodations to motions that occur at the sternoclavicular joint.

   3. Support. The articular capsule, thin and lax, is reinforced by several strong ligaments.

   a. The coracoclavicular ligament runs from the **coracoid process** of the scapula to the clavicle and is subdivided into **conoid** and **trapezoid ligaments**. These provide superior–inferior stability (see Fig. 6-1).

b. The acromioclavicular ligament bridges the joint and provides anterior–posterior stability.

c. Despite these ligamentous reinforcements, *acromioclavicular subluxation (shoulder separation)* is common, usually the result of downward displacement of the clavicle.

C. "Scapulothoracic joint"

1. Structure. It lies between the subscapularis muscle and the serratus anterior muscle. There is no articulation between the scapula and thoracic cage. However, it appears to function as a joint, although two of the movement pairs actually occur at the sternoclavicular joint.

2. Movement. The scapulothoracic joint has three degrees of freedom, which allow for considerable motion of the scapula on the posterolateral thoracic cage.
  a. Protraction/retraction (sternoclavicular)
  b. Elevation/depression (sternoclavicular)
  c. Rotation (acromioclavicular)

3. Support. This pseudojoint is supported by the clavicle and reinforced by the muscles that insert on the scapula from the axial skeleton, such as the rhomboids, serratus anterior, trapezius, and levator scapulae.

D. Glenohumeral (scapulohumeral) joint (see Fig. 6-1)

1. Structure. The glenoid fossa is the site of articulation with the head of the humerus (see Fig. 6-1). This shallow fossa is deepened slightly by a lip of fibrocartilage, the glenoid labrum.

2. Movement. As a ball-and-socket joint, it has three degrees of freedom.
  a. Flexion/extension
  b. Abduction/adduction
  c. Rotation

3. Ligamentous support. Its lax fibrous capsule is reinforced by tough articular ligaments.
  a. Superior, middle, and inferior glenohumeral ligaments run from the glenoid lip to the anatomic neck of the humerus.
  b. The coracohumeral ligament between the coracoid process and the humerus supports the dead weight of the free portion of the upper extremity.
  c. The coracoacromial ligament runs between the coracoid process and acromion process.
    (1) Together with the acromion process, this ligament forms the **coracoacromial arch**.
    (2) This arch buttresses the superior aspect of the glenohumeral joint to prevent superior displacement of the humerus.
    (3) This ligament transmits tensile force from the muscles that originate on the coracoid process to the acromion process and spine of the scapula.

4. Dynamic stability. This rather unstable joint is dynamically reinforced by muscle action (Figs. 6-2, 6-3, and 6-4).
  a. The rotator (musculotendinous) cuff (supraspinatus, infraspinatus, subscapularis, and teres minor muscles) acts as a dynamic "ligament," keeping the head of the humerus pressed into the glenoid fossa. Rotator cuff tears may limit shoulder mobility.
  b. The tendon of the long head of the biceps brachii muscle, passing over the humeral head en route to the supraglenoid tubercle of the scapula, also forces the humeral head medially into the joint.

5. Bursae. Several bursae lie between the various musculoskeletal components.
  a. The subacromial bursa separates the acromion process from the supraspinatus muscle (see Fig. 6-4).
    (1) This bursa is frequently the site of *subacromial bursitis* and *supraspinatus tendinitis* with calcium deposits.
    (2) Pain associated with subacromial bursitis, felt mainly during the initial stages of abduction and forward flexion, severely limits shoulder mobility.
  b. The subdeltoid bursa separates the deltoid muscle from the head of the humerus and the insertions of the rotator cuff muscles.
  c. Communications. The subacromial and subdeltoid bursae frequently communicate, but neither should communicate with the joint capsule. Such communication indicates a capsular tear.

6. Shoulder dislocation. Despite the musculotendinous cuff and other stabilizing features, the extreme mobility of the glenohumeral joint with the shallowness of the glenoid fossa result in loss of stability, which predisposes to dislocation of the humerus.

a. In *anterior dislocations*, the head of the humerus lies inferiorly to the coracoid process. The axillary nerve is sometimes injured. Dislocation stretches the anterior capsule and may avulse the glenoid labrum, which predisposes to recurrence.

b. In *posterior dislocations*, the humerus is displaced posteriorly.

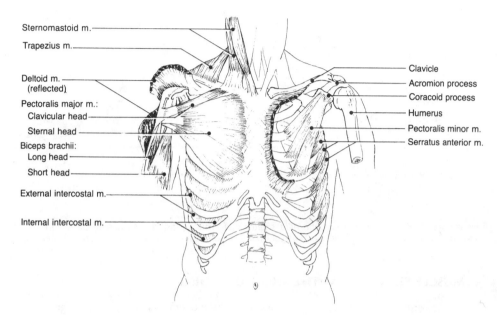

Figure 6-2. *Anterior pectoral musculature.* The right sternomastoid, trapezius, clavicular and sternal heads of the pectoralis major, and anterior portion of the deltoid, as well as the left pectoralis minor and serratus anterior, are depicted.

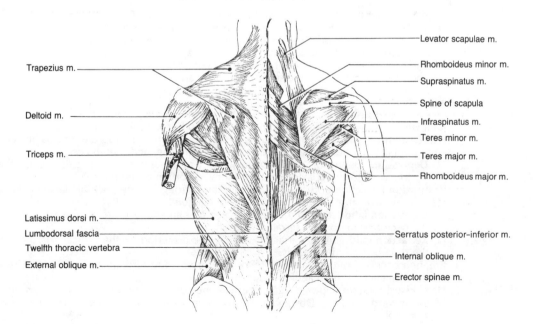

Figure 6-3. *Posterior pectoral musculature.* The left cervical and thoracic portions of the trapezius, the posterior and lateral portions of the deltoid, and the latissimus dorsi, as well as the right levator scapulae and major and minor rhomboids, are depicted.

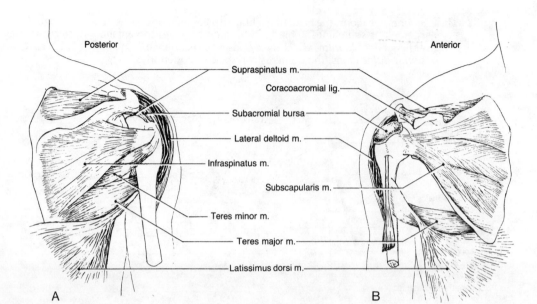

Posterior

Anterior

Supraspinatus m.

Coracoacromial lig.

Subacromial bursa

Lateral deltoid m.

Infraspinatus m.

Subscapularis m.

Teres minor m.

Teres major m.

Latissimus dorsi m.

A

B

Figure 6-4. *A, Posterior muscles of the right rotator cuff.* The supraspinatus, infraspinatus, and teres minor are shown. *B, Anterior muscles of the right rotator cuff.* The supraspinatus and subscapularis are shown.

V. MUSCLE FUNCTION AT THE SHOULDER JOINT

A. Movement of the pectoral girdle. Many of the movements of the pectoral girdle are accomplished by combinations of primitively dorsal and primitively ventral muscles that run between the axial skeleton and the bones of the pectoral girdle and arm (Table 6-1; see Figs 6-2 and 6-3).

1. **Shoulder movement** occurs at the sternoclavicular, acromioclavicular, and scapulothoracic joints (see Table 6-1).
   a. **Major posterior muscles** are arranged in two layers (see Fig. 6-3).
      (1) The superficial layer includes the **trapezius** and **latissimus dorsi**.
      (2) The deep layer includes the **levator scapulae** and the **major** and **minor rhomboids**.
   b. **Major anterior muscles** are also arranged in two layers (see Fig. 6-2).
      (1) The superficial layer includes the **pectoralis minor** and **sternomastoid**.
      (2) The deep layer includes the **serratus anterior** and **subclavius**.

2. **Group actions**
   a. **Elevation/depression** occurs through an anteroposterior axis through the sternoclavicular joint with a resultant vertical gliding at the scapulothoracic joint.
      (1) **Elevation.** Prime elevators, which pass superiorly to the anteroposterior axis, include the cervical portion of the trapezius, levator scapulae, and rhomboids.
      (2) **Depression.** Depressors, which pass inferiorly to the anteroposterior axis, include the pectoralis minor, subclavius, and the thoracic portion of the trapezius.
   b. **Protraction/retraction** occurs about a vertical axis through the sternoclavicular joint with a resultant horizontal gliding at the scapulothoracic joint.
      (1) **Protraction (abduction).** Protractors, which pass anteriorly to the vertical axis, include the pectoralis minor and serratus anterior.
      (2) **Retraction (adduction).** Retractors, which pass posteriorly to the vertical axis, include the rhomboids and trapezius.
   c. **Rotation** occurs about an anteroposterior axis through the acromioclavicular joint with a resultant rotation at the scapulothoracic joint.
      (1) **Upward rotation.** Upward rotator muscles include the trapezius and serratus anterior.
      (2) **Downward rotation.** Downward rotator muscles include the pectoralis minor and rhomboids.

3. **Group innervation.** Muscles acting on the scapula are innervated by the spinal accessory nerve, the dorsal scapular nerve, long thoracic nerve, and twigs from the brachial plexus (see Table 6-1).

Table 6-1. Muscles Acting on the Pectoral Girdle

Muscle	Origin	Insertion	Primary Action	Innervation
Anterior				
Sternomastoid	Mastoid process of cranium	Manubrium and proximal third of clavicle	Elevates sternum and clavicle; rotates the head	Spinal accessory n. (CN XI) and C2–C3
Subclavius	Costochondral junction of first rib	Middle third of clavicle	Depresses and protracts scapula	Nn. to subclavius (C5–C6, anterior)
Pectoralis minor	Outer surface of ribs 3–5	Coracoid process	Depression and protraction of scapula (elevates ribs if shoulder is fixed)	Medial pectoral n. (C8–T1, anterior)
Serratus anterior	Outer surface of ribs 1–9	Vertebral border of scapula	Protracts (abducts) and rotates scapula upward	Long thoracic n. (of Bell) [C5–C7, anterior]
Posterior				
Trapezius:				
Upper portion	Superior nuchal line, ligamentum nuchae, and spine of C7	Lateral third of clavicle and acromion process	Elevates and rotates scapula upward in elevation of arm	Spinal accessory n. (CN XI) and C3–C4
Lower portion	Spines of T1–T12	Spine of scapula	Depresses and rotates scapula upward	Spinal accessory n. (CN XI) and C2–C3
Levator scapulae	Transverse processes C1–C4	Superior portion of vertebral border of scapula	Elevates scapula	Nn. to levator scapular (C3–C4)
Rhomboids:				
Minor	Lower part of ligamentum nuchae and spines of C7–T1	Proximal portion of spine of scapula	Retract and elevate scapula	Dorsal scapula n. (C5, posterior)
Major	Spines of T2–T5	Vertebral border of scapula inferior to the spine		

B. Movement of the arm. Many of the movements at the glenohumeral joint are accomplished by combinations of primitively dorsal and primitively ventral muscles that run between the axial skeleton and the humerus (see Figs. 6-2, 6-3, and Table 6-1) as well as by muscles that run between the pectoral girdle and the humerus (Table 6-2; see Fig. 6-4).

  1. Muscles acting on the arm (see Table 6-2)
  a. Major posterior muscles are arranged in two layers (see Fig. 6-3).
   (1) The superficial layer includes the **deltoid** and **latissimus dorsi**.
   (2) The deep layer includes the **subscapularis, supraspinatus, infraspinatus, teres minor, teres major,** and **triceps brachii**.
  b. Major anterior muscles are also arranged in two layers (see Fig. 6-2).
   (1) The superficial layer includes the **pectoralis major** and **deltoid**.
   (2) The deep layer includes the **coracobrachialis** and **biceps brachii**.

  2. Group actions
  a. Abduction/adduction of the arm occurs about an anteroposterior axis through the glenohumeral joint.
   (1) Abduction. Arm abductors, which pass superiorly to the anteroposterior axis, include the supraspinatus (initial 15°) and the lateral part of the deltoid (10°–100°).
   (2) Adduction. Arm adductors, which pass inferiorly to the anteroposterior axis, include the pectoralis major, latissimus dorsi, coracobrachialis, teres major, and the anterior and posterior parts of the deltoid.

Table 6-2. Muscles Acting on the Arm

Muscle	Origin	Insertion	Primary Action	Innervation
		Anterior		
Pectoralis major:				
Clavicular head	Proximal third of clavicle	Lateral lip of bicipital groove (intertubercular sulcus)	Flexes, adducts, and medially rotates arm	Lateral pectoral n. (C5–C7, anterior)
Sternal head	Sternum and costal cartilages 1–7	Lateral lip of bicipital groove of humerus	Adducts, flexes, and medially rotates arm	Medial pectoral n. (C8–T1, anterior)
Coracobrachialis	Coracoid process of scapula	Midhumerus	Flexes, adducts, and medially rotates arm	Musculocutaneous n. (C5–C7, anterior)
Deltoid:				
Anterior part	Lateral third of clavicle	Deltoid tuberosity of humerus	Flexes, adducts, and medially rotates arm	Axillary n. (C5–C6, posterior)
Biceps brachii:				
Long head	Supraglenoid tubercle	Radial tuberosity and antebrachial fascia	Flexes shoulder and stabilizes shoulder joint	Musculocutaneous n. (C5–C7, anterior)
Short head	Coracoid process	Radial tuberosity and antebrachial fascia	Flexes shoulder	
		Posterior		
Deltoid:				
Lateral part	Acromion process of scapula	Deltoid tuberosity of humerus	Abducts arm between 10° and 100°	Axillary n. (C5–C6, posterior)
Posterior part	Spine of scapula	Deltoid tuberosity of humerus	Extends, adducts, and laterally rotates arm	
Supraspinatus	Supraspinous fossa of scapula	Greater tubercle of humerus	Elevates arm first 15° and stabilizes shoulder joint	Suprascapular n. (C5–C6, posterior)
Latissimus dorsi	Spinous processes of T6–L5, iliac crest, and ribs 10–12	Floor of bicipital groove of humerus	Extends, adducts, and medially rotates arm	Thoracodorsal n. (C6–C8, posterior)
Teres major	Inferior angle of scapula	Medial lip of bicipital groove of humerus	Extends, adducts, and medially rotates arm	Lower subscapular n. (C5–C6, posterior)
Triceps brachii:				
Long head	Infraglenoid tubercle	Olecranon process of ulnar	Extends and adducts arm	Radial n. (C6–C8, posterior)
Infraspinatus	Infraspinous fossa of scapula	Greater tubercle of humerus	Laterally rotates arm and stabilizes shoulder joint	Suprascapular n. (C5–C6, posterior)
Teres minor	Inferior angle of scapula	Greater tubercle of humerus	Laterally rotates arm and stabilizes shoulder joint	Axillary n. (C5–C6, posterior)
Subscapularis	Subscapular fossa	Lesser tubercle of humerus	Medially rotates arm and stabilizes shoulder joint	Upper and lower subscapular nn. (C5–C6, posterior)

      b. Flexion/extension of the arm occurs about a bilateral axis through the glenohumeral joint.

        (1) Flexion. Arm flexors, which pass anteriorly to the bilateral axis, include the anterior part of the deltoid, coracobrachialis, and biceps.

        (2) Extension. Arm extensors, which pass posteriorly to the bilateral axis, include the posterior part of the deltoid, latissimus dorsi, teres major, and long head of the triceps.

      c. Rotation of the arm occurs about a vertical axis through the glenohumeral joint.

        (1) Medial rotation. Medial rotators, which pass anteriorly to the vertical axis, include the subscapularis, teres major, pectoralis major, and the anterior part of the deltoid.

        (2) Lateral rotation. Lateral rotators, which pass posteriorly to the vertical axis, include the infraspinatus, teres minor, and the posterior part of the deltoid.

   3. Group innervation. The anterior muscles acting on the humerus are innervated largely by the lateral pectoral nerve, medial pectoral nerve, and the musculocutaneous nerve. The posterior muscles are innervated largely by the axillary nerve, suprascapular nerve, thoracodorsal nerve, and the upper and lower subscapular nerves (see Table 6-1).

C. Combined movements of the shoulder and arm

   1. Abduction of the arm. Approximately the first 100° occurs at the glenohumeral joint; the remaining 80° occurs by elevation and upward rotation of the pectoral girdle.

   2. Adduction of the arm. The pectoral girdle is depressed and rotated downward.

   3. Lateral rotation of the arm. The pectoral girdle is also retracted.

   4. Medial rotation of the arm. The pectoral girdle is also protracted.

VI. AXILLA

A. Configuration. The axilla is a space through which major neurovascular structures pass between the thorax and upper extremity. It is shaped like a pyramid.

   1. The apex is a triangular space limited by the first rib, the scapula, and the clavicle.

   2. The anterior wall is the **anterior axillary fold**, which overlies the pectoralis major and pectoralis minor muscles.

   3. The posterior wall is formed by the subscapularis muscle and the **posterior axillary fold**, which overlies the latissimus dorsi and teres major muscles.

   4. The broad medial wall is the serratus anterior muscle overlying the thoracic cage.

   5. The narrow lateral wall is the intertubercular (bicipital) groove.

   6. The base is formed by the skin and fascia of the axillary fossa (armpit).

B. Axillary vasculature (Fig. 6-5)

   1. Subclavian arteries

      a. Course. The **right subclavian artery** arises from the **brachiocephalic artery**; the **left subclavian artery** arises directly from the **aortic arch**. The subclavian arteries pass beneath the clavicles superiorly to the first rib between the insertions of the anterior and medial scalene muscles. On each side, the subclavian artery continues as the axillary artery once it passes the distal edge of the first rib.

      b. Branches. Each subclavian artery gives off several major distributing arteries in the base of the neck and shoulder.

        (1) The **vertebral artery** provides a primary supply to the brain.

        (2) The **internal thoracic (mammary) artery** supplies the anterior thoracic and abdominal walls.

        (3) The **thyrocervical artery (trunk)** supplies portions of the neck and shoulder. It usually gives off three significant branches.

          (a) The **inferior thyroid artery** supplies portions of the thyroid gland and larynx.

          (b) The **suprascapular artery** runs into the supraspinous fossa and passes through the **great scapular notch** superiorly to the **transverse scapular ligament** to reach the infraspinous fossa of the scapula. It anastomoses with the circumflex scapular artery and the descending (dorsal) scapular artery.

          (c) The **transverse cervical artery**, when present, crosses the posterior triangle and bifurcates into superficial and deep branches. When its branches arise separately (about 50% of the time), the transverse cervical artery does not exist.

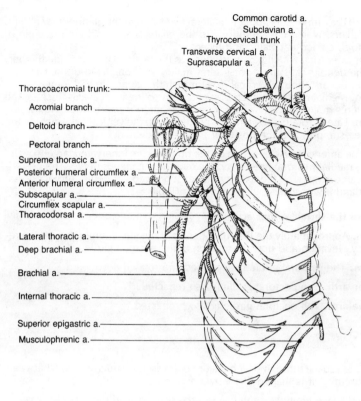

Common carotid a.
Subclavian a.
Thyrocervical trunk
Transverse cervical a.
Suprascapular a.

Thoracoacromial trunk:
Acromial branch
Deltoid branch
Pectoral branch
Supreme thoracic a.
Posterior humeral circumflex a.
Anterior humeral circumflex a.
Subscapular a.
Circumflex scapular a.
Thoracodorsal a.
Lateral thoracic a.
Deep brachial a.
Brachial a.
Internal thoracic a.
Superior epigastric a.
Musculophrenic a.

Figure 6-5. *Arterial supply to the right shoulder and proximal arm.*

      (i) The **superficial cervical branch** supplies the trapezius. This vessel may arise separately from the thyrocervical trunk.
     (ii) The **descending (dorsal) scapular branch** supplies the levator scapulae and rhomboids. This vessel may arise separately from the subclavian artery.
    (iii) The superficial cervical and descending (dorsal) scapular arteries anastomose with the suprascapular, subscapular, and circumflex scapular arteries in the vicinity of the scapular apex.
   (4) The **costocervical artery** passes back over the dome of the pleura to the neck of the first rib where it divides into deep cervical and superior intercostal arteries.

  2. Axillary artery
    a. Course. The axillary artery is the continuation of the subclavian artery from the lateral edge of the first rib. As the artery passes through the axilla, it is divided into three parts by the tendon of the overlying pectoralis minor muscle. Distally, the **axillary pulse** is palpable in the lateral wall of the axilla. The axillary artery becomes the brachial artery at the distal margin of the teres major muscle.
    b. Branches
      (1) The upper part of the axillary artery gives off the supreme (highest) thoracic artery, which supplies the more superficial subclavicular portion of the chest.
      (2) The middle part of the axillary artery gives off two branches.
        (a) The **thoracoacromial artery (trunk)** gives off four or five branches.
          (i) Pectoral branches descend to the pectoralis major and minor muscles.
         (ii) The **acromial branch** crosses to the deltoid muscle.
        (iii) A **clavicular branch** runs medially to the subclavius muscle.
        (iv) A **deltoid branch** supplies the deltoid and pectoral major muscles and accompanies the cephalic vein in the deltopectoral triangle.
        (b) The **lateral (long) thoracic artery** courses along the thoracic wall superficially to the serratus anterior muscle. It is a major supply to the mammary gland.
      (3) The lower part of the axillary artery gives off three branches. Here the anastomotic connections between the branches of the subclavian and axillary arteries around the shoulder joint are profuse and functional.
        (a) The **subscapular artery**, the largest branch, bifurcates.
          (i) The **circumflex scapular artery** curves around the lateral border of the scapula

and passes through the **triangular space** (teres major, teres minor, and long head of triceps brachii) to reach the dorsum of the scapula where it anastomoses with the suprascapular and dorsal (descending) scapular arteries.

  (ii) The **thoracodorsal artery** lies on the anterior surface of the latissimus dorsi muscle.

  (b) The **posterior humeral circumflex artery** accompanies the axillary nerve.
  (i) This vessel passes posteriorly to the humerus, through the **quadrangular space** (teres major, teres minor, long head of triceps brachii, and humerus).
  (ii) It anastomoses with the anterior humeral circumflex artery.

  (c) The **anterior humeral circumflex artery**
  (i) This small branch passes beneath the coracobrachialis muscle anteriorly to the humerus.
  (ii) It anastomoses with the posterior humeral circumflex artery.

3. Veins of the pectoral region
  a. Axillary vein
  (1) The axillary vein is formed by the joining of the **basilic vein** and one or more **brachial veins** near the the distal margin of the teres major muscle.
  (2) Usually lying medially to the axillary artery, it receives veins that correspond to the branches of the axillary artery with much variability.
  (3) It becomes the subclavian vein at the distal edge of the first rib.
  b. Cephalic vein
  (1) The cephalic vein lies in the **deltopectoral triangle**, together with the deltoid branch of the thoracoacromial artery.
  (2) While useful for insertion of intravenous lines, it is absent in approximately 10% of the population.
  (3) It terminates in the axillary vein just distal to the clavicle.

4. Lymphatic drainage
  a. Axillary lymph nodes are intercalated in the lymphatic drainage of the upper extremity and the posterior and anterior thoracic walls, including the major portion of each breast.
  b. Infection of the upper extremity or metastatic tumor from the breast may cause swollen axillary lymph nodes.

C. Brachial plexus

1. Development. The brachial plexus develops from the anterior primary rami of spinal nerves C4 through T1.

  a. As the embryonic somites migrate to form the extremities, they bring their own nerve supply, so that each dermatome and each myotome retain the original segmental innervation.
  b. With somite migration, some of the nerves come into close proximity and fuse in a particular pattern, forming a plexus.

2. Subdivisions (Fig. 6-6)
  a. Roots represent the anterior primary rami of the spinal nerves, and each innervates a specific dermatome and myotome. The roots lie between the anterior and middle scalene muscles.
  (1) Roots C3–C5 give branches to the **phrenic nerve**, which innervates the diaphragm.
  (2) Roots C5–C7 give rise to two nerves:
  (a) The **dorsal scapular nerve** (C5) innervates the rhomboid and levator scapulae muscles.
  (b) The **long thoracic nerve** (of Bell) [C5–C7] innervates the serratus anterior muscle.
  b. Trunks are formed by the joining of roots in the posterior cervical triangle.
  (1) **The upper trunk** is formed by the joining of roots C4–C6. It gives rise to two nerves.
  (a) The **suprascapular nerve** (C4–C6) passes through the suprascapular notch and beneath the transverse scapular ligament to innervate the supraspinatus and infraspinatus muscles.
  (b) The **subclavius nerve** (C5–C6) innervates the subclavius muscle.
  (2) **The middle trunk** is the continuation of root C7.
  (3) **The lower trunk** is formed by the joining of roots C8 and T1.
  c. Divisions are formed by bifurcation of the trunks deep to the clavicle.
  (1) **Posterior divisions** of the trunks are equivalent to the lateral perforating branches of the spinal nerves. These divisions innervate the primitive dorsal musculature (i.e., the extensors).
  (2) **Anterior divisions** of the trunks represent the continuations of the anterior primary rami. These divisions innervate the primitive ventral musculature (i.e., the flexors).

Netter pg 405

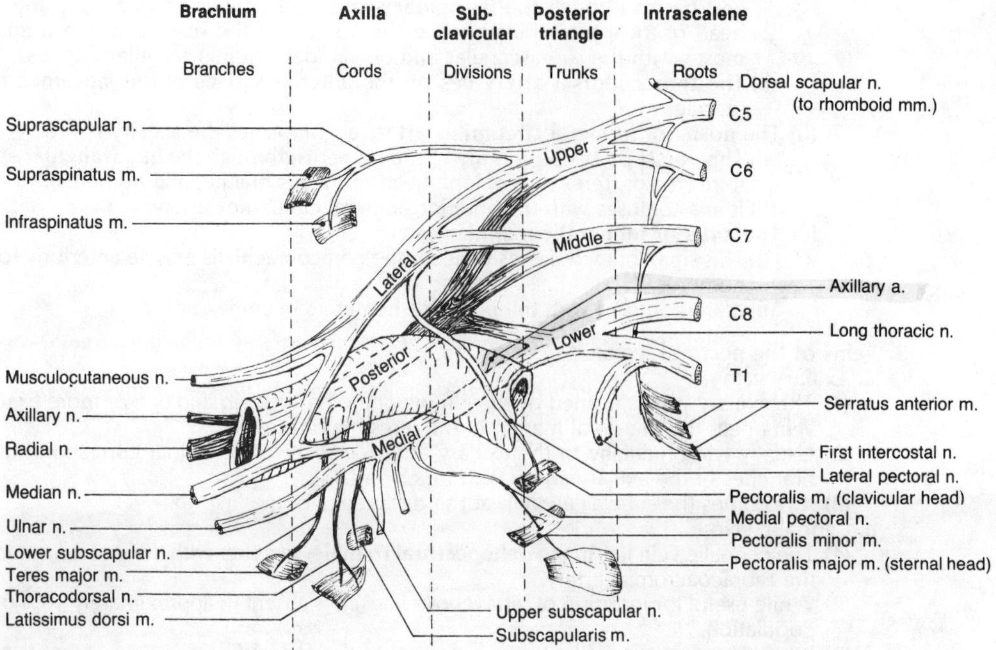

Figure 6-6. *Right brachial plexus.* The medial, lateral, and posterior cords are shown in relation to the axillary artery. The posterior division is *shaded*. The muscles innervated by branches of the roots and cords are indicated.

   d. Cords are formed by the joining of either the anterior divisions of the trunks or the posterior divisions of the trunks. They lie in the axilla beneath the pectoralis minor muscle and adjacent to the axillary artery.

     (1) The lateral cord (C4–C7) is formed by the joining of the anterior divisions of the upper and middle trunks and is named for its position relative to the axillary artery.

       (a) The **lateral pectoral nerve** (C5–C7) innervates the clavicular head of the pectoralis major muscle.

       (b) The **musculocutaneous nerve** (C4–C6) innervates the coracobrachialis, biceps brachii, and brachialis muscles before continuing as the **lateral antebrachial cutaneous nerve**.

       (c) The **median nerve** (C5–C7), to which both the medial cord and the lateral cord contribute, innervates the numerous flexor and pronator muscles of the forearm, wrist, and hand.

     (2) The medial cord (C8–T1) is the continuation of the anterior division of the lower trunk and is named for its position relative to the axillary artery.

       (a) The **medial pectoral nerve** (C8–T1) innervates the pectoralis minor and the sternal head of the pectoralis major muscles.

       (b) The **median nerve** (C6–C8), which is formed by the medial cord and the lateral cord, innervates numerous flexor and pronator muscles of the forearm, wrist, and hand.

       (c) The **ulnar nerve** (C8–T1) innervates numerous flexor muscles of the forearm, wrist, and hand.

       (d) The **medial antebrachial cutaneous nerve** (C8–T1) is sensory.

       (e) The **medial brachial cutaneous nerve** (T1) is sensory.

     (3) The posterior cord is formed by the fusion of the posterior divisions of the upper, middle, and lower trunks and is named for its position relative to the axillary artery.

       (a) The **upper subscapular nerve** (C5–C6) innervates the major portion of the subscapularis muscle.

       (b) The **thoracodorsal (middle subscapular) nerve** (C6–C8) innervates the latissimus dorsi muscle.

       (c) The **lower subscapular nerve** (C5–C6) innervates the teres major and a small portion of the subscapularis muscle.

       (d) The **axillary nerve** (C5–C6) innervates the teres minor muscle and all of the deltoid muscle. It passes through the quadrangular space.

       (e) The **radial nerve** (C5–T1) innervates all of the numerous extensors of the forearm, wrist, and hand.

3. **Lesions of the brachial plexus**
   a. **Injury to the roots** will produce not only complete paralysis of the upper extremity but a winging of the scapula due to paralysis of the long thoracic nerve. The most common cause of nerve root compression is cervical spondylosis in which the patient first complains of pain in the arm or shoulder in a dermatomal distribution.
   b. **Injury to the trunks and cords** produces distinct syndromes
      (1) **Upper trunk injury** (*Erb-Duchenne paralysis*) is the most common. Violent downward displacement of the arm, such as results from being thrown from a horse or motorcycle, may tear the 5th and 6th roots or the upper trunk.
         (a) It involves paralysis of all nerves innervated by the anterior and posterior divisions of the upper trunk.
         (b) The loss of abduction, radial flexion, and external rotation produces an upper extremity that hangs by the side in internal rotation, the ''waiter's tip'' position.
         (c) Involvement of the lateral pectoral nerve produces an inability to touch the opposite shoulder.
      (2) **Lower trunk injury** (*Klumpke's paralysis*) is less common. It is frequently caused by violent upward displacement of the arm, as in a difficult breach delivery, dislocation of the shoulder, apical tumors of the lung, a cervical rib, or scalene syndrome.
         (a) Paralysis of all nerves innervated by the anterior and posterior divisions of the lower trunk results.
         (b) There is a loss of ulnar flexion of the wrist and the use of many of the intrinsic muscles of the hand so that unopposed extension of the hand produces a ''claw-like'' position.
         (c) Involvement of the medial pectoral nerve produces an inability to adduct the arm in the lowered position against resistance.
      (3) **Posterior cord injuries**
         (a) Thoracodorsal nerve injury produces paralysis of the latissimus dorsi muscle.
         (b) Axillary nerve injury results in deltoid and teres minor paralysis (loss of shoulder abduction and weak external rotation) with loss of sensation over the deltoid muscle.
         (c) Injury to the posterior cord (*Saturday night palsy* or *crutch palsy*) or to the radial nerve results in loss of the extensors of the arm, wrist, and hand.
      (4) Injuries to the major distal nerves will be discussed subsequently.
   c. **Scalene syndrome** can be caused by spasm of the anterior and medial scalene muscles or by a cervical rib.
      (1) Spasms can compress portions of the brachial plexus, most commonly the lower trunk, causing pain along the medial border of the arm and atrophy of some of the small muscles of the hand.
      (2) Spasms can also compress the subclavian artery, causing ischemia of the arm.

the insertion of which is on the medial anterior border of the scapula).

7

Arm and Elbow Region

I. INTRODUCTION

A. Elbow region. This region consists of the distal part of the **arm (brachium)** and the proximal part of the **forearm (antebrachium).**

B. Bony landmarks (see Fig. 6-1; Fig. 7-1)

1. **The humerus,** including its head, greater tuberosity, bicipital (intertubercular) groove, shaft, lateral epicondyle, and medial epicondyle, is palpable.

2. **The ulna,** including its olecranon process and shaft, is palpable.

3. **The radius,** including its head and shaft, is palpable.

II. BONES OF THE ELBOW REGION

A. Distal humerus

1. **Characteristics.** The distal humerus flattens anteroposteriorly and broadens transversely into **medial** and **lateral supracondylar ridges,** which terminate in the **medial** and **lateral epicondyles,** all of which serve as muscle attachments.

2. **Articulations.** The distal end of the humerus has two articulations with the forearm at the elbow joint.
   a. **The trochlea** is a medial pulley-shaped articular surface that coapts the trochlear notch of the ulna to form the **humeroulnar joint** (see Fig. 7-1).
      (1) Anteriorly, proximal to the trochlea, the **coronoid fossa** receives the coronoid process of the ulna when the forearm is flexed.
      (2) Posteriorly, proximal to the trochlea, the **olecranon fossa** receives the olecranon process when the forearm is extended.
   b. **The capitulum** is an anterolateral hemispherical articular surface that coapts the head of the radius to form the **humeroradial joint** (see Fig. 7-1).

B. Proximal radius and ulna. These constitute the two bones of the forearm.

1. **Proximal ulna** (see Fig. 7-1)
   a. **Characteristics.** The proximal end of the ulna terminates in the thickened **olecranon process.**
      (1) This process provides attachment for the triceps muscle.
      (2) It is separated from the subcutaneous tissue by the **olecranon bursa,** which is prone to bursitis when subjected to repeated and prolonged pressure (*student's elbow*).
      (3) The triangular **coronoid process,** just distal to the trochlear notch, provides the insertion for the brachialis muscle.
      (4) The **radial notch,** lateral to the coronoid process, provides an articular surface for the proximal radioulnar joint.
      (5) The ulna narrows as it extends distally.
   b. **Articulations.** The deep **trochlear (semilunar) notch,** located on the anterior surface, provides an articular surface for the humeroulnar joint.

2. **Proximal radius** (see Fig. 7-1)
   a. **Characteristics**
      (1) A **cylindrical head** that is concave on the top is at the proximal end of the radius.
      (2) A **short** and **narrow neck** region lies distally to the head.

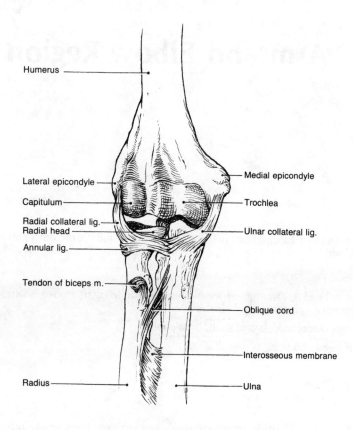

Humerus

Lateral epicondyle

Capitulum

Radial collateral lig.
Radial head

Annular lig.

Tendon of biceps m.

Radius

Medial epicondyle

Trochlea

Ulnar collateral lig.

Oblique cord

Interosseous membrane

Ulna

Figure 7-1. *Right elbow joint.* The elbow joint consists of the humeroulnar, humeroradial, and radioulnar joints and is reinforced by radial collateral and ulnar collateral ligaments and the annular ligament.

(3) The **radial (bicipital) tuberosity**, on the anteromedial aspect of the neck, receives the tendon of the biceps brachii muscle.

(4) The radius broadens as it extends distally.

b. Articulations

(1) The top of the head provides an articular surface for the humeroradial joint.

(2) The side of the head provides the articular surface for the radioulnar joint.

III. ARTICULATIONS

A. Elbow joint (see Figs. 3-2 and 7-1). The elbow region consists of three articulations between the humerus, radius, and ulna and the muscles that act across these joints to produce movement in the forearm.

1. The humeroulnar joint is formed by the trochlea of the humerus and the trochlear notch of the ulna.

a. This joint has one degree of freedom, permitting flexion/extension about a transverse axis through the trochlea.

b. It is reinforced by the **ulnar (medial) collateral ligament**, a strong, fan-shaped condensation of the fibrous joint capsule, which is composed of anterior, intermediate, and posterior fiber bundles. Sprain of this ligament permits abnormal abduction at the elbow joint.

c. The shape of the **trochlea** is variable among individuals and determines the angle (3°–29°) that the extended forearm makes with the arm, the **"carrying angle."** The carrying angle is usually greater in women than men—the anatomic reason that women tend to bowl with a curve.

2. The humeroradial joint is formed by the capitulum of the humerus and the head of the radius.

a. It has two degrees of freedom.

(1) Flexion/extension occurs about an axis through the trochlea.

(2) Pronation/supination (medial rotation/lateral rotation) occurs about an axis that passes through both centers of curvature of the proximal and distal radioulnar joints: The center of curvature of the proximal radioulnar joint is the head of the radius; that of the distal radioulnar joint is the head of the ulna.

b. It is reinforced by the **radial (lateral) collateral ligament**, a strong, fan-shaped condensation of the fibrous joint capsule, which is composed of anterior, intermediate, and posterior fiber bundles.

    (1) *Sprain* of this ligament will permit abnormal adduction at the elbow joint.

    (2) *Tennis elbow* seems to involve inflammation of this ligament, the periosteum about its insertion, or the small underlying bursa or a strain of the common extensor tendon.

3. **The proximal (superior) radioulnar joint** is formed by the side of the head of the radius and the radial notch of the ulna.

    a. It has one degree of freedom, permitting pronation/supination about an axis that passes through both the proximal and distal radioulnar joints.

    b. It is reinforced by two ligaments (see Fig. 7-1).

        (1) **The annular ligament** provides the major reinforcement.

            (a) Originating on the anterior lip of the radial notch of the ulna, it passes about the head and neck of the radius and inserts on the posterior lip of the radial notch of the ulna.

            (b) It permits rotation of the radius relative to the ulna.

        (2) **Interosseous membrane (septum)**

            (a) A broad sheet of strong connective tissue extends between the medial edge of the radius and the lateral edge of the ulna for nearly the entire length of the forearm.

            (b) Because the hand is attached primarily to the radius and because the humeroulnar joint is the most stable joint of the elbow, force is transmitted to the radius from the hand and through the interosseous membrane to the ulna and humerus.

            (c) The interosseous membrane also serves as attachment for the deep extrinsic flexor and extensor muscles of the hand.

            (d) The **oblique cord** of the interosseous membrane runs inferolaterally between the ulna, just distal to the radial notch, and the radius, just distal to the radial tuberosity.

    c. The proximal radioulnar joint functions in concert with the **distal (inferior) radioulnar joint** (see Fig. 8-1).

B. **Stability of the elbow joint**

    1. **Support.** In general, the elbow joint is very strong and stable. The coronoid process and the coaptation of the trochlea and the trochlear notch at the humeroulnar joint preclude dislocation except under extreme force or secondary to fracture. The humeroradial joint cannot withstand excessive traction in children because the radial head is undeveloped and can escape from the annular ligament, producing a *pulled elbow*.

    2. **Fractures.** The elbow joint is particularly susceptible to impacting fracture, including:

        a. Radial head fracture with or without fracture of the capitulum

        b. Fracture of the coronoid process with posterior dislocation of the elbow joint

        c. Fracture of the olecranon process or evulsion of the triceps brachii muscle insertion

IV. MUSCLE FUNCTION AT THE ELBOW JOINT

A. **Movement at the elbow joint.** Depending upon the lines of action, muscles in the arm produce movement at the glenohumeral joint, at the elbow joint, or both. Movement of the forearm at the elbow joint is accomplished by muscles that run from the pectoral girdle or the arm to the forearm or wrist (Table 7-1; Figs. 7-2 and 7-3).

    1. **The anterior (flexor) compartment** corresponds to primitively ventral musculature (see Table 7-1). Included are the **biceps brachii** and **brachialis**, which are innervated by the musculocutaneous nerve (see Fig. 7-2).

    2. **The posterior (extensor) compartment** corresponds to primitively dorsal musculature (see Table 7-1). Included are the **triceps brachii** and **anconeus**, which are innervated by the radial nerve (see Fig. 7-3).

B. **Group actions at the elbow joint**

    1. **Flexion/extension** (150°) occurs about an axis through the trochlea.

        a. **Flexion.** Flexor muscles include the biceps brachii (acting on the radius), brachialis (acting on the ulna), and brachioradialis when the forearm is semipronated. In addition, several of the wrist and hand flexor and extensor muscles that originate from the supracondylar ridges and epicondyles (see Table 9-1) serve as weak forearm flexors.

        b. **Extension.** Extensor muscles include the triceps brachii, the anconeus, and the brachioradialis when the forearm is fully supinated and nearly fully extended.

Table 7-1. Muscles Acting on the Forearm

Muscle	Origin	Insertion	Primary Action	Innervation
Flexor Compartment				
Biceps brachii:				
Long head	Supraglenoid tubercle of scapula	Radial tuberosity of the radius	Flex and supinate forearm at the elbow joint	Musculocutaneous n. (C5–C6, anterior)
Short head	Coracoid process of scapula			
Brachialis	Distal half of anterior surface of humerus	Coronoid process of ulna	Flexes forearm	Musculocutaneous n. (C5–C6, anterior)
Pronator teres:				
Long head	Medial epicondyle of humerus	Lateral suface of midradius	Pronate and weakly flex forearm	Median n. (C6–C7, anterior)
Short head	Proximal ulna			
Pronator quadratus	Anterior surface of ulna	Anterior surface of radius	Pronates forearm	Median n. (C6–C7, anterior)
Extensor Compartment				
Brachioradialis	Distal lateral surface of humerus	Styloid process of radius	Flexes, semipronates, and semisupinates forearm	Radial n. (C5–C6, posterior)
Triceps brachii:				
Long head	Infraglenoid tubercle of scapula	Olecranon process of ulna	Extend forearm at elbow joint	Radial n. (C7–C8, posterior)
Lateral head	Posterior surface of humerus			
Medial head	Posterior surface of humerus			
Anconeus	Posterior aspect of lateral epicondyle	Olecranon process	Extends forearm at elbow joint	Radial n. (C7–C8, posterior)
Supinator	Lateral epicondyle of humerus	Lateral aspect of midradius	Supinates forearm	Radial n. (C5–C7, posterior)

    2. Pronation/supination (160°) occurs about an axis that passes through both the capitulum of the humerus and the distal ulna as well as through the interosseous membrane.
        a. Pronation. Pronator muscles include the pronator teres (long and short heads) and the pronator quadratus, which is wrapped around the distal end of the ulna and acts by unwinding.
        b. Supination. Supinator muscles include the supinator, which wraps about the radius and acts by unwinding, and the biceps brachii, which inserts on the radial tuberosity. The strength of the biceps as a supinator accounts for the fact that it is easier to drive a screw into wood with the right hand than with the left hand—even for left-handed persons.

    3. Fractures
        a. In fracture of the upper third of the radius, the supinator muscles draw the proximal fragment laterally, while the pronator muscles draw the distal fragment medially.
        b. In fracture of the middle radius, there is less displacement than in proximal fractures because the action of the supinator is opposed by the pronator teres; however, the pronator quadratus still draws the distal fragment medially.

C. Group innervation

    1. Flexion of the forearm is mediated by the musculocutaneous, radial, and median nerves; thus, a lesion of any *one* of these nerves will only weaken, not abolish, flexion.

2. Extension is mediated by the radial nerve. Thus, radial nerve injury will produce an inability to extend the elbow.

3. Pronation is a function of the median nerve, although weak semipronation from a supinated position is possible through the brachioradialis muscle innervated by the radial nerve.

4. Supination is a function of the musculocutaneous and radial nerves; thus, a lesion of only one of these nerves will only weaken supination.

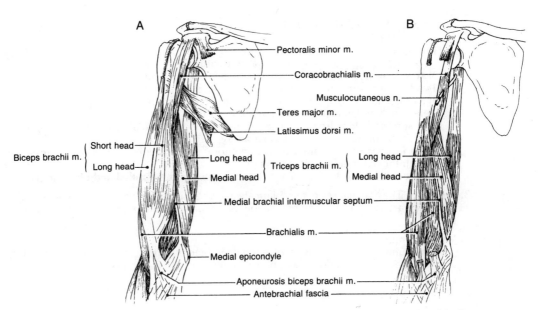

Figure 7-2. *A, Superficial muscles of the flexor compartment of the right arm. B, Deep muscles of the flexor compartment of the right arm.*

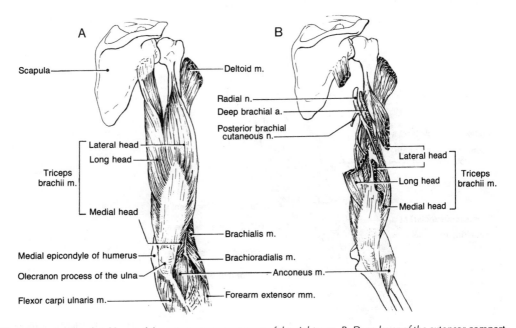

Figure 7-3. *A, Superficial layer of the extensor compartment of the right arm. B, Deep layer of the extensor compartment of the right arm.*

V. BRACHIAL VASCULATURE

A. Brachial artery (Fig. 7-4)

1. **Course.** The brachial artery, a continuation of the subclavian artery, begins at the distal edge of the teres major muscle. It descends rather superficially along the medial border of the arm. In the cubital fossa, it lies deeply to the bicipital aponeurosis, superficially to the brachialis muscle, and medially to the biceps brachii tendon. It is here that a **brachial pulse** may be palpated.

2. **Major branches**
   a. The **deep brachial (profunda brachii) artery** supplies the posterior compartment of the arm (see Fig. 7-3*B*).
      (1) Together with the radial nerve, this deep branch passes posteriorly to and then lies laterally to the humerus in the **radial (musculospiral) groove**.
      (2) A **recurrent branch** anastomoses with the posterior humeral circumflex artery.
      (3) A **lateral branch** anastomoses with the radial recurrent artery.
      (4) A **posterior branch** anastomoses with the recurrent interosseous artery.
   b. There is rich collateral circulation about the elbow joint.
      (1) A **superior ulnar branch** passes posteriorly to the medial epicondyle to anastomose with the posterior recurrent ulnar artery.
      (2) An **inferior ulnar branch** passes anteriorly to the medial epicondyle to anastomose with the anterior recurrent ulnar artery.
   c. In the cubital fossa, the brachial artery bifurcates into the **radial** and **ulnar arteries**.

3. **Variation.** The brachial artery may bifurcate anywhere in the brachium, resulting in a **superficial radial artery** (in 14% of the population) or a **superficial ulnar artery** (in 2% of the population).

4. **Vascular injury**
   a. A midhumeral fracture may involve the deep brachial artery and the radial nerve as they wind about the posterior aspect of the humerus in the radial groove.

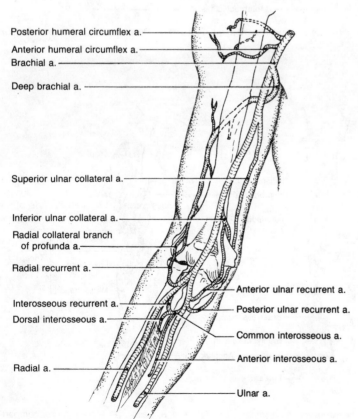

Posterior humeral circumflex a.
Anterior humeral circumflex a.
Brachial a.
Deep brachial a.
Superior ulnar collateral a.
Inferior ulnar collateral a.
Radial collateral branch of profunda a.
Radial recurrent a.
Interosseous recurrent a.
Dorsal interosseous a.
Radial a.
Anterior ulnar recurrent a.
Posterior ulnar recurrent a.
Common interosseous a.
Anterior interosseous a.
Ulnar a.

Figure 7-4. *Arteries of the arm and elbow region.*

      b. Supracondylar fractures of the humerus may sever the brachial vessels and injure the median nerve. Traction of the triceps brachii muscle draws the proximal portion of the ulnar posteriorly, while the brachialis muscle draws the distal humeral fragment anteriorly, jeopardizing the neurovascular bundle.

      c. Hemorrhage beneath the brachial fascia and antebrachial fascia may compress uninjured collateral blood vessels, thereby producing ischemia of the forearm and hand musculature with ultimate atrophy, known as *Volkmann's ischemic contracture.*

B. Venous return

    1. Deep veins of the brachium. The radial and ulnar veins of the forearm join to form usually two or three **brachial veins (venae comitantes brachiales)**, which anastomose freely about the brachial artery. These join with the basilic vein in the region of the teres major muscle to form the axillary vein.

    2. Superficial veins of the brachium lie in the subcutaneous tissue.

      a. Cephalic vein

        (1) This vein passes anteriorly to the lateral epicondyle along the anterior preaxial aspect of the brachium to the deltopectoral triangle, where it lies with the deltoid branch of the thoracoacromial artery.

        (2) It pierces the clavipectoral fascia and joins the axillary vein just distal to the first rib.

        (3) It is absent in about 10% of the population.

      b. Basilic vein

        (1) This vein passes anteriorly to the medial epicondyle.

        (2) It lies just medial to the biceps brachii along the brachium.

        (3) It penetrates the brachial fascia with the medial antebrachial cutaneous nerve.

        (4) It joins with the brachial veins near the teres major muscle to form the axillary vein.

      c. Median cubital vein

        (1) This connecting vein runs between the cephalic vein in the forearm through the cubital fossa to join the basilic vein in the arm.

        (2) It lies superficially to the bicipital aponeurosis.

        (3) It is subject to extreme variation.

C. Lymphatic drainage of the brachium

    1. Deep lymphatics accompany the brachial vein and drain through the axillary lymph nodes.

    2. Superficial lymphatics drain along the superficial veins.

      a. There are usually one or two supratrochlear lymph nodes in the distal brachium, just proximal to the medial epicondyle.

      b. The superficial drainage bypasses most of the axillary nodes, entering instead the subclavian nodes.

VI. BRACHIAL INNERVATION

A. Musculocutaneous nerve (C5–C6, anterior)

    1. Course. The musculocutaneous nerve originates from the lateral cord of the brachial plexus and contains contributions from the anterior divisions of roots C5–C6 (see Fig. 6-6). It lies between the two heads of the coracobrachialis muscle as it passes toward the lateral side of the brachium (Fig. 7-5). Just proximal to the lateral humeral epicondyle, it passes into the lateral aspect of the forearm as the lateral antebrachial cutaneous nerve.

    2. Distribution (see Fig. 7-5)

      a. Motor. This nerve innervates the flexor muscles of the arm (brachium), that is, the coracobrachialis, biceps brachii, and brachialis muscles.

      b. Sensory. The **lateral antebrachial cutaneous nerve** supplies the C6 dermatome along the preaxial (radial) side of the forearm.

    3. Musculocutaneous nerve injury in the brachium. A lesion involving the musculocutaneous nerve produces the inability to flex the forearm strongly, loss of the **biceps tendon reflex**, and loss of sensation along the lateral aspect of the forearm. Some weak elbow flexion is possible despite this injury because of the secondary flexor action of the brachioradialis (radial innervation) and the forearm muscles that originate on the medial humeral epicondyle (median and ulnar innervation).

B. Median nerve (C6–C8, anterior)

    1. Course. The median nerve originates as a lateral root from the lateral cord and a medial root

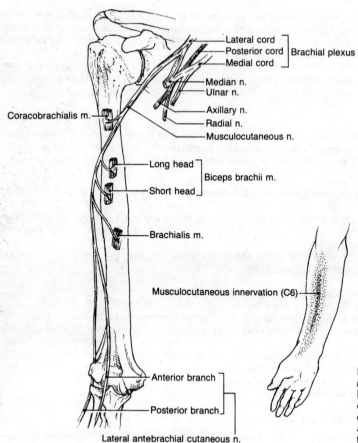

Lateral cord ⎤
Posterior cord ⎬ Brachial plexus
Medial cord ⎦

Median n.
Ulnar n.

Axillary n.
Radial n.
Musculocutaneous n.

Coracobrachialis m.

Long head ⎤
⎬ Biceps brachii m.
Short head ⎦

Brachialis m.

Musculocutaneous innervation (C6)

Anterior branch ⎤
⎬
Posterior branch ⎦

Lateral antebrachial cutaneous n.

Figure 7-5. *Musculocutaneous nerve in the arm.* The muscles innervated by branches of the nerve are indicated. Also indicated is the region of dermatomal innervation by the lateral antebrachial cutaneous nerve in the forearm.

from the medial cord. The roots join anteriorly to the third portion of the axillary artery and contain contributions from the anterior divisions of roots C6–C8 (see Fig. 6-6). The median nerve accompanies the brachial artery in the medial aspect of the brachium and gives off no branches in the brachium. It passes anteriorly to the medial epicondyle (see Fig. 9-6A) and through the cubital fossa deeply to the bicipital aponeurosis and usually (in 82% of the population) medially to the brachial artery.

2. **Distribution.** The median nerve has no motor function in the brachium. In the forearm, it supplies the pronator muscles and many of the flexors of the wrist and hand (see Fig. 9-6A). Because some of these muscles originate from the medial epicondyle, they can have a secondary flexor function at the elbow joint. This nerve has no sensory function in the brachium (see Chapter 9 IV).

3. **Median nerve injury.** A supracondylar fracture of the humerus may injure the median nerve and the brachial artery. Deficits will be in the forearm and hand. This nerve may be injured when drawing blood from the cubital fossa.

C. **Ulnar nerve** (C8–T1, anterior)

1. **Course.** The ulnar nerve is the direct continuation of the medial trunk, receiving contributions from the anterior divisions of roots C8–T1 (see Fig. 6-6). It descends rather superficially along the medial side of the brachium, lying first medially to the brachial artery then alongside the superior ulnar collateral artery. It passes posteriorly to the medial humeral epicondyle in the **ulnar groove** to enter the forearm (see Fig. 9-6B).

2. **Distribution.** The ulnar nerve gives off no branches in the brachium. In the forearm, it supplies a few of the flexors of the wrist and hand, some of which originate from the medial epicondyle and, therefore, have a secondary flexor function at the elbow joint (see Chapter 9 IV).

3. **Ulnar nerve injury.** Fracture of the medial epicondyle may produce nerve injury. Pressure to the nerve as it passes along the ulnar groove produces "funny bone" symptoms.

D. Medial brachial cutaneous nerve (C8–T1, anterior). This nerve originates from the medial cord of the brachial plexus and innervates the skin on medial side of the arm.

E. Axillary (humeral circumflex) nerve (C4–C5, posterior)

1. **Course.** The axillary nerve arises from the posterior cord of the brachial plexus, receiving contributions from the posterior divisions of roots C5–C6 (see Fig. 6-6). It runs laterally to the radial nerve and posteriorly to the axillary artery. It passes posteriorly to the humerus and anteriorly to the subscapularis muscle in company with the posterior humeral circumflex artery. It passes through the **quadrangular space** (teres minor, teres major, long head of the triceps, and surgical neck of the humerus) to reach the posterior aspect of the shoulder and brachium (see Fig. 7-6).

2. **Distribution** (Fig. 7-6)
   a. Motor. The axillary nerve supplies the teres minor and the deltoid muscles.
   b. Sensory. The nerve becomes superficial distally to the deltoid as the **superior lateral cutaneous nerve** of the arm, supplying the skin over the deltoid muscle.

3. **Axillary nerve injury.** Injury to this nerve results in deltoid paralysis with total inability to abduct the arm and severe impairment of flexion and extension at the glenohumeral joint.

F. Radial nerve (C5–T1, posterior)

1. **Course.** The radial nerve is the direct continuation of the posterior cord of the brachial plexus once the axillary nerve diverges, thus receiving contributions from the posterior divisions of roots C5–T1 (see Fig. 6-6). It runs posteriorly to the brachial artery. It passes posteriorly to the medial head of the triceps in the **musculospiral (radial) groove** of the humerus in company with the deep radial artery (see Fig. 7-3). It passes anteriorly to the lateral epicondyle between the brachialis and brachioradialis muscles to enter the forearm (see Fig. 8-4).

2. **Distribution** (see Fig. 7-6)
   a. Motor. Proximal to the musculospiral groove, medial muscular branches innervate the extensor muscles of the arm, that is, the long, medial, and lateral heads of the triceps brachii. In the vicinity of the lateral epicondyle, lateral muscular branches innervate the anconeus, brachioradialis, and extensor carpi radialis longus muscles.

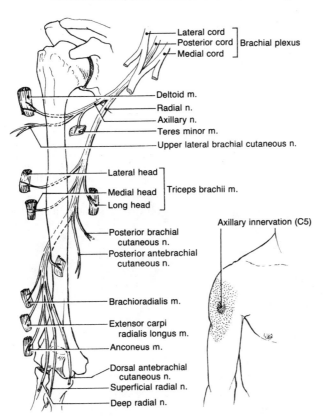

Lateral cord
Posterior cord ⎤ Brachial plexus
Medial cord ⎦

Deltoid m.
Radial n.
Axillary n.
Teres minor m.
Upper lateral brachial cutaneous n.

Lateral head ⎤
Medial head ⎥ Triceps brachii m.
Long head ⎦

Axillary innervation (C5)

Posterior brachial cutaneous n.
Posterior antebrachial cutaneous n.

Brachioradialis m.

Extensor carpi radialis longus m.

Anconeus m.

Dorsal antebrachial cutaneous n.
Superficial radial n.

Deep radial n.

Figure 7-6. *Radial and axillary nerves in the arm.* The muscles innervated by branches of these nerves are indicated. Also indicated is the region of dermatomal innervation by the axillary nerve in the arm.

b. Sensory. The **lateral brachial cutaneous nerve** supplies the preaxial border of the arm; the **posterior brachial cutaneous nerve** and **posterior antebrachial cutaneous nerve** supply the dorsal arm and forearm. These branches arise from the radial nerve in the vicinity of the musculospiral groove deep to the long head of the triceps muscle. In the vicinity of the lateral epicondyle, the superficial branch of the radial nerve arises and enters the forearm (see Chapter 8 VI).

3. **Radial nerve injury in the brachium.** Midhumeral fracture may damage the radial nerve as it lies in the musculospiral groove, causing paralysis of the extensor muscles of the wrist and hand. Because the medial muscular branches of the radial nerve to the triceps arise proximally to this groove, extension of the forearm usually is not affected by midhumeral fracture, and some supination remains possible by biceps brachii action.

8

Extensor Forearm and Posterior Wrist

I. INTRODUCTION

A. Organization. The primitively dorsal aspects of the forearm, wrist, and hand can be considered as a functional extensor unit.

B. Bony landmarks (see Fig. 7-1; Fig. 8-1)

  1. Ulna, including the shaft, head, and styloid process, is palpable.

  2. Radius, including the shaft, head, styloid process, and dorsal tubercle, is palpable.

  3. Carpal bones are all palpable.

II. BONES OF THE EXTENSOR FOREARM AND WRIST

A. Forearm. The **radius** and **ulna** are the bones of the forearm (antebrachium).

  1. Distal ulna (see Fig. 8-1)
    a. The ulna is strong at the elbow joint but becomes secondary at the wrist.
    b. At the distal end, it narrows and terminates as a slightly enlarged **head** and **styloid process**.
    c. Anterolaterally on the head, there is a surface that articulates with the ulnar notch of the radius, forming the **distal radioulnar joint**.
    d. The end of the ulna is separated from the lunate and triquetral bones by an **articular disk (triangular fibrocartilage)**.
      (1) The articular disk fills the gap between the ulna and variable portions of the triquetral and lunate bones.
      (2) This disk acts as a shock absorber, protecting the ulna and elbow joint from forces transmitted through the hand and wrist.

  2. Distal radius (see Fig. 8-1)
    a. The radius is supplemental at the elbow joint but becomes primary at the wrist. It receives most of the force transmitted from the hand and, as a consequence, is frequently fractured. **Fracture of the distal radius** (*Colles' fracture*) is surpassed in frequency only by fracture of the clavicle, ribs, and fingers.
      (1) The distal radius, composed of large amounts of cancellous bone within a thin rim of cortex, is the weakest point.
      (2) Anterior angulation of the proximal fragment may place the median nerve and radial artery in jeopardy.
    b. The distal radial shaft widens transversely. The posterior surface is grooved by the tendons of the extensor carpi radialis and the extensor pollicis longus muscles. Between these grooves is a prominent ridge, the **dorsal (radial) tubercle**.
    c. Laterally, it terminates in the **styloid process**, to which the brachioradialis muscle attaches.
    d. The **ulnar notch**, located on the medial surface of the distal head, coapts the head of the ulna and forms the **distal radioulnar joint**. The **interosseous membrane (septum)** reinforces the radioulnar joints and functions as a shock absorber. Forces transmitted to the radius from the hand are moderated by the membrane in their transmission to the ulna.
    e. The radius articulates with the proximal row of carpal bones. The slightly concave base of the radius comprises the **carpal articular surface**, which contacts the **scaphoid bone** (laterally) and the **lunate bone** (medially) to form the **radiocarpal joint** (see Fig. 9-4).

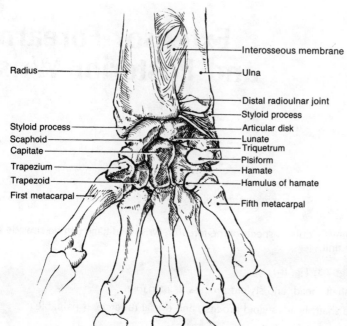

Radius

Styloid process
Scaphoid
Capitate
Trapezium
Trapezoid
First metacarpal

Interosseous membrane
Ulna
Distal radioulnar joint
Styloid process
Articular disk
Lunate
Triquetrum
Pisiform
Hamate
Hamulus of hamate
Fifth metacarpal

Figure 8-1. *Distal forearm and wrist.* Anterior view of the right forearm showing the radiocarpal, midcarpal, and carpometacarpal joints.

B. Wrist. Eight carpal bones, lying four in a row, comprise the skeleton of the wrist (corpus). The **transverse carpal arch** is formed by the nearly semicircular arrangement of the carpal bones. The transverse carpal ligament bridges the carpal tunnel.

1. **Proximal carpal bones** (see Fig. 8-1)
   a. **The scaphoid (navicular) bone** is the most lateral. It articulates with the radius proximally, with the lunate bone medially, and with the trapezium and trapezoid bones distally, thereby transmitting forces from the abducted hand to the radius. Because of its shape and location, the scaphoid is the carpal bone most prone to fracture.
      (1) Usually there is little displacement with fracture, and because the bony cortex of the scaphoid is very thin, fracture is easily missed on x-rays and can be misdiagnosed as a ligamentous sprain. If a second x-ray is obtained 10 days after fracture, following the bone resorption stage of healing, the diagnosis will be apparent.
      (2) Because the nutrient artery enters the scaphoid bone distally, fracture may destroy the blood supply to the proximal fragment and cause avascular necrosis.
   b. **The lunate bone** articulates with the radius and the triangular articular disk proximally, the scaphoid laterally, the triquetrum medially, and the capitate distally. It transmits forces from the adducted hand to the radius.
      (1) The lunate bone is not uncommonly dislocated anteriorly into the carpal tunnel, compressing the median nerve and causing the *carpal tunnel syndrome*.
      (2) The lunate bone receives its major blood supply through the anterior and posterior radiocarpal ligaments, and dislocation of the lunate can result in avascular necrosis.
   c. **The triquetral bone** (triquetrum) articulates with the articular disk proximally, the lunate bone laterally, the pisiform bone anteromedially, and the hamate bone distally.
   d. **The pisiform bone** is the most medial of the proximal row of carpal bones. It articulates only with the triquetral bone. Developmentally, it is a sesamoid bone, embedded in the tendon of the flexor carpi ulnaris muscle. Thus, the proximal carpal row can be considered to consist of three true carpal bones. It is the only proximal carpal bone with a muscular attachment.

2. **Distal carpal bones** (see Fig. 8-1)
   a. **The trapezium (greater multangular) bone** articulates with the scaphoid proximally, the trapezoid medially, and the first and second metacarpals distally. It is the most lateral of the distal row.
   b. **The trapezoid (lesser multangular) bone** articulates with the scaphoid proximally, the trapezium laterally, the capitate and the second metacarpal medially, and the second metacarpal distally.
   c. **The capitate bone** is the keystone of the **carpal arch**. It articulates with the scaphoid and

lunate bones proximally, the trapezoid laterally, the hamate medially, and the second, third, and fourth metacarpals distally. It transmits forces from the second, third, and fourth fingers to the proximal row of carpal bones. Lunate dislocation is accompanied by proximal displacement of the capitate, as a result of which the middle digit extends not much further than the second and fourth digits.

  d. **The hamate bone** is the most medial of the distal row of carpal bones. It articulates with the triquetrum proximally, the capitate laterally, and the fourth and fifth metacarpals distally. A hook-like process called the **hamulus**, which gives the hamate its name, projects toward the palm just distal to the pisiform bone.

III. TWO JOINTS OF THE WRIST

A. Radiocarpal joint

  1. **Structure.** This joint is located between the radial head and the scaphoid and lunate bones (see Figs. 8-1 and 9-4).

  2. **Movement.** The concave, ellipsoid articular surface of the distal end of the radius permits two degrees of freedom.

   a. **Abduction** (radial deviation)/**adduction** (ulnar deviation) occurs about an anteroposterior axis through the head of the capitate bone.

   (1) In abduction (15°), the scaphoid bone makes maximal contact with the radius, and the lunate contacts the articular disk.

   (2) In adduction (45°), the lunate bone makes maximal contact with the radius, and the triquetrum contacts the articular disk.

   b. **Flexion/extension** (170°) occurs about a transverse axis that passes between the lunate and capitate bones. The radiocarpal joint is most stable in full flexion where abduction/adduction is not possible.

   c. **Circumduction** of the wrist is possible because there are two degrees of freedom.

  3. **Support.** The radiocarpal joint is reinforced by several ligaments, but the high degree of mobility results in loss of strength.

   a. **The ulnar (medial) collateral ligament** connects the ulnar styloid process and the triquetrum.

   b. **The dorsal radiocarpal ligament** reinforces the dorsal side.

   c. **The radial (lateral) collateral ligament** connects the radial styloid process and the scaphoid bone.

   d. **The palmar radiocarpal ligament** reinforces the ventral side.

   e. **An ulnocarpal ligament** is on the palmar side.

   f. **The transverse carpal ligament**, from the triquetrum, pisiform, and hamulus medially to the scaphoid and trapezium laterally, bridges and maintains the **carpal arch** and forms the **carpal tunnel.**

B. Midcarpal joint

  1. **Structure.** This joint lies between the two rows of carpal bones (see Fig. 8-1).

  2. **Movement.** Small amounts of gliding movements of accommodation occur in the midcarpal joint during abduction/adduction, flexion/extension, and as the hand is flattened or hollowed.

  3. **Support.** The intercarpal joints are reinforced by numerous **dorsal intercarpal ligaments** and a palmar **carpal radiate ligament**.

IV. MUSCULATURE OF THE ANTEBRACHIAL EXTENSOR COMPARTMENT

A. Antebrachial fascia. The deep fascia of the forearm envelops the musculature of the forearm.

  1. **Organization.** Septa penetrate between the individual muscles and divide the arm into flexor and extensor compartments. Proximally, the **bicipital aponeurosis** of the biceps brachii muscle inserts into the fascial layer. The antebrachial fascia is the origin of some of the more superficial fascicles of the flexor and extensor muscles of the forearm.

  2. **In the posterior compartment**, the antebrachial fascia condenses at the wrist to form the **extensor retinaculum**, which is subdivided into six compartments by attachment in several places to bone (see Fig. 8-3). The tendons of the extrinsic extensor muscles of the wrist, hand,

and thumb pass through these compartments to gain access to the dorsum of the hand (see Fig. 8-3). On the dorsum of the hand, the antebrachial fascia and the extensor tendons fuse to form the **deep dorsal fascia**.

3. **The dorsal subcutaneous space** of the hand lies between the skin and the deep dorsal fascia. The loose connective tissue of this space accounts for the mobility of the skin on the dorsum of the hand.

4. **The subaponeurotic space** lies between the deep dorsal fascia and the deep fascia covering the dorsal interosseous muscles and metacarpal bones.

B. **Extensor compartment musculature.** Extensor muscles of the forearm are mostly "multijoint" muscles, acting across the elbow joint, the wrist joint, the carpal joints, and the metacarpophalangeal joints.

1. **Organization.** The extensor muscles of the wrist and digits originate on the lateral (preaxial) aspect of the distal arm and proximal forearm (primitively dorsal musculature).
   a. **The superficial group** (Figs. 8-2A and 8-3; Table 8-1) of extensor muscles of the wrist and hand originates from the lateral supracondylar ridge and the lateral epicondyle of the humerus, as well as from the proximal radius. They consist of the **brachioradialis** (which is usually a forearm flexor), **extensor carpi radialis longus** and **brevis**, **extensor digitorum communis**, **extensor digiti minimi**, and **extensor carpi ulnaris**.
   b. **The deep group** (Fig. 8-2B; see Fig. 8-3 and Table 8-1) of muscles originates from the mid-radius, the interosseous membrane, and the ulna. They consist of the **supinator** (which supinates the forearm), **abductor pollicis longus**, **extensor pollicis brevis**, **extensor pollicis longus**, and **extensor indicis proprius**.

2. **Group actions at the wrist.** The action of the muscles at the radiocarpal joint is determined by the location of the tendon relative to the two axes of rotation.
   a. **Flexion/extension** occurs about a transverse axis (see L–M in Fig. 9-4).
      (1) **Extensors of the wrist** are the muscles that pass dorsally to the transverse axis of the radiocarpal joint. The prime extensors are the extensor carpi radialis longus and the extensor carpi ulnaris.
      (2) **Flexors of the wrist** are the muscles that pass ventrally to this axis (see Chapter 9 II).
   b. **Abduction/adduction** occurs about an anteroposterior axis (see A–P in Fig. 9-4).
      (1) **Abductors of the wrist** are the muscles that pass on the radial side of the anteroposterior axis of the radiocarpal joint. The prime abductors are the extensor carpi radialis longus and flexor carpi radialis muscles.

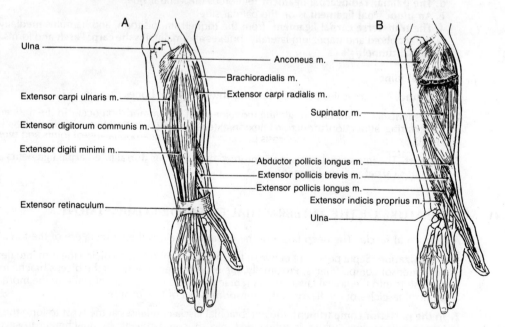

Figure 8-2. *A, Superficial layer of the extensor compartment of the right forearm. B, Deep layer of the extensor compartment of the right forearm.*

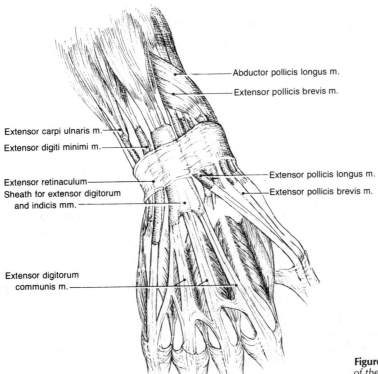

Abductor pollicis longus m.

Extensor pollicis brevis m.

Extensor carpi ulnaris m.

Extensor digiti minimi m.

Extensor pollicis longus m.

Extensor pollicis brevis m.

Extensor retinaculum

Sheath for extensor digitorum and indicis mm.

Extensor digitorum communis m.

Figure 8-3. *Extrinsic extensor tendons of the dorsum of the right hand.*

    (2) Adductors of the wrist are the muscles that pass on the ulnar side of this axis. The prime adductors are the extensor carpi ulnaris and flexor carpi ulnaris muscles.

    c. Dynamic stabilization of the wrist is performed by the actions of the extensor carpi radialis longus and brevis, the extensor carpi ulnaris, and the major flexors of the wrist.

    3. Group actions in the hand. Extension of the digits at the metacarpophalangeal joint (excluding the thumb) is accomplished by the extensor digitorum communis and the extensor digiti minimi of the superficial group as well as by the extensor indicis proprius of the deep group.

    a. Extensor tendons are held in place at the wrist by several **extensor retinacula** (see Fig. 8-3). To facilitate movement as the tendons pass beneath the extensor retinacula, the tendons are encased in **synovial sheaths**. These sheaths may become inflamed secondary to the widespread synovial inflammation that accompanies rheumatoid arthritis, resulting in *subacute tenosynovitis*. When inflammation invades the extensor tendons, they may rupture.

    b. The extrinsic extensor muscles, as well as the interosseous and lumbrical muscles, insert into the tendinous **extensor aponeurosis (dorsal expansion** or **hood)** of each digit (see Fig. 10-3). Each extensor aponeurosis passes over the metacarpophalangeal joints and then trifurcates.

    (1) The central portion of the extensor aponeurosis passes over the proximal interphalangeal joint and inserts into the base of the middle phalanx.

    (2) The two lateral bands pass over the proximal interphalangeal joint and the distal interphalangeal joint to insert into the base of the distal phalanx.

    c. The primary action of the extrinsic extensor muscles is to extend the metacarpophalangeal joints; secondarily, extrinsic extensors assist extension of the interphalangeal joints.

V. ANTEBRACHIAL VASCULATURE (see Chapter 9 III)

VI. INNERVATION OF THE POSTERIOR COMPARTMENT

    A. Radial nerve (C5–T1, posterior). Unlike the flexor side of the forearm, the extensor side of the forearm (Fig. 8-4) is innervated solely by the radial nerve.

Table 8-1. Forearm Muscles Acting on the Dorsal Side of the Wrist and Hand

Muscle	Origin	Insertion	Primary Action	Innervation
Superficial Group				
Brachioradialis	Distal lateral surface of humerus	Styloid process of radius	Flexes, semipronates, and semisupinates forearm	Superficial radial n. (C5–C6, posterior)
Extensor carpi radialis: Longus	Lateral epicondyle of humerus	Base of second metacarpal	Extends and abducts wrist	Superficial radial n. (C6–C7, posterior)
Brevis		Base of third metacarpal	Extends wrist	
Extensor carpi ulnaris	Lateral epicondyle of humerus	Base of fifth metacarpal	Extends and adducts wrist	Radial n. (C7–C8, posterior)
Extensor digitorum communis	Lateral epicondyle of humerus	Phalanges two and three of index, middle, and ring fingers	Extends MP joint and, when fist is clenched, extends wrist	Radial n. (C7–C8, posterior)
Extensor digiti minimi	Lateral epicondyle of humerus	All phalanges of fifth digit	Extends fifth digit	Radial n. (C7–C8, posterior)
Deep Group				
Abductor pollicis longus	Posterior interosseous membrane and ulna	Base of first metacarpal, laterally	Abducts thumb and wrist	Radial n. (C8–T1, posterior)
Extensor pollicis: Brevis	Posterior midshaft of radius and interosseous membrane	Base of first phalanx of thumb	Extend thumb and abduct wrist	Radial n. (C8–T1, posterior)
Longus	Posterior surface of interosseous membrane and posterior ulna	Base of second phalanx of thumb		
Extensor indicis proprius	Interosseous membrane and ulna	Phalanges two and three of index finger	Extends first digit and wrist	Radial n. (C8–T1, posterior)

1. **Course.** After leaving the musculospiral groove in the brachium, the radial nerve passes anteriorly to the lateral humeral epicondyle and between the brachialis and brachioradialis muscles (see Fig. 8-4). It sends branches to the anconeus, brachioradialis, and extensor carpi radialis longus muscles and divides into the superficial and deep branches before entering the forearm.

2. **Distribution**
   a. **Motor.** The **deep (posterior or dorsal interosseous) branch of the radial nerve** passes through the supinator muscle and winds around the radius to lie immediately dorsal to the interosseous membrane. It innervates the remaining extensor muscles of the posterior (extensor) compartment of the forearm (see Fig. 8-4). Often the **superficial branch** supplies the extensor carpi radialis brevis.
   b. **Sensory.** The **superficial branch of the radial nerve** usually arises at the level of the lateral epicondyle and supplies the lateral aspect of the dorsum of the wrist and hand with a region of exclusivity between the thumb and index finger (see Fig. 8-4). Unless it supplies the extensor carpi radialis brevis, this branch has no motor contribution.

B. **Radial nerve injury in the forearm.** Injuries involving the radial nerve produce signs that are dependent upon the level of the lesion.

1. **Injury proximal to the epicondyles**, such as by a supracondylar or epicondylar humeral fracture, involves both superficial and deep branches. The result is pronation of the hand (loss of

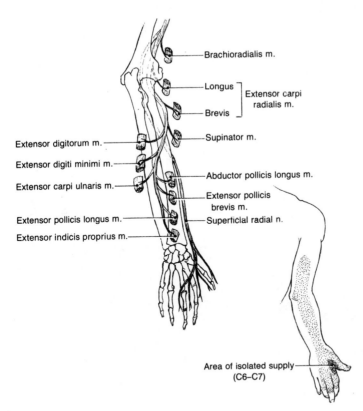

Brachioradialis m.

Longus ⎤
Brevis ⎦ Extensor carpi radialis m.

Supinator m.

Extensor digitorum m.

Extensor digiti minimi m.

Extensor carpi ulnaris m.

Abductor pollicis longus m.

Extensor pollicis brevis m.

Extensor pollicis longus m.

Superficial radial n.

Extensor indicis proprius m.

Area of isolated supply (C6–C7)

Figure 8-4. *Radial nerve in the forearm.* The muscles innervated by the branches of the radial nerve are indicated. Also depicted is the region of dermatomal innervation by the superficial branch of the radial nerve.

supination), wristdrop (inability to extend the wrist), inability to extend the digits and thumb, and loss of sensation to the dorsum of the hand and thumb.

2. **Injury distal to the epicondyles**, such as by fracture of the proximal third of the radius, usually involves only the deep branch and produces wristdrop with inability to extend the digits and thumb but without loss of supination or of sensation to the dorsum of the hand.

9

Flexor Forearm
and Anterior Wrist

I. INTRODUCTION

A. Organization. The **primitively ventral aspects** of the forearm, wrist, and hand can be considered a functional flexor unit.

B. Bony landmarks are in Chapter 8 I B. **Bones of the forearm** and **wrist** are in Chapter 8 II A, B.

II. MUSCULATURE OF THE ANTEBRACHIAL FLEXOR COMPARTMENT

A. Antebrachial fascia. The deep fascia of the forearm envelops the musculature of the forearm (see Chapter 8 IV A) and also divides the arm into flexor and extensor compartments.

 1. **At the wrist**, the antebrachial fascia condenses anteriorly to form the **flexor retinaculum**, which consists of two layers.
   a. **The volar carpal ligament** is superficial and separates the palmaris longus muscle from the underlying ulnar nerve and artery (see Fig. 9-4) and defines the tunnel of Guyon.
   b. **The transverse carpal ligament** (Fig. 9-1) is deeper and bridges the **carpal tunnel** (see Chapter 8 II B).
     (1) The **transverse carpal arch** is formed by the nearly semicircular arrangement of the carpal bones.
     (2) The transverse carpal ligament runs from the triquetrum, pisiform, and hamulus to the scaphoid and trapezium and maintains the concave carpal arch.
     (3) This ligament prevents the tendons of the extrinsic flexor muscles of the hand from bowstringing upon flexion.
     (4) Beneath it run the **median nerve**, the **flexor digitorum superficialis**, the **flexor digitorum profundus**, and the **flexor pollicis longus**.
 2. **In the hand**, the antebrachial fascia continues as the thickened **palmar aponeurosis**.

B. Flexor compartment musculature

 1. **Organization.** The flexor muscles of the wrist and hand originate from the postaxial (medial) aspect of the distal arm and proximal forearm and correspond to primitive ventral musculature. These are mostly "multijoint" muscles, acting across the elbow joint and the wrist joint, and, in some instances, the carpal joints, the metacarpal joints, and the interphalangeal joints.
   a. **The superficial group** (Fig. 9-2A; Table 9-1) includes the **pronator teres** (a pronator of the forearm), **flexor carpi radialis**, **palmaris longus** (missing in 13% of forearms), and **flexor carpi ulnaris**.
     (1) The superficial muscles originate from the medial supracondylar ridge and medial epicondyle of the humerus as well as from the anterior aspect of the proximal ulna.
     (2) These muscles constitute the primary flexors of the wrist.
   b. **The intermediate group** (see Fig. 9-2B and Table 9-1) consists of the **flexor digitorum superficialis**.
     (1) It originates from the medial epicondyle, proximal ulna, and proximal radius. The median nerve and ulnar artery pass between the humeroulnar and radial heads of this muscle. The flexor digitorum superficialis divides into four tendons, which pass beneath the flexor retinaculum. Each divides to insert into the base of the middle phalanx of the second through fifth digits.
     (2) This muscle is the primary flexor of the *proximal* interphalangeal joint.

79

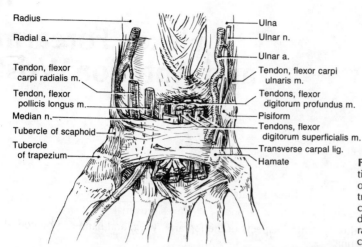

Radius

Radial a.

Tendon, flexor carpi radialis m.

Tendon, flexor pollicis longus m.

Median n.

Tubercle of scaphoid

Tubercle of trapezium

Ulna

Ulnar n.

Ulnar a.

Tendon, flexor carpi ulnaris m.

Tendons, flexor digitorum profundus m.

Pisiform

Tendons, flexor digitorum superficialis m.

Transverse carpal lig.

Hamate

Figure 9-1. *Carpal tunnel.* The relations of the extrinsic flexor tendons of the hand as they pass beneath the transverse carpal ligament are indicated. Also, the relations of the median and ulnar nerves, as well as the radial and ulnar arteries, are indicated.

    c. The deep group (Fig. 9-3*A*; see Table 9-1) consists of the **flexor digitorum profundus, flexor pollicis longus**, and **pronator quadratus**.

        (1) The deep muscles originate from the proximal ulna, the anterior surface of the interosseous membrane, and the anterior midradius. The tendons pass deeply to the flexor retinaculum (see Fig. 9-3*B*).

        (2) The flexor digitorum profundus passes through the split in the flexor digitorum superficialis tendons to insert into the base of the *distal* phalanx. It is the primary flexor of the distal interphalangeal joint.

        (3) The flexor pollicis longus inserts onto the distal phalanx of the thumb and is the primary flexor of the distal phalanx.

        (4) The pronator quadratus is a pronator of the forearm.

  2. Group actions at the wrist. The action of each muscle at the radiocarpal joint is determined by the location of the tendon relative to the two axes of rotation (Fig. 9-4).

    a. Flexion/extension occurs about a transverse axis (see L–M in Fig. 9-4).

        (1) Flexors of the wrist are the muscles of the anterior compartment, which pass ventrally to this axis. The prime flexors of the wrist are the flexor carpi radialis and flexor carpi ulnaris.

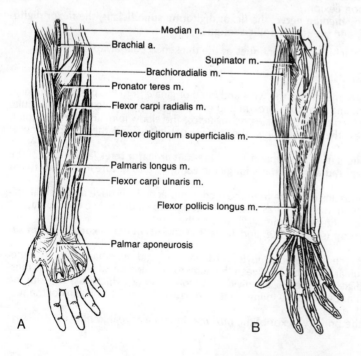

Median n.

Brachial a.

Supinator m.

Brachioradialis m.

Pronator teres m.

Flexor carpi radialis m.

Flexor digitorum superficialis m.

Palmaris longus m.

Flexor carpi ulnaris m.

Flexor pollicis longus m.

Palmar aponeurosis

A   B

Figure 9-2. *A, Superficial muscles of the flexor compartment of the right forearm. B, Intermediate layer of muscles of the flexor compartment of the right forearm.*

Table 9-1. Forearm Muscles Acting on the Palmar Side of the Wrist and Hand

Muscle	Origin	Insertion	Action	Innervation
Superficial Group				
Flexor carpi radialis	Medial epicondyle of humerus	Bases of second and third metacarpals	Flexes wrist and weakly flexes forearm	Median n. (C6–C7, anterior)
Palmaris longus	Medial epicondyle of humerus	Palmar aponeurosis	Flexes wrist	Median n. (C7–C8, anterior)
Flexor carpi ulnaris	Medial epicondyle of humerus	Pisiform bone and base of fifth metacarpal	Flexes wrist and weakly flexes forearm	Ulnar n. (C8–T1, anterior)
Intermediate Group				
Flexor digitorum superficialis	Medial epicondyle of humerus	Base of second phalanx of second through fifth digits	Flexes proximal interphalangeal joint, metacarpophalangeal joint, and wrist	Median n. (C7–T1, anterior)
Deep Group				
Flexor digitorum profundus	Anterior proximal ulna and interosseous membrane	Base of third phalanx of second through fifth digits	Flexes distal and proximal interphalangeal joint, metacarpophalangeal joint, and wrist	Heads for second and third digits: median n. (C7–T1, anterior) Heads for fourth and fifth digits: ulnar n. (C8–T1, anterior)
Flexor pollicis longus	Anterior mid-radius and interosseous membrane	Lateral aspect base of second phalanx of thumb	Flexes thumb, metacarpophalangeal joint, and wrist	Median n. (C7–T1, anterior)

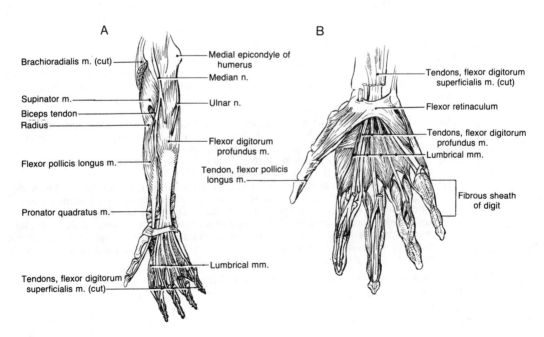

Figure 9-3. *A, Deep muscles of the flexor compartment of the right forearm. B, Extrinsic flexor tendons of the right hand.*

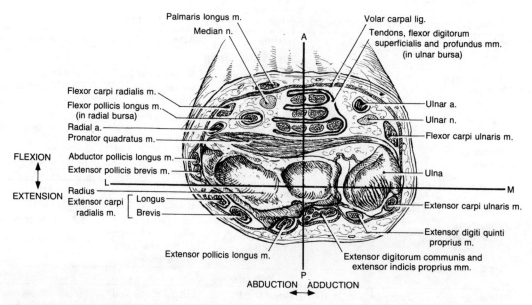

Palmaris longus m.
Median n.
A
Volar carpal lig.
Tendons, flexor digitorum superficialis and profundus mm. (in ulnar bursa)

Flexor carpi radialis m.
Flexor pollicis longus m. (in radial bursa)
Radial a.
Pronator quadratus m.

Ulnar a.
Ulnar n.
Flexor carpi ulnaris m.

FLEXION

EXTENSION

Abductor pollicis longus m.
Extensor pollicis brevis m.
L
Radius
Extensor carpi radialis m. { Longus / Brevis }

Ulna

M

Extensor carpi ulnaris m.

Extensor digiti quinti proprius m.

Extensor pollicis longus m.

Extensor digitorum communis and extensor indicis proprius mm.

P
ABDUCTION   ADDUCTION

Figure 9-4. *Extrinsic flexor and extensor tendons of the wrist.* The right wrist is in the anatomic position. The transverse axis (*L–M*) and anteroposterior axis (*A–P*) of the radiocarpal joint are indicated. Muscles that lie to the lateral side of the *A–P* axis abduct; those to the medial side adduct. Muscles that lie to the palmar side of the *L–M* axis flex; those to the dorsal side extend.

> (2) **Extensors of the wrist** are the muscles of the extensor compartment, which pass dorsally to the transverse axis of the radiocarpal joint.
> b. **Abduction/adduction** occurs about an anteroposterior axis (see A–P in Fig. 9-4).
>> (1) **Abductors of the wrist** are muscles that pass on the radial (lateral) side of the anteroposterior axis (see A–P in Fig. 9-4). The prime abductors of the wrist are the flexor carpi radialis and the extensor carpi radialis longus.
>> (2) **Adductors of the wrist** are muscles that pass on the ulnar (medial) side of this axis. The prime adductors of the wrist are the flexor carpi ulnaris and the extensor carpi ulnaris.
> c. The wrist is dynamically stabilized by the actions of the flexor carpi radialis, the flexor carpi ulnaris, and the major extensors of the wrist.

3. **Group actions on the fingers.** The flexor tendons are held in place at the wrist by the **flexor retinaculum** (**volar** and **transverse carpal ligaments**), which bridges the carpal arch to form the carpal tunnel (see Fig. 9-1 and 9-3*B*). To facilitate movement as the tendons pass beneath the retinacula, the tendons are encased in **synovial sheaths**. These tendon sheaths may be involved in generalized synovial inflammatory processes or by spread of infection from the fascial compartments of the palm or fingers.
> a. **Flexion/extension** occurs at the metacarpophalangeal joints and interphalangeal joints.
>> (1) Flexion of the digits (excluding the thumb) at the proximal interphalangeal joints is accomplished primarily by the flexor digitorum superficialis.
>> (2) Flexion of the digits (excluding the thumb) at the distal interphalangeal joints is accomplished primarily by the flexor digitorum profundus.
>> (3) Flexion at the metacarpophalangeal joint is accomplished in part by the flexor digitorum superficialis and profundus when the extensor digitorum communis is relaxed; otherwise, these muscles act in concert to stabilize the metacarpophalangeal joint.
> b. The tension-generating capacity of the long flexors of the digits is maximal when the wrist is extended, that is, in the stretched (''rest-length'') position. This explains why a child can be made to release an object or an assailant can be made to release a knife (not recommended) by flexing the wrist.

4. **Group innervation.** The muscles of the flexor compartment are innervated by the median and ulnar nerves.

III. ANTEBRACHIAL VASCULATURE

A. **Arterial supply.** The **radial artery** and **ulnar artery** arise from the bifurcation of the **brachial artery** in the cubital fossa.

1. Radial artery (Fig. 9-5)
  a. Course. This major vessel crosses the biceps brachii tendon deep to the bicipital aponeurosis, passes superficially to the pronator teres muscle, and descends along the anterior
  preaxial border of the forearm. The **radial pulse** may be palpated at the wrist between the
  tendons of the brachioradialis and the flexor carpi radialis muscles.
  b. Branches
    (1) The **radial recurrent branch** returns anteriorly to the lateral epicondyle to anastomose
       with the radial (anterior) collateral of the deep brachial artery. Numerous muscular
       branches are given off. Distally, the **palmar carpal branch** contributes to the anterior
       and posterior carpal networks (rete).
    (2) The **superficial palmar branch** provides the smaller radial contribution to the **super-
       ficial palmar arch**.
    (3) The main stem of the radial artery continues deeply to the volar (palmar) carpal liga-
       ment in the floor of the "anatomical snuff-box," toward the dorsal aspect of the hand.
       At the base of the first metacarpal, it passes between the two heads of the first dorsal in-
       terosseous muscle and becomes the major contributor to the **deep palmar arch**.

2. Ulnar artery (see Fig. 9-5)
  a. Course. This major vessel passes deeply to the pronator teres muscle and deviates toward
  the anterior postaxial border of the forearm. The ulnar artery continues deeply to the volar
  carpal ligament and superficially to the transverse carpal ligament as the major contributor
  to the **superficial palmar arch**. The **ulnar pulse** is palpable just to the radial side of the pisiform bone. On occasion, the ulnar artery may be very small or even absent.
  b. Branches
    (1) The **anterior ulnar recurrent branch** returns anteriorly to the medial epicondyle to
       anastomose with the inferior ulnar collateral of the brachial artery.
    (2) The **posterior ulnar recurrent branch** returns posteriorly to the medial epicondyle to
       anastomose with the superior ulnar collateral of the brachial artery.
    (3) The **common interosseous artery** bifurcates almost immediately.

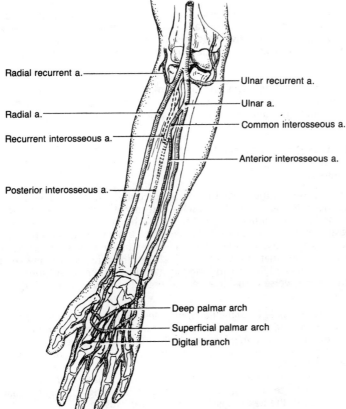

Radial recurrent a.—

Radial a.—

Recurrent interosseous a.—

Posterior interosseous a.—

Ulnar recurrent a.

Ulnar a.

Common interosseous a.

Anterior interosseous a.

Deep palmar arch

Superficial palmar arch

Digital branch

Figure 9-5. *Arterial supply of the forearm.*

(a) The **anterior interosseous artery** descends with the anterior interosseous nerve and passes through the interosseous membrane distally to anastomose with the posterior interosseous artery.

(b) The **posterior interosseous artery** returns through the interosseous membrane proximally.

  (i) The recurrent interosseous branch passes behind the elbow to anastomose with the posterior branch of the deep brachial artery.

  (ii) The posterior interosseous artery continues in the forearm and anastomoses with the anterior interosseous artery at the wrist.

(4) Near the pisiform bone, the ulnar artery gives off contributions to the anterior and posterior carpal networks (rete).

(5) At the wrist, the **deep palmar artery** arises and contributes to the deep palmar arch.

B. Venous return

1. **Superficial veins** are extremely variable, but generally they drain more of the distal forearm and hand than do the deep veins.

   a. **The cephalic vein** originates from a venous plexus on the radial (preaxial) side of the dorsum of the hand. It courses proximally, posterior to the styloid process of the radius. In the distal third of the forearm, it comes to lie anteriorly along the preaxial border to reach the elbow region. Just distal to the cubital fossa, it gives rise to the **median cubital vein**, which crosses the cubital fossa to join the basilic vein just superior to the medial epicondyle.

   b. **The basilic vein** originates from a venous plexus on the ulnar (postaxial) side of the dorsum of the hand. The basilic vein courses proximally, posterior to the ulna. In the middle third of the forearm, it comes to lie anteriorly along the postaxial border to reach the elbow region. Just below the medial epicondyle, it usually receives the **median (anterior) antebrachial vein**. Above the medial epicondyle, it receives the **median cubital vein**.

   c. **The median (anterior) antebrachial vein** is very variable, frequently consisting of a venous plexus. It drains the palm of the hand and anterior aspect of the forearm. It usually drains into the basilic vein, inferior to the medial epicondyle, but it may drain into the median cubital vein.

2. **Deep veins** accompany the named arteries often as venae commitantes.

3. **Clinical considerations. Phlebotomy**, or **venipuncture**, is easily accomplished on the superficial veins of the cubital fossa, but care must be taken to stay superficial to the bicipital aponeurosis. A snug, but not tight, tourniquet will occlude the superficial venous return, causing distension of the superficial veins. If a tourniquet is too tight, the arterial supply is occluded so that no venous return occurs.

C. Lymphatic drainage of the forearm. Superficial and deep lymphatics accompany the superficial and deep veins. A supratrochlear node lies just proximal to the medial epicondyle adjacent to the basilic vein. It receives drainage from the hypothenar aspect of the hand.

IV. INNERVATION OF THE ANTERIOR COMPARTMENT

A. Median nerve (C5–C7, anterior)

1. **Course.** After passing anteriorly to the medial humeral epicondyle in the cubital fossa, the median nerve lies anteriorly to the brachialis muscles and medially to the brachial artery, which is immediately medial to the tendon of the biceps brachii muscle (see Fig. 9-2A). The median nerve then passes between the humeral and ulnar heads of the pronator teres muscle and then between the humeroulnar and radial heads of the flexor digitorum superficialis muscle briefly in company with the ulnar artery (see Fig. 9-2B). At the wrist, it lies between the tendons of the palmaris longus and flexor carpi radialis muscles. It then passes through the carpal tunnel beneath the transverse carpal ligament to enter the hand (see Fig. 9-1).

2. **Distribution**

   a. **Motor.** The median nerve supplies all of the flexor muscles and pronator of the forearm except the flexor carpi ulnaris and the ulnar half of the flexor digitorum profundus.

   (1) **Muscular branches** arise in the cubital fossa, which innervate the superficial and intermediate groups of flexor muscles (i.e., pronator teres, flexor carpi radialis, palmaris longus, and flexor digitorum superficialis).

   (2) The **anterior interosseous branch** arises as the median nerve passes through the pronator muscle. It innervates the deep group (i.e., flexor pollicis longus, pronator quadratus, and the radial half of the flexor digitorum profundus muscles).

   b. Sensory. The **superficial palmar branch** arises just proximal to the carpal tunnel to supply
   the lateral proximal palmar surface of the hand.

3. **Median nerve injury in the forearm.** Injuries involving the median nerve produce signs that
   are dependent upon the level of the lesion.
   a. **Injury proximal to the epicondyles,** such as by a supracondylar or epicondylar humeral
      fracture, produces a supinated forearm, very weak flexion and abduction of the wrist, pa-
      ralysis of most of the muscles of the thenar side of the hand, and loss of sensation on the la-
      teral side of the palm.
   b. **Superficial injury at the wrist,** such as by superficial lacerations, may sever the superficial
      branch of the median nerve with loss of sensation on the thenar side without loss of sensa-
      tion along the major portions of the volar side of the first, second, and third fingers.
   c. **Carpal tunnel syndrome** results from compression of the median nerve within the carpal
      tunnel. It produces marked weakness of flexion and abduction of the thumb, inability to
      oppose the thumb, and inability to extend fully the first and second fingers, as well as loss
      of sensation along the volar aspects of the first, second, and third fingers. Because the su-
      perficial branch is not involved, there is no sensory loss along the radial side of the palm.
   d. **Deep injury at the wrist,** such as by deep lacerations, severs both the median nerve and its
      superficial branch, producing the deficits described above (IV A 3 b, c).

B. **Ulnar nerve** (C8–T1, anterior)

   1. **Course.** After passing posteriorly to the medial humeral epicondyle, the ulnar nerve enters
      the forearm by passing between the humeral and ulnar heads of the flexor carpi ulnaris mus-
      cle (Fig. 9-6*B*). It is joined by the ulnar artery as it descends in the midforearm. It passes later-
      ally to the pisiform bone under the carpal volar ligament and superficially to the transverse
      carpal ligament. In the hand, it divides into superficial and deep branches at the base of the
      hypothenar eminence.

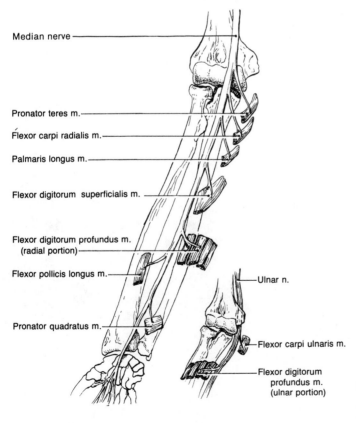

Median nerve

Pronator teres m.

Flexor carpi radialis m.

Palmaris longus m.

Flexor digitorum superficialis m.

Flexor digitorum profundus m.
   (radial portion)

Flexor pollicis longus m.

Ulnar n.

Pronator quadratus m.

Flexor carpi ulnaris m.

Flexor digitorum
   profundus m.
   (ulnar portion)

A   B

Figure 9-6. *A, Median nerve in the forearm.* The muscles innervated by branches of the median nerve are indicated. *B, Ulnar nerve in the forearm.* The muscles innervated by branches of the ulnar nerve are indicated.

2. Distribution

a. Motor. The ulnar nerve supplies only two flexor muscles in the forearm: the flexor carpi ulnaris and the ulnar half of the flexor digitorum profundus. These motor branches arise in the vicinity of the elbow.

b. Sensory. The **superficial palmar branch** arises in the middle of the forearm to supply the medial proximal palmar surface of the hand.

3. Ulnar nerve injury in the forearm. Injuries involving the ulnar nerve produce signs that are dependent upon the level of the lesion.

a. Injury in the vicinity of the medial epicondyle, such as by a supracondylar or epicondylar fracture, results in weakness of flexion and adduction of the wrist and paralysis of the hypothenar muscles, most of the deep muscles of the hand, and some of the muscles of the thenar side of the hand.

b. Injury to the ulnar nerve by fracture of the proximal ulna results in paralysis of the hypothenar muscles, most of the deep muscles of the hand, and some of the muscles of the thenar side of the hand. There is no loss of sensation along the medial side of the palm because the superficial branch is not involved.

C. Medial antebrachial cutaneous nerve (C8–T1, anterior). This nerve originates from the medial cord of the brachial plexus and innervates the skin on the medial side of the forearm. It generally accompanies the basilic vein.

I. INTRODUCTION

A. **Function.** The hand has a unique combination of mobility, dexterity, and sensitivity.

1. **The function of the upper extremity** is to position the hand to interact effectively with the environment.

2. **The function of the hand** is dependent upon extrinsic and intrinsic mobility, muscle strength, and sensation.
   a. **Stability and strength.** The hand has maximum stability and strength in the functional position, that is, when the wrist is partially extended and slightly abducted, when the digital joints are partially flexed and slightly abducted, and when the thumb is in partial abduction, flexion, and opposition.
   b. **Prehension.** The hand is characterized by several types of prehension.
      (1) **Digital prehension by terminal opposition**, for example, picking up a needle between the tips of the thumb and index finger
      (2) **Digital prehension by subterminal opposition**, for example, holding forceps between the thumb and index finger
      (3) **Digital prehension by subterminal three-point (chuck) opposition**, for example, holding an object between the thumb and the index and middle fingers. This is the most powerful and most stable type of digital prehension.
      (4) **Digital prehension by subterminal lateral opposition**, for example, holding a card between the thumb and the side of the index finger
      (5) **Lateral digital prehension**, for example, holding an object between any two fingers
      (6) **Digital palmar prehension**, for example, a grasp with opposition of the thumb, as in holding a baseball bat
      (7) **Palmar prehension**, for example, a grasp with the thumb adducted into the plane of the hand

B. **Bony landmarks**

1. **Metacarpals**

2. **Phalanges**

II. SKELETON OF THE HAND

A. **Bones of the wrist** (see Chapter 8 II B)

B. **Bones of the hand**

1. **Metacarpals.** Five metacarpal bones form the skeleton of the hand (see Fig. 8-1; Fig. 10-1).
   a. Each metacarpal bone has a proximal **base**, a **shaft**, and a distal **head**.
   b. The base of each metacarpal contacts the distal row of carpal bones to form the **carpometacarpal joint**.
   c. The head of each metacarpal contacts a proximal phalanx at a **metacarpophalangeal joint**.

2. **Phalanges.** Three rows of phalanges comprise the skeletons of the second through fifth digits; the thumb has only two phalanges.

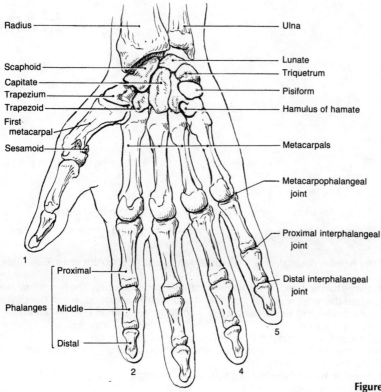

Radius

Ulna

Scaphoid

Lunate

Triquetrum

Capitate

Trapezium

Pisiform

Trapezoid

Hamulus of hamate

First metacarpal

Sesamoid

Metacarpals

Metacarpophalangeal joint

Proximal interphalangeal joint

1

Phalanges — Proximal

Distal interphalangeal joint

Middle

Distal

2

3

4

5

Figure 10-1. *Bones and joints of the right hand.*

C. Articulations

1. **Carpometacarpal joints** (see Fig. 10-1)
   a. **Second, third, fourth, and fifth carpometacarpal joints**
      (1) **Structure.** The joint surfaces are condyloid, permitting two degrees of freedom.
      (2) **Movement.** A slight amount of **flexion/extension** occurs, as well as some **abduction/adduction**. **Circumduction** at these joints allows the formation of the **transverse metacarpal arch** (the hollowing of the hand), which permits the fingers to meet.
      (3) **Support.** Each joint is reinforced by carpometacarpal ligaments.
   b. **First (thumb) carpometacarpal joint**
      (1) **Structure.** This joint is a saddle-shaped joint with two degrees of freedom between the trapezium of the first carpal row and the first metacarpal.
      (2) **Movement**
         (a) **Flexion/extension** (75°) of the thumb occurs about a complex axis. The plane of flexion/extension is approximately 60° to that of the hand.
            (i) **Flexion** brings the thumb ventrally to the plane of the hand toward the palm.
            (ii) **Extension** brings the thumb back into the plane of the hand.
         (b) **Abduction/adduction** of the thumb occurs about an axis perpendicular to the plane of the hand.
            (i) **Abduction** (15°) is movement of the extended thumb away from the index finger.
            (ii) **Adduction** (45°) brings the extended thumb against the index finger.
         (c) **Opposition** is a form of circumduction which, because of the sellar shape of the joint surfaces, involves a rotational movement of the thumb. Thus, opposition is flexion accompanied by medial rotation; **reposition** is extension and abduction accompanied by lateral rotation.

2. **Metacarpophalangeal joints** (see Fig. 10-1) represent the knuckles.
   a. **Structure.** These joints are condyloid with two degrees of freedom.
   b. **Movement**
      (1) **Flexion/extension** (100°) occurs about a transverse axis.

   (2) **Abduction/adduction** (as much as 30° for the index finger) occurs about an anteropos-
       terior axis with reference to the middle finger.
   (3) **Circumduction** (the combination of movements) is possible because there are two de-
       grees of freedom at the metacarpophalangeal joint.
   c. **Support. Metacarpophalangeal collateral ligaments** reinforce these joints. These ligaments
      are slack in the extended finger and taut in the flexed finger. The **deep transverse metacarpal
      ligament** interconnects the heads of the metacarpals.

3. **Interphalangeal joints** (see Fig. 10-1)
   a. **Structure.** These are hinge joints with one degree of freedom.
   b. **Movement.** Flexion/extension (90°) is about a transverse axis.
      (1) The **proximal interphalangeal joints** are between the proximal and middle phalanges.
      (2) The **distal interphalangeal joints** are between the middle and distal phalanges.
   c. **Support.** These joints are strongly reinforced by collateral ligaments, which are slacker
      when the finger is extended.

III. DORSAL HAND

A. **Fascia of the dorsal hand**

   1. **The antebrachial fascia** condenses at the wrist to form the **extensor retinaculum**.
      a. This retinaculum is subdivided into six compartments by attachment to bone.
      b. The tendons of the extrinsic extensor muscles of the hand pass through these compart-
         ments to gain access to the dorsum of the hand (see Fig. 8-3).
         (1) The tendons are enclosed in synovial sheaths beneath the extensor retinaculum.
         (2) The sheaths of the extensor digitorum communis terminate just distal to the extensor
             retinaculum, and the tendons lie directly on the metacarpal bones.
         (3) The tendons proceed to the individual digits, but there are variable interconnections
             that tend to limit the individual movements of the fingers.

   2. **Dorsal subcutaneous space.** The skin of the dorsum of the hand is very mobile over the un-
      derlying fascia, which is very vascular and contains a rich network of lymphatics.

   3. **Deep dorsal fascia** is formed by the fusion of the antebrachial fascia on the dorsum of the
      hand with the extensor tendons (which are unsheathed on the dorsum of the hand). It limits
      the dorsal subcutaneous space.

   4. **The dorsal subaponeurotic space** lies between the deep dorsal fascia or the extensor tendons
      and the deep fascia of the dorsal interossei and metacarpals. Laceration of the knuckles and
      inoculation with pyogenic organisms (such as occurs upon the violent meeting of fist with
      teeth) may result in infection of the subaponeurotic space. Otherwise, this space is not fre-
      quently involved in hand infections.

B. **Extrinsic extensor muscles** (see Chapter 8 IV B)

   1. **Organization.** This group of muscles on the dorsum of the hand includes many of those of the
      extensor compartment of the forearm. It consists of the tendons of the **extensor digitorum
      communis, extensor indicis proprius, extensor pollicis brevis, abductor pollicis longus**, and
      extensor pollicis longus muscles (Table 10-1; see Fig. 8-3 and Table 10-2).
      a. The subcutaneous tendons of the extensor digitorum communis lie in sheaths as they pass
         beneath the **extensor retinaculum**.
      b. The extensor digitorum communis divides into four slips; each slip inserts into the **extensor
         aponeurosis** on the dorsum of one of the second through fifth digits (Figs. 10-2 and 10-3).
         Each extensor aponeurosis passes over the metacarpophalangeal joint and then divides in-
         to one central and two lateral parts.
         (1) The central portion passes over the proximal interphalangeal joint to insert into the
             base of the middle phalanx.
         (2) The lateral bands pass over the proximal and distal interphalangeal joints to insert into
             the base of the distal phalanx.
      c. There are no intrinsic muscles on the dorsum of the hand.

   2. **Group actions.** The primary action of the extrinsic extensor muscles is to extend the metacar-
      pophalangeal joints. They also assist in the extension of the wrist.

   3. **Group innervation.** The extrinsic extensor muscles are innervated by the radial nerve.

Table 10-1. Muscles Acting on the Second through Fifth Digits

Muscle	Origin	Insertion	Action	Innervation
		Extrinsic Muscles		
Extensor digitorum communis	Lateral epicondyle of humerus	Phalanges two and three of index, middle, and ring fingers	Extends digits and wrist when fist is clenched	Radial nerve (C7–C8, posterior)
Extensor digiti minimi	Common extensor tendon	All phalanges of fifth digit	Extends fifth digit	Radial nerve (C7–C8, posterior)
Extensor indicis proprius	Interosseous membrane and ulna	Phalanges two and three of index finger	Extends index digit and wrist	Radial nerve (C8–T1, posterior)
Flexor digitorum superficialis	Medial epicondyle of humerus, proximal ulna, interosseous membrane, and oblique line of the radius	Base of second phalanx of each digit	Flexes proximal interphalangeal (PIP) joint, metacarpophalangeal (MP) joint, and wrist	Median nerve (C8–T1, anterior)
Flexor digitorum profundus	Anterior proximal ulna and interosseous membrane	Base of third phalanx of each digit	Flexes distal interphalangeal (DIP), PIP, and MP joints and wrist	Heads for second and third digits by median n. (C7–T1, anterior). Heads for fourth and fifth digits by ulnar n. (C7–T1, anterior)
		Intrinsic Muscles		
Dorsal interossei (4)	Medial side of first metacarpal; both sides of the second, third, and fourth metacarpal; and lateral side of fifth metacarpal	Tubercle of proximal phalanx and dorsal aponeurosis: laterally on second and third digits, medially on third and fourth digits	Abduct the second, third, and fourth digits from the midline of hand (third digit abducts to both sides). Flex MP joint and extend PIP and DIP joints	Ulnar nerve (C8–T1, anterior)
Palmar interossei (3)	Medial side of second and lateral side of fourth and fifth metacarpals	Tubercle of proximal phalanx and dorsal aponeurosis: medially on second digit, laterally on fourth and fifth digits	Adduct the second, fourth, and fifth digits to the midline of hand. Flex MP joint and extend PIP and DIP joints	Ulnar nerve (C8–T1, anterior)
Lumbricals 1 and 2	Tendons of the flexor digitorum profundus in the deep palm	Lateral side of dorsal expansion of second and third digits	Flex MP joint and extend PIP and DIP joints	Median nerve (C8–T1, anterior)
Lumbricals 3 and 4	Tendons of the flexor digitorum profundus in the deep palm	Lateral side of dorsal expansion of fourth and fifth digits	Flex MP joint and extend PIP and DIP joints	Ulnar nerve (C8–T1, anterior)
Palmaris brevis	Medial border of palmar aponeurosis	Skin over hypothenar region	Corrugates palmar skin	Ulnar nerve (C8–T1, anterior)
Abductor digiti minimi	Pisiform bone	Ulnar side of base of fifth proximal phalanx	Abduct the fifth digit	Ulnar nerve (C8–T1, anterior)

Table 10-1. Continued

Muscle	Origin	Insertion	Action	Innervation
Flexor digiti minimi brevis	Flexor retinaculum and hamulus	Ulnar side of base of fifth proximal phalanx	Flex fifth metacarpophalangeal joint	Ulnar nerve (C8–T1, anterior)
Opponens digiti minimi	Flexor retinaculum and hamulus	Ulnar side of fifth metacarpal	Flexion and opposition	Ulnar nerve (C8–T1, anterior)

IV. PALMAR HAND

   A. Fascia of the palm

   1. The subcutaneous fascia lies superficially to the palmar aponeurosis.
   a. Fibrous fasciculi connect the palmar skin to the palmar aponeurosis, greatly limiting independent movement of the palmar skin.
   b. The subcutaneous layer contains the **palmaris brevis** muscle and tendon as well as the **recurrent branch of the median nerve**.

   2. The antebrachial fascia condenses at the wrist to form the **flexor retinaculum**, which consists of the **volar carpal ligament** and the **transverse carpal ligament** (see Figs. 9-1 and 9-4). The antebrachial fascia then continues into the palm of the hand as the **deep palmar fascia**.
   a. Deep palmar fascia. The **thenar compartment** and the **hypothenar compartment** are formed when the deep fascia envelops the thenar and hypothenar muscle groups.
   b. Palmar aponeurosis. Between the thenar and hypothenar eminences, the deep palmar fascia thickens as the **palmar aponeurosis**.
   (1) Proximally, the palmar aponeurosis receives the insertion of the **palmaris longus muscle**, enabling that muscle to assist flexion of the hand.
   (2) Medially, beneath the hypothenar compartment, the palmar aponeurosis is attached along the length of the fifth metacarpal.

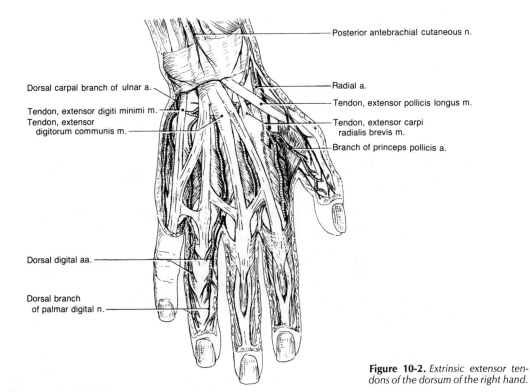

Posterior antebrachial cutaneous n.

Dorsal carpal branch of ulnar a.

Tendon, extensor digiti minimi m.
Tendon, extensor digitorum communis m.

Radial a.

Tendon, extensor pollicis longus m.

Tendon, extensor carpi radialis brevis m.

Branch of princeps pollicis a.

Dorsal digital aa.

Dorsal branch of palmar digital n.

Figure 10-2. *Extrinsic extensor tendons of the dorsum of the right hand.*

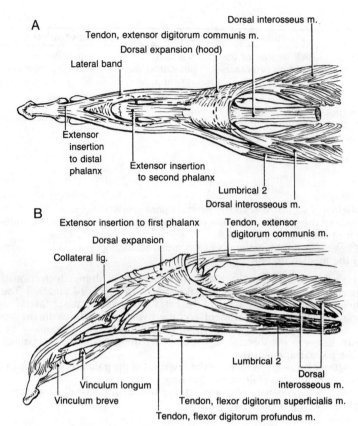

Figure 10-3. *Extensor aponeurosis of the third digit. A,* Dorsal view. *B,* Lateral view. The extensor digitorum communis, interossei, and lumbricals all insert into the extensor aponeurosis (dorsal hood).

(3) Laterally, beneath the thenar compartment, it is attached along the length of the first metacarpal.

(4) Centrally, the palmar aponeurosis attaches along the length of the third metacarpal, dividing the underlying palmar space.

(5) Distally, the palmar aponeurosis splits into four slips. Each slip has a superficial and a deep part.

    (a) The superficial part of each slip inserts into the skin at the base of a finger.

    (b) The deep part of each slip divides to pass along the sides of adjacent fingers to join the transverse metacarpal ligaments and the fibrous flexor sheaths to insert into the middle and proximal phalanges.

(6) In *Dupuytren's contracture*, there is fibrosis and shortening of the palmar aponeurosis, which is more severe toward the ulnar side.

    (a) Because the palmar aponeurosis inserts into the proximal and middle phalanges, shortening of the aponeurosis results in progressive flexion of the fourth and fifth digits.

    (b) A high correlation exists between *Dupuytren's contracture* and coronary artery disease, possibly the result of vasospasm produced by the effect of referred pain upon the sympathetic innervation of the vasculature within the T1 component of the ulnar nerve distribution.

c. The palmar (central) space lies beneath the palmar aponeurosis between the medial and lateral attachments. The attachment of the palmar aponeurosis to the third metacarpal divides the palmar space into a **thenar space** and a **midpalmar space.**

    (1) The **thenar space (bursa)** is lateral and contains the flexor pollicis longus tendon and the extrinsic flexor tendons of the index finger.

    (2) The **midpalmar space (bursa)** is medial and contains the extrinsic flexor tendons of the third, fourth, and fifth digits.

B. Palmar musculature. The palm can be divided into extrinsic and intrinsic musculature.

    1. Extrinsic flexor tendons include many muscles of the forearm flexor compartment (i.e., the palmaris longus, flexor digitorum superficialis, flexor digitorum profundus, and flexor pollicis longus (see Chapter 9 II B and Tables 9-1 and 10-1; Table 10-2). The tendons of these muscles lie in the carpal tunnel beneath the transverse carpal ligament (see Figs. 9-1 and 9-3B).

   a. The flexor digitorum superficialis divides into four tendons, which split near their insertions into the base of the middle phalanx of the second through fifth digits (see Fig. 10-3B).

     (1) Action. The flexor digitorum superficialis primarily flexes the proximal interphalangeal joints but also flexes the metacarpophalangeal joints and the wrist. The muscle is strongest when the wrist is extended.

     (2) Innervation is by the median nerve.

   b. The flexor digitorum profundus divides into four tendons, each of which passes through the split in the flexor superficialis tendon to insert into the base of the distal phalanx of the second through fifth digits (see Figs. 10-3B and 10-5).

     (1) Action. The flexor digitorum profundus primarily flexes the distal interphalangeal joints but also flexes the proximal interphalangeal joints, the metacarpophalangeal joints, and the wrist. The muscle is strongest when the wrist is extended.

     (2) Innervation. The median nerve innervates the lateral portion, which inserts on the second and third digits. The ulnar nerve innervates the medial portion, which inserts on the fourth and fifth digits.

   c. Fibrous flexor sheaths. A fibrous sheath extends along each digit to attach the flexor extrinsic tendons to the bone.

     (1) Annular ligaments are formed by condensations of these sheaths on each side of a joint.

     (2) Cruciate ligaments are formed across the joints by oblique fibers where the fibrous sheaths are thinner. They prevent the tendons from lifting away like a bowstring as the joint is flexed.

     (3) Vincula attach the superficial and deep flexors to the middle and distal phalanges, respectively, and provide routes for the vascular supply to each tendon (see Fig. 10-3). Inflammatory swelling of the tendon sheaths may compress the vincular blood supply, causing ischemic necrosis of the tendons.

   d. Synovial tendon sheaths enclose the flexor tendons as they approach and traverse the carpal tunnel. They extend variably into the hand.

     (1) The ulnar bursa surrounds the extrinsic digital flexor tendons as they pass beneath the transverse carpal ligament in the carpal tunnel.

       (a) This bursa is continuous with the synovial sheath of the tendon along the length of the fifth digit.

       (b) The more distal portions of the sheaths of the second, third, and fourth digits usually are not continuous with the ulnar bursa. Instead, each reforms at the distal end of the metacarpal and continues to the proximal end of the distal phalanx.

     (2) The radial bursa ensheaths the flexor pollicis longus tendon for its entire length. The radial bursa may or may not be continuous with the ulnar bursa.

     (3) An intermediate bursa, which is separate, occasionally invests the flexor tendon of the index finger. When present, an intermediate bursa may communicate with the radial bursa, the ulnar bursa, both, or neither.

     (4) Tenosynovitis

       (a) Infection within the synovial flexor sheath of the thumb and of the fifth digit will track proximally into and through the radial and ulnar bursae, respectively.

       (b) If the radial and ulnar bursae communicate, infection may track to the opposite side of the hand as a *horseshoe abscess*.

       (c) Inflammatory swelling of the bursae beneath the transverse carpal ligament produces *carpal tunnel syndrome*, whereby compression of the median nerve results in thenar paralysis with entrapment of the extrinsic flexor tendons.

2. Intrinsic flexor muscles (Fig. 10-4; see Fig. 10-5A)

   a. Dorsal interosseous muscles consist of four bipinnate muscles that originate from the first through fifth metacarpals. They insert into the extensor aponeuroses and the bases of the proximal phalanges on the lateral side of the second digit, onto both sides of the third digit, and onto the medial side of the fourth digit (see Figs. 10-3 and 10-4A).

     (1) Actions

       (a) Abduction. The phalangeal insertions are so arranged as to abduct the second and fourth digits from the midline of the hand and to abduct the third digit toward both the ulnar and radial sides of the hand.

       (b) Flexion. The insertions into the bases of the proximal phalanges result in flexion of the second, third, and fourth metacarpophalangeal joints.

       (c) Extension. By inserting into the extensor aponeuroses, they also extend the proximal and distal interphalangeal joints.

Table 10-2. Muscles Acting on the Thumb

Muscle	Origin	Insertion	Primary Action	Innervation
	**Extrinsic Muscles**				
Abductor pollicis longus	Posterior interosseous membrane and ulna	Base of first metacarpal, laterally	Abducts thumb and wrist	Radial n. (C8–T1, posterior)
Extensor pollicis brevis	Posterior midshaft of radius and interosseous membrane	Base of first phalanx	Extends thumb and abducts wrist	Radial n. (C8–T1, posterior)
Extensor pollicis longus	Posterior surface of interosseous membrane and posterior ulna	Base of second phalanx	Extends thumb and abducts wrist	Radial n. (C8–T1, posterior)
Flexor pollicis longus	Anterior mid-radius and interosseous membrane	Lateral aspect base of second phalanx	Flexes thumb, MP joint, and wrist	Median n. (C7–T1, anterior)
	**Intrinsic Muscles**				
Abductor pollicis brevis	Anterior surface of trapezium and scaphoid	Lateral aspect of base of first phalanx	Abducts thumb	Median n. (C8–T1, anterior)
Opponens pollicis	Trapezium	Anterolateral surface of first metacarpal	Medially rotates (opposes) thumb	Median n. (C8–T1, anterior)
Flexor pollicis brevis:				
Superficial head	Transverse carpal ligament and trapezium	Lateral side of base of first phalanx	Flex thumb	Median nerve (C8–T1, anterior)
Deep head	Lateral side of second metacarpal			Ulnar nerve (C8–T1, anterior)
Adductor pollicis:				
Oblique head	Anterior surface of capitate and second and third metacarpals	Medial side of base of first phalanx of thumb	Adduct thumb	Ulnar n. (C8–T1, anterior)
Transverse head	Distal half of third metacarpal			

   (2) **Innervation** of the dorsal interosseous muscles is by the ulnar nerve.
   b. **Palmar interosseous muscles** consist of three unipinnate muscles that originate from the second, fourth, and fifth metacarpals. They insert into the extensor aponeuroses and the bases of the proximal phalanges on the medial side of the second digit and onto the lateral sides of the fourth and fifth digits (see Fig. 10-4B).
      (1) **Actions**
         (a) **Adduction.** The phalangeal insertions are so arranged as to adduct the second and fourth digits toward the midline of the hand (i.e., they are antagonistic to the dorsal interossei) [mnemonic: PAD for **p**almar **ad**uct]. The third digit has no palmar interosseous muscle because the dorsal interossei give it full mobility.
         (b) **Flexion.** The insertions into the bases of the proximal phalanges result in flexion of the first, fourth, and fifth metacarpophalangeal joints (i.e., they are synergistic with the dorsal interossei).
         (c) **Extension.** By inserting into the opposite sides of the extensor aponeuroses from the dorsal interossei, they also extend the proximal and distal interphalangeal joints (i.e., they are synergistic with the dorsal interossei).

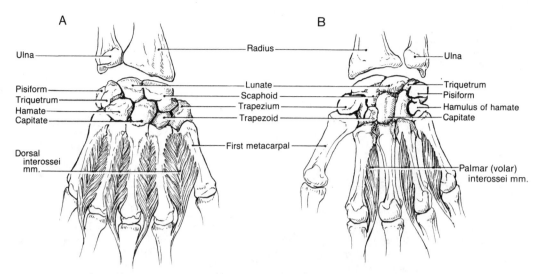

Figure 10-4. *A, Dorsal interossei of the right hand. B, Palmar interossei of the right hand.*

(2) Innervation of the palmar interosseous muscles is by the ulnar nerve.
   c. **The lumbrical muscles** originate from the flexor profundus tendons deep in the hand. They insert into the radial side of the extensor aponeuroses, which insert into the distal phalanges of the second through fifth digits (see Fig. 10-3*B*; Fig. 10-5*A*).
     (1) Actions
       (a) Flexion. Because they have longer lever arms than the interossei, the lumbricals are stronger flexors of the metacarpophalangeal joint (i.e., they are synergistic with the interossei).
       (b) Extension. By inserting into the extensor hood, they also extend the proximal and distal interphalangeal joints (i.e., they are synergistic with the interossei).
     (2) Innervation of the lumbrical muscles to the second and third digits is by the median nerve; those to the fourth and fifth digits is by the ulnar nerve.

C. Thenar musculature

   1. Organization. The thenar eminence comprises the intrinsic muscles of the thumb. The thenar muscles include the **abductor pollicis brevis**, the **opponens pollicis**, the **flexor pollicis brevis**, and the **adductor pollicis** (see Fig. 10-5 and Table 10-2).

   2. Actions. The actions of the thenar muscles are described by their names.
     a. **Flexion/extension**
       (1) Flexion is accomplished by the flexor pollicis brevis (intrinsic) and the flexor pollicis longus (extrinsic).
       (2) Extension is accomplished by the extensor pollicis longus and brevis (both extrinsic muscles).
     b. **Abduction/adduction**
       (1) Abduction is accomplished by the abductor pollicis brevis (intrinsic) and the abductor pollicis longus (extrinsic).
       (2) Adduction is a function of the adductor pollicis.
     c. **Opposition/reposition**
       (1) Opposition involves a medial rotation at the first carpometacarpal joint accomplished by the opponens pollicis muscle and assisted by the flexor pollicis brevis muscle. Opposition enables the pincer-like grasp that is characteristic of humans.
       (2) Reposition involves a lateral rotation accomplished by the synergistic actions of the abductors and extensors of the thumb.

   3. Innervation of the thenar muscles is by both median and ulnar nerves.
     a. **The recurrent branch of the median nerve** innervates the abductor pollicis brevis, opponens pollicis, and the superficial head of the flexor pollicis brevis muscle.
     b. **The deep terminal branch of the ulnar nerve** supplies the deep head of the flexor pollicis brevis muscle and the adductor pollicis muscle.

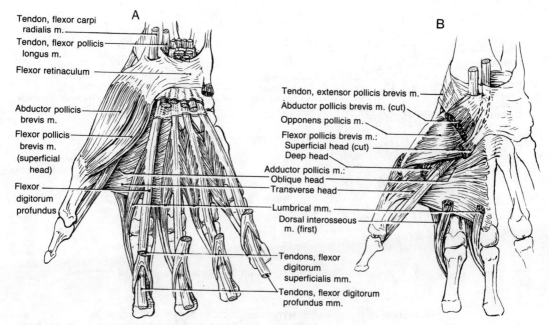

Tendon, flexor carpi radialis m.
Tendon, flexor pollicis longus m.
Flexor retinaculum
Abductor pollicis brevis m.
Flexor pollicis brevis m. (superficial head)
Flexor digitorum profundus

A

B

Tendon, extensor pollicis brevis m.
Abductor pollicis brevis m. (cut)
Opponens pollicis m.
Flexor pollicis brevis m.:
Superficial head (cut)
Deep head
Adductor pollicis m.:
Oblique head
Transverse head
Lumbrical mm.
Dorsal interosseous m. (first)
Tendons, flexor digitorum superficialis mm.
Tendons, flexor digitorum profundus mm.

Figure 10-5. *A, Lumbrical muscles. B, Thenar muscles.*

D. Hypothenar musculature

1. **Organization.** The hypothenar eminence comprises the intrinsic muscles of the fifth digit. The hypothenar muscles include the **abductor digiti minimi**, the **flexor digiti minimi brevis**, and the **opponens digiti minimi (quinti)** [see Table 10-1].

2. **Actions.** The actions of the hypothenar muscles are described by their names.

3. **Innervation** of the hypothenar muscles is by the ulnar nerve.

V. VASCULATURE OF THE HAND

A. Arterial supply

1. **Radial artery**
   a. **Course.** The radial artery passes laterally to the scaphoid (navicular) bone onto the dorsal side of the wrist between the tendons of the extensor pollicis longus and brevis muscles, the "anatomical snuff-box," where a **radial pulse** is palpable (see Fig. 9-5).
   b. **Branches**
      (1) The **dorsal carpal branch**, arising in the wrist, anastomoses with the dorsal carpal branch of the ulnar artery to supply the dorsum of the hand.
      (2) The **superficial palmar branch** of the radial artery arises at the level of the end of the radius. It passes through the thenar compartment and provides a minor contribution to the **superficial palmar arch** by anastomosing with the superficial palmar branch of the ulnar artery (Fig. 10-6).
      (3) The **deep palmar branch** is the continuation of the radial artery. It continues into the hand, diving through the first interosseous space to gain access to the thenar space. It gives off the **first metacarpal artery** to the thumb and lateral side of the index finger before it becomes the **deep palmar arch** and anastomoses with the deep palmar branch of the ulnar artery (see Fig. 10-6).
      (4) The **palmar metacarpal arteries** (three in number) arise from the deep palmar arch and join the common digital arteries of the superficial arch to form the **phalangeal arteries**, which run along the sides of the digits.

2. **Ulnar artery**
   a. **Course.** The ulnar artery continues into the hand on the medial side of the wrist, just lateral to the pisiform bone where the **ulnar pulse** is palpable (see Fig. 9-5).
   b. **Branches.** The ulnar artery gives off three major branches.

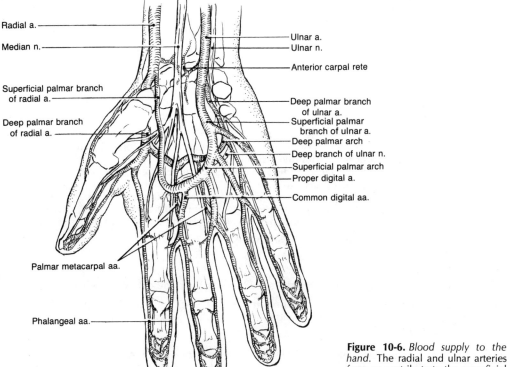

Figure 10-6. *Blood supply to the hand.* The radial and ulnar arteries form or contribute to the superficial and deep palmar arches.

Labels on figure:
Radial a.
Median n.
Superficial palmar branch of radial a.
Deep palmar branch of radial a.
Palmar metacarpal aa.
Phalangeal aa.
Ulnar a.
Ulnar n.
Anterior carpal rete
Deep palmar branch of ulnar a.
Superficial palmar branch of ulnar a.
Deep palmar arch
Deep branch of ulnar n.
Superficial palmar arch
Proper digital a.
Common digital aa.

(1) The **dorsal carpal branch**, arising in the wrist, anastomoses with the dorsal carpal branch of the radial artery to supply the dorsum of the hand.

(2) The **deep palmar branch** of the ulnar artery arises just distal to the lateral side of the pisiform bone and passes through the hypothenar compartment. This small branch makes a minor contribution to the **deep palmar arch**, which is formed by anastomosis with the deep palmar branch of the radial artery (see Fig. 10-6).

(3) The **superficial palmar branch** is the continuation of the ulnar artery. It lies just beneath the palmar aponeurosis and is the major contributor to the **superficial palmar arch**, which is formed by anastomosis with the superficial palmar branch of the radial artery (see Fig. 10-6).

(4) The **proper digital arteries** to the fifth digit and the four **common digital arteries** arise from the superficial palmar arch. Each common digital artery receives a common interosseous artery from the deep palmar arch before bifurcating into **phalangeal arteries**, which run along the sides of the fingers (see Fig. 10-6).

B. Venous return

1. The superficial veins provide a rich anastomotic network in the dorsal subcutaneous space.

a. The cephalic vein is formed by a coalescence of dorsal veins on the radial side of the hand. It winds from the posterior preaxial border of the forearm to lie along the ventral preaxial border.

b. The basilic vein is formed by a medial coalescence of dorsal veins on the ulnar side of the hand. It ascends nearly to the cubital fossa along the dorsal postaxial border.

2. Deep veins (venae comitantes) accompany the radial and ulnar arteries as well as other branches.

C. Lymphatic drainage

1. Superficial lymphatics. Most of the lymphatic drainage from the thenar compartment, hypothenar compartment, and digits is toward the dorsal subcutaneous space of the hand, explaining the extreme swelling of this region that accompanies infections of the digits or volar surface. The lymphatics drain along the cephalic and basilic veins and eventually drain through the clavipectoral and supraclavicular nodes.

2. **Deep lymphatics.** The thenar space, midpalmar space, and tendon sheaths drain through lymphatics that accompany the radial and ulnar vessels. The deep lymphatics drain through the axillary nodes.

VI. INNERVATION OF THE HAND

A. Radial nerve

1. **Course.** Only the **superficial branch of the radial nerve** reaches the hand (see Chapter 8 VI A 2 b).

2. **Distribution.** This superficial branch, containing contributions from the posterior roots of C6–C7, leaves the company of the radial artery and passes dorsally to the wrist where it divides into five **dorsal digital nerves** (see Fig. 8-4).
   a. **Motor.** While the radial nerve supplies the extrinsic extensor muscles of the hand, it supplies none of the intrinsic muscles.
   b. **Sensory.** The **dorsal digital nerves** supply the radial side of the dorsal hand from the level of the wrist to about the level of the proximal interphalangeal joints. There is a region of exclusivity over the web space between the thumb and index finger (see Fig. 8-4).

3. **Radial nerve injury** (see Chapter 8 VI B)

B. Median nerve

1. **Course.** The median nerve enters the hand through the carpal tunnel.

2. **Distribution.** Upon emerging from the transverse carpal ligament, it gives off the **recurrent branch.** The median nerve divides into four or five volar digital nerves, which run along the sides of the fingers.
   a. **Motor**
      (1) The median nerve supplies most of the extrinsic flexor muscles (see Chapter 9 IV A 2 a).
      (2) The **recurrent branch** supplies most of the thenar muscles, that is, the flexor pollicis brevis, opponens pollicis, and the superficial head of the flexor pollicis brevis (Fig. 10-7).
      (3) The **palmar (volar) digital branches** supply the lumbricals to the first and second fingers before becoming cutaneous (see Fig. 10-7).
   b. **Sensory**
      (1) The **palmar cutaneous branch** arises proximally to the flexor retinaculum, passes superficially to the flexor retinaculum, and supplies the radial side of the palm of the hand and the volar aspect of the thumb.

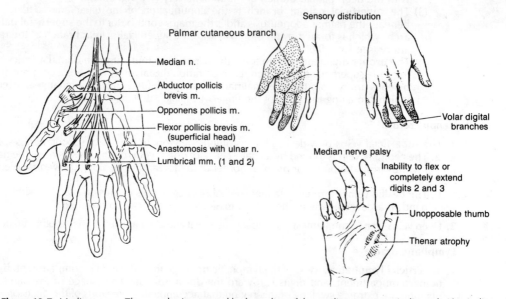

Figure 10-7. *Median nerve.* The muscles innervated by branches of the median nerve are indicated. Also indicated is the region of sensory distributions of the palmar cutaneous branch and the volar digital branches.

(2) The **palmar (volar) digital branches** supply the volar *and* dorsal surfaces distal to the proximal interphalangeal joints of the second and third digits with a variable portion of the fourth digit (see Figs. 10-2 and 10-7).

3. Median nerve injury
    a. Lesions of the median nerve in the vicinity of the elbow can produce paralysis of the flexor digitorum superficialis, the radial half of the flexor digitorum profundus, and flexor pollicis longus, as well as the muscles of the thenar compartment and lumbricals 1 and 2. This produces the *sign of benediction* in which the index and middle finger cannot be flexed and the thumb cannot be opposed. In addition, there is sensory loss over the radial side of the palm and of the digits lateral to the center of the ring finger.
    b. Compression of the median nerve in the carpal tunnel due to inflammation or displacement of the lunate bone can produce paralysis of the thenar compartment musculature (recurrent branch), loss of lumbricals 1 and 2 (volar digital nerves), and loss of sensation in the second and third digits with a portion of the fourth digit (volar digital nerves). However, there will be no loss of sensation over the proximal palm (palmar cutaneous branch).
    c. Laceration of the recurrent branch, probably the most commonly injured nerve, produces loss of thenar musculature (weakness of flexion and loss of opposition) but no sensory deficit.

C. Ulnar nerve

  1. Course. The ulnar nerve enters the hand with the ulnar artery.

  2. Distribution. The ulnar nerve gives off a **superficial terminal branch** in the vicinity of the hook of the hamate (against which it may become entrapped) and continues as the **deep terminal branch** (Fig. 10-8).
    a. Motor
        (1) The ulnar nerve supplies one and one-half extrinsic flexor muscles (flexor carpi ulnaris and the ulnar half of the flexor digitorum superficialis; see Chapter 9 IV B 2 a).
        (2) The **superficial terminal branch** supplies the palmaris brevis muscle before becoming cutaneous (see Fig. 10-8).
        (3) The **deep terminal branch** supplies the hypothenar musculature, lumbricals 3 and 4, the dorsal and palmar interossei, the adductor pollicis, and the deep head of the flexor pollicis brevis muscle (see Fig. 10-8).
    b. Sensory
        (1) The **palmar cutaneous branch** arises in the forearm and innervates the ulnar side of both the dorsum and palm of the hand (see Fig. 10-8).
        (2) The **dorsal branch** arises proximally to the wrist and supplies the dorsal aspects of the fourth and fifth digits (see Fig. 10-8).

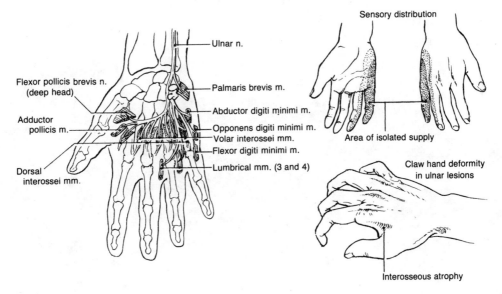

Figure 10-8. *Ulnar nerve.* The muscles innervated by branches of the ulnar nerve are indicated. Also indicated is the region of sensory distribution.

(3) The **superficial terminal branch** supplies the skin over the hypothenar eminence.

(4) **Volar digital branches** supply the volar side of the ulnar half of the fourth and fifth digits (see Fig. 10-8).

3. Ulnar nerve injury

a. Lesions of the ulnar nerve in the vicinity of the elbow will produce *claw hand*.

(1) The radial half of the flexor digitorum profundus, lumbricals 3 and 4, the dorsal and palmar interossei, and the hypothenar muscles will be paralyzed. When the metacarpophalangeal joints are extended, the proximal and distal interphalangeal joints cannot be extended because the interossei and half of the lumbricals are nonfunctional. The result is a "claw-like" posture. With time, there is muscular atrophy over the affected portion of the hand.

(2) Sensory deficit over the ulnar side of the dorsal and palmar aspects of the hand results.

b. Laceration of the ulnar nerve at the wrist leaves the innervation of the ulnar side of the flexor digitorum profundus intact but also results in a *claw hand*. There is loss of adduction of the thumb so that a piece of paper cannot be held between the side of the thumb and the index finger.

c. Coronary artery disease can refer pain to the T1 dermatomal components of the ulnar nerve.

STUDY QUESTIONS

Directions: Each question below contains five suggested answers. Choose the **one best** response to each question.

1. Which of the following statements concerning the teres major muscle, a contributor to the posterior axillary fold, is true?

(A) It is a major contributor to the stability of the posterior aspect of the shoulder joint
(B) It divides the axillary artery into three parts
(C) It inserts on the humerus just distal to the infraspinatus insertion
(D) It is active in adduction of the glenohumeral joint
(E) It is innervated by the same nerve that innervates the deltoid muscle

Questions 2–6

2. A new medical clerk is asked to place an intravenous line in the right arm of a patient. The clerk should know that the superficial veins of the arm are relatively consistent in their anatomic relationship and that all of the following statements concerning these veins are true EXCEPT

(A) at the level of the axilla, the basilic vein is joined by the cephalic vein to form the axillary vein
(B) the basilic vein runs along the medial aspect of the forearm
(C) the cephalic vein originates on the radial side of the dorsum of the hand
(D) the median cubital vein links the cephalic and basilic veins in the vicinity of the elbow joint
(E) the median cubital vein is separated from the brachial artery by the bicipital aponeurosis

3. In an attempt to distend the veins, the clerk applies a tight tourniquet approximately 2 inches proximally to the proposed site of venipuncture, but the veins do not distend when the patient's hand is vigorously flexed. The most probable reason for the failure of the veins to distend is

(A) that arterial flow to the foremarm is obstructed
(B) that deep venous return has not been obstructed
(C) that superficial venous return has not been obstructed
(D) that the venae comitantes are abnormally absent
(E) none of the above

4. After several attempts, the veins become visible and a vessel in the cubital fossa is penetrated. A small amount of blood is aspirated into the line to ensure proper placement. The blood is bright red, which makes it likely that the needle is in the

(A) brachial artery
(B) deep brachial artery
(C) cephalic vein
(D) median cubital vein
(E) venae comitantes

5. If the clerk changed the site of venipuncture to a slightly more medial one, the patient might experience pain radiating down the ventral surface of the forearm and hand, including the thumb, index finger, and middle finger; the nerve penetrated in this case would most likely be

(A) a branch of the musculocutaneous nerve
(B) the median antebrachial cutaneous nerve
(C) the median nerve
(D) the radial nerve
(E) the ulnar nerve

6. Upon successful placement of the intravenous line, the clerk is relieved but fails to notice an approximately 1.5 ml air bubble in the line entering the vein. The region at highest risk for air embolus is

(A) the brain
(B) the hand
(C) the heart
(D) the lungs
(E) none of the above

(end of group question)

7. Chronic traction on the ulnar nerve at the elbow may be relieved by a medial epicondyle osteotomy whereby the condyle is removed without disturbing the muscular attachments; the nerve is then relocated anteriorly to the humerus, and the medial epicondyle is reattached. Which of the following muscles originates from the medial epicondyle and is, therefore, directly involved in this procedure?

(A) Brachioradialis
(B) Extensor carpi ulnaris
(C) Flexor carpi radialis
(D) Flexor pollicis longus
(E) Supinator

8. All of the following structures pass deeply to the transverse carpal ligament EXCEPT the

(A) flexor digitorum superficialis tendon
(B) flexor digitorum profundus tendon
(C) flexor pollicis longus tendon
(D) median nerve
(E) ulnar artery

9. All of the following structures pass deeply to the flexor retinaculum EXCEPT the

(A) flexor digitorum profundus to the little finger
(B) flexor digitorum superficialis
(C) flexor pollicis longus
(D) median nerve
(E) ulnar artery

10. All of the following statements concerning the blood supply of the hand are true EXCEPT

(A) the arteries of the hand are accompanied by venae comitantes
(B) the blood supply reaches the long flexor tendons within the synovial sheaths through vincula
(C) the common digital arteries are branches of the deep palmar arch
(D) the deep and superficial palmar arches are supplied by the radial and ulnar arteries
(E) the dorsal carpal arch (rete) receives blood from the carpal branches of the radial, ulnar, and posterior interosseus arteries

11. The ulnar nerve innervates which of the following muscles of the thumb?

(A) Abductor pollicis brevis
(B) Abductor pollicis longus
(C) Deep head of the flexor pollicis brevis
(D) Opponens pollicis
(E) Superficial head of the flexor pollicis brevis

12. A large splinter deeply embedded adjacently to the nail on the hypothenar side of the ring finger develops into an abscess. Prior to lancing to remove the fragment and drain the abscess, the surgeon blocked with anesthetic the appropriate

(A) dorsal digital branch of the radial nerve
(B) dorsal digital branch of the ulnar nerve
(C) palmar digital branch of the radial nerve
(D) palmar digital branch of the ulnar nerve
(E) superficial branch of the radial nerve

Directions: Each question below contains four suggested answers of which **one or more** is correct. Choose the answer

A if **1, 2, and 3** are correct
B if **1 and 3** are correct
C if **2 and 4** are correct
D if **4** is correct
E if **1, 2, 3, and 4** are correct

13. With the upper limb adducted, hanging by the side and holding a heavy suitcase, support at the glenohumeral joint to prevent downward displacement is provided by the

(1) short head of the biceps brachii muscle
(2) coracohumeral ligament
(3) supraspinatus muscle
(4) coracoacromial ligament

14. Correct statements concerning the cephalic vein include which of the following?

(1) It accompanies the brachial artery
(2) It drains the radial side of the hand
(3) It is suitable for venipuncture anterior to the medial epicondyle
(4) It lies in the groove between the deltoid and pectoralis major muscles

15. The pectoralis minor muscle is a useful landmark structure that is characterized by which of the following statements?

(1) It is located at the level of the cords of the brachial plexus
(2) It is attached to the acromion process
(3) It divides the axillary artery into three portions
(4) It is innervated by the middle subscapular nerve

16. True statements concerning the pectoralis minor muscle include

(1) it attaches to the acromion process of the scapula
(2) it is an adductor and medial rotator of the humerus
(3) it is innervated by the middle subscapular (thoracodorsal) nerve
(4) it crosses the cords of the brachial plexus

17. Following complete severance of the musculocutaneous nerve, some weak flexion of the elbow is possible through contraction of the

(1) flexor carpi radialis
(2) brachioradialis
(3) flexor carpi ulnaris
(4) ulnar head of the pronator teres

18. Structures that are both medial to the biceps tendon and deep to the bicipital aponeurosis include the

(1) brachial artery
(2) deep (profunda) brachial artery
(3) median nerve
(4) median cubital vein

19. Correct statements concerning the subacromial bursa include which of the following?

(1) It frequently communicates with the subdeltoid bursa
(2) It normally communicates with the synovial cavity of the glenohumeral joint
(3) It underlies the coracoacromial ligament
(4) It permits movement between the scapula and the thoracic wall

20. Correct statements concerning the lunate bone include which of the following?

(1) It can compress the median nerve if displaced anteriorly
(2) It provides an attachment for the transverse carpal ligament
(3) It articulates with the fibrocartilage disk
(4) It is a component of the carpometacarpal joint

21. Which of the following statements concerning the scaphoid bone are true?

(1) It participates in the midcarpal joint
(2) It is the most susceptible of the carpal bones to fracture
(3) It receives an attachment for the transverse carpal ligament
(4) It articulates maximally with the radius in adduction

Questions 22–24

A tailor complains of difficulty picking up a needle and holding it while sewing. Physical examination reveals no sensory deficit. The patient is then tested for motor function of the second digit, as shown in the illustration below.

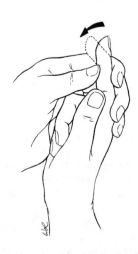

22. The muscles being tested in the illustration include the

(1) lumbrical 2
(2) flexor digitorum superficialis
(3) palmar interossei
(4) radial portion of the flexor digitorum profundus

SUMMARY OF DIRECTIONS

A	B	C	D	E
1, 2, 3 only	1, 3 only	2, 4 only	4 only	All are correct

---

23. If paralysis or paresis (weakness) of the terminal phalanx of the second digit were detected in the patient described, one might also expect to find

(1) paralysis or paresis of the fourth digit
(2) atrophy of the thenar eminence
(3) complete paralysis of the thumb
(4) weakness of pronation

24. Which of the following nerves supply muscles that produce antagonistic actions to that illustrated?

(1) Ulnar nerve
(2) Anterior interosseous nerve
(3) Median nerve
(4) Radial nerve

(end of group question)

25. In the distal portion of the upper extremity, a pulse may be palpated

(1) in the "anatomical snuff-box"
(2) at the radial side of the tendon of the flexor carpi radialis muscle
(3) between the tendons of the extensor pollicis longus and brevis muscles
(4) at the radial side of the pisiform bone

26. Locations where a pulse is readily palpable in the upper extremity include

(1) immediately lateral to the pisiform bone
(2) in the cubital fossa to the biceps brachii tendon
(3) between the tendons of the extensor pollicis longus and brevis
(4) against the humerus just distal to the pectoralis minor

27. Reposition of the thumb is accomplished by the

(1) abductor pollicis longus
(2) extensor pollicis brevis
(3) extensor pollicis longus
(4) adductor pollicis

28. An infection of the synovial sheath within the fifth digit (ulnar bursa) may

(1) spread distally into the thumb via the radial bursa
(2) spread proximally through the carpal tunnel into the distal forearm
(3) result in ischemic necrosis of the flexor tendons to the fifth finger
(4) spread distally into adjacent digits

29. Laceration of the recurrent branch of the median nerve will paralyze which of the following muscles of the thumb?

(1) Superficial head of the flexor pollicis brevis
(2) Abductor pollicis longus
(3) Abductor pollicis brevis
(4) Deep head of the flexor pollicis brevis

Directions: The groups of questions below consist of lettered choices followed by several numbered items. For each numbered item, select the **one** lettered choice with which it is **most closely** associated. Each lettered choice may be used once, more than once, or not at all. Choose the answer

- **A** if the item is associated with **(A) only**
- **B** if the item is associated with **(B) only**
- **C** if the item is associated with **both (A) and (B)**
- **D** if the item is associated with **neither (A) nor (B)**

Questions 30–34

For each muscle listed in the questions, select the appropriate action that is most apt to be associated with it.

(A) Flexes the metacarpophalangeal joint
(B) Extends the distal interphalangeal joint
(C) Both
(D) Neither

30. Palmaris longus

31. Dorsal interosseous muscles

32. Palmar interosseous muscles

33. Lumbrical muscles

34. Extensor digitorum communis

Questions 35–40

For each muscle listed in the questions, select the appropriate innervation.

(A) Ulnar nerve
(B) Median nerve
(C) Both
(D) Neither

35. Pronator quadratus muscle

36. Flexor digitorum profundus muscle

37. Abductor pollicis longus muscle

38. Lumbrical muscles

39. Dorsal interosseous muscles

40. Adductor pollicis muscle

ANSWERS AND EXPLANATIONS

1. The answer is D. [*Chapter 6 Table 6-2*] The teres major muscle, innervated by the lower subscapular nerve (C5–C6, posterior) inserts onto the medial lip of the bicipital groove of the humerus and, thus, is an adductor of the arm. The tendon of the pectoralis minor muscle, not the teres major muscle, divides the axillary artery into three parts.

2. The answer is A. [*Chapter 7 V C; Chapter 9 III B 1 a, b*] At the level of the axilla, the basilic vein and the brachial venae comitantes join to form the axillary vein. The cephalic vein joins the axillary vein just distal to the clavicle.

3. The answer is A. [*Chapter 5 III A 2 b; Chapter 7 V A 2, 4; Chapter 9 III B 3*] A tourniquet applied too tightly to the arm will compress the brachial artery and the collaterals, preventing blood flow into the forearm and, as a consequence, venous return.

4. The answer is A. [*Chapter 7 V A 2 a–c; Figure 7-4*] In the cubital fossa, the brachial artery contains bright red oxygenated blood and lies beneath the bicipital aponeurosis. The median cubital vein, containing darker deoxygenated blood, lies superficially to the bicipital aponeurosis. The deep brachial artery does not pass through the cubital fossa.

5. The answer is C. [*Chapter 7 VI B 1; Chapter 9 IV A 1, 2; Figure 9-6A*] The median nerve passes through the cubital fossa deeply to the bicipital aponeurosis, and just medial to the brachial artery.

6. The answer is D. [*Chapter 5 I A*] An embolus in the systemic venous system will pass through the right side of the heart and enter the pulmonary circulation where it will lodge in the small vessels of the lungs.

7. The answer is C. [*Chapter 8 Table 8-1; Chapter 9 Table 9-1*] The flexor carpi radialis along with the other superficial forearm flexors [i.e., the pronator teres, palmaris longus, flexor carpi ulnaris, and a portion of the flexor digitorum superficialis (the intermediate layer)] originates from the medial epicondyle. The superficial group of forearm extensors (i.e., the brachioradialis, extensor carpi radialis longus and brevis, extensor digitorum communis, and extensor carpi ulnaris) all originate from the lateral epicondyle as does the supinator of the deep forearm extensor group. The flexor pollicis longus originates from the radial shaft and interosseous membrane.

8. The answer is E. [*Chapter 9 II A 1 a, b*] The ulnar nerve and ulnar artery pass deeply to the volar carpal ligament but superficially to the transverse carpal ligament. The median nerve is the only major neurovascular structure that passes through the carpal tunnel.

9. The answer is E. [*Chapter 9 III A 2 a*] At the wrist, the ulnar artery lies between the tendons of the flexor carpi ulnaris muscle and the flexor digitorum superficialis muscle, where the ulnar pulse may be palpated. As the ulnar artery enters the hand, it lies superficially to the flexor retinaculum on the radial side of the pisiform bone.

10. The answer is C. [*Chapter 9 III A 1 b (1), 2 b (4); Chapter 10 VI A 1 b (2), (3), 2 b (2), (3)*] The deep palmar arch is supplied by the radial and ulnar arteries and gives off palmar metacarpal arteries. These arteries join the common digital arteries, which arise from the *superficial* palmar arch, before the latter bifurcate into phalangeal arteries.

11. The answer is C. [*Chapter 10 VII C 2 a (3)*] The thumb musculature is innervated by the ulnar, median, and radial nerves. The ulnar nerve innervates the deep head of the flexor pollicis brevis muscle as well as the adductor pollicis muscle. The opponens pollicis, abductor pollicis brevis, and superficial head of the flexor pollicis brevis are innervated by the median nerve. The abductor pollicis longus as well as the extensor pollicis longus and extensor pollicis brevis are innervated by the radial nerve.

12. The answer is B. [*Chapter 10 VII C 2 a, b*] The hypothenar (medial) side of the ring finger is generally innervated by the ulnar nerve. The dorsal digital branches innervate a region on the distal dorsum of the digits. The branches of the superficial radial nerve seldom extend beyond the proximal interphalangeal joints. The palmar digital branches innervate the volar surfaces of the digits.

13. The answer is A (1, 2, 3). [*Chapter 6 IV D 3 b, c; Table 6-2*] The dead weight of the arm is supported primarily by the coracohumeral ligament. In addition, the supraspinatus muscle, coracobrachialis, the short head of the biceps brachii muscle, and the long head of the triceps brachii all provide dynamic support at the glenohumeral joint. The coracoacromial ligament prevents *superior* displacement of the humerus at the glenohumeral joint.

14. The answer is C (2, 4). [*Chapter 7 V C 1; Chapter 9 III B 1 a; Chapter 10 VI C 1*] The cephalic vein drains the subcutaneous spaces of the radial (preaxial) side of the hand. It passes posteriorly to the styloid process, along the preaxial border of the forearm and anteriorly to the lateral epicondyle over which course it is readily accessible for venipuncture. Just distal to the cubital fossa it gives rise to the median cubital vein, which joins the basilic vein in the vicinity of the medial epicondyle. The cephalic vein continues along the preaxial border of the brachium, passing in the deltopectoral groove and piercing the clavipectoral fascia to enter the axillary vein.

15. The answer is B (1, 3). [*Chapter 6 VI B 2 a, b; Table 6-1*] As the pectoralis minor muscle courses toward its insertion on the coracoid process of the scapula, it divides the axillary artery into three regions. It is innervated by the medial pectoral nerve, a branch of the medial cord of the brachial plexus.

16. The answer is D (4). [*Chapter 6 VI C 2 d; Table 6-1*] The pectoralis minor muscle crosses the cords of the brachial plexus as it attaches to the coracoid process of the scapula. It, therefore, has no direct action on the humerus. It is innervated by the medial pectoral nerve of the medial cord of the brachial plexus.

17. The answer is A (1, 2, 3). [*Chapter 7 IV B 1 a, 3*] Aside from the brachialis and biceps brachii (both innervated by the musculocutaneous nerve), the brachioradialis (radial nerve), flexor carpi radialis (median nerve), and flexor carpi ulnaris (ulnar nerve) all take origin from the humerus and can weakly flex the elbow joint. The ulnar head of the pronator teres does not cross the humeroulnar joint.

18. The answer is B (1, 3). [*Chapter 7 V A 2 a, b; Chapter 9 IV A 1*] The brachial artery and median nerve are both deep to the bicipital aponeurosis and medial to the biceps brachii tendon. The medial cubital vein is superficial to the bicipital aponeurosis. The deep brachial artery anastomoses with the recurrent radial artery, which is lateral to the biceps brachii tendon.

19. The answer is B (1, 3). [*Chapter 6 IV D 5 a, c; Figure 6-4*] The subacromial and subdeltoid bursae may be separate, may communicate, or may form one large bursa. These bursae lie deeply to the deltoid muscle, extend inferiorly to the coracoacromial ligament, and lie superiorly to the supraspinatus muscle. Only in dislocations of the glenohumeral joint, when the joint capsule is torn, will the synovial cavity of the glenohumeral joint communicate with the subacromial or subdeltoid bursa.

20. The answer is B (1, 3). [*Chapter 8 II A 1 d, B 1 b; III A 2 a, 3 f*] The lunate bone articulates with the triangular fibrocartilage at the radiocarpal joint—much more so in abduction than in adduction. When displaced anteriorly, the lunate bone can impinge on the long flexor tendons of the digits and compress the median nerve, producing the carpal tunnel syndrome.

21. The answer is A (1, 2, 3). [*Chapter 8 II B 1 a; III A 3 f*] The scaphoid bone, articulating maximally with the radius in abduction, transmits the major forces from the hand to the forearm and is, therefore, the carpal bone most prone to fracture. The scaphoid provides a strong attachment for the transverse carpal ligament and participates in both the radiocarpal and midcarpal joints.

22. The answer is D (4). [*Chapter 9 II B 3 a, b; Chapter 10 Table 10-1*] The radial half of the flexor digitorum profundus flexes the terminal phalanx of the second and third digits. The flexor digitorum superficialis flexes the middle phalanx of the second through fifth digits. The lumbricals and interossei are extensors of the middle and distal phalanges.

23. The answer is C (2, 4). [*Chapter 9 IV A 2, B 2; Chapter 10 VII B 3 a*] The radial half of the flexor digitorum profundus, as well as the flexor digitorum superficialis and flexor pollicis longus, are innervated by branches of the median nerve in the forearm. A lesion that affected these muscles would most likely also affect the pronators teres and quadratus as well as some of the thenar muscles of the hand. Because the flexor digitorum profundus to the fourth and fifth digits is innervated by the ulnar nerve, rather than the median, paralysis of the fourth or fifth digit would not be expected to accompany paresis of the second digit; because some thumb muscles are innervated by the radial and ulnar nerves, rather than the median, injury of the median nerve would not cause complete paralysis of the thumb.

24. The answer is B (1, 3). [*Chapter 10 IV B 2 a, C 2 a; Table 10-1*] Extension of the terminal phalanges is accomplished primarily by the dorsal and palmar interossei, which are innervated by the ulnar nerve, as well as the lumbrical 2 innervated by the median nerve. The extensor digitorum communis and indicis proprius, which are innervated by the radial nerve, primarily extend the metacarpophalangeal joint.

25. The answer is E (all). [*Chapter 9 III A 1 a, 2 a; Chapter 10 VI A 1 a*] At the wrist, the radial pulse may be felt between the flexor carpi radialis and brachioradialis muscles, as well as in the "anatomical snuff-box'' (i.e., between the tendons of the extensor pollicis longus and brevis). The ulnar pulse is felt on the radial side of the insertion of the flexor carpi ulnaris into the pisiform bone.

26. The answer is E (all). [*Chapter 6 VI B 2 a; Chapter 7 V A 1; Chapter 10 VI A 1 a, 2 a*] The ulnar pulse is palpable immediately lateral to the pisiform bone. The brachial pulse is palpable in the cubital fossa just to the biceps brachii tendon. The radial pulse is palpable in the ''anatomical snuff-box'' between the tendons of the extensor pollicis longus and brevis muscles. The axillary pulse is palpable in the lateral axilla against the humerus just distal to the pectoralis minor.

27. The answer is A (1, 2, 3). [*Chapter 10 IV C 2; Table 10-2*] Reposition of the thumb, a combination of abduction and extension accompanied by lateral rotation, is accomplished primarily by the abductor pollicis longus and brevis muscles as well as by the extensor pollicis longus and brevis muscles. Opposition of the thumb is accomplished by flexion accompanied by medial rotation.

28. The answer is A (1, 2, 3). [*Chapter 10 IV B 1 d (1), (2), (4)*] An infection within the synovial sheath of the tendons of the fifth digit may compress the vincula and cause ischemic necrosis of the tendons. It also can spread along the synovial sheath through the carpal tunnel and into the distal forearm. If the ulnar and radial bursae communicate, infection may spread distally into the thumb. Infection of the fifth digit cannot, however, spread distally into the tendon sheaths of the second, third, and fourth digits.

29. The answer is B (1, 3). [*Chapter 10 IV C 3 a; VII B 3 c; Table 10-2*] The recurrent branch of the median nerve in the palm innervates the muscles of the thenar compartment, including the abductor pollicis brevis, opponens pollicis, and the superficial head of the flexor pollicis brevis muscle, except the deep head of the flexor pollicis brevis, which is innervated by the ulnar nerve. The abductor pollicis longus is innervated by a branch of the radial nerve in the forearm.

30–34. The answers are: 30-D, 31-C, 32-D, 33-C, 34-D. [*Chapter 9 Table 9-1; Chapter 10 Table 10-1*] The palmaris longus, inserting into the palmar aponeurosis, flexes the wrist. The lumbricals and interossei both flex the metacarpophalangeal joint and extend the proximal and distal interphalangeal joints. At the metacarpophalangeal joint, the line of action of lumbrical muscles is farther from the joint axis than the interossei, which makes them more effective. Since dorsal and palmar interossei and the lumbricals insert into the extensor expansion (hood), which passes dorsally to the proximal and distal interphalangeal joints, they act to extend these joints with extension being most effective when contraction of the extensor digitorum communis takes the slack out of the extensor hood. Even though the extensor digitorum communis extends the metacarpophalangeal joint, it cannot sufficiently take the slack out of the extensor hood to extend the proximal and distal interphalangeal joints by itself.

35–40. The answers are: 35-B, 36-C, 37-D, 38-C, 39-A, 40-A. [*Chapter 7 Table 7-1; Chapter 9 Table 9-1; Chapter 10 Tables 10-1 and 10-2*] Most of the muscles of the flexor compartment of the forearm, including the pronator quadratus, are innervated by the median nerve; the two exceptions are the flexor digitorum profundus, which receives innervation from both median and ulnar nerves, and the flexor carpi ulnaris, which receives ulnar innervation. All of the muscles of the extensor compartment of the forearm, including the abductor pollicis longus, are innervated by the radial nerve. In the hand, the median nerve generally innervates the thenar muscles as well as lumbricals 1 and 2. The ulnar nerve generally innervates the hypothenar muscles, both dorsal and ventral interossei muscles, the adductor pollicis muscle, and lumbricals 3 and 4. The flexor pollicis brevis usually receives innervation from both median and ulnar nerves.

Part III
Thoracic Region

Thoracic Cage

I. INTRODUCTION

A. Thorax

1. Boundaries. The thoracic region is delineated by the ribcage and diaphragm.
  a. The superior thoracic aperture (thoracic inlet) defines the upper boundary and communicates with the neck and upper extremities (see Fig. 11-2).
  b. The inferior thoracic aperture (thoracic outlet) is closed by the **diaphragm** (see Fig. 11-2).

2. Contents. The thoracic region includes the heart and lungs as well as neurovascular connections between the neck, upper extremities, and abdomen.

B. Pectoral region. Portions of the anterior chest wall provide attachments for muscles that insert onto the pectoral girdle and arms (see Chapter 6 V A 1 b, B 1 b and Fig. 6-2).

II. SURFACE LANDMARKS

A. Thoracic vertebrae. The **spinous process (spine)** of each of the 12 thoracic vertebrae are palpable. These landmarks are used to define thoracic levels by number (see Figs. 11-2 and 19-3).

B. Sternum

1. The jugular notch of the **manubrium** is prominent and approximates an anterior boundary between the neck and thorax.

2. The sternoclavicular joint is visible where the clavicle articulates with the manubrium (see Fig. 6-1).

3. The sternal angle (of Louis) is the junction between the manubrium and the body of the sternum (Fig. 11-1). It is located at the articulations of the 2nd ribs and (because the ribs slope downward) at the level of the 4th thoracic vertebra.

4. The body of the sternum is palpable for its entire length (see Fig. 11-1).

5. The xiphoid process, along with the adjacent costal margins, defines the **infrasternal angle (notch)**.

C. Ribs

1. The rib cage consists of 12 pairs of ribs.
  a. Seven pairs of **true ribs** (1–7) articulate with thoracic vertebrae and the sternum (see Fig. 11-1; Fig. 11-2).
  b. Five pairs of **false ribs** (8–12) articulate with thoracic vertebrae and the superjacent costal cartilage to form the anterior **costal margin**. The last two pairs of false ribs (11 and 12) articulate only with thoracic vertebrae and are sometimes called **floating ribs** (see Figs. 11-1 and 11-2).

2. The 12 pairs of ribs and the associated intercostal spaces are used to describe thoracic locations anteriorly.

D. Pectoral girdle. The clavicles and scapulae form the pectoral girdle, which is prominent and palpable (see Chapter 6 II B and Fig. 6-1).

1. The clavicles articulate with the manubrium at the **sternoclavicular joint**. Motion in this joint

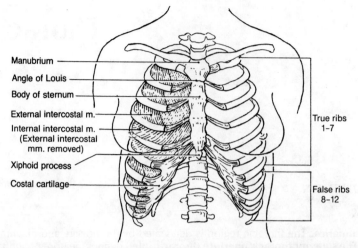

Figure 11-1. *Musculoskeletal thoracic cage.* The intrinsic thoracic musculature lies between the 12 pairs of ribs. The external intercostal muscles terminate at the costochondral junction, but the layer continues to the sternum as the external intercostal membrane. The internal intercostal layer is shown in the right 4th and 5th intercostal spaces.

is evident as the arm is circumducted. The **midclavicular line**, an important landmark, passes vertically through the center of the clavicles and approximately through the nipple in men and variably medial to the nipple in women (see Fig. 15-1).

2. **The scapula** is the most prominent dorsal landmark of the thorax and has several palpable features:
   a. The acromion process defines the superior surface of the shoulder.
   b. The coracoid process represents the third bone of the pectoral girdle that fuses with the scapulae.
   c. The spine of the scapula is obvious on the dorsal thoracic wall.
   d. The apex (lowest point) is usually located at the level of the 7th rib or 7th intercostal space.

E. **Mammary glands.** In the female, these are the most prominent superficial structures of the anterior thoracic wall (see Fig. 2-2). The **nipple** is located at the 4th intercostal space in men and variably lower in women.

F. **Internal thoracic organs.** While not directly palpable, during physical examination, the lungs and borders of the heart may be outlined by percussion and auscultation.

III. THORACIC CAGE

A. **Organization.** The bony elements of the thoracic cage consist of the **thoracic vertebrae**, **sternum**, and **ribs** (see Figs. 11-1, 11-2, and 19-3).

1. **Thoracic vertebrae.** Each of the 12 thoracic vertebrae has two main portions: the body and neural arch (see Fig. 19-3).
   a. The body of each vertebrae supports the mass of that portion superior to it. Thus, the bodies of successive vertebrae become more massive.
   (1) Each articulates with adjacent vertebrae through the **intervertebral disk**, which is an amphiarthrosis (a fibrocartilaginous joint with no capsule). The circumferential **annulus fibrosus** provides strength, while the enclosed gelatinous **nucleus pulposus** distributes pressure across the joint surfaces (see Fig. 19-5).
   (2) Each has facets for diarthrodial articulations with the heads of the immediately superior and inferior ribs (with some exceptions) near the junctions of the pedicles.
   b. The neural arch forms the **neural canal**, which contains the spinal cord. It is composed of pedicles and laminae (see Fig. 19-3).
   (1) Paired **pedicles** arise from the posterior aspect of the vertebral body and have facets for diarthrodial intervertebral articulation (see Fig. 19-3).
   (a) Facets on the **superior articular process** generally face posteriorly in the thorax.
   (b) Facets on the **inferior articular process** generally face anteriorly in the thorax.

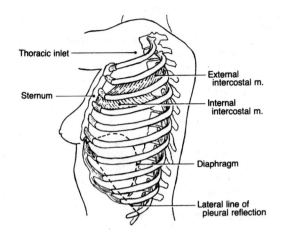

Thoracic inlet

Sternum

External intercostal m.

Internal intercostal m.

Diaphragm

Lateral line of pleural reflection

Figure 11-2. *Musculoskeletal thoracic cage.* Differences in the directionality and extent of the external and internal intercostal muscles are shown in the 2nd and 3rd intercostal spaces, respectively. Also depicted is the normal end-expiratory position of the diaphragm.

- **(2)** Paired **laminae** continue from the pedicles and fuse posteriorly in the midline (see Fig. 19-3).
- **(3)** A **spinous process** (spine) arises where the laminae meet (Fig. 19-3).
- **(4)** Paired **transverse processes** arise at the junction between the pedicles and the laminae and have articular facets for the tubercles of ribs 1–10 (see Fig. 19-3).
  - c. **The intervertebral canals (foramina)** transmit the **spinal nerves**. Each is formed by a deep notch in the inferior border of one vertebra and a shallow notch in the superior border of the subjacent vertebrae (see Fig. 19-5). Each is bounded by the intervertebral disk anteriorly and by the articulating zygapophyses posteriorly.

2. **Sternum.** Having developed as a segmental structure, the adult sternum retains segmentation that divides it into the **manubrium, corpus (or body)**, and **xiphoid process**. The articulation between the manubrium and corpus permits movement during respiration.
   - a. **Manubrium** (*L.* handle)
     - **(1) Structure.** The manubrium is somewhat shield-shaped. It is drawn upward by the action of the **sternocleidomastoid** and **scalene muscles** during extreme inspiratory effort.
     - **(2) Articulations.** It is anchored fairly firmly on each side by articulation with the clavicle at the **sternoclavicular joint** (see Fig. 6-1). It articulates with the synchondroses of the 1st ribs (see Fig. 6-1) and with the superior facets of the 2nd costal cartilages at the sternal angle, a useful landmark in counting ribs during physical examination (see Fig. 11-1). Finally, it articulates with the body of the sternum at the **manubriosternal synchondrosis** or **sternal angle (of Louis)** [see Fig. 11-1]. This joint acts as a hinge, allowing for anteroposterior expansion of the rib cage during inspiration.
   - b. **Corpus or body**
     - **(1) Structure.** This segment consists of four fused segments. It may be split (sternotomy) for surgical access to the middle mediastinum and superior mediastinum. Because there is very little fat over this bone, regardless of body build, it is a preferred site for bone marrow aspiration. It is sensitive to nociceptive stimulation; thus, firm rubbing of the sternum with the knuckles can rouse a stuporous patient.
     - **(2) Articulations.** In addition to the manubrium, the corpus articulates with the inferior facets of the 2nd costal cartilages at the sternal angle and with the 3rd–6th costal cartilages as well as with the xiphoid process.
   - c. **Xiphoid process**
     - **(1) Structure.** The xiphoid process is usually pointed but may be bifid. It can be palpated in the epigastrium in the infrasternal angle. Cartilaginous in the young, it calcifies during adolescence.
     - **(2) Articulation** with the sternum occurs at the **xiphisternal synchondrosis**, which usually fuses in the elderly.

3. **Ribs.** Twelve pairs of ribs form the main portion of the thoracic cage (see Fig. 11-1). Posteriorly, the ribs slope downward; anteriorly, they tend to slope upward slightly (see Fig. 11-2). The **true ribs** (1–7) become progressively longer; the **false ribs** (8–12) become progressively shorter.
   - a. **Structure**
     - **(1)** The **head**, an expanded portion, articulates with the vertebrae.
     - **(2)** The **neck**, the portion between the head and tubercle, is directed posterolaterally.
     - **(3)** The **tubercle** of the rib has a facet for articulation with the transverse process of the corresponding vertebra. Exceptions include the 11th and 12th ribs.

(4) The **shaft** of the rib has several features.
   (a) The **costal angle** is that region of the rib where it turns anteriorly.
   (b) The **costal groove** indicates the location of the **intercostal neurovascular bundle**. Although the neurovascular structures are protected within the costal groove, a fractured rib may injure the contained structures.
(5) The **costal cartilage** represents an unossified anterior portion of the rib. During inspiratory movements, the costal cartilages are bent upward and twisted. This flexion and torsion stores energy, which is subsequently released and used to assist expiratory effort (**extrinsic elastic recoil**).

b. Articulations
(1) The head articulates with the corresponding vertebral body and (except for the 1st rib) the immediately superjacent vertebral body via **inferior** and **superior articular facets**. Exceptions include the 1st, sometimes 10th, 11th, and 12th ribs, which articulate only with their corresponding thoracic vertebrae.
(2) The tubercle of ribs 1–10 articulates with the transverse process of the corresponding vertebra.
(3) The costal cartilages articulate with the manubrium and the body of the sternum in the following manner.
   (a) The 1st rib articulates with the manubrium below the sternoclavicular joint by a synchondrosis in which little or no movement is possible.
   (b) The 2nd costal cartilage articulates with the sternum at the sternal angle with one synovial joint on the manubrium and another on the body of the sternum.
   (c) The 3rd to the 6th or 7th costal cartilages articulate with the body of the sternum at the synovial joints.
   (d) The 7th or 8th to 10th costal cartilages articulate with the preceding costal cartilages at the synovial joints.
   (e) The 11th and 12th ribs (and 13th, if present), as floating ribs, end in the musculature of the abdominal wall.

c. Anomalies
(1) *Anomalous 13th ribs* sometimes exist.
   (a) A **cervical rib** may arise from the transverse process of vertebra C7 and usually articulates with the anterior third of the 1st rib; it may compress the lower trunk of the brachial plexus and the subclavian artery (misnamed the *"thoracic outlet"* syndrome).
   (b) A **lumbar ("gorilla") rib** may arise from vertebra L1.
(2) *Enlargement of the intercostal arteries*, such as occurs with coarctation of the aorta, results in erosion of the costal grooves, which is apparent on x-ray as scalloping of the inferior edges of the ribs.
(3) *Costochondritis* (inflammation of the synovial joints between the costal cartilages and sternum) is painful.
(4) *Calcification of costal cartilages* may occur in old age; the loss of resiliency affects ventilation and predisposes to fracture.
(5) *Deformities of or injuries to the thoracic cage* may impair ventilation of the lungs, which is the primary function of the thoracic cage.

B. Movement of the rib cage

1. Axis of rotation
a. Because there are two functional costovertebral joints between ribs 1–10 and the corresponding vertebra (one at the body and another at the transverse process), movement can occur about only one **axis of rotation** drawn through the centers of the two joints (see 1 and 2 in Fig. 11-3). Movement in either direction about one axis of rotation is described as **one degree of freedom**.
b. The defined axis of rotation extends diagonally across the midline and through the contralateral costochondral junction (see Fig. 11-3). Thus, the axes of opposite sides cross in the midline. Because the axes cross, the sternum elevates as the ribs rotate.

2. Group actions. External intercostal muscles on one side of the body, plus the contralateral interchondral portion of the internal intercostal muscles that have the same fiber direction *and* are on the same side of the axis of rotation (see A–B in Fig. 11-3) will act synergistically.

IV. MUSCULATURE OF THE THORAX

A. Extrinsic musculature. The extrinsic musculature of the thorax stabilizes and moves the pectoral girdle, upper limbs, and neck. Of these muscles, the **pectoralis major** (see Fig. 6-2); **pectoralis minor** (Fig. 11-4); **sternocleidomastoid** (a neck muscle) [see Fig. 11-4]; and the **anterior scalene**,

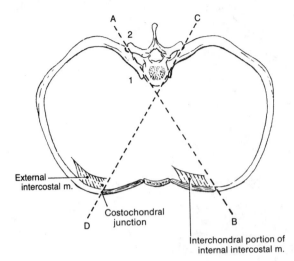

Figure 11-3. *Axis of rib rotation*. Rotation of the ribs occurs through two joints, one at the body of the vertebra (*1*) and the other at the transverse process (*2*), which define an axis of rotation (*AB*) that passes through the contralateral costochondral junction. Because the axes of the left and right ribs cross, the resultant movement of the shared arcs elevates the sternum. Because both the external intercostal muscle and the interchondral portion of the contralateral internal intercostal muscle have the same line of action and lie on the same side of the axis of rotation (*AB*), they have the same action.

middle scalene, and **posterior scalene** (also neck muscles) [see Fig. 27-4] are used to a significant degree as accessory muscles of respiration during extreme inspiratory effort (Table 11-1).

B. **Intrinsic musculature.** The intrinsic (body wall) musculature of the thorax and the abdomen is similar. There are three layers of muscles. On the anterior body wall, the outer layer runs down and medially, the intermediate layer runs down and laterally, and the deepest layer tends to run transversely. This "three-ply construction" provides excellent strength through bias bonding. All of the intrinsic muscles of the thorax and abdomen are used in respiratory effort.

1. **External intercostal muscles** constitute the outer layer of three thoracic musculature layers.
   a. **Attachments.** The external intercostals arise from the inferior border of the first 11 ribs and course diagonally and inferiorly to the superior border of the subjacent rib, extending from the edge of the vertebral column as far as the costochondral junction, where the muscle fibers cease, but the layer continues to the sternum as the **external intercostal membrane**.
   b. **Action.** These muscles elevate the ribs for inspiration.
      (1) Because of the directionality of the external intercostal muscles, each fascicle has a longer lever arm on the lower rib than on the upper rib. Hence, the lower rib is raised toward the upper when the muscles shorten. In Figure 11-5, points A and B are the axes of rotation of two adjacent ribs; line CD represents the line of action of the external intercostals; line AC is the lever arm of the muscle on the upper rib; and line BD is the lever arm of the lower rib. Because lever arm BD is greater than lever arm AC and because muscle tension (T) is equal at attachment points C and D, then (T) (BD) > (T) (AC). Thus, the lower rib will move toward the upper rib (D).
      (2) Each subsequent thoracic level is raised not only by the action of its external intercostals but also by the sum of the amounts of contraction within the external intercostals of the superior thoracic levels.
   c. **Innervation** is by intercostal nerves 1–11.

2. **Internal intercostal muscles**, the intermediate layer of the thoracic musculature with a fiber direction generally opposite to that of the external intercostals, can be divided into two functionally distinct muscle groups.
   a. **Internal intercostal portion**
      (1) **Attachments.** These arise from the superior border of ribs 2–12 and run diagonally and superiorly to the inferior border of the superjacent rib, extending from the costochondral junction as far as the angle of the rib where the muscle fibers cease, but the layer continues to the vertebral column as the **internal intercostal membrane**.
      (2) **Action.** This portion of the intercostal layer depresses (lowers) the ribs for forced expiration.
         (a) Because of the directionality of the internal intercostal muscles, each muscle fascicle has a longer lever arm on the upper rib than on the lower rib. Hence, the upper rib is drawn toward the lower rib. In Figure 11-5, points A and B are the axes of rotation of two adjacent ribs; line CD represents the line of action of the internal intercostals; line AC is the lever arm of the muscle on the upper rib; and line BD is the

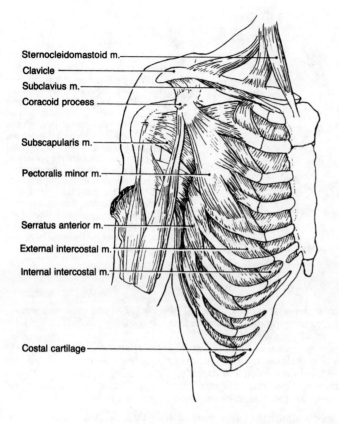

Sternocleidomastoid m.

Clavicle

Subclavius m.

Coracoid process

Subscapularis m.

Pectoralis minor m.

Serratus anterior m.

External intercostal m.

Internal intercostal m.

Costal cartilage

Figure 11-4. *Extrinsic thoracic musculature.* The pectoralis major muscle has been removed. In addition to their nonrespiratory actions, the pectoralis minor muscle elevates the ribs, and the sternocleidomastoid muscle elevates the manubrium.

lever arm of the lower rib. Because lever arm BD is less than lever arm AC and because muscle tension (T) is equal at attachment points C and D, then (T) (BD) < (T) (AC). Thus, the upper rib will move toward the lower rib (C).

    (b) Each subsequent thoracic level is lowered, not only by the action of its internal intercostal muscle proper, but also by the sum of the amounts of contraction within this muscle layer at lower levels.

  (3) Innervation is by intercostal nerves 1–11.

 b. Internal inter*chondral* portion

  (1) Attachments. These muscles run between the cartilaginous portions of ribs 1–10 from the sternum to the costochondral junction with the same directionality as the external intercostal muscle on the contralateral side.

  (2) Action. This portion of the internal intercostal layers elevates the ribs for inspiration. This paradox is explained by reference to Figure 11–3.

    (a) The axis of rotation of each rib passes approximately through the costochondral junction. Because this inter*chondral* portion of the internal intercostal muscle lies on the side of the rib axis of rotation opposite the internal intercostal proper, action in a similar direction will be antagonistic to the internal intercostals.

    (b) Because the internal inter*chondral* portion lies on the same side of the rib axis of rotation and has the same fiber directionality as the external intercostal of the opposite side of the chest, it will be synergistic with the external intercostals. Vectorial analysis illustrates these points.

  (3) Innervation is by intercostal nerves 1–11.

 c. Innermost intercostals, frequently poorly developed, are functionally part of the internal intercostal layer. The **intercostal neurovascular bundle** (intercostal artery, vein, or nerve) artificially divides the intermediate muscle layer into internal and innermost intercostals (see Fig. 11-6*B*).

3. Transverse thoracic (sternocostalis) muscles, frequently poorly developed, comprise the innermost layer of the thoracic musculature.

 a. Attachments are from the xiphoid process and the lower body of the sternum to ribs 2–6.

 b. Action is to lower the ribs in forced expiration.

 c. Innervation is by intercostal nerves 3–6.

Table 11-1. Muscles of Respiration—Inspiration and Expiration

Muscle	Origin	Insertion	Action	Innervation
Inspiration				
External inter-costals	Inferior border of ribs 1–11	Superior border of ribs 2–12	Elevate ribs	Nn. T1–T11
Interchondral portion of inter-nal intercostals	Inferior border of costal cartilages 1–9	Superior border of costal cartilages 2–10	Elevate ribs	Nn. T1–T9
Diaphragm	Costal margin	Central tendon	Lowers diaphragm	Phrenic n. (C3–C5) and twigs from T12–L2
Pectoralis major: Sternal head	Proximal third of clavicle, sternum, and costal carti-lages 1–6	Proximal shaft of humerus	Elevates ribs 1–6 when scapula is re-tracted and arm is extended	Medial pecto-ral n. (C5–C7)
Pectoralis minor	Ribs 2–5	Coracoid process	Elevates ribs 2–5 when scapula is re-tracted	Medial pecto-ral n. (C5–C7)
Sternomastoid	Mastoid process of cranium	Manubrium and proximal third of clavicle	Elevates sternum	Spinal acces-sory n. (CN XI)
Scalenes	Vertebrae C1–C6	Ribs 1–2	Elevate and stabi-lize ribs 1–2	Twigs from spinal nn. C3–C8
Expiration				
Internal inter-costals	Superior border ribs 2–12	Inferior border ribs 1–11	Depress ribs	Nn. T1–T11
Transverse tho-racic	Lower sternum and xiphoid pro-cess	Ribs 2–6	Depresses ribs 2–6	Nn. T3–T7
Rectus abdom-inis	Pubic arch	Costal cartilages 5–8, lower ster-num, and xiphoid process	Stretches and raises diaphragm	Nn. T5–L2
External oblique	Ribs 10–12 and iliac crest	Linea alba	Stretches and raises diaphragm	Nn. T7–L2
Internal oblique	Ribs 10–12 and thoracolumbar fascia	Linea alba	Stretches and raises diaphragm	Nn. T7–L12
Transverse ab-dominis	Ribs 5–12, thora-columbar fascia, and iliac crest	Linea alba	Stretches and raises diaphragm	Nn. T7–L2
Quadratus lumborum	Iliac crest	Rib 12	Stabilizes and lowers 12th rib	Twigs from nn. T12–L4

C. Diaphragm. The musculofibrous diaphragm separates the thoracic and abdominal cavities (see Fig. 11-2 and Fig. 18-4).

1. Structure
   a. The diaphragm is divided into several regions, reflecting its embryologic development.
      (1) The **costal portion** arises from the ribs that form the costal margin as far as the sternum.
      (2) The **sternal portion** arises from the xiphoid process and lower sternum.
      (3) The **crura (legs) of the lumbar portion** arise from the bodies of vertebrae L1–L3 (see Fig. 18-4).

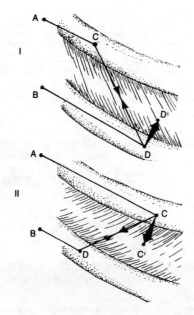

Figure 11-5. *Vectorial analysis of the intercostal muscle actions on the right side. I,* On contraction of the right external intercostal muscles, lever arm *BD* is longer than lever arm *AC,* resulting in elevation of the lower rib toward the upper rib. *II,* On contraction of the right internal intercostal muscles, lever arm *AC* is longer than lever arm *BD,* resulting in depression of the upper rib toward the lower rib.

 b. The central tendon is formed by the muscle fibers of the diaphragm, arching upward and converging (see Fig. 18-4). The right hemidiaphragm rises higher into the thoracic cavity than the left because of the liver in the upper right quadrant of the abdomen and the heart toward the left in the thoracic cavity.
 c. The aortic hiatus is formed by the left and right crura, which fuse over the aorta.
 d. The medial lumbocostal arch (medial arcuate ligament) is formed more laterally as the diaphragm arches from the vertebral bodies to the transverse process of L1. This arch bridges the uppermost origins of the psoas major muscle and the sympathetic chain.
 e. The lateral lumbocostal arch (lateral arcuate ligament) is formed as the diaphragm arches from the transverse process of L1 to the tip of the 12th rib. This arch bridges the **quadratus lumborum muscle**, which stablizes and depresses the 12th rib during expiration.

2. Apertures through the diaphragm
 a. The aortic hiatus, approximately in the midline, contains the aorta, azygos vein, and thoracic duct.
 b. The esophageal hiatus, located in the muscular portion on the left, contains the esophagus and the left and right vagus nerves; it is a frequent site of stomach herniation (hiatus or hiatal hernia).
 c. The vena caval foramen in the tendinous portion contains the inferior vena cava and branches of the right phrenic nerve.
 d. The splanchnic nerves and the hemiazygos vein pierce the crura.

3. Development. The diaphragm develops in the cervical region from at least six primordia, some of them paired, which include contributions from the septum transversum in the cervical region, pleuroperitoneal membranes, and the somatic wall. Failure of these primordia to fuse properly may result in anatomic defects through which herniation may occur.
 a. Incomplete closure of the pleuroperitoneal canal (usually on the left side) may allow herniation of abdominal contents into the thoracic cavity (*Bochdalek's hernia*).
 b. A foramen of Morgagni, the result of incomplete closure of the sternocostal triangle, is another potential, although rare, site of herniation.

4. Action. Since the muscle fibers of the diaphragm are radially arranged and insert into the central tendon, contraction of the diaphragm pulls the dome inferiorly into the abdomen, thereby increasing the thoracic volume during inspiration. Since the pericardial cavity and posterior mediastinum rest on the central tendon of the diaphragm, the diaphragm exerts a dynamic effect upon related structures. The position of the thoracic and abdominal organs varies with the respiratory cycle. Conversely, postural changes alter the ventilatory effectiveness of the diaphragm.

5. Innvervation. The motor and sensory innervation to the diaphragm is primarily by the phrenic nerves, which arise from levels C3–C5. However, there is some innervation from the lower thoracic and upper lumbar nerves to the crura. There is also some sensory innervation by intercostal nerves at the costal margin of the diaphragm.

    a. Voluntary control of the diaphragm is subject to override by the respiratory center of the brain stem.

    b. A phrenic nerve lesion may paralyze one hemidiaphragm.

    c. Hiccups are recurrent spasms of the diaphragm. Phrenicectomy is sometimes undertaken as a last resort to relieve chronic hiccups.

V. VASCULATURE OF THE THORACIC WALL

  A. Organization. The segmented skeleton of the thoracic cage is overlaid by segmentally innervated musculature (myotome) and skin (dermatome). On each side, each thoracic segment has a rib; a **neurovascular bundle** composed of a posterior intercostal artery and vein as well as the spinal nerve; and three layers of intercostal musculature interposed between the ribs (Fig. 11-6*A* and *B*).

    1. Posteriorly, the neurovascular bundle lies immediately behind the inferior edge of each rib in a slight depression, the **neurovascular (intercostal) groove** (Fig. 11-6*B*).

    2. Laterally, the neurovascular bundle gives off collateral branches, which move inferiorly across the intercostal space to lie immediately above the next rib anteriorly.

      a. Pleural taps (thoracentesis) usually are performed in the posterior or midaxillary line just above the superior edge of a rib.

      b. Posterior or lateral fractures of a rib may tear the associated vascular structures. If the parietal pleura that lines the pleural cavity is also torn by the sharp edges of a fractured rib, there may be bleeding into the pleural cavity (hemothorax).

  B. Arterial supply (see Fig. 11-6*A*)

    1. The aorta supplies the 3rd–11th **posterior intercostal arteries** and the **subcostal artery**.

    2. The internal thoracic (mammary) artery (a branch of the subclavian artery), supplies the:

      a. First through sixth anterior intercostal arteries, which anastomose with the highest and posterior intercostal arteries

      b. Superior epigastric artery, which anastomoses in the anterior abdominal wall with the **inferior epigastric artery** (a branch of the external iliac artery)

      c. Musculophrenic artery, which anastomoses with the 7th–10th posterior intercostal arteries

    3. The highest or superior intercostal artery arises from the **costocervical trunk** (a branch of the subclavian artery) to supply the 1st and 2nd posterior intercostal arteries.

    4. Pathways of collateral circulation

      a. In the event of *stenosis* or *coarctation of the aorta* (a congenital or slow-onset narrowing to ultimate occlusion), anastomotic connections between the posterior and anterior intercostals, as well as between the superior and inferior epigastric arteries, provide important collateral pathways.

      b. Hypertrophy of the intercostal arteries due to increased flow through collateral pathways results in erosion of the edges of the ribs in the vicinity of the neurovascular groove. This can be seen radiographically as notching or scalloping of the inferior edges of the ribs.

  C. Venous return. The veins generally parallel the arteries.

VI. NERVES OF THE THORACIC WALL

  A. Organization. Each of the 12 pairs of **thoracic nerves** is numbered after the vertebra just craniad to the intervertebral foramen through which each nerve passes; for example, nerve T4 emerges below vertebra T4.

    1. Dorsal sensory roots, which emerge from the dorsal aspect of the spinal cord, consist of afferent nerve fibers that mediate input from the sensory receptors of the body to the spinal cord (see Fig. 4-4).

      a. Afferent nerve classification

        (1) General somatic afferent (GSA) neurons from the body wall

        (2) General visceral afferent (GVA) neurons from the viscera

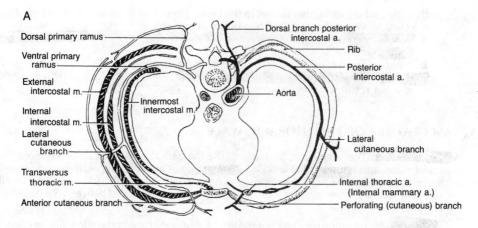

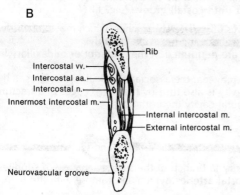

Figure 11-6. *Thoracic innervation and blood supply. A,* Branches of the right spinal nerve and left posterior intercostal artery. *B,* The relations of the intercostal neurovascular bundle to the rib and intrinsic thoracic musculature.

    b. The dorsal root ganglion within the intervertebral foramen contains the cell bodies of the sensory neurons.
    c. A dermatome is the region of skin supplied by the dorsal root of each spinal nerve (see Fig. 4-3). Because most areas of skin receive some innervation from adjacent spinal nerves (dermatomes overlap), the loss of one dorsal root or spinal nerve to a dermatome results in possible diminution rather than complete loss of sensation.

  2. Ventral (motor) roots emerge from the ventral aspect of the spinal cord (see Fig. 4-4); they consist of efferent nerve fibers that mediate output from the spinal cord to the effectors.
    a. Efferent nerve classification
      (1) General somatic efferent (GSE) neurons innervate the somatic musculature.
      (2) General visceral efferent (GVE) [autonomic] neurons innervate involuntary effectors.
    b. The ventral gray matter of the spinal cord (lateral and ventral horns) contains the cell bodies of the ventral root neurons.
    c. A myotome is a muscle group supplied by the fibers of the ventral root (see Fig. 4-3). Adjacent myotomes usually overlap so that loss of a single ventral root usually results in weakness rather than paralysis of a group of muscles.

  3. Spinal nerves are formed by the merger of dorsal and ventral roots in the intervertebral foramen (see Fig. 4-4). The spinal nerve is actually very short, dividing almost immediately into two major branches, the dorsal and ventral primary rami (see Fig. 11-6*A*).

B. Distribution (Fig. 11-6*A*)

  1. The dorsal primary ramus arches dorsally to innervate the posterior aspect of the dermatome and myotome (long muscles of the back).

  2. The ventral primary ramus continues ventrally as the intercostal nerve, which innervates the parietal pleura, intercostal muscles, and skin through the following branches.
    a. The lateral cutaneous branch penetrates the internal and external intercostal muscle layers at approximately the midaxillary line and subsequently bifurcates into **anterior** and

posterior branches of the lateral cutaneous nerve to supply the lateral aspect of the dermatome.

    b. The anterior penetrating branch penetrates the internal intercostal muscles and the external intercostal membrane just lateral to the sternum to supply the anterior aspect of the dermatome with some overlap across the midline.

    c. Intercostal nerve block is accomplished by injecting an anesthetic immediately beneath the inferior edge of a rib posteriorly.

    d. Nerve injury. Injury to an intercostal nerve, producing paralysis of intercostal musculature, is evident as a sucking in of the affected intercostal space upon inspiration and a bulging upon expiration.

VII. RESPIRATORY MECHANICS

A. Mechanical respiration. Ventilation can be divided into two categories—costal and diaphragmatic. Normally, ventilation is accomplished by a variable combination of costal and diaphragmatic contributions. In both types, air is moved into and out of the lungs by both muscular and mechanical forces.

  1. Intrathoracic (endothoracic, intrapleural) pressure (between lungs and thoracic wall) and **intrapulmonic pressure** (within the lungs) are measured or described relative to atmospheric pressure.

  2. The second law of thermodynamics states that flow (of air) must always be from higher energy (pressure) to lower.

  3. Boyle's law states that the product of pressure and volume is constant: $P_1V_1 = P_2V_2$. Since inspiration increases thoracic volume with a concomitant decrease in the intrapulmonic pressure, air will flow into the lungs (if the glottis is open) until the pressures equalize. Conversely, since expiration decreases the thoracic volume with a concomitant increase in intrapulmonic pressure, air must flow out of the lungs (if the glottis is open) until the pressures equalize.

B. Costal ventilation

  1. Normal (relaxed) costal inspiration lowers the intrathoracic and intrapulmonary pressures from about -4 cm H_2O to between -5 and -10 cm H_2O, producing the **tidal volume** (0.5–0.6 L). With a tidal volume of 0.5 L/breath and a respiratory rate of 14 breaths/minute, the lungs are ventilated with approximately 7 L of air/minute.

    a. Bucket-handle effects. The external intercostals and the interchondral portion of the internal intercostals increase the transverse diameter of the thorax by moving the ribs upward and outward (Fig. 11-7A).

    b. Pump-handle effect. Concomitantly, since the axes of rotation of the ribs on each side cross behind the sternum, elevation of the ribs moves the sternum forward and upward, thereby increasing the anteroposterior diameter of the thorax (Fig. 11-7B). The pump-handle action changes the sternal angle from about 160° in inspiration to about 180° in expiration.

    c. Since the interior surface of the rib is concave and since the rib rotates around its axis of rotation, there is a further increase in the transverse thoracic diameter when the ribs are elevated (Fig. 11-7C).

  2. Forced costal inspiration is required for ventilation in excess of tidal volume and contributes to the **inspiratory reserve volume** (about 3 L). This mechanism comes into action when respiratory requirements exceed 70–100 L/minute.

    a. In addition to the external intercostals, the pectoral muscles, which attach to the thoracic cage, assist with maximal elevation of the ribs to increase the transverse diameter of the thorax.

    b. The sternocleidomastoid and scalenes elevate the manubrium and 1st rib to increase the anteroposterior diameter of the thorax. (Recall the posture of a long-distance runner's neck and shoulders as he sprints for the finish line.)

  3. Relaxed costal expiration requires no muscular effort. Quiet tidal expiration is accomplished entirely by the elastic recoil of the costal cartilages and lungs.

    a. Extrinsic elastic recoil is provided by the costal cartilage.

      (1) Elevation of the ribs during inspiration deforms and twists the costal cartilage. When the external intercostals relax, this stored energy is released, rolling the ribs back to their resting (end-expiratory) position. Gravity also assists this mechanism.

A Frontal

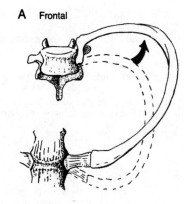

B Lateral

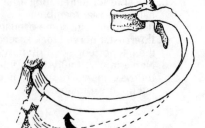

C Coronal section through ribs

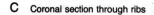

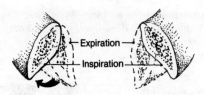

Expiration

Inspiration

Figure 11-7. *Respiratory movements of the ribs. A,* Bucket-handle motion of the left ribs whereby rotation about the axes produces elevation, which increases the transverse diameter of the thoracic cage. *B,* Pump-handle motion of the ribs whereby rotation about the crossed axes produces anterior elevation, which raises the sternum and increases the anteroposterior diameter of the thoracic cage. *C,* Because of the ellipsoid cross-sectional shape of the ribs, rotation also slightly increases the diameters of the rib cage.

(2) The importance of this mechanism decreases with age because calcification of the costal cartilages tends to immobilize the thoracic cage.
b. Intrinsic elastic recoil or **pulmonary compliance** is provided by the lungs.
  (1) The elastic fiber component of the interstitial tissue of the lung tends to cause the lung to shrink upon itself. Thus, the lung recoils after inspiratory stretching.
  (2) In conjunction with the intrinsic elasticity of the lung, the separation between the visceral pleura and parietal pleura by a thin flim of fluid produces a very high adhesive effect (surface tension). Thus, as the lungs deflate, they tend to pull the chest wall inward. Although this force, termed **intrathoracic pressure**, cannot be measured accurately, manometry shows that it may be between -3 cm H_2O and -5 cm H_2O (relative to atmospheric pressure) during relaxed tidal expiration and may reach values from -5 cm H_2O to -10 cm H_2O during relaxed tidal inspiration.
  (3) Diseases that reduce the compliance of the lung, such as *emphysema*, affect this aspect of ventilation.

4. Forced costal expiration necessitating muscular effort is required for air exchange greater than tidal volume. In addition to expelling the inspiratory reserve volume and the tidal volume, this mechanism provides **expiratory reserve volume** (about 1.2 L).
  a. Internal intercostals decrease the transverse and anteroposterior diameters of the thoracic cage, and intrathoracic pressure becomes positive (reaching as high as 300 mm Hg during a sneeze).
  b. The quadratus lumborum, running from the iliac crest to the 12th rib, lowers and stabilizes the 12th rib so that the thoracic cage can be depressed effectively.

C. Diaphragmatic ventilation

1. Diaphragmatic inspiration normally contributes to the **tidal volume** but also contributes to the **inspiratory reserve volume**.
  a. Contraction of the diaphragm lowers the level of the dome, thereby increasing the superoinferior thoracic diameter and the thoracic volume while decreasing the intrathoracic pressure. Vigorous diaphragmatic contraction may lower the dome by as much as 10 cm.
  b. During diaphragmatic contraction, the abdominal contents are displaced inferiorly and anteriorly, the movement being facilitated by relaxation of the abdominal wall muscles.

2. Diaphragmatic expiration contributes to **expiratory reserve volume**. The abdominal musculature comes into action when the respiratory requirements exceed 40–60 L/minute.

a. The abdominal musculature is antagonistic to the diaphragm. Since muscles accomplish work only when they contract, antagonistic muscles must contract to obtain work in the opposite direction.

b. Contraction of the abdominal muscles forces the abdominal viscera under the diaphragm (much like a fluid piston), thereby stretching the diaphragm into the thoracic cavity and decreasing thoracic volume to increase intrathoracic pressure.

3. Effects of gravity on ventilation. When a patient is in the sitting or standing position, the force of gravity acting on the abdominal and thoracic viscera tends to lower the diaphragm. With a patient in the horizontal position, the force of gravity on the abdominal viscera tends to stretch the diaphragm into the thoracic cavity (Fig. 11-8*A*). The plus and minus effects of gravity on ventilation explain the rationale for propping up a bedridden person with respiratory difficulty (Fig. 11-8*B*).

4. Normal balance between costal and diaphragmatic ventilation depends upon sex, body type, age, profession, state of health, and clothing. Children and elderly persons tend to breathe diaphragmatically, as do horn players and singers; obese people and women in advanced stages of pregnancy cannot effectively contract the diaphragm.

D. Approximate ventilation capacities

1. Tidal volume = 0.5–0.6 L.
  a. Tidal volume is limited to normal quiet inspiration and expiration.
  b. It is accomplished by the external intercostals and diaphragm with *return* by elastic recoil of the thoracic cage and lungs.

2. Inspiratory reserve volume = 3 L.
  a. Inspiratory reserve volume begins at the end of normal inspiration.
  b. It is accomplished by accessory muscles of inspiration in addition to the external intercostals and diaphragm with *return* by elastic recoil of the thoracic cage and lungs (assisted as necessary by expiratory muscles).

3. Expiratory reserve volume = 1.2 L.
  a. Expiratory reserve volume begins at end of normal expiration.
  b. It is accomplished by *in*ternal intercostals, the quadratus lumborum, and abdominal muscles with *return* accomplished by an expansionary elastic recoil of the thorax.

4. Vital capacity = 4.8 L (the sum of tidal volume and inspiratory and expiratory reserve volumes).

5. Anatomic dead space = 1.2 L (the nonrespiratory volume).

6. Total lung capacity = 6 L.

A

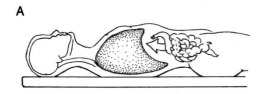

B

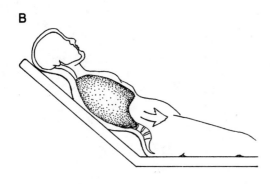

Figure 11-8. *Effects of posture on respiratory effort. A,* In a horizontal position, the weight of the abdominal viscera tends to stretch the diaphragm into the thoracic cavity. The patient must overcome this gravitational force on inspiration. *B,* A slight incline redistributes the abdominal viscera and facilitates inspiratory effort.

E. Clinical considerations

1. **Chest fractures**

   a. Fracture of several ribs in two locations (usually anteriorly and posterolaterally) produces "flail chest." The structural integrity of the thoracic cage is diminished with the loss of effective lever arms for the thoracic musculature. As the diaphragm is lowered, the flail portion of the wall cannot withstand the decrease in intrathoracic pressure and moves inward; conversely, as the abdominal muscles contract in expiration, the flail portion moves outward (*paradoxical respiratory movement*). The result is reduced ventilation and a decreased oxygenation/perfusion ratio. A positive-pressure respirator will correct the oxygenation/perfusion ratio.

   b. Fractured ribs may penetrate the thoracic wall or tear visceral pleura. Either circumstance can cause *pneumothorax*. Tearing associated blood vessels can result in *hemothorax*.

2. **A phrenic nerve lesion** may paralyze one hemidiaphragm.

   a. Paradoxical respiratory movements of a paralyzed hemidiaphragm result in poor oxygenation of blood in the corresponding lung.

      (1) During inspiration, expansion of the thoracic cavity with decreased intrathoracic pressure draws a paralyzed hemidiaphragm upward into the thorax, so that there is little or no increase in thoracic volume on the affected side.

      (2) During expiration, increased intrathoracic pressure draws a paralyzed hemidiaphragm inferiorly toward the abdomen so that there is little decrease in the thoracic volume on the affected side.

   b. Paralysis of one hemidiaphragm is evident on x-ray examination.

3. **Respirators**

   a. Negative-pressure devices, such as the Drinker ("iron lung"or tank) respirator, cyclically lower the extrathoracic and, therefore, the intrapulmonary pressure below atmospheric pressure, mimicking natural negative-pressure breathing.

   b. Positive-pressure devices cyclically elevate the atmospheric pressure above normal and above intrapulmonary pressure so that air is forced into the lungs.

I. INTRODUCTION

A. Basic principles. The vertebrate body consists of a somatic wall surrounding the coelom, within which the hollow viscera are suspended by mesenteries. Division of the common coelom produces the thoracic cavity and the peritoneal (abdominopelvic) cavity.

B. Subdivisions. The **thoracic cavity** contains the pleural cavities and the mediastinum.

1. The left and right pleural cavities are independent of each other and contain the **lungs**, each suspended by a mesentery.

2. The mediastinum separates the pleural cavities and is subdivided into four regions.
 a. The anterior mediastinum contains the inferior portion of the **thymus gland** and the sternopericardial ligaments.
 b. The middle mediastinum contains the **pericardial cavity** with its enclosed **heart, phrenic nerves**, and **lung roots**.
 c. The superior mediastinum contains the superior portion of the **thymus gland, aortic arch** and associated great vessels, **trachea** and **main stem bronchi, esophagus, superior vena cava** and its branches, and **autonomic nerves** to the heart and lungs.
 d. The posterior mediastinum contains the **esophagus, thoracic aorta, thoracic duct, inferior vena cava, azygos** and **hemiazygos veins, sympathetic chains**, and **vagus nerves (CN X)**.

II. PLEURAL CAVITIES

A. Parietal pleura. This layer of serous mesothelium lining the pleural cavities lies beneath the **endothoracic fascia**, which underlies the ribs, innermost intercostal muscles, and transverse thoracic muscle.

1. Topographic names describe the relation of the parietal pleura to adjacent structures, such as costal pleura, diaphragmatic pleura, mediastinal pleura, and cervical pleura (cupula of the pleura).

2. Lines of pleural reflection define the pleural cavities (Fig. 12-1).
 a. The anterior border of the right pleural cavity generally runs from the cupula (bounded by the base of the neck, clavicle, and trapezius muscle), through the right sternoclavicular joint, through the sternal angle in the midline, dropping in the midline to the level of the 6th costal cartilage or xiphoid process, where it begins to swing laterally to reach and then run just above the costal margin. The lower extent of this line of reflection is variable. A right subcostal incision carried into the sternocostal angle can nick the right pleural cavity, resulting in a *pneumothorax*.
 b. The anterior border of the left pleural cavity is similar, but it does not reach the midline; it deviates laterally at the level of the 4th rib to form the **cardiac notch** before diverging laterally at the level of the 6th or 7th costal cartilage.
 c. The inferior line of reflection of both pleural cavities moves progressively superior to the costal margin so that it is over the costochondral junction of the 8th rib and over the midshaft of the 11th rib.
 d. The inferior border of both pleural cavities runs horizontally across the 12th rib at midshaft to reach the vertebral column at about the 12th intervertebral level.
 (1) The inferior borders of the pleural cavities are variable.

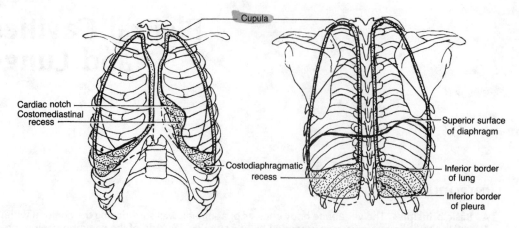

Figure 12-1. *Pleural cavities and lungs.* Because the lungs normally do not completely fill the pleural cavities, costo-diaphragmatic and costomediastinal recesses occur. The pleural recesses provide potential space for lung expansion on inspiration. The lines of pleural reflection vary between the extremes indicated by the *dashed lines*.

(2) A posterior abdominal incision inadvertently carried along the 12th rib runs a high probability of penetrating the pleural cavity and causing a pneumothorax.

e. **The posterior border of both pleural cavities** runs vertically between the 12th intervertebral level and the cupula.

3. **Pleural recesses** are regions of parietal pleura applied to parietal pleura where the lung fails to fill the pleural cavity. Pleural cavities are larger than pulmonary volumes (see Fig. 12-1). During deep inspiration, the lung parenchyma expands into, but never completely fills, these recesses. The pleural recesses are of considerable clinical importance.

a. **The left and right costodiaphragmatic recesses** are those parts of the pleural cavities between the costal pleura and the diaphragmatic pleura.

(1) When a patient is standing or sitting, the most dependent (lowest) part of the costodiaphragmatic recess is usually where the pleura passes over the 11th rib posterolaterally. In this position, pleural fluid accumulates in the costodiaphragmatic recess.

(2) Since the lung floats on pleural fluid, the fluid level can be determined by percussion.

(3) Blunting of the costodiaphragmatic recess will be visible radiographically with 400–500 ml of pleural fluid.

b. **The costomediastinal recess** is that part of the pleural cavity on the left side between the mediastinal pleura and the costal pleura. This recess is insignificant on the right side.

(1) The depth of the costomediastinal recess is variable.

(2) A needle inserted into the pericardial cavity or heart adjacent to the costomediastinal recess will avoid the pleural cavities and associated risk of pneumothorax.

4. **Pleural innervation**

a. **The costal pleura** is innervated by the **1st–12th intercostal nerves**.

b. **The peripheral diaphragmatic pleura** is innervated by the lower **intercostal nerves**, and pain from *irritation* may be *referred* to (appear as if originating from) the lateral thoracic and anterior abdominal walls because lower intercostal nerves innervate those regions.

c. **The central diaphragmatic pleura and mediastinal pleura** are innervated by the **phrenic nerves**; pain may be *referred* to the distribution of nerves C3–C5 (neck and shoulder).

5. **Clinical considerations.** Inflammation of the parietal pleura (*pleurisy*) is accompanied by pain (accentuated by respiratory movement), and a rasping sound (*friction rub*) may be heard on auscultation. If an effusion of fluid occurs, the friction rub disappears.

B. **Mesenteries (pulmonary ligaments).** A pulmonary ligament runs along the mediastinal surface of each lung from the **hilus** (the region where the vessels and air passages enter the lung) to the **base** (diaphragmatic surface).

1. **A mesentery** (frequently termed ligament) is formed by two layers of back-to-back mesothelium and supports a visceral structure from the body wall.

2. **The mesenteric pulmonary ligament** is formed where the parietal pleura of the anterior and posterior walls of the thoracic cavity is reflected off the mediastinal wall toward the lungs.

C. **Visceral pleura.** The mesothelium forming the pulmonary ligament diverges around the lungs to form the visceral pleura (pulmonary pleura). Thus, the parietal and visceral pleurae through the pulmonary ligament present a continuous surface.

1. **The intrathoracic cavity** is the potential space between these pleural layers. Introduction of air, blood, or fluid into the intrathoracic cavity results in *pneumothorax, hemothorax,* or *hydrothorax,* respectively.

2. **Pleural fluid,** normally a thin film, provides lubrication between the parietal and the visceral pleura, allowing the visceral surface to slide easily over the parietal surface. This thin fluid layer also provides tremendous surface tension between the parietal and the visceral pleural layers, which keeps the lung inflated. (For example, move a microscope slide over another; they can be pulled apart easily. However, place a drop of water between the slides, and repeat the procedure. There is less sliding friction, but the slides cannot be pulled apart.)

D. **Functional considerations**

1. **Normal conditions.** The pleural cavities are filled by the lungs with the exception of the costodiaphragmatic and costomediastinal recesses (see Fig. 12-1).
    a. The pleural recesses are only *potential spaces,* since parietal pleura is apposed to parietal pleura (i.e., costal pleura to diaphragmatic pleura in the costodiaphragmatic recess or to mediastinal pleura in the costomediastinal recess).
    b. The sizes of the pleural recesses vary with the respiratory cycle and with the degree of inspiratory and expiratory effort. As the thoracic cavities and lungs expand upon inspiration, a volume of lung tissue equivalent to the size of the inspiration moves into the pleural recesses. These changes in pulmonary volume between inspiration and expiration can be delineated by percussion.

2. **Pneumothorax**
    a. Negative intrathoracic (endothoracic) pressure (relative to atmospheric) and surface tension normally hold the lung surface against the thoracic wall. Thus, the lung remains inflated against its own intrinsic elasticity. A penetrating thoracic wound, the spontaneous rupture of a pulmonary bulla, a tear of abnormally fused pulmonary and parietal pleura, or the iatrogenic piercing of the pleural cavity, all of which allow entry of air into the pleural cavity, abolish the negative intrathoracic pressure and disrupt the surface tension. The unbalanced intrinsic elasticity results in collapse of the lung.
        (1) A **sucking pneumothorax** occurs if air enters and leaves the pleural cavity through the wound. This condition is accompanied by hyperexpansion of the chest wall on the normal side and a *mediastinal flutter* (a slight shift of the mediastinal mass toward the normal side during inspiration and toward the injured side during expiration).
        (2) A **tension pneumothorax** occurs if the wound acts as a check valve with air drawn into one pleural cavity during inspiration without expulsion of air from the pleural cavity during expiration. The resulting increase in intrathoracic pressure permanently forces the mediastinal contents toward the normal side (*mediastinal shift*), thereby reducing the vital capacity of the normal lung.
    b. Because blood continues to perfuse through the collapsed lung while little or no ventilation occurs in the collapsed lung, a significant portion of the cardiac output is not oxygenated. The result is *dyspnea* and even *cyanosis.* Mediastinal flutter, or a mediastinal shift, further reduces gas exchange by interfering with ventilation of the normal lung. Clearly, pneumothorax is a surgical emergency.
    c. Pneumothorax produces a hyperresonant percussive tone. Radiographically, the pleural cavity is markedly radiolucent with the free edge of the collapsed lung visible.

3. **Fluid in the pleural cavity**
    a. Fluid from pathologic processes (*hydrothorax* or *pyothorax,* if infected) or traumatic events (*hemothorax* or *chylothorax*) can collect in the costodiaphragmatic recesses when the person is erect.
        (1) Bleeding may result from injury to large or small thoracic vessels; chyle may issue from a torn thoracic duct; and fluid may result from direct or indirect pathologic processes (e.g., *pulmonary inflammation* and *congestive heart failure*).
        (2) Fluid levels may be ascertained by x-ray or by the level at which the percussive tympanic tone changes abruptly to a dull tone. As little as 500 ml of fluid may be seen on a thoracic x-ray film as blunting of the costodiaphragmatic recess.
        (3) Fluid accumulation reduces vital capacity of the lung and decreases the oxygenation-to-perfusion ratio.

b. Clinical considerations
  (1) Thoracentesis (pleural tap) usually is performed posteriorly to the midaxillary line with the patient in the sitting position. The fluid level is determined by percussion. At one or more intercostal spaces below the fluid level (but not below the 9th intercostal space), depending on the amount of pleural fluid, a needle is inserted immediately above the superior margin of a rib to avoid injury to the intercostal neurovascular bundle.
  (2) Chest tubes for the reduction of pneumothorax uncomplicated by fluid accumulation usually are inserted into the pleural cavity anteriorly through the 2nd intercostal space immediately superior to the 3rd rib. If fluids are or might be present, the chest tube is usually inserted in the posterior axillary line through the 5th or 6th intercostal space. Chest tubes are ordinarily maintained at -10 cm H_2O (relative to atmospheric pressure), approximating normal intrathoracic pressure.

III. LUNGS

A. Overview

1. The developing lung buds can be thought of as expanding into the pleural cavities, pushing the primitive mesothelium of the mediastinal pleura ahead of them. In this manner, the lung buds are covered by mesothelium (visceral pleura) with a connection maintained to the parietal pleura (pulmonary ligament) along the mediastinum.

2. Although the right lung is shorter because the liver is on the right side, it is wider because the heart is offset to the left. The right lung has a slightly larger capacity than the left.

B. External structure of the lungs. The lung surfaces are named to correspond to the related thoracic structures; for example, the diaphragmatic, costal, and mediastinal surfaces.

1. The apex lies in the **cupula**, which extends above the 1st rib into the base of the neck. Percussion and auscultation of the lung at the base of the neck and shoulder are performed routinely during thoracic examination.

2. The base is defined by the diaphragmatic surface.

3. The radix or root of the lung is on the mediastinal surface. Structures that enter or leave each lung at the **hilus** include the **main stem (primary) bronchus, pulmonary artery, pulmonary veins, bronchial lymphatics,** and **hilar lymph nodes.**

4. The pulmonary ligament, the mesentery of the lung, is formed by embryonic evagination of the lung buds into the thoracic cavity. It extends along the mediastinal surface from the hilus to the base and supports the lungs in the pleural cavity.

5. Lobes. Fissures of the visceral pleura divide the lungs into lobes (Fig. 12-2; Table 12-1).
  a. A **lobar (secondary) bronchus** supplies each lobe, accompanied by the lobar branch of the pulmonary artery and a lobar branch of a pulmonary vein.
  b. The **lingula** of the left lung, which corresponds to the middle lobe of the right lung, is formed by the more pronounced **cardiac incisure** in the left lung.
  c. The upper lobes lie more anteriorly, and the lower lobes extend posteriorly.

6. Bronchopulmonary segments are the functional units of the lungs.
  a. Structure. Each bronchopulmonary segment is characterized by:
    (1) A **segmental (tertiary) bronchus**, which lies centrally
    (2) A **segmental artery**, which accompanies the segmental bronchus
    (3) An **intersegmental septum** of connective tissue
    (4) Intersegmental veins, which form a peripheral venous plexus
  b. Distribution. There are 10 bronchopulmonary segments in the right lung and, depending upon the textbook, 8, 9, or even 10 described in the left lung (see Fig. 12-2; Table 12-2).

7. Clinical considerations
  a. A tumor of the upper lobe of either lung (Pancoast's tumor) may compress:
    (1) The subclavian or brachiocephalic vein, resulting in venous engorgement and edema of the face and arm on one side
    (2) A subclavian artery, resulting in diminished pulse in that extremity
    (3) A phrenic nerve, resulting in paralysis of a hemidiaphragm
    (4) A recurrent laryngeal nerve, resulting in vocal hoarseness
    (5) The sympathetic chain, resulting in Horner's syndrome
  b. Pathologic processes frequently are confined to a discrete bronchopulmonary segment. Surgical removal of a bronchopulmonary segment is feasible, but lobectomy is currently the more common procedure.

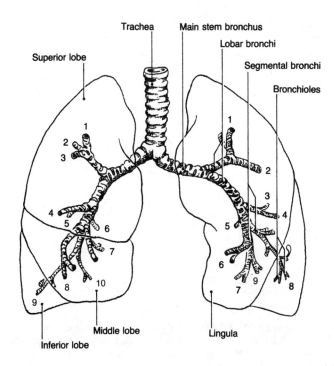

Figure 12-2. *Bronchial tree and lobes of the lung.* The right main stem bronchus divides into upper, middle, and lower lobar bronchi, which, in turn, give rise to a total of 10 segmental bronchi (*numbered*). The left main stem bronchus is more horizontal, divides into upper and lower lobar bronchi, and gives rise to 8–10 segmental bronchi (*numbered*).

IV. BRONCHIAL TREE

A. Trachea. The trachea is located in the superior mediastinum (see Fig. 12-2). It is approximately 12 cm long and 2 cm wide.

   1. The trachea is reinforced and kept patent by **C**-shaped rings (cartilages), which allow momentary expansion of the esophagus into the trachea during deglutition (swallowing).

   2. It bifurcates into main stem bronchi at the **carina** (septum) behind the sternal angle at level T5.

B. Main stem (primary) bronchi. Located posteriorly in the hilus behind the pulmonary artery (intermediate) and the pulmonary veins (anterior), each main stem bronchus supplies an entire lung.

   1. Right main stem bronchus (see Fig. 12-2)
   a. The right main stem bronchus is wider, shorter, and more vertical than the left main stem bronchus.
   b. It is the most probable resting place for large aspirated objects.
   c. It divides externally to the lung parenchyma into three lobar bronchi.

   2. Left main stem bronchus (see Fig. 12-2)
   a. The left main stem bronchus is narrower (because of smaller left lung volume), longer, and more horizontal than the right (because the heart is toward the left).
   b. It divides externally to the lung parenchyma into two lobar bronchi.

Table 12-1. Lobes and Fissures of the Lungs

Right Lung (3 lobes)		Left Lung (2 lobes)	
Anteriorly			
Upper lobe	⎫	Upper lobe	⎫
	Horizontal fissure		Oblique fissure
Middle lobe	⎬	Lower lobe	⎭
	Oblique fissure		
Lower lobe	⎭		
Posteriorly			
Upper lobe	⎫		
	Oblique fissure		
Lower lobe	⎭		

Table 12-2. Bronchopulmonary Segments of the Lungs

Right Lung	Left Lung	
Upper Lobe (3)	**Upper Lobe (4)**	
1. Apical	1. Apicoposterior	
2. Posterior	2. Anterior	
3. Anterior		
Middle Lobe (2)		
4. Medial	3. Superior lingular	
5. Lateral	4. Inferior lingular	
Lower Lobe (5)	**Lower Lobe (4–5)**	
6. Superior	5. Superior	
7. Basal medial	6. Basal medial	or Basal anteromedial
8. Basal anterior	7. Basal anterior	
9. Basal lateral	8. Basal lateral	
10. Basal posterior	9. Basal posterior	S MALP

C. Lobar (secondary) bronchi. Each bronchus supplies one lobe (see Fig. 12-2).

1. There are three lobar bronchi in the right lung and two in the left lung.

2. The **right lower lobar bronchus** is most vertical, most nearly continues the direction of the trachea, and is larger in diameter than the left; it is, therefore, the most common resting place for small aspirated objects. Such foreign bodies may block the airway and cause atelectasis (local collapse), inflammation, or even abscess.

3. The lobar bronchi divide into segmental bronchi.

D. Segmental (tertiary) bronchi. Each segmental bronchus supplies a bronchopulmonary segment.

1. The segmental bronchus enters the center of the segment accompanied by a segmental branch of the pulmonary artery.

2. A segmental bronchus may branch 6–18 times to produce 50–70 respiratory bronchioles, which become alveolar sacs and terminate in the alveoli, constituting the lung parenchyma.

3. Since the **superior segmental bronchi** of the left and right lower lobes are the most posterior when the body is supine, inhaled material (such as aspirated vomitus) easily enters the superior bronchopulmonary segments of the lower lobes. These segments, thus, are the ones most frequently involved in aspiration pneumonia (*Mendelson's syndrome*).

E. Lung. The pulmonary parenchyma represents the expanded terminal portion of the bronchial tree.

V. PULMONARY VASCULATURE

A. Arterial supply. Two separate systems of arterial vessels supply the lungs.

1. The pulmonary trunk arises from the **conus arteriosus** of the right ventricle of the heart; after approximately 5 cm, it bifurcates into the left and right pulmonary arteries (Fig. 12-3). **Pulmonary arteries** supply the respiratory (alveolar) parenchyma, carrying deoxygenated blood from the right ventricle.
    a. Right pulmonary artery (see Fig. 12-3)
        (1) The right pulmonary artery passes dorsally to the **ascending aorta** and the **superior vena cava.**
        (2) It crosses the right main stem bronchus to lie superiorly to that bronchus.
        (3) It lies posteriorly to the right superior pulmonary vein in the hilus.
        (4) It divides into three lobar, then segmental, arteries.
    b. Left pulmonary artery (see Fig. 12-3)
        (1) The left pulmonary artery passes anteriorly to the arch of the **descending aorta**, where it is connected to the arch of the aorta by the **ligamentum arteriosum** (a remnant of the **ductus arteriosus** or left 6th aortic arch) around which courses the recurrent laryngeal branch of the left vagus nerve.

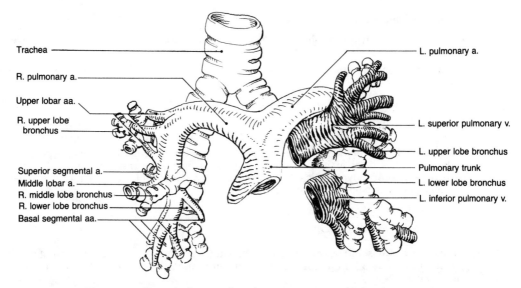

Figure 12-3. *Bronchial tree, right pulmonary arteries, and left pulmonary veins.*

(2) It crosses the left main stem bronchus to lie superiorly to that bronchus.
(3) It lies posteriorly to the left superior pulmonary vein in the hilus.
(4) It divides into two lobar, then eight to ten segmental, arteries.
- c. **Segmental arteries** enter the center of each bronchopulmonary segment with a segmental bronchus. The segmental arteries continue to subdivide to supply the capillary plexuses about each alveolus.
- d. **Clinical considerations. Pulmonary embolus** is a blocking body, such as a thrombus (thromboembolus), that is transported from one part of the systemic venous system (usually due to venous stasis or injury) or from the right side of the heart (usually due to valvular disease or myocardial infarct) to the pulmonary arterial circulation, where it becomes impacted. An embolus results in local **ischemia**. If the ischemia is severe, the tissues may undergo **necrosis**. The area of necrosis is termed an **infarct**. Pulmonary embolism is one of the greatest single causes of death in elderly injured or postoperative patients. A large embolus in a main stem or lobar artery may result in neurogenic shock and death within minutes. A small embolus may produce only mild pain, pleurisy, and perhaps hemoptysis before it resolves with healing. Supposedly, less than one-half of all pulmonary emboli are correctly diagnosed.

2. Bronchial arteries supply the conducting (nonrespiratory) portions of the lung with oxygenated blood.
- a. The left and right bronchial arteries arise from the descending aorta at the level of the tracheal bifurcation (T4–T5).
- b. During the course of pulmonary disease, bronchial arteries anastomose with the pulmonary circulation; these anastomoses may enlarge significantly to provide blood to the respiratory parenchyma.

B. Venous return (see Fig. 12-3). After transalveolar gaseous exchange, oxygenated blood returns to the left atrium via the pulmonary veins.

1. Venous capillaries arise from the alveolar capillary plexuses and coalesce peripherally in the connective tissue septa of the bronchopulmonary segments as the intersegmental veins.

2. The intersegmental veins combine to form the lobar veins.

3. The superior and middle lobar, or lingular, veins usually unite to form the superior pulmonary veins. The **inferior lobar veins** continue as the inferior pulmonary veins.

4. The left and right superior and inferior pulmonary veins generally enter the left atrium separately, but variations are common so that there may be three, four, or five openings into the left atrium.
- a. The left superior and inferior veins may unite to enter the left atrium as one vein.
- b. The right superior, middle, and inferior lobar veins may enter the left atrium separately.

C. **Pulmonary lymphatics**

1. **The pulmonary lymph nodes**, associated with the lobar bronchi, drain from the alveolar regions.

2. **Hilar, or bronchopulmonary, nodes** at the root of the lung receive lymph from the pulmonary nodes.

3. **Tracheobronchial nodes** are next along the main stem bronchi.

4. **The bronchomediastinal, or tracheal, nodes** unite with the **parasternal nodes** and drain into the subclavian veins.

5. **Clinical considerations. Pathways for metastatic spread of lung carcinoma** are provided by the lymphatics. If any of these lymph nodes become blocked by metastatic growth, there may be backup through lymphatic connections to the contralateral lung or to the celiac nodes in the epigastric region of the abdomen.

VI. PULMONARY INNERVATION (Fig. 12-4)

A. **Course and composition.** General visceral afferent (GVA) and general visceral efferent (GVE) innervation to the lungs are carried by branches from the thoracic sympathetic chain and the vagus nerve (CN X).

1. **Sympathetic nerves**
   a. The left and right sympathetic chains provide short splanchnic branches to the **anterior** and **posterior pulmonary plexuses**, which are located on the anterior and posterior aspects, respectively, of the main stem bronchi on each side.
   b. Sympathetic GVE neurons mediate vasoconstrictive function in the pulmonary vascular bed and secretomotor activity to the bronchial glands.

2. **Parasympathetic nerves**
   a. The left and right vagi in the mediastinum pass anteriorly to the main stem bronchi and, in this vicinity, give off twigs to the **anterior** and **posterior pulmonary plexuses.**
   b. Parasympathetic GVE neurons from the vagus innervate the bronchial smooth muscle. (Excessive stimulation produces the asthmatic syndrome.)
   c. The vagus nerves also carry respiratory reflex afferents (GVA).

B. **Regulation of respiration**

1. **Stretch reflexes.** The lungs contain stretch receptors innervated by neurons that run in the vagus nerve. When the lung is stretched, these GVA fibers inhibit the respiratory center in the brain stem (end-inspiratory reflex). There is similar GVA end-expiratory reflex.

2. **Pressure and chemical reflexes.** The **glossopharyngeal** (CN IX) and **vagus** (CN X) **nerves** innervate the **carotid** and **aortic bodies** and **sinuses**, respectively. These small structures located in conjunction with the branchial arches monitor changes in the partial pressure of oxygen and hemodynamic pressure, respectively. A decrease in PO_2 will increase inspiration through reflex mechanisms in the brain stem. An increase in blood pressure will decrease respiration through similar reflex mechanisms.

VII. FUNCTIONAL ANATOMY OF THE RESPIRATORY SYSTEM

A. **Basic principles.** The lungs are the organs through which ambient oxygen passes into the blood and ultimately reaches the intracellular mitochondria for oxidative metabolism. Similarly, the lungs are the organs through which the products of oxidative metabolism—carbon dioxide and some water vapor—are expelled to the atmosphere.

B. **Pulmonary function** depends on ventilation and respiration (alveolar gaseous exchange).

1. **Ventilation**, the mechanical displacement of gases, is effected by the muscles of the thoracic cage and abdominal wall, intrinsic elastic recoil of the thoracic cage, and intrinsic elastic recoil of the lungs.
   a. Ventilation moves gases through a series of airways with constantly decreasing diameters. Functionally, the airway is divided into conducting and respiratory portions.
      (1) The conducting portion consists of the trachea and bronchi as far as the terminal bronchiole. The air in the conductive portion at the end of inspiration does not exchange

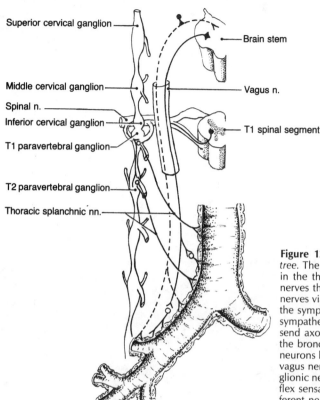

Superior cervical ganglion

Brain stem

Middle cervical ganglion

Vagus n.

Spinal n.

Inferior cervical ganglion

T1 spinal segment

T1 paravertebral ganglion

T2 paravertebral ganglion

Thoracic splanchnic nn.

Figure 12-4. *Autonomic innervation of the bronchial tree.* The first, or preganglionic, sympathetic neurons lie in the thoracic spinal cord, send axons to the spinal nerves through the ventral roots, and leave the spinal nerves via white rami communicantes to gain access to the sympathetic chain. The second, or postganglionic, sympathetic neurons lie in the sympathetic chain and send axons along thoracic splanchnic nerves to reach the bronchial tree. The preganglionic parasympathetic neurons lie in the brain stem and send axons along the vagus nerve (CN X) to the bronchial tree. The postganglionic neurons lie in the walls of the bronchial tree. Reflex sensation from the bronchial tree travels along afferent neurons (*dashed lines*) in the vagus nerve to the brain.

gases with the blood; similarly, the air in this portion at the end of expiration cannot be expelled (anatomic dead space).

  (2) The respiratory portion consists of respiratory bronchioles, alveolar ducts, alveolar sacs, and alveoli.

 b. Reduction of vital capacity (*flail chest, diaphragmatic paralysis, pneumothorax, pleural effusion, loss of intrinsic elasticity*) or obstruction of the airway (*mucus, asthma, neoplasm*) reduces ventilation, results in dyspnea, and may even produce cyanosis.

 2. Gaseous exchange occurs in the respiratory portion, primarily alveoli, whose total surface area is more than 70 m^2, or about the surface area of a tennis court. The alveoli are surrounded by capillary beds. **Alveolar respiration** is the exchange of gases across the **blood–air barrier**, which consists of the alveolar membrane, basement layer, and capillary endothelium, and is brought about by the differences in partial pressure between venous blood and alveolar air.

 a. Because gas exchange is very efficient, freshly oxygenated blood in the pulmonary veins normally reaches the same partial pressures as alveolar air. Ideally, alveolar ventilation is matched volume for volume by pulmonary vascular perfusion.

 b. However, excretion of CO_2 and absorption of O_2 may be reduced when the effective blood–air barrier is increased (**alveolar–capillary block**).

  (1) Gaseous diffusion across the alveolar membrane is a function of membrane thickness, surface area, partial pressure differentials, and the diffusion coefficients of the gases.

  (2) *Pulmonary fibrosis, pulmonary edema, bronchiectasis,* and *hyaline membrane disease* increase the **effective thickness** of the diffusion pathway, resulting in poor oxygenation. *Emphysema* reduces the **surface area** available for gas exchange, thereby reducing oxygen absorption. Finally, ineffective ventilation decreases pressure differentials so that the gas exchange becomes inefficient, and the blood perfusing the alveoli is incompletely oxygenated—in effect, an arteriovenous shunt.

13
Pericardial Cavity and Heart

I. MIDDLE MEDIASTINUM

A. Boundaries. The middle mediastinum is bounded by the left and right pleural cavities, diaphragm, anterior and posterior leaves of the fibrous pericardium, and superior boundary of the left and right pulmonary arteries.

B. Contents

1. Roots of the lungs

2. Ascending aorta

3. Pericardial cavity and heart

4. Pulmonary trunk and left and right pulmonary arteries

5. Left and right superior and inferior pulmonary veins

6. Superior and inferior venae cavae

7. Arch of the azygos vein

8. Left and right phrenic nerves

9. Lymph nodes

II. PERICARDIAL CAVITY

A. Characteristics. The pericardial cavity (sac) contains the heart and the roots of the great vessels (Fig. 13-1A).

1. It lies behind the sternum and the 2nd through 6th costal cartilages on the right side. On the left side, it extends inferolaterally beyond the costochondral junction of the 2nd through 4th ribs as far laterally as the midclavicular line.

2. The left and right **phrenic nerves** run along the lateral borders of the pericardial sac between the mediastinal pleura and the fibrous pericardium. It is for this reason that access to the pericardial cavity is usually gained by a longitudinal incision.

B. Structure. Two layers of pericardium comprise the pericardial sac.

1. **Fibrous pericardium** is an outer layer of tough, dense, irregularly arranged connective tissue, which is attached to the central tendon of the diaphragm and to the sternal periosteum by numerous small sternopericardial ligaments.
   a. It prevents transient overdistension of the heart but will stretch with time to accommodate chronic heart enlargement due to disease, such as congestive heart failure.
   b. It is continuous with the adventitia of the aorta, upper pulmonary trunk, superior and inferior venae cavae, and pulmonary veins.

2. **Serous pericardium** has two subdivisions.
   a. Parietal serous pericardium is a serous mesothelium lining the pericardial cavity.
   b. Visceral serous pericardium (epicardium) is a reflection of the mesothelial layer over the great vessels and surface of the heart to form the outermost layer of the heart wall. This layer can be thought of as being formed by invagination of the pericardial cavity by the developing heart.

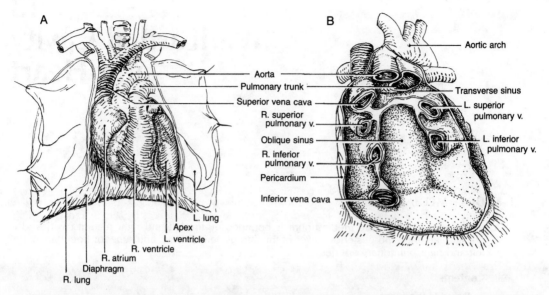

Figure 13-1. *A, Middle mediastinum*. The medial portions of the lungs have been retracted to show the location of the heart in the pericardial cavity. *B, Pericardial cavity*. With the heart removed, the line of pericardial reflection about the vessels that enter and leave the heart forms the oblique and transverse sinuses.

    c. Pericardial fluid is secreted from the serous mesothelium.
        (1) This keeps the surface layers moist and slippery, allowing nearly frictionless beating of the heart within the pericardial cavity.
        (2) It also provides surface tension between the pericardial layers with resultant expansion of the heart during inspiration. In conjunction with lowered inspiratory intrathoracic pressure, blood return to the right side of the heart is increased.

C. Pericardial sinuses. Reflection of the parietal pericardium around the great vessels to become the visceral pericardium results in the formation of sinuses.

    1. Oblique sinus (Fig. 13-1*B*)
        a. The oblique sinus represents a cul-de-sac in the pericardial cavity posterior to the heart.
        b. It results from reflection of the parietal pericardium around the inferior vena cava, right inferior and superior pulmonary veins, superior vena cava, and left superior and inferior pulmonary veins. Although not so named, this reflection is equivalent to a mesentery and suspends the heart in the pericardial cavity.

    2. Transverse sinus (see Fig. 13-1*B*)
        a. The transverse sinus is a passage behind the ascending aorta and pulmonary trunk.
        b. It is separated from the oblique sinus by the reflection of the pericardium (equivalent to a supporting mesentery) between the left and right superior pulmonary veins.
        c. It has clinical importance during cardiac surgery. A ligature may be passed through the transverse sinus and around the aorta and pulmonary trunk to control hemorrhage.

D. Clinical considerations

    1. Pericardial tamponade. An accumulation of fluid (blood or serous) in the pericardial sac will compress the heart and decrease diastolic capacity; this results in reduced cardiac output but an increased heart rate (indicated by a weak but rapid pulse) and increased venous pressure with associated jugular vein distension, pulsating liver, edema, and dyspnea resulting from pleural effusion.

    2. Pericardiocentesis and intracardiac injection
        a. Subcostal route. A needle may be safely introduced into the pericardial sac via the sternocostal angle adjacent to the xiphoid process. The needle is inserted on the left side adjacent to the xiphoid process, angling upward at about 45° and to the left. The risk of iatrogenic pneumothorax is small because the pleural cavities are avoided, and any fluid may be effectively drained if the patient is propped up slightly. Also avoided is the risk of puncturing the left anterior descending or the right marginal artery.

b. Parasternal route. Because of the cardiac notch in the pleura on the left side, a needle may be introduced into the pericardial cavity or heart through the left intercostal space immediately adjacent to the sternum with little chance of inducing a pneumothorax. Since the internal thoracic artery generally passes about 1 cm or more laterally to the margin of the sternum, this artery is avoided if the edge of the sternum is located with the tip of the needle before plunging into the pericardial sac. This is usually the preferred route for intracardiac injection.

3. Pericarditis. Inflammation of the parietal serous pericardium results in pain referred to (appearing as if originating from) the precordium and the epigastrium. A friction rub may be heard on auscultation.

III. HEART

A. Characteristics. The heart is located in the pericardial sac and covered with visceral pericardium (see Fig. 13-1). Normally, it weighs about 250 g in women and 300 g in men. The transverse diameter varies with inspiration and expiration but normally measures 8–9 cm in transverse diameter on a radiograph at end-maximum inspiration in the standing position. Generally, the transverse diameter should not be more than one-half the diameter of the chest.

B. Cardiac boundaries. Approximately one-third of the heart lies to the right of the midline and two-thirds to the left.

1. The right (acute) border delineates the superior vena cava, right atrium, and inferior vena cava.

2. The inferior border is delineated by the right ventricle.

3. The **apex** is the tip of the left ventricle. An **apical pulse** or **point of maximal impulse (PMI)** can be palpated and sometimes visualized in thin men at the 5th intercostal space just beneath the nipple.

4. The left (obtuse) border is formed by the left ventricle.

5. An upper border is indistinct radiographically.

C. Surface structure of the heart (Fig. 13-2)

1. The coronary (atrioventricular) sulcus completely encircles the heart and divides the atria from the ventricles. It marks the location of the annulus fibrosus.
    a. It has a nearly vertical orientation behind the sternum.
    b. It contains the **circumflex branch** of the **left coronary artery**, **right coronary artery**, **coronary sinus**, and **small cardiac vein**, which are all embedded in fat.

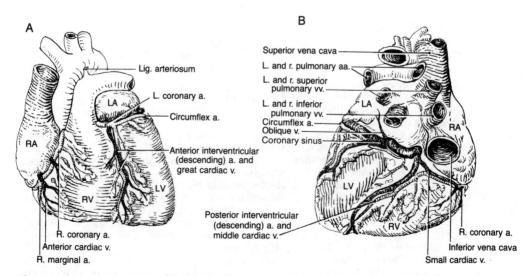

Figure 13-2. *A, Anterior aspect of the heart and coronary circulation. B, Posterior aspect of the heart and coronary circulation.*

2. **The anterior interventricular sulcus** marks the location of the **interventricular septum** separating the ventricles.
   a. It runs diagonally along the sternocostal surface close to the left border of the heart.
   b. It contains the **anterior interventricular (descending) artery** (a branch of the left coronary artery) and the **great cardiac vein**.

3. **The posterior interventricular sulcus** is the continuation of the anterior interventricular sulcus about the apex onto the diaphragmatic surface of the heart.
   a. It delineates the interventricular septum posteriorly.
   b. It contains the **posterior interventricular (descending) artery** (usually a branch of the right coronary artery) and the **middle cardiac vein**.

D. **Surface characteristics of the chambers of the heart.** The heart is divided into right and left sides, each of which is composed of two chambers, an atrium and a ventricle.

1. **The right atrium** forms the right border of the heart, part of the base, and part of the sternocostal surface.
   a. It gives off the right **auricular appendage**.
   b. It receives the superior vena cava, inferior vena cava, and coronary sinus.
   c. It discharges into the right ventricle through the tricuspid valve.

2. **The right ventricle** forms the largest portion of the sternocostal surface and a small part of the diaphragmatic surface.
   a. It projects to the left of the right atrium toward the apex.
   b. It receives blood from the right atrium.
   c. It discharges through the pulmonary semilunar valve into the pulmonary trunk just to the left of the aorta.

3. **The left atrium** forms the posterior surface (**base**) of the heart, lying entirely within the boundaries of the oblique sinus of the pericardial sac.
   a. It gives off the **left auricular appendage**.
   b. It receives two right pulmonary veins (sometimes three) and two left pulmonary veins (sometimes one).
   c. It discharges into the left ventricle through the mitral valve.

4. **The left ventricle** forms the left border, the **apex**, one-third of the sternocostal surface, and two-thirds of the diaphragmatic surface.
   a. It projects to the left of the left atrium toward the apex.
   b. It receives blood from the left atrium.
   c. It ejects blood into the aortic vestibule and through the aortic semilunar valve into the ascending aorta.

IV. CORONARY CIRCULATION

A. **Arterial circulation.** The myocardium and epicardium (visceral pericardium) are supplied by two main arteries, the **left** and **right coronary arteries**.

1. **Right coronary artery** (see Fig. 13-2; Fig. 13-3)
   a. **Course.** The right coronary artery arises from the **right aortic sinus** (of Valsalva) and runs in the coronary sulcus beneath the right auricular appendage and around the heart.
   b. **Major branches**
      (1) The **nodal branch**, which courses toward the superior vena cava, supplies the right atrium and the sinoatrial (S-A) node.
      (2) The **right marginal branch**, which passes along the inferior border toward the apex, supplies a portion of the right ventricle.
      (3) The **posterior interventricular (descending) branch**, which gives off a branch to the atrioventricular (A-V) node and anastomoses with the anterior descending branch of the left coronary artery on the diaphragmatic surface in the vicinity of the apex, supplies the posterior third of the interventricular septum and supplies the A-V node.
      (4) The **A-V nodal artery** arises from a deep loop of the right coronary artery immediately beyond the posterior interventricular branch.
      (5) The **right coronary artery** usually continues a few centimeters beyond the posterior interventricular sulcus to anastomose with terminal twigs of the **circumflex branch** of the left coronary artery.

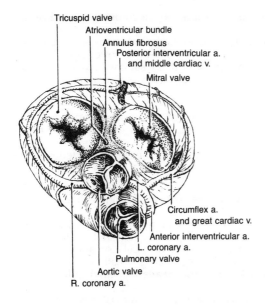

Tricuspid valve
Atrioventricular bundle
Annulus fibrosus
Posterior interventricular a.
and middle cardiac v.
Mitral valve

Circumflex a.
and great cardiac v.
Anterior interventricular a.
L. coronary a.
Pulmonary valve
Aortic valve
R. coronary a.

Figure 13-3. *Annulus fibrosus.* The left and right atria have been dissected away, showing the fibrous ring that supports the atrioventricular and semilunar valves and showing the origins of the left and right coronary arteries.

    c. Distribution. The right coronary artery generally supplies the right atrium (including the **S-A node**), the superior parts of the right ventricle, and the posterior third of the interventricular septum, including the **A-V node** and the **right branch of the A-V bundle** (of His). (Exception: In 10%–15% of the population, the posterior septum and A-V node are supplied by the left coronary artery.)

2. Left coronary artery (see Figs. 13-2 and 13-3)
    a. Course. The left coronary artery arises from the **left (left posterior) aortic sinus** (of Valsalva) and runs for a short, variable distance between the pulmonary trunk and aorta before bifurcating into the anterior interventricular and circumflex arteries.
    b. Major branches
        (1) The **anterior interventricular (descending) artery** enters the anterior interventricular sulcus.
            (a) It anastomoses with the posterior descending branch of the right coronary artery on the diaphragmatic surface of the heart in the vicinity of the apex.
            (b) It generally supplies the anterior aspects of the left and right ventricles and the anterior two-thirds of the interventricular septum, via septal branches including the **left branch of the A-V bundle** (of His) with some branches to the A-V node in about 40% of the population and total supply in 10%–15%.
        (2) The **circumflex artery** runs in the coronary sulcus toward the left border and around to the base of the heart.
            (a) It has as a major branch, the **left marginal artery**, which courses along the left border of the heart to supply the left ventricle.
            (b) It anastomoses with terminal twigs of the right coronary artery and gives rise to the posterior descending artery in 10%–15% of the population.
            (c) It generally supplies the posterior aspect of the left atrium and the superior portions of the left ventricle.

3. Variations in the coronary circulation
    a. Balanced coronary circulation is the coronary circulation described above.
        (1) It is present in 60%–65% of the population.
        (2) It is sometimes termed "right dominant" because the right coronary artery gives off the posterior descending branch, but this term does not distinguish this condition from right preponderant circulation.
    b. Left preponderant coronary circulation occurs when the circumflex branch of the *left* coronary artery gives rise to the posterior descending branch so that the entire interventricular septum, including the A-V node, is supplied by the left coronary artery.
        (1) It reduces possibilities for development of effective collateral circulation in the septal region and, therefore, lowers the chances of survival after septal infarct.
        (2) It is present in about 10%–15% of the population.

(3) It is sometimes termed "left dominant" because the left coronary artery gives off the posterior descending branch.

   c. Right preponderant coronary circulation occurs when the right coronary artery, in addition to supplying the posterior descending artery, crosses the posterior interventricular septum to reach as far as the left marginal artery so that a substantial portion of the diaphragmatic surface of the left ventricle is supplied by this extension of the right coronary artery.

   (1) It is present in 20%–25% of the population.

   (2) It is sometimes called "right dominant" because the right coronary artery gives off the posterior descending branch, but this term does not distinguish this condition from balanced circulation.

   d. Other variations frequently occur in the manner in which the coronary arteries arise from the aorta or even, lethally, from the pulmonary trunk.

4. Functional considerations

   a. Anastomoses. Normally, there are few anastomoses between left and right coronary arteries in the healthy, young individual.

   (1) In the normal healthy heart, most of the arteries are true end-arteries, which supply discrete volumes of myocardium with little overlap or few collaterals. Any anastomoses present are generally inadequate to maintain effective circulation in the event of sudden-onset occlusion of a major branch of the coronary circulation.

   (2) With slow-onset occlusion (atherosclerotic coronary artery disease), collateral circulation develops.

   (3) Obstruction in a coronary artery produces ischemia. Pain associated with ischemic cardiac tissue—**angina pectoris**—is *referred* to (appears as if originating from) the precordium, epigastrium, shoulder, and, frequently, the left arm. Ischemia may result in necrosis and hemorrhage (an infarct) with severe and life-threatening damage to the myocardium. If a patient survives the myocardial insult, the thrombus that forms over damaged endocardium may become dislodged and produce life-threatening emboli.

   b. Coronary perfusion. Because left ventricular systolic pressure and left ventricular transmural pressure are greater than or equal to aortic systolic pressure, blood flows through the coronary circulation of the *left* ventricle only during diastole. (Recall the second law of thermodynamics [Chapter 11 VII A 2].) Thus, the heart has a low margin of safety with respect to the amount of perfusion loss before ischemia results.

B. Venous circulation (see Fig. 13-2)

1. The coronary sinus receives most of the venous return from the epicardium and myocardium.

   a. Course. It is approximately 2 cm long and lies in the coronary sulcus, directed toward the diaphragmatic surface. It opens into the right atrium between the opening of the inferior vena cava and the right A-V valve.

   b. Tributaries

   (1) The **great cardiac vein** lies in the anterior interventricular groove.

   (a) It drains the anterior portion of the interventricular septum and the anterior aspects of both ventricles in the vicinity of the septum.

   (b) It accompanies the anterior descending (interventricular) branch of the left coronary artery.

   (2) The **middle cardiac vein** lies in the posterior interventricular groove.

   (a) It drains the posterior portion of the interventricular septum and the posterior aspects of both ventricles in the vicinity of the septum.

   (b) It accompanies the posterior descending (interventricular) branch of the right coronary artery.

   (3) The **small (lesser) cardiac vein** lies in the coronary sulcus to the right of the opening of the coronary sinus. It usually terminates as the **right marginal vein**, which occasionally opens separately into the right atrium.

   (a) It drains the marginal aspect of the right ventricle.

   (b) It accompanies the right coronary artery and its marginal branch.

   (4) The **oblique vein** (of Marshall), courses superolaterally across the posterior aspect of the left atrium. It is the embryologic remnant of the left common cardinal vein and may occasionally develop into a persistent left superior vena cava whether or not the right vena cava is present.

2. The anterior cardiac veins consist of one to several small veins that drain the anterior surface of the right ventricle.

   a. They course across the coronary sulcus anteriorly to the right coronary artery.

   b. They open separately into the right atrium.

3. The least cardiac (thebesian) veins or venae cordis minimae lie within the heart walls.
   a. They drain the endocardium and innermost layers of myocardium.
   b. They tend to empty directly into the atria with fewer draining into the ventricles.

V. CARDIAC WALL

A. Endocardium. The inner lining of the cardiac chambers is a thin, smooth endothelium.

B. Myocardium. This intermediate layer is composed of three layers of cardiac muscle, which are most definite and substantial in the ventricles.

   1. The myocardium varies in thickness between chambers, depending upon the functional requirements of each chamber.
      a. Each layer originates from the **annulus fibrosus** (fibrous ring); the layers are arranged in a spiral manner nearly perpendicular to each other. This produces a wringing maneuver to reduce the volume of the chambers during systole.
      b. Myocardial fibers do not cross the coronary sulcus.

   2. Excitable nodal tissue and the cardiac conducting system contained in the myocardium consist of the:
      a. S-A node, which initiates atrial systole
      b. A-V node, which is paced by the S-A node to initiate ventricular systole
      c. A-V bundle (of His), which transmits electrical activity throughout the ventricles

C. Epicardium. The serous visceral pericardium and a variable amount of fatty tissue form the outermost layer of the heart.

D. Fibrous ring (annulus fibrosus) [see Fig. 13-3]

   1. The annulus fibrosus is a layer of dense connective tissue, which is arranged in the A-V plane and demarcates the coronary sulcus. The annulus fibrosus forms the "skeleton" of the heart, which provides the site of origin for muscle layers of the atria and ventricles. It also surrounds and supports each valvular opening.

   2. The connective tissue stroma of the left and right A-V valves, as well as the pulmonary and aortic valves, is continuous with the fibrous ring.

   3. Only the A-V bundle passes through the fibrous ring.

VI. CHAMBERS OF THE HEART

A. Right atrium. The right atrium has the thinnest walls of the four chambers. The **crista terminalis** divides it into the sinus venarum and the atrium proper (Fig. 13-4).

   1. The sinus venarum is a smooth-walled region.
      a. It lies posteriorly and to the right of the crista terminalis, which represents the embryologic sinus venosus.
      b. It receives the:
         (1) Superior vena cava
         (2) Inferior vena cava, the valve of which (eustachian valve) is incompetent in the adult but in the fetus directs oxygenated blood from the inferior vena cava through the foramen ovale into the left atrium
         (3) Coronary sinus, the valve of which (thebesian valve) may reduce regurgitation of blood into the coronary sinus during atrial systole
         (4) Anterior cardiac veins

   2. The atrium proper comprises the anterior muscular portion of the atrium.
      a. It lies to the left of the **crista terminalis**, which is a muscular ridge running anteriorly along the atrial wall from just medial to the opening of the superior vena cava to just medial to the inferior vena cava. The crista terminalis gives origin to the **pectinate muscles**, which run across the atrial walls.
      b. The atrium contains the **right auricular appendage**, which has muscular walls and is a potential site for the formation of thrombi that, if dislodged, can result in pulmonary embolism.

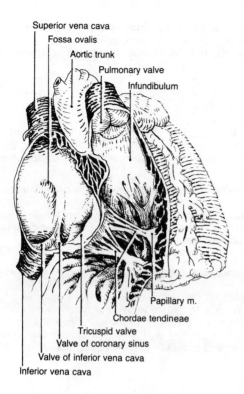

Superior vena cava
Fossa ovalis
Aortic trunk
Pulmonary valve
Infundibulum

Papillary m.
Chordae tendineae
Tricuspid valve
Valve of coronary sinus
Valve of inferior vena cava
Inferior vena cava

Figure 13-4. *Chambers of the right side of the heart.* The anterior heart wall has been dissected away, revealing the right atrium, tricuspid valve, right ventricle, and pulmonary valve.

    c. The **atrial septal wall** separates right and left atria.
        (1) The **fossa ovalis** is the embryologic remnant of the fetal **foramen ovale**, which connects the atria in the fetus and represents the **septum primum** (floor) and **septum secundum** (limbus or upper lateral edge). It may remain probe-patent in 10% of the population.
        (2) Atrial septal defects (ASDs) are most common in the vicinity of the fossa ovalis.
            (a) Septum primum defects are rare and usually accompanied by endocardial cushion defects involving valvular structure.
            (b) Septum secundum defects (the typical patent foramen ovale) comprise 10%–15% of all cardiac anomalies.
            (c) The left atrial pressure is normally slightly greater than the right atrial pressure so that there is typically a **left-to-right shunt** through an ASD.
                (i) Usually, there is no cyanosis associated with an ASD because oxygenated blood from the left side of the heart is shunted back to the right side.
                (ii) ASDs are usually compatible with a normal, healthy life unless extreme exercise, cardiac disease, or pulmonary disease alters normal chamber pressures, thereby inducing a **right-to-left shunt** with resultant cyanosis.
    d. The atrium proper empties into the right ventricle via the right A-V (tricuspid) valve.

B. Right ventricle. To overcome pulmonary vascular resistance, the right ventricle has a moderately thick wall with well-developed **trabeculae carneae.** It is separated into two regions, the **right ventricle proper** (for inflow) and the **infundibulum** (for outflow) by the **crista supraventricularis**, which lies between the right A-V valve and the pulmonary orifice (see Fig. 13-4).

    1. The right ventricle proper has rough, muscular walls with **trabeculae carneae** and anterior, posterior, and septal **papillary muscles.**
    a. The right A-V (tricuspid) valve transmits blood from the right atrium to the right ventricle.
        (1) It has three valve **leaflets** or **cusps:** anterior, posterior, and septal.
            (a) Each cusp has a connective tissue core that is attached to the annulus fibrosus and covered with endothelium.
            (b) Each is directed into the ventricle with connective tissues strands, **chordae tendineae,** running between the free edge and usually two papillary muscles.
            (c) Chordae tendineae are associated with a large **anterior papillary muscle,** a small **posterior papillary muscle,** and usually several very small **septal papillary muscles.**

(2) The tricuspid valve prevents regurgitation of blood into the atrium during ventricular systole.

   (a) During the isovolumic (pressure build-up) phase of ventricular systole, the papillary muscles tense to prevent eversion of the valve cusps.

   (b) Since the volume of the ventricle is reduced during the isotonic (blood ejection) phase of systole, shortening of the papillary muscles takes up any slack in the chordae tendineae to maintain the competence of the valvular closure.

b. **The interventricular septum** bulges into the right ventricle, giving the cavity a crescent shape in cross section.

   (1) It gives origin to the septal papillary muscle(s).

   (2) It is mostly muscular but has a small membranous upper portion.

   (a) The membranous portion is composed of connective tissue continuous with the annulus fibrosus.

   (b) It is the usual site of **ventricular septal defects** (VSDs). If small, a VSD may result in a left-to-right shunt; but in the presence of pulmonary stenosis, a VSD results in a **right-to-left shunt**, producing cyanosis and the ''blue-baby'' syndrome. A VSD is usually the principal factor in **tetralogy of Fallot**.

2. **The infundibulum** is smooth-walled and leads into the pulmonary orifice, which contains the **pulmonary semilunar valve**.

   a. The pulmonary valve consists of three semilunar cusps (demilunes): left, right, and anterior.

   (1) These cusps are attached peripherally to the annulus fibrosus.

   (2) Their free edges are strengthened with a rim of connective tissue, and a **nodule** in the center of the free edges completely closes the lumen. Their free edges are directed upward into the pulmonary trunk.

   b. The pulmonary semilunar valve prevents regurgitation of ejected blood from the pulmonary trunk into the ventricle during ventricular diastole.

3. **The pulmonary trunk** starts at the level of the pulmonary valves and lies to the left of the aortic outflow pathway for the first 3 cm.

   a. **Pulmonary sinuses** (of Valsalva), slight depressions in the walls of the pulmonary trunk, lie behind each valve leaflet.

   b. The pulmonary trunk bifurcates into the **right** and **left pulmonary arteries** at about 5 cm.

C. **Left atrium.** This chamber has slightly thicker walls than the right because of the increased effort required during atrial systole to overcome the elasticity of the extremely thick left ventricular walls (Fig. 13-5).

   1. The left atrium receives a variable number of right and left pulmonary veins—usually two from each side but frequently three on the right or one on the left.

   2. It contains the **fossa ovalis** on the septal wall, the remnant of the embryonic **foramen ovale**.

   3. It gives off the left **auricular appendage**, which has muscular walls and is a potential site for the formation of thrombi, which if dislodged can result in cerebral, systemic, or renal embolism. The left auricular appendage may be incised to provide access to the left atrium and mitral valve (i.e., mitral commissurotomy).

   4. It opens into the left ventricle via the left A-V valve.

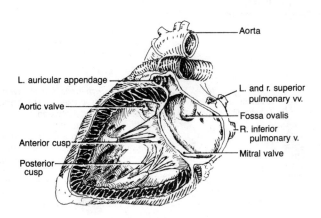

Figure 13-5. *Chambers of the left side of the heart.* The posterior heart wall has been dissected away revealing the left atrium, mitral valve, left ventricle, and aortic valve.

D. Left ventricle. To overcome the vascular resistance of the systemic vascular bed, the wall of the left ventricle is three times as thick as the wall of the right ventricle. The left ventricle is divided into the **left ventricle proper** and the **aortic vestibule** (see Fig. 13-5).

1. The left ventricle proper has the most muscular wall of all the heart chambers.

   a. The left A-V (bicuspid, mitral) valve transmits blood from the left atrium to the left ventricle proper (Fig. 13-6B).

   (1) This valve has a large **anterior cusp** and a small **posterior cusp** attached peripherally to the annulus fibrosus.

   (2) It has a greater number of **chordae tendineae** than the right A-V valve. The chordae run from the free edges of the cusps to a large **anterior papillary muscle** and a small **posterior papillary muscle**.

   (3) It prevents regurgitation of blood into the left atrium during ventricular systole in a manner similar to that described for the tricuspid valve.

   (a) *Mitral valve insufficiency* with transmission of the left ventricular systolic pressure to the left atrium and pulmonary vascular bed may lead to right heart failure (cor pulmonale).

   (b) *Mitral stenosis* will manifest on auscultation as late diastolic murmur; mitral insufficiency will be heard as a low-pitched, late systolic blowing murmur.

   b. The interventricular septum bulges into the right ventricle, giving the left ventricle a round cavity in cross section.

2. The aortic vestibule is located superiorly and to the right of the mitral valve. It leads into the **ascending aorta** and contains the **aortic valve**.

   a. The aortic valve consists of three semilunar cusps or demilunes, right (anterior), left (left posterior), and posterior (right posterior), which have anatomic characteristics similar to those of the pulmonary semilunar valves (see VI B 2).

   b. It prevents regurgitation of blood from the aorta into the ventricle during ventricular diastole.

   (1) A common anomaly is a bicuspid aortic valve, which is especially prone to stenosis with age.

   (2) *Aortic insufficiency* will manifest on auscultation as a diastolic murmur whereas aortic stenosis will tend to be heard as a high-pitched systolic murmur.

3. The ascending aorta starts at the level of the aortic valve (Fig. 13-6A).

   a. The aortic sinuses (of Valsalva), slight depressions in the walls of the aorta, lie behind the valve cusps.

   (1) The sinuses accommodate the volume of the open valve leaflets to reduce turbulent flow during ejection.

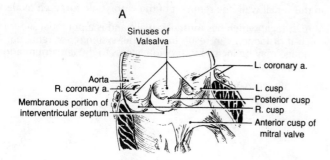

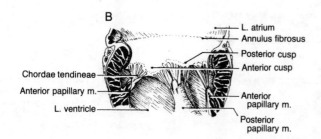

Figure 13-6. *Semilunar and atrioventricular valve structure. A,* The aorta has been opened between the left (posterior) and right semilunar valve cusps behind which are the ostia of the coronary arteries. *B,* The left ventricle has been opened, showing the anterior and posterior valve cusps, the chordae tendineae, and the anterior and posterior papillary muscles of the mitral valve.

       (2) They provide the origins of the **coronary arteries**.

          (a) The **right coronary artery** normally originates from the **right aortic sinus**.

          (b) The **left coronary artery** normally originates from the **left (left posterior) aortic sinus**.

     b. The elastic properties of the aorta accommodate the ejected blood volume and maintain the range of diastolic arterial pressure.

VII. CONDUCTING SYSTEM OF THE HEART

    A. Composition. Specialized cardiac muscle cells in certain regions of the myocardium have highly developed sensitivity and autorhythmicity.

       1. The S-A node is the autorhythmic pacemaker and initiates the contraction cycle with approximately 72 deplorizations a minute, which spread over both atria and to the A-V node (Fig. 13-7).

          a. Location. This node is about 7 mm x 2 mm x 1 mm and is located in the myocardium between the crista terminalis and the opening of the superior vena cava.

          b. Blood supply. It usually is supplied by the nodal branch of the right coronary artery.

          c. Innervation. It is influenced principally by the parasympathetic division of the autonomic nervous system, which slows the autorhythmicity.

       2. The A-V node is autorhythmic with approximately 40 depolarizations a minute. It is not connected to the S-A node by specialized conducting fibers but is stimulated by atrial depolarizations (see Fig. 13-7).

          a. Location. It is located in the right atrial floor near the interatrial septum, medially to the ostium of the coronary sinus and above the septal cusp of the tricuspid valve.

          b. Blood supply. It usually (in 85%–90% of cases) is supplied by the right coronary artery.

          c. Function. It has a slow conduction velocity (long latency), which allows atrial depolarization to spread over the entire right and left atria (0.09 second) before sending depolarization along the A-V bundle; thus, the atria normally contract before the ventricles.

          d. It gives rise to the A-V bundle.

       3. The A-V bundle (of His) arises from the A-V node and crosses the A-V ring (annulus fibrosus) [Fig. 13-7].

          a. Location. The A-V bundle descends along the posterior border of the membranous part of the interventricular septum to enter the muscular portion of the septum.

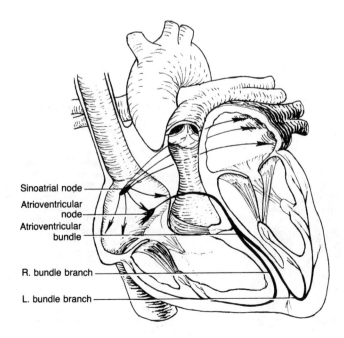

Sinoatrial node

Atrioventricular node

Atrioventricular bundle

R. bundle branch

L. bundle branch

Figure 13-7. *Conducting system of the heart.* Electrical activity initiated in the sinoatrial (S-A) node spreads throughout the atrial myocardium within 0.1 second, producing atrial systole and exciting the atrioventricular (A-V) node. From the A-V node, electrical activity enters the ventricles by the A-V bundle, travels along left and right bundle branches, and spreads over the ventricular myocardium to produce ventricular systole within 0.22 second.

    b. Branches. It divides into **left** and **right branches (bundles** or **crura)**; they descend into the interventricular septum and spread out into the walls of the ventricles, where they eventually become indistinguishable from the cardiac muscle fibers, which continue to conduct the impulse.

    c. Function. It consists of modified muscle cells (Purkinje fibers), which have a rapid conduction rate (low latency), so that the A-V bundle begins to depolarize at about 0.16 second after the S-A node and activity spreads throughout the ventricles by 0.22 second.

B. Clinical considerations

1. The contractile stimulus for any muscle is an electrical deplorization of its surface membrane. The heart has a large surface-to-volume ratio and consequently a large signal is generated that is detectable on the body surface by means of an electrocardiogram (ECG).

2. If a pathologic condition interrupts impulse propagation in the A-V bundle, heart block occurs with asynchronous beating of the atria and ventricles. If deplorization is initiated outside of the nodal system, the ventricular muscle may then fibrillate, resulting in no blood being pumped.

3. Bundles of Kent are occasional abnormal muscle bridges between the atria and the ventricles, which may cause ventricular preexcitation by bypassing the A-V node (Wolff-Parkinson-White syndrome).

VIII. CARDIAC INNERVATION

A. Motor control. The heart rate and ejection volume are controlled by the autonomic nervous system (Fig. 13-8).

1. Parasympathetic division. The **vagus nerve** sends fibers over the surface of the heart and to the nodal areas.

    a. These preganglionic fibers synapse with minute postganglionic fibers in the myocardium.

    b. Vagal activity slows the heart rate and reduces the stroke volume.

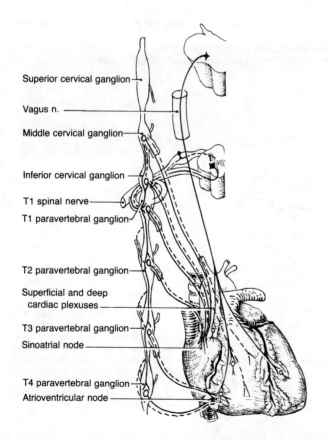

Superior cervical ganglion

Vagus n.

Middle cervical ganglion

Inferior cervical ganglion

T1 spinal nerve

T1 paravertebral ganglion

T2 paravertebral ganglion

Superficial and deep cardiac plexuses

T3 paravertebral ganglion

Sinoatrial node

T4 paravertebral ganglion

Atrioventricular node

Figure 13-8. *Autonomic innervation of the heart.* The first, or preganglionic, sympathetic neurons in the thoracic spinal cord (levels T1–T4) send axons through the ventral roots, spinal nerves, and white rami communicantes to gain access to the sympathetic chain. The second, or postsynaptic, neurons lie in the sympathetic chain ganglia and send axons either through the cervical cardiac accelerator nerves or through the thoracic splanchnic nerves to the heart. The preganglionic parasympathetic neurons lie in the brain stem and send axons along the vagus nerve (CN X) to the heart. The short postganglionic neurons lie in the walls of the heart. Axons conveying painful sensation from the heart travel along the sympathetic pathways to the upper thoracic segments of the spinal cord.

2. Sympathetic division. The **cardiac accelerator nerves** run in the neck to the heart from the superior, middle, and inferior ganglia of the cervical sympathetic chain. **Thoracic splanchnic nerves** run to the heart from ganglia T1–T3 of the thoracic sympathetic chain.
    a. These sympathetic fibers terminate in the vicinity of the S-A and A-V nodes.
    b. Sympathetic activity accelerates the heart rate and increases stroke volume.

B. Afferent sensation from the heart. Afferent nerves run along sympathetic pathways via both the cardiac accelerator nerves and the thoracic splanchnic nerves to reach levels T1–T4 of the spinal cord (see Fig. 13-8).

    1. Pain originating in the heart is referred to (perceived as coming from) the arm, shoulder, precordium, and sometimes epigastric region.

    2. The concept of referred pain must be understood to conduct and evaluate properly clinical signs and symptoms.
        a. Painful sensation mediated by visceral afferent fibers that enter the spinal cord at a particular level is referred to the somatic dermatome corresponding to that vertebral level.
        b. The mechanism of referred pain is not clear, but the most probable explanation involves a final common pathway in the spinal cord with cerebral interpretation toward the somatic rather than the visceral location.

IX. CARDIAC DYNAMICS

A. The cardiac cycle is approximately 0.8 second long (Fig. 13-9).

    1. Atrial systole (contraction)
        a. Atrial contraction is approximately 0.1 second long and produces pressures of approximately 5 mm Hg in the right atrium and 11 mm Hg in the left atrium.
        b. It pumps a small amount of blood into the ventricles. (Since the A-V valves are open during ventricular diastole and since ventricular filling is primarily passive, atrial systole results in a slight stretching of the ventricular muscle, thereby placing the heart muscle at the optimal point on the length-tension curve.) [Fig. 13-10*A* and *B*]

    2. Atrial diastole (relaxation)
        a. Atrial relaxation begins at 0.2 second into the cardiac cycle and is approximately 0.7 second long.
        b. It coincides with pressures of − 5 mm Hg in the right atrium and 0 mm Hg in the left atrium.

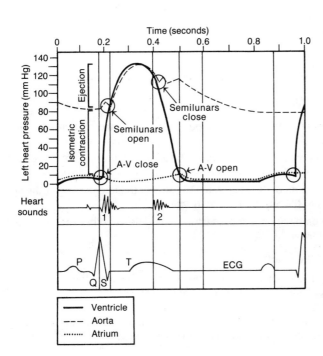

Figure 13-9. *Schematic diagram of the pressures developed during the cardiac cycle.* The atrioventricular valves close (S1, ''lub'') early in ventricular systole when ventricular pressures exceed atrial pressures. The semilunar valves close (S2, ''dup'') early in ventricular diastole when arterial pressures exceed ventricular pressures.

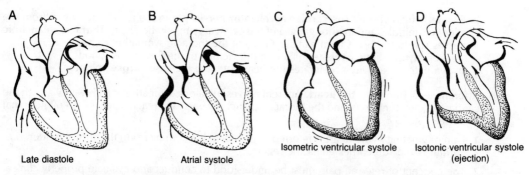

Late diastole   Atrial systole   Isometric ventricular systole   Isotonic ventricular systole (ejection)

Figure 13-10. *Cardiac cycle.* The cardiac cycle normally requires 0.8 second. *A,* Late diastole takes 0.4 second to 0.8 second. *B,* Atrial systole takes 0.0 second to 0.1 second. *C,* Isometric ventricular systole takes 0.1 second to 0.2 second. *D,* Isotonic ventricular systole takes 0.2 second to 0.4 second.

   3. Ventricular systole
- **a.** Ventricular contraction begins at approximately 0.1 second into the cardiac cycle and is approximately 0.3 second long.
- **b.** It produces pressures of approximately 23 mm Hg in the right ventricle and approximately 120 mm Hg in the left ventricle.
- **c.** It causes the A-V valves to close when ventricular pressure exceeds atrial pressure, thus producing the **S1 ("lub") heart sound** and beginning a period of isometric contraction during which pressure builds up (Fig. 13-10C).
- **d.** Ventricular systole causes the semilunar valves to open when ventricular pressure exceeds pulmonary diastolic pressure (10 mm Hg) and aortic diastolic pressure (80 mm Hg), thus beginning the period of isotonic contraction during which blood is ejected into the outflow trunks (Fig. 13-10D) and producing the maximum systolic pressure.

   4. Ventricular diastole
- **a.** Ventricular relaxation begins at approximately 0.4 second and has a 0.5-second duration.
- **b.** It results in approximately – 7 mm Hg pressure in the right ventricle and 0 mm Hg pressure in the left ventricle.
- **c.** It causes the semilunar valves to close when pulmonary (22 mm Hg) and aortic (120 mm Hg) pressures exceed end-systolic ventricular pressure, producing the **S2 ("dup") heart sound**. The dicrotic notch in the aortic pressure trace is probably due to elastic rebound when valve closure stops the momentary reverse flow of blood.
- **d.** Ventricular diastole coincides with rapid inflow of blood into the ventricular chambers caused by low intrathoracic pressure and the bulb-syringe effect (*vis a fronte*) of the ventricular walls (– 5 mm Hg right, 0 mm Hg left).

B. Ejection volume and blood pressure

   1. Ejection volume. The volume of blood ejected into the aorta during ventricular systole stretches that vessel and increases the elastic energy within its walls.
- **a.** This elastic pressure moves the blood from the aorta during diastole and maintains the range of diastolic pressure.
- **b.** With an ejection volume of 60–70 ml/beat and a pulse rate of 80 beats/minute, the heart will pump 5 L/minute.

   2. Systolic pressure can be a function of ejection volume. The greater the volume of blood ejected, the more the elastic walls of the aorta are stretched and the higher the systolic pressure.

   3. Diastolic pressure can be a function of the heart rate. The slower the rate, the lower the aortic diastolic pressure falls before the subsequent heart beat.

   4. Pulse pressure normally is a combination of these factors, in addition to other variables, such as peripheral vascular resistance and blood volume.

C. Projection of surface sounds of the heart

   1. Valve locations. The four heart valves not only are close together, but from the anterior–posterior viewpoint, the semilunar valves, as well as the A-V valves, are nearly superimposed behind the sternum.

   2. Conduction velocity. Because of the difference in the conduction velocity of sound through different types of tissues, the sound produced by each cycling valve is auscultated with maximal clarity and with minimal contribution from other valves over a distinct area of the thoracic wall (Fig. 13-11).

3. Auscultation areas
  a. Tricuspid valve sounds project to the midline and to the right side of the sternum at the 5th intercostal space.
  b. Mitral valve sounds project to the apex of the heart at the 5th intercostal space below the left nipple—the point of maximum impulse (PMI) or apical pulse.
  c. Aortic semilunar valve sounds project to the right of the sternum over the 2nd intercostal space and also can be auscultated in the neck over the carotid artery. Inspiration lowers endothoracic pressure and increases venous return to the right side of the heart. This increased filling can result in closure of the pulmonary valve momentarily after the aortic valve, producing a "split" S2.
  d. Pulmonary semilunar valve sounds project to the left of the sternum over the 2nd intercostal space. The murmur produced by a patent ductus arteriosus can best be auscultated just lateral to this point.

D. Valvular defects

  1. Valvular insufficiency. A valve is incompetent or insufficient if it permits backflow. Valvular defects can be congenital or can result from an infarct in the vicinity of a papillary muscle, rupture of the tendinous cords, or endocardial inflammatory processes.
  a. Tricuspid valve insufficiency leads to right-side heart failure in which right ventricular pressure is transmitted back to the venous system to produce distended neck veins, a pulsating liver, abdominal ascites, and edema.
  b. Mitral valve insufficiency leads to left-side heart failure in which left ventricular pressure is transmitted back to the lungs to produce pulmonary edema and pleural effusion (*hydrothorax*). This condition may, in turn, induce right-side heart failure.

  2. Valvular stenosis. A valve is **stenotic** if it restricts forward flow; stenosis can be either congenital or secondary to endocardial inflammatory processes.

  3. Auscultation. Both stenotic and incompetent valves produce murmurs that result from turbulent blood flow and can be localized by auscultation.

  4. Congestive heart failure. As the volume of blood flow through the heart changes, the heart compensates by increasing or decreasing its work (pressure-volume product). Shunts and valvular stenosis or incompetence increase the need for cardiac compensation. However, a point is reached where decompensation occurs with resultant **cardiac failure**. Stenosis may be amenable to simple surgical correction, but insufficiency usually requires correction by means of a prosthetic valve.

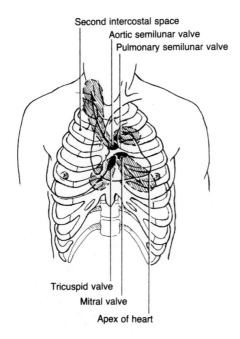

Second intercostal space
Aortic semilunar valve
Pulmonary semilunar valve

Tricuspid valve
Mitral valve
Apex of heart

Figure 13-11. *Surface projections of the heart sounds.* The atrioventricular and semilunar valves lie in the plane of the coronary sulcus and are nearly vertical behind the sternum. The mitral valve sound projects to the 5th intercostal space on the left, and the tricuspid valve is heard most distinctly in a region just to the right of the sternum in the 4th or 5th intercostal space. The aortic valve projects to the 2nd intercostal space at the right of the sternum and radiates into the neck. The pulmonic valve projects to the 2nd intercostal space at the left of the sternum.

X. FETAL AND EARLY POSTNATAL CIRCULATION

A. Fetal development. During fetal development, most of the blood passing through the heart must be shunted around the collapsed lungs (Fig. 13-12).

    1. The foramen ovale in the developing interatrial septum remains patent and shunts a portion of the blood to the systemic side.

    2. The ductus arteriosus (the 6th branchial arch on the left side) remains patent connecting the aorta (4th arch) and the left pulmonary artery. Blood entering the right ventricle flows into the aorta through this duct. Some blood does, however, flow through the lungs and is returned to the left atrium via the pulmonary veins.

B. Fetal blood flow (see Fig. 13-12)

    1. Ductus venosus. Fetal blood is charged with oxygen and nutriments in the placenta and returns to the fetus via the umbilical vein.

        a. Over one-half of the blood bypasses the liver through the ductus venosus and enters the inferior vena cava directly. The remaining blood from the umbilical vein drains through the portal vein and sinusoids of the liver before passing through the hepatic vein and joining the bypassed blood in the inferior vena cava.

        b. The blood flow from the umbilical vein is regulated by a muscular sphincter in the ductus venosus. When the sphincter relaxes, more blood passes through the ductus venosus; when the sphincter contracts, more blood is shunted to the portal vein and liver.

    2. Foramen ovale. After passing through the inferior vena cava, the blood enters the right atrium.

        a. Because the inferior vena cava also contains deoxygenated blood from the lower extremities, abdomen, and pelvis, the blood entering the right atrium is not quite as well oxygenated as that in the umbilical vein.

        b. The blood is guided by the valve of the inferior vena cava (eustachian valve) into the left atrium through the foramen ovale. The valve of the foramen ovale is located on the left side of the septum. The resistance to blood flow through the collapsed lungs makes the systolic pressure greater in the right atrium than in the left atrium, causing the blood to flow through the foramen ovale into the left atrium.

        c. In the left atrium, the well-oxygenated blood mixes with the small amount of deoxygenated blood returning via the pulmonary veins from the lungs. From the left atrium this blood flows successively through the left ventricle and ascending aorta.

        d. The coronary, carotid, and subclavian arteries are the first major arteries to receive blood from the ascending aorta. Hence, well-oxygenated and nutritional blood is directed to the heart, brain, head, neck, and upper extremities.

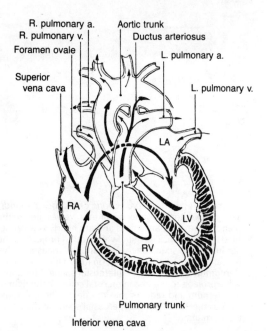

Figure 13-12. *Fetal circulation.* Oxygenated fetal blood, returning from the placenta through the inferior vena cava, passes predominantly through the foramen ovale into the left atrium, whence it passes through the left ventricle and supplies the vessels of the heart and head. Deoxygenated fetal blood, returning from the head through the superior vena cava, passes predominantly into the right ventricle and pulmonary trunk, whence most joins the aortic flow through the ductus arteriosus.

3. Ductus arteriosus. A small amount of blood entering the right atrium from the inferior vena cava is prevented from passing through the foramen ovale by the lower edge of the septum secundum, the so-called crista dividens, and, thus, remains in the right atrium.

   a. In the right atrium, it mixes with the deoxygenated blood returning from the head and upper extremities by way of the superior vena cava and from the heart via the coronary sinus. This deoxygenated blood is directed through the tricuspid valve into the right ventricle and then into the pulmonary trunk.

   b. Because the resistance in the pulmonary vessels during fetal life is high, the main portion of this deoxygenated blood does not pass through the pulmonary arteries to the lungs. Rather, it is diverted from the left pulmonary artery through the ductus arteriosus into the descending aorta, where it mixes with well-oxygenated blood from the aortic arch. This admixture occurs *after* the coronary and carotid arteries have been given off.

   c. Most of the mixed blood in the descending aorta passes to the placenta for reoxygenation; the remainder circulates through the lower part of the body.

C. Changes during the postnatal period

1. Physiologic changes at birth. The changes occurring in the vascular system at birth are related to the cessation of placental flow and the commencement of pulmonary respiration.

   a. Occlusion and ligation of the placental circulation cause an immediate fall in blood pressure in the inferior vena cava and right atrium.

   b. At birth, the amniotic fluid in the bronchial tree is expelled or suctioned and replaced by air; respiratory movements follow, and the lungs become aerated.

   c. Aeration of the lungs is associated with a drastic fall in pulmonary vascular resistance.

   d. The marked increase in pulmonary blood flow contributes to the drop in pressure in the right side of the heart and results in a rise in the blood pressure in the left atrium.

2. Closure of the foramen ovale. The differential blood pressure in the right atrium (lower) and the left atrium (higher) presses the valve of the foramen ovale (septum primum) against the atrial septum (septum secundum). Thus, the early functional closure of the foramen ovale is by pressure differential.

   a. During the first weeks of life, however, the closure is reversible, and crying creates a right-to-left shunt in newborns, accounting for cyanotic periods.

   b. Constant apposition gradually leads to fusion of the two septa after several months, resulting in the formation of the **fossa ovalis**. In 10%–15% of individuals, however, perfect anatomic closure may never be obtained.

3. Closure of the ductus arteriosus. The ductus arteriosus begins to close soon after birth, but during the first few postpartum days, a left-to-right shunt is not unusual, and a murmur produced by a patent ductus may be auscultated.

   a. Because of the angle at which the ductus joins the aorta and the resultant Bernoulli effect, the amount of backflow during this period is less than might be expected.

   b. Initial closure of the ductus arteriosus is by muscular contraction mediated by bradykinin, a substance released from the lungs during their initial inflation. As closure occurs, the amount of blood flowing into the lungs and left atrium increases.

   c. Complete anatomic obliteration by proliferation of the intima takes from 1–3 months. In the adult, the obliterated ductus arteriosus is called the **ligamentum arteriosum**.

4. Closure of the umbilical arteries is accomplished by contraction of the smooth muscles in the wall of the vessels and is probably caused by thermal and mechanical stimuli and a change in oxygen tension.

   a. Functionally, the arteries are closed a few minutes after birth, but the actual obliteration of the lumen by fibrous proliferation may take 2–3 months.

   b. The distal parts of the umbilical arteries then form the **medial umbilical ligaments**, while the proximal portions remain open as the superior vesical arteries.

5. Closure of the umbilical vein and ductus venosus occurs shortly after that of the umbilical arteries.

   a. After obliteration, the umbilical veins form the **ligamentum teres hepatis** in the lower margin of the falciform ligament.

   b. The ductus venosus, which courses from the ligamentum teres to the inferior vena cava, is also obliterated and forms the **ligamentum venosum**.

D. Congenital defects

1. Patent ductus arteriosus. Although the **ductus arteriosus** may remain patent throughout life with no major problems, it usually is ligated surgically when it is found to be open.

2. Septal defects

a. An **ASD**, such as a **patent foramen ovale**, frequently is compatible with normal life and activity because the shunting is from **left-to-right**.

b. A **VSD** may result in **tetralogy of Fallot**, in which:

(1) The VSD provides communication between ventricular chambers. The **right-to-left shunt** sends deoxygenated blood to the aorta, necessitating increased cardiac work to provide the same amount of oxygen to the body tissue with the possibility of decompensation and failure.

(2) The opening of the aorta overrides the VSD into the right ventricle.

(3) Distribution of hydrostatic pressure through the VSD from left ventricular systole results in hypertrophy of the right ventricular wall.

(4) Hypertrophy of the supraventricular crest produces a functional pulmonary stenosis whereby myocardial contraction occludes the pulmonary outflow tract.

3. Dextrocardia. In dextrocardia, the heart is normal, but in mirror image (0.02% of the population).

a. A very high incidence of situs inversus viscerum (reversed rotation of the gastrointestinal tract) occurs with dextrocardia.

b. Dextrocardia may be associated with either a right-sided or left-sided aorta.

E. Circulatory collapse (shock). The heart can pump only the blood that flows into it. In shock, there is venous pooling, and the blood fails to return to the heart. Blood pressure drops, and stroke volume decreases. The pulse becomes rapid but thready.

14
Superior Mediastinum and Posterior Mediastinum

I. SUPERIOR MEDIASTINUM

A. Overview. The principal contents of the superior mediastinum are the thymus gland and the vascular structures running to and from the heart. The veins tend to be anterior and toward the right, whereas the arteries tend to be located behind the veins and to the left. The trachea and esophagus are close to the midline (Fig. 14-1).

B. Thymus gland. Lying anteriorly in the superior mediastinum and anterior mediastinum beneath the manubrium and upper body of the sternum, the thymus gland extends laterally beneath the upper four costal cartilages, sometimes as far as the hilum of the lung on either side. It rises slightly into the base of the neck under the sternohyoid and sternothyroid muscles. It lies anteriorly to the aorta, brachiocephalic vein, and fibrous pericardium (see Fig. 14-1).

1. It varies in size with age.
    a. At birth, it weighs 10–15 g.
    b. Before puberty, it grows to 30–40 g.
    c. During late adult life, it regresses by involution and fatty atrophy to less than 10 g.

2. The thymus is a very important lymphoid organ associated with immunologic recognition.

3. Thymectomy is undertaken as a palliative measure for myasthenia gravis, an autoimmune disorder associated with the neuromuscular junctions.

C. Brachiocephalic veins. The left and right brachiocephalic veins receive the drainage from the head, neck, and upper extremities.

1. The right brachiocephalic vein and the longer, diagonal **left brachiocephalic vein** unite to form the superior vena cava, which continues the direction of the right brachiocephalic vein (see Fig. 14-1).

2. The superior vena cava receives the **azygos vein** before emptying into the right atrium.

3. Anomalies. The left brachiocephalic vein forms as an initial small shunt between the embryonic left and right anterior cardinal veins. As the shunt enlarges to become the left brachiocephalic vein, the left supracardinal vein normally regresses to form the oblique vein (of Marshall) of the left atrium. The left common cardinal vein into which the supracardinal and subcardinal veins empty in the embryo persists as the coronary sinus, which receives the oblique vein and empties into the right atrium. If this developmental sequence is reversed or modified, a single left superior vena cava can develop or, if the right vena cava persists, bilateral venae cavae may result.

D. Ascending aorta and aortic arch. The aorta emerges from the pericardial sac between the superior vena cava and the pulmonary trunk.

1. Course. The **ascending aorta** runs superiorly and toward the right before arching posteriorly as the **aortic arch** and then inferiorly as the **descending aorta** (see Fig. 14-4). The aortic arch produces the "aortic knob," a radiographic landmark on the left side of the mediastinum.

2. Branches of the ascending aorta and aortic arch include the:
    a. Left and right coronary arteries
    b. Brachiocephalic artery (on the right side only), which gives rise to the right common carotid and right subclavian arteries
    c. Left common carotid artery. However, the left common carotid artery sometimes arises from the brachiocephalic artery (especially in black people).
    d. Left subclavian artery

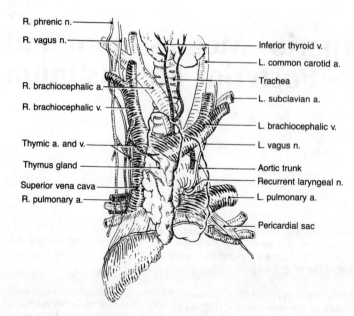

R. phrenic n.

R. vagus n.

R. brachiocephalic a.

R. brachiocephalic v.

Thymic a. and v.

Thymus gland

Superior vena cava

R. pulmonary a.

Inferior thyroid v.

L. common carotid a.

Trachea

L. subclavian a.

L. brachiocephalic v.

L. vagus n.

Aortic trunk

Recurrent laryngeal n.

L. pulmonary a.

Pericardial sac

Figure 14-1. *Anterior mediastinum and superior mediastinum.* The lower portion of the right half of the thymus gland is shown, and the remainder of the thymus is ghosted to show the major vessels of the anterior and superior mediastinum.

3. Development. The definitive aortic arch is derived from the artery of the 4th left branchial arch of the embryo. The right subclavian artery is derived from the artery of the 4th right branchial arch.

II. POSTERIOR MEDIASTINUM

A. Overview. By definition, the posterior mediastinum lies behind the pericardial sac; it is bounded by the mediastinal pleura on either side and by the diaphragm below. Superiorly, it is continuous with the superior mediastinum. It runs approximately from vertebral level T4–T12.

B. Descending aorta

1. Course. The descending aorta is located to the left within the mediastinum and bulges into the left pleural cavity, where it produces a distinctive radiopaque border on the right side of the mediastinum, superior to the heart shadow (see Fig. 14-4). It is about 3 cm in diameter. It enters the abdominal cavity through the aortic hiatus between the crura of the diaphragm.

2. Branches
a. The ligamentum arteriosum, a remnant of the embryonic **ductus arteriosus**, is derived from the artery of the 6th left branchial arch (the 6th arch degenerates completely on the right).
b. Intercostal arteries, a pair at each vertebral level, supply the thoracic wall and anastomose with the anterior intercostal branches of the internal thoracic (mammary) arteries.
c. Bronchial arteries supply the bronchial tree.
d. Esophageal branches supply the middle portion of the esophagus.

C. Esophagus

1. Relationships
a. The esophagus extends from the lower part of the laryngopharynx to the cardiac opening of the stomach. It is approximately 25–30 cm long in males (in the erect position) and a few centimeters shorter in females.
b. The esophagus lies anteriorly to the prevertebral fascia. High within the thorax, it lies in the midline behind the trachea. Inferior to the tracheal bifurcation, it deviates slightly to the right so that it lies slightly to the right of and slightly behind the aorta. Low in the thorax, it passes anteriorly to the aorta and pierces the diaphragm through the **esophageal hiatus**, which is about 2½ cm to the left of the midline in the muscular portion of the diaphragm (see Fig. 14-4). Below the diaphragm, it makes an abrupt turn to the left and enters the stomach (see Fig. 17-1).

2. Structure

a. The esophagus is the most muscular segment of the alimentary tract. Its upper quarter is composed primarily of striated muscle that initially is arranged irregularly but soon separates into inner circular and outer longitudinal layers.

(1) The **cricopharyngeal muscle** is the sphincter of the upper esophageal opening. It remains closed except during deglutition (swallowing), eructation (belching), and emesis (vomiting).

(2) Proceeding aborally, the quantity of smooth muscle increases and the striated fibers disappear.

(3) The intra-abdominal portion of the esophageal musculature acts as a diffuse sphincter (**cardiac sphincter**).

b. The **lumen** of the esophagus is small and irregular except during deglutition. It averages 1 cm in diameter orally and 2–3 cm aborally.

(1) The esophagus is normally flattened and the shape of the lumen is irregular because tension within the inner (circular) layer of the muscularis externa causes formation of longitudinal folds.

(2) Because of this folding, the esophagus is quite distensible, accommodating anything that passes through the epiglottis.

3. Esophageal vasculature (see Fig. 17-1)

a. Arterial supply is derived from the:

(1) Inferior thyroid arteries (branches of the subclavian arteries), which supply the cervical portion of the esophagus

(2) Esophageal branches of the aorta, which supply the middle portion

(3) Esophageal branches of the left gastric artery (a branch of the celiac trunk), which supply the lower thoracic and abdominal portions of the esophagus

b. Venous return is similar to the arterial supply.

(1) The veins of the upper and lower portions parallel the supplying arteries and drain into the **inferior thyroid veins** and the **left gastric vein**, respectively.

(2) The venous drainage of the middle portion is into the **azygos** and **hemiazygos veins**.

4. Esophageal function is discussed in association with the gastrointestinal tract (see Chapter 17 II A 4).

5. Clinical considerations are discussed in association with the gastrointestinal tract (see Chapter 17 II A 5).

D. Thoracic duct. Essentially all the lymph from the body below the diaphragm returns to the systemic circulation through the thoracic duct at a flow rate of 2–4 L of chyle a day.

1. Course

a. In the abdomen, the duct originates in the **cisterna chyli**, which is located at vertebral level T12 between the crura of the diaphragm and posterior to the aorta (Fig. 14-2).

b. In the thorax, the duct is located behind the esophagus and to the right of the aorta. It is very delicate and may occasionally be double or even triple.

c. In the superior mediastinum, the duct arches over the cupula of the left pleura to lie posteriorly to the left subclavian vein. It enters this vein at the angle formed by the juncture of the vein with the left internal jugular vein.

2. Branches

a. The intercostal nodes drain the thoracic wall and breasts.

b. The subclavian trunk drains the left arm, left breast, and left side of the neck.

c. The left jugular trunk drains the left side of the head and neck.

d. The left bronchomediastinal trunk drains the left bronchial tree.

3. Clinical consideration. Chyle from a ruptured or severed thoracic duct may drain into the right or left pleural cavity (chylothorax).

E. Thoracic lymph nodes

1. Preaortic nodes lie anteriorly to the aorta. These nodes drain the esophagus and empty into the thoracic duct.

2. Para-aortic nodes lie alongside the aorta (intercostal nodes). These nodes drain the posterior intercostal spaces (the anterior intercostal spaces are drained by parasternal nodes along the internal thoracic artery) and empty into the thoracic duct.

3. **Hilar nodes** are located around the root of the lung and within the lung parenchyma at the hilus. These nodes drain the lung and peripheral bronchial tree. Hilar nodes empty into the bronchomediastinal nodes.

4. **Bronchomediastinal nodes** lie adjacently to the bronchi and trachea. These nodes receive drainage from the hilar nodes. The bronchomediastinal nodes usually drain into the brachiocephalic veins on each side or, occasionally, into the thoracic duct.

F. **Venous drainage of the thorax**

1. **The azygos vein** lies to the right of the esophagus on the vertebral column (Fig. 14-3).
   a. It receives the right posterior intercostal veins, right bronchial veins, and hemiazygos veins.
   b. It enters the superior vena cava by passing to the superior mediastinum beneath the mediastinal pleura and arching anteriorly at vertebral level T4.

2. **The superior hemiazygos and inferior hemiazygos veins** may be separate or joined and are subject to extensive variation (see Fig. 14-3).
   a. **The superior (accessory) hemiazygos vein:**
      (1) Receives bronchial veins from the left lung
      (2) Receives the upper three to five left posterior intercostal veins
      (3) Either joins the inferior hemiazygos vein or enters the azygos vein
   b. **The (inferior) hemiazygos vein:**
      (1) Drains the lower eight or nine left posterior intercostal veins
      (2) May anastomose with the inferior vena cava or left adrenal vein or left renal vein in the abdomen
      (3) Joins the azygos vein, sometimes by more than one connection

G. **Thoracic sympathetic trunk**

1. **Course and composition.** Continuous along the length of the vertebral column, the sympathetic trunk lies across the necks of the ribs (see Fig. 14-4). It consists of a chain of interconnected **sympathetic** or **paravertebral ganglia**, 12 of which are in the thorax. Each thoracic ganglion has five roots: white ramus, gray ramus, superior connector, inferior connector, and splanchnic nerve.

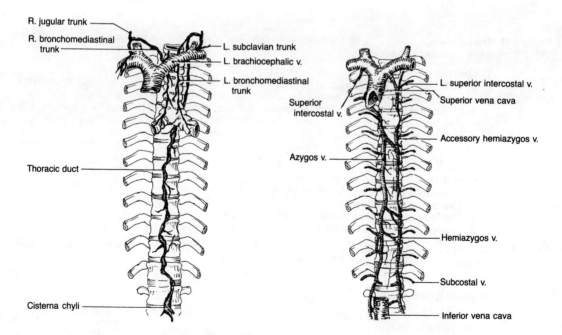

Figure 14-2. *Thoracic duct.* Lymph from the entire left side of the body as well as from the lower half of the right side is returned to the left brachiocephalic vein by the thoracic duct.

Figure 14-3. *Azygos system of veins.*

    a. White rami communicantes bring *presynaptic* (myelinated) general visceral efferent (GVE) neurons from the spinal nerve to or through the sympathetic chain. Those neurons passing through the ganglia without synapses may go either up or down the chain to synapse at an adjacent level, travel in splanchnic nerves to synapse in prevertebral ganglia, or synapse in the ganglion at the level where they enter the chain.

    b. Gray rami communicantes return *postsynaptic* (unmyelinated) GVE neurons to the spinal nerve for piloerection, sweating, and vasodilatory functions in the associated dermatome.

    c. The superior and inferior connectors carry presynaptic and postsynaptic neurons to adjacent or remote levels of the chain.

    d. Splanchnic nerves carry presynaptic neurons to prevertebral ganglia, such as the celiac or superior mesenteric ganglion, where they synapse. From the prevertebral ganglia, postsynaptic neurons innervate visceral structures. Other splanchnic nerves carry postsynaptic neurons from the sympathetic chain directly to visceral structures.

2. Distribution

    a. Ganglia T1–T3 contribute to the cervical chain since there are no white rami communicantes in the neck.

        (1) The **superior, middle** and **inferior cervical ganglia** give off cervical **cardiac (accelerator) nerves** to the heart, which are equivalent to splanchnic nerves.

        (2) All presynaptic sympathetic neurons to the head arise from these upper thoracic levels.

    b. Ganglia T1–T5 contribute **thoracic splanchnic nerves** to the heart and lungs.

    c. Ganglia T5–T9 contribute to the **greater splanchnic nerve**, which innervates the upper gastrointestinal tract.

    d. Ganglia T10–T11 contribute to the **lesser splanchnic nerve**, which innervates the midgut region.

    e. Ganglion T12 contributes to the **least splanchnic nerve**, which innervates the gonads, kidneys, and proximal ureters.

H. Vagus nerve (CN X)

1. Course and composition. The vagus nerve leaves the cranium as the right and left vagus nerves through the jugular foramen and passes through the neck and thorax into the abdomen to provide **parasympathetic innervation** to the cervical and thoracic viscera, foregut, and midgut. Because of the rotation of the gut within the abdomen, the left and right vagi become the anterior and posterior vagal trunks, respectively (Fig. 14-4).

2. Right vagus nerve (see Fig. 14-4)

    a. The right vagus nerve gives off the **right recurrent (inferior) laryngeal nerve** and arches inferiorly to and around the right subclavian artery, a remnant of the right 4th branchial arch (the right 5th and right 6th arches having degenerated completely) in the base of the neck.

    b. It gives off twigs to the **cardiac plexus**.

    c. It runs behind the right main stem bronchus and gives major contributions to the **anterior** and **posterior pulmonary plexuses**.

    d. It passes along the right surface of the esophagus to form an **esophageal plexus**, which re-forms as the **posterior vagal trunk** before passing through the esophageal hiatus of the diaphragm.

3. Left vagus nerve (see Fig. 14-4)

    a. The left vagus nerve gives off the **left recurrent (inferior) laryngeal nerve**, which courses around the ligamentum arteriosum (remnant of the 6th left branchial arch).

    b. Together with the recurrent laryngeal nerve, it gives off twigs to the **cardiac plexus**.

    c. It passes posteriorly to the root of the lung, where it gives off major contributions to the **posterior** and **anterior pulmonary plexuses**.

    d. It passes along the left side of the esophagus to form an **esophageal plexus**, which re-forms as the **anterior vagal trunk** just above the diaphragm.

4. Clinical considerations. The recurrent laryngeal nerves innervate the vocal musculature. Hilar or mediastinal tumors on the left side or an apical tumor on the right side, involving either of the respective nerves, will cause hoarseness, often the first sign of these tumors.

I. Functions of the autonomic nervous system in the thorax

1. Overview. The heart and lungs are innervated through the autonomic nervous system.

    a. The cardiac plexus is a region of intermingling sympathetic and parasympathetic (vagal) fibers located posteriorly to the arch of the aorta and anteriorly to the bifurcation of the pulmonary trunk. Its superficial and deep divisions are primarily associated with autonomic innervation to the heart (see Fig. 14-4).

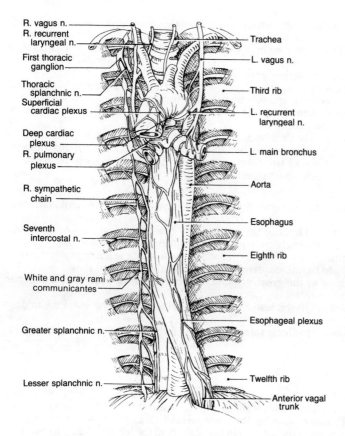

R. vagus n.
R. recurrent laryngeal n.
First thoracic ganglion
Thoracic splanchnic n.
Superficial cardiac plexus
Deep cardiac plexus
R. pulmonary plexus
R. sympathetic chain
Seventh intercostal n.
White and gray rami communicantes
Greater splanchnic n.
Lesser splanchnic n.

Trachea
L. vagus n.
Third rib
L. recurrent laryngeal n.
L. main bronchus
Aorta
Esophagus
Eighth rib
Esophageal plexus
Twelfth rib
Anterior vagal trunk

Figure 14-4. *Autonomic nerves of the thorax.* The sympathetic trunk and the vagus nerves bring sympathetic and parasympathetic innervation, respectively, to the thoracic viscera. Also shown are the aorta and esophagus.

    b. The pulmonary plexus is continuous with (sometimes not distinguished from) the cardiac plexus and lies about the roots of the lungs. Its anterior and posterior divisions are primarily associated with autonomic innervation to the lungs (see Fig. 14-4).

  2. Sympathetic pathways consist of two motor neurons, a myelinated preganglionic (presynaptic) neuron and an unmyelinated postganglionic (postsynaptic) neuron.
    a. Preganglionic fibers run from the spinal cord to the sympathetic chain of ganglia (see Figs. 12-4 and 13-8).
      (1) The **cell bodies of the presynaptic sympathetic neurons** are located in the **intermediolateral cell column** of spinal cord levels T1 through L2 or L3. The neurons associated with the heart and lungs arise only from levels T1 through T5. These axons pass through the ventral roots and ventral primary rami of spinal nerves T1 through T4 or T5 (see Fig. 4-5).
      (2) The **white rami communicantes** provide pathways for the myelinated presynaptic sympathetic neurons between the spinal nerves and the corresponding thoracic sympathetic ganglion (see Fig. 4-5).
      (3) Because there are no white rami communicantes in the cervical region, the presynaptic axons may travel up or down within the sympathetic chain before synapsing with postganglionic neurons in the three cervical ganglia or the upper four (sometimes five) thoracic ganglia. However, some presynaptic neurons may pass through the sympathetic chain to terminate in small ganglia (of Wrisberg) within the cardiac plexus.
    b. Postganglionic fibers run from the chain of sympathetic ganglia to the target organ (see Figs. 12-4 and 13-8).
      (1) The **cell bodies of postganglionic sympathetic neurons** to the heart and lungs have their cell bodies in the three cervical ganglia and upper four or five thoracic ganglia. However, some postsynaptic neurons may be located in small ganglia (of Wrisberg) within the cardiac plexus.
      (2) **Cervical cardiac nerves**, containing predominantly unmyelinated postsynaptic neurons, arise from the superior, middle, and inferior cervical ganglia and run deeply in the neck to the cardiac and pulmonary plexus to reach the target tissue.

       (3) Thoracic cardiac (splanchnic) nerves, containing predominantly unmyelinated post-synaptic neurons, arise from sympathetic ganglia T1 through T4 or T5 (variable on both sides) and run to the cardiac and pulmonary plexus to reach the target tissue (see Fig. 14-4).

       (4) Gray rami communicantes containing unmyelinated postsynaptic neurons leave the sympathetic ganglia and rejoin the corresponding spinal nerve to bring sympathetic innervation to the dermatomes (see Figs. 4-5 and 14-4).

  c. Effects of sympathetic innervation

     (1) Increases heart rate (tachycardia)

     (2) Increases stroke volume (ejection fraction)

     (3) Dilates coronary arteries

     (4) Produces piloerection, perspiration, and dilation or constriction of peripheral vessels in the dermatomes

3. Parasympathetic pathways also consist of two motor neurons, a myelinated preganglionic (presynaptic) neuron and an unmyelinated postganglionic (postsynaptic) neuron.

  a. Preganglionic fibers run from the brain to distal parasympathetic ganglia close to or within the target organ (see Figs. 12-4 and 13-8).

     (1) The **parasympathetic cell bodies** are located in the **dorsal vagal nuclei** of the brain. The myelinated presynaptic axons form cranial nerve X, the **vagus nerve**.

     (2) The **vagus nerve** leaves the cranium via the jugular foramen and courses deeply in the neck. At the base of the neck, it gives rise to the recurrent laryngeal nerve.

       (a) Numerous twigs arise from both the vagus nerves and the recurrent branch of each, which run to the cardiac plexus (see Fig. 14-4).

       (b) Some preganglionic vagal fibers synapse with neurons located in small ganglia (of Wrisberg) within the cardiac plexus and pulmonary plexus; most preganglionic vagal fibers synapse with neurons within the intrinsic cardiac ganglia.

  b. Postganglionic fibers arise from numerous small ganglia close to or within the target organ. They are unmyelinated and run to the tissue innervated (see Figs. 12-4 and 13-8).

  c. Effects of parasympathetic innervation

     (1) Decreases heart rate (bradycardia)

     (2) Decreases stroke volume (ejection fraction)

     (3) Produces bronchoconstriction

     (4) Increases bronchial secretion

4. Visceral afferents. Nerves mediating sensation from the thoracic viscera travel along autonomic pathways (see Figs. 12-4 and 13-8).

  a. Pain afferents follow sympathetic pathways.

     (1) Pain afferents from the heart follow the middle and inferior cervical cardiac nerves as well as the upper four or five thoracic cardiac nerves to reach the sympathetic chain (see Fig. 13-8).

     (2) White rami communicantes provide the pathway for pain afferents between the sympathetic chain and spinal nerves T1–T4. Because there are no white rami in the cervical region, the pain afferents arriving at the cervical ganglia descend to the T1 ganglion in order to reach a spinal nerve (see Fig. 4-5).

     (3) The **cell bodies** of the visceral afferent fibers are located in the **dorsal root ganglia** of the spinal nerves (see Figs. 4-5, 12-4, and 13-8).

     (4) Basis for referred pain

       (a) Painful stimuli mediated by visceral afferent neurons, which enter the spinal cord at a specific spinal level, are interpreted by the brain as originating from (referred to) the somatic dermatome corresponding to that particular spinal level.

       (b) As such, painful stimuli originating from the heart will be referred to the T1 through T4 dermatomes, which include the medial aspect of the upper arm and anterior thoracic wall as far as the level of the nipple.

       (c) Because there are no white rami communicantes in the cervical region, no visceral structures refer pain to the cervical dermatomes, except the diaphragm, which has a cervical origin.

  b. Reflex afferents follow parasympathetic pathways.

     (1) Afferents for respiratory and cardiac reflexes travel exclusively along the vagus nerve.

     (2) The **cell bodies** are located in the nodose (inferior) ganglion of the vagus nerve.

Part III Thoracic Region

STUDY QUESTIONS

Directions: Each question below contains five suggested answers. Choose the **one best** response to each question.

1. All of the following structures are located at the approximate level of the 4th or 5th thoracic vertebra EXCEPT the

(A) bifurcation of the trachea
(B) costosternal articulations of the 2nd ribs
(C) nipple in the male
(D) manubriosternal joint
(E) superior extent of the pericardial cavity

2. A characteristic of the intercostal neurovascular bundle that makes it particularly susceptible to injury from a fractured rib is that it lies

(A) behind the superior border of the rib
(B) beneath the inferior border of the rib
(C) between external and internal intercostal layers
(D) directly behind the midpoint of the rib
(E) halfway between two adjacent ribs

3. With a patient in a sitting position, there is percussive dullness below the level of the 5th rib in the right midaxillary line. Thoracentesis to remove the fluid is best accomplished here by inserting a needle

(A) adjacently to the sternum in the 2nd intercostal space
(B) adjacently to the sternum in the 5th intercostal space
(C) in the midaxillary line in the 5th intercostal space
(D) in the midaxillary line in the 10th intercostal space
(E) in none of the above spaces

4. In a hospital emergency room, a 23-year-old man states that he "inhaled" a peanut. On bronchoscopy, the peanut will most likely be located in the

(A) left lower lobar bronchus
(B) left main bronchus
(C) left superior segmental bronchus
(D) right lower lobar bronchus
(E) right superior segmental bronchus

5. Which of the following statements best characterizes bronchopulmonary segments, the functional units of the lungs?

(A) The bronchial tree and pulmonary vein are located centrally
(B) The pulmonary artery and pulmonary vein are distributed centrally
(C) The pulmonary artery and pulmonary vein are distributed peripherally
(D) The bronchopulmonary segments are supplied by a terminal bronchiole
(E) None of the above statements characterizes bronchopulmonary segments

6. Aspiration pneumonia is usually detectable by a distinct region of percussive dullness and a radiographic lightening of a particular bronchopulmonary segment. This bronchopulmonary segment is nearly always the

(A) left or right apical
(B) left or right superior
(C) left or right basal anterior
(D) left upper lingular
(E) left lower lingular

7. A blood clot breaking loose and entering the circulation from a leg vein is most likely to lodge and produce ischemia and local infarct in

(A) the brain
(B) the liver
(C) the lung
(D) the myocardium
(E) none of the above

8. Cancer of the right lung may metastasize to all of the following lymph nodes EXCEPT the

(A) left tracheobronchial
(B) right axillary
(C) right bronchomediastinal
(D) right hilar
(E) right tracheobronchial

9. The apex of the heart is normally located

(A) at the level of the 5th thoracic vertebra in the left midclavicular line
(B) at the level of the xiphoid process one finger breadth to the left of the midline
(C) deeply to the 3rd intercostal space in the left midclavicular line
(D) deeply to the 5th intercostal space in the left midclavicular line
(E) in none of the above places

10. Which of the following statements correctly describes the papillary muscles in the heart?

(A) They are rudimentary and have no major function
(B) They contract to close the atrioventricular (A-V) valves during ventricular systole (contraction)
(C) They contract to open the A-V valves during ventricular diastole (relaxation)
(D) They secure the chordae tendineae to the A-V valve leaflets
(E) None of the above

Questions 11–15

A man, aged 64, with severe angina pectoris is found by coronary angiography to have a 90% occlusion of the left coronary artery. A venous homograph is not possible because of severe venous varicosities; thus, the patient will be treated surgically by diverting his left internal thoracic artery into the left coronary artery distally to the occlusion.

11. Which of the following statements best describes the normal internal thoracic artery?

(A) It bifurcates into the inferior phrenic and superficial epigastric arteries
(B) It descends posteriorly to the sternum
(C) It is a branch of the axillary artery
(D) It is accompanied by the azygos vein on the right and hemiazygos vein on the left
(E) It provides significant blood supply to the mammary gland

12. The anterior interventricular artery is usually accompanied by the

(A) anterior cardiac vein
(B) coronary sinus
(C) middle cardiac vein
(D) oblique vein (of Marshall)
(E) none of the above

13. All of the following characteristics pertain to the left coronary artery EXCEPT

(A) it arises from the left aortic sinus (of Valsalva)
(B) it divides into the anterior descending and circumflex arteries
(C) it emerges from between the aortic and pulmonary trunks to enter the anterior interventricular sulcus
(D) it is seldom longer than 2 cm
(E) it passes to the left of the pulmonary trunk and to the right of the left auricular appendage

14. After a midline sternotomy, the pericardium is incised by a longitudinal incision. Had a horizontal incision been used, which of the following structures might have been jeopardized?

(A) Azygos vein
(B) Internal thoracic arteries
(C) Phrenic nerves
(D) Vagus nerves
(E) None of the above

15. Postoperatively, the chest wall region originally served by the internal thoracic artery receives blood flow from all of the following EXCEPT the

(A) left costocervical trunk
(B) left inferior epigastric artery
(C) left pericardiacophrenic artery
(D) left posterior intercostal arteries
(E) right internal thoracic artery

(end of group question)

Questions 16–19

A 36-year-old male office worker reports to the clinic complaining of weakness and shortness of breath that become marked after moderate exercise. The patient complains of a rapid, throbbing pulse on climbing two flights of stairs. Although the patient is not presently in pain, he admits to having occasional "heartburn" over the past year. Findings on physical examination include a weak and irregular pulse, a diastolic rumbling murmur attributed to the mitral valve, and rales (rattles) in the basal segments of the lungs.

16. The mitral valve is heard most distinctly on the

(A) left side, adjacent to the sternum in the 2nd intercostal space
(B) left side, adjacent to the sternum in the 5th intercostal space
(C) left side, in the midclavicular line in the 5th intercostal space
(D) right side, adjacent to the sternum in the 6th intercostal space
(E) right side, in the midclavicular line in the 6th intercostal space

17. Percussion reveals that the area of cardiac dullness extends two finger breadths to the right of the sternum. A chest x-ray confirms that the heart is enlarged with specific right ventricular hypertrophy and right atrial hypertrophy. The preliminary diagnosis is mitral valve stenosis. A sequela of mitral valve stenosis is hypertrophy of the right ventricle and right atrium, which is caused by all of the following EXCEPT

(A) impeded blood flow to the left ventricle
(B) rise in left atrial systolic pressure
(C) rise in pressure in the pulmonary vascular bed
(D) rise in right ventricular systolic and diastolic pressures
(E) fall in left ventricular pressure

18. A complication of mitral stenosis is the formation of thrombi on the walls of the enlarged left atrium. A thrombus dislodged from the left atrium may produce all of the following conditions EXCEPT

(A) a cerebral infarct
(B) a myocardial infarct
(C) renal necrosis and kidney failure
(D) gangrene in a lower extremity
(E) a pulmonary embolus

19. The patient is scheduled for open heart surgery to correct the valvular defect. The heart is approached by resection of the 5th rib on the left side. Commissurotomy, the surgical reestablishment of the divisions between the mitral valve cusps, occasionally can be accomplished by insertion of a finger through the valve. Toward this end, the valve may be approached best through the wall of the

(A) left atrium between the left and right superior pulmonary veins
(B) left atrium between the left superior and inferior pulmonary veins
(C) left auricular appendage
(D) left ventricle on the posterior surface
(E) right atrium and foramen ovale

(end of group question)

Directions: Each question below contains four suggested answers of which **one or more** is correct. Choose the answer

     A   if **1, 2, and 3** are correct
     B   if **1 and 3** are correct
     C   if **2 and 4** are correct
     D   if **4** is correct
     E   if **1, 2, 3, and 4** are correct

20. During physical examination, an elderly gentleman is observed to be breathing almost entirely with his diaphragm. However, there is no bulging of the intercostal space on expiration. One might expect that this respiratory pattern is the result of

(1) calcification of the costal cartilages
(2) paralysis of the thoracic musculature
(3) loss of extrinsic elastic recoil
(4) loss of intrinsic elastic recoil

21. Two elderly women, both diagnosed as having pleurisy, were comparing their illnesses. One had lateral thoracic pain, whereas the other had neck and shoulder pain. Assuming the diagnoses to be correct, the explanation for these pain distributions is that the pleura is innervated by

(1) thoracic splanchnic nerves
(2) intercostal nerves
(3) vagus nerves
(4) phrenic nerves

SUMMARY OF DIRECTIONS

A	B	C	D	E
1, 2, 3 only	1, 3 only	2, 4 only	4 only	All are correct

---

22. Elevation of the rib cage during inspiration may be aided by the

(1) sternocleidomastoid muscle
(2) pectoralis minor muscle
(3) anterior, middle, and posterior scalene muscles
(4) pectoralis major muscle

23. Respiratory mechanics involve coordinated activity of numerous muscles. Contraction of which of the following muscles contributes to forced diaphragmatic expiration?

(1) Transverse abdominis
(2) External oblique
(3) Rectus abdominis
(4) Internal oblique

24. Structures that are located in the anterior mediastinum include which of the following?

(1) Pericardial sac
(2) Trachea
(3) Heart
(4) Thymus

25. The movement of the ribs during thoracic inspiration involves

(1) movement at the costovertebral joints
(2) movement at the manubriosternal joint
(3) the crossing of the axes of rotation of the ribs
(4) increases in the transverse thoracic diameter

26. Characteristics of a bronchopulmonary segment include which of the following?

(1) It receives oxygen by the bronchial arteries
(2) It is supplied by a terminal bronchiole
(3) It is usually separated from adjacent segments by invaginations of visceral pleura
(4) It has vessels containing oxygenated blood located peripherally

27. A Pancoast's's tumor at the apex of the right lung may produce which of the following symptoms?

(1) Hoarseness of speech
(2) Loss of sympathetic supply to the right side of the head
(3) Paradoxical diaphragmatic movements
(4) Venous engorgement of the face and arm on the right side

28. During development, vessels that deliver blood to the left atrium include the

(1) pulmonary veins
(2) foramen ovale
(3) least cardiac veins (venae cordis minimae)
(4) ductus arteriosus

29. Blood is returned to the left side of the heart via which of the following vessels?

(1) Anterior cardiac veins
(2) Thebesian veins
(3) Coronary sinus
(4) Pulmonary veins

30. Myocardial infarction limited to the interventricular septum might be expected to produce which of the following problems?

(1) An aortic valve insufficiency
(2) A tricuspid valve incompetence
(3) A mitral valve incompetence
(4) Disturbance of cardiac impulse conduction

31. Circulatory changes that occur at birth normally include

(1) increased left atrial pressure
(2) increased blood flow through the lungs
(3) decreased right atrial pressure
(4) cessation of blood flow through the foramen ovale

32. Neurons that pass through the cardiac plexus include which of the following?

(1) Nonmyelinated postganglionic sympathetic fibers
(2) Visceral afferent fibers mediating reflexes
(3) Preganglionic parasympathetic fibers
(4) Visceral afferent fibers mediating pain

33. In angina pain of cardiac origin, the pain radiating across the precordium and perhaps down the arm to the wrist is mediated by increased activity in afferent fibers contained in the

(1) cervical cardiac nerves
(2) vagus nerves
(3) first four thoracic splanchnic nerves
(4) phrenic nerves

34. Structures that transit the diaphragm via the esophageal hiatus include the

(1) azygos vein
(2) thoracic duct
(3) hemiazygos vein
(4) right vagus nerve

Directions: The groups of questions below consist of lettered choices followed by several numbered items. For each numbered item, select the **one** lettered choice with which it is **most closely** associated. Each lettered choice may be used once, more than once, or not at all. Choose the answer

Questions 35–40

For each situation listed below, select the artery that is most likely to be associated with it.

(A) Right coronary artery
(B) Circumflex branch of left coronary artery
(C) Both
(D) Neither

35. Is found in the coronary sulcus

36. Arises from the left anterior aortic sinus

37. Supplies the anterior portion of the interventricular septum

38. Supplies the atrioventricular node

39. Supplies the sinoatrial node

40. Terminates as the posterior interventricular (descending) artery in most people

ANSWERS AND EXPLANATIONS

1. The answer is C. [*Chapter 11 II A 3, C 1 b, E; Chapter 12 IV A 2; Chapter 13 II A 1*] The bifurcation of the trachea, the costosternal articulations of the 2nd ribs, and the manubriosternal synchondrosis all occur at the approximate level of the 4th or 5th thoracic vertebrae. In addition, the pericardial cavity extends superiorly to that level. While the male nipple is located at the 4th intercostal space, anteriorly, it lies approximately at the level of the 7th thoracic vertebra.

2. The answer is B. [*Chapter 11 III A 3 a (4) (b); V A 1, 2 b*] Each intercostal neurovascular bundle lies beneath the inferior border of a rib between the internal intercostal muscle and the innermost intercostal muscle, which is a division of the internal intercostal. Fracture of a rib may injure the nerve or blood vessels. If a fragment of rib also tears the pleura, hemothorax may result.

3. The answer is E. [*Chapter 12 II D 3 b (1)*] Thoracentesis involves inserting a needle to withdraw fluid on which the lung floats. The presence of fluid in the thoracic cage reduces ventilation capacity. The needle is best inserted in the midaxillary line one or two ribs below the fluid level determined by percussion but not below the 9th intercostal space because of the proximity of the liver across the costodiaphragmatic recess.

4. The answer is D. [*Chapter 12 IV C 2*] Aspirated objects usually drop into the right main bronchus because it is more vertical than the left main bronchus and, thus, more nearly in line with the trachea. Small objects tend to lodge in the right lower lobar bronchus, which more nearly continues the direction of the right main bronchus.

5. The answer is E. [*Chapter 12 III B 6 a*] Bronchopulmonary segments are supplied by a segmental (tertiary) bronchus, which is distributed centrally along with a segmental branch of the pulmonary artery. The intersegmental veins lie in the connective tissue septa between the bronchopulmonary segments.

6. The answer is B. [*Chapter 12 IV D 3*] The left or right superior bronchopulmonary segment of the lower lobes are most dependent when a person is supine. Aspirated vomitus in the recumbent person drains into the nearly vertical superior segmental bronchi to produce a pneumonia, resulting in a region of percussive dullness just medial to the vertebral border of the scapula when the arm is elevated, and auscultatory rales.

7. The answer is C. [*Chapter 12 V A 1 d*] A blood clot moving in the venous system would pass through the right chambers of the heart and lodge in a branch of the pulmonary arterial circulation.

8. The answer is B. [*Chapter 12 V C 2–5*] The lymphatic drainage from the right lungs does not pass through the axillary lymph nodes but through the right and left tracheobronchial lymph nodes, the right bronchomediastinal lymph nodes, and the right hilar lymph nodes. If the bronchomediastinal lymph nodes become blocked by metastases, malignant cells may cross to the contralateral side.

9. The answer is D. [*Chapter 13 III B 3*] The apex of the heart, formed by the tip of the left ventricle, lies deeply to the 5th intercostal space in the left midclavicular line. It is at this point that the apical pulse may be palpated and the mitral valve sound auscultated with maximum discrimination.

10. The answer is E. [*Chapter 13 VI B 1 a (1) (c), (2) (a)*] The papillary muscles attach the chordae tendineae of the atrioventricular valve cusps to the walls of the ventricular chambers. Contraction of the papillary muscles prevents eversion of the valve cusps as the ventricular chambers decrease in volume during the second (ejection) phase of systole.

11. The answer is E. [*Chapter II V B 2 a–c*] The internal thoracic artery is the second branch of the subclavian artery, arising opposite the origin of each vertebral artery. It descends posteriorly to the costal cartilages and supplies the chest wall, including the mammary gland. It terminates by bifurcation into the musculophrenic and superior epigastric arteries.

12. The answer is E. [*Chapter 13 IV B 1 b (1), 2*] The great cardiac vein accompanies the anterior descending artery in the anterior interventricular sulcus. The anterior cardiac veins drain across the right coronary sulcus directly into the right atrium. The coronary sinus, into which the great cardiac vein drains, lies in the coronary sulcus and also receives the middle cardiac vein from the posterior interventricular sulcus.

13. The answer is C. [*Chapter 13 IV A 2 a, b*] The left coronary artery arises from behind the left aortic cusp and courses between the pulmonary trunk and left auricular appendage for only 1 or 2 cm before bifurcating into the circumflex artery and anterior interventricular artery.

14. The answer is C. [*Chapter 12 VI A 2; Chapter 13 I B 8; Chapter 14 II F 1 a*] The phrenic nerves lie anteriorly along the lateral boundaries of the middle mediastinum between the fibrous pericardium and the mediastinal pleura. The left and right vagus nerves, which pass adjacently to the root of the each lung, as well as the azygos vein, are components of the posterior mediastinum.

15. The answer is C. [*Chapter 11 V C 1, 2*] The internal thoracic artery and its musculophrenic branch anastomose with the posterior intercostal arteries, which arise from the costocervical trunk (the superior intercostal artery) and the dorsal aorta. The superior epigastric branch anastomoses with the inferior epigastric artery. There are also anastomoses across the midline with the contralateral internal thoracic artery. The pericardiacophrenic artery supplies the pericardium and diaphragm and does not supply the chest wall significantly.

16. The answer is C. [*Chapter 13 IX C 3 b*] The sound of mitral valve closure projects to the apex of the heart and is heard most distinctly in the left 5th intercostal space in the midclavicular line. In men, this is usually just below the left nipple.

17. The answer is E. [*Chapter 13 XI A 1–4*] Mitral valve stenosis, whereby blood flow to the left ventricle is impeded, results in elevated left atrial systolic pressure and consequent elevation of pulmonary pressure. This requires the right side of the heart to work more to overcome the pulmonary pressure with consequent hypertrophy. The elevated pulmonary pressure produces pulmonary effusion and results in rales.

18. The answer is E. [*Chapter 13 VI A 2 b, C 3*] Thrombi dislodged from the left atrium can result in cerebral, coronary, renal, or systemic embolism. Thrombi dislodged from the right atrium produce pulmonary embolism.

19. The answer is C. [*Chapter 13 VI C 3; Figure 13–5*] The best approach to mitral commissurotomy is to gain access to the left atrium through the left auricular appendage and thence direct the finger through the mitral valve into the ventricle.

20. The answer is B (1, 3). [*Chapter 11 III A 3 a (5), c (4); VII C 4*] In the elderly, frequently there is progressive calcification of the costal cartilages with resultant loss of thoracic cage elasticity that inhibits or even precludes thoracic respiratory movement. The person compensates by using the diaphragm for inspiration and the abdominal musculature for expiration. Paralysis of the thoracic musculature results in bulging of the intercostal space during expiration.

21. The answer is C (2, 4). [*Chapter 12 II A 4 b, c*] The parietal pleura is innervated by branches of the intercostal nerves, and parietal pleuritic pain is localized to the region of irritation (lateral thoracic and anterior abdominal walls). The diaphragmatic pleura is innervated in large part by the phrenic nerves, arising from C3–C5; thus, diaphragmatic pleuritic pain is referred to the midcervical dermatomes (neck and shoulder).

22. The answer is E (all). [*Chapter 11 IV A; Table 11–1*] The pectoralis major and minor muscles, as well as the scalene muscles, attach to the ribs and can be used to elevate the ribs during exertional inspiration. The sternocleidomastoid muscle, attached to the manubrium, elevates the sternum, thereby increasing the anteroposterior thoracic diameter during exertional inspiration.

23. The answer is E (all). [*Chapter 11 VII C 2; Table 11-1; Chapter 15 IV B 4*] The muscles of the abdominal wall (including the external oblique, internal oblique, transverse abdominis, and rectus abdominis) are antagonistic to the diaphragm and participate in forced expiration.

24. The answer is D (4). [*Chapter 12 I B 2 a*] The anterior mediastinum contains the inferior portion of the thymus gland as well as sternopericardial ligaments and connective tissue. The superior portion of the thymus gland and the trachea are located in the superior mediastinum. The pericardial cavity and heart occupy the middle mediastinum.

25. The answer is E (all). [*Chapter 11 III B 1, 2; VII B 1 a, b*] The inspiratory action of the external intercostal muscles and the interchondral portion of the internal intercostal muscles results in upward rotation of the ribs about an axis described by the costovertebral joints. Because the axes of rotation cross anteriorly, there is resultant upward movement of the sternum and a flexion at the manubriosternal joint. The anteroposterior diameter of the thoracic cage is thereby increased.

26. The answer is D (4). [*Chapter 12 III B 6 a; VII B 2 a*] The bronchopulmonary segment, as the basic functional unit of the lungs, receives a central segmental artery and a segmental bronchus, which divides 10–20 times before reaching the terminal bronchiole. The intersegmental septa contain intersegmental veins, which return oxygenated blood to the lobar veins.

27. The answer is E (all). [*Chapter 12 III B 7 a*] A tumor at the apex of the right lung may involve compression of the recurrent laryngeal nerve with hoarseness of speech; compression of the sympathetic chain with loss of sympathetic supply to the right side of the head (Horner's syndrome); compression of the phrenic nerve with paradoxical diaphragmatic movements; compression of the right brachiocephalic vein with venous engorgement of the face and arm on the right side; and compression of the subclavian artery with diminished pulses in the right upper extremity.

28. The answer is A (1, 2, 3). [*Chapter 13 IV B 3; VI C 1, 2; X B 2, 3*] The pulmonary veins, foramen ovale, and the least cardiac veins (venae cordis minimae) of the left atrial wall all deliver blood to the left atrium in the developing fetus. The ductus arteriosus shunts blood from the left pulmonary artery to the aorta.

29. The answer is C (2, 4). [*Chapter 13 IV B 3*] Blood is returned to the left side of the heart from the lungs via the pulmonary veins and from the innermost layers of the left atrial wall via the thebesian veins. The anterior cardiac veins, coronary sinus, superior vena cava, and inferior vena cava, as well as the thebesian veins, return blood to the right side of the heart.

30. The answer is C (2, 4). [*Chapter 13 VI B 1 a; VII B 2; XI A 1 a*] A myocardial infarction limited to the interventricular septum may produce tricuspid valve incompetence as a result of necrosis of the septal papillary muscles as well as irregularities in impulse conduction as a result of involvement of the atrioventricular bundle. The mitral valve, having no septal cusp or septal papillary muscles, is not directly affected nor is the aortic valve within the aortic outflow tract.

31. The answer is E (all). [*Chapter 13 X C 1–3*] At birth, the lungs expand with a concomitant increase in the pulmonary blood flow, which both reduces the pressure in the right atrium and increases the pressure in the left atrium. This pressure reversal causes the foramen ovale to close, preventing a reverse flow of blood to the right atrium.

32. The answer is E (all). [*Chapter 13 Figure 13–8; Chapter 14 II I 2–4*] The cardiac plexus contains postganglionic sympathetic neurons from the sympathetic chain, preganglionic parasympathetic neurons from the vagus nerve, and afferent neurons for both pain and reflexes.

33. The answer is B (1, 3). [*Chapter 13 VIII B; Chapter 14 II 1 4 a*] Pain originating in the heart travels along visceral afferent nerve fibers, which take two pathways to reach the spinal cord. Some cardiac afferents run along the middle and inferior cardiac cervical (accelerator) nerves to reach the sympathetic chain and then travel inferiorly to the level of T1 or T2. Other cardiac afferents travel along the upper thoracic splanchnic nerves to reach the sympathetic chain. The cardiac afferent fibers leave the sympathetic chain through the white rami communicantes and travel along the spinal nerve to reach the corresponding level of the spinal cord. Pain of cardiac origin is referred to (appears as if originating from) dermatomes T1–T4.

34. The answer is D (4). [*Chapter 11 IV C 2 a, b*] The left and right vagus nerves along with the esophagus pass through the diaphragm via the esophageal hiatus. The azygos vein, the hemiazygos vein, and the thoracic duct pass through the aortic hiatus.

35–40. The answers are: 35-C, 36-D, 37-D, 38-A, 39-A, 40-A. [*Chapter 13 IV A 1, 2*] Both the right coronary artery and the circumflex branch of the left coronary artery lie in the coronary sulcus. The left coronary artery arises from the left posterior aortic sinus, whereas the right coronary artery arises from the right aortic sinus. The right coronary artery usually supplies both the sinoatrial node and the atrioventricular node and usually terminates as the posterior interventricular (descending) artery. The anterior portion of the interventricular septum is normally supplied by the anterior interventricular (descending) branch of the left coronary artery.

Part IV
Abdominal Region

I. INTRODUCTION

A. Abdomen. The abdominal region is below the **diaphragm** and above the bones and ligaments of the **pelvic brim**.

B. Surface landmarks of the abdomen (Fig. 15-1A)

1. **Hard tissue landmarks** include the **xiphoid process** of the sternum, **costal margin**, **iliac crest** of the pelvis, **anterior–superior iliac spine**, **pubic tubercle**, **inguinal ligament** between the anterior–superior iliac spine and the pubic tubercle, and **pubic crest**.

2. **Soft tissue landmarks** include the **linea semilunaris**, just lateral to the rectus abdominis muscle, **tendinous inscriptions** of the anterior sheath of the rectus abdominis muscle, **linea alba** in the midline, and **umbilicus**.

C. Abdominal subdivisions. References to **quadrants** and **regions** are used to describe the location of anatomic structures within the abdomen as well as to describe symptoms and the results of physical examinations pertaining to this area.

1. **Abdominal quadrants.** The abdomen is divided by vertical and horizontal planes through the umbilicus into four quadrants (Fig. 15-1B).
   a. Left and right upper quadrants
   b. Left and right lower quadrants

2. **Abdominal regions.** The abdomen is divided into nine regions by the horizontal **subcostal plane** (beneath the lowest extension of the thoracic cage) or the **transpyloric plane** (halfway between the jugular notch and the pubis), and the horizontal **intertubercular plane** (through the iliac tubercles) as well as the two vertical **midclavicular planes** or **midinguinal planes**.
   a. Left and right hypochondriac regions underlie the rib cage.
   b. Epigastric region
   c. Left and right lateral (posterolateral **lumbar** and anterolateral **flank**) **regions**
   d. Umbilical region
   e. Left and right inguinal (iliac) regions
   f. Pubic (hypogastric) region

D. Internal structures

1. **Numerous abdominal organs and structures may be palpated** through the anterior abdominal wall (see Fig. 15-1B).
   a. The aorta can be outlined both above and below the umbilicus from its pulse; the aortic pulse even may be seen in thin individuals. An abdominal aortic aneurysm can be discerned upon palpation.
   b. The liver usually is palpable in the right upper quadrant, depending upon a person's build. A palpable liver may or may not indicate a pathologic condition. In persons with hepatomegaly, it may extend to the iliac crest.
   c. The gallbladder, if it is enlarged or contains choleliths (gallstones), may be palpated in the right hypochondriac region at the junction of the linea semilunaris and the costal margin.
   d. The stomach is palpable only if engorged.
   e. The spleen normally is not palpable. When it is enlarged, however, it may be felt in the left hypochondriac region and may even be palpable in the lower quadrants.
   f. The kidneys (only lower poles) usually can be palpated in the lateral regions.
   g. The ascending colon usually is palpable only if distended by gas or chyme.

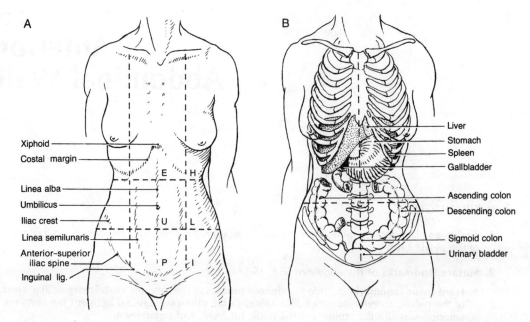

A

Xiphoid
Costal margin

Linea alba
Umbilicus
Iliac crest
Linea semilunaris
Anterior–superior
iliac spine
Inguinal lig.

E H

U L

P I

B

Liver
Stomach
Spleen
Gallbladder

Ascending colon
Descending colon

Sigmoid colon
Urinary bladder

Figure. 15-1. *A, Abdomen divided by regions.* Left and right hypochondriac (*H*), lateral (*L*), and inguinal (*I*) regions with medial epigastric (*E*), umbilical (*U*), and pubic (*P*) regions. Prominent landmarks and surface features are indicated. *B, Abdomen divided by quadrants.* Left and right upper and lower quadrants are depicted with the principal underlying structures.

> **h. The descending colon and sigmoid colon**, containing solidifying feces, usually are palpable in the lower left quadrant.
>
> **i. The urinary bladder**, when distended, is palpable in the pubic (hypogastric) region.
>
> **j. The uterus** can be palpated bimanually in the pubic region with the cervix being supported by the fingers of one hand.
>
> **k. The ovaries**, if enlarged, may be palpated bimanually in the inguinal regions.

2. Several abdominal organs may be percussed or auscultated through the abdominal wall.

> **a. An empty stomach** upon percussion will produce a tympanic tone in the left upper quadrant in an upright patient and in the epigastric region in a supine patient.
>
> **b. The bowel** normally produces tympanic tones upon percussion; hypertympanic tones are elicited when the bowel is distended by gas.
>
> **c. Bowel sounds** produced by peristalsis (**borborygmi**) normally are evident on auscultation; the frequency of borborygmi is an indication of intestinal motility.
>
> **d. The abdominal aorta and common iliac arteries** may be auscultated for pulse sounds and pathologic bruits.

II. INTEGUMENT

A. Overview. Review Chapter 2 I.

B. Dermis. Beneath the **epidermis**, the connective tissue fibers of the dermis have a prevailing directionality.

1. Cleavage lines (of Langer) are of clinical importance. They define the prevailing arrangement of connective tissue fibers and are evident as crease lines in the skin.

> **a.** Langer's lines are nearly horizontal along the anterior abdominal wall (see Fig. 2-1).
>
> **b.** An incision made across the prevailing direction of the connective tissue fibers tends to gape, resulting in a prominent scar upon healing.
>
> **c.** An incision made parallel to Langer's lines severs fewer connective tissue bundles, gapes less, and heals with a more cosmetic scar.

2. Elasticity. The dermis resists tearing, shearing, and localized pressure but is flexible enough to permit some stretch. Skin that is stretched for long periods, as in pregnancy or in obese persons, can be damaged by rupture of the connective tissue fascicles with subsequent scar formation called striae or "stretch marks." Striae appear perpendicular to Langer's lines.

III. FASCIAE OF THE ANTERIOR ABDOMINAL WALL

A. Superficial fascia (subdermis or hypodermis). The superficial fascia contains the superficial blood vessels, lymphatics, and nerves. It has two distinct layers.

1. **The superficial layer of the superficial fascia (of Camper)** underlies the epidermis and is predominantly an adipose layer, containing most of the fat of the subdermis.
   a. Camper's fascia continues over the chest as the superficial layer of the superficial thoracic fascia.
   b. It crosses the inguinal ligament to merge with the superficial fascia of the thigh.
   c. It also continues over the pubis as the superficial layer (**of Cruveilhier**) of the superficial perineal fascia.

2. **The deep layer of the superficial fascia (of Scarpa)** is a fibrous layer that will hold sutures.
   a. Scarpa's fascia continues over the chest as the deep layer of superficial thoracic fascia.
   b. It passes into the upper thigh, where it is attached to the fascia lata just below and parallel to the inguinal ligament.
   c. It also continues over the pubis as the deep layer (**of Colles**) of the superficial perineal fascia.

3. **Regional differences.** Camper's and Scarpa's layers of the superficial fascia fuse over the penis and the scrotum.
   a. The reflection of superficial fascia over the penis, as the superficial fascia of the penis, has no apparent adipose contribution from Camper's fascia.
   b. The reflection over the scrotum, the **dartos layer**, contains smooth muscle.

B. Deep fascia

1. **Location.** By definition, the deep fascia is the investing fascia of the musculature, aponeuroses and large neurovascular structures. It cannot be separated easily from the underlying epimysium of muscle, perineurium of nerve, periosteum of bone, and perichondrium of cartilage.

2. **Regional extensions**
   a. The deep fascia continues over the spermatic cord as the **external spermatic fascia.**
   b. It passes over the pubis and perineal musculature as the deep perineal fascia (**of Gallaudet**).
   c. It also extends over the penis as the deep penile fascia (**of Buck**).

C. Transversalis fascia. This is an inner layer of connective tissue that enwraps the abdominal cavity and supports the peritoneum. Loose extraperitoneal connective tissue contains varying amounts of fat and separates the transversalis fascia from the peritoneum.

IV. MUSCULATURE OF THE ANTERIOR ABDOMINAL WALL

A. Intrinsic abdominal musculature. Arranged in three layers, the outer layer runs superolaterally–inferomedially; the intermediate layer tends to run inferolaterally–superomedially; and the deepest layer tends to run transversely. As in the thorax, this three-ply bias bond construction provides maximum strength.

1. **The external oblique muscle**, the outer layer of the abdominal wall musculature, has fibers that run from superolateral origins to inferomedial insertions. The muscle layer becomes aponeurotic (broadly tendinous) at approximately the midclavicular line (Fig. 15-2; Table 15-1).
   a. Attachments. From the inferior borders of ribs 5–12, this superficial muscle layer runs diagonally inferiorly and medially.
      (1) It inserts into the iliac crest as far as the anterior–superior iliac spine.
      (2) It also inserts into the pubic tubercle and pubic crest, and the linea alba by decussating with fibers of the contralateral aponeurosis.
   b. The inferior lumbar trigone (of Petit) is formed by the free posterior border of the external oblique muscle (between the rib cage and iliac crest), the iliac crest, and the lateral border of the latissimus dorsi muscle.
   c. The inguinal ligament (of Poupart) is formed by the lower margin of the external oblique aponeurosis (between the **anterior iliac spine** and the **pubic tubercle**).
   d. The superficial (external) inguinal ring, although so termed, is actually a triangular defect in the aponeurosis of the external oblique muscle.
      (1) The aponeurosis of the external oblique muscle splits into two crura.
         (a) The **lateral crus** inserts onto the pubic tubercle with some tendinous fibers reflected to the **superior pubic ramus** as the **lacunar ligament (of Gimbernat)**.

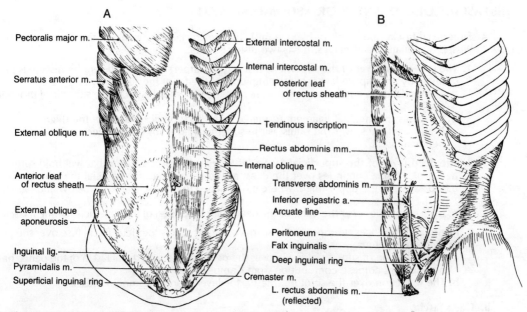

Figure 15-2. *Abdominal musculature. A,* The right external oblique and left internal oblique muscles are depicted. The anterior leaf of the rectus sheath is shown on the right side and dissected away on the left side to show the left rectus abdominis muscle. *B,* The left transverse abdominis and the right rectus abdominis muscles are depicted. The left rectus abdominis has been removed to show the posterior leaf of the rectus sheath and arcuate line.

      (b) The **medial crus** inserts into the **pubic symphysis.**
      (c) Intercrural fibers strengthen the apex of the superficial inguinal ring.
    (2) The superficial inguinal ring transmits the **spermatic cord** (in the male) or the **round ligament of the uterus** (in the female).

2. The internal oblique muscle is the intermediate layer of the abdominal wall musculature. Fibers of the internal oblique muscle and aponeurosis originate inferolaterally and insert superomedially, usually running perpendicular to those of the external oblique. However, in the inguinal and hypogastric regions, they run parallel to the fibers of the external oblique aponeurosis and provide muscular support for the abdominal wall in these regions. The intermediate muscle layer becomes aponeurotic at approximately the linea semilunaris (see Fig. 15-2 and Table 15-1).

    a. Attachments. From the thoracolumbar (thoracodorsal) fascia, transverse processes of the lumbar vertebrae, iliac crest, and inguinal ligament, this intermediate muscle layer inserts into the:
      (1) Inferior borders of ribs 10, 11, and 12
      (2) Linea alba by decussating with fibers of the contralateral aponeurosis
      (3) Pubic symphysis
    b. The floor of the inferior lumbar trigone is formed by the internal oblique muscle.
    c. The **cremaster muscle** in the male is derived as the spermatic cord passes through the internal oblique muscle. Thus, the **inguinal canal** passes diagonally through and is formed in part by the internal oblique muscle.

3. The transverse abdominal muscle constitutes the innermost layer of the abdominal wall musculature. Fibers of the transverse abdominal muscle and aponeurosis tend to run horizontally across the abdominal wall. At approximately the linea semilunaris, the muscle fibers cease and the layer becomes aponeurotic (see Fig. 15-2 and Table 15-1).

    a. Attachments. From the inferior borders of costal cartilages 5–10, ribs 11 and 12, lumbodorsal fascia, and iliac crest, this innermost muscle layer inserts into the:
      (1) Linea alba by decussating with fibers of the contralateral aponeurosis
      (2) Pubic crest, thereby reinforcing the inferomedial portion of the inguinal region
    b. Falx inguinalis. The lower free edge of the transverse abdominal muscle and aponeurosis (falx inguinalis) is usually (in 95% of males) craniad to the inguinal canal and takes no part in forming the spermatic cord.

4. The rectus abdominis muscle (and the occasional **pyramidalis muscle**) represents the same

Table 15-1. Muscles of the Anterior Abdominal Wall

Muscle	Origin	Insertion	Action	Innervation
External oblique	Inferior border of ribs 5–12	Iliac crest, pubic crest, and linea alba	Compresses abdomen and rotates vertebral column	Nn. T7–T12, iliohypogastric n. (L1–L2)
Internal oblique	Thoracodorsal fascia, iliac crest, and inguinal ligament	Inferior borders of ribs 10–12, linea alba, and pubic crest	Compresses abdomen and rotates vertebral column	Nn. T7–T12, iliohypogastric and ilioinguinal nn. (L1–L2)
Transverse abdominis	Ribs 10–12, thoracodorsal fascia, and iliac crest	Linea alba and pubic crest	Compresses abdomen	Nn. T8–T12, iliohypogastric and ilioinguinal nn. (L1–L2)
Rectus abdominis	Costal cartilages 5–8 and xiphoid process	Pubic crest and symphysis	Flexes vertebral column and compresses abdomen	Nn. T5–T12, iliohypogastric and ilioinguinal nn. (L1–L2)

layer as the occasional sternalis muscle in the thorax and the pubococcygeus muscle of the pelvis (see Fig. 15-2 and Table 15-1).

  a. Attachments. From costal cartilages 5–7 and the xiphoid process, this innermost muscle layer inserts into the pubic crest.

  b. The linea semilunaris (of Spigelius) is defined by the lateral margins of the rectus abdominis muscle.

  c. The linea alba in the midline is defined by the medial margins of the rectus abdominis muscle.

  d. Tendinous inscriptions (three or four) lie between the anterior leaf of the rectus sheath and the rectus abdominis muscle craniad to the umbilicus. The rectus sheath is formed by the aponeuroses of the muscles of the lateral abdominal wall.

  e. The pyramidalis muscle (frequently undeveloped, absent, or unilateral) lies anteriorly to the lower rectus abdominis muscle between the linea alba and the pubic crest.

B. Actions of the muscles of the anterior abdominal wall. Depending upon the type of motion, the muscles of the abdominal wall can be antagonistic to or synergistic with each other (see Table 15-1).

  1. Flexion of the vertebral column and pelvis by contraction of the external and internal oblique muscles, bilaterally, as well as the rectus abdominis muscle, which is the most powerful abdominal flexor

  2. Abduction of the trunk by ipsilateral contraction of the external and internal obliques, acting synergistically on one side

  3. Rotation of the trunk by contraction of the external oblique muscle on one side with the internal oblique of the contralateral side. The fibers of the external oblique on one side have the same prevailing direction as those of the internal oblique on the contralateral side.

  4. Respiration. Abdominal expiration is accomplished by contraction of the abdominal musculature as a unit, which forces the abdominal viscera (as a fluid piston) against the inferior surface of the diaphragm and causes the diaphragm to stretch and elevate into the thoracic cavity, thereby decreasing thoracic volume.

  5. Fixation of the abdominal wall

  a. Contraction of the abdominal musculature in concert with the thoracic musculature and with the glottis closed will result in increased thoracoabdominal pressure, the **Valsalva maneuver**.

  b. Fixation of the thoracoabdominal wall is utilized in lifting heavy weights, coughing, sneezing, micturition, parturition, defecation, eructation, and emesis. In combination with weakness of the abdominal wall, herniation may result from excessive thoracoabdominal fixation.

C. Rectus sheath. The rectus sheath is formed by fusion of the aponeurosis of the internal oblique, external oblique, and transverse abdominis muscles. It invests the rectus abdominis muscle (see Fig. 15-2).

1. Above the arcuate line (of Douglas)
  a. The anterior leaf of the rectus sheath is formed by fusion of the external oblique aponeurosis *and* the anterior portion of the split internal oblique aponeurosis. Two or three **tendinous inscriptions** (remnants of segmentation) attach the anterior leaf of the rectus sheath to the rectus abdominis muscle and frequently are visible surface features superior to the umbilicus.
  b. The posterior leaf of the rectus sheath is formed by fusion of the posterior portion of the split internal oblique aponeurosis *and* the transverse abdominis aponeurosis. There are no tendinous inscriptions on the posterior surface of the rectus sheath.

2. At the arcuate line
  a. One or two inches below the umbilical level, all of the aponeurotic layers of the lateral abdominal muscles pass *anteriorly* to the rectus abdominis muscle. The lower free edge of the posterior rectus sheath formed by this transition is the arcuate line.
  b. If that portion of the posterior rectus sheath contributed by the internal oblique aponeurosis passes anteriorly to the rectus abdominis, superior (craniad) to that portion contributed by the transverse abdominis aponeurosis, there will be two arcuate lines.
  c. The junction of the arcuate line with linea semilunaris is the site of a *lateral ventral (spigelian) hernia* into the rectus sheath.

3. Below the arcuate line
  a. The anterior leaf of the rectus sheath is formed by fusion of the aponeuroses of all three lateral abdominal muscles.
  b. The transversalis fascia lies against the rectus abdominis muscle since there is no posterior leaf of the sheath below the arcuate line.

V. ABDOMINAL WALL VASCULATURE

A. Arterial supply

1. Epigastric arterial system (Fig. 15-3)
  a. The superior epigastric artery:
    (1) Arises from the bifurcation of the **internal thoracic artery**
    (2) Enters the rectus sheath beneath the costal margin
    (3) Anastomoses with the inferior epigastric artery within the rectus abdominis muscle
  b. The inferior epigastric artery:
    (1) Arises from the **external iliac artery**
    (2) Lies along the medial margin of the **deep inguinal ring** as it passes obliquely upward from its origin and enters the rectus sheath by passing superficially to the arcuate line
    (3) Anastomoses with the superior epigastric artery within the rectus abdominis muscle
  c. Clinical considerations. Anastomoses in the epigastric arterial system can provide important shunts to supply blood to the lower part of the body in conditions such as coarctation of the aorta.

2. Other major anterior abdominal arteries (see Fig. 15-3)
  a. The musculophrenic artery:
    (1) Arises as the other bifurcating branch of the internal thoracic (internal mammary) artery
    (2) Courses along the costal margin
    (3) Supplies numerous branches to the anterior abdominal wall and to the diaphragm
  b. Intercostal arteries 9 through 12:
    (1) Arise from the **thoracic aorta** and extend ventromedially beyond the costal margin into the abdominal wall
    (2) Enter the rectus sheath laterally
    (3) Anastomose with the superior and inferior epigastric arteries
  c. The deep circumflex iliac artery:
    (1) Arises from the **external iliac artery**
    (2) Courses deeply to the inguinal ligament toward the anterior–superior iliac spine
    (3) Supplies the deep inguinal region
    (4) Anastomoses with the lower intercostal arteries
  d. Superficial epigastric arteries:
    (1) Arise from the **femoral artery**
    (2) Supply the superficial inguinal and pubic regions
    (3) Anastomose with lower intercostal arteries

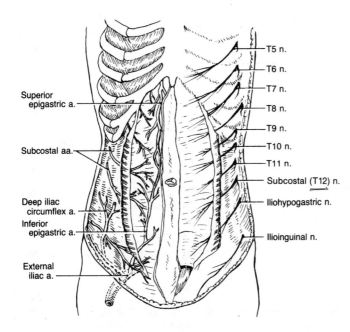

Superior
epigastric a.

Subcostal aa.

Deep iliac
circumflex a.

Inferior
epigastric a.

External
iliac a.

T5 n.
T6 n.
T7 n.
T8 n.
T9 n.
T10 n.
T11 n.
Subcostal (T12) n.
Iliohypogastric n.
Ilioinguinal n.

Figure 15-3. *Abdominal vasculature and innervation.* The vascular anastomoses between the superior epigastric, musculophrenic, intercostal, and inferior epigastric arteries are indicated. The 10th intercostal nerve supplies the umbilical region.

B. Venous drainage of the anterior abdominal wall

1. Superficial venous drainage of the anterior abdominal wall is toward the following veins:
a. Superior and inferior epigastric veins
b. Axillary vein via the thoracic and thoracoepigastric veins
c. Femoral vein via the superficial branches of the great saphenous vein
d. Hepatic portal vein via the paraumbilical veins
(1) Pathologic obstruction of the passage of blood from the gastrointestinal tract through the liver results in retrograde flow of blood to the systemic venous return of the body wall, via enlarged paraumbilical anastomoses and is termed **caput medusae**.
(2) Although caput medusae may not be readily apparent in cases of portal hypertension, it nearly always is demonstrable in a photograph taken with infrared film.

2. Deep venous drainage of the anterior wall is along veins that parallel the arterial supply previously noted.

C. Lymphatic drainage. Abdominal lymphatic vessels are divided into superficial and deep pathways.

1. Above the umbilicus
a. Superficial lymphatics, draining the subcutaneous tissues, drain primarily to the axillary nodes with some drainage to the parasternal nodes.
b. Deep lymphatics, draining muscles and investing fascia, drain primarily to the parasternal nodes.

2. Below the umbilicus
a. Superficial lymphatics drain toward the superficial inguinal nodes.
b. Deep lymphatics drain through the external iliac nodes to the para-aortic nodes.

VI. NERVES OF THE ANTERIOR ABDOMINAL WALL

A. Intercostal nerves T7–T12. The **ventral primary rami** of the spinal nerves supply the dermatomes and myotomes of the abdominal wall (see Fig. 15-3).

1. Course. They continue beyond the costal margin into the abdominal wall between the internal oblique and the transverse abdominis muscles. They penetrate the rectus sheath laterally.

2. Distribution. They supply the skin, muscles, and parietal peritoneum.

      a. The lateral cutaneous branches arise approximately at the anterior axillary line. They pierce the abdominal musculature in the midaxillary line to reach the dermis, bifurcating into anterior and posterior branches.

      b. The anterior cutaneous branches arise as terminal branches of the anterior rami. Each enters the rectus sheath laterally and exits from the rectus sheath anteriorly to reach the dermis. They bifurcate into medial and lateral branches.

    3. Landmarks. T4 and T10 supply dermatomes that contain, respectively, the nipple and the umbilicus.

  B. Lumbar nerves L1 and L2. The anterior primary rami of T12–L4 contribute to the **lumbar plexus** (see Fig. 18-6).

    1. The lumbar plexus has anterior and posterior divisions.

    2. The anterior divisions of T12, L1, and L2 form the **iliohypogastric**, **ilioinguinal**, and **genitofemoral nerves**, which supply the pubic (hypogastric), inguinal, and anterior perineal regions as well as the anterior proximal thigh and spermatic cord.

VII. CLINICAL CONSIDERATIONS RELATING TO THE ANTERIOR ABDOMINAL WALL

  A. Access to the peritoneal cavity

    1. Via the rectus sheath

      a. An incision into the anterior leaf of the rectus sheath medially has the advantage that the rectus abdominis muscle may be retracted laterally without compromising either its vascular or nerve supply; a second incision through the posterior leaf provides access to the peritoneal cavity. When the incisions in the posterior and anterior leaves are sutured, the undamaged rectus abdominis muscle will support the incisions well.

      b. If the rectus abdominis muscle is split longitudinally, the nerves supplying the muscle medially to the incision will be severed, resulting in paralysis of the medial portion of the muscle and a weakened abdominal wall disposed to ventral herniation.

      c. If the rectus muscle is transsected, the arterial supply can be embarrassed. Moreover, reanastomosis of a transsected muscle (compared to trying to sew two paint brushes together) may come apart upon Valsalva fixation (i.e., a cough or sneeze).

    2. Via the linea alba

      a. The linea alba is a relatively avascular region, hence its name *alba*. A midline incision through this structure is relatively bloodless but may not heal well without leaving a large, unsightly roll of tissue.

      b. A midline incision may predispose to midline herniation, one of the most common postsurgical complications.

    3. Via the ventrolateral abdominal wall

      a. Although many transverse incisions can take advantage of Langer's cleavage lines, they also may transsect the nerve supply to the abdominal musculature unless appropriate precautions are taken.

      b. The muscle splitting approach (of McBurney) for appendectomy in the right lower quadrant uses Langer's lines for the superficial incision, then splits and retracts the external oblique muscle, splits and retracts the internal oblique muscle (locating and retracting the iliohypogastric nerve), splits and retracts the transverse abdominis muscle, and, finally, incises the peritoneum. Incisions in each layer close naturally with few, if any, sutures and provide a strong repair. Valsalva fixation (i.e., a cough or sneeze) tends to close a split muscle but to tear the sutures from a transsected muscle.

  B. Abdominal hernias

    1. Epigastric herniation is a midline protrusion of fat or abdominal viscera through the linea alba (usually above the umbilicus). It is a frequent postoperative complication but rare otherwise.

    2. Umbilical herniation is a protrusion of abdominal viscera through the umbilical ring.

      a. Congenital umbilical hernia, which has an embryologic basis, occurs in about 1 in 50 births. It may be apparent at birth, (omphalocele), or it may occur spontaneously in infants and children.

      b. Acquired umbilical hernia usually occurs in adult life.

        (1) It is frequently the result of repeated Valsalva fixation, pregnancy, ascites, or obesity along with a weakened abdominal wall.

       (2) Repair of umbilical herniation may be complicated by portal hypertension, which causes the paraumbilical anastomoses to be greatly enlarged and bleed profusely if divided.

   3. Semilunar (spigelian) herniation is a ventrolateral herniation that occurs most frequently at the junction of the arcuate line and the linea semilunaris but may occur along the linea semilunaris inferior to the arcuate line.

   4. Diaphragmatic herniation involves upward displacement of abdominal viscera or fat into the thorax; it is more common on the left side because of the location of the esophageal hiatus.

     a. The **esophageal hiatus** through which the esophagus passes into the abdomen is a common site for herniation of the stomach (*hiatus hernia*).

     b. The diaphragm develops from at least six primordia, some of them paired, which include contributions from the septum transversum in the cervical region, pleuroperitoneal membranes, and somatic wall. Failure of these primordia to fuse properly results in defects through which herniation may occur.

       (1) Incomplete closure of the pleuroperitoneal canal (usually on the left side) may be a site of herniation of abdominal contents into the thoracic cavity (*Bochdalek's hernia*).

       (2) Incomplete closure of the sternocostal triangle (**foramen of Morgagni**) is another potential, although rare, cause of herniation.

VIII. INGUINAL REGION

   A. Introduction

     1. Descent of the testes. The inguinal region is best described with reference to descent of the testes.

      a. The gonads develop retroperitoneally from the urogenital ridge in the region of the kidneys. They migrate retroperitoneally within the transversalis fascia toward the scrotum during embryonic development, apparently guided by the **gubernaculum testis**, a ligamentous structure that runs between the lower pole of each gonad to each labial-scrotal fold.

       (1) By the third embryonic month, the gonads beneath the peritoneum have moved into the false pelvis.

         (a) In females, migration continues into the deep pelvis.

         (b) In males, a peritoneal pouch, the **processus vaginalis**, evaginates into the developing scrotum.

       (2) By the seventh embryonic month, the testes reach the abdominal end of the inguinal canal (retroperitoneally).

       (3) Between the seventh and eighth embryonic months, the testes migrate through the inguinal canal *behind* the processus vaginalis (retroperitoneally) into the scrotum to form the spermatic cord.

       (4) Before birth, the testes reach their definitive position in the scrotum behind the tunica vaginalis, the remnant of the processus vaginalis.

       (5) The processus vaginalis may close before birth. However, it may remain patent, thereby predisposing to congenital indirect hernia during the early years of life.

      b. The testes retain their original nerve and blood supply; this explains the course of the spermatic vessels and nerves.

      c. The ovaries migrate into the deep pelvis, also trailing their vessels and nerves.

     2. Cryptorchidism is a failure of descent of the testes, which remain at the deep ring or within the inguinal canal. Cryptorchidism may correct spontaneously during the first year of life; if it persists, surgery is indicated. Failure to correct undescended testes before puberty results in sterility and, if bilateral, failure of the secondary sex characteristics to develop.

     3. Inguinal hernia. The inguinal region is a potential site of **abdominal herniation**.

      a. Passage of the testes through the ventral abdominal wall produces a weakened area, which may predispose to herniation with danger of strangulation, necrosis of the bowel loop, and embarrassment of the testicular blood supply.

      b. A persistent processus vaginalis (**funicular process**) may result in an **indirect congenital hernia** in which a loop of gut passes into and may even continue through the patent extension of the abdominal cavity into the scrotum.

   B. Inguinal canal. The anterior abdominal wall is breached by the inguinal canal from the superficial inguinal ring to the deep inguinal ring.

     1. The superficial (external) inguinal ring is a triangular interruption in the external oblique aponeurosis (Fig. 15-4*A*).

a. **Structure.** It is formed by the splitting of the aponeurosis of the external oblique muscle into two crura.

(1) The **lateral crus** inserts onto the **pubic tubercle** with some fibers reflected to the **superior pubic ramus** as the **lacunar ligament** (of Gimbernat).

(2) The **medial crus** inserts into the **pubis symphysis**.

(3) **Intercrural fibers** strengthen the apex of the superficial inguinal ring.

(4) The superficial inguinal ring may be palpated (but not normally entered) by inversion
of the scrotum (preferably with the little finger) directed along the vas deferens toward
the pubic tubercle.

b. **Contents.** The superficial ring transmits the **spermatic cord** in males and the **round ligament of the uterus** in females.

(1) The deep fascia (investing fascia of the external oblique muscle) continues over the
spermatic cord as the **external spermatic fascia**.

(2) At the superficial ring (the point of exit of both **direct** and **indirect inguinal hernias**), a
small hernia will bulge slightly during Valsalva fixation, for example, a cough, whereas
a large hernia will be palpable or even visually apparent.

2. **The deep (internal) inguinal ring** is a partial interruption in the transversalis fascia (Fig.
15-4*B*).

a. **Structure.** It is covered by peritoneum (unless there is a patent funicular process).

b. **Contents.** The vas deferens and the testicular neurovascular bundle pass through the deep
inguinal ring.

(1) Some of the transversalis fascia is drawn into the spermatic cord during descent of the
testes, forming the **internal spermatic fascia**.

(2) The deep ring is the entry site of an **indirect inguinal hernia**.

3. **The walls of the inguinal canal** are formed by the muscular, aponeurotic, and fascial layers of
the abdominal wall.

a. **The anterior wall** of the inguinal canal is formed by the **external oblique aponeurosis** superior to the **intercrural fibers** and the fibers of the **internal oblique muscle** (Fig. 15-5).

(1) The fibers of the external oblique muscle rarely extend as far as the inguinal canal.

(2) Contraction of the internal oblique muscle occludes and strengthens the inguinal
canal.

(3) Because the free edge of the transverse abdominis muscle lies superiorly to the deep
ring, it seldom contributes (less than 5% of the time) to the anterior wall of the inguinal
canal or to the cremaster muscle.

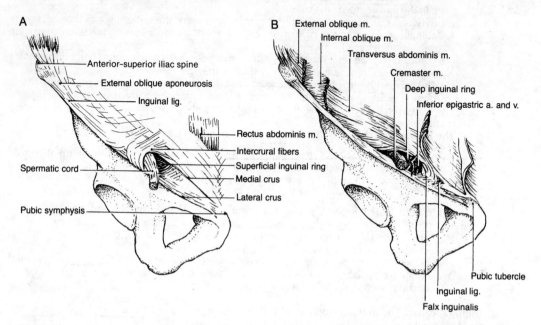

Figure 15-4. *A. Superficial inguinal ring.* The defect in the external oblique aponeurosis forms the superficial inguinal ring. The internal oblique muscle contributes the cremaster layer of the spermatic cord. *B, Deep inguinal ring.*
The deep ring is the location of the fused processus vaginalis. The inguinal canal passes beneath the inferior free
edge of the transverse abdominis muscle and aponeurosis (falx inguinalis).

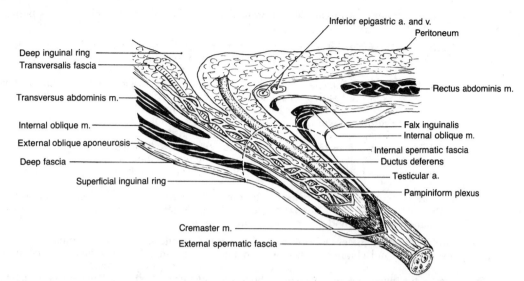

Figure 15-5. *Spermatic cord.* The deep fascia of the external oblique layer contributes the external spermatic fascia, the internal oblique layer contributes the cremaster muscle, the transverse abdominis layer usually makes no contribution, and the transversalis fascia contributes the internal spermatic fascia.

   b. The superior wall (roof) of the inguinal canal is formed by the **falx inguinalis**, which is the arcing free edge of the aponeurosis of the transverse abdominis muscle (see Fig. 15-4B).
      (1) The falx inguinalis (*falx, L. sickle*) runs from the lateral portion of the inguinal ligament to the rectus sheath and pubic ramus, leaving a weak area inferolaterally to it.
      (2) Many texts and atlas figures mistakenly and unfortunately refer to the falx inguinalis as the "conjoined tendon." The falx inguinalis actually is the lower, curving portion of the transverse abdominis aponeurosis. The conjoined tendon, on the other hand, results from the fusion (conjoining) of the aponeuroses of the internal oblique muscle and the transverse abdominis muscle as they form the anterior wall of the rectus sheath. Less than 5% of the time does a true conjoined tendon extend more than 1 cm laterally to the linea semilunaris. Therefore, a true conjoined tendon should be regarded as the rarity it is.
   c. The inferior wall (floor) of the inguinal canal is formed by the **inguinal ligament** (of Poupart) and the **lateral crus** of the superficial ring.
      (1) The inguinal ligament runs from the anterior–superior iliac spine to the pubic tubercle (see Fig. 15-4B).
      (2) The floor also receives a contribution from the lacunar ligament, which is formed by fibers reflected from the inguinal ligament posteriorly and runs superiorly to join the pectinate ligament along the pectinate line of the superior pubic ramus (see Figs. 15-4B and 15-6).
   d. The posterior wall of the inguinal canal is formed by fibers of the internal oblique muscle, the **transversalis fascia** and the **interfoveolar ligament** (of Hesselbach), a variable condensation of transversalis fascia of the inferior epigastric vessels (see Fig. 15-4B).

C. Spermatic cord

   1. Embryologic development (see Fig. 15-5)
      a. The spermatic cord begins as an evagination of the peritoneal cavity through the anterior abdominal wall, the **processus vaginalis**. Normally, the processus vaginalis persists from the third to the ninth months.
      b. As the processus vaginalis and the testes descend through the abdominal wall in the inguinal region, contributions from some layers of the abdominal wall are incorporated into the spermatic cord. These contributions include:
         (1) A peritoneal remnant of the obliterated processus vaginalis or, if patent, the **funicular process**
         (2) The **internal spermatic fascia**, derived from the transversalis fascia
         (3) The **cremaster muscle**, derived from the internal oblique muscle
         (4) The **external spermatic fascia**, derived from the deep (investing) fascia of the external oblique muscle with no other contribution from the external oblique layer, since the aponeurosis splits to form the superficial inguinal ring

(5) The **superficial fascia**, derived from Camper's and Scarpa's layers of the superficial fascia

2. Contents in the male. Because the testes remain attached to their original excretory ducts, vascular supply, and nerve supply, the contents of the spermatic cord include (see Fig. 15-5):

a. The vas deferens, the efferent duct of the testes

b. The testicular arteries, each of which arises from the aorta just caudal to the renal arteries

c. The arteries of the vas deferens, each of which arises from the internal iliac artery in the deep pelvis and is homologous to the uterine artery in females

d. The pampiniform plexus of veins, which coalesces into the **testicular veins**, which drain into the inferior vena cava (right) and the renal vein (left)

(1) This venous plexus provides a countercurrent heat exchange mechanism, which keeps the testes 1°–3° F cooler than the core body temperature.

(2) Varicosities in the pampiniform plexus of veins occur on the left side in about 85% of cases.

e. Testicular nerves, which include autonomic nerves to and visceral afferent nerves from the testes, as well as the **genital branch** of the **genitofemoral nerve**, which innervates the cremaster muscle, and the sensory **ilioinguinal nerve**

3. Contents in the female. Only the **round ligament of the uterus**, which is a remnant of the gubernaculum, and a branch of the **ilioinguinal nerve** pass through the inguinal canal (of Nuck).

D. Nerves of the inguinal region arise from the **lumbar plexus** on each side. They pass through the psoas major muscle and anterior to the quadratus lumborum muscle, posteriorly. They course between the transverse abdominis and internal oblique muscles, laterally. They become superficial anteriorly.

1. Iliohypogastric nerve (T12–L1, anterior) [see Fig. 18-6]

a. This nerve is distributed to the skin of the hypogastric region just superior to the pubic crest and pubic symphysis.

b. It usually can be observed penetrating the external oblique aponeurosis 1–2 cm superomedially to the superficial inguinal ring.

c. It provides afferent (sensory) and one of the several efferent (motor) limbs of the **abdominal reflex**. Stroking the skin of the abdomen parallel to the inguinal ligament initiates a rippling of the rectus abdominis muscle or flank muscles of the abdomen.

2. Ilioinguinal nerve (L1, anterior)

a. This nerve accompanies the spermatic cord or round ligament of the uterus through the superficial inguinal ring.

b. It is distributed to the medial aspect of the thigh, the base of the penis or mons, and the scrotum or labia majora.

3. Genitofemoral nerve (L1 and L2, anterior)

a. This nerve lies on the anterior surface of the psoas major muscle.

b. Its **femoral branch**, which passes beneath the inguinal ligament, is distributed to the anterolateral aspect of the thigh and provides the afferent limb of the **cremaster reflex**, initiated by stroking the medial thigh.

c. Its **genital branch**, which lies laterally or posterolaterally to the spermatic cord as it passes through the superficial inguinal ring, is distributed to the cremaster muscle and scrotum, and provides the efferent limb of the **cremaster reflex**, whereby a testis is elevated within the scrotum.

E. Inguinal triangle (of Hesselbach)

1. Boundaries. The **inguinal triangle** (Fig. 15-6) lies in the inferomedial inguinal region at the base of the medial inguinal fossa where it is bounded by the:

a. Linea semilunaris (the lateral edge of the rectus sheath), medially

b. Inguinal ligament (of Poupart), inferolaterally

c. Inferior epigastric artery, laterally

2. Structure. The inguinal triangle is an area of potential weakness and, thus, is often the site of a **direct inguinal hernia**.

a. The abdominal wall contributes little or no muscular support to the inguinal triangle.

b. Anteriorly, the inguinal triangle is covered by the external oblique aponeurosis. It is weakened by a defect in the anterior abdominal wall, the superficial inguinal ring.

c. Behind the superficial ring, the reinforcement of the inguinal triangle is provided mainly by the **falx inguinalis** and almost insignificantly by the **interfoveolar ligament**. The extent of support derived from the falx inguinalis varies among individuals, depending on the extent of its lateral insertion upon the superior pubic ramus.

d. Toward the base of the inguinal triangle, support is provided only by the weak transversalis fascia and the presence of the spermatic cord within the superficial ring.

3. **Clinical considerations relating to inguinal hernias.** Fully 97% of all hernias in males and 50% in females are inguinal. Inguinal hernias, of which there are two types, occur superiorly to the inguinal ligament.

   a. Indirect inguinal hernias
      (1) Indirect herniation is with*in* the inguinal canal.
         (a) The herniated viscus or fat lies with*in* the spermatic cord.
         (b) The hernia passes through the deep inguinal ring; therefore, it must pass *laterally* to the inferior epigastric artery to enter the inguinal canal; this places its origin, by definition, laterally to the inguinal triangle and in the lateral inguinal fossa.
         (c) The hernia exits the abdominal wall through the superficial inguinal ring.
         (d) The hernia is directed by the spermatic cord toward the scrotum.
         (e) The risk of strangulation and infarct is high because the hernia passes through the muscular inguinal canal.
      (2) **A congenital (funicular) indirect inguinal hernia** occurs through a patent funicular process, usually during the first year or so of life.
      (3) **An acquired indirect inguinal hernia** occurs through a weakened area behind the remnant of the fused processus vaginalis, usually in middle age or later.

   b. Direct inguinal hernia
      (1) Direct herniation through the inguinal triangle (of Hesselbach), pushes the peritoneum and transversalis fascia through the superficial inguinal ring.
         (a) The herniated viscus or fat lies *adjacent to* (not within) the spermatic cord.
         (b) The hernia does *not* enter the deep inguinal ring and does *not* pass through the inguinal canal; therefore, it passes *medially* to the inferior epigastric artery and by definition, through the inguinal triangle at the the base of the medial inguinal fossa.
         (c) The hernia exits the abdominal wall *directly* through the superficial inguinal ring.
         (d) Direct hernias seldom enter the scrotum.
         (e) The risk of strangulation in direct inguinal herniation is low because passage is not through muscular layers or strong tendinous confines.
      (2) Direct inguinal hernias are always acquired.
         (a) They are most common in middle-aged males and less common in females (although they have been seen more often in recent years).
         (b) Although these hernias are acquired, genetic factors predispose to the weakness or strength of the abdominal wall in the inguinal triangle (i.e., the lateral extent of the attachment of the falx inguinalis along the superior pubic ramus).

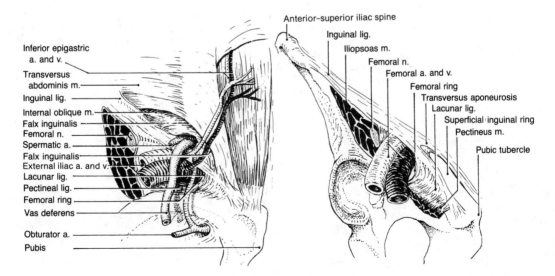

Figure 15-6. *Inguinal triangle.* The boundaries of the inguinal triangle (viewed from behind) are the lateral border of the rectus abdominis muscle, the inguinal ligament, and the inferior epigastric artery.

Figure 15-7. *Femoral ring.* The lateral muscular compartment contains the iliopsoas muscle and the femoral nerve. The medial vascular compartment contains the femoral vein, the femoral artery, and the femoral ring through which lymphatics pass.

    c. All inguinal hernias, **in contrast to femoral hernias**:
        (1) Lie superiorly to the inguinal ligament
        (2) Lie superiorly and medially to the pubic tubercle to which the inguinal ligament and lateral crus of the superficial ring both attach

F. Femoral triangle

   1. Boundaries. This region is bounded by the inguinal ligament, the sartorius muscle, and the adductor longus muscle.

   2. Compartments. The area between the inguinal ligament and the superior pubic ramus is divided into two compartments (Fig. 15-7).
     a. The muscular compartment contains the:
       (1) Iliopsoas muscle, a flexor of the thigh at the hip joint
       (2) Femoral nerve, which is derived from the anterior division of the lumbar plexus and innervates the anterior thigh
     b. The vascular compartment contains the:
       (1) Femoral artery, supplying the greater portion of the thigh and leg
       (2) Femoral vein
       (3) Femoral ring, containing lymphatics
       (4) Lacunar ligament

   3. Femoral ring. The femoral artery, femoral vein, and **femoral canal** (the extension of the femoral ring) are wrapped in the **femoral sheath**, which is a continuation of the transveralis fascia leaving the abdominal cavity along with the femoral vessels.
     a. Boundaries of the femoral ring include (see Figs. 15-6 and 15-7) the:
       (1) Femoral vein, laterally
       (2) Inguinal ligament (of Poupart), anteriorly, which is the inferior border of the external oblique aponeurosis between the anterior–superior iliac spine and the pubic tubercle
       (3) Lacunar ligament (of Gimbernat), medially, which is a reflection of the inguinal ligament posteriorly to the pectinate line of the superior pubic ramus
       (4) Pectineal ligament (of Cooper), posteriorly, which is an extension of the lacunar ligament along the pectinate line of the superior pubic ramus
     b. Contents of the femoral ring consist of lymphatics from the legs and perineum, which gain access to the peritoneal cavity via the femoral ring.

   4. Clinical considerations relating to femoral hernias
     a. Fully 34% of all hernias in females are femoral, but these hernias are rare in males.
     b. Femoral hernias enter the femoral ring and femoral canal inferior to the inguinal ligament. Because of the ligamentous boundaries, there is a high potential for strangulation and necrosis of a herniated loop of bowel.
     c. These hernias may run medially to, laterally to, or even bifurcate to run on both sides of an **aberrant obturator artery**, which arises from the inferior epigastric artery (see Fig. 15-6). This situation can present a danger for the unaware during surgical repair of a femoral hernia.
     d. Femoral hernias, **in contrast to inguinal hernias**:
       (1) Lie inferiorly to the inguinal ligament
       (2) Lie inferiorly and laterally to the pubic tubercle to which the inguinal ligament attaches
       (3) Usually lie in the thigh and present through the fossa ovalis of the great saphenous vein

I. INTRODUCTION

A. Basic principles. The **peritoneal (abdominopelvic) cavity** is the largest cavity of the body.

1. **The abdominal cavity** lies within the abdomen proper.
   a. It extends from the inferior surface of the musculotendinous diaphragm to the brim of the minor (deep) pelvis.
   b. It contains most of the gastrointestinal tract, the accessory digestive glands, the kidneys, and the major vessels.

2. **The pelvic cavity** lies within the true pelvis.
   a. It extends from the minor (deep) pelvis to the pelvic diaphragm (levator ani muscle) and is about the size of a clenched fist.
   b. It contains loops of small bowel, the sigmoid colon, the terminal portion of the gastrointestinal tract (i.e., the rectum and anal canal), and the organs of the urogenital system with associated adnexa.

B. Peritoneum

1. **The parietal peritoneum** is a continuous layer of mesothelial serosa that lines the abdominopelvic cavity (labeled somatic mesoderm layer in Fig. 16-3B).
   a. In males, the peritoneal cavity is completely enclosed by peritoneum.
   b. In females, it is perforated by the ostia of the uterine tubes. This pathway enables infectious organisms to enter the peritoneal cavity from the external environment.

2. **The visceral peritoneum** is a continuous layer of mesothelial serosa that covers surfaces of the abdominal and pelvic viscera (labeled splanchnic mesoderm layer in Fig. 16-3B).

3. **Mesenteries** are formed by the reflection of mesothelium between the parietal peritoneum and visceral peritoneum (labeled dorsal and ventral mesenteries in Fig. 16-3B).
   a. As the parietal peritoneum from one side leaves the body wall in the midline, it fuses with its contralateral counterpart. These back-to-back mesothelial layers form mesentery or a so-called ligament.
   b. As the mesentery reaches the viscera, the layers split and enclose the organ as visceral peritoneum.
   c. The mesenteries not only support the viscera but also provide pathways for the associated neurovascular structures and, occasionally, ducts.

II. DEVELOPMENT OF THE PERITONEAL CAVITY

A. Early differentiation. The **embryonic disk** develops during the third week. From the fourth to the eighth weeks, differential development produces tissues and organs.

B. The primitive gut is divided into three regions.

1. **The head fold, lateral folds, and tail fold** incorporate within the developing embryo a portion of the **upper pole** of the primitive **yolk sac** to form the **primitive gut** (Figs. 16-1 and 16-2).
   a. The foregut is formed by the **head fold**.
   (1) It is closed by the buccopharyngeal membrane while a blind tube and eventually opens through the **stomodeum**.
   (2) It is supplied by the **celiac artery**.

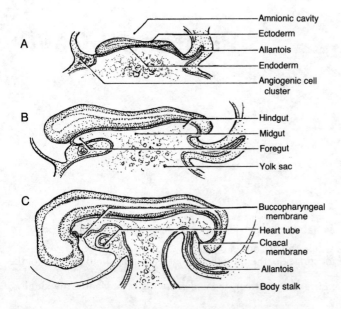

Figure 16-1. *Early differentiation of the gastrointestinal tract. A*, The germ disk at 19 days lies flat on the yolk mass. *B*, By 21 days, head and tail folds form the early foregut and hindgut, respectively. *C*, By 24 days, the head and tail folds have deepened, and lateral folds have come together to form the midgut.

    b. The midgut is formed largely by **lateral folds**.
        (1) It retains a connection with the yolk sac by the **omphalomesenteric (vitelline) duct**, which normally ultimately degenerates but which may persist in whole or part, resulting in an **ileal diverticulum** (of Meckel).
        (2) It is supplied by the **superior mesenteric artery**.
    c. The hindgut is formed by the **tail fold**.
        (1) Although the hindgut is a blind tube closed by the cloacal membrane, it eventually opens through the proctodeum.
        (2) It is supplied by the **inferior mesenteric artery**.
  2. The enteric tube is formed by the ventral fusion of the splanchnopleure (viscera primordium) as the two lateral folds meet with the head and tail folds at the umbilical stalk; the body wall is formed by the fusion of the somatopleure (body wall primordium) [see Fig. 16-3].

C. The coelomic cavity (coelom) is the space enclosed between the splanchnopleure (the layer that forms the viscera) and the somatopleure (the layer that forms the body wall) [see Fig. 16-3].

  1. The visceral pleura and visceral peritoneum, serous mesothelia, cover the outer surfaces of the splanchnopleure.

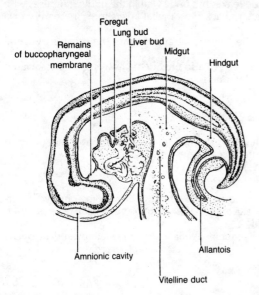

Figure 16-2. *Later differentiation of the gastrointestinal tract.* During the fourth week, flexion occurs with development of the body stalk. The midgut retains a connection with the yolk sac through the vitelline duct.

2. **The parietal pleura and parietal peritoneum**, serous mesothelia, cover the inner surfaces of the somatopleure.

3. **The dorsal and ventral mesenteries**, suspending the gastrointestinal tract within the coelom, divide the coelomic cavity into left and right halves (see Fig. 16-3*B*).

D. Mesenteries. Back-to-back layers of parietal peritoneum reflect off the body wall to become mesenteries which, in turn, become visceral peritoneum (Fig. 16-3).

1. **The ventral mesentery** largely degenerates, except for the cephalad and caudal portions. The liver develops *within* the cephalad portion of the ventral mesentery, dividing it into the lesser omentum and the falciform ligament (Fig. 16-4).
   a. **The lesser omentum**, between the stomach and the liver, has two named portions (see Fig. 16-5*C*).
      (1) The **gastrohepatic ligament** runs between the lesser curvature of the stomach and the liver.
      (2) The **hepatoduodenal ligament** runs between the superior duodenum and the liver.
         (a) This portion of the lesser omentum contains the hepatic artery, hepatic portal vein, and common bile duct.
         (b) The inferior free edge of the ventral mesentery forms the **omental (epiploic) foramen** (of Winslow).
   b. **The coronary ligaments** attach the liver to the diaphragm and provide continuity between the falciform ligament and the lesser omentum (see Fig. 17-7).
      (1) **Left** and **right anterior** as well as **left** and **right posterior coronary ligaments** provide continuity between the visceral peritoneum covering the liver and the parietal peritoneum lining the inferior surface of the diaphragm.
      (2) Nonapposition of the anterior and posterior leaves of the coronary ligaments results in an area between the liver and the diaphragm not covered by mesothelium, the **bare area** (see Fig. 17-7).
      (3) Lateral extensions of the anterior and posterior leaflets of the coronary ligaments form the **left** and **right triangular ligaments**.
   c. **The falciform ligament** (*falx, L. sickle*) attaches the liver to the ventral body wall.
      (1) The left and right leaflets of the anterior coronary ligament meet anteriorly to form the falciform ligament.

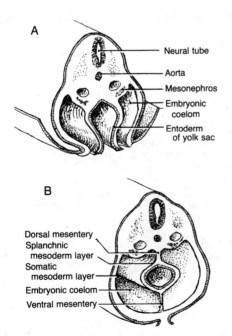

Figure 16-3. *Formation of the dorsal and ventral mesenteries. A,* About 21 days, the lateral folds delineate the intraembryonic coelom and the midgut. *B,* By 28 days, dorsal and ventral mesenteries become defined.

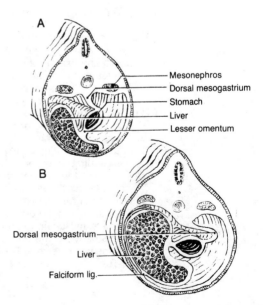

Figure 16-4. *Rotation of the stomach. A,* During the fifth week, the stomach rotates so that the primitive left side becomes the anterior surface. *B,* By the sixth week, the greater omentum begins to form, and the liver develops toward the right. The ventral mesentery caudal to the falciform ligament begins to degenerate.

(2) The falciform ligament extends between the ventral surface of the liver and the ventral abdominal wall as far inferiorly as the umbilicus.

(3) It contains the **round ligament of the liver (ligamentum teres)**, which is the remnant of the obliterated **umbilical vein(s)**. It may remain patent (rare in adults).

d. The median umbilical fold is a caudal remnant of the ventral mesentery.

(1) It runs from the apex of the urinary bladder to the umbilicus.

(2) It contains the **urachus**, which is the remnant of the fused allantoic stalk.

(a) A completely patent urachus (rare) will result in seepage of urine from the umbilicus, while isolated patency may produce a urachal cyst.

(b) Partial patency, extending from the apex of the bladder, may complicate surgery in the umbilical and pubic (hypogastric) regions.

2. The dorsal mesentery is the primary support of the gastrointestinal tract and provides passage for the blood vessels that supply the viscera (see Fig. 16-4).

a. The dorsal mesogastrium is divided into several ligaments.

(1) The **gastrophrenic ligament** runs between the greater curvature of the stomach and the diaphragmatic peritoneum.

(2) The **gastrosplenic ligament** runs between the stomach and the spleen, and the **splenorenal (lienorenal, phrenicolienal) ligament** continues between the spleen and the dorsal body wall superior to the left kidney.

(3) The **greater omentum** runs between the dorsal body wall and the greater curvature of the stomach (see Fig. 16-6).

b. The dorsal mesoduodenum is an embryonic structure only.

(1) Except for a minute portion along the superior duodenum, this mesentery becomes apposed and fixed to the peritoneum of the dorsal body wall, making the duodenum *secondarily retroperitoneal*.

(2) It contains the head and body of the pancreas, so that those portions of that organ also become *secondarily retroperitoneal*.

(3) The region of fusion (fascia of Toldt) may be incised surgically, thereby mobilizing the duodenum and pancreas together with their shared blood supplies.

c. The dorsal mesointestine is divided into several regions, each of which supports a segment of bowel.

(1) The mesentery proper supports most of the small intestine (see Fig. 16-5C).

(a) This ligament runs from the dorsal body wall to the jejunum and ileum.

(b) It is fan-shaped, so that a root about 9 inches long supports about 20 feet of small bowel.

(c) The positions in which it supports the various loops of small bowel are extremely variable.

(2) The ascending mesocolon is an embryonic structure only.

(a) Except for a small portion along the inferior cecum, this mesentery becomes apposed and fixed to the peritoneum of the dorsal body wall, making the ascending colon *secondarily retroperitoneal*.

(b) The line of fusion may be incised surgically, thereby reestablishing the mesentery and mobilizing the ascending colon with its blood supply.

(3) The transverse mesocolon supports the transverse colon.

(a) It runs between the dorsal body wall and the transverse colon.

(b) It suspends the transverse colon in greatly variable positions (see Fig. 17-15).

(c) It fuses with the greater omentum to form the gastrocolic ligament (see Fig. 16-6).

(4) The descending mesocolon is an embryonic structure only.

(a) This mesentery becomes apposed and fixed to the peritoneum of the dorsal body wall, making the descending colon *secondarily retroperitoneal*.

(b) The line of fusion may be incised surgically, thereby reestablishing the mesentery and mobilizing the descending colon with its blood supply.

(5) The sigmoid mesocolon supports the sigmoid colon (see Fig. 16-5C).

(a) It runs between the dorsal body wall and the sigmoid colon.

(b) It suspends the sigmoid colon in variable positions within the deep pelvis (see Fig. 17-16).

III. ROTATION OF THE FOREGUT

A. Early stage. The definitive position of the gut in adults is due to rotation during development. The gastric anlage develops as a dilatation of the foregut during the fifth week (see Figs. 16-4 and 16-5).

1. There is a 90° rotation to the right around the longitudinal axis of the gut.

2. The left side of the stomach becomes anterior; the left vagus nerve now innervates the anterior surface while the right vagus nerve now innervates the posterior surface.

B. Later stages

1. Stomach. During rotation, the primitive dorsal surface of the stomach undergoes accelerated growth to produce the greater curvature.
  a. Differential growth pulls the mesogastrium to the left (see Fig. 16-4).
    (1) This growth produces the greater omentum.
    (2) A portion of the right coelom becomes entrapped behind the stomach to form the **omental bursa (lesser sac)**, which communicates with the rest of the coelomic cavity (greater sac) via the **omental (epiploic) foramen** (of Winslow) [see Fig. 16-7].
  b. Because of the concomitant growth of the intestines, the pylorus is moved craniad and to the right, and the stomach assumes the definitive adult position.
  c. The **spleen** develops *within* the dorsal mesogastrium.
    (1) The position of the spleen divides the dorsal mesogastrium into **gastrosplenic** and **splenorenal (phrenicolienal) ligaments**.
    (2) The lienorenal ligament is continuous with the peritoneum of the posterior abdominal wall and contains the tail of the developing pancreas.
  d. The dorsal mesogastrium continues to grow, doubling upon itself.
    (1) The omental bursa (lesser sac) initially extends into the space between the folds of the dorsal mesogastrium.
    (2) Fusion of the mesothelial leaflets of the dorsal mesogastrium on the lesser sac side forms the **greater omentum**.
    (3) Fat develops in the greater omentum.

2. Duodenum. During rotation, the **duodenum** assumes a C-shape.
  a. It is moved posteriorly and to the right because of gastric rotation.
  b. During the sixth week, it comes to lie against the dorsal body wall and (except for the first portion) becomes secondarily retroperitoneal as the dorsal mesoduodenum fuses with the peritoneum of the dorsal body wall.
  c. Upon fusion of the dorsal mesoduodenum to the posterior body wall, the pancreas (with the exception of the terminal portion of the tail) becomes secondarily retroperitoneal.
  d. During the second month, the duodenum passes through a solid stage and reopens somewhat later. Incomplete recanalization results in **pyloric** or **duodenal stenosis**, which must be surgically corrected.

IV. ROTATION OF THE MIDGUT

A. First stage (Fig. 16-5A)

1. Rapid growth of the midgut during the fifth week results in the formation of the **primary intestinal loop**, which is connected to the yolk sac via the **omphalomesenteric vitelline duct**.
  a. The cranial limb of the primary intestinal loop develops into the distal duodenum, the jejunum, and the ileum as far as the omphalomesenteric duct.
  b. The caudal limb of the primary intestinal loop develops into the terminal ileum, cecum and appendix, ascending colon, and most of the transverse colon.
  c. The omphalomesenteric duct, or its remnant in 3% of adults (Meckel's diverticulum), represents the boundary between the cranial and the caudal limbs of the primary intestinal loop.

2. A physiologic **umbilical herniation** of the primary intestinal loop normally occurs.
  a. Because of rapid hepatic and intestinal growth, the abdominal cavity becomes temporarily too small for the developing gut.
  b. The primary intestinal loop herniates into the umbilical stalk during the sixth week and reaches its maximal extent during the ninth week.

3. The primary intestinal loop rotates 90° counterclockwise around an axis provided by the superior mesenteric artery, the artery of the midgut.
  a. The cranial limb comes to lie on the right and the caudal limb on the left within the umbilical stalk. This positioning is probably the result of extremely rapid growth of the cranial limb during this phase.
  b. The cranial limb forms secondary loops and coils within the umbilical stalk.
  c. A clockwise rotation at this stage produces **situs inversus viscerum** (with a high probability of visceral reversal in the thorax). Situs inversus occurs in about 1 in 10,000 births.

4. Rotation of the stomach and duodenal fixation to the dorsal body wall occur simultaneously with this stage of intestinal rotation.

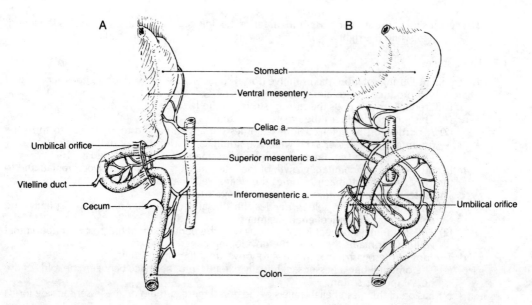

A

Stomach

Ventral mesentery

Celiac a.

Aorta

Superior mesenteric a.

Umbilical orifice

Inferomesenteric a.

Vitelline duct

Cecum

Colon

B

Umbilical orifice

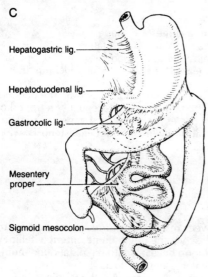

C

Hepatogastric lig.

Hepatoduodenal lig.

Gastrocolic lig.

Mesentery proper

Sigmoid mesocolon

Figure 16-5. *Rotation of the gut. A, Stage I: umbilical herniation.* During the sixth week, the rapidly growing intestine herniates into the umbilical stalk. *B, Stage II: reduction of the umbilical herniation.* During the tenth week, the intraembryonic coils of intestine are rapidly withdrawn into the coelomic cavity. The cranial loop passes posteriorly to the caudal loop so that an effective rotation results. *C, Stage III: fixation of the gastrointestinal tract.* Once the intestines have returned to the body cavity, their mesenteries fuse with the parietal peritoneum in several regions, making those portions of the tract appear retroperitoneal, thus the term *secondarily retroperitoneal.*

B. Second stage (Fig. 16-5*B*)

  1. During the tenth week, the umbilical herniation is completely withdrawn.
    a. The abdominal cavity has become larger as a result of the slower growth of the liver.
    b. The proximal jejunum is withdrawn first.
      (1) It passes posteriorly to the superior mesenteric artery and the caudal limb of the primary intestinal loop, coming to lie in the upper left quadrant. This explains the position of the superior mesenteric artery anterior to the terminal duodenum.
      (2) With progressive reduction of the umbilical herniation, successive loops lie progressively to the right as they enter the abdominal cavity, and secondary intestinal loops form.

  2. The caudal limb of the primary intestinal loop is withdrawn last.
    a. The cecum, therefore, tends to lie to the right below the liver.
    b. The caudal limb of the primary intestinal loop now lies anteriorly to the former cranial limb.

  3. Withdrawal of the physiologic herniation results in additional 180° rotation counterclockwise. Thus, the total rotation from first and second stages is 270° counterclockwise about an axis formed by the superior mesenteric artery.

C. Third stage (Fig. 16-5C)

1. The cecal region of the caudal limb of the primary intestinal loop moves inferiorly into the lower right quadrant, and the appendix develops.

2. The ascending mesocolon and the descending mesocolon fuse with the peritoneum of the dorsal body wall.

   a. The ascending and descending colons, thus, become secondarily retroperitoneal (i.e., once supported in the peritoneal cavity by the mesentery but adherent to the dorsal body wall in adults).

   b. The dorsal mesogastrium grows markedly and fuses with itself, obliterating that portion of the omental bursa that had been contained between the two redundant leaflets and becoming the greater omentum (Fig. 16-6).

      (1) The greater omentum, thus, contains four layers of mesothelium, two of which are apposed and fused.

      (2) The inferior surface of the greater omentum also fuses with the superior surface of the transverse mesocolon, forming the **gastrocolic ligament**, which, thus, is composed of six layers of mesothelium. The gastrocolic ligament divides the peritoneal cavity into supracolic and infracolic compartments.

D. Clinical considerations. Anomalies of rotation predispose toward strangulation, intestinal obstruction, ischemic necrosis, and infarct. Although major developmental anomalies of gastrointestinal tract rotation occasionally are incidental findings, they usually manifest clinically in neonates.

1. **Nonrotation of the gut**

   a. The primary intestinal loop may return to the abdominal cavity without rotation.

   b. The jejunum and ileum may lie on the right, and the duodenum may remain peritoneal. The small intestine may twist around the superior mesenteric artery (**volvulus**), obstructing the intestine or compressing its blood supply and resulting in ischemic necrosis and infarction of the small bowel.

   c. The ascending colon and descending colon may lie on the left side so that there is, in effect, no transverse colon.

      (1) The descending colon may become fixed.

      (2) Peritoneal bands or adhesions may develop because of partial fixation and lead to intestinal obstruction.

      (3) The floating ascending colon may undergo volvulus.

2. **Reversed rotation (situs inversus viscerum)**

   a. Situs inversus is rare, occurring in about 1 in 10,000 births.

   b. The external appearance is normal except that in males the right testis is lower. The visceral positions are a mirror image of normal.

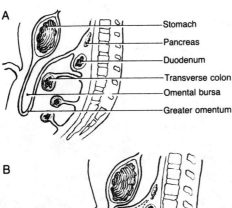

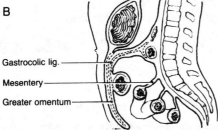

Figure 16-6. *Formation of ligaments. A,* The omental bursa (lesser sac) extends between the redundant leaflets of the greater omentum. *B,* Fusion of the redundant leaflets of the greater omentum and subsequent fusion with the transverse mesocolon form the gastrocolic ligament.

c. This anomaly is caused by clockwise withdrawal of the primary intestinal loop into the abdominal cavity.

d. Situs inversus can seriously confuse the unaware.

　　(1) In one extensive study, 55% of situs inversus cases were recognized *prior* to surgery, 32% were recognized *during* surgery, and 13% were not recognized until *after* surgery.

　　(2) In this same study, 14% of cases had a correct diagnosis and appropriate incision, 45% had an incorrect diagnosis, and 31% had an inappropriate incision; in 10%, the abdomen was closed immediately because the surgeons became confused.

3. Malrotation of the gut

a. There are various stages or degrees of incomplete rotation. Depending on the degree of nonrotation and subsequent peritoneal fixation, volvulus or strangulation may occur.

b. Nonfixation occurs most frequently at the cecum, and always to a variable extent in the region of the sigmoid colon, so that the normal length of the peritoneal portion of the sigmoid colon varies widely.

c. Malrotation may produce **paraduodenal hernias**, which are readily detected on x-rays.

　　(1) Left paraduodenal hernia. As the cranial limb of the primary intestinal loop withdraws from the umbilical herniation into the abdominal cavity, it may pass beneath the inferior mesenteric vein and thereby pass between the descending mesocolon and the parietal peritoneum. This passage prevents complete fusion of the descending mesocolon to the parietal peritoneum and forms an extensive left paraduodenal fossa, containing loops of small bowel.

　　(2) Right paraduodenal hernia. Similarly, in instances of nonrotation where the small intestine lies to the right, the ascending colon may move properly to the right and partially fixate to the parietal peritoneum, entrapping the small intestine beneath the ascending mesocolon and forming an extensive right paraduodenal fossa.

4. Omphalocele and umbilical hernia

a. Omphalocele is an anterior abdominal wall defect, covered only by peritoneum and perhaps a layer of amnion, though which viscera may protrude.

b. An umbilical hernia results from incomplete reduction of the physiologic umbilical hernia during the 10th week with retention of part of the primary intestinal loop within the umbilical cord.

　　(1) This anomaly occurs in approximately 2% of infants.

　　(2) A very small umbilical hernia may not be immediately recognized, and a loop of small intestine may be divided when the umbilical cord is sectioned and tied.

5. Meckel's diverticulum is a persistent remnant of the vitelline (omphalomesenteric) duct that is present in 3% of adults.

a. It may remain attached to the umbilicus as either a fibrous cord or a vitelline fistula.

b. Epithelium-lined cysts may persist along a fibrous connector.

c. Vitelline vessels may persist with attachment at the umbilicus and anatomoses with somatic vessels of the anterior abdominal wall.

d. Volvulus may occur around any persistent connection between the ileum and umbilicus with possible obstruction or strangulation.

e. Even if Meckel's diverticulum is not attached to the ventral body wall, there may be complications, such as inflammation of the diverticulum and peptic ulcers of the ileum, because Meckel's diverticula frequently contain gastric mucosae that secrete acid into the normally alkaline ileum.

V. PERITONEUM

A. The parietal peritoneum is the innermost layer of the abdominal wall, completely covering the inside wall of the abdominal cavity.

1. It is composed of mesothelial serous membrane.

2. It is supported by extraperitoneal connective tissue (transversalis fascia and fat) in which are embedded the great vessels, kidneys, and ureters.

3. Innervated by twigs from the lower intercostal nerves and lumbar nerves, it is very sensitive to painful stimuli.

B. The visceral peritoneum is the reflection of the peritoneal layer over the viscera.

1. Histologically, it is composed of a mesothelial serous membrane, the tunica serosa.

2. The serosa comprises the external surface of much of the gastrointestinal tract.

    3. Innervated by sphanchnic nerves, it is relatively insensitive to pain.

  C. Mesenteries are continuations between the parietal peritoneum and the visceral peritoneum. Mesenteries are formed by back-to-back layers of reflected peritoneum. They provide pathways to the viscera for blood vessels and nerves from the body walls as well as support for the ducts from the digestive glands (see Table 16-1).

    1. Primitive dorsal mesentery
      a. The primitive dorsal mesentery persists throughout embryonic development and adult life.
      b. It suspends the entire abdominal gut from the dorsal body wall (even in regions where the gut becomes secondarily retroperitoneal).
      c. It has as named portions: the **dorsal mesogastrium (gastrophrenic, gastrosplenic, phreni-colienal ligaments), greater omentum, mesentery proper, mesoappendix, transverse mesocolon**, and **sigmoid mesocolon**.

    2. Primitive ventral mesentery
      a. The primitive ventral mesentery largely degenerates during development, but where persistent, it suspends the viscera from the ventral body wall.
      b. It has as named portions: the **lesser omentum (gastrohepatic ligament** and **hepatoduodenal ligament), coronary ligament, left** and **right triangular ligaments, falciform ligament**, and **median umbilical fold**.

  D. Peritoneal fluid. Small amounts of serous fluid normally exude through the peritoneal membranes.

    1. This serous fluid provides a lubricating film for the parietal and visceral surfaces and facilitates free mobility of the viscera with minimal friction.

    2. During inflammatory processes involving the peritoneum or other pathologic conditions, large quantities of fluid (ascites) may collect in the abdominal cavity.

  E. Omental bursa (lesser sac)

    1. The omental bursa, representing the primitive right coelomic cavity, is isolated from the **greater sac** by rotation of the stomach as well as by elevation and fixation of the duodenum.

    2. It lies posteriorly to the stomach and lesser omentum.

    3. The **epiploic foramen** (of Winslow), which provides access to the lesser sac, is bounded by:
      a. The **caudate lobe** of the liver, superiorly
      b. The parietal peritoneum over the **inferior vena cava**, posteriorly
      c. The **superior duodenum**, inferiorly
      d. The **hepatoduodenal ligament**, anteriorly, which contains the **common bile duct, hepatic artery**, and **hepatic portal vein**

    4. The epiploic foramen is a potential site for intra-abdominal herniation. If a loop of bowel passes through the epiploic foramen and becomes incarcerated, none of the boundaries of the foramen can be safely incised. Instead, the bowel must be deflated with a needle, and the loop of gut can then be withdrawn easily.

VI. PERITONEAL RELATIONSHIPS

  A. Peritoneal fixation. The definitive positions and attachments of the gastrointestinal tract are the result of rotation of the gut. Certain regions of the gut and associated mesentery become applied to the peritoneum covering the dorsal body wall and appear to lose the supporting mesentery (i.e., they become **secondarily retroperitoneal**). Secondarily retroperitoneal portions of the gut may be sugically mobilized along with the supporting mesentery and its contained blood vessels and nerve supply by incising the fusion between the mesentery and the parietal peritoneum (Table 16-1).

  B. Peritoneal subdivisions. Peritoneal reflections as mesenteries divide the abdominal cavity into sacs, compartments, and several blind fossae or recesses. The peritoneal spaces, recesses, and fossae are potential sites for infection.

    1. Sacs and compartments. The dorsal mesogastrium and lesser omentum separate the greater and lesser sacs. The gastrocolic ligament, the fusion of the greater omentum and transverse mesocolon, divides the greater sac.
      a. The supracolic compartment lies craniad to the gastrocolic ligament.
      b. The infracolic compartment lies inferiorly to this ligament.

Table 16-1. Summary of Peritoneal Relationships

Organ	Classification	Support
Esophagus:		
Thoracic	Retroperitoneal	(Adventitia)
Abdominal	Peritoneal	Lesser omentum and gastrophrenic lig.
Stomach	Peritoneal	Dorsal mesogastrium (gastrophrenic, gastrosplenic, and phrenicolienal ligs., and greater omentum) and ventral mesogastrium (gastrohepatic lig.)
Liver	Peritoneal	Lesser omentum (gastrohepatic and hepaticoduodenal ligs.) and falciform, coronary, and triangular ligs.
Pancreas:		
Head, body, and most of tail	Secondarily retroperitoneal	(No mesentery)
Tip of tail	Peritoneal	Lienorenal lig.
Duodenum:		
First part	Peritoneal	Hepatoduodenal lig. and greater omentum
Second, third, and fourth parts	Secondarily retroperitoneal	(No mesentery)
Jejunum	Peritoneal	Mesentery proper
Ileum	Peritoneal	Mesentery proper
Cecum:		
Terminal part	Peritoneal	Ileocecal fold of mesentery proper
Body	Secondarily retroperitoneal	(No mesentery)
Appendix	Peritoneal	Mesoappendix
Ascending colon	Secondarily retroperitoneal	(No mesentery)
Transverse colon	Peritoneal	Transverse mesocolon and gastrocolic lig.
Descending colon	Secondarily retroperitoneal	(No mesentery)
Sigmoid colon	Peritoneal	Sigmoid mesocolon
Rectum	Retroperitoneal	(Adventitia)

2. **Named recesses, spaces, and fossae** (Figs. 16-7 and 16-8)
    a. **The subphrenic (suprahepatic) recess** is located anteriorly and superiorly to the liver, beneath the diaphragm.
        (1) It is divided into **right** and **left subphrenic recesses** by the falciform ligament.
        (2) It is the second most frequently infected abdominal space.
            (a) Adhesions between hepatic and diaphragmatic peritoneum may result in abscess formation.
            (b) A pulmonary abscess may erode across the diaphragm, invading the abdominal cavity and involving the suprahepatic recess. This type of infection was a frequent cause of mortality before antibiotics and still is a serious condition.
    b. **The infrahepatic recess** is located inferiorly to the liver within the lesser sac.
    c. **The right subhepatic recess** and the **hepatorenal recess** form the **pouch of Morison**.
        (1) The pouch of Morison is located inferiorly and posteriorly to the liver in the supracolic compartment.
        (2) These spaces communicate with (see Fig. 16-8):
            (a) The **lesser sac** via the **omental foramen**
            (b) The **right paracolic gutter** and, hence, the pelvic cavity
            (c) The **subphrenic recess**
        (3) When a person is supine, the hepatorenal recess becomes the lowest portion of the abdominal cavity and will collect fluid.
        (4) The pouch of Morison is probably the most frequently infected abdominal space.
            (a) If a patient has symptoms of an abdominal infection but no local abdominal signs, an abscess in the pouch of Morison will probably be the correct diagnosis.

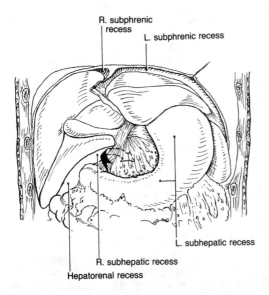

R. subphrenic recess
L. subphrenic recess

L. subhepatic recess

R. subhepatic recess

Hepatorenal recess

Figure 16-7. *Recesses of the supracolic compartment.* The named recesses are indicated. The *arrow* passes through the epiploic foramen into the omental bursa (lesser sac).

      (b) The primary causes of infection include appendicitis, a perforated duodenal or gastric ulcer, a perforated gallbladder or biliary tree, a liver abscess, and surgical procedures.
  d. Paraduodenal fossae
    (1) The **superior, inferior, left,** and **right paraduodenal fossae** occur in association with the **terminal duodenum,** the **inferior mesenteric vein,** and the **left colic artery.**
    (2) Herniation into a paraduodenal fossa may occur as a developmental anomaly.
  e. Fossae of the cecal region (see Fig. 17-14)
    (1) The **superior ileocecal fossa** is bounded by the **mesentery proper** and the **superior ileocecal (vascular) fold**.
    (2) The **inferior ileocecal fossa** is bounded by the **mesoappendix** and the **inferior ileocecal** (bloodless) **fold**.
    (3) The **retrocecal fossa** lies beneath the cecum, is variable in extent, is formed by adhesion of the cecum to the posterior parietal peritoneum, and frequently contains the appendix.
  f. Paracolic recesses (gutters) [see Fig. 16-8]
    (1) The **right colic recess (gutter)** lies laterally to the ascending colon.
      (a) It communicates with the supracolic compartment, the pouch of Morison, and the pelvic cavity.

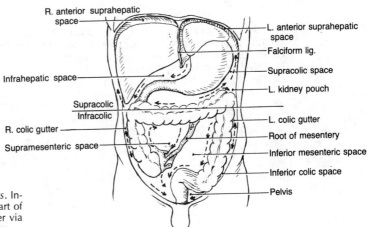

R. anterior suprahepatic space

L. anterior suprahepatic space

Falciform lig.

Supracolic space

L. kidney pouch

Infrahepatic space

Supracolic
Infracolic

L. colic gutter

R. colic gutter

Root of mesentery

Supramesenteric space

Inferior mesenteric space

Inferior colic space

Pelvis

Figure 16-8. *Drainage pathways.* Infection can spread from one part of the peritoneal cavity to another via the recesses, spaces, and gutters.

(b) It provides a route for the spread of infection between the pelvis and the upper abdominal region.

(2) The **left colic recess (gutter)** lies laterally to the descending colon.

(a) It communicates with the supracolic and infracolic compartments and the pelvic cavity.

(b) It seldom becomes infected.

g. **The infrasigmoid (parasigmoid) fossa**, which lies beneath the sigmoid colon, at the juncture of the sigmoid mesocolon and parietal peritoneum, is a potential site for intra-abdominal herniation of the sigmoid colon or small bowel.

h. **Abdominal wall fossae**

(1) The **lateral inguinal fossae** lie laterally to the **lateral umbilical folds**, which are formed by the underlying inferior epigastric arteries.

(a) The deep inguinal rings lie in this fossa.

(b) These are the sites where *in*direct inguinal herniation begins.

(2) The **medial inguinal fossae** lie medially to the lateral umbilical folds and laterally to the **medial umbilical folds**, which are formed by the underlying obliterated umbilical arteries.

(a) This nearly coincides with the inguinal triangle (of Hesselbach).

(b) These are the sites where direct inguinal herniation begins.

(3) The **supravesical fossae** lie medially to the medial umbilical folds and laterally to the **median (middle) umbilical fold**, which is formed by the underlying urachus.

(a) These fossae are superior to the bladder and posterior to the rectus abdominis muscles.

(b) These are the sites of supravesicular herniation.

C. Clinical considerations

1. Paracentesis

a. Inflammatory exudate tends to collect in the subphrenic and hepatorenal recesses, as well as in the pelvic cavity, when a patient who has a peritoneal infection is supine.

b. To arrive at a diagnosis, the fluid level may be percussed along the lateral abdominal walls and seen on x-rays.

c. Treatment includes puncture of the abdominal wall to withdraw fluid. A cannula inserted in the flank will pass through the skin, superficial fascia, deep fascia, aponeurosis of the external oblique muscle, internal oblique muscle, transverse abdominis muscle, transversalis fascia and extraperitoneal fat, and the parietal peritoneum.

2. Peritonitis, the "acute abdomen"

a. Inflammation of the parietal and visceral peritoneum produces different symptoms.

(1) The **parietal peritoneum** is innervated by somatic nerves of the anterior body wall. An inflamed parietal peritoneum is exquisitely sensitive to palpation or stretching in an area that is rather *discretely localized* on the abdominal wall. The abdominal musculature will tense—*guarding* or *splinting*—to produce a rigid abdomen and thereby minimize the pain.

(2) The **visceral peritoneum** is innervated by visceral nerves, which travel along autonomic pathways. An inflamed visceral peritoneum results in diffuse, crampy or colicky abdominal pain, which may be *referred* to specific dermatomes on the abdominal wall (see Table 17-2).

b. Infection may gain access to the peritoneal cavity by means of:

(1) Traumatic or surgical penetration of the abdominal wall

(2) Perforation of the gastrointestinal tract

(3) Hematogenous or lymphatic spread

(4) The uterine tubes, which normally open into the peritoneal cavity

I. INTRODUCTION

A. General principles

1. **Location.** The major portion of the gastrointestinal (GI) tract, as well as many of the associated digestive glands, is contained within the abdominopelvic cavity.

2. **Function.** The principal functions of the GI tract are digestion and nutrient absorption.

3. **Innervation.** The motor innervation of the GI tract is via the autonomic nervous system, the pathways of which are shared by visceral afferent nerves.

B. Developmental considerations. The division, locations, and mesenteric attachments of the portions of the GI tract may be described best with reference to their embryonic development.

1. **Divisions.** The GI tract can be divided by two methods.
   a. **The common organization** is by morphologically distinct organs, that is, the esophagus, stomach, small intestine, large intestine, and accessory glands.
   b. **A functional organization** of the abdominopelvic portion of the GI tract groups organs according to common similarities in the embryonic blood supply (see Chapter 16 II B), innervation, and principal function, that is, the foregut, midgut, and hindgut.
      (1) The **foregut** consists of the esophagus, stomach, and duodenum with the associated liver and pancreas.
         (a) The foregut is supplied largely by the **celiac artery**.
         (b) It receives parasympathetic innervation from the **vagus nerve** (CN X) and sympathetic innervation from the **greater splanchnic nerve** (T5–T9).
         (c) It functions as a conduit (esophagus), as a reservoir (stomach), and in digestion (stomach, duodenum, liver, and pancreas).
      (2) The **midgut** consists of the jejunum, ileum, ascending colon (including the cecum and appendix), and the transverse colon.
         (a) The midgut is supplied primarily by the **superior mesenteric artery**.
         (b) It receives parasympathetic innervation from the **vagus nerve** (CN X) and sympathetic innervation from the **lesser splanchnic nerve** (T10–T11).
         (c) It functions primarily in absorption of nutrients (jejunum and ileum) and water (ascending and transverse colons).
      (3) The **hindgut** consists of the descending colon, sigmoid colon, and rectum.
         (a) The hindgut is supplied primarily by the **inferior mesenteric artery**.
         (b) It receives parasympathetic innervation from the **pelvic splanchnic nerves** (S2–S4) and sympathetic innervation primarily from the **lumbar splanchnic nerves** (L1–L2).
         (c) It functions primarily in fecal transport (descending colon), storage (sigmoid colon), and evacuation (rectum).

2. **Adult position of the abdominal viscera.** The definitive position of regions of the GI tract are determined during development as the gut elongates rapidly, rotates, and portions become adherent to the peritoneal wall (see Chapter 16 IV).

3. **Mesenteric attachments.** The existence of persisting mesenteries and localized peritoneal fusion subsequent to rotation of the gut anchor portions of the GI tract in certain locations. Existing mesenteries limit movement of organs and provide support for the viscera as well as pathways for neurovascular structures.

II. FOREGUT

A. Esophagus

1. **Esophageal relationships.** The esophagus is approximately 25–30 cm long in males (in the erect position) and a few centimeters shorter in females.
   a. **In the thorax**, the esophagus extends from the lower part of the laryngopharynx and passes through the **esophageal hiatus** of the musculotendinous diaphragm (see Fig. 14-4). The relationships of the thoracic portion of the esophagus are discussed in Chapter 14 II C.
   b. **In the abdomen**, the esophagus makes an abrupt turn to the left from the esophageal hiatus of the diaphragm and enters the cardiac opening of the stomach (Fig. 17-1).
      (1) The esophageal hiatus is slightly to the left of the midline in the muscular portion of the diaphragm (see Fig. 18-4).
      (2) The hiatus is a frequent site of herniation of the stomach into the thoracic cavity (*hiatus hernia*).

2. **Esophageal vasculature**
   a. **Arterial supply** for the upper, middle, and lower portions is by respective branches from the **inferior thyroid arteries**, by the midline **esophageal arteries** and **bronchial arteries**, and by branches from the **left gastric artery** (see Fig. 17-1).
   b. **Venous return** generally parallels the arterial supply except that the long middle portion drains into the **azygos vein** and the **hemiazygos vein** (see Fig. 14-3). The terminal portion drains into the hepatic portal system via the **left gastric vein**.
   c. **Lymphatic drainage** parallels the blood vessels, an important consideration in the spread of esophageal carcinoma to cervical, mediastinal, and celiac nodes.

3. **Esophageal innervation**
   a. **Parasympathetic innervation**
      (1) The **cell bodies of the parasympathetic presynaptic neurons** lie in the **dorsal motor nucleus of CN X**. The presynaptic axons leave the brain in the vagus nerve (CN X), exiting the skull via the jugular foramen.
         (a) In the thorax, the **left** and **right vagus nerves** form the **esophageal plexus**, which reunites distally to form the vagal trunks. Due to rotation of the stomach, the left and right vagus nerves change their relative positions to become **anterior** and **posterior vagal trunks**, respectively (see Fig. 14-4). Both vagal trunks pass through the esophageal hiatus to enter the abdomen.
         (b) The parasympathetic presynaptic neurons, which leave the vagus nerves, synapse with the parasympathetic postsynaptic neurons in ganglia located between the longitudinal and circular muscle layers.
      (2) The **cell bodies of the parasympathetic postsynaptic neurons** lie in the **myenteric plexuses** (of Auerbach) and the **submucosal plexuses** (of Meissner). Dysfunction of the parasympathetic pathway at the level of the myenteric plexus results in *achalasia* (*cardiospasm*), which is characterized by *dysphagia* (inability to swallow) and a progressively dilated and tortuous esophagus.
   b. **Sympathetic innervation**
      (1) The **cell bodies of the sympathetic presynaptic neurons** to the esophagus lie in the lateral horns of the central gray matter of the thoracic spinal cord and exit with the ventral roots of the spinal nerves (see Figs. 4-5 and 12-4).
         (a) Reaching the sympathetic chain via an associated **white rami**, the sympathetic presynaptic neurons emerge from the chain ganglia as thoracic splanchnic nerves.
         (b) Thoracic splanchnic nerves from T1–T4 supply the thoracic portion of the esophagus. The splanchnic outflow from T5 contributes to the greater splanchnic nerve and supplies the abdominal portion of the esophagus.
      (2) The **cell bodies of the sympathetic postsynaptic neurons** lie either in the sympathetic chain (paravertebral ganglia) or in prevertebral ganglia.
         (a) The postsynaptic sympathetic neurons coursing in the thoracic splanchnic nerves (T1–T4) synapse in the **sympathetic chain**.
         (b) The neurons coursing in the T5 component of the greater splanchnic nerve to the terminal esophagus synapse in a prevertebral ganglion, in this case, the **celiac ganglion** (see Fig. 4-5).
   c. **Afferent innervation**
      (1) **Sensory pathways.** Afferent fibers mediating painful sensation course along the thoracic splanchnics and greater splanchnic nerve to the sympathetic chain. From there, the afferents travel along a white ramus communicans to the respective spinal nerves (see Fig. 4-5).

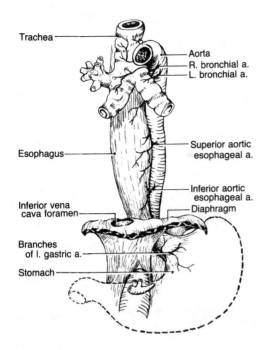

Trachea

Aorta
R. bronchial a.
L. bronchial a.

Esophagus

Superior aortic
esophageal a.

Inferior aortic
esophageal a.
Diaphragm

Inferior vena
cava foramen

Branches
of l. gastric a.

Stomach

Figure 17-1. *Esophagus.* The esophagus passes through the esophageal hiatus of the diaphragm to gain access to the abdomen where it joins the stomach. The blood supply is by branches of the aorta and left gastric artery.

(2) Referred esophageal pain. The majority of painful sensation from the esophagus originates in the terminal segments and is mediated via the splanchnic nerves associated with T4 and T5. The esophageal pain is, therefore, referred to (appears as if originating from) the inferior thoracic and epigastric regions. The differential diagnosis involving the pain of coronary artery insufficiency and the pain of esophageal origin is critical.

4. Esophageal function. The esophagus serves solely for the transport of food **deglutition**.
   a. Upper esophagus. The **cricopharyngeus muscle** is the sphincter of the upper esophageal opening (see Fig. 33-7). It remains closed except during deglutition (swallowing), eructation (belching), and emesis (vomiting).
   b. The midesophagus propels a bolus of food aborally by peristaltic activity.
     (1) A solid bolus entering the esophagus initiates an involuntary peristaltic wave, which reaches the cardiac sphincter in 5–6 seconds. The cardiac sphincter, normally closed, relaxes to allow the bolus to pass into the stomach.
     (2) A fluid bolus outruns the peristaltic wave due to gravity but is stopped by the cardiac sphincter until this sphincter is inhibited by the approach of the peristaltic wave. Thus, only one sound is heard when a solid bolus is swallowed, whereas there are two separate sounds with a fluid bolus.
   c. Lower esophagus. The **cardiac sphincter** is located at the cardioesophageal junction (Fig. 17-2). The smooth muscle of this junction is continuous with that of the stomach, and the muscle is not considered a true structural sphincter. Nevertheless, it is a functional sphincter, intrinsic to the esophagus and not dependent upon external constriction by the crura of the esophageal hiatus of the diaphragm (as suggested by some textbooks).
     (1) There is considerable evidence that the sphincteric barrier is, in part, due to valve-like folds of mucosa maintained by the tonic activity of the muscularis mucosae. This barrier and the diffuse tonic activity of the muscularis externa as well as the fairly constant tone within the muscle layers of the cardiac region of the stomach constitute the physiologic closing mechanism.
     (2) The cardiac sphincter, thus, maintains a broad zone of elevated intramural (luminal) pressure at the aboral end of the esophagus. The pressure within the stomach is always greater than that in the midesophagus. Without the zone of increased transmural pressure, the gastric contents would reflux into the esophagus continuously. Because of peristaltic inhibition, the cardiac sphincter can be opened by a pressure of 2–7 cm of water within the esophagus, but eructation will not occur until the intragastric pressure exceeds 25 cm of water. When a series of swallows is made in rapid succession (as in "chugging"), the cardiac sphincter may remain inhibited so that fluid passes directly into the stomach unassisted by peristalsis.

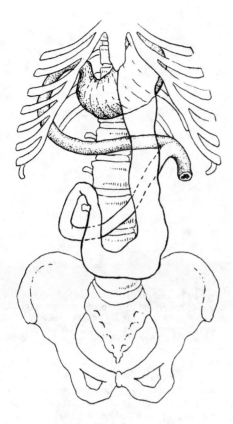

Figure 17-2. *Location of the stomach.* The location, size, and shape of the stomach are variable as indicated.

5. **Clinical considerations**
   a. **Esophagitis**
      (1) Occasionally, gastric contents reflux into the esophagus. The acidic peptic chyme burns and inflames the unprotected stratified squamous epithelium of the esophageal mucosa, producing *regurgitative esophagitis*—the uncomfortable sensation commonly known as "heartburn."
      (2) Repeated reflux can result in peptic ulcers in the esophageal mucosa, which can cause *dysphagia* and serious bleeding.
   b. **Esophageal varices**
      (1) The submucosa of the esophagus contains an extensive plexus of large blood vessels.
      (2) The veins of the midesophagus, which drain into the systemic circulation via the azygos and the hemiazygos veins, anastomose with the veins of the lower esophagus, which anastomose with the veins of the hepatic portal system via the left gastric vein. These anastomoses provide an important but potentially dangerous shunt in the event of *portal hypertension* and result in the development of *esophageal varices* (see Fig. 17-21).
         (a) These tortuous and dilated veins directly beneath the mucosa protrude into the esophageal lumen, usually in the distal third of the esophagus, where they are subject to mechanical trauma during deglutition, emesis, or the passage of diagnostic instrumentation.
         (b) *Esophageal varices* produce no symptoms until they rupture, causing massive hematemesis. Among patients with advanced cirrhosis of the liver, over half of the deaths result from rupture of an esophageal varix.
   c. **Hiatus hernia**
      (1) Herniation of the stomach through the esophageal hiatus produces a sac-like dilation above the diaphragm.
         (a) In 90% of *hiatus hernia* cases, the esophagus ends above the diaphragm ("hour glass stomach" or *sliding hiatus hernia*).
         (b) In the remaining 10%, the cardiac region of the stomach dissects alongside the esophagus through a defect in the esophageal hiatus to produce an intrathoracic sac (*paraesophageal hernia*).
      (2) *Hiatus hernia* may produce incompetence of the cardiac sphincter, resulting in *regurgitative esophagitis*, which may ultimately lead to peptic ulceration of the esophagus.

B. Stomach

1. **Gastric position and relationships.** The stomach is located in the left hypochondriac and epigastric regions of the abdomen (see Fig. 15-1). Because it is suspended by mesenteries, it is a mobile and easily displaced organ with no fixed position.

   a. **Empty**, the stomach is almost tubular or J-shaped, except for the bulge of the fundus; it may be almost entirely under the rib cage (see Fig. 17-2).

   b. **Full**, because it is very distensible and can accommodate more than 2 L, the stomach may pendulate as far as the pelvis (see Fig. 17-2).

2. **External structure of the stomach.** The stomach is composed of two sides, two curvatures, and two orifices.

   a. **The greater curvature** represents the primitive dorsal surface (Fig. 17-3). It receives ligamentous support from the primitive dorsal mesentery (see Fig. 16-4).

      (1) The **greater omentum** is a redundant portion of the dorsal mesogastrium.

      (2) The **gastrophrenic ligament** is the portion of the dorsal mesogastrium between the fundic region and the dorsal body wall in the vicinity of the diaphragmatic crura.

      (3) The **gastrosplenic ligament** is the portion of the dorsal mesogastrium between the greater curvature of the stomach and the spleen.

      (4) The **splenorenal ligament** is a continuation of the dorsal mesogastrium between the spleen and the dorsal body wall in the vicinity of the left kidney.

   b. **The lesser curvature** represents the primitive ventral surface (see Fig. 17-3). It receives ligamentous support from the derivatives of the primitive ventral mesentery, which becomes the **gastrohepatic (hepatogastric) ligament** of the **lesser omentum** (see Figs. 16-4 and 16-5).

   c. **The cardiac and pyloric sphincters** define the oral and aboral ends.

   d. **Divisions.** The stomach is divided into four regions (see Fig. 17-3).

      (1) The **cardia** is the region located in the immediate vicinity of the esophagus.

         (a) The stomach receives the esophagus at the cardiac opening, which lies at the superior junction of the greater and lesser curvatures.

         (b) The **cardiac notch** is the pronounced incisure between the esophagus and the fundus.

      (2) The **fundus** is located cephalad to the level of the esophageal juncture. It usually contains a variable amount of air.

         (a) Air in the fundus, evident on an x-ray film, is a useful radiographic landmark.

         (b) A tympanic note may be percussed from the fundic air bubble.

      (3) The **corpus (body)** constitutes the major portion of the stomach.

      (4) The **pylorus**, located distally, is subdivided into three regions.

         (a) The **pyloric antrum** begins as a slight dilation, which produces the **angular incisure** in the lesser curvature.

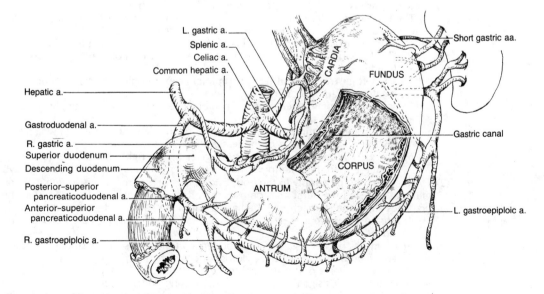

Figure 17-3. *Stomach.* The regions of the stomach are indicated. The window in the body of the stomach shows the rugae and the location of the gastric canal. The blood supply to the stomach is by the three branches of the celiac artery.

(b) The **pyloric canal** is 2–3 cm long.

(c) The **pyloric sphincter** is neither demarcated from the circular muscle of the adjacent regions nor are there any discernible physiologic differences.

(i) A pressure gradient of only 2–4 cm of water is sufficient to move chyme through the pylorus, indicating that it is not a true sphincter. Also, unlike a true sphincter, the pylorus contracts synergistically with peristalsis and functions as a unit with the pyloric end of the stomach.

(ii) While the pylorus does not control the rate of stomach emptying, it does control the size of the particles that enter the duodenum. A kernel of corn is about the limit.

(iii) The pyloric sphincter may be abnormally thickened at birth (*congenital hypertrophic pyloric stenosis*), a condition that requires immediate surgical correction (see Chapter 16 III B 2 d).

3. Internal structure of the stomach

a. At the gastroesophageal juncture, the whitish stratified squamous epithelium of the esophagus changes abruptly to pink simple columnar epithelium.

b. Rugae are formed by the folding of the gastric mucosa (see Fig. 17-3).

(1) Permanent rugae form the **gastric canal (gutter)** in the region of the lesser curvature. These direct fluids downward to the pylorus.

(a) Unlike the rugae in the rest of the stomach, those of the gastric canal do not diminish as the stomach fills. These rugae are usually evident on upper GI x-ray films with barium contrast.

(b) The gastric canal is especially vulnerable to burns from accidental ingestion of caustic substances.

(2) Gastric rugae in regions away from the gastric canal are temporary high folds of mucosa. These are produced by muscular tension in the muscularis mucosae, which pleats the mucosa and underlying submucosa.

(a) These temporary rugae run primarily in a longitudinal direction.

(b) They provide for stomach expansion, becoming flattened as the stomach fills.

c. The gastric mucosa contains small permanent furrows, **gastric sulci**, that divide the surface into **gastric (mamillated) areas**.

(1) Gastric areas do not flatten out with stomach distension. However, within gastric areas, the surface of the distended stomach appears smooth.

(2) Gastric glands (pits) open onto the surface of the gastric areas.

d. Muscularis. Three muscle layers comprise the muscularis externa, but their predominant directions are difficult to characterize.

4. Vasculature of the stomach

a. Arterial supply

(1) The **celiac artery** supplies the stomach through its three branches.

(a) The **left gastric artery** supplies the superior regions of the lesser curvature (body, cardia, and fundus) as well as the lower esophagus (see Fig. 17-3).

(b) The **splenic artery** gives off two sets of branches to the stomach (see Fig. 17-3).

(i) Short gastric arteries supply the left superior portions of the greater curvature (fundus and body).

(ii) The **left gastro-omental (gastroepiploic) artery** supplies the left portion of the greater curvature (body).

(c) The **common hepatic artery** gives off two branches, which, in turn, supply the stomach (see Fig. 17-3).

(i) The **proper hepatic artery**, in turn, gives off the **right gastric artery**, which supplies the inferior region of the lesser curvature (pylorus).

(ii) The **gastroduodenal artery**, in turn, gives off the **right gastro-omental artery**, which supplies the inferior portion of the greater curvature (pylorus and body).

(2) The **collateral blood supply** to the body and fundus of the stomach is so rich in extramural as well as intramural anastomoses that, from a surgical standpoint, any of the branches along the stomach may be ligated with little risk of ischemia and necrosis.

(a) Anastomoses between gastric, esophageal, and splenic arteries are constant and abundant.

(b) Anastomoses between gastric and duodenal arteries are scant, resulting in the so-called "**bloodless line**" at the pyloroduodenal junction.

b. Venous return. The veins of the stomach generally parallel the arterial supply, but they diverge from the arteries significantly to join the hepatic portal system.

(1) The anastomoses between the **left gastric (coronary) vein** and the esophageal veins are important portal–systemic shunts and may lead to *esophageal varices*.

(2) Anastomoses between the esophageal veins and both the left gastroepiploic vein and short gastric veins may also become engorged if the splenic vein becomes occluded, such as by *pancreatic carcinoma*.

  c. Lymphatic drainage. The routes of lymphatic drainage from the stomach generally follow the arteries and are so named, although alternative names are frequently used.

5. Stomach function. Functions of the stomach include trituration, formation of chyme, acid enzymatic digestion and some absorption, as well as serving as a reservoir.

  a. Storage. The upper half of the stomach serves as a reservoir for ingested food and expands passively as it is filled.

    (1) Persistent tonic contracture in the region of the cardia and fundus aids in the establishment of the cardiac sphincter, and serves to deliver food to the lower and more motile regions of the stomach.

    (2) The gastric glands of both the fundus and the body contain cells that secrete hydrochloric acid; the mucosal glands of the cardia and pylorus are mostly devoid of acid-secreting cells.

      (a) In the upper regions of the stomach, boluses of food are bathed in gastric juice, which converts the surface of a bolus into a liquid mixture, chyme.

      (b) Pepsin, an enzyme secreted by gastric glands and activated at a low pH, converts proteins to polypeptides.

  b. Triturition. In the lower half of the stomach, chyme is thoroughly mixed with the gastric secretions.

    (1) The muscle layers increase in thickness from the cardia to the pylorus, and peristaltic activity increases in intensity along this gradient. Peristaltic waves, which occur at about 20-second intervals, begin as ring-like contractions about midway in the body of the stomach and propagate toward the pylorus with increasing vigor.

    (2) Once formed, chyme is moved toward and through the pylorus.

    (3) While the pylorus is devoid of acid-secreting cells, it contains most of the secretin-producing cells. This hormone influences acid secretion in the rest of the stomach.

  c. Peristalsis. Factors controlling motility are both intrinsic and extrinsic.

    (1) Gastric peristalsis is the result of intrinsic rhythm within enteric neurons.

      (a) Vagal activity accelerates peristalsis.

      (b) The introduction of acid chyme into the duodenum releases the hormone enterogastrone, which inhibits gastric peristalsis, the enterogastric reflex.

    (2) After an initial adjustment period, the stomach empties its contents at a relatively constant rate.

  d. Absorption. Some substances, such as alcohol, enter directly through the gastric mucosa.

6. Clinical considerations

  a. Gastritis. Excessive vagal activity may produce *gastritis* by excessively stimulating the acid-secreting glands. Substances that irritate the gastric mucosa, such as aspirin and steroids, can also produce gastritis.

  b. Peptic ulceration. Peptic ulcers, due to a variety of causes, occur in the non–acid-secreting regions of the upper GI tract, such as the duodenum (primarily), the antral region of the stomach along the lesser curvature (less frequently), and the lower esophagus (more rarely). The mucosa may be inspected with a fiberoptic gastroscope.

    (1) Peptic ulcers may produce severe bleeding, obstruction from edema or scarring, and peritonitis from perforation. Erosion into a subjacent blood vessel results in massive hematemesis or intra-abdominal bleeding with complicating peritonitis.

    (2) Pain from *peptic ulceration* of the lower esophagus, stomach, or superior duodenum is referred to the distributions of the 5th and 6th dermatomes, which include the epigastric and hypochondriac regions.

    (3) Ulceration may be treated with drugs that block acid secretion. Alternatively, selective vagotomy (section of the gastric branches of the anterior and posterior vagal trunks) reduces peptic secretion and may be combined with pyloroplasty.

  c. Hemigastrectomy

    (1) Surgical treatment of duodenal and gastric ulceration frequently involves resection of approximately one-third of the distal stomach and, because a collateral blood supply is inconsistent, the superior duodenum (Billroth I gastrectomy).

    (2) Although the distal portion of the stomach produces little acid, it does produce most of the **secretin** that induces acid secretion in the fundus and body. Resection of the lower portion not only removes the ulcers but also reduces acid secretion in the upper regions and largely preserves the capacity of the stomach as a reservoir.

  d. Malignant metastases. The venous and lymphatic drainage of the stomach is such that malignant cancer at this site can spread to other organs and regions.

(1) Metastases to the liver may occur via either the portal vein or by reverse flow along the lymphatics.

(2) Metastases to any other part of the body may occur via the thoracic duct and circulatory system.

C. Duodenum

1. **External structure of the duodenum.** While the small intestine is arbitrarily divided into three parts, the morphologic and physiologic differences between duodenum and jejunum, as well as between jejunum and ileum, are slight. These differences occur by slow transitions over considerable distances.

   a. **Length and position.** The duodenum is the shortest portion of the small intestine. It is about 12 finger breadths (*duodeni*, L. twelve) or 25 cm (10 inches) from the pyloric sphincter to the duodenojejunal flexure. While the position and size are variable (see Fig. 17-2), it loops in a **C**-shape to the right (see Fig. 17-3).

   b. **Divisions.** The duodenum is divided into four parts.

   (1) The **superior duodenum**, the first part, is 3–5 cm (1.5–2 inches) long.

   (a) This portion is peritoneal, supported by the **hepatoduodenal ligament**, which is a portion of the ventral mesentery and lesser omentum.

   (b) Internally, the mucosa lacks circular folds and, because of its characteristic smooth appearance on x-ray, it is termed the **duodenal cap.**

   (c) It receives chyme through the pylorus. It is relatively quiet.

   (2) The **descending duodenum**, the second part, is 8–10 cm (3.5–4 inches) long.

   (a) This portion is secondarily retroperitoneal. The transverse colon lies anteriorly, the right kidney, posteriorly, and the right lobe of the liver, superiorly.

   (b) Internally, circular folds of mucosa (**plicae circulares**) make their appearance, increasing in size and complexity toward the jejunum.

   (c) This segment receives the **hepatopancreatic ampulla** and the **secondary pancreatic duct** (of Santorini), if present.

   (i) Usually the common bile duct and the primary pancreatic duct join to form the hepatopancreatic ampulla (see Fig. 17-8). However, numerous variations are common (see Fig. 17-11).

   (ii) Where the hepatopancreatic ampulla enters the duodenum, a small protrusion occurs, the **duodenal papilla** (see Fig. 17-11).

   (iii) The **hepatopancreatic sphincter** (of Oddi) surrounds the duodenal papilla.

   (d) The activity of the descending duodenum is similar to a churning or milling action with true peristaltic activity becoming evident only in the distal duodenum.

   (3) The **inferior duodenum** comprises the third and fourth parts.

   (a) The **transverse portion**, the third part, is 2.5–5 cm (1–2 inches) long.

   (i) This horizontal segment is secondarily retroperitoneal.

   (ii) As a result of intestinal rotation during development, the **superior mesenteric vessels** are located anteriorly to the transverse portion. With wasting diseases or after severe dieting, the superior mesenteric vascular bundle may compress the duodenum and produce intestinal obstruction (*superior mesenteric artery syndrome*).

   (b) The **ascending portion**, the fourth part, is 2.5–5 cm long.

   (i) This segment, also secondarily retroperitoneal, terminates at the duodenojejunal flexure.

   (ii) The **suspensory ligament (of Treitz)**, a surgical landmark, holds the ascending portion in place (Fig. 17-4). It runs from the fourth duodenal segment to the right crus of the diaphragm. This ligament is composed of a tendon insinuated between slips of smooth muscle from the muscularis externa of the duodenum and fascicles of striated muscle from the diaphragmatic crus.

   (iii) At the **duodenojejunal flexure**, the small intestine becomes the peritoneal jejunum, supported by the **mesentery proper** (see Fig. 16-5).

2. **Function of the duodenum.** As the terminal portion of the foregut, the function of the duodenum is still primarily digestion. It is here that chyme is mixed with the products of the liver and pancreas as well as with enzymes secreted by the duodenum. It functions more in absorption toward the distal portions, which is indicated by the increase in the amount and complexity of the plicae circulares.

3. **Vasculature of the duodenum**

   a. **Arterial supply** to the duodenum is from two sources. The celiac artery supplies the proximal duodenum via the right gastric, gastroduodenal, and superior pancreaticoduodenal arteries; the superior mesenteric artery supplies the distal duodenum via the inferior pancreaticoduodenal artery (see Figs. 17-4 and 17-10).

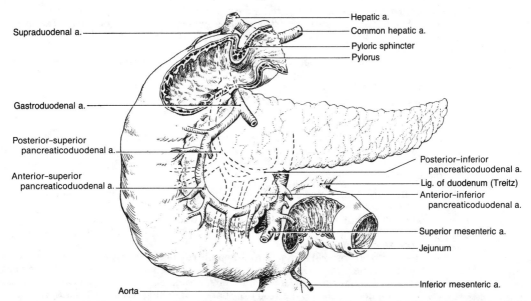

Figure 17-4. *Duodenum and pancreas.* The window in the superior portion of the duodenum shows the pyloric sphincter and beginnings of the plica circularis. The blood supply to the duodenum is by branches of the celiac and superior mesenteric arteries. The head of the pancreas is intimately associated with the duodenum, sharing a common blood supply.

(1) The blood supply to the first (superior) portion is sparse, sometimes to the point of being precarious with few, if any, extramural anastomoses between the branches of the right gastric and gastroduodenal arteries. Because of this, the superior duodenum is removed when the distal portion of the stomach is resected.

(a) The **supraduodenal artery** usually arises from the gastroduodenal artery but may arise from the right gastric artery or common hepatic artery. It is absent in 30% of individuals.

(b) **Retroduodenal arteries** are variable in number and location. They may arise from the gastroduodenal artery, the right gastroepiploic artery, superior pancreatic artery, or any combination thereof.

(2) Blood is profusely supplied to the descending and inferior portions of the duodenum by anastomosing arcades from the **superior pancreaticoduodenal artery** (a branch of the gastroduodenal artery) and the **inferior pancreaticoduodenal artery** (a branch of the superior mesenteric artery).

b. Venous drainage of the duodenum is into the hepatic portal vein by way of both the common hepatic vein and the superior mesenteric vein. Because the duodenum is mostly secondarily retroperitoneal, there are variable anastomotic connections (**veins of Retzius**) with the systemic veins of the body wall.

4. Clinical considerations

a. Mobilization of the duodenum. Since the blood supply is from the medial side, an incision of the peritoneum (fascia of Toldt) along the right edge of the descending duodenum mobilizes this viscus as well as the head of the pancreas. This reestablishes the former dorsal mesentery with its contained blood supply.

b. Metastatic routes. Since the venous and lymphatic channels may anastomose with those of the dorsal body wall, carcinoma of the duodenum and pancreas frequently has a poor prognosis.

c. Duodenal ulcers. Peptic ulceration in the duodenal cap is four times more frequent than gastric ulcers. Chronic duodenal ulcers may require a Billroth I gastroduodenal resection.

D. Liver

1. Developmental considerations. The liver develops from the **hepatic diverticulum**, which penetrates the ventral mesogastrium during the third week.

a. By the tenth week, the liver has assumed transient hematopoietic function.

b. The developing liver is an important metabolic center in the umbilical venous channel and represents 10% of the embryonic weight.

2. Hepatic relations. The liver is located almost under the rib cage (Fig. 15-1).

 a. Right hypochondriac and epigastric regions are largely occupied by liver. The liver extends superiorly as far as the 5th rib anteriorly on the right side and the 5th intercostal space on the left.

  (1) Normally, the right lobe extends just beneath the right costal margin where it may or may not be palpable, depending upon the consistency. The examiner's fingers are placed adjacently to the costal margin just lateral to the rectus abdominis muscle.

  (2) Because of its mesenteric attachments, the liver moves with the diaphragm during respiration. As the patient exhales, the liver might be felt as it slides down beneath the examiner's fingertips.

 b. Subphrenic space. Hepatic cysts and subphrenic abscesses can erode through the diaphragm into the pleural cavity; conversely, pulmonary abscesses can erode through the diaphragm into the subphrenic space and involve the liver. Before antibiotics, this was commonly fatal.

3. Hepatic structure. Morphologically, the liver appears to be one of the simplest organs; functionally, it is one of the most complex.

 a. A compound tubular gland, the liver secretes up to a liter of bile a day.

  (1) It is the largest gland in the body, averaging 1200–1600 g with the larger end of the range seen in males.

  (2) The liver is enclosed in a thin, fibrous **hepatic capsule** (of Glisson) that lies just beneath the visceral peritoneum. From the hepatic capsule, septa project inward into the hepatic parenchyma. Pain associated with *acute hepatomegaly* and *transient venous congestion* (*"runners stitch"*) is the result of sudden stretching of Glisson's capsule.

 b. Lobation

  (1) Anatomic lobation. In addition to the very large **right lobe** and much smaller **left lobe**, the liver also has two rudimentary lobes, the **quadrate lobe** or so-called medial segment of the right lobe and the **caudate lobe** (Figs. 17-5 and 17-6).

  (2) Functional lobation. Functionally, the liver is nearly evenly divided.

   (a) Right and **left lobes** have separate biliary drainage and vascular supplies (see Fig. 17-6).

   (b) The **quadrate lobe** is functionally part of the left lobe because it receives blood from the left hepatic artery and drains the hepatic duct. This is counter to its alternate name and location to the right of the segmental fissure, which is marked by the falciform ligament.

   (c) The **caudate lobe**, receiving blood from the left and right hepatic arteries and secreting bile into both left and right hepatic ducts, is functionally part of both the left and right lobes.

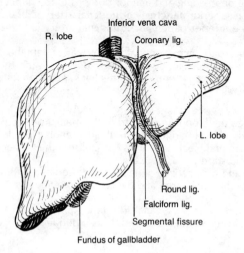

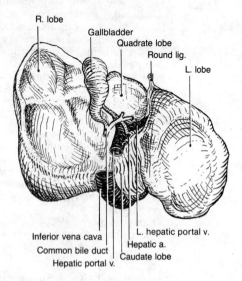

Figure 17-5. *Liver.* The anterior surface of the liver is depicted with the right and left lobes separated by the falciform ligament.

Figure 17-6. *Liver.* The inferior surface of the liver is shown with the left, right, quadrate, and caudate lobes as well as the gallbladder located between the right and quadrate lobes.

4. Mesenteric attachments of the liver

a. The ventral mesogastrium, within which the liver develops, is divided by the liver into two principal ligaments (see Fig. 16-4).

  (1) The **lesser omentum** runs between the liver and the lesser curvature of the stomach as well as the first part of the duodenum. It is, therefore, subdivided into the **hepatoduodenal ligament** and the **gastrohepatic (hepatogastric) ligament**.

  (2) The **falciform ligament** is a continuation of the former ventral mesentery from the anterior surface of the liver to the ventral body wall as far inferiorly as the umbilicus (Fig. 17-7).

    (a) The **round ligament (ligamentum teres)** of the liver, which is the remnant of the obliterated umbilical vein, is in the inferior free edge of this ligament.

    (b) Patency may be reestablished in the round ligament by cannulation through the umbilicus to measure portal venous pressure or to sample portal blood.

b. The coronary ligaments are formed by reflection of ventral mesentery between the superior pole of the liver and the inferior surface of the diaphragm (see Fig. 17-7B); that is, they provide a path of continuity between the lesser omentum and the falciform ligament over the top of the liver (see Fig. 17-7A).

  (1) As the liver enlarges, it forces the two leaflets of the ventral mesogastrium apart superiorly, leaving a **bare area** where neither the diaphragm nor the liver is covered by peritoneum.

  (2) The boundaries of the bare area on the left side are the falciform ligament, the left anterior leaf of the coronary ligament, the left triangular ligament where the left anterior and posterior coronary leaflets unite, the posterior left leaf of the coronary ligament, and the gastrohepatic portion of the lesser omentum. The boundaries are complementary on the right side (see Fig. 17-7).

  (3) The **hepatic veins** enter the **inferior vena cava** as the latter passes through the bare area.

5. Blood supply to the liver. The liver receives arterial blood from the hepatic artery and venous blood rich in nutrients from the hepatic portal vein (see Figs. 17-6, 17-9, and 17-21).

a. Hepatic arterial supply. The liver is an extremely vascular organ, bleeding profusely when ruptured by blunt trauma or sections in surgery.

  (1) The **common hepatic artery**, one of the three branches of the celiac artery, becomes the proper hepatic artery once the gastroduodenal artery is given off.

  (2) The **proper hepatic artery** ascends in the hepatoduodenal ligament (the right edge of the lesser omentum) and divides into the left and right hepatic arteries.

    (a) The **right hepatic artery** supplies the right lobe and the right half of the caudate lobe.

    (b) The **left hepatic artery** supplies the left and quadrate lobes as well as the left half of the caudate lobe.

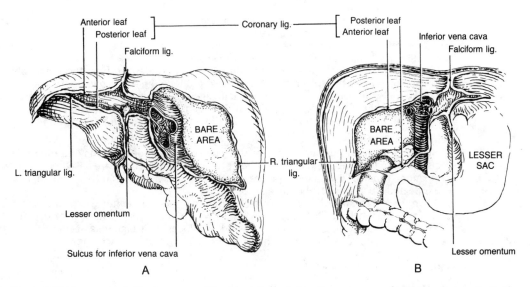

Figure 17-7. *Bare areas. A,* The bare area of the liver is bounded by the coronary and triangular ligaments. *B,* The bare area of the inferior surface of the diaphragm where the ligaments become parietal peritoneum is indicated.

(3) Extrahepatic variation
- **(a)** The usual textbook representation of the hepatic blood supply is actually rather uncommon, because 60%–70% of individuals have some variation.
  - **(i)** The level of bifurcation into left and right hepatic arteries is variable.
  - **(ii)** The right hepatic artery may arise (15%) from the superior mesenteric artery via the inferior pancreaticoduodenal artery.
  - **(iii)** The left hepatic artery may arise (25%) from the left gastric artery. This anomaly has a high incidence of covariance with a right hepatic artery from the superior mesenteric artery.
- **(b)** Lack of familiarity with the many possible variations in the course of the right hepatic and cystic arteries may result in profuse and unexpected hemorrhage during surgery. Hemorrhage from the hepatic vascular supply may be controlled by compressing the hepatic pedicle between the thumb and a forefinger inserted into the epiploic foramen.

(4) Extrahepatic occlusion. Hepatic arteries and accessory hepatic arteries are not collateral arteries because they supply discrete regions of the liver parenchyma. Sudden occulusion of these vessels can produce ischemic necrosis of a region of the liver.
- **(a)** Ligation of the hepatic portal vein may be tolerated for up to 30 minutes.
- **(b)** Depending on the location, ligation of the hepatic arterial supply may be tolerated.
  - **(i)** The common hepatic artery may be permanently ligated because there are ample anastomotic connections between vessels supplying the stomach and the terminal branches of the common hepatic artery. Thus, the liver will continue to be perfused.
  - **(ii)** The proper hepatic artery may be permanently ligated only proximally to the right gastric artery.
- **(c)** If an occlusion develops over a prolonged period, adequate intrahepatic and extrahepatic collateral circulation will develop.

b. Hepatic venous supply is by the **hepatic portal vein**. A portal system is a venous system that begins as venous capillaries, such as in the stomach and intestine, and terminates in venous sinusoids, such as in the liver.

c. Hepatic venous drainage usually is by three **hepatic veins**, which drain into the inferior vena cava.

6. Hepatic function. The liver is primarily a metabolic center.
- **a. Intrahepatic perfusion.** The liver has a uniform structure, consisting of anastomosing sheets of cells that are separated by intervening venous sinusoids. All venous blood from the GI tract percolates through the liver sinusoids.
  - **(1)** The liver cells are supplied by blood from the hepatic portal vein (nutrient-rich, unoxygenated, sometimes toxic) and the hepatic artery (oxygenated).
  - **(2)** Because of this uncommon vascular arrangement, a lower than usual oxygen content bathes the liver cells. This may explain, in part, the high susceptibility of the liver to damage and disease.
  - **(3)** The sinusoids drain into the central veins of the lobules and ultimately into the **hepatic veins**, which join the inferior vena cava.
- **b. Bile secretion.** The liver secretes up to a liter of bile a day into the biliary tree. Among the constituents of bile are bile pigments and bile salts.
  - **(1) Bile pigments**, including bilirubin and biliverdin, are derived from the breakdown of hemoglobin in the spleen and liver. These produce the distinctive color of feces.
  - **(2) Bile salts,** formed from cholesterol, aid in digestion by emulsification of fats, thereby facilitating their absorption by the intestinal mucosa. Most of the bile salts are reabsorbed by the mucosa and returned to the liver by the portal veins for resecretion.

7. Clinical considerations
- **a. Hepatic resection.** Although the liver is a vital organ, portions of it may be surgically excised without irreparable damage to the body. Hepatic resections are based on functional lobation. The liver has tremendous regenerative power, which can be detrimental when undirected, such as in cirrhosis.
- **b. Biliary overload and obstruction.** *Icterus (jaundice)* occurs either when the liver overloads beyond its capacity, such as by excessive hemolysis of blood or liver damage, or when the biliary tree is obstructed. In both instances, bile pigments accumulate in the blood, giving the skin and eyes a characteristic yellow coloration accompanied by itching.

E. Hepatic tree and gallbladder

1. Hepatic tree
- **a. Internal structure of the hepatic tree.** Hepatocytes secrete bile into bile canaliculi.

(1) Bile canaliculi, which traverse the sheets of hepatocytes, unite to form intrahepatic bile ducts. These lie in portal triads along with branches of the hepatic portal veins and hepatic arteries.

(2) Bile ductules converge to form the left and right hepatic ducts. This pattern defines the functional segments of the liver.

b. External structure of the hepatic tree. The hepatic tree consists of the left and right hepatic ducts, cystic duct, common hepatic duct, common bile duct, and gallbladder (Fig. 17-8).

    (1) **Right** and **left hepatic ducts** leave the respective lobes of the liver and join to form the common hepatic duct (see Fig. 17-8).

        (a) The **left hepatic duct** drains the left and quadrate lobes of the liver as well as the left side of the caudate lobe.

        (b) The **right hepatic duct** drains the right lobe and the right half of the caudate lobe.

    (2) The **cystic duct** to the gallbladder arises from the common hepatic duct (see Fig. 17-8).

        (a) The length of the cystic duct is variable but is usually 1–2 cm long.

        (b) The spiral fold (valve of Heister) maintains the patency of the cystic duct.

        (c) The course of the cystic duct is very variable.

    (3) The **common hepatic duct** and the cystic duct converge to form the common bile duct (see Fig. 17-8).

    (4) The **common bile duct**, which is about 10 cm long and 1 cm in diameter, empties into the duodenum.

        (a) It contributes to the **hepatic pedicle**, which is located in the free edge of the hepatoduodenal ligament and forms the anterior boundary of the epiploic foramen (see Fig. 17-9).

            (i) The **common bile duct** lies anteriorly to the right within the hepatic pedicle.

            (ii) The **hepatic artery** lies anteriorly to the left.

            (iii) The **hepatic portal vein** lies posteriorly.

        (b) The common bile duct passes posteriorly to the superior and descending duodenum. It usually is embedded in pancreatic parenchyma but is easily dissected free.

        (c) It enters the descending duodenum about 3–8 cm beyond the pyloric sphincter (see Fig. 17-10).

            (i) The intramural portion of the common bile duct is its narrowest part.

            (ii) The **hepatopancreatic ampulla** and **duodenal papilla** (of Vater) are formed by the confluence of the common bile duct and the pancreatic duct (see Fig. 17-11).

            (iii) The duodenal papilla is very variable. In about 65% of individuals, bile from the common duct and pancreatic juice from the pancreatic duct are mixed in a true hepatopancreatic ampulla. However, the two ducts may open into the duodenal lumen separately, each having its own papilla (see Figs. 17-11*B* and 17-11*C*).

            (iv) Blockage of the hepatopancreatic ampulla by a bile calculus, muscle spasm, or edema may precipitate *pancreatitis* as a result of bile reflux along the pancreatic duct.

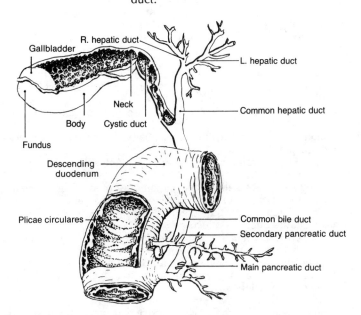

Figure 17-8. *Biliary tree, gallbladder, and descending duodenum. The divisions of the biliary system are indicated, as is the pancreatic drainage.*

(d) The **hepatopancreatic sphincter** (of Oddi) normally keeps the ostium of the hepa-topancreatic ampulla closed. It is divided into three parts.

    (i)The **sphincter choledochus** is a well-developed portion around the termination of the common bile duct.

    (ii)The **papillary sphincter** surrounds the hepatopancreatic ampulla.

    (iii)The **pancreatic sphincter** is a poorly developed portion around the termination of the pancreatic duct.

(e) Vagal activity (and possible cholecystokinin) results in relaxation of the hepatopan-creatic sphincter.

c. Arterial supply to the hepatic tree. The common bile duct is supplied by a rich anasto-motic network from the right hepatic, proper hepatic, retroduodenal, and posterior pan-creaticoduodenal arteries.

  (1) The twigs, variable and inconstant, course in the adventitia of the bile duct.

  (2) During surgical manipulation of the common duct, care must be taken to prevent isola-tion of the blood supply, as necrosis and infarct may result.

2. Gallbladder

a. Relations of the gallbladder. The gallbladder is 6–10 cm long with a capacity of about 45 ml. It is located in a shallow fossa on the inferior surface of the liver between the right and quadrate lobes (see Fig. 17-6).

  (1) It may be embedded within the liver parenchyma (rare); it may be adherent to the liver by peritoneum (most common); it may be surrounded by peritoneum (floating); or it may have a short mesentery to the inferior surface of the liver.

  (2) It lies adjacently and may be adherent to the superior duodenum and transverse colon, anteriorly.

    (a) A large gallstone may erode through the adhered common walls into the bowel.

    (b) To relieve common duct obstruction, the gallbladder may be anastomosed surgi-cally to the duodenum.

  (3) The gallbladder usually cannot be palpated unless it contains stones, whereupon it may be felt in the right upper quadrant at the angle between the costal margin and the rectus abdominis muscle (see Fig. 15-1).

b. External structure of the gallbladder. The gallbladder is divided into three parts: the **fun-dus**, the **body**, and the **neck** (see Fig. 17-8). The neck may contain a pendulous sacculation (**Hartmann's pouch**).

  (1) During surgery, gentle traction on Hartmann's pouch causes the cystic duct to stand out.

  (2) Hartmann's pouch may be adherent to the cystic duct, or even to the common bile duct, and may be accidentally included in a ligature passed around the cystic duct.

c. Vasculature of the gallbladder

  (1) The **cystic artery** normally arises as a single stem from the right hepatic artery within the **cystohepatic triangle** (of Calot). It subsequently bifurcates into **anterior** and **poste-rior cystic arteries** (Fig. 17-9). Cystic arteries are quite variable; they may arise from the right, left, or even proper hepatic arteries in any combination.

  (2) **Cystic veins** may drain into the right hepatic vein or, more usually, drain directly into the liver.

d. Gallbladder function

  (1) The spinchter choledochus and papillary sphincter are normally closed so that bile, which is secreted continuously by the liver, refluxes through the cystic duct into the gallbladder.

  (2) The gallbladder stores and concentrates bile by mucosal absorption of electrolytes and water.

  (3) The ability of the gallbladder to concentrate bile forms the basis for radiographic visu-alization of that organ (cholangiography).

    (a) When radiopaque, iodinated compounds are administered either orally or in-travenously, they are excreted by the liver.

    (b) As the bile is concentrated in the gallbladder, the amount of radiopaque dye in-creases so that the entire organ, and often the biliary tree, is visualized.

  (4) A spurt of fat through the pylorus into the duodenum causes the release of cholecysto-kinin by intestinal glands in the duodenal mucosa.

    (a) This hormone induces contraction of the gallbladder musculature, causing the fun-dus to rise and the body to narrow.

    (b) The resulting rise in biliary pressure overcomes the sphincters of the lower biliary tree, in which the muscular tone may have been decreased slightly by vagal activi-ty, and bile spurts into the duodenum.

e. Clinical considerations

  (1) Duct variation. The numerous variations in the biliary tree provide difficulties in surgery.

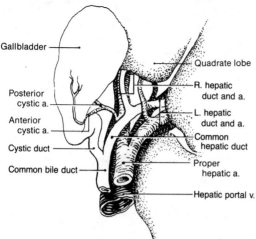

Figure 17-9. *Vascular supply to the liver and gallbladder.*

(a) Although procedures on the biliary tree are routine, complications in cholecystec-tomy (removal) and cholecystostomy (drainage) are suprisingly frequent.

(b) Inadvertent injury to the biliary tree with leakage of bile into the peritoneal cavity rapidly produces *bile peritonitis.* Subsequent edema or scar tissue may cause ob-struction with pain and jaundice.

(2) **Arterial variation** in the branches of the hepatic artery within the cystohepatic triangle (of Calot) constitutes the major hazard in biliary surgery (see Fig. 17-9).

(a) The boundaries of the cystohepatic triangle include the following:

(i) The **common hepatic duct** on the left

(ii) The **cystic duct** and **gallbladder** on the right

(iii) The **liver** on the superior border

(b) The cystohepatic triangle contains most structures of importance in cholecystec-tomy:

(i) 90% of all **cystic arteries**

(ii) 82% of all **right hepatic arteries**

(iii) 95% of all **accessory right hepatic arteries**

(iv) 91% of all **accessory bile ducts**

(c) Structures in the cystohepatic triangle and in the hepatoduodenal ligament should never be clamped or divided until dissected free and positively identified.

(3) **Gallstones (choleliths).** Overconcentration of the bile, perhaps due to biliary stasis, re-sults in the precipitation of bile salts and pigments, leading to concretions.

(a) Gallstones may be composed purely of cholesterol, which may calcify and be visi-ble in an ordinary radiograph. They may be of purely bile pigments, which appear as a filling defect in a cholangiogram. They may be mixed in concentric layers and calcified.

(b) In women, it is not uncommon to find more than one population of similarly sized stones; each population represents a different pregnancy.

(c) Gallstones are most common in overweight, multiparous women over the age of 40—the so-called "4-F individual" (fat, fertile, female, and forty).

(d) Small stones pass through the ducts and sphincter without incident. Stones too large to enter the cystic duct irritate the cystic mucosa (cholecystitis) and may result in ulceration with possible erosion of the wall of the gallbladder and subsequent peritonitis.

(e) Stones of the appropriate size may enter and lodge within the duct system, espe-cially at points of narrowing, such as the point where the duct penetrates the duo-denal wall, the sphincter choledochus, and the ostium of the duodenal papilla.

(i) If the stone lodges in the cystic duct, there is cholecystitis without jaundice, be-cause the bile produced by the liver continues to drain into the duodenum.

(ii) If the stone lodges in the common bile duct, there is jaundice as well as chole-cystitis.

(f) Biliary obstruction is painful, especially when fatty chyme elicits the release of cho-lecystokinin, which elevates biliary pressure ("*gallbladder attack*").

(i) Pain associated with cholecystitis reaches the spinal cord, levels T5–T6, via visceral afferent fibers traveling with the greater splanchnic nerves. Biliary pain is referred to the 5th and 6th thoracic dermatomes, which include the upper quadrants of the abdomen and epigastrium.

(ii) When the peritoneum becomes involved, the pain becomes localized in the right upper quadrant (right hypochondriac region).

(4) Anomalies of the gallbladder include diverticula, duplication, and congenital absence, as well as ductal duplication, atresia, and stenosis.

F. Pancreas

1. Developmental considerations. The pancreas develops from two intestinal diverticula (one in the dorsal mesentery and the other in the ventral mesentery) in association with the hepatic diverticulum.

a. This accounts for the primary and secondary pancreatic ducts.

b. Due to rotation of the gut, the head of the pancreas lies on the right side and the head and body become secondarily retroperitoneal as the duodenum fuses to the dorsal wall peritoneum.

2. External structure of the pancreas

a. Location. The pancreas lies in the epigastric and left hypochondriac regions. Except for the tail, the pancreas is secondarily retroperitoneal.

(1) The anterior surface of the pancreas is covered by the posterior wall of the lesser sac (omental bursa). The posterior surface is adherent to the parietal peritoneum of the dorsal body wall by the fascia of Toldt. Incising the fascia of Toldt and reestablishing the mesentery allows mobilization of the duodenum and pancreas while maintaining the body supply.

(2) Because of its deep position, the pancreas is difficult to palpate. Usually, by the time a tumor is large enough to be palpable, it has metastasized widely and, therefore, is inoperable.

b. Divisions. The pancreas is divisible into the head, body, and tail.

(1) The **head** is located within the curve of the duodenum (see Fig. 17-4).

(a) The initimate association between the head of the pancreas and the duodenum results from the shared mesentery and the shared superior and inferior pancreaticoduodenal arteries (see Fig. 17-4).

(b) The **uncinate process** is a portion of the head that projects from the left posterior to the superior mesenteric vessels.

(2) The **body** lies posteriorly to the stomach.

(a) The blood supply to the body is derived in part from the superior pancreaticoduodenal vessels and variable branches from the splenic artery (Fig. 17-10).

(b) A perforated gastric ulcer may result in adhesions between the stomach and pancreas with subsequent erosion into the pancreatic parenchyma and resultant *pancreatitis. Hemorrhagic pancreatitis* may be rapidly fatal.

(3) The **tail** projects into the **splenorenal ligament** toward the hilus of the spleen (see Fig. 17-10). It is, therefore, peritoneal.

(a) The blood supply to the tail is extremely variable. The splenic artery, usually embedded in the parenchyma of the tail, gives off a variable **great pancreatic artery** and often several **caudal pancreatic arteries** (see Fig. 17-10).

(b) The tail contains the majority of the **pancreatic islets**, which is a major consideration in pancreatic resection.

3. Internal structure of the pancreas. Two pancreatic ducts drain the pancreas.

a. The primary pancreatic duct (of Wirsung) is formed during development by the joining of the distal portion of the primitive dorsal duct to the proximal portion of the primitive ventral duct.

(1) The primary duct courses the entire length of the pancreas (see Fig. 17-10).

(2) It usually joins the common bile duct, which also develops in the ventral mesentery, to form the **hepatopancreatic ampulla (duodenal papilla)** [Fig. 17-11A].

b. The accessory (secondary) pancreatic duct (of Santorini) represents the proximal portion of the primitive dorsal duct.

(1) This duct drains a small portion of the head and body.

(2) This duct usually enters the duodenum separately at the **accessory duodenal papilla** about 2 cm above the primary duodenal papilla (see Fig. 17-10).

c. Ductal variations

(1) The embryonic pattern persists in about 10% of individuals.

(2) The distal end of the secondary duct may remain connected to the primary duct (see Fig. 17-10).

(3) The primary pancreatic duct may not fuse with the common bile duct so that each opens separately into the duodenum (see Fig. 17-11B).

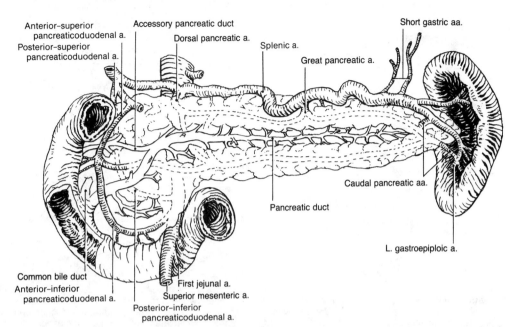

Figure 17-10. *Pancreas and spleen.* The primary and secondary (if present) pancreatic ducts empty into the descending portion of the duodenum. The pancreas receives its blood supply from branches of the celiac and superior mesenteric arteries. The spleen receives a profuse blood supply from the splenic branch of the celiac artery.

4. Pancreatic vasculature

a. Arterial supply to the pancreas is from both the celiac and superior mesenteric distributions.

(1) The **great pancreatic artery**, **caudal pancreatic arteries**, and an occasional **dorsal pancreatic artery** all arise from the splenic artery.

(2) The **superior pancreaticoduodenal artery**, arising from the gastroduodenal artery, bifurcates into anterior and posterior branches, which supply the head of the pancreas and the duodenum.

(3) The **inferior pancreaticoduodenal artery**, arising from the superior mesenteric artery, also bifurcates into anterior and posterior branches, which anastomose with those from the superior pancreaticoduodenal artery.

b. Pancreatic veins generally drain into the hepatic portal system through the superior mesenteric vein as well as the common hepatic and splenic veins. Because the pancreas is mostly secondarily retroperitoneal, there are variable anastomotic connections (**veins of Retzius**) with the systemic veins of the posterior body wall.

5. Pancreatic function

a. Endocrine function. The **pancreatic islets (of Langerhans)** comprise the endocrine portion of the pancreas. The islet cells secrete insulin and glucagon into the circulatory system.

b. Exocrine function. As a compound tubuloacinar exocrine gland, the pancreas secretes 500–1200 ml of pancreatic juice a day into the duodenum.

(1) This colorless, viscous fluid is alkaline due to a high bicarbonate content. It maintains the alkaline pH of the duodenum.

(2) Pancreatic juice contains three classes of powerful digestive enzymes: trypsin, amylase, and lipase, all of which are activated by an alkaline pH.

(3) The secretion of pancreatic juice is regulated by local chemical stimuli as well as by parasympathetic innervation.

6. Clinical considerations

a. Pancreatic obstruction. Because the pancreatic and common bile ducts join and enter the duodenum via a common ostium in 65% of individuals, spasm of the hepatopancreatic sphincter or blockage of the duct at the duodenal papilla by a cholelith not only results in biliary stasis and but also may cause reflux of bile into the pancreas. This situation causes *acute pancreatitis (chemical autolytic pancreatic necrosis)*.

b. Carcinoma of the pancreas is not uncommon.

(1) The lymphatic drainage of the pancreas is diffuse, in part due to its secondary retroperitoneal position. The prognosis for recovery from pancreatic cancer is usually poor because metastases may spread widely by retroperitoneal channels.

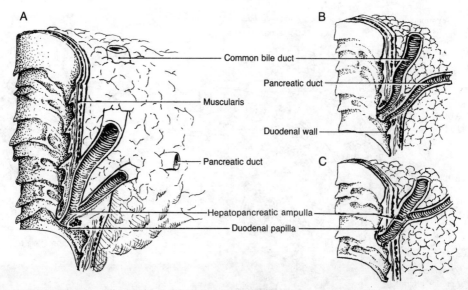

A   B

Common bile duct

Pancreatic duct

Muscularis

Duodenal wall

Pancreatic duct   C

Hepatopancreatic ampulla

Duodenal papilla

Figure 17-11. *Duodenal papilla. A,* The usual configuration whereby the common bile duct and the primary pancreatic duct join to enter the duodenum through a hepatopancreatic duct. *B,* The common bile duct and primary pancreatic duct may enter the duodenum separately. *C,* A more proximal joining of the biliary and pancreatic ducts produces an unusually long hepatopancreatic ampulla.

       (2) Fully 80% of pancreatic carcinomas are located in the head.
          (a) Jaundice with no indication of biliary pathology or hepatic dysfunction may result from compression of the common bile duct by a pancreatic tumor. This is often a life-saving signal, as pancreatic tumors are frequently silent until it is too late for surgery.
          (b) Pancreatic carcinoma frequently invades the splenic vein and even the portal vein with symptoms of portal hypertension, such as esophageal varices, without indications of liver disease. By this time, metastases have usually spread widely by hematogenous routes.
     c. Anomalies. Aside from the commonly occurring **ectopic pancreatic tissue** (possible in the stomach, small intestine, gallbladder, or spleen), the pancreas may completely surround the descending duodenum (**annular pancreas**) with constriction or even obstruction of this portion of the small bowel.

G. Spleen

   1. Splenic relations. The spleen is located in the left hypochondriac region (upper left quadrant) adjacent to the stomach. It is superior to the left colic flexure and anterior to the left kidney (see Fig. 15-1). The upper half is usually anterior to ribs 11 and 12, while the lower pole may extend to the level of L2.
     a. Support. Arising within the dorsal mesogastrium, its supports include the **gastrosplenic ligament** and the **splenorenal ligament**.
       (1) The tail of the pancreas may reach into the lienorenal ligament as far as the hilum of the spleen.
       (2) Variable in position, it may prolapse into the false pelvis, especially in *splenomegaly*.
     b. Variation. One to ten accessory spleens in the dorsal mesogastrium are common.

   2. Splenic structure and function
     a. The size of the normal spleen varies within wide limits, but it is usually about 12 cm x 7 cm x 3 cm and weighs 150–200 g.
     b. The spleen is an intensely vascular tissue. It is hematopoietic in the fetus and later functions in blood cell destruction. Once thought to be the seat of bad humor ("...to vent one's spleen...") and sadness, caused by an excess of "black bile" (melancholy), it is not a vital organ in the adult.
       (1) It is composed of lymphoid tissue separated by blood-filled sinuses. The lymphocytes participate in immunologic activity by producing opsonins and antibodies to blood-borne antigens.
       (2) As a reticuloendothelial structure situated in the hepatic portal system, the spleen removes particulate matter and cellular residue from the blood.

(3) Due to its vascularity and capacity to store blood, as well as its location in the hepatic portal system, the spleen hemorrhages profusely when ruptured, such as by a broken 11th or 12th left rib. Internal exsanguination may occur quickly.

3. **Splenic vasculature.** The spleen is supplied profusely by the **splenic artery** (see Fig. 17-10). It is drained by the **splenic vein**, a major branch of the hepatic portal vein (see Fig. 17-21).

4. **Clinical considerations**
   a. **Splenomegaly.** The spleen is palpable only when enlarged, which may be a sign of chronic infection (e.g., mononucleosis), blood dyscrasias, lipid-storage disease, or lymphosarcoma.
   b. **Splenectomy.** When therapeutic splenectomy is indicated, all accessory splenic tissue must be located and excised.

III. MIDGUT

A. **General principles**

1. **Extent.** The midgut consists of the jejunum, ileum, ascending colon (including the cecum and appendix), and the transverse colon.

2. **Vasculature.** It is supplied primarily by the **superior mesenteric artery**.

3. **Innervation.** It receives parasympathetic innervation from the vagus nerve (CN X) and sympathetic innervation from the lesser splanchnic nerve (T10–T11).

4. **Function.** The midgut portion of the small intestine functions:
   a. In absorption of nutrients and water from digested chyme
   b. In the conservation of fluid, a process that converts chyme into semisolid feces

B. **Jejunum and ileum**

1. **Location and support of the jejunum and ileum.** The small bowel is supported by the mesentery proper. The jejunal loops tend to be located in the left lateral region of the abdominal cavity while the ileal loops tend to be located in the pelvic cavity.
   a. **The mesentery proper** originates from a root or radix (line of peritoneal attachment) that is 15–20 cm (6–9 inches) long and runs diagonally down the posterior body wall from upper left to lower right (see Fig. 16-8). From the radix, the mesentery fans out to support more than 7 m of small bowel.
      (1) Except at their terminations, the jejunum and ileum are highly mobile. Their relationships, as seen on x-ray, may change from day to day.
      (2) Excessive twisting (volvulus) of the small intestine may compress the blood supply that runs within the mesentery, possibly resulting in ischemia (*intestinal angina*) and infarct.
   b. **Paraduodenal recesses (fossae)** are formed by small folds of peritoneum at the duodeno-jejunal flexure where the small intestine changes from secondarily retroperitoneal to peritoneal.
      (1) The left and right paraduodenal fossae are potential sites for intra-abdominal herniation (*paraduodenal hernia*) during development as the result of malrotation of the gut.
      (2) The openings to the paraduodenal fossae are bounded by the duodenum (medially) and by the inferior mesenteric and left colic veins (laterally).

2. **External structure of the jejunum and ileum.** The peritoneal portion of the small bowel averages 7 m (22 feet), varying from 5–11 m (15–34 feet). Because the jejunum and ileum share a common blood supply, innervation, mesenteric support, and function and because there is no distinct demarcation between them, they are best discussed as a unit.
   a. **Divisions.** The **jejunum** constitutes the proximal three-fifths of the peritoneal portion of the small intestine; the **ileum** constitutes the distal three-fifths. Because the distinctive characteristics of the jejunum gradually become those of the ileum, the one-fifth in common cannot be ascribed to either segment.
   b. **Distinguishing characteristics.** Although the differences between the jejunum and ileum are subtle, it is relatively easy to distinguish between the proximal jejunum and the distal ileum (Fig. 17-12 and Table 17-1). Differentiation is important when resecting the small bowel; inattention to this detail is can be fatal.
      (1) The size and number of arterial arcades and arteriae rectae, as well as the relative quantity of mesenteric fat between these vessels, are the more obvious distinguishing characteristics for surgery. The jejunum has larger arcades with fewer loops, longer straight vessels, and less mesenteric fat (see Fig. 17-12).
      (2) Radiographically with barium contrast, the differences are apparent by the diameter and the appearance of the plicae circulares. The jejunum has a wider lumen with a feathery appearance due to the abundant, tall, and branched plica circularis (see Table 17-1).

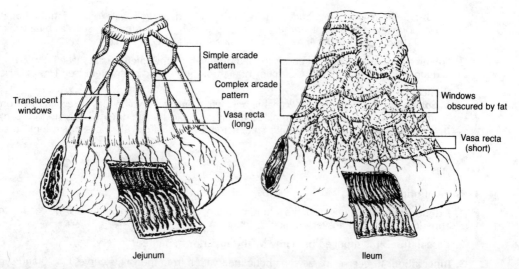

Figure 17-12. *Jejunum and ileum.* Several differences between the proximal jejunum and distal ileum are indicated.

3. **Internal structure of the jejunum and ileum.** The function of the small intestine is absorption, accomplished through structural modification that increases the surface area.
    a. **Length.** The 5–8 m length of the small intestine with a diameter of 2–4 cm provides a substantial surface area for absorption.
    b. **Plicae circulares**, circular folds, substantially increase the mucosal surface area.
    c. **Intestinal villi**, surface projections on the plicae circulares, tremendously increase the folded surface area.
    d. **Microvilli**, cytologic modifications of the luminal surface of the mucosal cells, increase the absorptive surface area by magnitudes.

4. **Vasculature of jejunum and ileum**
    a. **Arterial supply** to the jejunum and ileum is from the **jejunal** and **ileal branches** of the **superior mesenteric artery.**
        (1) **Jejunal** and **ileal arteries** (up to 20) branch to form **arterial arcades** within the mesentery proper (Fig. 17-13).
            (a) The arcades provide abundant collateral circulation, which is especially important during peristaltic movements when some of the arteries may be constricted.
            (b) The arcades of the ileum are smaller and more numerous than found in the jejunum (see Fig. 17-12).
        (2) **Straight arteries (arteriae rectae** or **vasa recta)** run from the arcades directly to the viscus wall (see Fig. 17-12) where they give off anterior and posterior branches to the sides of the viscus.
            (a) Adjacent vasa recta usually anastomose within the intestinal wall on the mesenteric side. On the antimesenteric side, the vessels are true **end-arteries**, which are, by definition, nonanastomotic and supply exclusive regions of tissue.
                (i) Intramural anastomoses on the mesenteric side of the small intestine are sufficient to sustain several centimeters of intestinal tissue in the event of embarrassment of the direct blood supply.

Table 17-1. Distinguishing Characteristics of the Jejunum and Ileum

Characteristic	Jejunum	Ileum
Diameter	2–4 cm	2.5–3 cm
Wall	Thick, heavy	Thin, light
Vascularity	Greater	Less
Color	Deeper red	Paler pink
Arteriae rectae	Tall	Short
Arcades	Large loops, few	Short loops, many
Fat in mesentery	Less	More
Plicae circulares	Well developed	Rudimentary
Peyer's patches	Few	Many

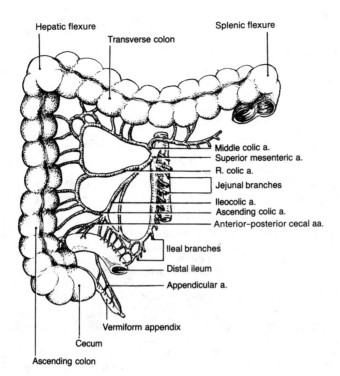

Figure 17-13. *Superior mesenteric arterial distribution.*

   (ii) If an end-artery to the antimesenteric side becomes occluded, local infarct results.
  (b) The straight arteries of the ileum are shorter and more numerous than those of the jejunum.
 (3) Ileal twigs from the ileocolic branch of the superior mesenteric artery supply the terminal ileum.
 b. Venous return of the jejunum and ileum is to the hepatic portal system via the **superior mesenteric vein**.
 c. Lymphatic drainage. Lymphatics of the small intestine absorb and transport triglycerides as well as lymph to the circulation via the thoracic duct. Lymphatics of this region are of minor clinical importance because the jejunum and ileum are noted for a low incidence of carcinoma.

5. Function of the small intestine
 a. Intestinal motility propels chyme through the gut.
  (1) Segmentation occurs at the rate of about 10 constrictions a minute and corresponds to the frequency of **borborygmi** (the sounds auscultated with a stethoscope applied to the abdominal wall).
   (a) Intestinal movements are classified as segmental when columns of chyme are divided into segments by nonpropagating contraction of the circular muscle.
   (b) These contractions are the result of reflexes initiated by intestinal wall distension.
   (c) Subsequent contractions in adjacent regions both divide a segment of chyme and reunite previously divided adjacent segments.
  (2) Pendular movements are slight waves of contraction that pass bidirectionally and rapidly along short segments of the gut at a rate of 12–13 a minute.
   (a) Both circular and longitudinal muscle layers are involved in pendular movements. These movements mix chyme with intestinal and pancreatic digestive juices and bile as well as assist absorption by continuously bringing chyme into contact with the surface mucosa.
   (b) These movements have an important massaging influence on the blood and lymphatic channels of the small intestine.
  (3) Peristalsis, a unidirectional intestinal movement, occurs in the jejunum, ileum, and large intestine. While not very forceful in the jejunum, peristalsis becomes very pronounced in the ileum. This can be appreciated on a barium contrast x-ray.
   (a) One form of peristaltic wave is a relatively slow (1–2 cm/second) advancing contraction that proceeds aborally over relatively short distances (4–5 cm) at a time.

(b) The second form is **peristaltic rush**, which sweeps along the intestine, without pause, for great distances. This usually occurs after meals, especially the first of the day.

(4) Control of GI motility

(a) Enteric reflexes. Segmental and pendular movements are probably enteric in origin.

(i) Many of the nerve cells of the myenteric plexuses are intrinsic and involve local reflexes.

(ii) Enteric reflexes may explain the **Bayliss-Starling law of the gut**, whereby a bolus of chyme, exerting transverse pressure on the intestinal wall, results in increased muscular tone immediately oral to the bolus and relaxation in the adjacent aboral region. This reflex mechanism ensures unidirectional oral to aboral flow of intestinal contents.

(b) Autonomic influences affect persistalsis

(i) The sympathetic (adrenergic) neurons from the celiac and superior mesenteric ganglia synapse on the enteric neurons of the myenteric plexuses. Sympathetic activity inhibits the enteric neurons, thereby slowing peristalsis.

(ii) Parasympathetic (cholinergic) neurons of the myenteric plexuses are stimulated by the vagus nerve. Parasympathetic stimulation increases peristalsis, probably by facilitating the enteric neurons and the enteric reflexes.

b. Exocrine function. The digestive functions of the small intestine require alkaline secretion (pancreatic juice) as well as the production of proteolytic enzymes by intestinal glands.

c. Absorption. The mucosa of the small intestine absorbs carbohydrates, triglycerides and fatty acids, amino acids, vitamins, electrolytes, and water.

6. Clinical considerations

a. Ileojejunal resection. Excision of up to one-third of the small intestine is compatible with normal life. Survival is possible with as little as 18 inches.

(1) The stomach may be anastomosed to the jejunum (gastrojejunostomy) when the duodenum must be bypassed or resected (Whipple's procedure).

(2) As a radical weight control measure, a segment of small intestine may be bypassed to reduce the absorptive area.

b. Ileal (Meckel's) diverticulum is a persistent remnant of the omphalomesenteric (vitelline) duct of the embryo.

(1) Present in about 3% of the population, Meckel's diverticulum is located about 3 feet from the ileocecal junction (on the antimesenteric side of the ileum) and is usually less than 3 inches long.

(2) Meckel's diverticulum may contain gastric mucosa with oxyntic (acid-secreting) cells. Peptic ulceration of adjacent ileal mucosa is a frequent complication. Associated edema may produce intestinal obstruction.

(3) An ileal diverticulum may become inflamed with symptoms similar to those produced by an inflamed appendix, and require surgical intervention.

(4) Rarely, the omphalomesenteric duct remains a patent connection with the umbilicus so that chyme may be discharged to the exterior. The connection to the umbilicus may persist with an obliterated lumen; however, the gut may revolve about this structure (volvulus) with resultant obstruction of the intestinal lumen, occlusion of blood vessels, and subsequent gangrene.

(5) During appendectomy and other surgical procedures, the terminal ileum is routinely explored for a possible ileal diverticulum. If found, it is resected as a prophylactic measure.

c. Paralytic ileus. Inflammation or trauma (even surgical manipulation) to abdominal or related organs, such as the kidneys or gonads, usually leads to inhibition of the small bowel, *intestinal shut-down (paralytic ileus)*. The reemergence of borborygmi is anxiously awaited postoperatively as the sign of reestablished bowel function, an indication that feeding by mouth may be resumed.

C. Ascending colon. This portion of the large bowel consists of three parts: cecum, appendix, and ascending colon proper (see Fig. 17-13).

1. Overview. The colon or large bowel is characterized by its capacity, distensibility, the length of time that it retains its contents, and the special arrangement of musculature within the tunica muscularis. These characteristics are related to the role of the large intestine in the formation, transport, and evacuation of the feces.

a. The principal function of the colon is the conversion of fluid by the conversion of liquid chyme into semisolid feces. The mucosa of the colon is characterized by a mixture of absorptive cells and mucous cells, which provide lubrication to enable the forming stool to be moved toward the rectum.

b. Structure of the colon
   (1) The large bowel is approximately 2 m (5 feet) long, varying between 1 and 3 m (3 and 7 feet).
   (2) Fat-filled tags (**appendices epiploicae**) are scattered over its surface (Fig. 17-14).
   (3) The outer longitudinal layer of the muscularis externa is incomplete, being restricted to three longitudinal bands, the **teniae coli** (see Fig. 17-14).
      (a) The teniae coli, about 1 cm wide, are most obvious on the cecum and ascending colon.
      (b) Their position is somewhat variable along the transverse colon, becoming diffuse along the descending colon, rather indistinct along the sigmoid colon, and fusing to form a complete outer layer over both the appendix and rectum.
   (4) Muscle tone in the teniae coli results in sacculations (**haustra**) along the large intestine.
c. Divisions. The large bowel is divided according to its mesenteric attachments, or lack thereof, into ascending, transverse, descending, sigmoid, and rectal regions. The ascending and transverse portions are part of the midgut (see Fig. 17-13); the descending, sigmoid, and rectal portions constitute the hindgut (see Fig. 17-17).

2. The cecum, a dilated pendulous sac inferior to the ileocecal juncture, is the first part of the ascending colon (see Fig. 17-14).
a. Ileocecal juncture. The ileum terminates by entering the cecum posteromedially with some variation (see Fig. 17-14).
   (1) The **ileal papilla** is the conical projection at the point of entry.
   (2) The **ileocecal valve** is formed by two folds that border the ileocecal ostium (see Fig. 17-14).
      (a) While the ileocecal valve impedes regurgitation of cecal contents into the ileum, it is largely incompetent and of minor mechanical importance. Backflow of chyme is prevented primarily by the direction of peristalsis.
      (b) A barium enema that fills the colon completely also penetrates the terminal ileum to a variable extent.
   (3) Intussusception. The terminal ileum is vulnerable to herniation into the cecum (*intussusception*). The resultant functional obstruction necessitates immediate surgery.
b. Mesenteric support of the cecum
   (1) The **superior ileocecal (vascular) fold**, an extension of the mesentery proper, supports the pendulous position of the cecum, thereby making it a peritoneal structure (see Fig.

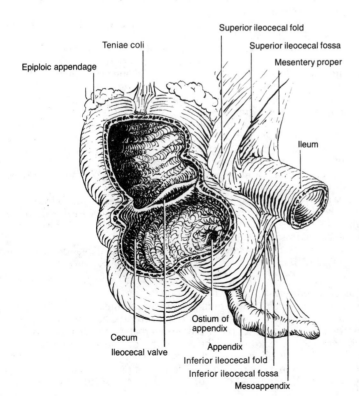

Figure 17-14. *Cecum and appendix.* The mesenteries and mesenteric folds supporting the cecum and appendix are shown as is the internal structure of the cecum.

17-14). The extent of this mesentery is variable, but it always contains the **anterior cecal artery**.

(2) Cecal fossae. The more superior portions of the cecum, along with the rest of the ascending colon, are secondarily retroperitoneal. At the line of fusion between the primitive ascending mesocolon and the parietal peritoneum, peritoneal folds define three cecal fossae (see Fig. 17-14).

(a) The **superior ileocecal fossa** lies between the mesentery proper and the superior ileocecal fold (see Fig. 17-14).

(b) The **inferior ileocecal fossa** lies between the mesoappendix and the inferior ileocecal (bloodless) fold (see Fig. 17-14).

(c) The **rectocecal fossa** is formed by the incomplete fusion of the visceral peritoneum of the cecum and the parietal peritoneum of the posterior abdominal wall. The appendix is frequently found coiled within this fossa.

(3) Mobilization of the cecum. The cecum may be mobilized with its blood supply by incision along its lateral juncture with the parietal peritoneum.

c. Cecal teniae coli converge toward the appendix and provide useful landmarks for the location of the appendix during surgery, which sometimes can be exasperatingly difficult.

d. Arterial supply to the cecum is by the **ileocolic artery**, which is the terminal branch of the superior mesenteric artery (see Fig. 17-13). This artery has four sets of branches.

(1) The **ascending colic artery** supplies the basal portion of the ascending colon (see Fig. 17-13).

(2) Anterior and **posterior cecal arteries** supply the cecum. The anterior cecal artery runs in the superior ileocecal fold (see Fig. 17-13).

(3) The **appendicular artery** supplies the appendix (see Fig. 17-13).

(4) Ileal branches of the ileocolic artery provide the major supply to the terminal ileum (see Fig. 17-13)

(a) Any anastomoses between the last ileal arcade of the superior mesenteric artery and the ileal branches of the ileocolic artery are weak, if present at all.

(b) In cecal resection with excision of the ileocolic artery, it is customary to remove the terminal ileum up to the first substantial arcade to avoid the possibility of ischemic necrosis.

3. Vermiform appendix. A narrow, hollow, muscular viscus that arises from the posteromedial aspect of the cecum about 2–3 cm below the ileocecal orifice, the appendix represents the tip of the cecum, which fails to enlarge (see Fig. 17-14).

a. Structure of the appendix. The appendix averages 9 cm (3.5 inches) in length but may vary between 0.5 cm and 25 cm (0.15 and 10 inches).

(1) A complete longitudinal muscular coat gives the appendix a smooth appearance.

(2) Within the mucosa and submucosa are large accumulations of lymphatic tissue.

(3) The region of the appendix adjacent to the cecum has a thickened muscular coat and a slightly narrower lumen. The luminal diameter varies with age.

(a) The lumen tends to be rather wide (6–8 mm) in infants and young children.

(b) It is often entirely obliterated by middle-age.

(c) The lumen is dangerously narrow in adolescents and young adults. During this period, it may be occluded by a fecalith or even by edema and swollen lymphatic tissue associated with mild inflammation. This explains the high frequency of acute and chronic *appendicitis* in teenagers and young adults.

b. Support and vasculature. The **mesoappendix** suspends the appendix from the dorsal body wall, making this organ a peritoneal structure (see Fig. 17-14).

(1) Because it is supported by the mesoappendix, the position of the appendix is extremely variable. At operation, the appendix may be found in the retrocecal fossa (65%), in the iliac fossa (31%), in the right paracolic recess (2%), in the superior ileocolic fossa (1%), or in the inferior ileocolic fossa (1%).

(2) The mesoappendix contains the **appendicular artery**, a terminal branch of the ileocolic artery (see Fig. 17-13).

c. Appendicitis

(1) The pathogenesis of *appendicitis* is poorly understood.

(a) If the lumen is blocked, local bacteria multiply rapidly and their toxins produce inflammation, nausea, and fever. The resultant distension of the appendix from gas or edema compromises the blood supply, which exacerbates the inflammatory process. Local ischemia with breakdown of the ischemic blood vessels produces intramural hemorrhage. This situation is a surgical emergency. Delay results in *gangrenous appendicitis* with danger of rupture and complicating peritonitis.

(b) Because the appendix is innervated by neurons of the lesser splanchnic nerve, initial colicky pain is referred to the 10th and 11th dermatomes, which include the periumbilical region. However, this is not pathognomonic.

    (c) As the adjacent parietal peritoneum becomes inflamed, pain becomes exquisitely localized to the lower right quadrant, and the overlying musculature exhibits the reflex spasm of the acute abdomen (guarding, splinting). This sign, following the initial paraumbilical pain and nausea, confirms the diagnosis. In *situs inversus*, peritoneal pain will be on the left.

  (2) The usual route for surgical access to the appendix is at McBurney's point, about 5 cm along a line from the anterior–superior iliac spine to the umbilicus. Surgeons frequently use McBurney's muscle-splitting incision for appendectomy (see Chapter 15 VII A 3 b).

4. The ascending colon proper is located along the right side of the abdominal cavity (see Fig. 15-1). It is 15–20 cm long (about 8 inches).

  a. Structure and relations of the ascending colon

    (1) Support. The ascending colon proper is fused to the posterior abdominal wall, making it secondarily retroperitoneal.

      (a) The relations of the ascending colon proper are somewhat variable, depending on the degree to which it is adherent to the peritoneum.

      (b) Incomplete fixation may occur as a developmental anomaly, a situation that predisposes to volvulus.

      (c) It may be mobilized and its primitive mesentery with its contained blood supply re-established by incising along the line of peritoneal fusion.

    (2) The **right paracolic gutter** is defined by the ascending colon. This recess communicates with the supracolic compartment, above, and the infracolic compartment, below (see Fig. 16-8). Infection may track between the pelvic cavity and the subhepatic recess along the right paracolic gutter.

    (3) The **right colic flexure** of the ascending colon marks the transition between ascending colon and transverse colon.

  b. Vasculature of the ascending colon

    (1) Arterial supply is by the middle colic, right colic, ileocolic branches of the superior mesenteric artery.

      (a) The **ascending colic branch** of the ileocolic artery supplies the most inferior portion of the ascending colon proper (see Fig. 17-13).

      (b) The **right colic artery**, through its ascending and descending branches, usually supplies the major portion of the ascending colon proper (see Fig. 17-13). This artery is the most variable branch of the superior mesenteric artery and is absent in 18% of individuals.

      (c) The **middle colic artery** bifurcates into left and right branches. The right branch may supply a variable portion of the ascending colon in the vicinity of the hepatic flexure.

      (d) The **marginal artery** (of Drummond) lies in the mesentery adjacent to the colon (see Fig. 17-20). It is an important vascular channel formed by the anastomoses of various branches of the superior mesenteric artery.

        (i) The ascending colic branch of the ileocolic artery anastomoses with the descending branch of the right colic artery.

        (ii) The ascending and descending branches of the right colic artery anastomose.

        (iii) The ascending branch of the right colic artery anastomoses with the right branch of the middle colic artery.

    (2) Venous return is via the **superior mesenteric vein** to the hepatic portal vein.

D. Transverse colon. This second and longest segment of the large bowel begins at the hepatic flexure and traverses the peritoneal cavity to the splenic flexure (see Fig. 17-15). It is 45–50 cm (18–20 inches) long and is suspended by mesentery (transverse mesocolon), which makes this segment of bowel peritoneal.

  1. Structure and relations of the transverse colon. The position of the transverse colon is variable and depends upon its degree of fullness (Figs. 17-15 and 17-16).

    a. The **hepatic (right colic) flexure**, which is located anteriorly to the lower pole of the kidney and the inferior segment of the duodenum, anchors the proximal end of the transverse colon to the secondarily retroperitoneal ascending colon (see Fig. 17-13).

    b. The **splenic (left colic) flexure**, which is located anteriorly to the pole of the kidney and inferiorly to the spleen, anchors the distal end to the secondarily retroperitoneal descending colon (see Fig. 15-1).

    c. The **gastrocolic ligament** is formed by the fusion of the transverse mesocolon with the inferior surface of the greater omentum (see Fig. 16-6). This ligament suspends the transverse

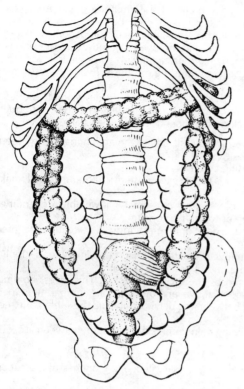

Figure 17-15. *Variation in the position of the transverse colon.*

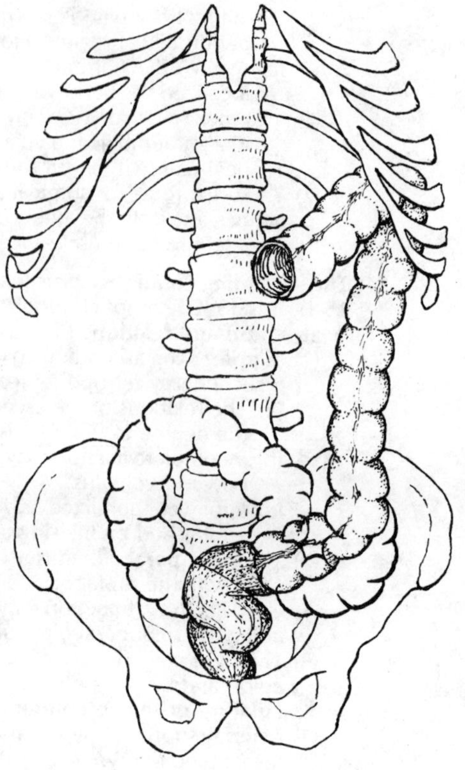

Figure 17-16. *Variation in the length and location of the sigmoid colon.*

colon between the two flexures. This ligament divides the peritoneal cavity into supracolic and infracolic compartments (see Fig. 16-8).

2. Vasculature of the transverse colon

  a. Arterial supply is by the **middle colic artery** (see Fig. 17-13).

    (1) The middle colic artery is the first major branch of the superior mesenteric artery after the inferior pancreaticoduodenal artery; it lies within the transverse mesocolon.

    (2) Its branches contribute to the **marginal artery** (see Fig. 17-20).

      (a) The right branch anastomoses with the ascending branch of the right colic artery or, if missing, the ascending branch of the ileocolic artery.

      (b) The left branch usually anastomoses with the ascending branch of the left colic artery from the inferior mesenteric artery. However, anastomoses may be poor or even absent in the vicinity of the splenic flexure—a danger area.

  b. Venous return is to the **superior mesenteric vein** and hepatic portal vein.

3. Innervation of the transverse colon. This segment of bowel is the most caudal portion of the GI tract to be supplied with parasympathetic innervation from the **vagus nerve**. The sympathetic innervation is via the **lesser splanchnic nerves** from spinal levels T10 and T11 via the superior mesenteric ganglia.

IV. HINDGUT

  A. General principles

    1. Extent. The hindgut consists of the descending colon, sigmoid colon, and rectum. The common features of the large bowel are discussed in the section on the ascending colon (see III C 1).

    2. Vasculature. The hindgut is supplied primarily by the **inferior mesenteric artery**.

    3. Innervation. It receives parasympathetic innervation from the **pelvic splanchnic nerves** (S2–S3) and sympathetic innervation from the **lumbar splanchnic nerves** (L1–L2).

    4. Function. The hindgut portion of the large intestine functions to conserve fluid, a process that solidifies the feces, and to transport, store, and evacuate the feces.

B. Descending (left) colon. The initial segment of hindgut, is located along the left side of the abdominal cavity (see Fig. 15-1). It is about 25 cm long (10 inches).

    1. Structure and relations of the descending colon

        a. Support. The descending colon is fused to the posterior abdominal wall from the splenic flexure to the pelvic brim, making this segment secondarily retroperitoneal. It may be mobilized, along with its blood supply, in the primitive descending mesocolon, by incising the line of peritoneal adhesion.

        b. The left paracolic gutter is defined by the descending colon. This recess communicates between the pelvic fossa and the supracolic compartment (see Fig. 16-8).

        c. External structure. Because the bulk of the feces decreases with the absorption of water, the diameter of the descending colon is smaller than either the ascending or transverse segments of large bowel. Because the luminal contents are more solid, the descending segment is more prone to *diverticulosis*. If a fecalith should become entrapped in a diverticulum, inflammation ensues (*diverticulitis*).

    2. Vasculature of the descending colon

        a. Arterial supply is by the **left colic branch** of the **inferior mesenteric artery** (Fig. 17-17).

        (1) The ascending branch of the left colic artery anastomoses inconsistently with the left branch of the middle colic artery.

        (2) The descending branch anastomoses with the arcades of the sigmoid arteries (see Fig. 17-17).

        (3) These anastomoses contribute to the **marginal artery** (of Drummond) [see Fig. 17-20].

        b. Venous return is via the **inferior mesenteric vein** to the splenic vein or superior mesenteric vein.

    3. Innervation of the descending colon

        a. Parasympathetic innervation to the descending colon is from the **pelvic splanchnic nerves (nervi erigentes)** from spinal levels S2–S4.

        b. Sympathetic innervation is from the **lumbar splanchnic nerves** (L1–L2).

        c. Afferent innervation from the descending colon reaches the L1–L2 levels of the spinal cord by traveling the lumbar splanchnic pathways. Pain originating within the descending colon is referred to the L1–L2 dermatomes of the inguinal region and thigh.

C. Sigmoid colon. This segment of the large bowel begins where the colon again becomes peritoneal, usually at the pelvic brim. It ranges in length from 25–40 cm (10–15 inches).

    1. Structure and relations of the sigmoid colon. The structure is similar to the rest of the large bowel except that the mucosa is composed primarily of mucous cells.

        a. Support. The sigmoid colon is peritoneal, suspended from the posterior abdominopelvic wall by the **sigmoid mesocolon.**

        (1) Its position is extremely variable; a large loop may occupy the deep pelvis.

        (2) The **intersigmoid recess**, an occasional site for intra-abdominal herniation, is located at the juncture of the inferior surface of the sigmoid mesocolon and the dorsal body wall.

        b. Relations. Anteriorly, the sigmoid colon is related to the urinary bladder in males and to the posterior surface of the uterus and upper part of the vagina in females. Posteriorly, it is related to the external iliac vessels and the sacrum. The sigmoid colon continues into the rectum at the third sacral level.

    2. The primary function of the sigmoid colon is storage of feces.

    3. Vasculature of the sigmoid colon

        a. Arterial supply is by sigmoid arteries and the rectosigmoid artery that arise from the inferior mesenteric artery.

        (1) Sigmoid arteries, usually four in number, anastomose profusely to form arcades.

            (a) The first sigmoid artery anastomoses with the descending branch of the left colic artery to contribute to the **marginal artery.**

            (b) The last rectal arcade may or may not anastomose with the rectosigmoid artery.

        (2) The **rectosigmoid artery**, the terminal branch of the inferior mesenteric artery, supplies the terminal sigmoid colon and the superior rectum.

            (a) Because the rectosigmoid artery may or may not anastomose with the last rectal arcade, this region is known as the **critical point of Sudeck**. Adequate anastomoses between these two vessels occur in only 52% of individuals, a point of major concern in colorectal resection.

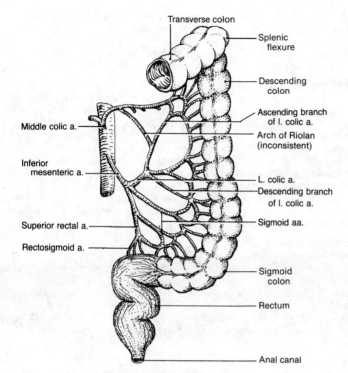

Transverse colon

Splenic
flexure

Descending
colon

Ascending branch
of l. colic a.

Arch of Riolan
(inconsistent)

Middle colic a.

Inferior
mesenteric a.

L. colic a.

Descending branch
of l. colic a.

Superior rectal a.

Sigmoid aa.

Rectosigmoid a.

Sigmoid
colon

Rectum

Anal canal

Figure 17-17. *Inferior mesenteric arterial distribution.* Anastomoses between the middle colic and left colic arteries via an arch of Riolan are inconsistent.

    (b) The rectosigmoid artery may or may not anastomose with the superior rectal (hemorrhoidal) arteries, branches of the internal iliac arteries.

  b. Venous return is via the inferior mesenteric vein to the splenic vein or superior mesenteric vein.

  c. Lymphatic drainage of the terminal portion of the large bowel is of considerable importance due to the high frequency of colonic carcinoma.

    (1) In general, the lymphatic drainage follows the arteries. However, there are regional nodes along the lines of the left colic vein and both the inferior mesenteric artery and vein.

      (a) The major lymph drainage from the descending and sigmoid colons is via the left colic nodes, which drain, in turn, into the inferior mesenteric nodes, the para-aortic nodes, and the thoracic duct.

      (b) Lymph from the descending colon may drain into the celiac nodes via the lymphatic channels along the inferior mesenteric vein.

      (c) Drainage may also be to the deep pelvic nodes via lymphatics associated with the superior hemorrhoidal vessels.

    (2) Because the drainage pathways are variable, carcinoma of the descending colon, sigmoid colon, and rectum (the most frequent sites in males) not only may spread to the liver hematogenously but also may metastasize widely to the lymph nodes of the abdomen and pelvis.

  4. Innervation of the sigmoid colon is the same as for the descending colon.

  5. Clinical considerations

    a. Diverticulosis. The sigmoid colon is the most common site for diverticula, which are small saccular protrusions of intestinal mucosa through the colonic wall. These usually occur at the margins of the teniae coli adjacent to penetrating blood vessels.

    (1) Diverticula result from a combination of factors, such as an incomplete outer muscular layer, the penetration of the intestinal wall by arteriae rectae with resultant weak spots in the muscularis, and the more solid composition of the feces.

    (2) The prevalence of diverticula increases with age, associated with increased intraluminal pressure brought about by straining at stool with chronic constipation.

    b. Diverticulitis. Diverticula may become occluded by fecaliths and may then become inflamed (*diverticulitis*). Ulcerative diverticulitis with bleeding into the intestinal lumen is a frequent cause of anemia in the elderly. An inflamed diverticulum may perforate, resulting in severe intra-abdominal bleeding and peritonitis. Surgical resection of the involved portion of bowel is frequently the treatment of choice.

c. **Hirschsprung's disease**
 (1) Congenital agenesis of the myenteric plexus in the terminal portion of the sigmoid colon precludes peristaltic activity and results in functional blockage with toxic megacolon. This usually manifests as chronic constipation in neonates or young children.
 (2) Bowel function may be restored surgically by an abdominopelvic pull-through procedure with resection of the aganglionic portion of bowel and anastomoses of normal bowel to the region of the anal sphincter.

D. **Rectum.** The part of the large intestine between the sigmoid colon and the anal canal comprises the rectum.

   1. **External structure of the rectum.** The upper limit of the rectum is where the sigmoid mesocolon disappears; the lower limit is the pelvic floor (see Fig. 17-17; Fig 17-18).
      a. **Support.** The rectum, a retroperitoneal structure, is about 13 cm (5 inches) long.
      b. **The ampulla**, which lies just above the pelvic floor, is the widest part of the rectum and is capable of considerable distension.
         (1) Because feces usually are stored in the sigmoid colon, the ampulla usually is empty.
         (2) Movement of feces into the ampulla generates the sensation of rectal fullness and the urge to defecate.
      c. **Function.** The rectum pierces the pelvic diaphragm (levator ani muscle) to become the anal canal.
         (1) The **rectal sling** is formed by the **puborectal muscle**, which is the most medial portion of the **pubococcygeal muscle** of the **levator ani** group (see Chapter 20 II A 2).
            (a) The puborectal muscle fibers loop posteriorly to the rectum with some fibers attaching to the rectum.
            (b) Tension in this muscle cants the rectum forward.
         (2) The rectal sling is the single most important factor in fecal continence.

   2. **Internal structure of the rectum.** Three transverse **rectal folds** (Houston's valves) are formed by the inner three layers of the intestinal wall within the rectum.
      a. **The inferior rectal fold** projects from the left side about 2 cm (1 inch) above the anal canal and can be palpated digitally.
      b. **The middle rectal fold** projects from the right about 2 cm (1 inch) above the inferior fold.
      c. **The superior rectal fold** projects from the left about 2 cm (1 inch) above the middle fold.

   3. **Vasculature of the rectum.** The arterial supply, venous return, and lymphatic drainage of the rectum are of considerable importance.

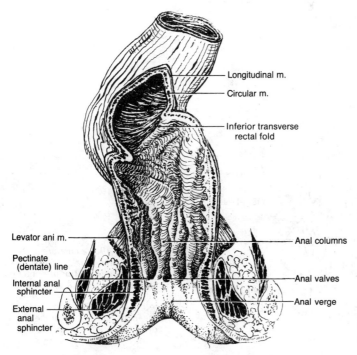

Longitudinal m.

Circular m.

Inferior transverse rectal fold

Levator ani m.

Pectinate (dentate) line

Internal anal sphincter

External anal sphincter

Anal columns

Anal valves

Anal verge

Figure 17-18. *Rectum and anal canal. The levator ani muscle marks the boundary between the rectum and the anal canal.*

a. Arterial supply
  (1) The rectum receives branches from the inferior mesenteric artey via the **superior rectal (hemorrhoidal) artery**, from the internal iliac artery via **middle rectal (hemorrhoidal) arteries** as well as from the internal pudendal arteries via **inferior rectal (hemorrhoidal) arteries.**
  (2) There are abundant anastomotic connections in the submucosa between tributaries of these arteries (Fig. 17-19).
b. Venous return is along similarly named vessels that have especially rich anastomotic connections.
c. Lymphatic drainage. Because the lymph vessels of the rectum parallel the arteries, metastatic carcinoma of the rectum may be disseminated widely within the abdomen and pelvis as well as the inguinal nodes. In addition, hematogenous spread to the liver occurs frequently.

4. Clinical considerations
  a. Carcinoma. The high incidence of rectal, colonic, and prostatic carcinoma makes digital and sigmoidoscopic examination important in the postmiddle-aged male.
    (1) Most rectal cancers can be discovered by digital palpation.
    (2) Carcinomatous spread from the rectum posteriorly may involve the sacral plexus with pain distribution down the leg and over the perineum. Anterior spread may involve pelvic viscera, such as the prostate, bladder, uterus, or vagina.
  b. Prolapse of the rectum through the anus is often related to damage of the levator ani muscle, usually the result of obstetric trauma.
  c. Hemorrhoids
    (1) Portal hypertension (accompanying cirrhosis) or local compression of the inferior mesenteric artery (because of chronic constipation with dilation of the sigmoid colon or because of the presence of the fetus during the later stages of pregnancy) results in the shunting of blood from the rectum and anal canal to the systemic venous system.
    (2) This shunting results in variceal dilation of the submucosal anal and perianal venous plexuses (hemorrhoids).
    (3) Hemorrhoidal varices may rupture during the evacuation of feces, producing considerable blood loss.
    (4) Unlike esophageal varices, hemorrhoidal varices are amenable to treatment by ligature or cautery.

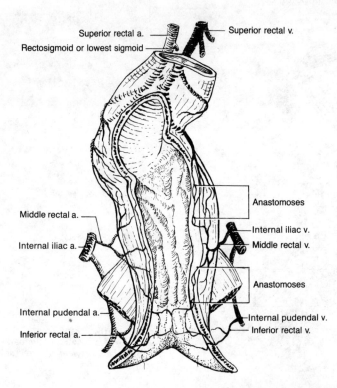

Superior rectal a.
Superior rectal v.
Rectosigmoid or lowest sigmoid
Anastomoses
Middle rectal a.
Internal iliac v.
Middle rectal v.
Internal iliac a.
Anastomoses
Internal pudendal a.
Internal pudendal v.
Inferior rectal v.
Inferior rectal a.

Figure 17-19. *Vasculature of the rectum and anal canal.* The arterial supply from the superior mesenteric artery (superior rectal artery) and the internal iliac arteries (middle and inferior rectal arteries) is depicted on the right side. The parallel venous drainage is shown on the opposite side.

V. VASCULAR SUPPLY TO THE GI TRACT

A. Celiac artery (trunk, axis) [see Fig. 17-3]

   1. Origin and course. The celiac artery leaves the aorta in the midventral line just caudal to the aortic hiatus of the diaphragm. It supplies that portion of the abdominal GI tract derived from the primitive foregut.

   2. Branches. Only 0.5–3 cm long, the celiac artery trifurcates almost immediately.

     a. The left gastric artery supplies the right superior region of the stomach as well as the abdominal portion of the esophagus (see Fig. 17-3). The left gastric artery anastomoses with the right gastric artery (from the common hepatic artery) along the lesser curvature.

       (1) It gives off the **esophageal branch**, which anastomoses with the midesophageal branches of the aorta. This artery provides the pathway for the **posterior vagus nerve** to reach the **celiac plexus**.

       (2) It may give rise to an aberrant **left hepatic artery**.

     b. The splenic artery supplies the spleen and the entire left side of the stomach as well as portions of the pancreas and greater omentum (see Figs. 17-3 and 17-10).

       (1) Short gastric arteries supply the left superior portion of the stomach and anastomose with the arteries of the esophagus.

       (2) The **left gastro-omental (gastroepiploic) artery** supplies the left inferior region of the stomach and portions of the greater omentum; it anastomoses with the right gastro-omental artery along the greater curvature.

       (3) Several **caudal pancreatic arteries** supply the tail of the pancreas.

     c. The common hepatic artery supplies the liver and gallbladder as well as portions of the stomach, duodenum, and pancreas (see Figs. 17-3 and 17-4).

       (1) The **right gastric artery** supplies the right inferior region of the stomach and anastomoses with the left gastric artery along the lesser curvature.

       (2) The **gastroduodenal artery** gives off two major branches.

         (a) The **right gastro-omental (gastroepiploic) artery** supplies the right inferior region of the stomach and a portion of the greater omentum. It anastomoses with the left gastro-omental artery (from the splenic artery) along the greater curvature.

         (b) The **superior pancreaticoduodenal artery** supplies the upper portion of the duodenum and the head of the pancreas.

           (i) This artery divides into anterior–superior and posterior–superior branches.

           (ii) It anastomoses with the anterior–inferior and posterior–inferior branches of the inferior pancreaticoduodenal arteries which, in turn, arise from the superior mesenteric artery.

       (3) The **proper hepatic artery** is the continuation of the common hepatic artery (see Fig. 17-9).

         (a) The **left hepatic artery** supplies the left lobe, the quadrate lobe, and the left half of the caudate lobe.

         (b) The **right hepatic artery** supplies the right lobe, the right half of the caudate lobe, and the gallbladder via the cystic artery.

           (i) The **cystic artery** is usually a branch of the right hepatic artery.

           (ii) The cystic artery bifurcates into the **anterior cystic artery** and the **posterior cystic artery**.

B. Superior mesenteric artery (see Figs. 17-10 and 17-13)

   1. Origin and course. The superior mesenteric artery leaves the aorta in the midventral line a few centimeters caudally to the celiac artery. It gives off numerous branches that supply that portion of the GI tract derived from the primitive midgut.

   2. Branches

     a. The inferior pancreaticoduodenal artery divides into **anterior** and **posterior** branches, which anastomose with anterior and posterior branches of the superior pancreaticoduodenal artery and supply the descending and inferior portions of the duodenum as well as the head of the pancreas.

     b. Jejunal and ileal branches provide abundant collateral circulation to the peritoneal portion of the small bowel.

     c. The ileocolic artery is the terminal branch of the superior mesenteric artery. Its branches include the **anterior** and **posterior cecal arteries**, the **appendicular artery**, the **ascending colic artery**, and the **ileal branches**.

     d. The right colic artery, which is frequently missing, supplies the ascending colon with great variation.

     e. The middle colic artery supplies the transverse colon.

3. Clinical considerations
  a. Intestinal angina. Stenosis of the vascular supply may produce intestinal angina (colicky abdominal pain referred to the umbilical region).
  b. Superior mesenteric infarct. Due to the pattern of blood supply, occlusion of the superior mesenteric artery by a thrombus results in infarction of the small intestine and of the large intestine as far as the vicinity of the splenic flexure.

C. Inferior mesenteric artery (see Fig. 17-17)

1. Origin and course.
The inferior mesenteric artery originates from the aorta distally to the gonadal arteries and supplies that portion of the GI tract derived from the primitive hindgut.

2. Branches
  a. The left colic artery supplies the descending colon; the ascending branch anastomoses with the left branch of the middle colic artery and the descending branch anastomoses with the sigmoid arteries.
  b. Sigmoidal arteries supply the sigmoid colon and anastomose with the descending branch of the left colic artery.
  c. Rectosigmoid arteries supply the terminal sigmoid colon and superior rectum, variably anastomosing with sigmoid and superior hemorrhoidal arteries at the critical point (of Sudeck).
  d. The superior hemorrhoidal (rectal) artery supplies the upper rectum; this artery has poor anastomotic connections with sigmoid arteries.

3. Clinical considerations.
Stenosis or occlusion of the inferior mesenteric artery results in ischemia and possible infarct of the descending colon and sigmoid colon. Colicky abdominal pain will be referred to the inguinal regions and anterior thigh.

D. Marginal artery (Fig. 17-20)

1. Origin and course.
Anastomoses between branches of the **superior mesenteric** and **inferior mesenteric arteries** result in the formation of a continuous arterial trunk, the **marginal artery** (of Drummond). The marginal artery lies in the mesentery, close to the border of the large intestine, and runs from the ileocecal valve to the rectosigmoid junction.

2. Clinical considerations.
There are several so-called ''critical points'' at which the marginal artery may not provide adequate collateral supply.

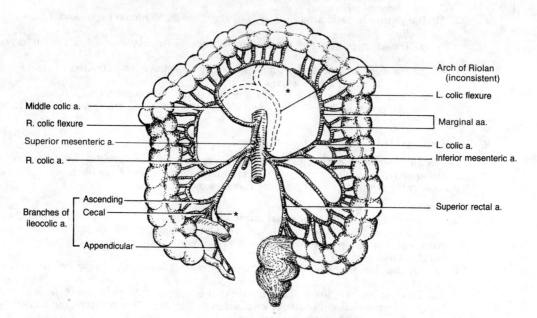

Figure 17-20. *Marginal artery.* Anastomoses occur between the branches of the superior mesenteric arteries as well as between the branches of the inferior mesenteric arteries. Inconsistent anastomoses occur in the regions of the ileocecal junction and the splenic flexure (*).

    a. Anastomoses between the ileal branches of the superior mesenteric artery and the ileal branches of the ileocolic artery seldom provide adequate collateral circulation in the ileocecal region.

    b. There may be few anastomotic connections between the left branch of the middle colic artery and the ascending branch of the left colic artery in the vicinity of the splenic flexure.

       (1) Occlusion or ligature of one of these vessels may result in ischemic necrosis of a segment of large bowel.

       (2) The arch of Riolan occasionally provides a strong anastomotic connection between the inferior mesenteric artery and the left branch of the middle colic artery (see Fig. 17-17).

    c. Anastomotic connection between the lowest sigmoid branch and the upper rectal branch of the superior rectal (hemorrhoidal) artery is sometimes by a very small vessel incapable of providing adequate collateral circulation in the vicinity of Sudeck's critical point.

VI. HEPATIC PORTAL SYSTEM

  A. Venous return from the GI tract. The hepatic portal system begins as the venous capillaries of the GI tract and ends as the venous sinusoids in the liver; this system delivers nutrient-rich (sometimes toxic) blood to the liver.

  B. Hepatic portal vein

    1. Course. The hepatic portal vein lies in the **hepatoduodenal ligament** just posterior and slightly to the left of the common bile duct and the proper hepatic artery (see Fig. 17-9).

    2. Distribution. This vein receives three major veins and numerous smaller veins (Fig. 17-21).

    a. The superior mesenteric vein drains the small intestine and large intestine as far as the splenic flexure.

       (1) The superior mesenteric vein generally parallels and drains the regions supplied by the superior mesenteric artery.

       (2) It is the largest contributor to the hepatic portal vein.

    b. The inferior mesenteric vein drains the descending and sigmoid colons.

       (1) This vein generally parallels and drains the regions supplied by the inferior mesenteric artery.

       (2) It usually enters the superior mesenteric vein but may enter the splenic vein.

    c. The splenic vein drains the spleen as well as portions of the stomach and pancreas.

       (1) The splenic vein generally parallels the splenic artery and may be embedded in pancreatic parenchyma.

       (2) This vein joins the superior mesenteric vein to form the hepatic portal vein.

    d. Celiac veins. Small veins corresponding to the named branches of the celiac artery drain directly into the hepatic portal vein.

  C. Hepatic veins

    1. Origin and course. The hepatic sinusoids coalesce into interlobular veins, which come together to form the hepatic veins.

    2. Variations. The hepatic veins may join before emptying into the inferior vena cava, or they may enter separately. There are usually three hepatic veins, draining the left, right, and quadrate lobes of the liver.

  D. Clinical considerations

    1. Portal hypertension. Because the hepatic portal system has no valves, blood need not flow toward the liver. Liver disease (such as cirrhosis) or compression of a vein (as in pregnancy or constipation) results in blood shunting through the anastomotic connections in the systemic venous system.

    a. Esophageal varices

       (1) The branches of the **left gastric vein** anastomose profusely with the **azygos**, as well as the **hemiazygos, veins.**

       (2) With portal hypertension, increased shunting through the esophageal anastomoses results in *esophageal varices* within the mucosa of the esophagus.

       (a) A varix may rupture when a large bolus is swallowed, upon emesis, or upon the passage of diagnostic instrumentation.

       (b) Uncontrollable and frequently fatal hemorrhage results from variceal rupture. Hematemesis associated with esophageal varices is the most common terminal event in alcohol-related liver disease.

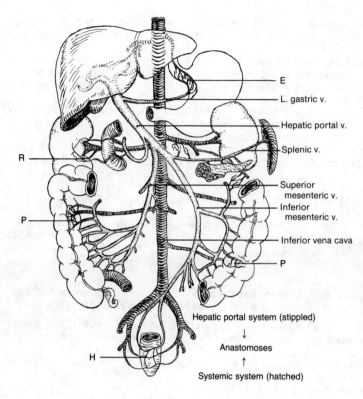

E

L. gastric v.

Hepatic portal v.

Splenic v.

R

Superior
mesenteric v.

Inferior
mesenteric v.

P

Inferior vena cava

P

Hepatic portal system (stippled)

↓

Anastomoses

↑

Systemic system (hatched)

H

Figure 17-21. *Hepatic portal system.* Most of the blood returns from the gut through the liver. Anastomoses occur in several regions: *E,* esophageal veins; *H,* hemorrhoidal veins; *P,* veins of Retzius between secondarily retroperitoneal structures and the veins of the abdominal wall; and *R,* veins between both the duodenum and renal veins and as well as between the spleen and renal veins. Varix development is possible in any of these anastomoses.

b. Caput medusae
(1) The **periumbilical veins** in the falciform ligament (and potentially even a recannulized **ligamentum teres**) anastomose abundantly with **superior** and **inferior epigastric veins** and other veins of the anterior abdominal wall.
(2) Although the grossly dilated veins of a caput medusae appear to be rare, this condition is always present in portal hypertension and can be demonstrated by infrared photography.
(3) These periumbilical anastomoses may bleed profusely during midline surgical approaches through the abdominal wall.
c. Hemorrhoids
(1) The **superior rectal vein** anastomoses profusely with the **middle rectal vein**, which is a branch of the **internal iliac vein**. In addition, the **middle rectal vein** anastomoses with the **inferior rectal vein**, which is a branch of the **internal pudendal vein**.
(2) Increased shunting through these anastomotic channels results in varices within the submucosa of the terminal rectum (**internal hemorrhoids**) and of the anal canal (**external hemorrhoids**).
d. Veins of Retzius
(1) The veins of a secondarily retroperitoneal structure may anastomose with the veins of the dorsal body wall, forming veins of Retzius.
(2) These varices usually are silent and produce few problems. However, they may bleed profusely when a secondarily retroperitoneal structure is mobilized surgically.

2. Surgical correction. The life-threatening **sequelae of portal hypertension** are amenable to surgical correction to provide routes for visceral blood to bypass the obstructed liver. Although shunts reduce the dangers inherent in varices, toxic blood shunted around the obstructed liver has a long-term effect on the central nervous system, causing such disorders as the encephalopathy that results from alcohol-related liver disease.
a. Surgical creation of a portacaval fistula produces a side-to-side anastomosis between the hepatic portal vein and the inferior vena cava, across the epiploic foramen.
b. A splenorenal anastomosis is also a common procedure. Splenectomy is followed by end-to-end anastomosis of the splenic vein to the left renal vein.

VII. LYMPHATIC DRAINAGE OF THE GUT

A. Lymphatics

1. **The lymphatic channels** generally follow the vascular system.

2. **Lymph nodes** are interposed in the lymphatic channels. These nodes are of great importance due to lymphatic spread of carcinoma of the GI tract.

B. Celiac nodes

1. **Location.** Celiac nodes are located around the celiac axis. There are four drainage areas about the stomach.

   a. Left gastric nodes
   - (1) These are located along the lesser curvature of the stomach.
   - (2) The left gastric nodes generally drain along left gastric vessels to the celiac nodes.
   - (3) Because they communicate with esophageal lymphatics, the left gastric nodes provide a potential route for spread of carcinoma from the stomach to the esophagus.

   b. Gastroepiploic nodes
   - (1) This group is located along the inferior portion of the greater curvature, pyloric antrum, and pyloric canal.
   - (2) The drainage is generally along the right gastroepiploic vessels toward the pyloric and celiac nodes.
   - (3) This pathway gains significance from the high prevalence of gastric carcinoma (the most common carcinoma in the region) and the high prevalence of duodenal and pancreatic carcinomas.

   c. Splenic nodes
   - (1) These are located along the lateral portion of the greater curvature and fundus.
   - (2) They generally drain along left gastroepiploic and splenic vessels to the celiac nodes.

   d. Right gastric nodes
   - (1) This group of nodes is located along the superior region of the pyloric antrum and the pyloric canal.
   - (2) The drainage is along the right gastric vessels to the celiac and hepatic nodes.

2. **Drainage course.** The celiac nodes drain into the **intestinal lymph trunk**, which, in turn, drains the cisterna chyli.

C. Superior mesenteric nodes

1. **Location.** Superior mesenteric nodes are located along the root of the superior mesenteric artery. These nodes receive drainage from the intermediate nodes, which are intercalated in lymph pathways along the named branches of the superior mesenteric artery.

   a. Mesenteric nodes are associated with the small intestine.

   b. Ileocolic nodes are associated with the cecum and appendix.

   c. Right colic nodes are associated with the ascending colon.

   d. Middle colic nodes are associated with the transverse colon.

2. **Drainage course.** Superior mesenteric nodes also receive drainage from the duodenum and pancreas. There are numerous anastomotic connections with the celiac and inferior mesenteric nodes. The superior mesenteric nodes and the celiac nodes give rise to the **intestinal trunk**, which drains into the **cisterna chyli**.

D. Inferior mesenteric nodes

1. **Location.** Intermediate nodes intercalated in lymph pathways along the named vessels include the following:

   a. Left colic nodes are associated with the descending colon and the sigmoid colon.

   b. Superior rectal nodes are associated with the rectum.
   - (1) There are profuse anastomotic connections with:
     - (a) **Middle rectal nodes**, which drain along the internal iliac vessels to the hypogastric nodes.
     - (b) **Inferior rectal nodes**, which drain along the external iliac vessels to the inguinal and hypogastric nodes.
   - (2) Carcinoma of the rectum may metastasize along abdominal as well as pelvic lymph channels, resulting in wide dissemination of the tumor.

2. Drainage course. Inferior mesenteric nodes drain into the superior mesenteric nodes. There is also considerable communication with the celiac nodes (from lymphatics, which follow the inferior mesenteric vein rather than the inferior mesenteric artery) as well as with the middle colic nodes.

E. Cisterna chyli

1. Location. The cisterna chyli receives the celiac and intestinal trunks and drains the **hypogastric nodes** from the systemic lymphatics of the lower body wall, lower extremities, and pelvis.

2. Drainage course. It drains into the **thoracic duct**, which empties into the left subclavian vein at the base of the neck.

VIII. INNERVATION OF THE GI TRACT

A. The autonomic nervous system, which innervates the GI tract, is subdivided into two divisions.

1. Parasympathetic (craniosacral) division (Fig. 17-22)

    a. The vagus nerve (CN X) brings parasympathetic presynaptic neurons to the GI tract almost as far as the splenic flexure of the transverse colon.

    (1) The vagus nerve reaches abdominal viscera via the esophageal plexus. Left and right vagal plexuses consolidate into anterior and posterior vagal trunks as they pass through the diaphragm at the esophageal hiatus.

    (a) The **anterior vagal trunk** (left vagus nerve) courses along the anterior surface of the abdominal esophagus.

      (i) Hepatic branches course through the lesser omentum to the **hepatic plexus** on the hepatic artery to innervate the liver, gallbladder, and bile duct. Some fibers course retrogradely along the hepatic artery to join the **celiac plexus** and innervate portions of the duodenum and the sphincter of Oddi.

      (ii) Gastric branches continue along the anterior surface of the stomach as far as the pylorus to innervate the smooth muscle and glands of the stomach.

      (iii) The **celiac branch** courses along the left gastric artery, contributing to and coursing within the **celiac plexus** and the **superior mesenteric plexus**. It supplies the small and large intestines approximately as far as the splenic flexure by coursing along the arteries.

    (b) The **posterior vagal trunk** (the right vagus nerve) courses along the posterior surface of the esophagus and stomach.

      (i) Gastric branches continue along the posterior surface of the stomach, not quite reaching the pylorus, to innervate smooth muscle and glands of the stomach.

      (ii) The **celiac branch** courses along the left gastric artery, contributing to and coursing through the **celiac plexus** and **superior mesenteric plexus**. It supplies the small and large intestines approximately as far as the splenic flexure by coursings along the arteries.

      (iii) The posterior vagal trunk does not appear to give rise to any hepatic branches.

    (2) Functional considerations

    (a) The vagus nerve is composed of *pre*ganglionic (*pre*synaptic) neurons.

      (i) These neurons synapse with *post*ganglionic (*post*synaptic) neurons within the enteric plexuses.

      (ii) Parasympathetic presynaptic neurons use acetylcholine as the neurotransmitter (cholinergic transmission).

    (b) The short *post*ganglionic neurons are located in the enteric plexuses within the walls of the gut. They synapse with effectors, such as smooth muscle cells and gland cells. Parasympathetic postsynaptic neurons are also cholinergic.

    (c) The parasympathetic division functions primarily to promote digestion.

      (i) Vagal activity increases acid, alkaline, and enzymatic secretion from the various regions of the GI tract.

      (ii) It enhances peristaltic activity and decreases sphincteric tone.

    (3) Clinical considerations

    (a) Selective vagotomy (section of the gastric branches of the anterior and posterior vagal trunks) reduces peptic secretion. This procedure is sometimes used as treatment for chronic gastric and duodenal ulcers.

    (b) Total vagotomy (section of the anterior and posterior vagi on the surface of the abdominal esophagus) has other sequelae, such as a decreased rate of stomach emptying, gallbladder dilation, biliary stasis with predisposition to stone formation, and dumping syndrome.

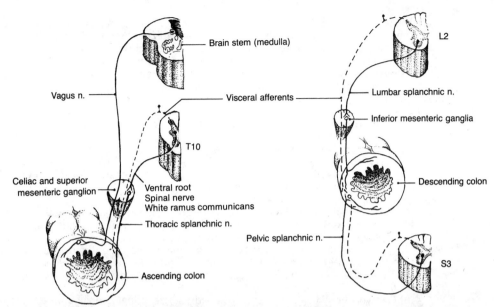

Figure 17-22. *Innervation of the gastrointestinal tract.* Preganglionic sympathetic neurons (*solid cell body*) in the thoracic spinal cord (levels T5–L2) send axons through the ventral roots, spinal nerves, white rami communicantes, the sympathetic chain, and splanchnic nerves to gain access to prevertebral ganglia. The postsynaptic sympathetic neurons (*open cell body*) of the prevertebral ganglia send their axons to the effectors in the gut walls. The cranial presynaptic parasympathetic neurons (*solid cell body*) in the brain send axons along the vagus nerve (CN X) to the enteric ganglia within the walls of the viscera as far as the splenic flexure. The sacral presynaptic parasympathetic neurons (*solid cell body*) in the spinal cord (levels S2–S4) send axons through the spinal nerves and pelvic splanchnic nerves to distribute to the enteric ganglia of the lower colon. The short parasympathetic postganglionic neurons (*open*) of the enteric ganglia send axons to the effectors. Afferent axons (*dashed*), which convey painful stimuli from the gut, travel along the autonomic pathways to both thoracolumbar and pelvic segments of the spinal cord.

    b. The nervi erigentes (sacral nerve, pelvic splanchnic nerves) represent the sacral portion of the parasympathetic division of the autonomic nervous system.

        (1) The nervi erigentes arise from S2, S3, and S4 (see Fig. 17-22) and innervate the lower gastrointestinal tract approximately between the splenic flexure and the pectinate line of the anal canal.

        (2) These nerves contribute to the **lateral pelvic plexus** and run in the adventitia of the rectum.

            (a) Some fibers course retrogradely along the rectum, sigmoid colon, and descending colon.

            (b) Other fibers course retrogradely in the **hypogastric plexus** as far as the branches of the inferior mesenteric artery, which they follow to the descending bowel.

        (3) The *pre*ganglionic neurons of the pelvic splanchnic nerves synapse with *post*ganglionic fibers in enteric plexuses, using cholinergic neurotransmission.

        (4) The short *post*synaptic neurons have their cell bodies in enteric plexuses.

            (a) These *post*ganglionic neurons synapse with effectors, such as smooth muscle cells and secretory cells.

            (b) The parasympathetic *post*synaptic neurons are also cholinergic.

        (5) Clinical considerations

            (a) *Aganglionosis (Hirschsprung's disease)* is a genetic defect wherein the parasympathetic ganglia do not develop in the walls of the descending colon.

            (b) In this condition, there is loss of initiation of propulsive function within the sigmoid colon—a functional blockage—with ensuing megacolon. With extreme stretching of the colon, ischemia with abdominal signs results. *Toxic megacolon* is usually treated surgically.

  2. Symptatic (thoracolumbar) division (see Fig. 17-22)

    a. *Pre*ganglionic sympathetic fibers arise from levels T1–L2, or sometimes, L3. All sympathetic *pre*ganglionic neurons use acetylcholine as their neurotransmitter (cholinergic transmission).

        (1) The **greater splanchnic nerve** arises from T5–T9 and synapses with postganglionic fibers in the **celiac ganglion.**

(2) The **lesser splanchnic nerve** arises from T10 and T11 and generally synapses with post-ganglionic fibers either in the sympathetic chain or in the **superior mesenteric ganglion**.

(3) The **least splanchnic nerve** arises from T12, may be incorporated into the lesser splanchnic nerve, and synapses with postganglionic fibers in the **aorticorenal ganglion**.

(4) **Lumbar splanchnic nerves** arise from L1–L2 (occasionally L3) and synapse with *post*-ganglionic fibers in small ganglia located in the **aortic plexus**.

b. *Post*ganglionic sympathetic neurons, which innervate the GI tract, lie in the prevertebral ganglia.

(1) The **prevertebral ganglia** are paired bilaterally and are variable in size—sometimes fused on either side to a variable extent.

(a) **Celiac ganglia** are located in the **celiac plexus** and supply sympathetic nerves to that portion of the viscera associated with the celiac artery.

(b) **Superior mesenteric ganglia** are located in the **celiac** or **superior mesenteric plexuses** and supply sympathetic nerves to that portion of the viscera associated with the superior mesenteric artery.

(c) **Aorticorenal ganglia** lie in the **celiac** or **aortic plexuses** and supply the kidneys.

(d) The **ganglia of the aortic plexus** (small ganglia within the adventitia of the aorta) supply viscera associated with the inferior mesenteric artery.

(2) The **postganglionic sympathetic fibers** contribute to the various named plexuses and follow the arteries to the effectors, usually the musculature of blood vessels.

(a) These fibers are generally long, compared with postganglionic parasympathetic fibers.

(b) *Post*synaptic sympathetic neurons are adrenergic, using norepinephrine as the neurotransmitter.

(3) **Adrenal medulla**

(a) The cells of the adrenal medulla are embryonic equivalents of sympathetic *post*ganglionic neurons; both are derived from neural crest tissue.

(b) The adrenal medulla secretes adrenaline and norepinephrine into the bloodstream, whereby they act on distant effector sites over a longer time period.

(c) It receives preganglionic innervation from greater, lesser, and least splanchnic nerves.

(4) **Functional aspects**

(a) Activity within the sympathetic division results in constriction of the blood vessels of the GI tract.

(b) Sympathectomy has no appreciable effect on the gut.

(i) It has been used to promote GI vasodilation in cases of extreme hypertension.

(ii) Sympathectomy has also been used to relieve intractable pain, because the visceral afferents travel along the sympathetic pathways.

B. Visceral afferent nerves

1. The visceral afferent nerves are not considered part of the autonomic nervous system, although they do share the pathways.

2. As a general rule, vagal pathways carry reflex afferents, whereas splanchnic pathways convey pain afferents.

   a. Afferent reflex neurons travel along with the vagus fibers to the brain (the gastric filling reflex), as well as along pelvic splanchnics to the spinal cord (the micturition reflex).

   b. Pain and pressure afferents travel along the thoracic splanchnics (sympathetic), lumbar splanchnics (sympathetic), and pelvic splanchnics (parasympathetic) to the spinal cord.

   (1) This arrangement explains the basis for **referred pain**.

   (a) Nociceptive input from the viscera at any thoracic or lumbar level of the spinal cord cannot be distinguished from that of somatic origin.

   (b) The pain is referred to the somatic dermatome associated with the particular splanchnic nerve.

   (2) Knowing approximate visceral innervation patterns is essential to understanding referred pain and making accurate diagnoses.

3. Because the pain fibers run along the sympathetic pathways, sympathectomy is undertaken to relieve visceral pain, as in untreatable carcinoma.

4. A summary of referred pain patterns is found in Table 17-2.

Table 17-2. Summary of Visceral Afferent Pathways

Organ	Afferent Pathway	Vertebral Level	Referred To
Heart	Middle inferior cardiac nerves Thoracic splanchnics	Cervical ganglia, thence T1–T2 T1–T4	Postaxial arm, high thorax, and mid-thorax postaxial arm
Esophagus Stomach Gallbladder Liver Bile duct Superior duodenum	Celiac plexus and greater splanchnic nerve	T5–T9	Low thorax and epigastric region
Inferior duodenum Jejunum Ileum Appendix Ascending colon Transverse colon	Superior mesenteric plexus and lesser splanchnic nerve	T10–T11	Umbilical region
Kidneys High ureters Gonads	Aorticorenal plexus, least splanchnic nerve, and upper lumbar splanchnic nerves	T12–L1	Lumbar and inguinal regions (unilaterally)
Descending colon Sigmoid colon Midureter Uterus	Aortic plexus and lumbar splanchnics	L1–L2	Pubic and inguinal regions, anterior scrotum or labia, and thigh
Cervix Bladder Low ureters Rectum Superior anal canal	Plevic plexus and pelvic splanchnics	S2–S4	Leg, foot, and perineum

Kidneys and
Posterior Abdominal Wall

I. KIDNEYS

A. Introduction

1. **Location.** The paired kidneys are located posteriorly in the abdominal cavity.
   a. The kidneys are retroperitoneal, embedded in a considerable amount of perinephric and paranephric fat.
   b. About the size of a clenched fist, each kidney normally weighs 150 ± 25 g in men and 135 ± 20 g in women.

2. **Relations.** Each kidney lies on the ventral surface of the quadratus lumborum muscle, just lateral to the psoas muscle and vertebral column. Both are in contact posterosuperiorly with the inferior surface of the diaphragm and are capped anteromedially by an adrenal gland.
   a. **The right kidney** is 2–8 cm (1–2 inches) lower than the left, due to the presence of the liver on the right side. It is usually related to the 12th rib, posteriorly and is related to the liver, duodenum, and hepatic flexure of the colon, anteriorly.
   b. **The left kidney** is related to the 11th and 12th ribs, posteriorly and is related to the pancreas, spleen, and splenic flexure of the colon, anteriorly.

B. Renal (perirenal) fascia (of Gerota; false capsule)

1. **Structure.** A discrete fascial layer surrounds each kidney (see Fig. 18-5).
   a. **Origin.** This layer arises from the prevertebral fascia and passes anteriorly to the vertebral column.
   b. **Divisions.** This fascial layer splits into two leaflets, which lie anteriorly and posteriorly to the kidney.
      (1) Laterally and superiorly to each kidney, the leaflets fuse, thereby partially enclosing each kidney with its associated adrenal gland.
      (2) The renal fascia divides the fat associated with the kidney into two distinct regions.
         (a) **Perirenal (perinephric) fat** is contained within the renal fascia.
         (b) **Pararenal (paranephric) fat** is external to the renal fascia; it is found most abundantly posterolaterally.
      (3) Although the renal fascia fuses around the renal vessels, it is open inferiorly along each ureter, forming the **periureteral sheath**.
         (a) Infection or renal abscess may be limited by the renal fascia but may track between the leaflets of this fascia into the pelvis.
         (b) Air injected into the retroperitoneal tissue of the pelvis will rise within the renal fascia to surround the kidney (insufflation), enabling radiographic visualization of the kidneys and adrenal glands.

2. **Support**
   a. **Mobilization.** Normally, kidneys do not exceed a 9-cm (3–4-inch) excursion when an individual changes position from supine to erect.
   b. **Position** of the kidneys is maintained primarily by deposits of paranephric and perinephric fat.
      (1) This fat (similar to periorbital and ischiorectal fat) is very slow to be reabsorbed in wasting disease or acute starvation.
      (2) Renal vessels provide some support, although it is uncertain how much.
   c. **Nephroptosis** (floating kidney)
      (1) This condition occurs frequently among truck drivers, horseback riders, and motorcyclists.
      (2) An aberrant inferior polar artery may compress the ureter with resultant *hydronephrosis*.
      (3) The pain of nephroptosis is due, apparently, to traction on the renal vessels.

C. Renal capsule

1. **Structure.** The renal capsule, the true fibrous capsule of the kidney, is stripped readily from a normal kidney but is adherent in certain pathologic conditions in which scarring has occurred.

2. **Function.** The renal capsule provides a barrier against the spread of infection.

D. Renal sinus

1. **Location.** The renal sinus, the cavity of the kidney, opens medially at the **hilus** (Fig. 18-1).

2. **Contents.** Most structures enter and leave the kidney at the hilus.
   a. **The renal artery** branches extensively within the sinus to supply **renal segments**.
   b. **The renal vein** leaves the kidney at the renal hilus.
   c. **The renal pelvis**, the funnel-shaped proximal end of the ureter, is continuous with the ureter proper distally and with the major calyces within the kidney.
   d. **Perirenal fat** fills the spaces between the various structures within the renal sinus.

E. Internal structure of the kidney (see Fig. 18-1)

1. **Divisions.** The renal parenchyma is divisible into two parts.
   a. **The cortex** is composed primarily of renal corpuscles, proximal convoluted tubules, and distal convoluted tubules. It also extends between the renal pyramids as **renal columns** (of Bertini).
   b. **The medulla** is formed by 9–14 **renal pyramids** (of Malpighi). It consists primarily of straight tubules, Henle's loops, and collecting ducts.

2. **Subdivisions.** The kidney is subdivided into lobes.
   a. Each **renal pyramid** with the overlying cortex and adjacent regions of renal columns forms the basic functional unit of the kidney (the lobe). Fusion of lobes (pyramids) at the apices forms 8–18 renal papillae.
   b. Each **renal papilla** opens into the cup-shaped end of a minor calyx, which tends to be arranged either anteriorly or posteriorly.

3. **Collecting system**
   a. **Minor calyces** fuse and join to form 2-4 major calyces.
   b. **Major calyces** join to form the funnel-shaped renal pelvis.
   c. **The renal pelvis** narrows to form the ureter.

F. Renal vasculature (Fig. 18-2)

1. **Renal arteries.** The blood supply to the kidneys is profuse with the kidneys receiving approximately one-fifth of the cardiac output.
   a. **Origin and course.** The left and right renal arteries arise from lateral aspects of the aorta, usually between L1 and L2 just below the origin of the superior mesenteric artery. The right renal artery passes posteriorly to the inferior vena cava (see Fig. 18-2).

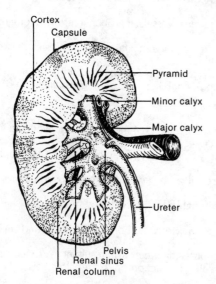

Figure 18-1. *Right kidney.* The division of the renal parenchyma into the cortex and medulla is evident. Renal pyramids empty into calyces, which coalesce into the renal pelvis and ureter.

b. Extrarenal branches. Left and right renal arteries give rise to:
    (1) Inferior suprarenal arteries to the adrenal glands
    (2) Numerous twigs to the ureters
    (3) Occasional **gonadal arteries** (15%)
    (4) Occasional **inferior phrenic arteries** (10%)
c. Intrarenal branches. Renal arteries divide into **segmental arteries** within the renal sinus.
    (1) Renal vascular segments. This division defines five renal segments.
        (a) Superior (apical)
        (b) Anterior–superior (upper)
        (c) Anterior–inferior (middle)
        (d) Inferior (lower)
        (e) Posterior
    (2) Renal divisions. The distributions of the anterior–superior and anterior–inferior branches, as distinct from the posterior branch, divide the kidney into **anterior** and **posterior divisions.**
        (a) There are few anastomoses between the anterior segments and the posterior segment.
          (i) A longitudinal incision along the **avascular line** (Brödel's white line) produces minimal damage to the blood supply. This approach is used for removal of renal (staghorn) calculi.
          (ii) The apical and lower segments have a more variable blood supply.
        (b) Ligation of a segmental artery results in necrosis of the entire segment.
        (c) Although the possibility of segmental resection for segmentally localized kidney disease is appealing, it seems yet to be of marginal practicality.
    (3) Interlobar arteries arise from each segmental artery (see Fig. 18-2).
        (a) Each interlobar artery enters a renal column.
        (b) Each gives off **arcuate arteries**, which run across the bases of the pyramids between the cortex and medulla.
          (i) Arcuate arteries give off **interlobular arteries**, which course between the medullary rays.
          (ii) Interlobular arteries supply the **afferent glomerular arterioles.**
d. Vascular variations. Aberrant or supernumerary segmental arteries are common (32%). They are derived from the fetal lobation pattern with failure of the renal arterial segments to fuse into a single renal artery.
    (1) About half of all aberrant segmental arteries are hilar (arising from the renal artery) and the other half, polar (arising from the aorta).
        (a) Polar arteries, arising directly from the aorta, tend to be larger than hilar arteries, which arise from the renal vessels. **Lower polar arteries** typically pass anteriorly to the ureter. Compression of the ureter by an inferior polar artery produces *hydronephrosis.*
        (b) Supernumerary renal arteries, cannot be considered collaterals because each supplies one segment. Ligation almost always produces necrosis of a renal segment.
    (2) Angiographic studies prior to renal surgery are usually advisable. Lack of awareness of arterial variations can result in overwhelming hemorrhage at surgery. If torn, a polar artery usually separates where it arises from the aorta, leaving no stump to clamp and making control exasperatingly difficult.

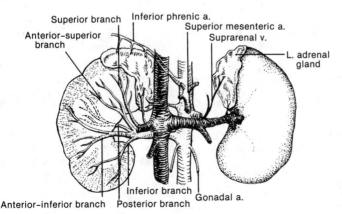

Superior branch — Inferior phrenic a.
Anterior–superior branch — Superior mesenteric a. — Suprarenal v. — L. adrenal gland
Anterior–inferior branch — Inferior branch — Posterior branch — Gonadal a.

Figure 18-2. *Blood supply and venous drainage of the kidneys and adrenal glands.* The five segmental arteries of the right kidney are shown as are the three arteries supplying the right adrenal gland. The venous drainage of the left kidney and adrenal gland is depicted on the opposite side.

2. Renal veins (see Fig. 18-2)
   a. The right renal vein enters the inferior vena cava at a lower point than the left renal vein. It usually has no significant tributaries.
   b. The left renal vein is longer than the right renal vein and passes anteriorly to the aorta. It receives the left gonadal vein, left suprarenal vein, and left inferior phrenic vein, and communicates with the azygos vein.
   c. Variations
      (1) Multiple renal veins are less common (14%) than supernumerary arteries (32%).
      (2) The most common variation is doubling of a renal vein (6%).

G. Renal innervation

   1. Motor pathways. Sympathetic preganglionic nerves from T12–L2 pass through the sympathetic chain ganglia and run in the least splanchnic nerve and lumbar splanchnic nerves.
   a. Postganglionic cells lie in aorticorenal ganglia and course within the renal plexus to the kidneys.
   b. Sympathetic nerves principally supply the renal vasculature.
      (1) With excessive sympathetic stimulation, urine production ceases.
      (2) Sympathectomy, interrupting input to the aorticorenal ganglion, produces renal vasodilation and transient renal diuresis.

   2. Afferent pathways. Visceral afferent nerves from the kidneys reach the spinal cord via the least splanchnic nerve (T12) and the lumbar splanchnic nerves (L1–L2) along the corresponding white rami communicantes. Pain derived from the kidney, its vascular supply, and upper ureter is referred to the T12–L2 dermatomal distribution (lumbar and inguinal regions as well as the anterosuperior thigh).

H. Renal function

   1. Electrolyte balance. One of the functions of the kidney is the conservation of minerals and the maintenance of ionic balance in the body fluids.
   a. Each kidney contains 10^6 or more nephrons, which comprise the basic morphologic and physiologic units of the kidney.
      (1) An ultrafiltrate of plasma (provisional urine) exudes through the glomerular capillaries into Bowman's capsule, the beginning of the nephron.
      (2) Along the proximal tubule, select substances are absorbed from the provisional urine, resulting in osmotic uptake of water. In addition, some substances are secreted into the provisional urine by the tubular epithelium.
      (3) In the distal tubule, sodium and water may or may not be absorbed, depending on the circulating concentrations of antidiuretic hormone and aldosterone.
      (4) The fluid that remains passes into the collecting ducts as urine.
      (5) About 500 collecting ducts, distributed within 9–14 pyramids in the normal adult kidney, terminate in 7–9 minor calyces.
   b. Although approximately 150–200 L of provisional urine are produced each day, reabsorption results in only 1–2 L of urine to be voided by micturition.

   2. Blood pressure regulation. The kidneys also produce vasoactive substances, which control blood pressure.

I. Clinical considerations

   1. Systemic hypertension. A kidney with a stenotic or occluded renal artery, or one that is nonfunctional with respect to production of urine, will produce an overabundance of angiotensin with resultant systemic hypertension.

   2. Intravenous pyelography. The renal outflow tract can be visualized radiographically by intravenous pyelography (IVP). Contrast material, injected intravenously, is excreted by the kidney and concentrated in the outflow tract where it appears radiopaque. X-rays taken at intervals demonstrate this process.

   3. Kidney stones. Urinary components (calcium compounds and urea) may salt out in the calyces, and concretions (**nephroliths**) may develop. Usually, nephroliths are small enough to pass through the ureter and be eliminated. However, they may become large enough to lodge within and obstruct a ureter. Large staghorn concretions may completely occlude a calyx.

II. URETERS

A. Origin and course of the ureters. The ureters are the excretory ducts between the kidneys and the urinary bladder (Fig. 18-3).

1. **Within the kidney**, the **renal pelvis**, lying in the renal sinus, narrows at the ureteropelvic junction to form the ureter.

2. **In the abdomen**, each ureter descends retroperitoneally within the **periureteral sheath**, which is an extension of the renal fascia, anterior to the psoas muscle and dives over the brim of the deep pelvis.

3. **In the pelvis**, each ureter courses retroperitoneally in the endopelvic fascia anterior to the sacrum.
   a. After crossing the common iliac vessels, each ureter receives vascular twigs from the various iliac arteries.
   b. At this location, the ureter may be affected by aneurysms of the iliac artery.
   c. Because this is one of the narrowest parts of the ureter, calculi may lodge here with excruciating pain referred to the inguinal and pubic regions, the thigh, and the anterior scrotum or labia (L1–L2).

B. Relations of the ureters

1. **The right ureter** has the following relationships in the abdomen:
   a. It is posterior to the descending portion of the duodenum.
   b. It lies posteriorly to the root of the mesentery proper, containing the ileocolic and right colic vessels.
   c. It is posterior to the right gonadal vessels, which contribute important vascular twigs to the right ureter.

2. **The left ureter** has the following relationships in the abdomen:
   a. It lies posteriorly to the left colic vessels.
   b. It is adjacent to the left gonadal vessels, which contribute important vascular twigs to the left ureter.
   c. It is posterior to the sigmoid mesocolon, containing the sigmoidal arteries and superior rectal artery

3. **Both ureters** have the following relationships within the deep pelvis:
   a. In males, the ureters pass inferiorly to the vas deferens.
   b. In females, the ureters pass inferiorly to the cardinal ligaments and uterine vessels, where they may be inadvertently clamped or sectioned along with uterine vessels.
   c. The ureters converge to enter the bladder posteroinferiorly.
   d. The diameter of the ureter decreases as the ureter passes through the bladder wall, providing another point at which a nephrolith may become lodged.

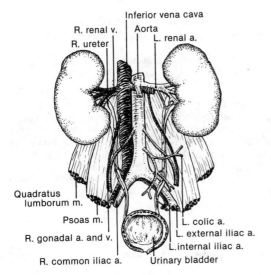

Figure 18-3. *Ureters and urinary bladder.* The blood supply to the ureters is from numerous sources.

C. Ureteral vasculature

1. **Arterial supply.** The blood supply to the ureters is very diffuse and variable. Each ureter is serviced by small arterial twigs from the renal arteries, aorta, small arteries of the posterior abdominal wall, gonadal arteries, and common and internal iliac arteries as well as the inferior vesical arteries (see Figs. 18-3 and 22-1).

2. **Mobilization** of a ureter, or even traction upon a ureter during surgery, should be avoided, because interruption of the delicate blood supply results in ischemic necrosis of the ureter.

D. Innervation of the ureters

1. **Motor pathways**
   a. **The least splanchnic nerve** from T12, with synapses in the aorticorenal ganglion, supplies the sympathetic innervation to the renal pelvis.
   b. **The lumbar splanchnic nerves** from L1–L2 supply sympathetic innervation to the abdominal and pelvic portions of the ureter.
   c. **The pelvic splanchnic nerves** from S2–S4 supply parasympathetic innervation to the entire ureter.

2. **Visceral afferent nerve fibers** from the various regions of each ureter travel to the spinal cord via the closest splanchnic nerve. Pain originating from specific regions of the ureter is referred accordingly (see Table 17-2).
   a. Upper ureteral obstruction and distension cause pain to be referred to the lumbar region (T12 and L1).
   b. Middle ureteral obstruction and distension cause pain to be referred toward inguinal and pubic regions as well as to the anterior scrotum or mons pubis and to the superoanterior thigh (L1 and L2).
   c. Lower ureteral obstruction and distension cause pain to be referred to the perineum and sometimes to the posterior thigh and leg (S2–S4).

E. Clinical considerations

1. **Urination.** The ureters display intrinsic peristaltic activity.
   a. The ureters contain a thick muscular wall, composed of circular and longitudinal layers, which moves the urine toward the bladder by waves of contraction (**urination**) that occur one to six times a minute.
   b. Evidence of normal peristaltic activity is seen on an intravenous pyelogram.

2. **Nephroliths**
   a. The ureter narrows at the junction between the renal pelvis and the ureter proper, at the point where the ureter crosses the pelvic brim, and again where the ureter passes through the wall of the bladder.
      (1) A stone may lodge or pass slowly through the ureter in these three regions.
      (2) There is ureteral distension (hydronephrosis) proximal to the stone.
   b. The distension causes excruciating pain (renal colic), which is purported to be among the most intense pains experienced (see Table 17-2).
   c. When a kidney stone obstructs a ureter, hydronephrosis results. If the stone fails to move despite copious imbibition, hydronephrosis may produce kidney damage and, thus, require treatment.
      (1) Instrumentation can be passed through the urethra and urinary bladder into the obstructed ureter to fragment the stone.
      (2) Alternatively, the ureter may be approached by surgical incision.

3. **Anomalies.** Renal and ureteral anomalies are rather frequent (4%).
   a. Anomalies in renal development include **polycystic kidneys**, in which the collecting tubules fail to join a calyx; **horseshoe kidney**, in which the two kidneys are joined across the midline; **lobated kidneys**, which maintain their fetal pattern of segmentation; and **ectopic kidneys**, which may be located in the pelvis or both on the same side of the body.
   b. There are numerous branching patterns possible with respect to the ureter.
   c. Dural ureters have a slower flow of urine than normal ureters and are, therefore, more susceptible to infection, which can spread to the kidneys from the urinary bladder.

III. ADRENAL (SUPRARENAL) GLANDS

A. Location. The **adrenal glands** lie anteromedially to the kidneys between the kidneys and the diaphragm.

B. Internal structure of the adrenal glands. The **adrenal parenchyma** is divided into two zones.

    1. The adrenal cortex is the outer zone. It is derived from the mesoderm of the embryonic urogenital ridge, and it produces three classes of steroid hormones.
        a. Mineralocorticoids (aldosterone), necessary for salt metabolism
            (1) Deficiency of mineralocorticoids results in *Addison's disease.*
            (2) Adrenal hyperplasia with excessive mineralocorticoid secretion results in *Conn's syndrome.*
        b. Glucocorticoids (such as cortisone)
            (1) These affect metabolism, especially that of the connective tissues.
            (2) Adrenal hyperplasia with excessive glucocorticoid production leads to *Cushing's syndrome;* insufficiency leads to *Addison's disease.*
        c. Male and female sex hormones
            (1) Hyperplasia with excessive secretion of sex hormones produces the *adrenogenital syndrome.* Depending upon the hormone produced, the result is either masculinization of females or feminization of males.
            (2) Sex hormones produced by the adrenals are metabolized in the liver; therefore, liver dysfunction in males (often associated with alcohol-related liver disease) results in *gynecomastia* and the development of a feminine escutcheon.

    2. Adrenal medulla is the inner zone. It is derived from embryonic neural crest tissue, which also gives rise to sympathetic ganglia and consists primarily of chromaffin (pheochrome) cells.
        a. The adrenal medulla is innervated variably by the greater, lesser, least, and lumbar splanchnic nerves.
        b. It secretes epinephrine into the bloodstream in response to cholinergic stimulation.

B. Adrenal vasculature

    1. Adrenal arteries arise from three sources (see Fig. 18-2).
        a. Superior suprarenal arteries (one or more) arise from each inferior phrenic artery.
        b. Middle suprarenal arteries arise from the aorta on each side.
        c. Inferior suprarenal arteries (one or more) arise from each renal artery.

    2. One suprarenal vein (see Fig. 18-2) exits at the hilus of each adrenal gland.
        a. The right adrenal vein usually drains into the inferior vena cava.
        b. The left adrenal vein usually drains into the left renal vein.

C. Clinical considerations

    1. Pheochromocytoma is a tumor of the adrenal medulla. Excessive bursts of epinephrine and norepinephrine, which result in paroxysms of hypertension, may be released from these tumors upon sympathetic activation or even abdominal palpation.

    2. Adrenalectomy. Great care must be taken during adrenalectomy. The suprarenal vein must be ligated as soon as practical before manipulation of the gland so that catecholamines do not escape into the circulation. The right adrenal gland is more difficult to approach surgically than the left because it is in part posterior to the inferior vena cava.

    3. Ectopic adrenal tissue
        a. Ectopic adrenal tissue usually consists of both cortex and medulla.
        b. Pheochromocytomas may occur anywhere along the embryonic urogenital ridge.

IV. POSTERIOR ABDOMINAL WALL

A. Structure of the posterior abdominal wall. The posterior abdominal wall is formed by the diaphragm, the bilateral quadratus lumborum muscle, the iliopsoas muscle, and the thoracolumbar fascia (Table 18-1).

    1. The diaphragm is a dome-shaped muscular sheet separating the thoracic from the abdominal cavity (Fig. 18-4; see Table 18-1). The upper surface is covered by parietal pleura and pericardium; the lower surface is covered by parietal peritoneum.
        a. Attachments
            (1) The sternal origin is the posterior surface of the xiphoid process.
            (2) The costal origin comprises the inner surfaces of the costal margin (costal cartilages 7–9 as well as the tips of ribs 10–12).
            (3) The lumbar origins are two **crura** arising from the lateral aspects of the upper three or four lumbar vertebrae.

Table 18-1. Muscles of the Posterior Abdominal Wall

Muscle	Origin	Insertion	Action	Innervation
Diaphragm	Costal margin, L1–L4	Central tendon	Lower diaphragm	Phrenic n. (C3–C5) and twigs from nn. T12–L2
Iliopsoas	Iliac fossa and transverse processes T12–L4	Lesser trochanter of femur	Flexes vertebral column and thigh	Twigs from nn. T12–L4
Quadratus lumborum	Iliac crest, transverse processes L4–L5	Rib 12	Stabilizes and lowers 12th rib; abducts vertebral column	Twigs from nn. T12–L4

  (a) The **lateral arcuate ligament** is formed by the free edge of the diaphragm between its attachments at the tip of the 12th rib and the transverse process of L1. Beneath it runs the quadratus lumborum muscle.
  (b) The **medial arcuate ligament** is formed by the free edge of the diaphragm between the transverse process of L1 and vertebral bodies of L1 and L2. Beneath it runs the most cephalad portion of the psoas muscle.
  (4) The muscle fascicles insert into the **central tendon** of the diaphragm.
  b. Action. Diaphragmatic contraction pulls the central tendon caudally, thereby increasing thoracic volume and decreasing intrathoracic pressure.
  c. Weak areas
  (1) The **sternocostal triangle** (foramen of Morgagni) is located between the sternal and costal portions.
    (a) This area transmits superior epigastric vessels.
    (b) It is a potential site for herniation (usually acquired).
  (2) The **lumbocostal triangle** (foramen of Bochdalek) is located between costal and lumbar portions.
    (a) This area may be nonmuscular with only two layers of serous membrane separating the thoracic from the abdominal cavity.
    (b) It is a potential site for herniation of a peritoneal sac of the abdominal viscera.
    (c) Although the stomach is the most commonly herniated viscus, nearly every abdominal organ (except the sigmoid colon) has been been reported herniated into the thorax.
    (d) Diaphragmatic hernia is 8–10 times more common on the left than on the right.
  d. Hiatuses
  (1) The **aortic hiatus** is located in approximately the midline between the diaphragmatic crura and transmits the aorta, azygos vein, thoracic duct, and occasionally, splanchnic nerves.
  (2) The **caval hiatus** (foramen of the inferior vena cava) is located to the right of the midline and transmits the inferior vena cava.

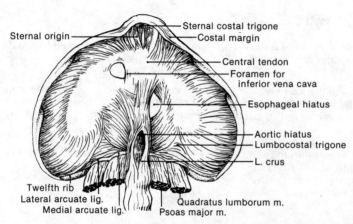

Figure 18-4. *Diaphragm.* The inferior surface of the diaphragm with the three hiatuses is depicted.

    (3) The **esophageal hiatus** transmits the esophagus and vagal trunks.
        (a) The esophageal hiatus is located slightly left of the midline in the muscular portion of the diaphragm.
        (b) It is the site of *hiatus hernia*.
            (i) Hiatus hernia is relatively common (1%), accounting for 98% of all diaphragmatic hernias.
            (ii) The cardiac end of the stomach may herniate through this hiatus into the thorax (sliding hiatus hernia).
    (4) The splanchnic nerves and the hemiazygos vein usually pierce the crura.
  e. Innervation of the diaphragm
    (1) The diaphragm develops in the cervical region along with the heart. Motor innervation (except for the crural regions) is provided by the phrenic nerves (C3–C5).
    (2) Sensory innervation is dual.
        (a) Sensation from the larger central region is carried by phrenic nerves to segments C3–C5 of the spinal cord. Irritation of the diaphragmatic peritoneum (such as by free air or blood) is referred to the shoulder region.
        (b) Sensation from the peripheral region is supplied by intercostal nerves. Pain from the periphery of the diaphragm is referred to the thoracic and abdominal wall.

2. The quadratus lumborum muscle is the most lateral muscle of the posterior abdominal wall (see Fig. 18-4 and Table 18-1; Fig. 18-5).
  a. Attachments. This muscle originates from the iliac crest and transverse processes of vertebrae L3–L5. It passes beneath the lateral arcuate ligament of the diaphragm to insert into the inferior border of the 12th rib and the transverse processes of L1.
  b. Actions of the quadratus lumborum muscle include lateral flexion of the vertebral column as well as fixation of the 12th rib during inspiration and depression of the 12th rib during forced expiration.

3. The iliopsoas muscle has two separate muscular heads (see Fig. 23-4 and Table 18-1).
  a. Attachments
    (1) The **psoas muscle** originates from lateral aspects and transverse processes of vertebral bodies and discs T12–L4 (see Figs. 18-4 and 18-5).
    (2) The **iliac muscle** originates from the iliac fossa.
    (3) Iliac and psoas heads fuse inferiorly to form the **iliopsoas muscle**, which inserts into the lesser trochanter of the femur.
  b. Actions. This muscle acts to flex the pelvis and the thigh at the hip joint.
  c. Relations of the psoas muscle
    (1) The lumbar plexus passes through the psoas muscle.
    (2) The kidneys lie anterolaterally to the psoas muscle.
    (3) The ureters lie anteriorly to and cross the psoas muscle to enter the deep pelvis.
    (4) The sympathetic chain is located anteromedially to the psoas muscle.
    (5) The common iliac vessels are medial to the psoas muscle.
  d. Psoas fascia
    (1) The psoas muscle is covered with a dense fascia derived in part from the lumbodorsal fascia (see Fig. 18-5).
    (2) Infections associated with the psoas muscle or vertebral column may be limited by the psoas fascia and track into the thigh.

4. Lumbodorsal (thoracodorsal) fascia (see Fig. 18-5)
  a. Origin. The posterior abdominal wall lateral to the quadratus lumborum muscle consists largely of **lumbodorsal fascia**, which is formed by the fusion of the posterior aponeuroses of the internal oblique and transverse muscles of the abdomen.
  b. Divisions. This fascia splits into three leaves that form two muscular compartments.
    (1) The posterior leaf inserts into the vertebral spinous processes and provides the site of origin for the latissimus dorsi muscle.
    (2) The middle leaf inserts into the tip of the vertebral transverse processes. With the posterior leaf, it encloses the erector muscle of the spine.
    (3) The anterior leaf inserts midway along the transverse processes.
        (a) With the middle leaf, it encloses the quadratus lumborum muscle.
        (b) The psoas fascia arises about midway along the middle leaf.
  c. The lumbar triangle (of Petit) is bounded by the posterolateral free edge of the external oblique muscle, the anterolateral border of the latissimus dorsi muscle, and the superior aspect of the iliac crest.
    (1) The floor of the trigone (triangle) consists of the lumbodorsal fascia and, sometimes, fibers of the internal oblique muscle.
    (2) Since the abdominal wall is thinner here, this trigone may be the site of a lumbar hernia. It is used in surgical approaches to the kidneys and ureters.

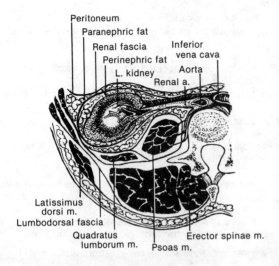

Peritoneum
Paranephric fat
Renal fascia
Perinephric fat
L. kidney
Renal a.
Inferior vena cava
Aorta
Latissimus dorsi m.
Lumbodorsal fascia
Quadratus lumborum m.
Psoas m.
Erector spinae m.

Figure 18-5. *Posterior abdominal wall and renal fascia.*

B. Posterior abdominal vessels

1. Aorta (see Fig. 18-3)

 a. Course. The aorta enters the abdomen via the aortic hiatus between the diaphragmatic crura. It is located on the anterior aspect of the vertebral column, deviating slightly to the left during passage toward the bifurcation.

 b. Relations

  (1) The aorta lies to the left of the inferior vena cava.

  (2) It is associated with the celiac and superior mesenteric plexuses and the included ganglia as well as with the aortic, inferior mesenteric, and hypogastric plexuses.

 c. Branches. Before bifurcating into the common iliac arteries, branches of the abdominal aorta include the inferior phrenic arteries, the celiac artery, the superior mesenteric artery, renal arteries, gonadal arteries, numerous twigs to the ureters, the inferior mesenteric artery, lumbar arteries, and the middle sacral artery.

 d. Clinical considerations

  (1) Aneurysms of the aorta commonly involve the origins of the aortic branches. A large aneurysm may erode into an adjacent vertebra and produce back pain as the principal symptom.

  (2) Coarctation (occlusion) of the aorta superior to the renal arteries is usually fatal, because the kidneys require far more blood than can flow through collateral vessels. Slow occlusion below the renal arteries results in progressive development of collateral pathways.

   (a) A major shunt between the internal thoracic artery and the inferior epigastric artery will develop with enlargement of the intercostal and abdominal arteries.

   (b) The marginal artery of Drummond may hypertrophy if occlusion occurs between the origins of the superior and inferior mesenteric arteries.

2. Inferior vena cava (see Fig. 18-3)

 a. Course. The inferior vena cava lies to the right of the aorta along the anterior aspect of the vertebral column. It enters the thorax via the foramen of the inferior vena cava.

 b. Branches. This vein receives the common iliac veins, lumbar veins, right gonadal vein, renal veins, right adrenal vein, right inferior phrenic vein, and hepatic veins.

 c. Variations. The inferior vena cava develops from three distinct embryologic venous pathways that fuse and atrophy in various places.

  (1) There are numerous collateral pathways and variations.

  (2) The vein may be absent, its drainage functions served by an enlarged azygos system entering the thorax via the aortic hiatus in the diaphragm.

C. Lymphatic channels

1. Common iliac nodes are adjacent to the common iliac vessels.

 a. They receive drainage from the lower limbs, the perineum, and the gluteal region via the **inguinal nodes**.

 b. Three nodes drain into the aortic nodes.

2. **Aortic (lumbar) nodes** lie adjacently to the aorta and form the lumbar lymphatic trunk, often one on each side of the aorta.
   a. They receive drainage from the posterior abdominal wall and iliac nodes.
   b. They drain into the cisterna chyli.

3. **The cisterna chyli** lies posteriorly and a little to the right of the aorta, usually between the crura of the diaphragm at the aortic hiatus.
   a. It receives drainage from the intestinal lymphatic trunk and from the celiac and superior mesenteric nodes as well as from the aortic (lumbar) lymphatic trunks.
   b. It represents the lower expanded portion of the thoracic duct.
      (1) The thoracic duct courses through the posterior mediastinum between the aorta and the vertebral column.
      (2) It empties into the left subclavian vein near the junction of this vein with the internal jugular vein.

D. Nerves of the posterior abdominal wall

1. **Sympathetic nerve** (Fig. 18-6)
   a. **Paired sympathetic trunks** run in a groove between the psoas muscle and the vertebral column.
      (1) Each sympathetic chain consists of ganglia (approximately one for each vertebral level) and interconnecting nerve fibers.
      (2) The five lumbar ganglia may be fused to varying extents.
   b. **Rami communicantes** provide pathways between the spinal cord and the lumbar ganglia.
      (1) Only thoracic ganglia (T1–T12) and lumbar ganglia L1–L2 (occasionally L3) receive **white rami** from the spinal cord (see Fig. 18-6).
         (a) White rami contain myelinated presynaptic (preganglionic) nerve fibers as well as visceral afferent fibers.

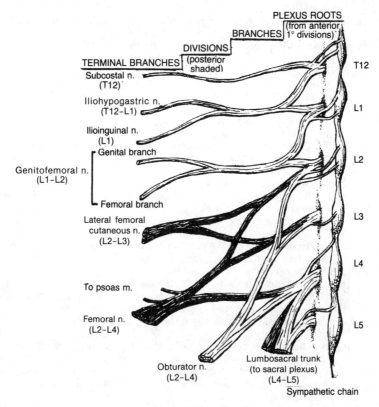

Figure 18-6. *Lumbar plexus.* The lumbar plexus is formed by the anterior primary rami of the lumbar spinal nerves. The anterior divisions of the lumbar plexus are *light*, and the posterior divisions are *dark*. The lumbosacral trunk from L4–L5 joins the sacral plexus. Spinal nerves T12, L1, and L2 have both white and gray rami communicantes. Spinal nerves L3, L4, and L5 have only gray rami communicantes.

(b) The sympathetic nerves of the white rami synapse in either the paravertebral or the prevertebral ganglia.

(2) Every sympathetic ganglion gives off a **gray ramus**.

(a) Gray rami contain unmyelinated postsynaptic (postganglionic) nerve fibers.

(b) These postganglionic sympathetic fibers return to their respective spinal nerves to subserve peripheral autonomic functions, such as vasoconstriction, piloerection, and perspiration.

c. Lumbar splanchnic nerves, arising from ganglia L1–L2, innervate portions of the gastrointestinal tract, kidneys, and ureters.

(1) Each lumbar splanchnic nerve is composed of presynaptic and postsynaptic nerve fibers and also carries visceral afferent nerves.

(2) The lumbar splanchnic nerves join the **aortic plexus**, which:

(a) Communicates superiorly with the superior mesenteric and celiac plexuses

(b) Communicates inferiorly with the hypogastric plexus, which conveys sympathetic pathways into the pelvis for urinary and sexual function

(c) Contains small sympathetic ganglia, which provide postsynaptic fibers to the descending colon, sigmoid colon, and ureters

(3) Sympathetic nerves to the viscera are primarily vasoconstrictors.

(4) Lumbar splanchnic pathways and the associated white rami communicantes convey to the lumbar region of the spinal cord **afferent pain fibers** from the descending colon and sigmoid colon as well as from the upper and middle portions of the ureters. Pain originating from these structures is referred to lumbar, inguinal, and pubic regions as well as to the anterosuperior portion of the thigh (L1–L2).

2. Somatic nerves of the posterior abdominal wall arise from the **lumbar plexus**.

a. The lumbar plexus is the upper portion of the **lumbosacral plexus**. This plexus of somatic nerves is formed from spinal nerves as the extremities develop from the segmental dermatomes and myotomes (see Chapter 4 III C).

(1) The posterior division of a somatic plexus is equivalent to the lateral branch of a spinal nerve (see Fig. 4-2).

(2) The anterior division of a somatic plexus is equivalent to the continuation of a spinal nerve distal to the lateral branch (see Fig. 4-2).

b. Major nerves of the lumbar plexus innervate the abdominal musculature and convey sensation from the skin and parietal peritoneum (see Fig. 18-6).

(1) Iliohypogastric nerve (T12–L1, anterior)

(a) This nerve emerges from the lateral side of the psoas muscle to run within the musculature of the abdominal wall.

(b) Its **iliac branch** is sensory to the upper gluteal region.

(c) Its **hypogastric branch** is motor to the internal oblique and transverse muscles of the abdomen and sensory to the pubic (hypogastric) region.

(d) The iliohypogastric nerve supplies afferent and efferent limbs for the **abdominal reflex**, whereby stroking the lower abdominal wall produces a rippling of the underlying abdominal muscles.

(2) Ilioinguinal nerve (L1, anterior)

(a) The ilioinguinal nerve emerges from the lateral border of the psoas muscle and runs parallel to, and sometimes anastomoses with, the iliohypogastric nerve. Its terminal branch accompanies the spermatic cord through the superficial inguinal ring.

(b) It is motor to the internal oblique and transverse muscles of the abdomen and sensory to the upper medial part of the thigh, the root of the penis or mons pubis, and the anterior scrotum or superior portion of the labia majora.

(3) Genitofemoral nerve (L1–L2, anterior)

(a) This nerve emerges anteriorly to the psoas muscle and runs along its anterior surface.

(b) Its **genital branch** (the external spermatic nerve) passes through the inguinal canal. This branch is motor to the cremaster muscle (the efferent limb of the **cremaster reflex** in men) and sensory to the anterior scrotum or superior portion of the labia majora.

(c) Its **femoral branch** (the lumboinguinal nerve) passes into the thigh beneath the inguinal ligament, medial to the psoas muscle. This branch is sensory to the anterior-superior aspect of the thigh and provides the afferent limb of the **cremaster reflex** in men, whereby stroking the anteromedial thigh produces elevation of the testes within the scrotum.

(4) Lateral femoral cutaneous nerve (L2–L3, posterior)

(a) The lateral cutaneous nerve emerges laterally to the psoas muscle, runs across the iliac fossa, and passes beneath the inguinal ligament close to the anterior–superior iliac spine.

(b) It is sensory to the lateral aspect of the thigh.

(5) Femoral nerve (L2–L4, posterior)
  (a) The femoral nerve emerges from the lateral border of the psoas muscle just before the inguinal ligament and enters the thigh by passing beneath the inguinal ligament.
  (b) It is motor to the anterior aspect of the thigh, sensory to the anterior and medial aspects of the thigh, and, by its saphenous branch, sensory to the anterior and medial aspects of the leg. This nerve provides both afferent and efferent limbs of the **knee-jerk reflex**, whereby transient stretch of the patellar tendon produces brief contraction of the quadriceps femoris muscle.

(6) Obturator nerve (L2–L4, anterior)
  (a) The obturator nerve emerges from the medial border of the psoas muscle and enters the thigh through the obturator foramen.
  (b) It is motor, and occasionally sensory, to the medial aspect of the thigh.

(7) Lumbosacral trunk nerve (L4–L5, anterior and posterior)
  (a) This large connector joins the lumbar plexus with the sacral plexus.
  (b) The posterior division contributes to the common peroneal portion of the sciatic nerve. In general, this nerve is sensory and motor to the anterior leg and dorsum of the foot.
  (c) The anterior division contributes to the tibial portion of the sciatic nerve. In general, this nerve is sensory and motor to the posterior thigh, posterior leg, and plantar aspect of the foot.

Part IV Abdominal Region

STUDY QUESTIONS

Directions: Each question below contains five suggested answers. Choose the **one best** response to each question.

1. Which of the following statements concerning the superficial inguinal ring is most accurate?

(A) It is formed in part by the falx inguinalis
(B) It is formed in part by the rectus sheath
(C) It is a perforation in the internal oblique muscle
(D) It is a perforation in the transversalis fascia
(E) None of the above statements is true

2. Stimulation of the cremaster muscle draws the testis up from the scrotum toward the superficial inguinal ring. These afferent impulses are carried centrally by

(A) an anterior primary ramus
(B) the iliohypogastric nerve
(C) the subcostal nerve
(D) the vagus nerve
(E) a white ramus communicans

3. The medial umbilical folds are created by peritoneum overlying the

(A) falciform ligament
(B) inferior epigastric arteries
(C) lateral borders of the rectus sheath
(D) obliterated umbilical arteries
(E) urachus

Questions 4–11

A 43-year-old advertising executive is brought to the emergency room with searing pain to the midback and "coffee grounds" hematemesis. The present illness began shortly after eating a heavy lunch, which included two drinks. He relates a history of periodic attacks of heartburn, nausea, and pain (pointing to his epigastric region). These attacks are often relieved by antacids or upon eating. He smokes two packs of cigarettes a day and admits to moderate social use of alcohol. An EKG is normal. Upon examination the patient exhibits abdominal rigidity, but no region elicits particularly marked tenderness upon palpation.

4. All of the following structures refer pain via the greater splanchnic nerve to the epigastric region EXCEPT the

(A) abdominal esophagus
(B) descending duodenum
(C) gallbladder
(D) ileum
(E) stomach

5. The sharp searing midback pain is explained by stimulation of neurons traveling along the

(A) greater splanchnic nerve
(B) lesser splanchnic nerve
(C) least splanchnic nerve
(D) lumbar splanchnic nerves
(E) spinal nerves (T12–L2)

6. Perforation of a peptic ulcer is suspected, and the patient is prepared for immediate surgery. Under general anesthesia, the skin and anterior wall of the rectus sheath are incised in a paramidline incision from the xiphoid process to the umbilicus. The rectus abdominis muscle is retracted laterally, and the posterior wall of the rectus sheath is incised. As a result of the incision and manipulation of the rectus abdominis muscle, which of the following results might be expected?

(A) Ischemic necrosis of the rectus abdominis muscle *superior* to the umbilicus
(B) Ischemic necrosis of the rectus abdominis muscle *inferior* to the umbilicus
(C) Paralysis of the rectus abdominis muscle in the region *medial* to the incision of the sheath
(D) Paralysis of the rectus abdominis muscle in the region *lateral* to the incision of the sheath
(E) None of the above results can be expected

7. Upon exploration of the abdominal cavity, a slight amount of blood and other fluid is observed in the pouch of Morison and aspirated. The pouch of Morison (subhepatic and hepaticorenal recesses) is directly continuous with

(A) the inferior mesenteric space
(B) the infracolic compartment
(C) the left paracolic gutter
(D) the lesser sac
(E) none of the above

8. No sign of perforation was seen on the anterior aspects of the stomach or duodenum. An incision was made in the membrane between the stomach and liver, avoiding the blood vessels of the lesser curvature as well as those of the portal triad. This incision passed through

(A) the falciform ligament
(B) the gastrohepatic ligament
(C) the greater omentum
(D) the hepatoduodenal ligament
(E) none of the above

9. The lesser sac was aspirated and a ¼ cm perforation was seen on the superior–posterior aspect of the pyloric antrum. The ulcer had eroded a small arterial branch, which was bleeding. The artery involved in this ulceration is most likely a branch of

(A) the esophageal artery
(B) the left gastric artery
(C) the left gastroepiploic artery
(D) a short gastric artery
(E) none of the above arteries

10. The surgeon required more exposure to resect the ulcer. He extended the previous opening in the lesser omentum toward the esophagus whereupon the field suddenly filled with bright red blood. The surgeon quickly gained control of the bleeding. Both ends of the artery were clamped and ligated. Within a few moments he noticed that a portion of the left lobe of the liver blanched, indicating ischemia. The ligated artery most likely was the

(A) left gastric artery
(B) common hepatic artery
(C) right gastric artery
(D) right hepatic artery
(E) left gastroepiploic artery

11. The gastric ulcer was successfully resected. The surgical pathology examination was negative for malignancy. The postoperative course was stormy with spiking temperatures, but the patient gradually improved. The reason that the patient did not succumb to liver necrosis in this instance is that abundant anastomoses exist between the distal portion of the left gastric artery and all of the following EXCEPT the

(A) esophageal arteries
(B) right gastric artery
(C) short gastric arteries
(D) right hepatic artery
(E) splenic arteries

(end of group question)

Questions 12–20

A 34-year-old man came into the emergency room with excruciating abdominal pain in the left flank and inguinal region. A urine sample was positive for blood. While the patient was waiting for x-rays, he complained that the pain was worse and had moved into the anterior and medial aspects of the left thigh as well as to the pubic region. The "KUB" (kidneys, ureters, and bladder) flat plate indicated a possible calculus in the lower left quadrant.

12. The initial pain in the flank and inguinal region resultant from a nephrolith is due to afferent pain fibers, which travel via the

(A) greater splanchnic nerve
(B) iliohypogastric nerve
(C) least splanchnic nerve
(D) lumbar splanchnics
(E) vagus nerve

13. The subsequent renal colic referred to the medial and anterior aspects of the left thigh as well as to the pubic region is the result of afferent pain fibers, which travel via the

(A) genitofemoral nerve
(B) ilioinguinal nerve
(C) lesser splanchnic nerve
(D) lumbar splanchnic nerves
(E) vagus nerve

14. The patient was admitted for observation and possible surgery. The pain remained in the lateral thigh and pubic regions and was unremitting. From this particular pattern of renal colic and the anatomy of the renal system, it is suspected that a stone has lodged

(A) at the junction of the renal pelvis and ureter
(B) at midureter as it passes beneath the gonadal vessels
(C) at the pelvic brim
(D) in the intramural portion of the ureter where it penetrates the bladder
(E) in the urethra

15. The patient is scheduled for surgery. Under general anesthesia, the peritoneal cavity is opened by incision of the skin, fascia, and musculature of the abdominal wall from the iliac crest to the pubis above and parallel to the left inguinal ligament. All of the following layers are severed in the anterior abdominal incision EXCEPT

(A) Camper's fascia
(B) deep fascia
(C) renal fascia
(D) Scarpa's fascia
(E) transversalis fascia

16. Two large nerves are encountered and preserved in the process of making the abdominal incision. The location of these nerves is most apt to be

(A) between the internal oblique and external oblique muscles
(B) between the transverse abdominis and internal oblique muscles
(C) in the superficial fascia
(D) in the transversalis fascia
(E) within the rectus sheath

17. The surgeon mobilized the descending bowel from the posterior abdominal wall. He gently lifted the bowel and separated its mesentery from the peritoneum covering the dorsal body wall. This reestablished mesentery contained the

(A) left lumbar arteries
(B) left renal artery
(C) left testicular artery
(D) middle colic artery
(E) none of the above arteries

18. With the descending colon and its mesentery drawn to the midline, the ureter was visible beneath the peritoneum. The stone was palpated and removed through a small longitudinal incision. Care was taken to preserve the blood supply to the ureter, which is obtained from all of the following vessels EXCEPT the

(A) common iliac artery
(B) gonadal artery
(C) inferior mesenteric artery
(D) inferior vesical arteries
(E) renal artery

19. The surgeon notes that the blood supply to the kidneys appears normal. All of the following statements concerning the blood supply in the vicinity of the kidneys are correct EXCEPT the

(A) left gonadal artery usually originates from the left renal artery
(B) left gonadal vein usually drains into the left renal vein
(C) left renal vein passes anteriorly to the aorta
(D) right adrenal vein drains directly into the inferior vena cava
(E) right renal artery passes posteriorly to the inferior vena cava

20. The abdominal incision was closed in layers. The patient was allowed up on the first postoperative day and was discharged on the tenth day. As a result of the location and direction of the incision, healing might be expected to be accompanied by

(A) paralysis of a portion of the rectus abdominis muscle
(B) minimal scarring
(C) ischemia to the rectus abdominis muscle
(D) significant weakness of a portion of the lateral abdominal wall
(E) none of the above results

(end of group question)

Directions: Each question below contains four suggested answers of which **one or more** is correct. Choose the answer

- A if **1, 2, and 3** are correct
- B if **1 and 3** are correct
- C if **2 and 4** are correct
- D if **4** is correct
- E if **1, 2, 3, and 4** are correct

21. Structures that might be encountered within the layers of the abdominal wall while performing an appendectomy at McBurney's point (midway between the anterior–superior iliac spine and the umbilicus), using the McBurney muscle-splitting incision, include the

(1) inferior epigastric artery
(2) lateral femoral cutaneous nerve
(3) aberrant obturator artery
(4) iliohypogastric nerve

22. Statements that characterize Meckel's diverticulum include which of the following?

(1) It is located within 3 feet of the ileocecal valve
(2) It is usually lined by gastric mucosa
(3) It is present in about 3% of the population
(4) It is a remnant of the urachus

23. Considering the incidence of lymphatic metastatic carcinoma from the rectum, knowledge of its lymphatic drainage is important. Near the middle rectal valve, the lymphatic drainage is to the

(1) inferior mesenteric nodes
(2) internal iliac nodes
(3) superficial inguinal nodes
(4) superior mesenteric nodes

24. After radical esophagectomy (total removal of esophagus), a functional reanastomosis can be made between the pharynx and stomach, using a living segment of transverse colon brought up into the thorax and neck on a mesenteric pedicle containing the original blood vessels and nerves. Important considerations in performing this operation would be to

(1) anastomose the hepatic flexure end of the transverse colon to the stomach and the splenic flexure end to the pharynx
(2) take care not to sever the pelvic splanchnic nerves
(3) preserve the esophageal venous plexus
(4) preserve the middle colic vessels to the transverse colon

Directions: The groups of questions below consist of lettered choices followed by several numbered items. For each numbered item, select the **one** lettered choice with which it is **most closely** associated. Each lettered choice may be used once, more than once, or not at all. Choose the answer

- A if the item is associated with **(A) only**
- B if the item is associated with **(B) only**
- C if the item is associated with **both (A) and (B)**
- D if the item is associated with **neither (A) nor (B)**

Questions 25–29

For each description listed below, select the type of hernia most apt to be associated with it.

(A) Direct inguinal hernia
(B) Indirect inguinal hernia
(C) Both
(D) Neither

25. Presents through the superficial inguinal ring

26. Lies laterally to the inferior epigastric artery

27. Is found within the spermatic cord

28. Seldom enters the scrotum

29. Is bounded by the inguinal, lacuna, and pectineal ligaments as well as the femoral vein

Directions: The groups of questions below consist of lettered choices followed by several numbered items. For each numbered item select the **one** lettered choice with which it is **most** closely associated. Each lettered choice may be used once, more than once, or not at all.

Questions 30–35

For each inflammatory condition named below, select the dermatomes to which pain is *initially* referred.

(A) C3–C5
(B) T1–T4
(C) T5–T9
(D) T10–T11
(E) L1–L2

30. Cholecystitis

31. Diverticulitis of the sigmoid colon

32. Free air irritating the inferior surface of the diaphragm

33. Gastritis

34. Intussusception of the ileum at the ileocecal valve

35. Ischemic volvulus of the transverse colon

Questions 36–40

For each area named below, select the artery that is most likely to supply it.

(A) Celiac trunk
(B) Inferior mesenteric artery
(C) Inferior phrenic arteries
(D) Superior mesenteric artery
(E) None of the above

36. A portion of the adrenal gland

37. The appendix

38. The tail of the pancreas

39. A portion of the left ureter

40. A portion of the rectum

ANSWERS AND EXPLANATIONS

1. The answer is E. [*Chapter 15 VIII B 1*] The superficial inguinal ring is an opening in the external oblique aponeurosis. The superficial ring transmits the spermatic cord in males and the round ligament of the uterus in females. In males, the internal oblique muscle contributes the cremaster muscle. The deep ring is a perforation in the transversalis fascia.

2. The answer is A. [*Chapter 15 VIII D 3 b, c*] The afferent limb of the cremaster reflex is carried by the femoral branch of the genitofemoral nerve. This is the continuation of an anterior primary ramus of the L1–L2 spinal nerves.

3. The answer is D. [*Chapter 16 VI B 2 h (2)*] The medial umbilical folds are formed by peritoneum draping over the umbilical ligaments (obliterated umbilical arteries). The lateral folds overlie the inferior epigastric arteries, while the median umbilical fold in the midline is formed by the underlying urachus.

4. The answer is D. [*Chapter 17 II B 6 b (2); VIII B; Table 17-2*] The visceral afferent neurons from the lower esophagus, stomach, duodenum, gallbladder, and bile duct all travel through the celiac plexus. These neurons gain access to the greater splanchnic nerve, pass through the ganglia of the sympathetic chain (levels 5–9), reach the associated spinal nerves via white rami communicantes, and enter the spinal cord via the dorsal root. The visceral afferents from the terminal duodenum, jejunum, ileum, ascending colon, and transverse colon travel through the superior mesenteric plexus then along the lesser splanchnic nerve.

5. The answer is E. [*Chapter 16 VI C 2 a (1)*] Sharp localized pain is an indication of peritonitis. The parietal peritoneum is innervated by twigs from the anterior primary rami of the spinal nerves.

6. The answer is E. [*Chapter 15 VII A 1 a–c*] Because the blood supply enters the rectus abdominis muscle from the superior and inferior ends and because the innervation enters this muscle from the lateral aspect, the incision described will compromise neither the vascular supply nor the innervation.

7. The answer is D. [*Chapter 16 VI B 2 c (2) (a)*] The pouch of Morison communicates with the lesser sac via the epiploic foramen, the supracolic compartment, the right paracolic gutter, and the subphrenic space. It does not communicate directly with the infracolic compartment or the left paracolic gutter.

8. The answer is B. [*Chapter 16 II D 1 a; V E*] The lesser omentum runs between the foregut and the liver. The portion between the stomach and liver, the gastrohepatic ligament, is relatively avascular. The portion between the duodenum and liver, hepatoduodenal ligament, contains the portal triad, which includes the major blood supply to the liver.

9. The answer is E. [*Chapter 17 II B 4 a (1) (c); V A 2 c (1)*] The most probable arteries involved in pyloric peptic ulceration are the right gastric, common hepatic, gastroduodenal, and right gastroepiploic arteries. In addition, ulceration of the posterior wall of the body of the stomach may involve the splenic artery with profuse bleeding.

10. The answer is A. [*Chapter 17 II D 5 a (3) (a) (iii), (4)*] An aberrant left hepatic artery frequently arises from the left gastric artery. Since the arteries supplying the liver are end-arteries, ligation of an aberrant artery or (as in this case) its supplying artery usually will produce ischemia of a portion of the liver.

11. The answer is D. [*Chapter 17 II B 4 a (2), D 1 b (2) (a), 5 a (4)*] There are few anastomoses between the left and right lobes of the liver, thus precluding anastomotic flow from the right hepatic artery. Abundant anastomoses among the left gastric artery, the esophageal arteries, right gastric artery, and short gastric branches of the splenic artery hypertrophy so that the aberrant left hepatic artery receives sufficient blood to supply the left lobe of the liver.

12. The answer is C. [*Chapter 17 Table 17-2; Chapter 18 II D 2 a*] The renal calyces, renal pelvis, and most of the proximal portion of the ureters generally receive afferent innervation from spinal segment T12 via the least splanchnic nerve. Since the somatic portions of spinal segment T12 supply the lumbar and inguinal regions of the abdominal wall, pain is referred accordingly.

13. The answer is D. [*Chapter 17 Table 17–2; Chapter 18 II D 2 b*] The middle section of each abdominal ureter generally receives innervation from spinal segments L1–L2. As such, the afferent pathway is via the lumbar splanchnic nerves. Since the somatic portions of spinal segments L1 and L2 supply the pubic region and anterior aspects of the scrotum or labia majora, as well as the anterior and medial portions of the thigh, pain is referred accordingly.

14. The answer is C. [*Chapter 18 II A 2, 3 c*] The ureters are narrowest at the pelvic brim where they cross the iliac vessels and enter the deep pelvis. Thus, the probability of a nephrolith lodging in a ureter at this point is high and is confirmed by the pattern of referred pain.

15. The answer is C. [*Chapter 15 III; VII A 3 b; Chapter 16 VI C 1 c; Chapter 18 II A 2*] An incision through the anterior abdominal wall passes through Camper's fascia (the superficial layer of the superficial fascia), Scarpa's fascia (the deep layer of the superficial fascia), the deep fascia investing the musculature, and the transversalis fascia before reaching the peritoneum. Gerota's fascia or the renal fascia comprises the false capsule of the kidney. After gaining access to the peritoneal cavity, the periureteral sheath, an extension of the renal fascia (of Gerota), is incised to gain direct access to the ureter.

16. The answer is B. [*Chapter 15 VI A 1, B 2; Chapter 18 IV D 2 b (1) (a), (2) (a)*] The nerves of the anterior abdominal wall are located between the internal oblique and transverse abdominis layers and their aponeurotic extensions. The two large nerves encountered in the incision would be the iliohypogastric and the ilioinguinal nerves.

17. The answer is E. [*Chapter 16 VI A; Chapter 18 II B 2*] During development, the mesentery of the descending colon becomes adherent to the dorsal body wall of the peritoneal cavity anterior to the left ureter and the left gonadal neurovascular bundle. As such, the left colic artery and its sigmoidal and superior rectal branches lie in the former descending mesocolon anterior to the left ureter.

18. The answer is C. [*Chapter 18 II C 1*] The ureter obtains its blood supply from twigs of numerous vessels, including the renal, gonadal, common iliac, internal iliac, and inferior vesical arteries. While the left colic artery comes into close proximity to the ureter, there are normally no anastomoses between this vessel and those of the ureter.

19. The answer is A. [*Chapter 18 I F 1, 2; III B 2*] The left gonadal vein usually drains into the left renal vein, but both gonadal arteries usually arise directly from the abdominal aorta just distal to the renal arteries. The abdominal aorta lies to the left of the inferior vena cava, and the renal veins usually lie anteriorly to the renal arteries. Therefore, the left renal vein passes anteriorly to the aorta while the right renal artery passes posteriorly to the inferior vena cava. The right adrenal artery drains directly into the inferior vena cava while the left suprarenal artery drains into the left renal vein.

20. The answer is B. [*Chapter 2 I D*] In the lower abdominal wall, the cleavage lines run in a direction nearly parallel to the inguinal ligament. An incision approximating this direction gapes less and produces minimal scarring.

21. The answer is D (4). [*Chapter 15 VI A 2 b; Chapter 17 III C 3 c (2)*] The ilioinguinal nerve lies between the internal oblique and transverse abdominis muscles just superior to the inguinal ligament. At the level of McBurney's point, the inferior epigastric artery has entered the rectus sheath. The lateral femoral cutaneous nerve, lying on the iliacus muscle in the iliac fossa, escapes from the pelvis by passing beneath the inguinal ligament close to the anterior–superior iliac spine. An aberrant obturator artery crosses the femoral ring where it may be encountered during repair of a femoral hernia.

22. The answer is A (1, 2, 3). [*Chapter 17 III B 6 b*] An ileal diverticulum (of Meckel) is a remnant of the vitelline duct, which connects the midgut to the yolk sac during early development. The urachus is a remnant of the duct, which connects the bladder to the allantois. A ''rule of threes'' applies to Meckel's diverticulum: it is usually 3 inches long, within 3 feet of the ileocecal valve, and present in about 3% of the population. It is usually lined with gastric mucosa, the secretions of which often produce ileal ulceration.

23. The answer is A (1, 2, 3). [*Chapter 17 IV C 3 c, D 3 c; VII D 1 b*] The lymphatic drainage from the midrectum is along the inferior mesenteric, internal iliac, and external iliac nodes. In general, the lymphatic drainage follows the arteries. Thus, carcinoma of the rectum may disseminate widely within the abdomen and pelvis as well as to the inguinal nodes.

24. The answer is D (4). [*Chapter 17 III D 2 a; V B 2 e, D 2 b (1)*] The transverse colon receives its blood supply by the middle colic branch of the superior mesenteric artery and its parasympathetic innervation by the vagus nerve, which sends axons along the perivascular plexus. Failure to preserve the vascular pedicle would result in denervation and necrosis.

25–29. The answers are: 25-C, 26-B, 27-B, 28-A, 29-D. [*Chapter 15 VIII E 3 a, F 4 a, c*] Direct inguinal herniation occurs through the inguinal triangle medially to the inferior epigastric vessels; indirect inguinal herniation occurs through the deep inguinal ring laterally to the inferior epigastric vessels. Due to the path of an indirect hernia through the inguinal canal, the indirect hernia lies within the cremaster muscle and, thus, is within the spermatic cord and, unlike the direct hernia, directed toward the scrotum. Both direct and indirect inguinal hernias present through the superficial inguinal ring. Femoral hernias, in contrast to inguinal hernias, pass through the femoral ring, which is bounded by the inguinal, lacunar, and pectineal ligaments as well as the femoral vein.

30–35. The answers are: 30-C, 31-E, 32-A, 33-C, 34-D, 35-D. [*Chapter 17 VIII B; Table 17-2; Chapter 18 IV A 1 e (2)*] Pain arising from visceral structures is referred to the dermatomes corresponding to the spinal nerves that convey the pain afferents to the spinal cord. The peritoneum of the diaphragm is innervated primarily by the phrenic nerves, which arise from C3–C5. Afferents from the stomach, duodenum, and biliary tree travel along the greater splanchnic nerves to T5–T9 spinal levels. The small intestine and large intestine as far as the splenic flexure send afferent nerves to spinal levels T10–T11 along the lesser splanchnic nerve. Afferents from the descending colon and sigmoid colon travel along the lumbar splanchnic nerves to the L1 and L2 spinal levels.

36–40. The answers are: 36-B, 37-D, 38-A, 39-E, 40-B. [*Chapter 17 I B 1 b (1) (a), (2) (a), (3) (a); II F 2 b (3); III C 2 d (3); IV D 3 a (1); Chapter 18 II C 1; III B 1*] The celiac trunk supplies the primitive foregut through its three major branches. The left gastric artery supplies the terminal esophagus and stomach; the common hepatic artery supplies the stomach, liver, gallbladder, duodenum, and the head of the pancreas; and the splenic artery supplies the body and tail of the pancreas, stomach, and spleen. The superior mesenteric artery supplies the head of the pancreas and the bowel from the duodenum to the splenic flexure. The inferior mesenteric artery supplies the large bowel from the splenic flexure to the midrectum. The adrenal glands are supplied by the inferior phrenic and renal arteries as well as by twigs directly from the aorta. The ureters receive blood from numerous sources, none of which include the celiac, superior, or inferior mesenteric arteries.

Part V
Back

I. VERTEBRAL COLUMN

A. General structure

1. Vertebrae. The vertebral column is composed of 33 bones (vertebrae), divided into cervical, thoracic, lumbar, sacral, and coccygeal regions. Vertebrae differ in appearance according to region.

2. Intervertebral disks. There are 23–24 intervertebral disks.
  a. Location. The disks form **amphiarthroses (symphyses)** between the vertebral bodies.
    (1) In the cervical region, there are no disks between the occiput and the atlas nor between the atlas and the axis.
    (2) Remnants of disks occur within the sacrum and coccyx.
  b. Structure. The disks vary in shape, closely corresponding to the associated vertebral bodies, and permit a limited amount of movement between adjacent vertebrae.

3. Intervertebral ligaments. The vertebral column is stabilized by ligaments, which run between the vertebral bodies, vertebral arches, transverse processes, and spinous processes.

B. Spinal curvatures

1. Normal curvatures are due almost entirely to the shape of the intervertebral disks rather than the vertebral bodies (Fig. 19-1).
  a. The sacral curvature (S1 to coccyx) is concave anteriorly.
    (1) This curve is characteristic of the fetus and lower animal forms.
    (2) It is considered a primary curvature.
  b. The lumbar curvature (T12–L5) is concave posteriorly.
    (1) This curve appears between the ages of 12 and 18 months.
    (2) It tends to be more pronounced in females than in males and presents a greater potential for instability than other regions.
    (3) It represents the major adaptation to upright posture.
  c. The thoracic curvature (T2–T12) is concave anteriorly.
    (1) This curve is characteristic of the fetus and lower animal forms.
    (2) It is considered a primary curvature.
  d. The cervical curvature (C2–T2) is concave posteriorly.
    (1) This curve appears between the late intrauterine period and 3–4 months after birth.
    (2) It represents an adaptation to quadripedal posture.

2. Abnormal curvatures
  a. Kyphosis ("hunchback") is an exaggerated thoracic curvature that usually results from congenital deformities, pathologic erosion, or traumatic collapse of one or more vertebral bodies.
  b. Lordosis ("swayback") is a greater than normal lumbar curvature. Most commonly, it is due to compression of the posterior portion of the lumbar intervertebral disks as compensation for a forward shift in the center of gravity, such as that accompanying the late stages of pregnancy or occurring in obesity.
  c. Scoliosis is a lateral curvature. The causes of scoliosis include congenital malformation, pathologic erosion, and unilateral weakness or paralysis of vertebral musculature. Most commonly, it is idiopathic (of unknown causes). A compensatory curve in the opposite direction in another region of the spine usually accompanies scoliosis.
    (1) As scoliosis progresses, there is a concomitant rotation toward the convex side. This deformity progressively interferes with mechanical respiration.
    (2) Scoliosis is amenable to surgical correction, preferably in the late teens and definitely prior to the age of 30.

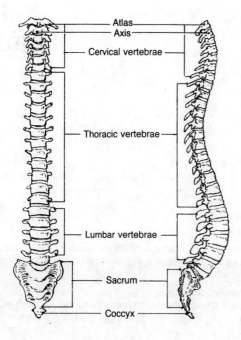

Atlas
Axis
Cervical vertebrae

Thoracic vertebrae

Lumbar vertebrae

Sacrum

Coccyx

Figure 19-1. *Vertebral column.* The anterior and lateral aspects of the spine illustrate the major differences between the vertebra of the six regions. The primary sacral and thoracic curvatures as well as the secondary lumbar and cervical curvatures are evident in the lateral view.

C. Structure of typical vertebra (see Fig. 19-3)

1. **The body** forms the major supportive portion of the vertebra.
   a. **Structure.** The body is approximately cylindrical. As an adaptation to upright posture, each subsequent body must support a greater proportion of the mass of an individual; therefore, bodies increase in bulk toward the sacrum.
   b. **Articulations**
      (1) **Amphiarthroses.** The bodies of most vertebrae articulate with the superjacent and subjacent bodies through intervertebral disks that form symphyses.
      (2) **Diarthroses.** The superior and inferior articular processes of most vertebrae articulate with the superjacent and subjacent vertebrae by synovial joints.

2. **The vertebral (neural) arch** forms the posterior portion of the vertebra.
   a. **Pedicles** arise posterolaterally from the vertebral body and form the foot processes of the vertebral arch.
   b. **Laminae** are supported by the pedicles and fuse in the midline to form the roof of the vertebral arch.

3. **Vertebral processes**
   a. **The spinous process** arises from the vertebral arch where the laminae fuse. It serves as a lever for the attachment of vertebral musculature and ligaments.
   b. **Transverse processes** arise on each side of the vertebral arch where the pedicles and laminae fuse. They serve as levers for the attachment of vertebral musculature and ligaments as well as points of articulation for the necks of ribs 2–10 in the thoracic region.
   c. **Costal processes** arise on each side of the vertebral body anteriorly to the pedicle. They form the ribs in the thoracic region and contribute to the transverse processes of the cervical vertebrae as well as to the alae (wings) of the sacrum.
   d. **Superior articular processes** (zygapophyses) are located on the superior surface of the pedicles.
      (1) These form diarthrodial articulations with inferior articular processes of the vertebra immediately above.
      (2) These processes generally face posteriorly in the cervical region, posteromedially in the thoracic region, medially in the lumbar region, and posteriorly in the sacrum.
   e. **Inferior articular processes** (zygapophyses) arise from the inferior surfaces of the pedicles.
      (1) These form diarthrodial articulations with the superior articular processes of the vertebra immediately below.
      (2) These processes generally face anteriorly in the cervical region, anterolaterally in the thoracic region, and laterally in the lumbar region, but become anterior by the fifth lumbar vertebra.

4. Foramina

　a. The vertebral foramen is formed by the posterior surface of the vertebral body and the vertebral arch.

　　(1) Successive foramina form the **vertebral (neural) canal**.

　　(2) The vertebral foramina contain the spinal cord and its meningeal coverings as well as the associated vessels embedded in adipose tissue.

　b. *Inter*vertebral foramina are located bilaterally between adjacent vertebrae.

　　(1) The intervertebral foramen on each side is formed by a deep notch in the inferior surface of each pedicle of one vertebra and a shallow notch in the superior surface of each pedicle of the vertebra immediately below.

　　(2) These foramina transmit the spinal nerves as they exit the vertebral canal.

D. Regional variation in vertebral characteristics

1. Cervical vertebrae (7)

　a. The atlas is the first vertebra (Fig. 19-2).

　　(1) Structure. This vertebra, lacking a body and a spinous process, is composed of an anterior arch, the vertebral arch, and paired transverse processes.

　　(2) Articulations

　　　(a) The atlas articulates superiorly with the occiput of the skull (**atlanto-occipital joint**).

　　　　(i) Two large concave facets receive the occipital condyles.

　　　　(ii) The atlanto-occipital joint functions in flexion/extension ($\pm 15°$).

　　　　(iii) There is no intervertebral disk at the atlanto-occipital joint, which accounts for the extensive range of movement there.

　　　(b) The atlas articulates inferiorly with the axis.

　b. The axis is the second vertebra (see Fig. 19-2).

　　(1) Structure. The second vertebra is similar to the typical cervical vertebrae except that it has a superior projection, the **dens**, and large flat superior articular surfaces, which are nearly horizontal. The **dens (odontoid process)** represents the body of the atlas, which has become separated from the first vertebra and fused with the second.

　　(2) Articulations

　　　(a) The axis articulates superiorly with the atlas (**atlantoaxial joint**) both at the dens, which projects into the atlas, and at the zygapophyses.

　　　　(i) The dens, stabilized within the atlas by several ligaments, serves as a pivot about which rotation occurs. Fracture and posterior dislocation of the dens may crush the spinal cord at the level of the first cervical vertebra with subsequent death from paralysis of all respiratory musculature.

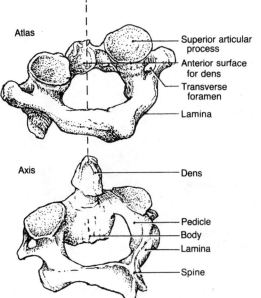

Figure 19-2. *Atlas and axis.* The atlas articulates with the occipital condyles of the cranium and has no body or spinous process. The odontoid process (dens) of the axis, about which rotation occurs, represents the misplaced body of the atlas.

(ii) The atlantoaxial joint functions primarily in rotation (± 25°).

(iii) There is no intervertebral disk at the atlantoaxial joint, which accounts for the extensive range of rotational movement there.

(b) The axis articulates inferiorly with the third cervical vertebra through an intervertebral disk and at the zygapophyses.

c. Typical cervical vertebrae (Fig. 19-3) are relatively small but increase in size toward the thoracic region.

(1) Structure. They have a characteristic **transverse foramen** in each transverse process.

(a) The transverse foramen represents a hiatus that results from incomplete fusion of the costal and transverse processes.

(b) Except for the seventh cervical vertebra, the transverse foramen transmits the **vertebral artery**, which provides collateral circulation to the brain.

(c) The cervical spinous processes angle sharply downward and are frequently bifid. The spinous process of the seventh cervical vertebra, the **vertebra prominens**, is nearly horizontal and usually not bifid; it is a most useful surface landmark.

(2) Articulations. The cervical articular processes have a characteristic orientation.

(a) These articular facets are cup-shaped and lie in a more-or-less transverse plane.

(b) The superior articular surfaces generally face posteriorly, whereas the inferior articular surfaces generally face anteriorly.

(c) These joints have three degrees of freedom, allowing rotation, flexion/extension, and lateral flexion (bending).

(d) The orientation of and the great mobility at the cervical zygapophyses predispose to dislocation.

d. Summary of movements over the cervical region.

(1) Flexion: 40°/**extension:** 90°, primarily at the atlanto-occipital joint

(2) Lateral flexion (abduction): 35°–40°, primarily between the second through seventh cervical vertebrae

(3) Rotation: ± 45°–50°, half of which occurs at the atlantoaxial joint

(4) Circumduction. The combination of flexion/extension and lateral flexion permits circumduction of the head.

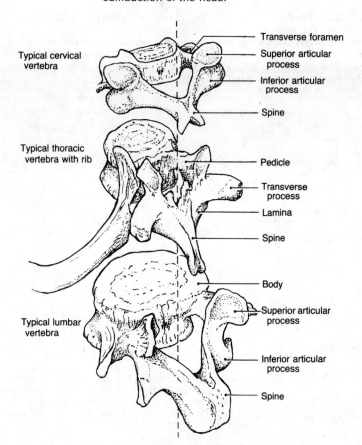

Typical cervical vertebra — Transverse foramen, Superior articular process, Inferior articular process, Spine

Typical thoracic vertebra with rib — Pedicle, Transverse process, Lamina, Spine

Typical lumbar vertebra — Body, Superior articular process, Inferior articular process, Spine

Figure 19-3. *Typical cervical, thoracic, and lumbar vertebrae.*

2. Thoracic vertebrae (12) are larger than the cervical vertebrae and increase in size toward the lumbar region (see Fig. 19-3).

a. Structure

(1) The thoracic spinous processes are very long and oriented caudally.

(2) The costal processes develop into ribs.

(3) The bodies and transverse processes have articular surfaces for the ribs (diarthrodial joints).

  (a) The upper 10 thoracic vertebrae have articular facets on the transverse processes for the necks of the ribs as well as articular demifacets on the bodies near the origins of the pedicles for the heads of the associated ribs.

  (b) The eleventh and twelfth vertebrae have costal articular facets on the bodies only.

  (c) The second through ninth ribs articulate with superior and inferior vertebrae at the level of the intervertebral disk.

   (i) The bodies of the first and tenth through twelfth thoracic vertebrae have complete facets for the articulation of the associated ribs.

   (ii) The second through ninth thoracic vertebrae have demifacets on the bodies for articulations with the associated ribs.

b. Articulations. The thoracic zygapophyses have a characteristic orientation.

(1) The paired superior articular processes face posterolaterally, whereas the inferior articular processes face anteromedially.

(2) The articular surfaces lie approximately on an arc of a circle, the center of which is located in the vertebral body, thereby allowing rotation.

  (a) Over the 12 thoracic vertebrae, there is a total of $\pm 35°$ of rotation.

  (b) There is some flexion (40°) and extension (20°).

  (c) There is a small amount of lateral flexion (20°).

(3) The movements of the thoracic vertebral column are limited by the ribs and their attachment to the sternum.

3. Lumbar vertebrae (5) are very large and heavy, becoming progressively larger toward the sacral region (see Fig. 19-3).

a. Structure. The spinous processes are stubby and project posteriorly. This allows considerable extension as well as access to the intervertebral canal with a needle (**lumbar puncture**).

b. Articulations. The lumbar articular processes have a characteristic orientation.

(1) They lie on nearly parasagittal planes. The articular surfaces lie on an arc of a circle, the center of which is located near the tips of the spinous processes.

(2) The superior articular surfaces face medially, whereas the inferior articular surfaces face laterally. By the level of the fifth lumbar vertebra, however, the inferior facets become directed anteriorly.

  (a) This orientation limits rotation over the lumbar region to $\pm 5°$.

  (b) It, however, allows considerable flexion (60°)/extension (35°).

  (c) There is a moderate amount of lateral flexion (20°).

  (d) The combination of flexion/extension and lateral flexion allows a considerable amount of circumduction of the trunk.

  (e) The articular facets for the fifth lumbar vertebra are directed anteriorly and engage the posteriorly directed superior processes of the first sacral vertebra.

4. Sacral vertebrae (5) fuse to form the sacrum (Fig. 19-4).

a. Structure

(1) Four horizontal lines of fusion are evident on the anterior surface.

  (a) There may be a rudimentary disk between the first and second sacral vertebrae.

  (b) Rarely, there may be sacralization of the fifth lumbar vertebra whereby that vertebra becomes partially or completely fused with the sacrum.

(2) The fused spinous processes form the **median sacral crest**.

(3) The fused transverse processes together with fused costal processes form the **alae** or sacroiliac articular processes. This fusion occurs around the spinal nerves to form sacral foramina.

  (a) The **posterior sacral foramina** transmit the dorsal primary rami of the sacral spinal nerves.

  (b) The **anterior sacral foramina** transmit the ventral primary rami of the sacral spinal nerves.

(4) The lamina of the fifth sacral vertebra fail to form, resulting in the **sacral hiatus**. The pedicles of the fifth sacral vertebra form the **sacral cornua**, important landmarks used in locating the sacral hiatus for the administration of *caudal anesthesia*.

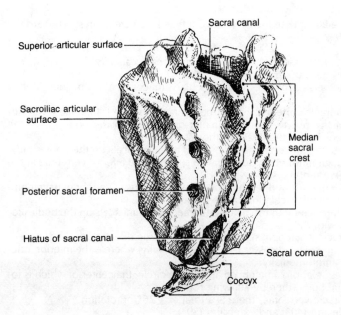

Superior articular surface

Sacral canal

Sacroiliac articular surface

Median sacral crest

Posterior sacral foramen

Hiatus of sacral canal

Sacral cornua

Coccyx

Figure 19-4. *Sacrum and coccyx.*

b. Articulations
   (1) The articular facets for the first sacral vertebra are directed posteriorly, thereby preventing the lumbar vertebrae from slipping forward on the anteriorly slanted body of the sacrum.
   (2) The sacrum articulates with the first coccygeal vertebrae to form the **sacrococcygeal joint**.

5. Coccygeal vertebrae (see Fig. 19-4)
   a. Structure
      (1) The four or five coccygeal vertebrae fuse to form the coccyx, but there may be a disk between the first and second coccygeal vertebrae.
      (2) The first (and sometimes the second) coccygeal vertebra has rudimentary pedicles that form the coccygeal cornua.
      (3) The second through fourth coccygeal vertebrae represent rudimentary vertebral bodies.
   b. Articulations. The first coccygeal vertebra articulates with the sacrum. The sacrococcygeal joint usually has a small intervertebral disk but may fuse in some individuals.

6. Clinical considerations
   a. Compression fractures of the vertebral bodies may result from pathology or trauma. The internal collapse of the vertebral body not only may result in kyphosis or scoliosis but also may reduce the size of the intervertebral foramen and increase the possibility of spinal nerve compression.
   b. Fracture of the pedicles dissociates a vertebral body from the stabilizing influences of the associated zygapophyses. Such fractures predispose to misalignment of the vertebral column with risk of compression of the spinal cord as well as damage to the exiting spinal nerves.
      (1) Because the vertebral canal is relatively wide in the cervical region, a small displacement may not compress the spinal cord.
      (2) Because the vertebral canal is relatively narrow in the thoracolumbar region, paraplegia is a common sequel to a thoracic fracture dislocation.
   c. Spondylolisthesis
      (1) The large angle (140°) between the axes of the fifth lumbar vertebra and the sacrum at the lumbosacral joint and the considerable angle (41°) of the body of the first sacral vertebra from the horizontal impose enormous forces upon the superior articular processes of the first sacral vertebra.
      (2) Spondylolisthesis is an anterior displacement of the vertebral column, usually at the lumbosacral joint. This may be due to:
         (a) Congenital defect of the lamina of L5 (**spondylolysis**), which is found in 5% of the population
         (b) Fracture of the pedicles or inferior articular processes of the fifth lumbar vertebra
         (c) Fracture of the lamina or superior articular processes of the sacrum

(3) Since the center of gravity normally passes through the lumbosacral joint, gravity causes the vertebral column to slide forward on the intervertebral disk at the lumbosacral joint with the anterior longitudinal ligament supporting the entire vertebral structure.

(4) Spondylolisthesis usually involves compression of spinal nerves S1 and S2, producing symptoms of sciatica.

d. Coccygodynia. Fracture of the fused coccygeal joints or of a fused sacrococcygeal joint or arthritis in these joints causes a great deal of pain.

E. Intervertebral disks

1. External structure. The disks conform to surfaces of apposed vertebral bodies (see Fig. 19-5).
  a. The intervertebral disks comprise about 25% of the total length of the vertebral column.
  b. In the cervical region, the disks are small and thin, but by the lumbar region, the disks have become progressively larger and thicker.
  c. Differences in thickness anteriorly and posteriorly define the secondary vertebral curvatures (i.e., they are thicker anteriorly in the lumbar and cervical regions).

2. Internal structure. Each disk is composed of an outer fibrocartilaginous portion and an inner gelatinous central portion.
  a. The annulus fibrosus forms the outer portion of the intervertebral disk.
    (1) This structure is composed of lamelliform connective tissue bands. Each band is oriented in a different direction, providing considerable strength.
    (2) The annulus fibrosus is firmly attached to and joins adjacent vertebrae, forming a symphysis type of amphiarthrosis.
    (3) It supports the central nucleus pulposus.
  b. The nucleus pulposus forms the central portion of the intervertebral disk.
    (1) This structure is a remnant of the embryologic **notochord**.
    (2) It is surrounded and supported by the annulus fibrosus.
    (3) The nucleus pulposus is composed of mucinous tissue with fibrous and mucopolysaccharide complexes, which contribute to its high osmotic pressure.
    (4) The extremely high water content of the disk (70%–80%) accounts for the fact that a person may be as much as 1 cm taller upon rising in the morning than at the end of a strenuous day.

3. Functions. The intervertebral disks bind the vertebrae together (**amphiarthroses**) and permit limited movement between adjacent vertebrae. The nucleus pulposus functions as a noncompressible but deformable pad, which distributes forces uniformly over the entire surface of the vertebra regardless of the degree of flexion/extension, rotation, or lateral flexion.

4. Clinical considerations
  a. Disk degeneration
    (1) Degeneration is associated with chronic dehydration of the nucleus pulposus.
    (2) Loss of water and mucopolysaccharide leads to a narrowing of intervertebral space and reduces the capacity of the disk to act as a cushion between vertebrae.
    (3) Narrowing of the intervertebral disks results in diminished stature. It also decreases the size of the intervertebral foramina, thereby increasing the possibility of spinal nerve compression.
    (4) Occasionally, degeneration arthroses produce progressive calcification at the superior and inferior margins of the vertebral bodies to form **osteophytes**. These may compress the spinal nerves, especially in the cervical region with resultant arm pain, or cause soft tissue irritation, especially in the lumbar region with resultant back pain.
  b. Disk herniation (*"slipped disk"*)
    (1) Herniation is the prolapse (extrusion) of the nucleus pulposus through the annulus fibrosus.
    (2) Because of the lumbar curvature and the considerable mass of the body superior to this region, herniation usually occurs between the fourth and fifth lumbar vertebrae or between the fifth lumbar and first sacral vertebrae. Disk herniation also occurs in the cervical region (5%–10%).
    (3) Rarely, prolapse may occur into the superjacent or subjacent vertebral bodies (Schmorl's node).
    (4) Most commonly, herniation is directed posterolaterally where the annulus fibrosus is not reinforced by the posterior longitudinal ligament. The presence of the anterior longitudinal ligament and the posterior longitudinal ligament reinforces the underlying annulus fibrosus.
    (5) Posterolateral prolapse impinges upon the spinal nerve of the next lower vertebral level, causing symptoms associated with the dermatomic and myotomic distributions

of that nerve. This is true in *both* the cervical and the lumbar regions, but for different reasons.

(a) In the cervical region, there are eight cervical nerves, but seven vertebrae; thus, herniation of the disk between the fourth and fifth cervical vertebrae impinges upon spinal nerve C5, which exits through the intervertebral foramen formed by the fourth and fifth cervical vertebrae (see III B 1).

(b) In the lumbar region, the deeply notched pedicles of the vertebrae are located high on the body, and the intervertebral disks are thicker; thus, herniation of the disk between the fourth and fifth lumbar vertebrae does not harm spinal nerve L4, which exits just superior to the disk. However, the herniation compresses spinal nerve L5 as it descends across the disk between the fourth and fifth lumbar vertebrae to its intervertebral foramen, producing the symptoms of **sciatica**. If the herniation is more substantial and posteriorly directed, it may involve one or more spinal nerves below the level of herniation; if very large, it may involve some spinal nerves on the contralateral side as well.

(c) In addition to the pain referred to the distribution of the compressed spinal nerve, local pain produced from the stretched annulus fibrosus initiates painful spasm of the back muscles.

F. Ligaments of the vertebral column (Fig. 19-5)

1. **The supraspinous ligament** connects spinous processes at the tips and forms the **ligamentum nuchae** in the cervical region.

2. **The interspinous ligaments** run between spinous processes.
   a. These ligaments limit the range of motion of the vertebrae.
   b. Small tears in these ligaments, the result of hyperextension trauma, may be the basis for ''whiplash'' injury.

3. **The ligamentum flavum** (*flavum, L.* yellow) stretches between adjacent laminae. Except for a midline gap, the paired ligaments, together with the vertebral arch, complete the posterior aspect of the vertebral canal.
   a. Unlike the other ligaments, the ligamentum flavum is composed of elastic tissue.
   b. Upon traumatic hyperextension of the neck, the capacity of the ligamentum flavum to take up slack may be exceeded, and the resultant buckling may injure the spinal cord.

4. **The anterior longitudinal ligament** connects the vertebral bodies and intervertebral disks anteriorly.
   a. This ligament is a very broad band and runs from the sacrum to the occipital bone.
   b. It resists the gravitational tendency toward increased lordosis.
   c. It reinforces the annulus fibrosus anteriorly and directs herniation posteriorly.
   d. It can be used to splint fractured vertebrae when the trunk is cast in extension.

5. **The posterior longitudinal ligament** connects the vertebral bodies and intervertebral disks posteriorly.
   a. This ligament is denticulate in shape, wider posterior to the vertebral body and narrower posterior to the intervertebral disk.

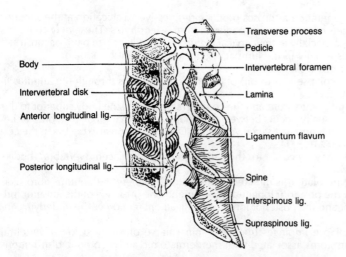

Figure 19-5. *Ligaments of the vertebral column.*

b. It resists the gravitational tendency toward increased kyphosis.

c. It supports the annulus fibrosus posteriorly and directs disk herniation posterolaterally.

II. SOFT TISSUES OF THE BACK

A. Posterior muscles of the vertebral column

1. **Organization.** The intrinsic muscles of the back are arranged in three layers. The most super-ficial layer runs upward and obliquely outward, the middle layer runs parallel to the vertebral column, and the innermost layer runs upward and obliquely inward. The actions of these muscles include extension, rotation, and lateral flexion.

   a. **The spinotransverse group**, representing the superficial layer, originates from spinous processes and the nuchal ligament, runs superiorly and obliquely laterally, and inserts onto either transverse processes or the base of the skull (Fig. 19-6)

      (1) **Divisions**

         (a) The **splenius capitis muscle** runs from the spines of the seventh cervical through the fourth thoracic vertebrae and the nuchal ligament to insert into the superior nuchal line and mastoid process of the skull.

         (b) The **splenius cervicis muscle** runs from the spines of the third through sixth thoracic vertebrae to the transverse processes of the second through fourth cervical vertebrae.

      (2) **Actions.** These muscles, acting unilaterally, rotate the head and neck toward the same side and, acting bilaterally, elevate the head and extend the neck.

   b. **The sacrospinalis group (erector spinae)**, representing the middle layer, originates from the sacrum, iliac crests, and spinous processes of the lumbar and lower thoracic vertebrae; runs parallel to the vertebral column; and inserts into the ribs and transverse processes (see Fig. 19-6).

      (1) **Divisions**

         (a) The **iliocostalis muscles** comprise the most lateral portion of the erector spinae muscle.

            (i) The iliocostalis lumborum muscle runs from the sacrum and iliac crests to the inferior borders of the lower six ribs.

            (ii) The iliocostalis thoracis muscle runs from the superior borders of the lower six ribs to the inferior borders of the upper six ribs.

            (iii) The iliocostalis cervicis muscle runs from the superior borders of the upper six ribs to the inferior borders of the transverse processes of the fourth through sixth cervical vertebrae.

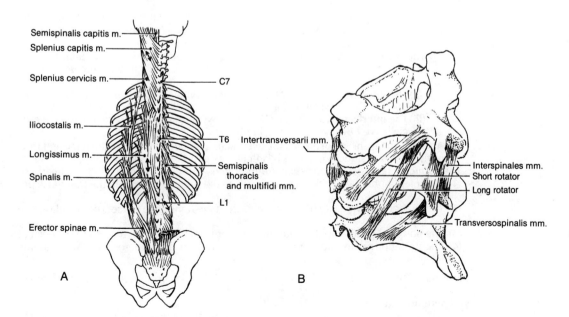

Figure 19-6. *Muscles of the back. A,* Long muscles. *B,* Short muscles.

    (b) The **longissimus muscle** comprises the middle portion of the erector spinae muscle (see Fig. 19-6).

        (i) The longissimus thoracis muscle runs from the sacrum to the inferior borders of the transverse processes and associated ribs of the lower four thoracic vertebrae.

        (ii) The longissimus cervicis muscle runs from the superior borders of the transverse processes of the upper six thoracic vertebrae to the inferior borders of the transverse processes of the second through sixth cervical vertebrae.

        (iii) The longissimus capitis muscle runs from the superior borders of the second through sixth cervical vertebrae to the mastoid process of the skull.

    (c) The **spinalis muscle** comprises the medial portion of the erector spinae muscle and is poorly developed.

        (i) The spinalis thoracis muscle, which is the most highly developed portion, runs between spinous processes.

        (ii) The spinalis cervicis muscle is inconsistently present.

        (iii) The spinalis capitis muscle is usually considered the medial portion of the semispinalis capitis muscle.

  (2) Actions. The sacrospinalis group of muscles, acting unilaterally, flexes the vertebral column to the same side and, acting bilaterally, extends the vertebral column.

c. The transversospinal group (transversospinalis), representing the innermost layer, originates from transverse processes, runs superiorly and obliquely medial, and inserts into spinous processes (see Fig. 19-6). The deeper muscles traverse shorter distances.

  (1) Divisions

    (a) The **semispinalis muscle** lies beneath the erector spinae muscle and generally passes over five or more vertebrae (see Fig. 19-6).

        (i) The semispinalis thoracis muscle runs from the transverse processes of the lower seven thoracic vertebrae to the spines of the upper four thoracic vertebrae.

        (ii) The semispinalis cervicis muscle runs from the transverse processes of the upper six thoracic vertebrae to the spines of the second through fifth cervical vertebrae.

        (iii) The semispinalis capitis muscle overlies the semispinalis cervicis muscle and runs from the transverse processes of the fourth cervical through the fourth thoracic vertebrae to the inferior nuchal line of the skull.

        (iv) The medial portion of this muscle, originating from spinous processes, is termed the spinalis capitis muscle.

    (b) The **multifidus muscles (multifidi)** lie deep to the semispinalis muscle and generally pass over two to three vertebrae (see Fig. 19-6).

        (i) Except for their shorter bundles, they are nearly indistinguishable from the semispinalis muscle.

        (ii) The multifidus muscles are best developed in the lumbar and cervical regions.

    (c) The **long rotators** run from the transverse processes to the spinous processes two vertebrae above.

    (d) The **segmental muscles** run between adjacent vertebrae (see Fig. 19-6).

        (i) The **short rotators** run between the transverse and spinous processes of adjacent vertebrae.

        (ii) The **interspinales muscles** run between spinous processes and are best developed in the cervical region.

        (iii) The **intertransversarii muscles** run between transverse processes and are especially well developed in the lumbar and cervical regions.

  (2) Actions. The **transversospinalis muscles**, acting unilaterally, rotate the neck and trunk to the *opposite* side and, acting bilaterally, extend the vertebral column.

d. The suboccipital muscles (a special group of deep muscles) act to extend the head.

  (1) The inferior oblique capitis muscle runs from the spine of the axis to the transverse process of the atlas.

  (2) The superior oblique capitis muscle runs from the transverse process of the atlas to the occiput.

  (3) The posterior rectus capitis major muscle runs from the spine of the axis to the occiput.

  (4) The posterior rectus capitis minor muscle runs from the spine of the atlas to the occiput.

2. Innervation. The **posterior muscles** are innervated by dorsal primary rami of the spinal nerves.

B. Anterior vertebral muscles

1. The scalene muscles run from transverse processes of cervical vertebrae to the first and second ribs (see Chapter 27 II C 2 a).

    2. The longus colli muscle runs from the fourth through sixth cervical vertebrae to the occiput (see Chapter 27 II C 2 b).

    3. The lateral rectus capitis muscle runs between the transverse process of the atlas and the jugular process of the occiput (see Chapter 27 II C 2 b).

    4. The quadratus lumborum muscle runs from the iliac crests to the inferior borders of the twelfth ribs (see Chapter 18 IV A 2).

C. Functional considerations. In addition to major movements of the vertebral column, the normal function of the vertebral musculature is the making of fine adjustments to keep the center of gravity of the body over the first sacral vertebra.

    1. When a person stands erect and balanced with a 10-kg (22-pound) weight in each hand, 20 kg is distributed evenly over each vertebral disk.

    2. If, instead, the 20 kg is held 20 cm anteriorly to the vertebral column by flexing the elbows, the forward shift in the center of gravity requires contraction of the back muscles in order to keep the individual erect.

        a. The muscles that insert into the tips of the spinous processes (i.e., those having the maximum mechanical advantage because of a 2-cm lever arm) exert 200 kg of counterforce (20 kg x 20 cm/2 cm = 200 kg).

        b. This counterforce, pulling downward on one side of the vertebra, is additive to the weight of the object lifted. Thus, the total force distributed over each intervertebral disk is 220 kg.

    3. If, in addition, the individual supports the 20-kg weight with extended arms and with the trunk flexed forward so that the weight extends 50 cm anteriorly to the fifth lumbar vertebra, the back muscles must generate 500 kg of counterforce (20 kg x 50 cm/2 cm = 500 kg).

        a. This muscular pull downward on the vertebral spine is additive to the weight of the object lifted, so that 520 kg must be distributed over each lumbar intervertebral disk.

        b. Since a disk may rupture when forces exceed 500–800 kg, the individual is clearly placed in physical jeopardy.

III. SPINAL CORD

A. General structure. The spinal cord is an extension of the brain and a component of the central nervous system.

    1. Vertebral (neural) canal. The spinal cord and its meningeal coverings lie within the osseofibrous vertebral canal (see Fig. 19-5).

        a. Structure. At the vertebral level, the vertebral canal is bounded by the vertebral body and the neural arch, which consists of the pedicles and laminae. At the intervertebral levels, it is bounded by the intervertebral disks and the ligamenta flava, which run between successive laminae on either side of the midline.

        b. The epidural space between the boundaries of the vertebral canal and the dura mater is filled with loose fatty connective tissue and contains the extensive **vertebral venous plexus**. This is the space into which anesthetic is infused for epidural anesthesia.

    2. External structure of the spinal cord. The spinal cord extends from the **medulla oblongata** at the level of the first cervical vertebra and terminates as the **conus medullaris** at the level of the intervertebral disk between the first and second lumbar vertebrae.

        a. Fetal location. Until the third fetal month, the spinal cord is as long as the vertebral canal, and both extend to the level of the fourth sacral vertebra.

        b. Law of descent. After the fourth month, the vertebral column outgrows the spinal cord so that the cord ultimately ends at the upper lumbar levels. However, the spinal roots proceed through the dural sac to the appropriate intervertebral foramina (Fig. 19-7). The resultant descending distribution of spinal roots form the **cauda equina** (L. horse's tail).

        c. Postnatal location. In most individuals, the spinal cord terminates between the first and second lumbar vertebrae (see Fig. 19-7). Because the spinal cord extends below the second lumbar vertebra in 1% of the population (especially in short-statured individuals), neither lumbar puncture nor spinal (intrathecal) anesthesia should be attempted above the level of the fourth lumbar vertebra.

    3. Spinal meninges. Three layers envelop the brain and spinal cord: dura mater, the arachnoid layer, and pia mater. The dura is sometimes referred to as the **pachymeninx** (G. thick membrane), while the combined arachnoid and pia can be referred to as the **leptomeninges** (G. delicate membranes).

        a. The dura mater (L. hard mother) is the outermost meningeal layer. It is a dense fibrous membrane that forms a sheath about the central nervous system (Fig. 19-8).

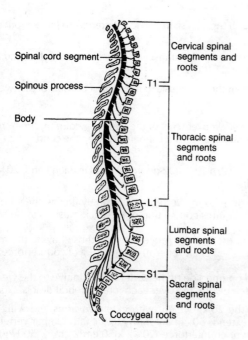

Spinal cord segment

Spinous process

Body

T1

L1

S1

Coccygeal roots

Cervical spinal segments and roots

Thoracic spinal segments and roots

Lumbar spinal segments and roots

Sacral spinal segments and roots

Figure 19-7. *Schematic diagram of the spinal cord and roots.* The spinal cord usually terminates just caudally to the L1 vertebral body below which the spinal roots form the cauda equina as each root courses to its respective intervertebral foramen.

    (1) This layer begins at the foramen magnum, where the dura mater is no longer fused with the periosteum of the cranial vault. It extends to the level of the second sacral vertebra, where it forms the **coccygeal ligament**, which covers the pial filum terminale.

    (2) This layer evaginates into each intervertebral foramen to surround the spinal nerve and fuse with the periosteum of the vertebrae. It becomes continuous with the epineurium, the connective tissue covering of the spinal nerve.

    (3) The dura defines the **epidural space** and the **subdural space**.

  b. The arachnoid (L. spiderlike) is the intermediate meningeal layer (see Fig. 19-8).

    (1) Subdural space. Because the arachnoid is loosely adherent to the dura, there is a potential space between the dura and arachnoid layers (**subdural space**), which does not contain cerebrospinal fluid. However, blood or pus may extravasate within this space.

    (2) Subarachnoid space. The arachnoid is attached to the underlying pia by numerous **arachnoid trabeculae**. The considerable cavity between the arachnoid and the pial layers defines this subarachnoid space. This space contains the cerebrospinal fluid, which suspends the central nervous system and nerve roots. Large blood vessels pass within this space.

      (a) The **dural sac** (a misnomer) is that portion of the subarachnoid space between the conus medullaris (approximately vertebra L1) and the point at which the coccygeal ligament begins (about vertebra S2). By definition, the dural sac contains only spinal roots suspended in cerebrospinal fluid.

      (b) It is the dural sac into which a needle is inserted in the lumbar puncture (spinal tap) and for induction of spinal anesthesia. The suspended roots usually are displaced by the needle and, therefore, are not harmed.

  c. The pia mater (L. tender mother) is the innermost meningeal layer. It intimately invests the brain and continues through the foramen magnum to intimately invest the spinal cord. It contains the plexus of small blood vessels that supply the neural tissue.

    (1) The **denticulate ligament** supports the entire spinal cord so that it is located in the center of the subarachnoid space. It consists of 21 pairs of lateral pial reflections, which pass through the arachnoid to attach to the dura. In this manner, the spinal cord is tethered in the cerebrospinal fluid.

    (2) The **filum terminale** is an extension of the pia beyond the tip of the spinal cord (conus medullaris). At the distal end of the dural sac, it is covered by the dura, thereby contributing to the **coccygeal ligament**.

 4. Internal structure of the spinal cord

  a. Gray matter. Groups of nerve cell bodies, organized into nuclei and laminae, form the gray matter. These are arranged into dorsal horn, lateral horn, and ventral horn as well as a central gray area (Fig. 19-9).

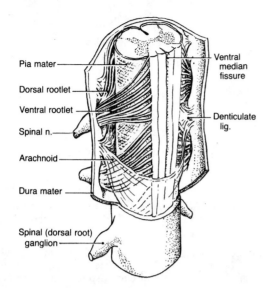

Figure 19-8. *Spinal cord and meninges.*

(1) The **laminae of the dorsal horn** are sensory, comprising the second-order neurons of the spinothalamic sensory pathways and simple reflex arcs (see Figs. 4-4 and 4-5).

(2) The **nuclei of the lateral horn** comprise the preganglionic neurons for the sympathetic division of the autonomic nervous system. The axons leave via the ventral roots between levels T1–L2 and synapse with postganglionic neurons in a peripheral ganglion (see Fig. 4-5).

(3) The **nuclei of the ventral horn** comprise the lower motor neurons for voluntary activity (alpha motoneurons) and smaller neurons that regulate muscle tone (gamma motoneurons). The outflow tracts of these neurons leave via ventral roots (see Fig. 4-4).

(4) The **central gray area** contains a multitude of small interneurons, many of which are associated with reflexes and some part of the spinoreticular tract.

b. **White matter.** Bundles of nerve axons organized into **tracts** form the white matter. There are numerous descending (motor) and ascending (sensory) tracts in the spinal cord, some of which cross in the anterior commissure to the contralateral side. These tracts define three columns: the dorsal, lateral, and ventral funiculi (see Fig. 19-9; Fig. 19-10).

(1) **Motor (descending) tracts** are composed of **upper motor neurons**; two of these tracts originate primarily in the motor cortex (precentral gyrus) of the brain and descend in the brain stem and spinal cord. Because the descending pathways exert their influence upon the activity of **lower motor neurons**, they remain entirely within the central nervous system. The corticobulbar tract (in the brain stem) and the lateral corticospinal (pyramidal) tract (in the lateral funiculus of the spinal cord) are major descending (motor) tracts (see Fig. 19-10).

(a) The **corticobulbar tract** extends through the brain stem and mediates voluntary movements in the head and neck. Its axons influence lower motor neuron activity in the nuclei of the cranial nerves: some ipsilateral, some contralateral, and some a mixture of both.

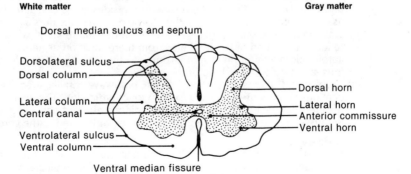

Figure 19-9. *Cross section of the spinal cord with characteristic regions of white matter and gray matter.* (Reprinted with permission from DeMyer W: *Neuroanatomy.* New York, John Wiley & Sons, 1988, p. 101.)

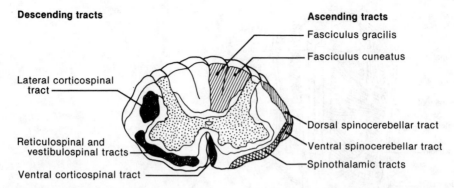

Descending tracts

Ascending tracts

Fasciculus gracilis

Fasciculus cuneatus

Lateral corticospinal tract

Dorsal spinocerebellar tract

Ventral spinocerebellar tract

Reticulospinal and vestibulospinal tracts

Spinothalamic tracts

Ventral corticospinal tract

Figure 19-10. *Cross section of the spinal cord, showing the location of the major ascending and descending tracts.* (Reprinted with permission from DeMyer W: *Neuroanatomy.* New York, John Wiley & Sons, 1988, p. 110.)

(b) The **corticospinal tract** mostly decussates (crosses to the opposite side) in the lower brain stem so that each side of the brain controls the contralateral side of the body. The axons descend in the spinal cord as the **lateral corticospinal tract** (see Fig. 19-10) and terminate on interneurons and lower motor neurons of the gray matter ventral horn. This neuronal pathway mediates voluntary movement below the neck.

(c) "**Extrapyramidal pathways**," such as the vestibulospinal and reticulospinal tracts, influence the lower motor neuron activity and, thereby, affect muscular tone (see Fig. 19-10). These pathways also control visceral sympathetic function.

(2) Sensory (ascending) tracts convey different sensory modalities to the brain stem. Some are crossed, whereas others ascend on the ipsilateral side of the spinal cord to cross in the brain stem. The posterior column pathways, the spinothalamic tracts, and the spinoreticulothalamic tracts are three principal ascending pathways (see Fig. 19-10).

(a) The **dorsal columns (fasciculus cuneatus and fasciculus gracilis)** convey touch and proprioception via a three-neuron pathway. The cell body of the first neuron in this pathway resides in the dorsal root ganglion, and its central process enters the dorsal funiculus. These axons ascend ipsilaterally to the nuclei gracilis (for the lower extremity) or cuneatus (for the upper extremity) in the medulla (see Fig. 19-10). From there, second-order neurons decussate to reach the ventral posterolateral nucleus of the thalamus. The third neuron projects to the sensory cortex (postcentral gyrus) of the brain.

(b) The **lateral spinothalamic tract** conveys sharp pain and temperature while the **anterior spinothalamic tract** conveys light touch. Each tract is a three-neuron pathway. The cell body of the first neuron resides in the dorsal root ganglion, and its central process terminates in the dorsal horn of the central gray matter. The second neuron decussates through the anterior commissure (see Fig. 19-9) to the contralateral ventral column (see Fig. 19-10) and ascends to the midbrain where synapse occurs with the third neuron in the ventral posterolateral nucleus of the thalamus. The third neuron projects to the sensory cortex (postcentral gyrus) of the brain.

(c) The **spinoreticulothalamic tract** conveys dull and diffuse pain via a multineuron pathway. The cell body of the first neuron resides in the dorsal root ganglion, and its central process terminates on the second neuron in the dorsal horn of the central gray matter. The second neurons ascend in the anterolateral white matter to the reticular formation of the medulla. From there, high-order neurons project to the autonomic nuclei of the hypothalamus, to the intralaminar nucleus of the thalamus, and to the limbic lobe of the cortex.

(d) Crossed and uncrossed **spinocerebellar pathways** convey subconscious proprioception and pressure to the cerebellum for coordination. The uncrossed dorsal spinocerebellar tract and the cross ventral spinocerebellar tracts run in the lateral column (see Fig. 19-10) and enter the pons to be distributed to the cerebellar cortex.

5. Clinical considerations

a. Paralysis. A lesion on one side involving the corticospinal tract in the cortex or brain stem above the pyramidal decussation produces a contralateral hemiplegia; a lesion below the decussation produces ipsilateral hemiplegia. Bilateral involvement of the pyramidal tracts in the spinal cord produces paraplegia or quadriplegia, depending on the level of the lesion.

    b. Syringomyelia. A cavitating lesion of the central canal involves the spinothalamic tracts for pain and temperature as the fibers decussate in the anterior commissure. The patient usually observes bilateral finger burns with little or no pain.

    c. Tabes dorsalis. Involvement of the dorsal columns with loss of proprioception and sense of position produces a characteristic shuffling gait.

    d. Ipsilateral spinal cord crush injury (Brown-Séquard syndrome). Paralysis on the ipsilateral side, loss of the sensation of pain and temperature on the contralateral side below the level of the injury, and bilateral anesthesia at the level of the injury result.

    e. Spinal cord transsection. Complete paralysis and anesthesia below the injury result.

B. Nerve roots (see Fig. 19-7)

  1. Spinal nerves. There are 31 pairs of **spinal nerves**: eight cervical (C1 through C7 above each respective cervical vertebra and C8 below the seventh cervical vertebra), twelve thoracic, five lumbar, five sacral, and one coccygeal (each below the respective vertebra).

    a. Rootlets. Each dorsal and ventral root is formed by six to eight rootlets.

      (1) The sensory rootlets arise in linear fashion from the posterolateral sulci and the motor rootlets arise from the ventrolateral sulci.

      (2) The regions of the spinal cord, which contribute rootlets to specific spinal nerves, are termed **spinal segments (levels)**.

    b. Roots. Spinal nerves are formed by a dorsal (sensory) and a ventral (motor) root.

  2. Law of descent. As a result of unequal growth of the neural canal and spinal cord, there is a progressive descending obliquity of the nerve roots, forming the **cauda equina**.

    a. Cervical roots. Only in the cervical region do the segments of the spinal cord correspond in position to the level of the corresponding vertebrae.

    b. Thoracic, lumbar, and sacral roots. Below the cervical region, each spinal nerve from a given cord segment travels inferiorly (sometimes several inches as do the sacral rootlets) before exiting at the appropriate intervertebral foramen.

    c. Cauda equina. The nerve roots of the cauda equina float in cerebrospinal fluid. Therefore, a needle introduced into the dural sac will displace the roots without damage.

  3. Primary rami. The spinal nerve leaves the vertebral canal through the intervertebral foramen. Each spinal nerve bifurcates into a dorsal primary ramus and a ventral primary ramus (see Fig. 4-2).

    a. Dorsal primary rami supply the dermatomes and myotomes of the median portion of the back.

    b. Ventral primary rami supply the dermatomes and myotomes of the lateral and anterior portions of the body as well as the extremities in their entirety.

C. Vasculature of the spinal cord

  1. Arterial supply to the spinal cord is mainly by branches of the vertebral, deep cervical, intercostal and lumbar arteries. These vessels give off spinal branches.

    a. Vertebral artery branches. The spinal branches of the vertebral artery give off ascending and descending branches, which run in the pia mater and fuse to form the midline **anterior spinal artery** and two **posterior spinal arteries**.

    b. Intervertebral arteries. The intervertebral branches of the deep cervical, intercostal, and lumbar vessels bifurcate into anterior and posterior branches.

      (1) Radicular arteries. Most of the anterior and posterior intervertebral branches supply the dorsal and ventral roots and terminate in the pia associated with these roots as radicular arteries.

      (2) Medullary arteries. In the lower cervical and upper lumbar regions, six to eight larger anterior and posterior intervertebral branches reach the spinal cord where they course in the pia mater as medullary arteries.

        (a) Some anterior medullary arteries reach the anterior median fissure of the spinal cord and divide into ascending and descending branches. These anastomose with each other to continue the **anterior spinal artery**, which supplies the ventromedial aspects of the spinal cord.

        (b) Some posterior medullary arteries reach the posterolateral sulcus. Their ascending and descending branches anastomose to continue the paired **posterior spinal arteries**, which supply the dorsal and lateral aspects of the spinal cord.

  2. Venous return of the spinal cord forms a plexus in the pia mater.

    a. Two **median longitudinal veins**, two **anterolateral longitudinal veins**, and two **posterolateral longitudinal veins** are usually distinguishable. These veins drain into the vertebral venous plexus.

    b. The **vertebral venous plexus**, which lies in the fatty connective tissue of the epidural space, drains into **intervertebral veins**, which generally accompany the radicular arteries. The vertebral venous plexus also communicates superiorly with the venous sinuses of the cranium and inferiorly with the venous sinuses of the deep pelvis. Thus, the vertebral venous plexus offers not only a collateral pathway for venous drainage from the pelvis but also a potential pathway for the metastatic spread of carcinoma of the pelvic viscera.

D. Clinical considerations

  1. Lumbar puncture (spinal tap) is used to obtain samples of cerebrospinal fluid for laboratory analysis.

    a. Lumbar puncture is facilitated by the fact that the spinal cord rarely extends below the third lumbar vertebra.

    b. The spinous processes of the lower lumbar vertebrae project posteriorly, thereby providing access to the vertebral canal. When a patient is placed on the side in a decubitus position, thereby maximally flexing the vertebral column, the space between the spinous processes of adjacent lumbar vertebrae is widened further.

    c. A line tangential to the highest points of the iliac crests passes through the level of the fourth lumbar vertebra or the interspace between the fourth and fifth lumbar vertebrae.

    d. Local anesthetic is applied over the interspace between the fourth and fifth vertebrae, and a spinal needle is inserted through skin, superficial fascia, supraspinous ligament, interspinous ligament, between the paired flaval ligaments, and through the epidural space, the dura, and the arachnoid to enter the subarachnoid space, which contains cerebrospinal fluid.

  2. Spinal (intrathecal) anesthesia is used to block the roots of the spinal nerves *within* the dural sac (subarachnoid space).

    a. The procedure is the same as that for lumbar puncture.

    b. The number of segmental levels anesthetized is controlled by the amount of anesthetic injected and the position of the patient. Thus, the operating table must not be tilted so that the patient's head is below the horizontal during spinal anesthesia.

  3. Epidural anesthesia is used to block the spinal nerves in the epidural space in the lumbar region (i.e., external to the dural sac).

    a. The procedure is similar to that for lumbar puncture except that the dura is not penetrated.

    b. The number of nerves anesthetized above or below the injection site is controlled by the amount of anesthetic injected into the connective tissue of the epidural space.

  4. Caudal (sacral) anesthesia is used to block the spinal nerves in the epidural space (i.e., external to the dural sac).

    a. A needle inserted between sacral cornua penetrates the sacrococcygeal ligament to gain access to the sacral hiatus. The anesthetic is infiltrated into the epidural space around the dural sac.

    b. The degree of ascent of anesthesia can be controlled (e.g., for parturition) so that afferent pain fibers from the perineum and cervix running in the sacral nerves can be blocked (saddle block), leaving unaffected the lumbar fibers that control the abdominal musculature, as well as those involved in reflex uterine contractions.

Part V Back

STUDY QUESTIONS

Directions: Each question below contains five suggested answers. Choose the **one best** response to each question.

1. Which of the following structures pass through the posterior sacral foramina?

(A) Anterior primary rami
(B) Filum terminale
(C) Sensory rootlets
(D) Spinal nerves
(E) None of the above structures

2. The vertebral venous plexus is found in

(A) the dural sac
(B) the epidural space
(C) the subarachnoid space
(D) the subdural space
(E) none of the above

3. All of the following statements pertain to the pia mater EXCEPT

(A) it forms one of the boundaries of the subarachnoid space
(B) it forms the filum terminale
(C) it is adherent to the spinal cord
(D) it supports the anterior and posterior spinal arteries
(E) it terminates at the foramen magnum

4. The dura mater terminates caudally as the

(A) coccygeal ligament
(B) denticulate ligament
(C) dural sac
(D) filum terminale
(E) conus medullaris

5. In the adult vertebral column, the spinal cord usually terminates between

(A) T12 and L1
(B) L1 and L2
(C) L2 and L3
(D) L3 and L4
(E) L4 and L5

6. A spinal segment is defined as that region of the spinal cord that

(A) corresponds to a collection of nerves passing up or down within the white matter
(B) corresponds to the region of the vertebral column (i.e., cervical, thoracic, lumbar, and sacral) to which spinal nerves are sent
(C) sends rootlets to a particular spinal nerve
(D) underlies the neural arch of a particular vertebra in the adult
(E) none of the above

Directions: Each question below contains four suggested answers of which **one or more** is correct. Choose the answer

A   if **1, 2, and 3** are correct
B   if **1 and 3** are correct
C   if **2 and 4** are correct
D   if **4** is correct
E   if **1, 2, 3, and 4** are correct

7. Characteristics of the atlas include

(1) an odontoid process
(2) flat inferior articular facets
(3) a prominent spinous process
(4) bowl-shaped superior articular facets

8. Which of the following structures are formed by or receive a contribution from the costal processes of the developing vertebrae?

(1) Transverse processes of the cervical vertebrae
(2) Ribs of the thoracic region
(3) Alar plates of the sacrum
(4) Sacral cornua

9. Support for the nucleus pulposus is provided by which of the following structures?

(1) Anterior longitudinal ligament
(2) Annulus fibrosus
(3) Posterior longitudinal ligament
(4) Ligamentum flavum

10. The most common herniation of the nucleus pulposus of the L4–L5 intervertebral disk will

(1) impinge on the fifth lumbar nerve
(2) impinge on the fourth lumbar nerve
(3) be directed posterolaterally
(4) be covered by a layer of the posterior longitudinal ligament

11. The transversospinalis group includes which of the following intrinsic back muscles?

(1) Long and short rotators
(2) Semispinalis
(3) Multifidi
(4) Splenius cervicis

12. Large volumes of cerebrospinal fluid are found in which of the following spaces?

(1) Epidural space
(2) Dural sac
(3) Subdural space
(4) Subarachnoid space

Directions: The group of questions below consists of lettered choices followed by several numbered items. For each numbered item, select the **one** lettered choice with which it is **most closely** associated. Each lettered choice may be used once, more than once, or not at all. Choose the answer

A   if the item is associated with **(A) only**
B   if the item is associated with **(B) only**
C   if the item is associated with **both (A) and (B)**
D   if the item is associated with **neither (A) nor (B)**

Questions 13–17

For each joint or vertebral region, select the principal movements permitted from the list below.

(A) Flexion/extension
(B) Rotation
(C) Both
(D) Neither

13. Atlantoaxial joint

14. Atlanto-occipital joint

15. Thoracic region

16. Lumbar region

17. Sacral region

Directions: The group of questions below consists of lettered choices followed by several numbered items. For each numbered item select the **one** lettered choice with which it is **most closely** associated. Each lettered choice may be used once, more than once, or not at all.

Questions 18–20

For the procedure named in each question, select the most appropriate landmark structure from the list below.

(A) Iliac crests
(B) Posterior–superior iliac spines
(C) Spinous process of T12
(D) Sacral cornua
(E) Coccyx

18. Induction of caudal anesthesia

19. Obtaining cerebrospinal fluid by means of a spinal tap

20. Induction of spinal (intrathecal) anesthesia

ANSWERS AND EXPLANATIONS

1. The answer is E. [*Chapter 19 I D 4 a (3) (a)*] The sacral spinal nerves leave the vertebral canal via intervertebral foramina, which bifurcate almost immediately into anterior and posterior foramina. The dorsal primary rami pass through the posterior sacral foramina to innervate the back muscles and provide sensation to the dorsal portion of the dermatomes.

2. The answer is B. [*Chapter 19 III A 1 b, 3 a (3), b*] The vertebral venous plexus lies in the loose fatty connective tissue of the epidural space. It receives drainage from the spinal vertebral plexus and drains through the intervertebral veins. It provides a collateral route for venous drainage of the pelvis and, as such, is a pathway for the spread of pelvic malignancies.

3. The answer is E. [*Chapter 19 III A 3 b (2), c, C 1 a*] The pia mater, the innermost layer of the meninges, tightly invests the spinal cord and supports the arterial supply to the spinal cord. The space between the arachnoid and pia constitutes the subarachnoid space. The pia continues beyond the termination of the spinal cord as the filum terminale; it also continues through the foramen magnum as the pia of the brain.

4. The answer is A. [*Chapter 19 III A 3 a (1)*] From the dural sac, the dura mater extends to the second sacral vertebra as the coccygeal ligament, covering the pial filum terminale. The denticulate ligament supports the entire spinal cord in the center of the subarachnoid space.

5. The answer is B. [*Chapter 19 III A 2*] In most individuals, the spinal cord extends from the medulla oblongata at the level of the first cervical vertebra and terminates as the conus medullaris at the level of the first lumbar intervertebral disk (between vertebrae L1 and L2). In only 1% of the population does the conus medullaris extend beyond the L2 level.

6. The answer is C. [*Chapter 19 III B 1 a (2), 2 a, b; Figure 19-8*] A spinal segment is defined as that region of the spinal cord that sends rootlets to a particular spinal nerve. Because the vertebral column grows postnatally while the spinal cord does not, most of the vertebrae are displaced from the corresponding spinal segments in the adult, which accounts for the cauda equina.

7. The answer is C (2, 4). [*Chapter 19 I D 1 a, b*] The atlas, the first cervical vertebra, has large concave superior facets, which articulate with the occipital condyles. Considerable flexion/extension occurs at the atlanto-occipital joint. The inferior facets, which articulate with the axis, are rather flat, permitting considerable rotation at the atlantoaxial joint. The atlas has neither a body nor a prominent spinous process, the former being represented by the odontoid process of the axis.

8. The answer is A (1, 2, 3). [*Chapter 19 I D 1 c (1) (a), 2 a (2), 4 a (3)*] The costal processes of the developing vertebrae participate in the formation of the transverse processes of the cervical vertebrae, the ribs, and the alar plates of the sacrum. The sacral cornua are the pedicles of the fifth sacral vertebra where the lamina have failed to form.

9. The answer is A (1, 2, 3). [*Chapter 19 I E 2 a (3), F 4 c, 5 c*] The annulus fibrosus forms the outer portion of the intervertebral disk and provides the principal support for the more gelatinous nucleus pulposus. Additional support is provided by the anterior and posterior longitudinal ligaments. Disk herniation occurs most frequently in the posterolateral aspect of the disk where these ligaments do not meet.

10. The answer is B (1, 3). [*Chapter 19 I E 4 b (5), F 5*] The combination of the anterior and posterior longitudinal ligaments is deficient posterolaterally. Herniation of the nucleus pulposus through this weak area will be in a posterolateral direction. Spinal nerve L4 is not jeopardized because the intervertebral notch in the inferior border of the pedicle of L4 is cut so high that spinal nerve L4 exits superiorly to the disk. However, the herniated nucleus pulposus impinges on the next lower spinal nerve, L5 in this case, as it courses posterolaterally past the herniated disk toward the next intervertebral foramen.

11. The answer is A (1, 2, 3). [*Chapter 19 II A 1 a, c*] While all of the muscles listed are primarily rotators of the vertebral column, the splenius cervicis and splenius capitis muscles are members of the more superficial spinotransverse group. They originate on the spinous processes and diverge superolaterally to insert into transverse processes. The semispinalis, multifidi, and rotators comprise the deep transversospinalis group. These muscles originate on transverse processes and converge superomedially onto spinous processes.

12. The answer is C (2, 4). [*Chapter 19 III A 1 b, 3 b (2) (a)*] Cerebrospinal fluid is contained in the sub-arachnoid space. Below the termination of the conus medullaris, the subarachnoid space is termed the dural sac. The subdural space is only a potential space, and the epidural space contains the vertebral venous plexus.

13–17. The answers are: 13-B, 14-A, 15-C, 16-A, 17-D. [*Chapter 19 I D 1 a, b, 2 b, 3 b, 4 b*] In the cervical region, most of the flexion/extension occurs at the atlanto-occipital joint, while most of the rotation occurs at the atlantoaxial joint. In the thoracic region, there is a moderate amount of both rotation and flexion/extension, largely possible due to the elasticity of the costal cartilages. The principal movement in the lumbar region is flexion/extension because the zygapophyseal facets are oriented in anterior–posterior planes. The fused sacral vertebrae permit no movement.

18–20. The answers are: 18-D, 19-A, 20-A. [*Chapter 19 I D 4 a (4); III D 1*] The sacral cornua (rudimentary pedicles of the fifth sacral vertebra) are important landmarks for the location of the sacral hiatus through which caudal (sacral epidural) anesthesia is induced. Because the spinal cord always ends by the L3 level and because the spinous process of L4 is nearly horizontal, the interspace between L4–L5 is the preferred site for a spinal tap to sample cerebrospinal fluid as well as for the induction of spinal (intrathecal) anesthesia. The intertubercular plane through the iliac crests, which passes through the body and spinous process of L4 or through the fourth intervertebral disk, provides the best landmark for these procedures. The spinous process of the twelfth thoracic vertebra is most difficult to locate and is, therefore, not a landmark. The posterior–superior iliac spines lie at the L1–S1 level.

Part VI
Pelvic Region

20
Pelvis

I. OSSEOUS PELVIS

A. General structure. The pelvis is formed by the bilateral **hip (pelvic) bones**, the **sacrum**, and the **coccyx**.

1. **The hip (pelvic or coxal) bone** on each side of the pelvis is formed by the fusion of three separate bones—the **ilium**, **ischium**, and **pubis**. These join at the **acetabulum** (*L.* vinegar cruet), which is the fossa of the hip joint.

   a. **The ilium** is the most superior portion of the coxal bone (Fig. 20-1).

   (1) A flared **alar plate** provides attachment for back and thigh musculature, whereas the **iliac fossa**, the concave inner surface, provides attachment for the iliacus muscle.

   (2) The **auricular surface**, the posterior portion of the ala, articulates with the sacrum at the **sacroiliac joint**.

   (3) The **iliac crest** at the free superior edge of the ala terminates at the antero- and postero-superior iliac spines and provides attachment for back and abdominal musculature as well as several named ligaments.

   (4) The **greater sciatic notch** of the ilium provides an aperture through which neurovascular as well as muscular structures pass from the deep pelvis into the thigh.

   (a) The **piriformis muscle** originates from the anterior aspect of the sacrum, passes through the greater sciatic notch, and inserts at the base of the greater trochanter of the femur.

   (b) The **suprapiriformis recess** is that portion of the **greater sciatic foramen** superior to the piriformis muscle. It contains the **superior gluteal neurovascular bundle**.

   (c) The **infrapiriformis recess** is that portion of the greater sciatic foramen inferior to the piriformis muscle. It contains the **inferior gluteal neurovascular bundle**, the **sciatic nerve**, and the **pudendal neurovascular bundle**. A portion of the **sciatic nerve** occasionally may pass through the piriformis muscle, which may compress the nerve to produce *sciatica*.

   (5) The **linea terminalis** describes the pelvic inlet and provides the boundary between the greater pelvis and the deep pelvis.

   b. **The ischium** is the most inferior portion of the coxal bone (see Fig. 20-1).

   (1) The **ischial ramus** fuses with the inferior pubic ramus.

   (2) The **ischial tuberosity**, a palpable landmark, provides an attachment for muscles of the posterior thigh as well as for the **sacrotuberous ligament**.

   (a) In a person who is standing, the gluteus maximus muscle on each side covers the ischial tuberosity.

   (b) In a person who is sitting, the gluteus maximus muscles move laterally so that one sits on the ischial tuberosities.

   (3) The **ischial spine** provides attachment for the **sacrospinous ligament**, which bridges the greater sciatic notch just superior to the sacrotuberous ligament to form the greater sciatic foramen.

   (4) The **lesser sciatic notch** of the ischium lies between the ischial tuberosity and the ischial spine. It provides an aperture through which neurovascular as well as musculotendinous structures pass from the deep pelvis to the thigh and perineum.

   (a) The lesser sciatic notch with the sacrospinous and sacrotuberous ligaments forms the **lesser sciatic foramen**.

   (b) The **obturator internus muscle** originates on the internal aspect of the ilium and pubis, exits the pelvis through the lesser sciatic foramen, and inserts at the base of the greater trochanter of the femur.

   (c) The **pudendal neurovascular bundle** arises in the deep pelvis, exits through the greater sciatic foramen, and passes through the lesser sciatic foramen to reach the perineum.

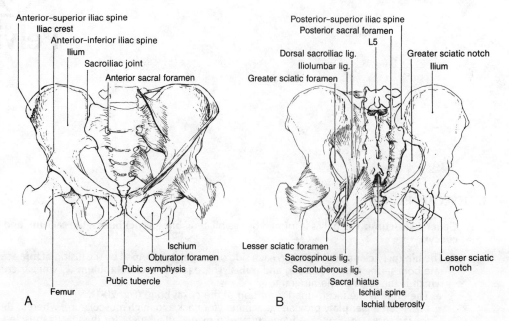

Figure 20-1. *Pelvis. A,* Anterior aspect of the female pelvis. The anterior ligaments supporting the left lumbosacral joint are included. *B,* Posterior aspect of the female pelvis. The posterior ligaments supporting the left lumbosacral joint are included.

   c. The pubis is the most anterior portion of the coxal bone (see Fig. 20-1A).
     (1) The **superior pubic ramus** fuses with the ilium.
     (2) The **inferior pubic ramus** fuses with the ischial ramus.
     (3) Anteriorly, left and right pubic bones articulate at the **pubic symphysis**.
       (a) The pubic symphysis, a true amphiarthrosis, is composed of hyaline cartilage artic-
        ular surfaces separated by a fibrocartilage disk.
       (b) This joint relaxes under the influence of progesterone in the weeks prior to parturi-
        tion, allowing the birth canal to widen.

  2. The sacrum forms the posterior portion of the pelvis (see Figs. 19-4 and 20-1).
   a. Structure. The five sacral vertebrae fuse to form the sacrum, and four lines of fusion are
    evident on the anterior surface.
    (1) The fused spinous processes form the **medial sacral crest**.
    (2) The fused transverse processes together with fused costal processes form the **alae**, or
     sacroiliac articular processes.
   b. Sacral foramina. The **posterior sacral foramina** transmit the dorsal primary rami of the sa-
    cral spinal nerves and the **anterior sacral foramina** transmit the ventral primary rami of the
    sacral spinal nerves.
   c. Sacral hiatus. The laminae of the fifth sacral vertebra fail to form, resulting in the sacral hia-
    tus. The pedicles form the **sacral cornua**, which are important landmarks used in locating
    the sacral hiatus for the administration of *caudal anesthesia*.

  3. The sacroiliac joint is formed by the articulation of the alar plates of the sacrum with the ilium
   (see Fig. 20-1B).
   a. Structure. Because there is a synovial cavity within it, the sacroiliac joint is a diarthrosis.
    (1) Although there is slight mobility at this joint, movement is greatest in the pregnant fe-
     male in the weeks prior to parturition.
    (2) Movement at this joint decreases with age; ankylosis usually is complete by age 50.
   b. Ligamentous support. As a result of ligamentous reinforcement, this joint is probably
    among the strongest diarthrodial joints in the body. These ligaments oppose the rotational
    effect of gravity at the sacroiliac joint.
    (1) The **sacrotuberous ligament** runs from the sacrum to the ischial tuberosity. It strongly
     reinforces the sacroiliac joint and transmits directly to the sacrum the forces generated
     by the hamstring muscles of the posterior thigh as well as the gluteus maximus.
    (2) The **sacrospinous ligament** runs from the sacrum to the ischial spine. It closes the
     greater sciatic notch and bounds the lesser sciatic foramen. The boundaries of the les-
     ser sciatic foramen include the sacrotuberous ligament, lesser sciatic notch, and sacro-
     spinous ligament.

(3) The **iliolumbar ligament** runs between the iliac crest and the transverse processes of the fifth lumbar vertebra.

(4) The **dorsal sacroiliac ligaments** run between the posterosuperior iliac spine and the dorsum of the sacrum.

(5) The **ventral sacroiliac ligament** runs across the sacroiliac joint on the anterior surface of the pelvis.

(6) The **interosseous ligaments** run between the sacroiliac articular surfaces. The tendency of the sacrum to rotate forward tightens these ligaments, locking the joint.

   c. **"Low back" pain** is often misdiagnosed as "sacroiliac sprain" or "lumbago." There have been essentially no documented instances of sacroiliac sprain. Low back pain is usually of muscular origin, the result of either arthritis of the lumbar zygapophyses or a herniated disk at the level of L4 or L5.

4. **The coccyx**, or **tail bone**, forms the inferoposterior portion of the pelvis.
   a. **Structure.** The four or five coccygeal vertebrae fuse to form the coccyx, but there may be a disk between the first and second coccygeal vertebrae.
      (1) The first coccygeal vertebra articulates with the sacrum. The sacrococcygeal joint usually has a small intervertebral disk but may fuse in some individuals.
      (2) The first (and sometimes second) coccygeal vertebra has rudimentary pedicles, which form the coccygeal cornua.
      (3) The second through fourth coccygeal vertebrae represent rudimentary vertebral bodies.
   b. **Sex differences.** In the male, the coccyx tends to be directed more anteriorly toward the pubis, whereas in the female, the coccyx tends to be slightly more vertical, thereby providing a larger pelvic outlet.
   c. **Coccygodynia.** Fracture of the fused coccygeal joints or of a fused sacrococcygeal joint, or arthritis in these joints, causes the painful condition, *coccygodynia*.

B. **Birth canal.** The bony pelvis is made up of the sacrum and pelvic girdle; the enclosed pelvic cavity is an extension of the abdominal cavity. This region contains portions of the gastrointestinal tract; with the exceptions of a gravid uterus and the fundus of the full urinary bladder, no pelvic organs lie in the major pelvis.

1. **The major (greater or false) pelvis** is that portion of the bony pelvis between the iliac crests and the pelvic brim.
   a. Laterally, the **alar plates** (wings) of the ilia form the walls of the major pelvis.
   b. Posteriorly, the major pelvis is bounded by the lumbar vertebrae, L3, L4, and L5.
   c. Anteriorly, the major pelvis is bounded by the anterior abdominal wall.

2. **The minor (lesser, deep, or true) pelvis** lies inferiorly to the pelvic brim.
   a. **The pelvic inlet** demarcates the minor from the major pelvis.
      (1) **Boundaries.** The pelvic inlet is defined by the sacral promontory and the linea terminalis of the innominate bone. The **linea terminalis** includes the pubic crest, iliopectineal line, and the arcuate line of the ilium.
      (2) **Measurements of the female pelvic inlet**
         (a) **Conjugate diameters**
            (i) The **true conjugate** (about 10 cm) is the anteroposterior diameter from the sacral promontory to the superior margin of the pubic symphysis. This can be measured only radiographically in a lateral projection.
            (ii) The **diagonal conjugate** (about 10.5 cm) is measured from the sacral promontory to the inferior margin of the pubic symphysis. This is easily noted on pelvic examination.
            (iii) The **obstetric conjugate** is the least anteroposterior diameter from the sacral promontory to a point a few millimeters below the superior margin of the pubic symphysis.
         (b) The **oblique diameter** (about 12.5 cm) is measured from the sacroiliac joint to the contralateral iliopectineal eminence.
         (c) The **transverse diameter** (13.5 cm) is the widest distance across the pelvic brim.
   b. **The pelvic cavity** is defined by the minor pelvis, forms the birth canal, and contains portions of the gastrointestinal tract (loops of ileum, sigmoid colon, rectum, and anal canal) as well as the lower portion of the urinary tract and certain reproductive organs.
   c. **The pelvic outlet** is closed by the pelvic diaphragm and covered by the perineum.
      (1) **Boundaries.** The pelvic outlet is defined by the coccyx, ischial tuberosities, inferior pubic ramus, and pubic symphysis. It is closed by the pelvic diaphragm and reinforced by the urogenital diaphragm.
      (2) **Measurements of the female pelvic outlet**
         (a) The **transverse diameter** is between the ischial tuberosities. It is approximately as long as a clenched fist is wide.

(b) The **transverse midplane diameter** (about 10.5 cm) is measured between the ischial spines. If this diameter is less than 9.5 cm, there is nearly a 50% chance that surgical intervention will be necessary during labor.

(c) The **anteroposterior (sagittal) diameter** (about 13.5 cm) is measured from the lower margin of the pubic symphysis to the sacrococcygeal joint.

(3) Pelvimetry, although controversial, can be a valuable diagnostic tool in cases of cephalopelvic disproportion.

3. Characterization of the pelvis

a. Inlet shape

(1) The normal male pelvis is **android** with a heart-shape due to the prominent sacrum.

(2) The normal female inlet shape is **gynecoid**.

(3) An **anthropoid pelvis** has a lengthened anteroposterior diameter and a shortened transverse diameter.

(4) A **platypoid pelvis** is oval with a shortened anteroposterior diameter and a lengthened transverse diameter.

(5) A pelvis may be **asymmetric** due to scoliosis or congenital malformation.

b. Sex differences

(1) The bones of the female pelvis are thinner than those of the male, the pelvic cavity is less funnel-shaped, the surface for articulation of the sacrum with L5 is smaller, and the subpubic angle is greater (almost 90°).

(2) A summary of sexual differences is found in Table 20-1.

II. PELVIC MUSCULATURE

A. Organization. The **pelvic floor (pelvic diaphragm)** is composed of the **coccygeus** and **levator ani muscles** (Fig. 20-2).

1. The coccygeus (ischiococcygeus) muscle forms a small portion of the pelvic floor and is a rudimentary "tail-wagger." This simple muscle originates from the ischial spine and inserts onto the lateral borders of the fourth sacral to the second coccygeal vertebrae as well as along the **sacrospinous ligament**. This ligament may be regarded as a degenerated portion of the coccygeus muscle.

2. The levator ani muscle forms most of the pelvic floor (Table 20-2).

a. Subdivisions. The complex levator ani muscle is divided into three parts.

(1) The **iliococcygeus muscle** is a rudimentary "tail-wagger." It arises laterally from the **ischial spine** and the **arcus tendineus (white line)** of the **obturator fascia**. It inserts onto the coccyx.

(2) The **pubococcygeus muscle** is another rudimentary "tail-wagger." It arises from the arcus tendineus, laterally, and the **pubic arch**, anteriorly. It inserts primarily onto the coccyx and into the anococcygeal raphe.

(3) The **puborectalis muscle** is the most medial portion of the levator ani muscle.

(a) Arising from the pubic arch medially to the pubococcygeus muscle, the fibers loop around the rectum to form the **rectal sling**, the principal mechanism for fecal continence.

(b) Some fibers of the puborectalis muscle loop posteriorly to the vagina in the female and posteriorly to the prostate in the male to insert onto the central tendon of the perineum. These fibers are considered by some to be separate muscles, the **pubovaginalis** and **puboprostaticus**.

b. The urogenital hiatus is a midline defect in the levator ani muscle through which pass the rectum, the urethra, and in the female, the vagina. The urogenital hiatus is reinforced inferiorly by the **urogenital diaphragm** in the perineum.

Table 20-1. Differences between the Male and Female Pelvis

Characteristic	Male	Female
Size	Smaller	Larger
Greater sciatic notch	Narrower	Wider
Inferior pubic angle	Narrower	Wider
False pelvis	Tall	Flaired
Pelvic inlet	Android	Gynecoid
Pelvic outlet	Smaller	Larger

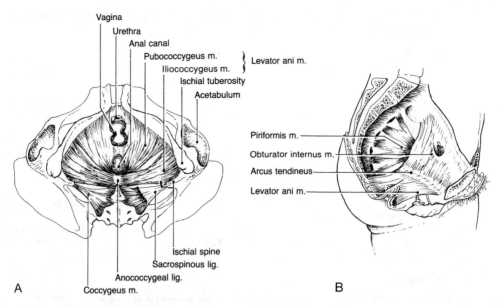

Figure 20-2. *Pelvic floor. A,* Inferior view of the female pelvis showing the muscles of the pelvic diaphragm. *B,* Sagittal section of the female pelvis showing the muscles of the pelvic diaphragm.

    3. Innervation. The muscles of the pelvic floor are innervated by twigs from the sacral plexus.

    4. Group actions
        a. Urinary continence. The pelvic floor supports the urinary bladder and participates in urinary continence.
        b. Fecal continence. The rectal sling, by drawing the rectum anteriorly, is the principal component for the maintenance of fecal continence.
        c. Uterine support. In the female, the pelvic floor stabilizes the vagina and indirectly participates in the support of the uterus.

  B. Other muscles. The iliacus, obturator internus, and piriformis originate within the pelvis but are associated with the lower extremity and act across the hip joint (see Fig. 20-2*B*).

III. PELVIC FASCIAE AND PERITONEUM

  A. Visceral (endopelvic) fascia

    1. Location. The endopelvic fascia, a continuation of the extraperitoneal connective tissue layer, ensheathes viscera and forms septa between the organs.

    2. Pelvic ligaments
        a. The pelvic ligaments are thickened continuities between the parietal and visceral fasciae; in some locations, they are condensations of the neurovascular sheaths.
        b. These condensations of endopelvic fascia support pelvic viscera and provide pathways for the associated neurovascular bundles.

Table 20-2. Muscles of the Pelvic Floor

Muscle	Origin	Insertion	Action	Innervation
Levator ani: Coccygeus	Ischial spine	Vertebrae S4–Cx2	Supports the pelvic viscera	Nn. S4–Cx2
Iliococcygeus	Arcus tendineus	Coccyx	Supports the pelvic viscera	Nn. S5–Cx2
Pubococcygeus	Pubic arch and arcus tendineus	Coccyx	Supports the pelvic viscera	Nn. S4–Cx1
Puborectalis (rectal sling)	Pubic arch	Midline and rectum	Rectal continence	Nn. S3–S5

c. The pelvic ligaments are especially important in the support of the uterus and, to some extent, the urinary bladder (see Chapter 22 VI B).

B. Parietal fascia

1. **Location.** The parietal fascia is a continuation of the transversalis fascia into the pelvis. As investing fascia, it covers the piriformis and obturator internus muscle.

2. **Subdivisions.** This fascial layer, which attaches to the arcuate line of the pubis and ilium, thickens over the obturator internus muscle to form the **arcus tendineus**, the origin of the levator ani muscle.
   a. At the arcus tendineus, a diaphragmatic fascial layer splits to cover both superior and inferior surfaces of the levator ani muscle as the **superior** and **inferior fasciae of the pelvic diaphragm**.
      (1) The superior fascia of the pelvic diaphragm reflects onto the pelvic viscera as **visceral fascia**.
      (2) The inferior fascia of the pelvic diaphragm contributes to the **superior fascia** of the **urogenital diaphragm**.
   b. Inferior to the arcus tendineus, the parietal fascial layer splits around the pudendal neurovascular bundle to form the **pudendal canal** (of Alcock) along the medial surface of the obturator internus muscle.

C. Peritoneum

1. **Parietal peritoneum** overlies portions of the pelvic wall and endopelvic fasciae.

2. **Visceral peritoneum** reflects onto the pelvic viscera, forming mesenteries in the process.

IV. PELVIC VASCULATURE

A. Arterial supply

1. **Common iliac arteries** (Fig. 20-3)
   a. Origin and course. The left and right common iliac arteries arise with the bifurcation of the **aorta** at the level of the fourth lumbar vertebra. The common iliac vessels bifurcate at the level of the fifth lumbar intervertebral disk to form **internal** and **external iliac** vessels.
   b. Distribution. With the exception of small branches to the psoas muscle, twigs to the ureters, and an occasional iliolumbar artery or accessory inferior renal artery, the common iliac arteries give rise to no major branches.

2. **External iliac artery** (see Fig. 20-3)
   a. Origin and course. The external iliac artery and vein course in the greater pelvis just above the pelvic brim. The external iliac artery exits the pelvis beneath the inguinal ligament and becomes the **femoral artery**.
   b. Distribution. The external iliac artery gives off the **inferior epigastric artery** and **deep iliac circumflex artery** in the vicinity of the inguinal ligament.

3. **Internal iliac artery** (see Fig. 20-3)
   a. Origin and course. The internal iliac artery dives into the deep pelvis.
   b. Distribution. Each internal iliac artery supplies the walls of the pelvis, the pelvic viscera, and the perineum. The branches of the internal iliac vessels are extremely variable in site of origin, arrangement, number, and position.
      (1) The **posterior trunk of the internal iliac artery** provides only somatic branches.
         (a) The **iliolumbar artery** courses upward behind the external iliac vessels as far as the medial border of the psoas muscle, where it divides into an iliac branch and a lumbar branch. The iliolumbar artery is especially vulnerable to tearing in posterior pelvic fractures with resultant severe hemorrhage.
            (i) The **lumbar branch** supplies the psoas and quadratus lumborum muscles as well as the lower lumbar vertebra and dural sac.
            (ii) The **iliac branch** supplies the iliacus muscle. Its anastomoses with the distributions of the superior gluteal artery, iliac circumflex artery, and lateral femoral circumflex artery provide collateral circulation about the hip.
         (b) The **lateral sacral artery** supplies the region of the sacrum.
         (c) The **superior gluteal artery** is the largest branch of the internal iliac artery.
            (i) This artery leaves the pelvic cavity via the **suprapiriform portion of the greater sciatic foramen**.
            (ii) It supplies portions of the gluteus maximus, gluteus medius, and gluteus minimus muscles and sends twigs to the hip joint.

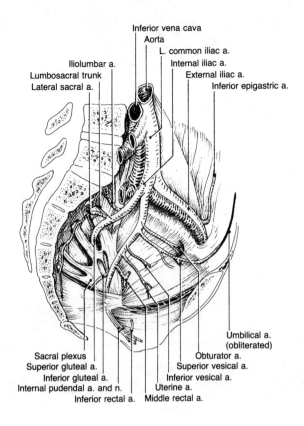

Iliolumbar a.
Lumbosacral trunk
Lateral sacral a.

Inferior vena cava
Aorta
L. common iliac a.
Internal iliac a.
External iliac a.
Inferior epigastric a.

Umbilical a.
(obliterated)
Obturator a.
Superior vesical a.
Inferior vesical a.
Uterine a.
Middle rectal a.

Sacral plexus
Superior gluteal a.
Inferior gluteal a.
Internal pudendal a. and n.
Inferior rectal a.

Figure 20-3. *Vasculature of the pelvis.*

(2) The **anterior trunk of the internal iliac artery** supplies both somatic and visceral structures.

(a) The **inferior gluteal artery** supplies somatic structures.

(i) Inside the pelvic cavity, this vessel supplies portions of the coccygeus, piriformis, and levator ani muscles.

(ii) This artery leaves the pelvic cavity via the **infrapiriform portion of the greater sciatic foramen** to supply portions of the gluteus maximus muscle, the lateral rotator muscles, and the hip joint.

(b) The **internal pudendal artery** supplies the somatic structures of the perineum.

(i) This vessel exits the pelvis via the infrapiriform portion of the greater sciatic foramen.

(ii) It enters the ischiorectal fossa of the perineum via the **lesser sciatic foramen**.

(iii) The branches of the internal pudendal artery are discussed in Chapter 21 IV E 1 a.

(c) The **obturator artery** supplies somatic structures in the anteromedial thigh.

(i) This vessel runs ventrally along the pelvic wall, medially to the obturator fascia, and leaves the pelvic cavity via the obturator canal to supply the adductor muscles and the artery of the ligamentum teres of the femoral head.

(ii) Within the pelvic cavity, this vessel provides twigs to the iliacus muscle.

(iii) The obturator artery may arise as an aberrant obturator artery from the inferior epigastric artery (30%). Such an anomalous artery may complicate surgery in the region of the femoral ring.

(d) The **umbilical artery** retains a lumen for a short distance beyond the internal iliac artery and gives off one or two vesical branches to the urinary bladder.

(i) The **superior vesical artery** supplies a number of small branches to the cranial portion of the urinary bladder.

(ii) The **middle vesical arteries** are variable and, when present, supply the fundus of the urinary bladder.

(iii) The **medial umbilical ligaments** are the obliterated remnants of the **umbilical artery**.

(e) The **uterine artery** in the female is homologous to the **deferential artery** in the male. In the female, the **uterine artery** courses medially on the superior surface of the levator ani toward the cervix, passes superiorly to the ureters in the **transverse**

cervical (cardinal) ligament, and ascends within the broad ligament to reach the uterus. This artery hypertrophies greatly during pregnancy.

   (i) **Tubal branches** supply the oviducts and anastomose with the distribution of the ovarian arteries.

   (ii) **Vaginal branches** supply the inner portions of the vagina and anastomose with the vaginal branches of the internal pudendal artery.

(f) The **deferential artery** in the male supplies the vas deferens and epididymis.

   (i) This vessel anastomoses with the spermatic artery about the testes.

   (ii) Usually, it arises separately but it may arise from the umbilical artery or the inferior vesical artery.

(g) **Middle rectal (hemorrhoidal) artery**

   (i) The middle rectal artery is distributed to the rectum.

   (ii) It anastomoses with the inferior mesenteric artery via the superior rectal artery and with the internal pudendal artery via the inferior rectal artery.

(h) **Inferior vesical arteries** (one or two) supply the neck of the urinary bladder, and in the male, portions of the prostate and the seminal vesicles. They may give rise to the deferential artery.

B. Pelvic venous return

1. **Course.** The small veins of the pelvic cavity generally follow the arteries but with even greater variation. These drain into the **internal iliac vein**, which reaches the inferior vena cava via the **common iliac vein**.

2. **Venous anastomoses.** There are several areas of important venous anastomoses within the pelvic cavity.

   a. **The rectal venous plexus** lies about the rectum.

      (1) Abundant anastomoses exist among superior rectal, middle rectal, and inferior rectal (hemorrhoidal) veins.

      (2) These anastomoses connect the hepatic portal system with systemic drainage.

   b. **The vesical venous plexus** lies about the base of the bladder and, in the male, the prostate.

      (1) This venous plexus receives drainage from the penis or clitoris via the deep dorsal vein and from the urinary bladder, as well as the vas deferens, seminal vesicle, and prostate in the male.

      (2) Because drainage from this plexus is diffuse and without valves, cancer of the prostate may spread via several hematogenous routes with the potential for metastatic rests along the courses.

         (a) The principal drainage is into the internal iliac vein.

         (b) Collateral drainage occurs into the **vertebral venous plexus** within the neural canal, the obturator veins (which frequently drain into the external iliac veins) and the rectal plexus.

C. Pelvic lymphatic drainage

1. **Course.** The lymphatic drainage of the pelvis, which is extremely variable, is divided into two groups of nodes.

   a. **The external nodes** lie near the pelvic brim along the external and common iliac vessels. They receive lymph from the superficial and deep inguinal nodes, draining the lower extremity and portions of the external genitalia.

   b. **The internal nodes** lie within the deep pelvis along the internal iliac vessels. These drain the pelvic floor and pelvic viscera.

2. **Lymphatic anastomoses.** Anastomoses occur widely with the abdominal lymphatics around the rectum and with the superficial lymphatic drainage of the perineum. The lymphatic drainage is of major surgical concern in procedures involving carcinoma of the pelvic viscera.

V. LUMBOSACRAL PLEXUS

A. Organization. The lumbosacral plexus provides the somatic innervation to the pelvis and lower extremities.

1. **Lumbosacral trunk.** The **lumbar plexus** (T12–L4) lies in the posterior abdominal wall and iliac fossa (see Chapter 18 IV D 2 a). The **lumbosacral trunk** (L4–L5, anterior and posterior divisions) contributes to the **sacral plexus** and joins the lumbar plexus with the sacral plexus.

   a. **The anterior division** of the lumbosacral trunk contributes the **tibial** portion of the **sciatic nerve**. In general, this nerve is sensory and motor to the posterior aspects of the thigh and leg as well as to the plantar surface of the foot.

b. **The posterior division** of the lumbosacral trunk contributes the **superior** and **inferior gluteal nerves** as well as the **common peroneal** portion of the sciatic nerve. In general, these nerves are sensory and motor to the posterior hip as well as to the anterior leg and to the dorsum of the foot.

2. **The sacral plexus** (L4–S3) lies in the minor pelvis and supplies the gluteal region, posterior thigh, leg, and foot as well as the perineum.

3. **The pelvic splanchnic nerves** (nervi erigentes) arise from spinal nerves S2–S3 or S3–S4.
   a. These consist of parasympathetic preganglionic fibers.
   b. They also contain visceral afferent fibers from specific portions of the pelvic viscera.

4. **The coccygeal plexus** (S4–Cx1) supplies the terminal portion of the vertebral column.

B. **Major branches of the sacral plexus** (Fig. 20-4)

1. **Superior gluteal nerve** (L4–S1, posterior)
   a. This nerve exits the pelvis via the suprapiriform portion of the greater sciatic foramen.
   b. It supplies the gluteus medius and gluteus minimus muscles.
   c. Severance of this nerve results in ''abductor lurch,'' a rolling (Trendelenburg) gait due to loss of abductive power in the hip.

2. **Inferior gluteal nerve** (L5–S2, posterior)
   a. This nerve exits via the infrapiriform region of the greater sciatic foramen.
   b. It supplies the gluteus maximus muscle.
   c. Paralysis results in difficulty climbing stairs or rising from a chair.

3. **The common peroneal nerve** (L4–S2, posterior) forms the posterior component of the sciatic nerve.
   a. This nerve exits via the infrapiriform region of the greater sciatic foramen.
   b. It innervates the anterior aspect of the leg and the dorsum of the foot.
   c. Paralysis results in the inability to dorsiflex the foot (foot drop) and to evert the foot.

4. **The tibial nerve** (L4–S3, anterior) forms the anterior component of the sciatic nerve.
   a. This nerve exits from the infrapiriform region of the greater sciatic foramen.
   b. It innervates the posterior region of the thigh and leg and the plantar surface of the foot.
   c. Paralysis is indicated by an inability to stand on the toes and loss of the Achilles tendon reflex.

5. **Pudendal nerve** (S2–S4, anterior)
   a. This nerve exits the pelvis via the infrapiriform region of the greater sciatic foramen. It then enters the ischiorectal fossa of the perineum via the lesser sciatic foramen. It comes to lie in the **pudendal canal (of Alcock)**, a condensation of the obturator fascia on the medial surface of the obturator muscle.
   b. Branches of the pudendal nerve supply the perineum.
      (1) The **inferior rectal nerve** is motor to the external anal sphincter and sensory to the inferior portion of the anal canal and the anal triangle of the perineum.

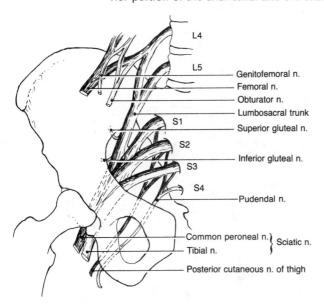

L4
L5
Genitofemoral n.
Femoral n.
Obturator n.
Lumbosacral trunk
S1
Superior gluteal n.
S2
Inferior gluteal n.
S3
S4
Pudendal n.
Common peroneal n. } Sciatic n.
Tibial n.
Posterior cutaneous n. of thigh

Figure 20-4. *Sacral plexus.* The anterior divisions (*light*) and posterior divisions (*shaded*) of the sacral plexus are indicated.

    (2) The **perineal nerve** is motor to the muscles of the urogenital diaphragm and external genitalia and sensory to the urogenital triangle of the perineum, including the external genitalia.

  c. The pudendal nerve may be anesthetized by *pudendal block*.

    (1) With the patient in the lithotomy position, the ischial tuberosities are palpated and, after cutaneous anesthesia, the needle is passed just medial to the tuberosity to a depth of about 2 cm.

    (2) Alternatively, the transvaginal route involves palpation of the ischial spine per vagina; the needle is inserted through the lateral vaginal wall to a point approximately 1 cm medial to the ischial spine and 1 cm below the sacrospinous ligament.

I. PERINEUM

A. General structure

1. **Perineal boundaries.** The perineum is defined as the area between the **ischial tuberosities**, extending from the **pubis** to the **coccyx**.

2. **Perineal function.** The perineum covers the **pelvic outlet** and is pierced by anal and urogenital canals, which incorporate sphincteric mechanisms.

B. Divisions. The perineum is divisible by the **transverse diameter** (a line connecting the ischial tuberosities) into the **anal triangle** posteriorly and the **urogenital triangle** anteriorly (Fig. 21-1). Both perineal triangles share the levator ani muscle, the same innervation, and the same blood supply. However, each perineal triangle has special musculature.

1. **Anal triangle.** The principal structures of the anal triangle are the anal canal and the anus, the terminal portion of the gastrointestinal tract.

2. **Urogenital triangle.** The principal structures of the urogenital triangle are the external genitalia. Although the differences between the superficial structures of the urogenital triangle of the male and female are obvious, the structures have homologous counterparts in both sexes.

II. DEVELOPMENT OF THE EXTERNAL GENITALIA

A. Indifferent stage (weeks 3–6)

1. **Cloacal folds** form on either side of the **cloacal membrane**, which separates the lumen of the hindgut from the perineal region.

2. **The genital tubercle** is formed anteriorly by fusion of the cloacal folds.

3. **The urorectal septum** divides the cloaca into the **urogenital sinus**, anteriorly, and the **anal canal**, posteriorly, by the sixth week.
   a. This transverse septum divides the cloacal membrane into the **urogenital membrane** and the **anal membrane**. The cloacal sphincter is similarly divided, becoming the **urogenital diaphragm** and the **external anal sphincter**.
   b. The fusion of the urogenital septum with the cloacal membrane forms the **central tendon** of the perineum (**perineal body**) and divides the cloacal folds into **urethral folds** and **anal folds**.

4. **Genital swellings (labial–scrotal folds)** develop laterally to the urethral folds.

5. **The urogenital and anal membranes** degenerate shortly afterward, opening the alimentary and urogenital canals to the perineum.

B. Definitive external genitalia of the male

1. **The phallus** is produced by rapid elongation of the genital tubercle.
   a. During this growth process, the urethral folds are drawn anteriorly, producing the **urogenital (urethral) groove**.
   b. By the twelfth week, the urethral folds fuse over the urethral groove to form the **penile urethra**.
   c. During the sixteenth week, the **glans penis** differentiates and the **external urethral meatus** connects with the penile urethra at the **fossa navicularis**.

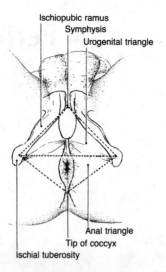

Ischiopubic ramus
Symphysis
Urogenital triangle

Anal triangle
Tip of coccyx
Ischial tuberosity

Figure 21-1. *Perineum.* The boundaries and landmarks of the perineum are indicated. The perineum is divided into anal and urogenital triangles by the transverse diameter of the pelvic outlet.

   d. The skin of the body of the penis grows over the glans penis as the **prepuce**. Initially, the prepuce fuses to the glans, but it becomes separated by birth or shortly thereafter.

   2. The scrotal folds migrate somewhat posteriorly and fuse, forming the scrotal septum; each fold forms one-half of the definitive **scrotum**.

   3. The testes, or male gonads, develop in the cranial region of the posterior abdominal wall (see Fig. 22-4A).
   a. Toward the end of the ninth week, the testes begin a retroperitoneal migration toward the iliac fossa.
      (1) A fascial condensation, the **gubernaculum testis**, runs from the inferior pole of each testis, through the inguinal canals, to the respective scrotal fold.
      (2) The gubernaculum normally fails to lengthen during the subsequent period of rapid body growth, which results in a shift of the gonads into the major pelvis.
   b. By 13 weeks, the testes lie in the inguinal region but retain the original neurovascular connections.
   c. At about this time, an evagination of the coelom, the **processus vaginalis (vaginal process)**, penetrates the abdominal wall of the inguinal region and enters the scrotal fold, drawing with it some of the layers of the abdominal wall.
   d. During the seventh month, the testes commence their descent through the inguinal canal, posteriorly to the vaginal process (see Fig. 22-4B).
   e. In the scrotum, the vaginal process in the vicinity of the testes forms the visceral and parietal layers of the **tunica vaginalis**.
   f. The neck of the vaginal process normally fuses by birth or shortly thereafter.

   4. Congenital anomalies of the penis include *hypospadias* (incomplete fusion of the penile urethra with the opening on the ventral surface), *epispadias* (with an opening on the dorsal surface), *micropenis*, congenital absence, and doubling of the glans.

C. Definitive external genitalia of the female

   1. The clitoris is formed by a slight elongation of the genital tubercle.
   a. The urethral folds do not fuse but develop into the **labia minora**.
   b. The urogenital groove remains as the **vestibule**.

   2. Labia majora develop from the genital swellings.

   3. Ovaries, the female gonads, develop in the cranial portion of the posterior abdominal cavity (see Fig. 22-4A). Subsequent to the ninth week, the ovaries migrate into the minor (deep) pelvis (see Fig. 22-4C).
   a. A fascial condensation runs from the inferior pole of each ovary, through the inguinal canals, to the respective labial fold.
   b. This ligament normally fails to lengthen during the subsequent period of rapid body growth, which results in a shift of the gonads into the major pelvis.
   c. A portion of this ligament is incorporated into the developing uterine wall, which, thus, divides the ligament into a cranial **ovarian ligament** and a caudal **round ligament of the ovary**.

d. By 13 weeks, the ovaries lie in the minor pelvis but retain the original neurovascular connections that form the **suspensory ligament of the ovary**.

e. At about this time, an evagination of the coelom, the **processus vaginalis (canal of Nuck)**, penetrates the abdominal wall of the inguinal region and enters the labial fold.

f. The neck of the processus vaginalis normally fuses by birth or shortly thereafter.

III. ANAL TRIANGLE

A. Ischioanal (ischiorectal) fossa. This is the region about the anal canal (Fig. 21-2).

1. Boundaries of the ischioanal fossa include the **levator ani muscle** (superiorly), the **obturator internus muscle** (anterolaterally), and the sacrotuberous ligament and the overlying **gluteus maximus muscle** (posterolaterally).

a. An **anterior horn** extends forward into the urogenital triangle on each side superiorly to the **urogenital diaphragm**.

b. The ischioanal fossae communicate contralaterally posterior to the anal canal.

2. Contents

a. The considerable **ischioanal fat** is separated by stringy connective tissue fibers and septa.

   (1) This fatty tissue is a counterpart of the superficial layer of superficial fascia (Camper's fascia of the abdominal wall and Cruveilhier's fascia of the urogenital triangle of the perineum).

   (2) Tension in the gluteus maximus muscles compresses the fat of the ischioanal fossa around the anal canal, contributing to fecal continence; however, the fat allows dilation of the anal canal during defecation when the muscles are relaxed.

b. **Inferior rectal (hemorrhoidal) vessels** and **nerves** pass through the ischioanal fossae (see Figs. 21-8 and 21-13).

   (1) These are branches of the **internal pudendal vessels** and the **pudendal nerve**.

   (2) They supply and innervate the **external anal sphincter**.

   (3) Inadvertent severing of the nerves in the ischioanal fossa results in *sphincteric incontinence* during brief peristaltic waves.

3. Abscesses in the ischioanal fossa may become large or may even extend to the contralateral side around the anus (*"horseshoe" abscess*).

B. Anal canal (see Fig. 17-18)

1. External structure. The anal canal is that part of the alimentary canal inferior to the pelvic diaphragm.

a. The anal canal is directed posteriorly and inferiorly from the **rectum** to the **anal verge**, is 3–4 cm long, and is open to the exterior via the **anus**.

b. The most medial fibers of the **pubococcygeus muscle** (the puborectalis or rectal sling) form the **perineal flexure** ("carrying angle") of the rectum (see Figs. 20-2A and 22-1). Thus, the anal canal is directed downward and backward from the **rectum** to the **anal verge** (see Fig. 17-18).

2. Internal structure (see Fig. 17-18)

a. The anal canal can be divided into an upper two-thirds and a lower one-third, representing a division between the visceral and somatic portions.

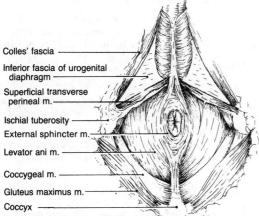

Colles' fascia
Inferior fascia of urogenital diaphragm
Superficial transverse perineal m.
Ischial tuberosity
External sphincter m.
Levator ani m.
Coccygeal m.
Gluteus maximus m.
Coccyx

Figure 21-2. *Anal triangle.* The fat of the ischiorectal fossa has been removed to show the underlying pelvic floor and external anal sphincter.

(1) The upper two-thirds of the anal canal belong to the intestine with respect to the mucosa, blood supply, and autonomic innervation.

(2) Only the lower one-third belongs to the perineum with respect to mucosa, blood supply, and somatic innervation.

b. The mucosa of the upper two-thirds is thrown into longitudinal folds about 1 cm long, the **anal columns** (of Morgagni). The bases of the anal columns are joined together by small semilunar folds of tissue, the **anal valves** (see Fig. 17-18).

(1) The bases of the anal columns and intervening anal valves define the **pectinate (dentate) line**, which approximates the division between visceral and somatic portions of the anal canal.

(a) Superior to the pectinate line, the epithelium is insensitive to touch, whereas below it, the epithelium is extremely sensitive.

(b) Venous anastomoses between the portal system draining the rectum and the systemic system draining the anal canal are potential sites for varicosities (hemorrhoids).

(2) Between the bases of the anal columns, behind the anal valves, are small blind sacs—the **anal sinuses (crypts)**—into which the **anal glands** open.

(3) The anal glands are rudimentary circumanal glands.

c. The mucosal lining of the lower portion of the anal canal is squamous epithelium.

3. Anal sphincters (see Figs. 17-18 and 21-2)

a. The anal canal is always closed except during the passage of feces or flatus. The sphincteric mechanism has a remarkable ability to distinguish between and separate flatus from feces.

(1) The **internal anal sphincter** is a tube of involuntary muscle, which encloses the entire anal canal. It is a continuation of the circular muscular layer of the intestine.

(a) The intrinsic tone of the involuntary internal anal sphincter maintains this closure.

(b) This sphincter is inhibited by the approach of a peristaltic wave and, thus, is not a major factor in fecal continence.

(2) The **external anal sphincter** is striated muscle that is under voluntary control via the rectal branches of the pudendal nerve.

(a) The external sphincter is a complex structure composed of three distinct layers (see Fig. 21-2).

(i) The **subcutaneous portion**, also known as the **corrugator ani muscle**, is a remnant of the panniculus carnosus, a primitive superficial muscle that also gives rise to the platysma muscle about the lower face and neck.

(ii) The **superficial portion** arises from the coccyx, passes to either side of the anus, and inserts onto the perineal body.

(iii) The **deep portion** completely surrounds the anal canal.

(b) The external anal sphincter can maintain a voluntary tonic contracture for about 10–20 seconds, a duration that provides the voluntary contraction necessary to counter the passage of a peristaltic wave when the internal anal sphincter relaxes.

b. Continued anal continence is a function of the **rectal sling**, a portion of the levator ani muscle.

C. Vasculature of the anal triangle

1. Arterial supply. The perineum is largely supplied by the **internal pudendal artery** (see Figs. 21-8 and 21-13).

a. Its **inferior hemorrhoidal (rectal) branches** supply the anal triangle.

b. These vessels form anastomotic connections with the middle hemorrhoidal (rectal) artery.

2. Venous return. The perineum is drained in large part by the **internal pudendal vein**.

a. The **inferior hemorrhoidal (rectal) veins** parallel the inferior hemorrhoidal arteries.

b. These veins form extensive anastomotic connections with the **middle rectal (hemorrhoidal) veins**.

D. Innervation of the anal triangle

1. The inferior rectal branch of the pudendal nerve is motor to the external anal sphincter and sensory to the inferior portion of the anal canal and the integument of the anal triangle (see Figs. 21-8 and 21-13).

2. Anal incontinence. Severance of the inferior rectal branch results in paralysis of the external anal sphincter and may result in momentary embarrassments as peristaltic waves reach the terminal portion of the gastrointestinal tract.

E. Functional considerations

1. Fecal continence. When the feces stored within the sigmoid colon are suddenly moved into the rectum, dilatation of the rectal ampulla occurs and the urge to evacuate the rectum is perceived. Defecation is prevented primarily by the **puborectalis muscle (rectal sling)**, which cants the lower part of the rectum forward (**carrying angle**), effectively kinking the lumen. Additionally, when a person is standing erect, the mass of fat within the ischiorectal fossa is compressed against the anal canal anteriorly and laterally by the location and tonic activity of the buttocks. Occasional contraction of the external anal sphincters and gluteus maximus muscles is required at this point to prevent defecation.

2. **Defecation.** When the time and place are propitious, compression of the anal canal by the ischiorectal fat is released by suitable anatomic positioning; the puborectalis muscle is relaxed, thereby allowing the rectum to straighten and descend slightly; and the anal sphincters are relaxed, the internal, according to the Bayliss-Starling law, and the external, voluntarily. The muscular movements of the terminal portion of the alimentary canal, assisted by gravity, evacuate the rectum. This may be accompanied by the Valsalva maneuver, which increases the intra-abdominal pressure and facilitates the expulsion of feces.

3. **Restoration of fecal continence.** After passage of each fecal mass, the puborectalis muscle re-establishes the carrying angle, and the anal sphincters contract, thereby restoring anal continence.

F. **Clinical considerations**

1. **Anal fistulas.** The anal glands empty into the base of the anal crypts behind the anal valves. These glands are prone to infection and may give rise to fistulous tracts that require surgical correction.

2. **Anal hemorrhoids.** Because there are rich anastomoses between the hepatic portal system and the systemic venous drainage, portal hypertension often results in *hemorrhoidal varices*.
   a. **Internal hemorrhoids.** Hemorrhoids above the **pectinate line** are classified as internal.
      (1) The innervation above this line is similar to that of the rest of the gut; there are pressure receptors but no definitive pain receptors (i.e., there is an urge to defecate when feces enter the rectum).
      (2) Internal hemorrhoids are painless because the epithelium is innervated by visceral afferents.
      (3) Hemorrhoids in this region are likely to be large, silent, and dangerous and may be noticed only when they become large enough to prolapse through the anus or bleed profusely. Such bleeding may be sufficient to produce anemia.
   b. **External hemorrhoids.** Hemorrhoids below the pectinate line are termed external.
      (1) Conversely, the anal canal below the pectinate line is innervated somatically via the rectal (hemorrhoidal) branches of the pudendal nerve.
      (2) These hemorrhoids are painful because the epithelium is innervated by rectal branches of the pudendal nerve.
      (3) External hemorrhoids are so exquisitely sensitive that even small hemorrhoids in this region demand attention.

3. **Ischioanal abscess.** Incisions into the ischioanal fossa to drain abscesses must avoid the inferior rectal neurovascular bundle.

IV. MALE UROGENITAL TRIANGLE

A. **External genitalia**

1. **Penis**
   a. **Orientation.** Because the penis is suspended along the abdominal wall in lower animals, the anterior surface in humans is termed the dorsal surface.
   b. **External structure** (Fig. 21-3)
      (1) **Composition.** The penis is composed of three bodies of vascular **erectile tissue** (Fig. 21-4). These begin as the **radix** or **root** of the penis in the superficial perineal pouch. The **body** of the penis is formed by the fusion of these three vascular erectile structures as they leave the perineum (Fig. 21-5A).
         (a) The central **corpus spongiosum (corpus cavernosum urethrae)** is ventrally situated (see Figs. 21-4A and 21-5A).
            (i) It begins as the **bulb** of the corpus spongiosum, which is located in the superficial perineal pouch (Fig. 21-6).
            (ii) It terminates as the **glans penis** (Fig. 21-5A). Proximally, the glans flares to define the **coronary sulcus** of the penis.
            (iii) It contains the **penile urethra** (see Figs. 21-4A and 21-5A).
         (b) The bilateral **corpora cavernosa** are situated dorsally (see Figs. 21-4A and 21-5A).
            (i) Each corpus cavernosum begins as a **crus**, which is attached along an ischiopubic ramus in the superficial perineal pouch (see Fig. 21-6).

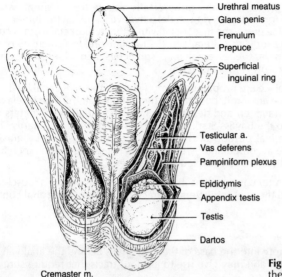

Urethral meatus
Glans penis
Frenulum
Prepuce
Superficial inguinal ring

Testicular a.
Vas deferens
Pampiniform plexus

Epididymis
Appendix testis

Testis

Dartos

Cremaster m.

Figure 21-3. *Male external genitalia.* The contents of the left scrotum and spermatic cord are indicated.

(ii) These erectile bodies terminate immediately proximal to the glans penis (see Fig. 21-5A)

(2) Epithelium and fascia. The penis is covered by pigmented and relatively hairless skin.

(a) The **prepuce** or **foreskin** is an extension of the integument over the glans penis (see Figs. 21-3 and 21-5A). The amount is variable; often not covering the glans but usually redundant over the tip of the glans.

(i) The foreskin is attached to the ventral raphe of the glans by the **frenulum** (see Fig. 21-3).

(ii) *Circumcision* involves removal of redundant foreskin to a variable extent.

(b) The **superficial fascia of the penis** is an extension of the superficial fascia of the abdominal wall over the body of the penis.

(i) The superficial layer of the superficial fascia loses its fat and fuses with the deep layer of the superficial fascia to form the **superficial fascia of the penis** (see Fig. 21-5A). It is rather loosely attached to the underlying tunica albuginea so that the overlying skin is quite mobile.

(ii) The superficial fascia of the penis does not extend over the glans penis, which is covered by mucous epithelium tightly bound to the underlying tunica albuginea.

(c) The **deep fascia (of Buck) of the penis** invests the muscles at the root of the penis and extends distally as far as the glans, even though the muscle layer terminates where the penis becomes pendulous (see Figs. 21-4 and 21-5A).

(3) Support. The **suspensory ligament** arises from the linea alba and inserts onto the deep fascia (of Buck) of the penis (see Fig. 21-5A).

c. Internal structure

(1) Erectile tissue

(a) The corpora cavernosa and corpus spongiosum are composed of vascular sinusoids. The filling of these sinusoids with arterial blood results in erection of the penis.

(b) Each erectile structure is surrounded by a distinctive connective layer, the **tunica albuginea**.

(i) The tunica albuginea that surrounds the corpora cavernosa is very dense and not particularly elastic. When the penis becomes tumescent, this layer greatly impedes venous return, which results in the extreme turgidity characteristic of the corpora cavernosa.

(ii) Because the tunica albuginea that surrounds the central corpus spongiosum is less dense, this erectile tissue and its terminal glans penis does not become excessively turgid upon erection. Therefore, the penile urethra is never occluded, permitting the passage of the ejaculate.

(2) The **penile urethra** extends along the entire penis within the corpus spongiosum (see Fig. 21-5A). This is homologous to the urogenital sinus (vestibule) in the female.

(a) This portion of the urethra begins in the superficial perineal pouch just as the membranous urethra penetrates the urogenital diaphragm (see Fig. 21-5A).

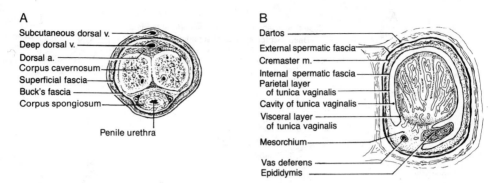

Figure 21-4. *A,* Cross section of the penis, showing the fascial layers and erectile bodies. *B,* Cross section through the scrotum and testis.

    (b) After about 2.5 cm, it receives the ducts of the **bulbourethral glands (of Cowper)** [see Fig. 21-5*A*]. More distally, mucous **urethral glands** (of Littré) line the urethra.

    (c) Terminally, the penile urethra widens within the glans penis, forming the **fossa navicularis**, which terminates as the slit-like **urethral meatus**.

    (d) Introduction of instruments into the male urethra must be accomplished carefully with the penis straightened into an approximately erect posture.

  d. Vasculature of the penis

    (1) Arterial supply. The erectile tissue of the penis is supplied by the **internal pudendal arteries**. The superficial skin of the penis is supplied by both the internal pudendal arteries as well as the external pudendal arteries, which arise from each femoral artery.

    (2) Venous return is via the **dorsal vein** of the penis to the prostatic venous plexus.

  e. Innervation of the penis is by terminal branches of the **pudendal nerve**.

2. Scrotum

  a. Orientation. The scrotum develops from the **labial–scrotal folds**, which fuse in the male. The scrotum is suspended from the male perineum and contains the testes.

  b. Structure

    (1) Epithelium and fascia

      (a) The skin of the scrotum is thin with very little fat, an important factor in maintaining a lower testicular temperature.

      (b) The **dartos layer** or **tunic** is formed by the fusion of the superficial and deep layers of the superficial fascia (see Fig. 21-3).

        (i) The dartos layer contains smooth muscle, which also functions in temperature regulation.

        (ii) Although core body temperature is 37°C, scrotal temperature is 33.9°C. When below this temperature, the scrotum contracts to bring the testes into close contact with the body to conserve heat; when too warm, the scrotum relaxes and becomes pendulous to dissipate heat.

      (c) *Transillumination of the scrotum* is used to define the testicular outline and distinguish between fluid (i.e., hydrocele) and tissue (i.e., hernia or tumor).

    (2) Innervation. It is innervated by the **ilioinguinal**, **genitofemoral**, and **dorsal nerves** of the penis.

3. Spermatic cord

  a. Orientation. The spermatic cord is formed during the descent of the testes. It provides the route of neurovascular and ductal communication between the scrotum and the abdominopelvic cavity.

  b. Structure. The spermatic cord has several layers, most of which correspond to those of the abdominal wall (see Fig. 15-5).

    (1) The **external spermatic fascia** forms the outermost layer of the spermatic cord.

      (a) The external spermatic fascia is derived from the deep (investing) fascia of the **external oblique muscle**.

      (b) Because the superficial inguinal ring is a defect in the external oblique aponeurosis, there is no direct contribution from that muscle.

    (2) The **cremaster muscle** and **fascia** form the intermediate layer.

      (a) The cremaster muscle is derived from the **internal oblique muscle** and its investing fascia.

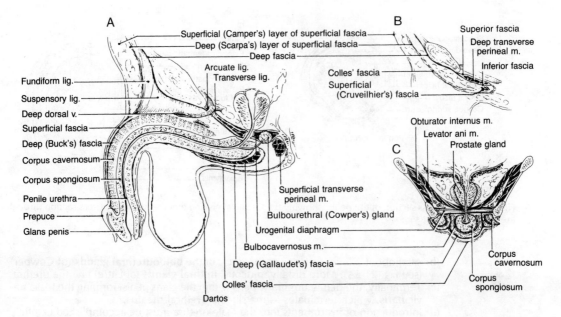

Figure 21-5. *Perineal fascia in the male. A,* Midsagittal section. *B,* Parasagittal section. *C,* Coronal section.

> **(b)** This layer is innervated by the **genital branch of the genitofemoral nerve**, which provides the motor limb for the *cremaster reflex* (L1–L2), the elevation of the testes within the scrotum when the inner thigh is stroked.
>
> **(3)** The **internal spermatic fascia** is the innermost layer.
> > **(a)** This layer is derived from the **transversalis fascia**.
> > **(b)** It contains the testicular neurovascular bundle and the vas deferens (see Fig. 21-3) with its accompanying deferential artery.
> > > **(i)** Each **testicular artery** arises from the abdominal aorta (Fig. 18-3).
> > > **(ii)** As the veins leave the testes, they form the anastomotic **pampiniform plexus** of veins, which surrounds each testicular artery (see Fig. 21-3). This plexus serves as a countercurrent heat exchanger, cooling the arterial blood before it reaches the testes. Before leaving the spermatic cord, each plexus coalesces into a **testicular vein**. The right testicular vein drains into the inferior vena cava; the left testicular vein drains into the left renal vein.
> > > **(iii)** The **artery of the vas deferens** (deferential artery) anastomoses with the testicular artery to provide some collateral circulation to the testes and scrotum.
> > > **(iv)** The **vas deferens** rises from the inferior pole of the testes and passes into the spermatic cord (see Fig. 21-3).
> > **(c) Clinical considerations**
> > > **(i)** *Varices* of the pampiniform plexus are very common (80% of all males). Fully 90% of all varices are on the left side because local venous hypertension results from compression of the left testicular vein by the sigmoid colon, which contains stored fecal matter. Because varices reduce the efficiency of the heat exchange mechanism, fertility may be decreased.
> > > **(ii)** *Vasectomy* is generally performed by section of the vas at the superolateral aspect of the scrotum.
>
> **(4)** The **tunica vaginalis** is a remnant of the **processus vaginalis**.
> > **(a)** The **processus vaginalis** is a fetal evagination of the peritoneal cavity into the scrotum. It is usually occluded postnatally.
> > > **(i)** A patent processus vaginalis predisposes to *indirect (congenital) inguinal hernia*.
> > > **(ii)** Partial occlusion of a processus vaginalis can result in fluid accumulation (*hydrocele processus vaginalis*), which may not be distinguishable from a hernia or incompletely descended testes until surgery is performed.
> > **(b)** The **tunica vaginalis** consists of two layers of peritoneum with an intervening potential space that represents the detached portion of the processus vaginalis within the scrotum (see Fig. 21-4B). It is anterior and lateral to the testes.
> > > **(i)** The **parietal layer** of the tunica vaginalis is equivalent to the parietal peritoneum and lines the outer wall of the potential space (see Fig. 21-4B).

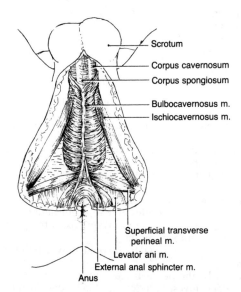

Scrotum

Corpus cavernosum

Corpus spongiosum

Bulbocavernosus m.

Ischiocavernosus m.

Superficial transverse
perineal m.

Levator ani m.

External anal sphincter m.

Anus

Figure 21-6. *Superficial perineal pouch in the male.*

> **(ii)** The **visceral layer** of the tunica vaginalis is equivalent to visceral peritoneum
> and overlies the testes (see Fig. 21-4*B*).
>
> **(iii)** The **mesorchium** or mesentery of the testes is formed by the reflection of the
> parietal layer to become the visceral layer (see Fig. 21-4*B*).
>
> **(iv)** Although it is accepted that the testes develop retroperitoneally and descend in-
> to the scrotal sacs retroperitoneally, there is some debate as to whether the de-
> scended testes are peritoneal or retroperitoneal with resultant confusion in ter-
> minology. The presence of a mesorchium supporting the testes justifies the clas-
> sification of the postnatal descended testes as a peritoneal structure.

B. Fascia of the male urogenital triangle (see Fig. 21-5). The underlying muscular and fascial structures support and contribute to the function of the external genitalia as well as reinforce the abdominal cavity.

1. **The superficial layer of the superficial perineal fascia** (of Cruveilhier) is a fatty layer just beneath the dermis.
   a. This layer is continuous with Camper's fascia of the abdominal wall (see Fig. 21-5*B*).
   b. It passes across the perineum to become the fat of the ischioanal fossa.

2. **The deep layer of the superficial perineal fascia (of Colles)** is a membranous layer that retains sutures.
   a. This layer is continuous with Scarpa's fascia of the abdominal wall (see Fig. 21-5*B*).
   b. It is attached as follows:
   (1) Laterally, it is attached to the ischiopubic rami (see Fig. 21-5*C*).
   (2) Anteriorly, it is continuous with Scarpa's fascia of the abdominal wall and fuses with
   the superficial layer of the superficial fascia, which passes over the penis as the **superficial fascia of the penis** (see Fig. 21-5*A*).
   (3) Medially, it inserts into the adventitia of the urethra.
   (4) Inferiorly, if fuses with the superficial layer of the superficial fascia and passes over the
   scrotum, where it contributes to the **dartos layer**.
   (5) Posteriorly, it passes inferiorly (superficially) to the **superficial transverse perineal
   muscle** to the posterior edge of the urogenital triangle; then it runs superiorly to the
   posterior edge of the **deep transverse perineal muscle**.
   (a) The midline fibers join the **central tendon (perineal body)**.
   (b) The more lateral fibers merge with the inferior and superior leaflets of the fascia of
   the urogenital diaphragm, which run anteriorly toward the pubis, enclosing the
   deep transverse perineal muscle. Thus, these two leaflets define the **deep perineal
   pouch** (see Fig. 21-7).
   (i) The superficial (inferior) layer is termed the **inferior fascia of the urogenital diaphragm**, or the **perineal membrane** (see Fig. 21-5*B*).
   (ii) The deep (superior) layer is termed the **superior fascia of the urogenital diaphragm** (see Fig. 21-5*B*).

3. The deep or investing fascia (of Gallaudet) covers the muscles of the superficial pouch (see Fig. 21-5C). It is continuous with Buck's fascia.

C. **Superficial perineal pouch (space) in the male** (see Fig. 21-6)

1. **Boundaries.** The superficial perineal pouch is defined as that region between Colles' fascia and the inferior fascia of the urogenital diaphragm (perineal membrane).

2. **The contents** of the superficial perineal pouch include several structures.
   a. **Superficial transverse perineal muscle** (see Fig. 21-6; Table 21-1)
      (1) This muscle arises on each side from the ischial tuberosities and inserts into the perineal body (central tendon).
      (2) Although it is a very small and weak muscle, which may be absent unilaterally or bilaterally, it acts to stabilize the perineum.
      (3) It is a palpable surgical landmark that delineates the posterior extent of the superficial pouch.
   b. **Ischiocavernosus muscles and crura** of the penis (see Fig. 21-6)
      (1) These structures arise from the ischial tuberosities and adjacent rami of the ischia.
      (2) On each side, a crus composed of erectile tissue forms a radix of the penis.
      (3) Each crus is covered by an ischiocavernosus muscle.
         (a) These muscles insert onto the penis just distal to the point at which the crura join to form the body of the penis.
         (b) These muscles are innervated by the perineal branches of the pudendal nerve.
         (c) They are covered by the deep investing fascia (of Gallaudet) in the superficial pouch, which becomes the deep fascia (of Buck) on the penile shaft.
   c. **Bulbocavernosus (bulbospongiosus) muscles and bulb** of the penis (see Fig. 21-6)
      (1) The bulb of the **corpus spongiosum** in the male is composed of erectile tissue.
      (2) The bulbospongiosus muscles overlie the bulb of the penis.
         (a) The muscles arise from the central tendon, posteriorly, and from the **median raphe**.
         (b) Each inserts into the deep fascia of the penis just after the crura fuse.
         (c) They are covered by the deep investing fascia (of Gallaudet) in the superficial pouch, which becomes the deep fascia (of Buck) on the penile shaft.
         (d) These muscles are innervated by the perineal branches of the pudendal nerve.
         (e) Once termed the compressor urethrae, the bulbospongiosus muscle functions to expel the final drops of urine from the penile urethra. Because the muscle does not extend beyond the very base of the penile shaft, it is rather ineffectual in this function.

3. **Perforations of the penile urethra** produce extravasation of urine into the superficial pouch. From this space the urine may extravasate along the abdominal wall (beneath Scarpa's fascia), into the scrotum (under the dartos tunic), and onto the penis (superficial to Buck's fascia).

Table 21-1. Muscles of the Perineum

Muscle	Origin	Insertion	Action	Innervation
Superficial transverse perineal	Ischial tuberosity	Central perineal body	Stabilizes perineum	Pudendal n. (S2–S4)
Deep transverse perineal	Ischiopubic rami	Midline raphe and central perineal body (vaginal adventitia)	Reinforces pelvic floor (stabilizes vagina)	Pudendal n. (S2–S4)
External urethral sphincter	Circumurethral	Circumurethral	Urinary continence	Pudendal n. (S2–S4)
Ischiocavernosus	Ischiopubic rami	Body of penis or clitoris	. . .	Pudendal n. (S2–S4)
Bulbocavernosus	Central body and midline raphe	Body of penis or clitoris	. . .	Pudendal n. (S2–S4)
External anal sphincter	Circumanal	Circumanal	Fecal continence	Pudendal n. (S2–S4)

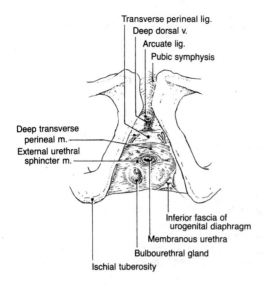

Transverse perineal lig.
Deep dorsal v.
Arcuate lig.
Pubic symphysis

Deep transverse
perineal m.
External urethral
sphincter m.

Inferior fascia of
urogenital diaphragm
Membranous urethra
Bulbourethral gland
Ischial tuberosity

Figure 21-7. *Deep perineal pouch in the male.*

D. Deep perineal space (pouch) in the male

1. **Boundaries.** The deep perineal pouch is between the superior and inferior fascial planes of the deep transverse perineal muscle (see Fig. 21-5B).

2. **Contents.** In the male, the deep perineal space contains the deep transverse perineal muscle, the external urethral sphincter, and the bulbourethral glands (Fig. 21-7).

    a. **The deep transverse perineal muscle** originates along the ischiopubic rami and inserts into the central raphe of the urogenital diaphragm and the central tendon of the perineum (see Fig. 21-7 and Table 21-1).

      (1) The deep transverse perineal muscle and its associated superior and inferior fascial layers form the **urogenital diaphragm** (see Fig. 21-5C).

      (2) The urogenital diaphragm lies inferiorly to the **urogenital hiatus** of the **levator ani muscle** and supports this potentially weak region of the pelvic floor (see Fig. 20-2).

    b. **The external urethral sphincter** is a modified portion of the deep transverse perineal muscle (see Fig. 21-7).

      (1) Where the membranous urethra passes through the urogenital diaphragm, muscle fascicles of the deep transverse perineal muscle completely encircle the urethra.

      (2) The external urethral sphincter is voluntary, innervated by twigs from the perineal branch of the pudendal nerve.

      (3) This muscle is a prime factor in urinary continence once the desire to void is perceived.

    c. **The paired bulbourethral glands** (of Cowper) are embedded in the deep transverse perineal muscle (see Fig. 21-5). The ducts drain into the base of the penile urethra (see Fig. 21-7).

    d. **The inferior fascia of the urogenital diaphragm (perineal membrane)** bounds the deep pouch inferiorly; the **superior fascia of the urogenital diaphragm** bounds the deep pouch superiorly (see Fig. 21-5B).

      (1) The merging of the inferior and superior fascial layers of the urogenital diaphragm anteriorly to the free edge of the deep transverse perineal muscle forms the **transverse perineal ligament** (see Fig. 21-7).

      (2) The **arcuate ligament** lies behind the pubic symphysis and is the most anterior component of this layer. The **deep dorsal vein** of the penis runs through the hiatus between the transverse and arcuate ligaments to reach the prostatic venous plexus (see Fig. 21-7).

      (3) The attachments of the superior and inferior fascial layers of the urogenital diaphragm to the ischiopubic rami limit the deep pouch laterally.

      (4) The superior and inferior fascial layers of the urogenital diaphragm also fuse at the posterior free edge of the deep transverse perineal muscle. Along this line of fusion, the deep layer of perineal fascia (of Colles) attaches to define the posterior limit of the superficial perineal pouch (see Figs. 21-5A and B).

3. **Perforation of the urethra** superior to the urogenital diaphragm, resulting from trauma or disease, may produce intrapelvic, intra-abdominal, intraperitoneal, or extraperitonal urinary extravasations.

E. Vasculature of the male urogenital triangle

1. **Arterial supply**
   a. **The internal pudendal artery** supplies most of the perineum (Fig. 21-8).
      (1) This artery leaves the pelvic cavity via the infrapiriform portion of the **greater sciatic foramen**.
      (2) It then passes through the **lesser sciatic foramen** to enter the ischiorectal fossa of the perineum, where it courses in the **pudendal canal** (of Alcock) on the medial side of the obturator internus muscle.
      (3) The internal pudendal arteries give rise to the:
         (a) **Inferior rectal (hemorrhoidal) arteries**
         (b) **Superficial perineal arteries**, including:
            (i) The **transverse perineal branch**, which supplies the superficial and deep perineal pouches
            (ii) The **posterior scrotal** or **labial branch**
         (c) **Deep perineal arteries**, which supply blood to the erectile tissues of the penis
            (i) The **dorsal arteries** of the penis lie beneath the deep fascia (of Buck) and external to the tunica albuginea.
            (ii) The **deep (central) arteries**, located within the corpora cavernosa, supply those erectile tissues.
            (iii) The **bulbar arteries** of the bulb of the penis supply erectile tissue of the corpus spongiosum and glans penis.
   b. **The external pudendal arteries**, which arise from the femoral arteries, supply the anterior superficial portion of the perineum and the subcutaneous tissue of the penis.

2. **Venous return of the male urogenital triangle**
   a. The veins of the perineum generally parallel the internal pudendal artery.
   b. The **deep dorsal vein** of the penis penetrates the urogenital diaphragm between the **arcuate** and **transverse ligaments** (see Fig. 21-7) and drains into the **prostatic venous plexus**.
   c. The **superficial dorsal veins** of the penis and the veins of the anterior scrotum drain into the **external pudendal veins**, which enter the femoral vein.

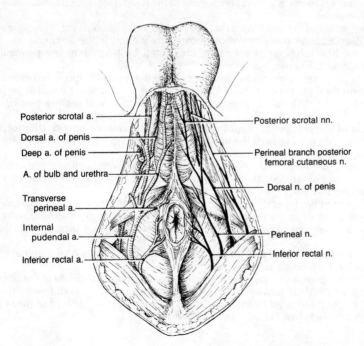

Posterior scrotal a.
Dorsal a. of penis
Deep a. of penis
A. of bulb and urethra
Transverse perineal a.
Internal pudendal a.
Inferior rectal a.

Posterior scrotal nn.
Perineal branch posterior femoral cutaneous n.
Dorsal n. of penis
Perineal n.
Inferior rectal n.

Figure 21-8. *Vasculature and innervation of the male perineum.*

F. Innervation of the male urogenital triangle

    1. Pudendal nerve

        a. The perineal branch of the pudendal nerve is a mixed nerve, containing the general somatic afferent (GSA), general somatic efferent (GSE), and sympathetic (GVE) nerves as components.

            (1) The perineal branches of the pudendal nerve are named according to the arteries that they accompany.

            (2) On the dorsum of the penis, the dorsal nerve of the penis (the terminal branch of the pudendal nerve) lies between the deep penile fascia (of Buck) and the tunica albuginea. It supplies the dorsum and glans of the penis.

        b. Saddle block. The perineum may be blocked by anesthesia of the pudendal nerve as it courses along the pudendal canal. A needle is introduced just medial to the ischial tuberosity and directed toward the ischial spine.

    2. The ilioinguinal, genitofemoral, and posterior femoral cutaneous nerves also send twigs to the perineum.

G. Functional considerations

    1. Erection is a vascular condition. Parasympathetics from the nervi erigentes (S2–S4) reach the penis via the lateral pelvic plexus and prostatic plexus, crossing the urogenital diaphragm with the deep dorsal vein. Stimulation relaxes the arterioles supplying the erectile tissue, thereby increasing the blood flow. The resultant tumescence compresses the veins that drain the erectile tissue as they pass through the tunica albuginea, thereby impeding venous return and producing full erection.

    2. Ejaculation is a sympathetic event. Sympathetics from the lumbar chain (L1–L2) follow the hypogastric plexus to reach the lateral pelvic plexus and prostatic plexus. Stimulation produces vigorous contraction of the muscular vas deferens, seminal vesicles, and prostate gland with expulsion of the seminal ejaculate.

V. FEMALE UROGENITAL TRIANGLE

A. External genitalia

    1. The vulva is the most obvious feature of the female urogenital triangle (Fig. 21-9).

        a. Orientation. The vulva consists of the genital labia and the enclosed urogenital sinus (vestibule). These structures are homologous to the scrotum and the penile urethra, respectively, in the male.

        b. Structure

            (1) The **labia majora** are two folds of hirsute skin, each supported by underlying fat pads (see Fig. 21-9). The labia majora with the anterior and posterior commissures define the boundaries of the **pudendal cleft**.

                (a) The **anterior commissure** is formed by the blending of the labia majora over the pubic symphysis.

                (b) The **posterior commissure**, a fold of tissue anterior to the anus, connects the labia majora posteriorly.

                (c) The labia majora develop from the **labial–scrotal folds**, which do not fuse in the female.

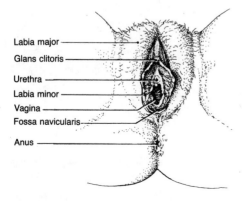

Labia major
Glans clitoris
Urethra
Labia minor
Vagina
Fossa navicularis
Anus

Figure 21-9. *Female external genitalia.*

(2) The **labia minora** are two small, hairless, sometimes pendulous, folds located medially to the labia majora (see Fig. 21-9). The labia minora enclose the **vestibule (urogenital sinus)**.

    (a) Anteriorly, each labium minus divides into two parts.

        (i) Superior to the clitoris, the lateral portions fuse to form the **prepuce of the clitoris**.

        (ii) Inferior to the clitoris, the medial portions fuse to form the **frenulum of the clitoris**.

    (b) Posteriorly, even though the labia minora appear to blend with the labia majora, there is a subtle fold, the **fourchette** (frenulum of the labia), which connects the labia minora posteriorly to the vaginal introitus.

    (c) The labia minora develop from the urethral folds. (In contrast to the female, these folds fuse over the urogenital sinus in the developing male to form the penile urethra.)

(3) The **vestibule**, or **urogenital sinus**, is bounded by the labia minora, frenulum of the clitoris, and fourchette.

    (a) The **external urethral ostium** is located anterosuperiorly to the vaginal opening, approximately 2 cm inferiorly to the clitoris (see Fig. 21-9).

    (b) The **vaginal introitus**, the predominant feature of the vestibule, is covered incompletely by the **hymen**.

        (i) The hymen is a remnant of the urogenital membrane.

        (ii) An imperforate hymen, which is rare, must be surgically opened prior to or at the time of the first menses.

        (iii) The hymen is usually stretched, and sometimes torn, during the first coitus. It often survives partially intact until the first vaginal childbirth.

    (c) The **fossa navicularis** (vestibular fossa of the vagina) is the region between the vaginal introitus and the fourchette (see Fig. 21-9).

    (d) The **greater vestibular glands (of Bartholin)** are located on either side of the vaginal introitus (Fig. 21-10).

        (i) These mucous glands are located in the superficial perineal pouch.

        (ii) They are homologous to the bulbourethral glands (of Cowper) in the male.

        (iii) They may become infected, resulting in *Bartholin's cyst*.

    (e) The **lesser vestibular glands** are numerous, small glands, which secrete mucus directly into the vestibule.

    (f) The **paraurethral glands (of Skene)** open on either side of the external urethral ostium.

        (i) These glands are prone to infection by pathogenic organisms.

        (ii) They are homologous to the prostate gland in the male.

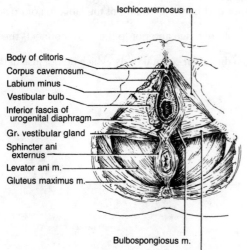

Ischiocavernosus m.

Body of clitoris
Corpus cavernosum
Labium minus
Vestibular bulb
Inferior fascia of urogenital diaphragm
Gr. vestibular gland
Sphincter ani externus
Levator ani m.
Gluteus maximus m.

Bulbospongiosus m.
Superficial transverse perineal m.

Figure 21-10. *Superficial perineal pouch in the female.*

2. Clitoris

a. Orientation. The clitoris is homologous to the penis in most respects.

b. Structure. The clitoris is composed of three bodies of vascular **erectile tissue** (see Fig. 21-10). These begin as the **radix** or **root** of the clitoris in the superficial perineal pouch. The filling of these sinusoids with arterial blood results in erection of the clitoris.

  (1) The **corpora cavernosa**, which are bilateral, fuse to form the body of the clitoris (see Fig. 21-10).

  (a) Each corpus cavernosum begins in the superficial perineal pouch as a **crus**, which is attached along an ischiopubic ramus and covered by an **ischiocavernosus muscle** (see Fig. 21-10).

  (b) These erectile bodies terminate immediately proximal to the glans clitoris.

  (2) The **vestibular bulbs**, which are also bilateral, fuse to form the glans clitoris (see Fig. 21-10).

  (a) The vestibular bulbs are located in the superficial perineal pouch where they lie deeply to the labia minora and flank the vaginal introitus (see Fig. 21-10). Each is covered by a **bulbocavernosus muscle**.

  (b) Anteriorly, the thread-like commissures of the vestibular bulbs fuse and terminate as the **glans clitoris** (see Fig. 21-9).

  (3) The **prepuce** or **foreskin** is an extension of the labia minora over the glans clitoris. The amount is variable but usually amply covers the glans.

  (4) The **suspensory ligament** arises from the pubic symphysis and inserts into the deep fascia of the clitoris. Like the male, the distal portion of the phallus is pendulous, but in the female, this portion is largely embedded in the tissue of the mons pubis, which limits its mobility.

c. Vasculature of the vulva and clitoris

  (1) **Arterial supply.** The erectile tissue of the clitoris is supplied by the internal pudendal arteries. The superficial skin of the labia and vulva is supplied by both the internal pudendal arteries, which arise from each internal iliac artery, and the external pudendal arteries, which arise from each femoral artery.

  (2) **Venous return** from the clitoris is via the dorsal vein to the vesicle venous plexus at the base of the urinary bladder. Return from the vulva is largely via the internal pudendal vein.

d. Innervation of the vulva and clitoris is by terminal branches of the pudendal nerve. There are numerous sensory nerve endings converging on the glans clitoris.

B. Fascia of the female urogenital triangle. The underlying muscular and fascial structures support and contribute to the function of the external genitalia as well as reinforce the abdominal cavity.

1. Superficial layer of the superficial fascia (of Cruveilhier)

  a. This fatty layer is continuous with Camper's layer of the abdominal wall as well as the fat of the ischioanal fossa. It gives form to the mons pubis and labia majora.

  b. The fat is lost as this layer extends into the labia minora.

2. Deep layer of the superficial fascia (of Colles)

  a. This membranous layer is continuous with Scarpa's fascia of the anterior abdominal wall.

  b. Relations of Colles' fascia in the female

  (1) Laterally, it is attached to the ischiopubic rami.

  (2) Anteriorly, it is continuous with Scarpa's fascia.

  (3) Medially, it inserts into the adventitia of the urethra and vagina.

  (4) Inferiorly, it passes deeply to the labia majora.

  (5) Posteriorly, it delimits the urogenital triangle by passing superficially to the **superficial transverse perineal muscle**.

  (a) The midline fibers insert into the adventitia of the vagina and the **central tendon of the perineum**.

  (b) The more lateral fibers reach the posterior edge of the **deep transverse perineal muscle** before merging with the fascia of the urogenital diaphragm.

  (i) The fascia over the inferior surface of the deep transverse perineal muscle forms the **inferior** or **external fascia of the urogenital diaphragm** (see Fig. 21-10).

  (ii) The fascia over the superior surface of the deep transverse perineal muscle forms the **superior** or **internal fascia of the urogenital diaphragm**.

  (iii) These two layers fuse anteriorly to the deep pouch as the **transverse perineal ligament**.

3. The deep, or investing, fascia (of Gallaudet) covers the muscles of the superficial pouch (see Figs. 21-5*A* and 21-5*C*).

C. Superficial perineal space (pouch) in the female (see Fig. 21-10)

   1. Boundaries. Colles' fascia bounds the superficial pouch inferiorly; the inferior fascia of the urogenital diaphragm bounds this space superiorly. Anteriorly, the superficial perineal space is continuous with the potential space in the anterior abdominal wall between the deep layer of the superficial fascia (of Scarpa) and the deep investing fascia.

   2. Contents of the superficial pouch include the roots of the external genitalia, the crura of the clitoris, and the vestibular bulbs with the associated muscles, nerves, and vessels.

   a. Corpora cavernosa are paired bodies of vascular erectile tissue lying along the ischiopubic rami.

   (1) These fuse to form the body of the clitoris.

   (2) They are covered by the ischiocavernosus muscles and are homologous to the male corpora cavernosa.

   b. Vestibular bulbs are paired masses of vascular erectile tissue on either side of the vaginal introitus under the labia minora.

   (1) These are covered by the bulbospongiosus muscles and are homologous to the fused halves of the bulb of the penis in the male.

   (2) They tend to spread the labia minora and open the vaginal introitus, when turgid.

   c. Ischiocavernosus muscles overlie the clitoral crura.

   (1) Their origin is the medial side of the ischial tuberosities and ischial rami.

   (2) They insert into the margin of the pubic arch and into each crus of the clitoris.

   d. Bulbospongiosus (bulbocavernosus) muscles overlie the vestibular bulbs.

   (1) Their origin is the anterior portion of the central tendon.

   (2) The medial fascicles attach to the deep fascia of the dorsum of the clitoris; the lateral fascicles attach to the inferior fascia of the urogenital diaphragm.

   e. Superficial transverse perineal muscle (see Table 21-1)

   (1) It originates along the anterior portions of the ischial tuberosities and inserts into the central tendon of the perineum (perineal body).

   (2) Its action is to stabilize the central tendon of the perineum.

   f. Greater vestibular (vulvovaginal) glands (of Bartholin)

   (1) The ducts of these mucous glands open into the vestibule.

   (2) These glands are homologous to the bulbourethral glands in the male.

   g. The central tendon (perineal body) is a common point of attachment of connective tissue and muscles of the anal and urogenital triangles (see Fig. 21-12). As such, it has been considered the most important structure of the perineum and must be sutured if torn or incised.

   3. Clinical considerations. Tearing of the vascular vestibular bulbs during difficult parturition will produce considerable bleeding into the superficial pouch with extravasation into the labia majora and along the mons pubis and abdominal wall beneath Scarpa's fascia.

D. Deep perineal space (pouch) in the female

   1. Boundaries. The deep perineal pouch is delineated by the superior and inferior fascial layers of the deep transverse perineal muscle (see Fig. 21-5*B*).

   2. Contents. In the female, the deep perineal space contains the deep transverse perineal muscle and the external urethral sphincter (Fig. 21-11).

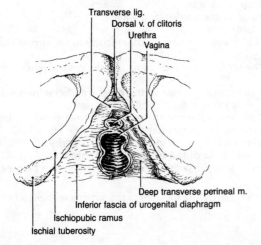

Transverse lig.
Dorsal v. of clitoris
Urethra
Vagina

Deep transverse perineal m.
Inferior fascia of urogenital diaphragm
Ischiopubic ramus
Ischial tuberosity

Figure 21-11. *Deep perineal pouch in the female.*

    a. The deep transverse perineal muscle originates along the ischiopubic rami and inserts into the central raphe of the urogenital diaphragm, the adventitia of the vaginal wall, and the central tendon of the perineum (see Fig. 21-11 and Table 21-1). The vagina passes through a substantial portion of this muscle.

        (1) The deep transverse perineal muscle and its associated superior and inferior fascial layers form the **urogenital diaphragm** (see Fig. 21-5C).

        (2) The urogenital diaphragm lies inferiorly to the **urogenital hiatus** of the **levator ani muscle** and supports this potentially weak region of the pelvic floor (see Figs. 20-2 and 21-10).

    b. The sphincter urethrae (external urethral sphincter) is a modified portion of the deep transverse perineal muscle (see Fig. 21-11).

        (1) Where the urethra passes through the urogenital diaphragm, muscle fascicles of the deep transverse perineal muscle arch anteriorly to the urethra but do not pass posteriorly to the urethra as in the male, because at this level, the urethra is embedded in the adventitia of the anterior wall of the vagina (Fig. 21-12). This anatomic arrangement predisposes toward *urinary stress incontinence*, especially after trauma associated with difficult parturition.

        (2) The external urethral sphincter is voluntary, innervated by twigs from the perineal branch of the pudendal nerve.

        (3) This muscle is a prime factor in urinary continence once the desire to void is perceived.

    c. The inferior fascia of the urogenital diaphragm (perineal membrane) bounds the deep pouch inferiorly; the **superior fascia of the urogenital diaphragm** bounds the deep pouch superiorly (see Fig. 21-5B).

        (1) The merging of the inferior and superior fascial layers of the urogenital diaphragm anterior to the free edge of the deep transverse perineal muscle forms the **transverse perineal ligament** (see Fig. 21-7).

        (2) The **arcuate ligament** lies behind the pubic symphysis and is the most anterior component of this layer. The **deep dorsal vein** of the clitoris runs through the hiatus between the transverse and arcuate ligaments to reach the prostatic venous plexus (see Fig. 21-11).

        (3) The attachments of the superior and inferior fascial layers of the urogenital diaphragm to the ischiopubic rami limit the deep pouch laterally.

        (4) The superior and inferior fascial layers of the urogenital diaphragm also fuse at the posterior free edge of the deep transverse perineal muscle. Along this line of fusion, the deep layer of perineal fascia (of Colles) attaches to define the posterior limit of the superficial perineal pouch (see Figs. 21-5A and B)

    d. The dorsal artery and dorsal nerve of the clitoris, the terminal branches of the internal pudendal artery and pudendal nerve, course through the deep perineal pouch.

E. Vasculature of the female urogenital triangle (Fig. 21-13)

    1. Arterial supply

        a. The internal pudendal artery supplies most of the external genitalia.

        (1) The **transverse perineal branch** runs parallel to the superficial transverse perineal muscle and gives off the **bulbar branches**, which supply the vestibule and erectile tissues of the vestibular bulbs.

        (2) The **posterior labial branches** supply the posterior portion of the vulva.

        (3) The **deep (central) arteries** of the clitoris, located in each corpus cavernosum, supply the erectile tissues of those bodies.

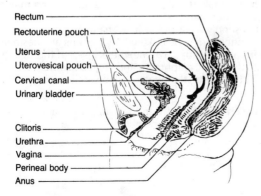

Rectum —
Rectouterine pouch —
Uterus —
Uterovesical pouch —
Cervical canal —
Urinary bladder —

Clitoris —
Urethra —
Vagina —
Perineal body —
Anus —

Figure 21-12. *Midsagittal section through the female genitalia.*

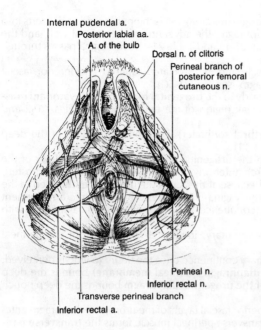

Internal pudendal a.
Posterior labial aa.
A. of the bulb
Dorsal n. of clitoris
Perineal branch of posterior femoral cutaneous n.

Perineal n.
Inferior rectal n.
Transverse perineal branch
Inferior rectal a.

Figure 21-13. *Vasculature and innervation of the female perineum.*

    (4) The **dorsal artery** of the clitoris, which lies between the deep fascia of the clitoris and the tunica albuginea, supplies the superficial aspects of the clitoris.

    b. The external pudendal branches of each femoral artery supply the anterior aspects of the mons pubis, anterior portions of the labia majora, and subcutaneous tissue of the clitoris.

2. Venous return

    a. With one exception, the venous drainage of the female urogenital triangle generally parallels the arterial supply.

    b. The **deep dorsal vein** of the clitoris passes between the arcuate and transverse ligaments of the urogenital diaphragm to drain into the vaginal venous plexus (see Fig. 21-11).

3. Lymphatic drainage pathways of the perineum are of considerable clinical importance because of the high prevalence of cervical carcinoma.

    a. The superficial lymphatics of the clitoris drain toward the superficial inguinal nodes, whereas the deep clitoral vessels drain into the internal iliac nodes.

    b. The lymphatics of the vulva and inferior vagina drain largely into the superficial and deep inguinal nodes.

    c. The middle vagina generally drains toward the internal iliac nodes.

    d. The upper vagina generally drains toward the internal iliac and common iliac nodes.

F. Innervation of the female urogenital triangle (see Fig. 21-13)

    1. The pudendal nerve supplies most of the perineum via perineal branches.

        a. The pudendal nerve arises from spinal segments S2–S4. The **perineal branch** of the pudendal nerve gives rise to:

            (1) The **posterior labial branches**, which supply the skin of the labia majora and minora

            (2) The **dorsal nerve** of the clitoris, which supplies that structure

            (3) The **inferior rectal branches**, which supply most of the anal triangle

        b. Pudendal block. Pain fibers from the perineum may be blocked by injecting an anesthetic in the vicinity of the pudendal nerve in the perineum. The needle is inserted through the posterolateral vaginal wall, just beneath the pelvic diaphragm and angled toward the ischial tuberosity. Within a centimeter of the tuberosity, the needle will be in the vicinity of the **pudendal canal.**

    2. Ilioinguinal and genitofemoral nerves, which arise from L1 and L2, supply the anterior part of each labium as well as the corresponding half of the mons pubis.

    3. Perineal branches of the posterior femoral cutaneous nerve arise from spinal segments S2–S4 and supply the posterolateral region of the urogenital triangle.

G. Clinical considerations

1. Bartholin's cysts result from infection of the greater vestibular glands.

2. Episiotomy may be performed, in which the perineum is sectioned prior to parturition to prevent uncontrolled tearing. Suturing an incision is preferable to repairing a ragged tear.

a. Midline episiotomy

    (1) The incision is into the posterior vaginal wall and carried posteriorly in the midline through the fossa navicularis to divide the central tendon.

    (2) The incision does not extend into the deep fibers of the external anal sphincter.

    (3) The advantage of the midline episiotomy is that it is relatively bloodless and painless because no major vessels or nerves are transected. However, this incision provides a limited expansion of the birth canal with a slight possibility of tearing the anal sphincters.

b. Mediolateral episiotomy

    (1) The incision into the posterior vaginal wall is carried posterolaterally into the ischioanal fossa.

    (2) This incision incises skin over the ischioanal fossa, bulbospongiosus muscle, superficial transverse perineal muscle, fascia and muscle of the urogenital diaphragm, and transverse perineal branches of the internal pudendal artery and pudendal nerve.

    (3) The advantage of the mediolateral episiotomy is that it allows greater expansion of the birth canal into the ischioanal fossa. However, there is an increased risk of infection due to contamination of the ischioanal fossa, and the incision is more difficult to close layer by layer.

I. PELVIC PORTION OF THE GASTROINTESTINAL TRACT: THE RECTUM

A. External structure. The rectum is that part of the large intestine, about 13 cm (5 in) long, between the sigmoid colon and the anal canal (see Fig. 17-18).

1. **The rectum** is a retroperitoneal structure located immediately superior to the pelvic diaphragm.
   a. The upper limit of the rectum is where the sigmoid mesocolon disappears.
   b. The **ampulla**, which lies just above the pelvic floor, is the widest part of the rectum and is capable of considerable distension.
      (1) Because the feces are stored in the sigmoid colon, the ampulla is usually empty.
      (2) Movement of feces into the rectal ampulla from the sigmoid colon generates the sensation of rectal fullness and the urge to defecate.
   c. The rectum pierces the pelvic diaphragm (the levator ani muscle) to become the anal canal.

2. **The rectal sling** is formed by the **puborectalis muscle**, which is the most medial portion of the **pubococcygeus muscle** of the **levator ani** group (Fig. 22-1). The rectal sling is the single most important factor in fecal continence.
   a. The puborectalis muscle fibers loop posteriorly to the rectum, and some fibers attach to the rectum.
   b. Tension in this muscle cants the rectum forward, establishing fecal continence.
   c. Prolapse of the rectum through the anus is often related to damage of the levator ani muscle, usually the result of obstetric trauma.

B. Internal structure. Within the rectum are three transverse **rectal folds** (plicae transversae, valves of Houston) formed by the inner three layers of the intestinal wall.

1. The high incidence of rectal, colonic, and prostatic carcinoma makes digital and sigmoidoscopic examination important in the post–middle-aged male. Most rectal neoplasms can be detected by digital palpation.

2. Carcinomatous spread posteriorly from the rectum may involve the sacral plexus with pain distribution down the leg. Anterior spread may involve the pelvic viscera, such as the prostate, bladder, uterus, or vagina.

C. Vasculature of the rectum. The arterial supply, venous return, and lymphatic drainage of the rectum are of considerable importance.

1. **Arterial supply.** The rectum receives branches from the inferior mesenteric artery via the **superior rectal (hemorrhoidal) artery**, from the internal iliac artery via the **middle rectal (hemorrhoidal) arteries**, and from the internal pudendal arteries via the **inferior rectal (hemorrhoidal) arteries**. Anastomotic connections are abundant in the submucosa between tributaries of these arteries (see Fig. 17-20).

2. **Venous return** from the rectum is along similarly named vessels between which anastomoses are especially rich (see Fig. 17-20).
   a. *Portal hypertension* (accompanying cirrhosis) or local compression of the inferior mesenteric vein (because of chronic constipation with dilation of the sigmoid colon or because of the presence of the fetus during the later stages of pregnancy) results in the shunting of blood from the rectum and anal canal into the systemic venous system.
   b. This shunting results in *hemorrhoids*, which are variceal dilatations of the submucosal anal and perianal venous plexuses. Hemorrhoidal varices may rupture during the evacuation of

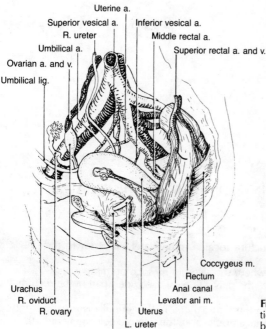

Uterine a.

Superior vesical a.

Inferior vesical a.

R. ureter

Middle rectal a.

Umbilical a.

Superior rectal a. and v.

Ovarian a. and v.

Umbilical lig.

Coccygeus m.

Rectum

Anal canal

Urachus

R. oviduct

Levator ani m.

R. ovary

Uterus

L. ureter

Bladder

Figure 22-1. *Female pelvic viscera.* A parasagittal section indicates the relationships between the urinary bladder, uterus, and rectum as well as the major pelvic vasculature.

feces and produce considerable blood loss. Chronically bleeding hemorrhoids may lead to anemia. Hemorrhoids are amenable to treatment by ligature or cautery.

3. **Lymphatic drainage** of the rectum parallels the various arterial pathways. Therefore, metastatic carcinoma of the rectum may be widely disseminated within the abdomen and pelvis as well as to the inguinal nodes. Also, hematogenous spread to the liver frequently occurs.

II. PELVIC PORTION OF THE URINARY SYSTEM

A. Ureters

1. **Course.** The ureters convey urine from the kidneys to the urinary bladder.
   a. To reach the urinary bladder the ureters pass anteriorly to the psoas muscles and common iliac vessels to enter the minor (deep) pelvis (see Fig. 18-3). Here the ureters narrow somewhat, and a nephrolith may lodge.
   b. They run along the posterolateral wall of the deep pelvis in which they lie extraperitoneally in the endopelvic fascia.
   c. In the male, they pass posteriorly to the **vas deferens** (see Fig. 22-2).
   d. In the female, they pass beneath the **cardinal ligament** and **uterine vessels** (see Fig. 22-1).
   e. They converge to enter the urinary bladder posteroinferiorly, where they define the upper lateral limits of the **trigone** (see Fig. 22-7).
      (1) Because the bladder is muscular, the intramural portion of each ureter is narrow. This is another point at which nephroliths may lodge.
      (2) The oblique course of each ureter through the bladder wall functions as a check valve to prevent reflux of urine from the urinary bladder into a ureter.

2. **Blood supply to the ureters**
   a. The diffuse blood supply to the ureters is from the aorta, the renal arteries, the gonadal arteries, the common and internal iliac arteries, as well as from the inferior vesical arteries.
   b. Anastomoses may be weak so that inadvertent interruption of a ureteral twig may lead to necrosis and rupture of a ureteral segment about 1 week postoperatively. This may become evident as a fistulous tract, such as a ureterovaginal fistula.

3. **Innervation of the ureters**
   a. The proximal portion of each ureter is innervated by autonomic nerves that arise from T12 and run in the least splanchnic nerve. As such, pain from this region refers to the lumbar region.
   b. In the vicinity of the pelvic brim, the ureter is innervated by autonomic nerves that arise

from L1 and L2 and run in the lumbar splanchnic nerves. As such, pain from this region refers to the inguinal and pubic regions as well as to the lateral and anterior aspects of the thigh.

    c. The terminal portion of the ureter is innervated by autonomic nerves that arise from S2–S4 and run in the pelvic splanchnic nerves. As such, pain from this portion refers to the perineum as well as the posterior thigh and leg.

4. Clinical considerations. The pelvic ureters (aside from the common bile duct) are the structures most often abused during abdominal and pelvic surgical procedures.

    a. Injuries must be recognized and repaired at once to prevent urine extravasation and possible subsequent peritonitis.

    b. The female ureters are frequently injured due to inadvertent clamping, ligation, or sectioning along with the ovarian vessels or the uterine vessels.

    c. Careful identification and subsequent avoidance of the ureters is a major objective of abdominopelvic surgery.

B. Urinary bladder

1. External structure. The urinary bladder is located immediately behind the pubic symphysis and the superior pubic rami (see Figs. 22-2 and 22-8).

    a. The fundus of the urinary bladder is extraperitoneal.

      (1) This region expands freely, rising above the pubic crest.

      (2) It normally accommodates 250–300 ml but may accommodate up to 500 ml.

    b. The base of the urinary bladder rests on the pelvic floor.

      (1) Internally, the base of the urinary bladder has a relatively unexpandable portion, the **trigone**, which the ureters enter and the urethra leaves.

      (2) The base is associated with the prostate gland in the male.

    c. The anterior surface of the urinary bladder defines the **retropubic space**, a potential space filled with endopelvic fascia.

      (1) The retropubic surgical approach to the prostate is through this space.

      (2) Urethral trauma superior to the urogenital diaphragm or rupture of the urinary bladder results in extravasation of urine into the retropubic space.

    d. The posterior surface is covered by peritoneum.

      (1) In the female, the posterior surface of the urinary bladder is related to the **vesicouterine pouch** and the **vesicovaginal septum** (see Figs. 22-1 and 22-8).

      (2) In the male, the posterior surface is related to the **rectovesical pouch, rectovesical septum** (Denonvilliers' fascia), and **seminal vesicles** (Fig. 22-2). The rectovesical septum is in part formed by embryonic fusion of the inferior portion of the rectovesical pouch.

    e. Support of the the urinary bladder is by several pelvic structures and condensations of endopelvic fascia.

      (1) It is supported by the **urogenital diaphragm** and is stabilized by the penetration of the urogenital diaphragm by the urethra (see Figs. 22-2 and 22-8).

      (2) There is a supportive condensation of **endopelvic fascia** at the base of the urinary bladder.

      (3) The **medial umbilical ligaments**, remnants of the **umbilical arteries**, support the urinary bladder anterolaterally.

      (4) The **median umbilical ligament** supports the bladder superiorly.

        (a) The **urachus** forms from the allantoic duct (between the cloaca and the placental allantois), which obliterates (see Fig. 22-1).

        (b) The urinary bladder forms in the base of the urachus.

2. Internal structure

    a. The walls of the urinary bladder are composed of a substantial layer of smooth muscle, the **detrusor muscle**.

    b. The mucosal lining is composed of **transitional epithelium.**

      (1) The transitional epithelium, unique to the urinary system, becomes attenuated as the bladder fills and allows considerable stretching.

      (2) This layer is thicker and thrown into folds in the empty bladder.

    c. The **trigone** is a triangular area at the base of the bladder (see Fig. 22-7).

      (1) It is relatively indistensible and has few mucosal folds when the bladder is empty.

      (2) It is defined posterosuperiorly by the ostia of the ureters (about 5 cm apart) and inferiorly by the urethra.

      (3) The ostia of the ureters are protected by the **ureterovesical valves** (of Sampson).

        (a) The angular penetration of the ureters through the walls of the detrusor muscle within the trigone results in a valve-like action.

        (b) Urine in the filling bladder exerts pressure against the bladder walls, thereby compressing the anterior aspect of the intramural portion of the ureters to prevent reflux of urine from the bladder into the ureters.

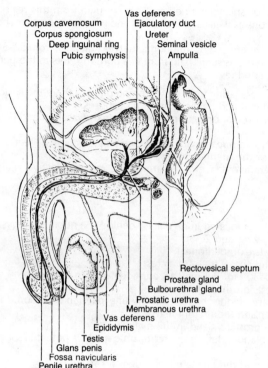

Corpus cavernosum
Corpus spongiosum
Deep inguinal ring
Pubic symphysis
Vas deferens
Ejaculatory duct
Ureter
Seminal vesicle
Ampulla

Rectovesical septum
Prostate gland
Bulbourethral gland
Prostatic urethra
Membranous urethra
Vas deferens
Epididymis
Testis
Glans penis
Fossa navicularis
Penile urethra

Figure 22-2. *Male pelvic viscera.* A midsagittal section depicts the lower portion of the male genitourinary tract.

3. **Vasculature of the urinary bladder** is somewhat variable.
   a. **Arterial supply**
      (1) The **superior vesical arteries** arise from the patent portion of the umbilical arteries and supply the cranial portion of the bladder.
      (2) The **middle vesical arteries** are inconsistent and, when present, may arise from a variety of sources to supply the fundus.
      (3) The **inferior vesical arteries** usually arise from the internal iliac arteries, but may arise from the deferential artery, and supply the base and trigonal region.
   b. **Venous return.** The veins tend to drain into the **prostatic venous plexus** at the base of the bladder.

4. **Innervation of the bladder** (see Fig. 22-3)
   a. The motor innervation to the bladder is via the nervi erigentes.
   b. The afferent pathways from the bladder are controversial.
      (1) The sensory perception for bladder fullness appears to travel largely via afferent neurons that lie along the sympathetic pathways (hypogastric nerve) to spinal segments T12–L2. Also, as the bladder fills to the extreme, urine continues to fill the ureters; this transient hydronephrosis causes additional pain in the lowest thoracic and upper lumbar dermatomes.
      (2) The afferent limb of the bladder-emptying reflex appears to travel via afferent neurons that lie along the parasympathetic pathways (pelvic nerves and nervi erigentes) to spinal segments S2–S4. This is also the pathway for the efferent limb of the bladder-emptying reflex.

5. **Functional considerations**
   a. **Urination** is the process of urine moving down the ureters by peristaltic activity to fill the urinary bladder.
      (1) Upon filling, the urinary bladder rises in the pelvic cavity.
         (a) Afferent reflex impulses from the stretch receptors of the urinary bladder reach segments S2–S4 of the spinal cord via the pelvic nerve.
         (b) Autonomic efferent reflex impulses leave segments S2–S4 of the spinal cord at the same level and travel by the pelvic nerves to the detrusor muscle. Stretch in the detrusor muscle elicits a reflex contraction of that muscle, enhancing the urge to void.

(c) In the normal average male, bladder content is first noted when it reaches 100–150 ml.
(2) There does not appear to be any physiologic or morphologic internal urethral sphincter to close the opening of the prostatic urethra. The initial urinary sphincteric function has been debated intensely. It seems that physiologic sphincteric action in the empty to partially full bladder is due to the longitudinal folding of the mucosa in the region of the trigone. During ejaculation, contraction of the smooth muscle in the prostate gland and the peristaltic contraction of the prostate urethra function as an internal urethral sphincter to prevent retrograde ejaculation into the bladder.
(3) Upon filling, the mucosa of the trigone area around the ureteral outlet is stretched, thereby opening the prostatic urethra so that urinary continence of the partially full to full urinary bladder is maintained by action of the **external urethral sphincter**.
   (a) The external urethral sphincter is formed about the membranous urethra from the adjacent circularly arranged fibers of the deep transverse perineal muscle and is under voluntary control by the deep perineal branches of the pudendal nerve.
   (b) In addition, the most medial fibers of the **puborectalis muscle** of the pelvic floor contribute to the support of the bladder and may contribute to urinary continence by raising the bladder, thereby attenuating the urethral lumen.
   (c) In the male, distension becomes uncomfortable at a volume of 350–400 ml, beyond which painful sensations are experienced. If unrelieved, involuntary micturition occurs at about 500 ml.
   (d) In the female, these values are considerably lower because the bladder is smaller and the external urinary sphincter is incomplete as a result of the penetration of the urogenital diaphragm by the vagina and the intimate association between the urethra and vagina.

b. Micturition is the process of emptying the urinary bladder through the urethra. It is a reflex action, initiated by the stretching of the detrusor muscle.
   (1) When the puborectalis muscle is relaxed in anticipation of micturition, the bladder descends slightly; this process relieves the stretching that attenuates the lumen of the initial segment of the urethra.
   (2) When the external urethral sphincter is allowed to relax, the bladder commences to void, often assisted by a Valsalva maneuver, which raises the intra-abdominal pressure.
   (3) At the termination of micturition, strong contractions of the external urethral sphincter, pelvic floor, and in the male, bulbocavernosus muscles expel the residual urine from the urethra, and continence is restored.
   (4) The detrusor muscle relaxes with the absence of stretch.
   (5) Thus, the bladder is under the influence of the autonomic nervous system (probably the parasympathetic division) via reflex arcs along the pelvic nerve with overriding voluntary control of the external urethral sphincter via the pudendal nerve.

6. Clinical considerations
   a. Transurethral cystoscopy is performed to inspect the mucosa of the urinary bladder and prostate. It may be combined with procedures to remove ureteral calculi and bladder stones. It is also an integral part of the transurethral approach to prostatectomy.
   b. Patent urachus
      (1) A completely patent urachus (very rare) allows reflux of urine through the umbilicus.
      (2) Approximately 33% of all individuals have to some extent a patent lumen of the urachus. Severance of such a patent urachus by a transverse suprapubic incision can result in urine extravasation into the peritoneal cavity.
   c. Pelvic fractures may damage the urinary bladder, urethra, rectum, uterus, vagina, nerves, and, especially blood vessels, with a potential for severe internal hemorrhage.
      (1) The urinary bladder is most vulnerable in pelvic fracture, particularly when distended with resulting intrapelvic and subperitoneal extravasation of urine.
      (2) Following severe pelvic fractures, immediate surgical repairs to soft tissues may be essential.

C. Urethra

1. The male urethra is divided into prostatic, membranous, and penile segments (see Fig. 22-2).
   a. The prostatic urethra is the portion that begins at the vesical neck at the apex of trigone, extends through the prostate gland, and terminates at the superior fascia of the urogenital diaphragm. This portion of urethra has the thickest walls.
      (1) The **urethral crest (crista urethralis)** is located on the posterior wall of the prostatic urethra.
         (a) The numerous **prostatic ducts** open individually into the prostatic urethra on either side of the urethral crest.
         (b) The **colliculus seminalis (verumontanum)** is the widest portion of the urethral

crest. The **prostatic utricle (uterus masculinus)**, an invagination about 5 mm deep, lies in the midline of the verumontanum.

(i) The prostatic utricle is thought by some to represent the terminal remnant of the fused terminal portion of the paramesonephric (müllerian) ducts, which form the uterus and oviducts in the female.

(ii) This structure is prone to occasional cyst formation.

(c) **Ejaculatory ducts** enter the prostatic urethra through the urethral crest immediately lateral to the prostatic utricle (see Figs. 22-2 and 22-7).

(2) Injury to the prostatic urethra or the membranous urethra above the urogenital diaphragm may result in intrapelvic, extraperitoneal extravasation of urine.

b. **The membranous urethra** begins at the superior fascia of the urogenital diaphragm, extends through the deep transverse perineal muscle, and terminates at the inferior fascia of the urogenital diaphragm. This short portion of the urethra (1–2 cm long) is relatively thin-walled.

(1) Except for the external urethral meatus, this is the narrowest and least distensible portion of the urethra. This seems to be due to the tone of the surrounding **external urethral sphincter**.

(2) Due to its narrowness, delicate walls, and the sharp angle between the penile and membranous portions, this segment of the urethra is most likely to be ruptured by trauma or passage of instrumentation.

(a) If damage occurs above the urogenital diaphragm, extravasation of urine is intrapelvic and extraperitoneal.

(b) If damage is below this diaphragm, extravasation is within the superficial perineal space with possible extension into the scrotum beneath the dartos layer, the penis superficial to Buck's fascia, and the anterior abdominal wall beneath Scarpa's fascia.

c. **The penile (spongy, cavernous) urethra** begins at the inferior fascia of the urogenital diaphragm. It extends into the bulb of the penis, where it is surrounded by **corpus spongiosum (corpus cavernosum urethrae)**, and terminates at the external urethral meatus of the glans penis (see Fig. 22-2).

(1) It varies in length among individuals and with the state of erection.

(2) The ducts of the **bulbourethral glands** (of Cowper) enter the penile urethra just below the urogenital diaphragm.

(3) The penile urethra enlarges into the **fossa navicularis** just prior to termination at the **external urethral meatus**.

(4) Damage to the penile urethra may result in extravasation of urine.

(a) If the deep penile fascia (of Buck) remains intact, extravasation occurs deep to this fascial layer and is limited to the penis.

(b) If the deep penile fascia is damaged, extravasation of urine occurs between the deep and superficial penile fascial layers with possible extension beneath Colles' fascia into the superficial perineal space, beneath the dartos fascia into the scrotum, and beneath Scarpa's fascia in the anterior abdominal wall.

2. **The female urethra** is approximately 3.75 cm in length (see Fig. 22-8). It corresponds to the prostatic and membranous portions of the male urethra.

a. It is fused to the adventitia of the anterior vaginal wall.

(1) Because of this physical relationship, the fibers of the deep transverse perineal muscle do not pass posteriorly to the urethra; thus, the external urethral sphincter in the female is incomplete. This explains, in part, the much higher incidence of stress incontinence among women, especially if the urogenital diaphragm is damaged during parturition or if urethrocele (herniation of the urethra into the vagina) develops.

(2) The intimate relationship of the urethra to the vagina predisposes the urethra to injuries and subsequent cystocele or urethrocele associated with difficult parturition.

b. The **urethral glands** open into the urethra; the **paraurethral** glands (of Skene) open into the vestibule adjacent to the urethral orifice.

(1) These glands are homologous to the prostate.

(2) They tend to be especially prone to infection.

c. The urethra opens into the **vestibule** between the **labia minora**.

III. INNERVATION OF THE PELVIC VISCERA

A. **Sympathetic innervation to the pelvic viscera** arises from spinal segments T12–L2 (Fig. 22-3).

1. **Presynaptic pathways.** White rami communicantes carry the preganglionic sympathetic motor neurons from the lowest thoracic and upper lumbar spinal nerves to the sympathetic chain of paravertebral ganglia.

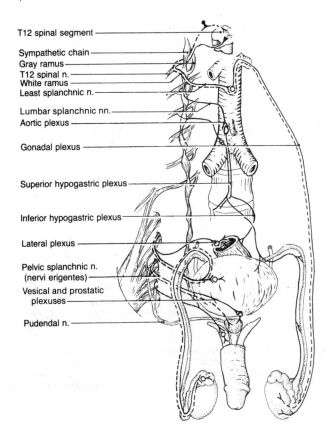

T12 spinal segment

Sympathetic chain
Gray ramus
T12 spinal n.
White ramus
Least splanchnic n.

Lumbar splanchnic nn.
Aortic plexus

Gonadal plexus

Superior hypogastric plexus

Inferior hypogastric plexus

Lateral plexus

Pelvic splanchnic n.
(nervi erigentes)
Vesical and prostatic
  plexuses

Pudendal n.

Figure 22-3. *Innervation of the male bladder, reproductive tract, and genitalia.* The sympathetic pathways arise from the lower thoracic and upper lumbar spinal levels (there are no white rami below L2) and reach the pelvic viscera via thoracic and lumbar splanchnic nerves and then the hypogastric plexuses. The parasympathetic pathways arise from the midsacral spinal levels and reach the pelvic viscera via the pelvic splanchnic nerves. Visceral afferent fibers (*dashed*) from the pelvic viscera travel specifically along either one or the other autonomic pathways, producing specific patterns of referred pain. The pudendal nerve provides somatic innervation to and from the perineum.

2. **Sympathetic ganglia.** Some of these neurons may synapse within the paravertebral ganglia before leaving via the lumbar splanchnic nerves to join the **aortic plexus**. Those neurons that do not synapse in the paravertebral ganglia seem to synapse in small ganglia located along the aortic plexus.

3. **Postsynaptic pathway.** After synapse, the axons run inferiorly in the **aortic plexus**.
   a. **The superior hypogastric plexus** is the continuation of the aortic plexus over the bifurcation of the aorta. This pathway brings sympathetic nerves into the pelvis.
   b. **The inferior hypogastric plexuses (left and right hypogastric nerves)** are formed by the bifurcation of the superior hypogastric plexus at the pelvic brim.
      (1) These sympathetic pathways run along the anterior surface of the sacrum toward the rectum.
      (2) They provide the sympathetic contribution to the **pelvic plexus (lateral pelvic plexus)** on the walls of the rectum.

B. **Parasympathetic innervation to the pelvic viscera** arises from spinal segments S2–S4 (see Fig. 22-3).

   1. **Presynaptic parasympathetic neurons** arise from spinal segments S2 and S3 *or* S3 and S4 (seldom S2, S3, *and* S4) and leave the spinal nerves in the vicinity of the anterior sacral foramina to form the **pelvic nerves (nervi erigentes, pelvic splanchnic nerves).**

   2. These parasympathetic **pelvic splanchnic nerves** run along the lower sacrum to provide the parasympathetic contribution to the pelvic plexus (lateral pelvic plexus).

C. **Pelvic plexus (lateral pelvic plexus)**

   1. **Location.** This autonomic plexus on the lateral walls of the rectum contains postsynaptic sympathetic neurons from the hypogastric plexus, presynaptic parasympathetic neurons from the nervi erigentes, small autonomic ganglia, and visceral afferent neurons.

   2. **Divisions.** The lateral plexus gives rise to several autonomic plexuses.
      a. **The rectal plexus** provides autonomic innervation to the rectum and anal canal as well as

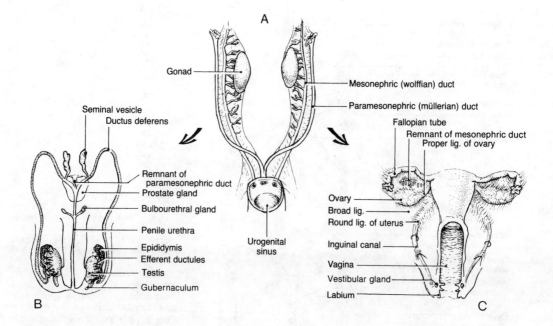

Figure 22-4. *Development of the gonads and genital ducts. A,* In the indifferent stage both mesonephric and paramesonephric ducts lead to the urogenital sinus. The gonads are located high in the abdomen. *B,* In the developing male, the mesonephric ducts develop into the vas deferens and epididymis. The testes descend into the scrotum. *C,* In the female, the paramesonephric ducts partially fuse to form the uterus, while the unfused portions form the oviducts. The ovaries descend into the deep pelvis.

parasympathetic innervation in a retrograde direction to the large colon as far as the splenic flexure.

    b. The uterovaginal or **prostatic plexus** gives rise to the **cavernous plexuses**, which convey parasympathetic nerves to the erectile tissues as well as sympathetic nerves to the glands associated with the reproductive tract.

    c. The vesical plexus is associated with urinary bladder function.

  3. Clinical consideration. Surgical interference with the pelvic plexus, as for example in rectal resection, results in varying degrees of dysfunction of the pelvic and abdominal structures innervated.

IV. DEVELOPMENT OF THE UROGENITAL DUCTS

  A. Early development of the genitourinary system. The development of the kidneys proceeds through three distinct pairs of excretory organs that develop sequentially in the abdominal cavity from the cranial to the caudal regions of the urogenital ridge (Fig. 22-4A).

    1. The pronephros develops by the fourth week but is never completely functional.

    2. The mesonephros develops late in the fourth week.

      a. Rows of parallel nephrons drain into the **mesonephric (wolffian) duct**, which empties into the cloaca.

      b. By the sixth week, the mesonephros consists of elongated organs projecting into the coelomic cavity on either side of the midline of the posterior abdominal wall.

      c. By the eighth week, most of the mesonephros has degenerated, but the mesonephric duct remains.

    3. The metanephros, the definitive kidney, develops during the fifth week and is fully functional by the eighth week. It develops from two primordia:

      a. A **metanephric diverticulum** arises from the posterior wall of the caudal region of the mesonephric duct.

      b. The **metanephric mass** arises from the posterior region of the urogenital ridge. The growth of the metanephric duct into the metanephric mass results in the formation of several kidney lobes.

B. During the indifferent stage (weeks 6–8), there are no apparent differences between the developing male and female (see Fig. 22-4*A*).

  1. Genital ducts. By the sixth week, two pairs of genital ducts have formed.

    a. Mesonephric (wolffian) ducts are urinary tracts exiting from the primitive mesonephros and draining into the unpartitioned cloaca. In the male, the **mesonephric ducts** form the genital tract, and the paramesonephric ducts largely degenerate.

    b. Paramesonephric (müllerian) ducts arise parallel to the mesonephric ducts. In the female, the **paramesonephric ducts** form the oviducts and uterus, and the mesonephric ducts largely degenerate.

  2. The urinary bladder is formed by partition of the cloaca by the urorectal septum into an anterior **urogenital sinus** and a posterior **anorectal canal.**

    a. The most cranial portion of the urogenital sinus forms the urinary bladder.

      (1) The cranial portion of the urogenital sinus is continuous with the **allantois**, a diverticulum of the hindgut into the placenta.

      (2) When the lumen of the allantoic duct is obliterated, it becomes the cord-like **urachus**, connecting the superior pole of the urinary bladder to the umbilicus and forming the **median umbilical ligament.**

    b. The middle portion of the urogenital sinus forms the **urethra** in the female, which is equivalent to the membranous and prostatic portions of the urethra in the male.

    c. The inferior portion of the urogenital sinus remains as the definitive urogenital sinus, which persists as the **vestibule** in the female and forms the **penile urethra** in the male on fusion of the urethral folds.

  3. The mesonephric ducts enter the urogenital sinus with the partition of the cloaca by the urorectal septum. As the walls of the terminal region of the mesonephric ducts become incorporated into the developing urinary bladder, the metanephric diverticula separate from the mesonephric ducts so that the former enter the urinary bladder, whereas the latter enter the urethra.

C. Development of the male genital tract (see Fig. 22-4*B*)

  1. The testes develop from the mesonephric (urogenital) ridge.

    a. Seminiferous cords develop into **seminiferous tubules, straight tubules,** and the **rete testis.**

    b. Between the seminiferous cords, interstitial mesenchyme differentiates into hormone-producing **interstitial cells** (of Leydig).

  2. The mesonephric ducts develop under the influence of the fetal testicular hormones.

    a. The greater part of the mesonephric duct forms the **vas deferens.**

    b. A diverticulum toward the caudal end of each mesonephric duct forms each **seminal vesicle.**

    c. Distal to the seminal vesicles, each mesonephric duct becomes the **ejaculatory duct,** which joins the prostatic urethra.

    d. As the mesonephros degenerates, some mesonephric tubules are incorporated into the developing genital duct system.

      (1) Some of the mesonephric tubules become the **efferent tubules,** which open into the portion of the mesonephric duct that develops into the epididymis.

      (2) Some of these mesonephric tubules fail to connect with the epididymis, yet fail to degenerate. These become the vestigial **paradidymis** and the **appendix epididymis.**

      (3) Most of the more cranial portions of the mesonephric duct degenerate except for a small cranial portion that remains at the **appendix testis.**

    e. The paramesonephric ducts degenerate except for the **prostatic utricle,** which may be a site of cyst formation (müllerian cyst).

  3. The prostatic, membranous, and penile portions of the urethra form the terminal portion of the genital tract in the male.

    a. The prostate develops as five groups of diverticula from the cranial portion of the urethra.

    b. The **bulbourethral glands** develop as diverticula of the membranous urethra.

D. Development of the female genital tract (see Fig. 22-4*C*)

  1. The ovaries develop from the mesonephric (urogenital) ridge.

    a. Cortical cords break up into the **primary follicles.**

    b. All 2 million primary follicles are formed during the early embryonic period.

    c. The developing ovaries descend toward the deep pelvis.

  2. The paramesonephric ducts develop as the female genital tract in the absence of testicular hormone.

    a. The cranial terminus of each paramesonephric duct opens directly into the peritoneal cavity as the **ostium** of the oviduct.

    b. The paramesonephric ducts fuse caudally to form the uterovaginal primordium. As the ovaries descend into the minor pelvis, the paramesonephric fusion extends cranially. Thus, the original paired vaginas and paired uteri become a single midline tube.

       (1) This fusion brings the developing oviducts toward the midline, lifting them away from the pelvic walls.

       (2) This elevation forms the **broad ligament**, which divides the inferior aspect of the pelvic cavity into an anterior **uterovesical pouch** and a posterior **uterorectal pouch.**

       (3) These fused ducts form the corpus of the uterus and the cervix.

    c. Two groups of vestigial mesonephric tubules remain in the vicinity of the ovary, the epoophoron and the paroophoron. These can be sites of cyst formation.

    d. The mesonephric ducts degenerate except for the most cranial and most caudal portions, which persist as the appendix vesiculosa and the Gartner's ducts, respectively. These can be sites of cyst formation.

  3. Vagina

    a. The vagina develops along the posterior wall of the urogenital sinus from two solid sinovaginal bulbs that fuse in the midline.

    b. The termination of the fused paramesonephric ducts (uterovaginal primordium) joins the sinovaginal bulbs.

    c. The vaginal lumen is formed by cavitation of the sinovaginal bulbs, except for the most inferior portion, which persists as the **hymen**; the hymen separates the vaginal cavity from the vestibule until the perinatal period.

V. MALE REPRODUCTIVE TRACT

  A. Testes

    1. External structure (Fig. 22-5)

      a. Each testis is approximately 4.5 cm x 3 cm x 2.7 cm.

      b. Each testis is surrounded by the **tunica albuginea**, a tough connective tissue layer. On the medial, anterior, lateral, and inferior aspects, the testis is covered by the visceral layer of the **tunica vaginalis** (visceral peritoneum). Within the scrotum, the left testis is usually lower.

    2. Internal structure. Each testis is divided into approximately 300 lobules, each of which contains approximately 1–4 **seminiferous tubules**. As these tubules are approximately 75 cm long, there are approximately 750 m of tubules per testis.

      a. Sixty-one day cycles of spermatogenesis (development of spermatids) and subsequent spermiogenesis (maturation of spermatids into spermatozoa) occur in a wave-like manner along the seminiferous tubules.

      b. The seminiferous tubules lead into about 25 **straight tubules**, which coalesce in the mediastinum as the **rete testes**. From the rete, 15–20 **efferent ductules** pierce the tunica albuginea to enter the head of the **epididymis**.

    3. Vasculature of the testes

      a. Arterial supply (see Figs. 18-3 and 21-3)

        (1) Each **testicular artery** arises from the aorta below the respective renal arteries and courses anteriorly to the psoas muscle and through the deep inguinal ring to enter the spermatic cord. It enters the testis via the mesorchium.

        (2) There is a rich collateral blood supply to the testes.

          (a) The **artery** of the **ductus deferens** arises from the internal iliac artery and is homologous to the uterine artery.

          (b) The **cremaster artery** arises as a branch of the inferior epigastric artery.

          (c) The **external pudendal arteries** arise as branches of the femoral arteries.

      b. Venous return (see Figs. 18-3 and 21-3)

        (1) The **testicular (internal spermatic) veins** provide the principal drainage of the testes.

          (a) The 8–10 veins of the **pampiniform plexus** in the lower end of the spermatic cord unite to form each testicular vein. The pampiniform plexus functions as a **countercurrent heat exchanger**. This plexus is frequently (80% of males) affected by **varicocele**.

            (i) Fully 90% of all varices are on the left side. This is apparently due to local venous hypertension, resulting from compression of the left testicular vein by the sigmoid colon, which contains stored feces.

            (ii) Varicocele thrombosis is often painful, affects temperature regulation, and may result in decreased sperm viability.

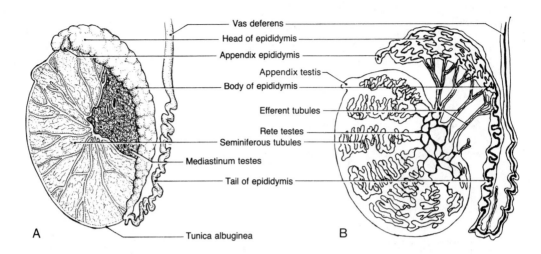

Figure 22-5. *Testis. A*, This parasagittal section depicts the testicular parenchyma and the undissected epididymis. *B*, The duct system is schematized.

(iii) The best resolution is through surgical ligation of the testicular vein superior to the deep inguinal ring so that the blood returns via collaterals. This procedure alleviates the back pressure, resulting from the hydrostatic columns of venous blood within the abdomen, and encourages collateral venous return.

(b) The right testicular vein drains into the inferior vena cava, whereas the left testicular vein drains into the left renal vein. Contrary to wide acceptance, this is not the cause of varicocele.

(2) Collateral venous drainage is via the **external pudendal veins**, **posterior scrotal veins**, **cremaster veins**, and **veins of the vas deferens**.

c. **Lymphatic drainage.** The lymphatics from the testes parallel the testicular veins to drain directly into the para-aortic lymph nodes. This explains in part why testicular seminoma disseminates widely and rapidly in the midabdominal retroperitoneum.

4. **Innervation of the testes** (see Fig. 22-3)
   a. The motor role of the autonomic nerves to the testes is uncertain.
      (1) Sympathetic nerves from the aortic and renal plexuses innervate the testes.
      (2) Parasympathetic nerves from the prostatic plexus supply the vas deferens and epididymis.
   b. Afferent nerves from the testis (see Fig. 22-3).
      (1) Afferent fibers travel to the spinal cord along the sympathetic pathways, parallel to the testicular vessels.
      (2) These pass through the aorticorenal plexus, thence along the lesser and least splanchnic nerves to spinal segments T10–T12 as well as along the lumbar splanchnic nerves to spinal segments L1–L2.
      (3) Pain originating in the testes is referred to the middle and lower abdominal wall.

5. **Testicular torsion**
   a. Rotation (torsion) of the testes may occur from 90°–360° about the spermatic cord.
   b. Precipitating factors seem to be strong contractions of the cremaster muscle and the presence of a long mesorchium between the parietal and visceral layers of the tunica vaginalis, the "mesentery of the testes." It is most common in the adolescent.
   c. Torsion is usually external rotation. Aside from trauma, testicular torsion represents the only urologic emergency.
      (1) Compression of testicular vessels leads to ischemic necrosis of the testes in 6 hours.
      (2) Failure to recognize and intervene immediately has resulted in loss of the testis in about 80% of all cases, subsequent atrophy of the testis in 10% more, and fertile resolution in only 10%.
      (3) In the emergency room, the twisted testis may be medially rotated by gentle external pressure. At subsequent operation, both sides will be sewn into position (orchiopexy) because the other testis is very valuable.

B. Epididymis

1. **Structure.** The highly convoluted **efferent ductules** open into the single duct, forming the epididymis (see Fig. 22-5).

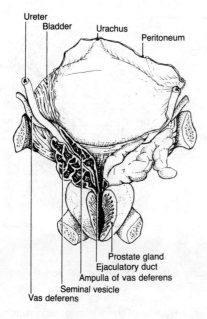

Ureter
Bladder
Urachus
Peritoneum

Prostate gland
Ejaculatory duct
Ampulla of vas deferens
Seminal vesicle
Vas deferens

Figure 22-6. *Accessory glands of the male.* The posterior aspect of the urinary bladder, seminal vesicles, and prostate gland is indicated with the terminal portion of the left vas deferens, the left seminal vesicle, and the prostate gland all dissected open.

    a. The epididymis represents a persistent portion of the **mesonephric (wolffian) duct** (see Fig. 22-4).
    b. The epididymis is about 6 m long and compactly coiled to form the **head** and **body** of the epididymis. These segments comprise the sites of sperm maturation and storage.
    c. In the **tail** of the epididymis, the duct becomes thicker and forms the **vas deferens.**
    d. The epididymis is partially covered by the visceral layer of the tunica vaginalis and is adherent to the superior pole of the testis.

  2. Function. Maturation and storage of spermatozoa occurs in the epididymis.

  3. Blood supply to the epididymis is predominantly by the **deferential artery.**

  4. Innervation of the epididymis
    a. The parasympathetic and visceral innervation of the epididymis is by the **nervi erigentes.**
    b. Afferent nerves from the epididymis and vas deferens travel along the vas deferens to the prostatic plexus, thence to the lateral pelvic plexuses, nervi erigentes, and sacral nerves to reach spinal segments S2–S4. Pain originating in the epididymis, such as that accompanying epididymitis, is referred to the distribution of S2–S4.

C. Vas deferens

  1. Structure. The vas deferens represents the **mesonephric (wolffian) duct** (see Fig. 22-4). It is a direct continuation of the duct of the epididymis. It has a very muscular wall, which accounts for the cord-like composition.
    a. The external course of the vas deferens
      (1) From the inferior pole of the testis, the vas deferens ascends and passes into the center of the spermatic cord, which is the usual site for vasectomy.
      (2) It transits the inguinal canal and enters the abdominal cavity by passing through the deep inguinal ring located laterally to the inferior epigastric artery.
      (3) The vas deferens may be double on occasion, an important factor if vasectomy is to be successful.
    b. The internal course of the vas deferens
      (1) In the abdominal cavity, the vas deferens lies within the transversalis fascia.
      (2) At the deep inguinal ring, it immediately diverges from the **testicular neurovascular bundle** and passes extraperitoneally over the **external iliac vessels** and **obturator neurovascular bundle** to enter the deep pelvis.
      (3) In the pelvic cavity, the vas deferens lies within endopelvic fascia.
      (4) It passes anterosuperiorly to the **ureters** to converge toward the prostatic urethra at the base of the urinary bladder.
    c. Ampulla (Fig. 22-6)
      (1) The ampulla of the vas deferens is an enlarged, sacculated portion near the base of the **prostate gland.**

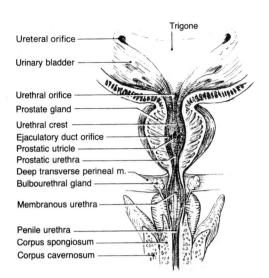

Figure 22-7. *Male urethra.*

(2) The **seminal vesicles** join the vas deferens where the ampulla becomes the ejaculatory duct.

  d. Ejaculatory duct (see Fig. 22-6; Fig. 22-7)

    (1) The ejaculatory duct is the narrow (0.5 mm), thin-walled terminal intraprostate portion of the vas deferens.

    (2) It passes through the parenchyma of the prostate gland.

    (3) It joins the **prostatic urethra** by an opening on the **colliculus seminalis (verumontanum)** on either side of the **prostatic utricle.**

2. Blood supply to the vas deferens is by the artery of the vas deferens (deferential artery), the homologue of the uterine artery in the female.

3. Innervation of the vas deferens is primarily by parasympathetic fibers from the lateral pelvic plexus.

D. Seminal vesicles (see Fig. 22-6)

  1. Structure. Each seminal vesicle is a dilated convoluted tube about 4 cm long (9 cm if straightened). Located at the base of the urinary bladder, posteriorly, they may be palpated per rectum if swollen and inflamed.

  2. Function. The seminal vesicles are secretory glands.

    a. They contribute seminal fluid, which contains fructose and choline. Fructose, which is not produced anywhere else in the body, provides a forensic determination for the occurrence of rape. However, choline crystals form the preferred basis for the determination of the presence of semen (Florence's test).

    b. These glands do *not* function for sperm storage.

E. Prostate gland

  1. Structure

    a. The prostate gland lies between the base of the urinary bladder and upper surface of the levator ani and deep transverse perineal muscles (see Fig. 22-7).

    (1) The anterior surface is adjacent to the **retropubic space** (of Retzius).

    (2) The posterior surface is adjacent to the seminal vesicles and the **rectovesical septum (prostatoperitoneal membrane, Denonvilliers' fascia).** It is palpable per rectum.

    b. The prostate has a true capsule and a false capsule (prostatic fascia) formed by endopelvic fascia.

    c. The prostate gland forms as five diverticula of the prostatic urethra; the prostatic parenchyma is subdivided into five lobes.

    (1) The **left** and **right lateral lobes** are extensive and include what was formerly termed an anterior lobe.

    (2) The **left** and **right posterior lobes** also include the apex of the gland.

      (a) These lobes are the lobes most predisposed to *malignant transformation*.

      (b) Malignant cells are found in nearly 50% of the prostate glands of all males over the age of 50 years. This is the most common malignancy in the adult male. Hard malignant nodes may be palpated rectally.

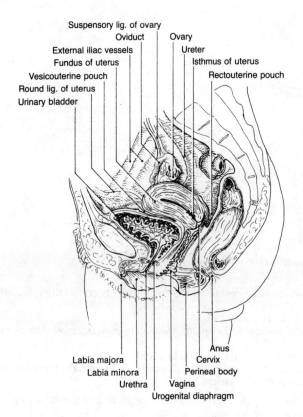

Suspensory lig. of ovary
Oviduct
Ovary
External iliac vessels
Ureter
Fundus of uterus
Isthmus of uterus
Vesicouterine pouch
Rectouterine pouch
Round lig. of uterus
Urinary bladder

Anus
Labia majora
Cervix
Labia minora
Perineal body
Urethra
Vagina
Urogenital diaphragm

Figure 22-8. *Female pelvic viscera.* This midsagittal section depicts the lower portion of the female urinary tract and the genital tract.

(c) Metastatic prostatic carcinoma represents the third most common cause of death from neoplastic diseases.

(3) The **medial lobe** surrounds the prostatic urethra.

    (a) This lobe is predisposed to *benign prostatic "hypertrophy"* (middle lobe hypertrophy, Albarrán's lobe, BPH). This condition is actually *proliferative hyperplasia* of the periurethral glands associated with the median lobe.

    (b) Middle lobe hypertrophy results in obstruction of the urethra and visceral neck of the urinary bladder.

    (c) This condition begins at approximately the age of 45 and occurs in 80% of all men by 80 years of age. However, only about 10% require surgical treatment.

2. Function. The prostate is a compound tubuloalveolar gland.

  a. It secrets 0.5–2 ml of fluid a day, which contributes 10%–20% of the seminal fluid of the ejaculate.

  b. This fluid contains citric acid, acid phosphatase, prostaglandins, and fibrinogen.

3. Vasculature of the prostate. An extensive prostatic venous plexus lies between the dense prostatic capsule and a layer of visceral pelvic fascia (prostatic fascia).

  a. Prostatectomy may result in extensive bleeding.

  b. The prostatic plexus drains into the internal iliac veins as well as into the vertebral venous plexus. This may explain the wide metastatic spread of prostatic cancer to the vertebral column and brain.

F. Clinical considerations

1. Digital examination. Valuable clinical information can be obtained by digital examination per anus. Palpable in the male are the membranous part of a catheterized urethra, posterior and lateral lobes of the prostate gland, rectovesical fossa, seminal vesicles if enlarged, bladder when distended, bulbourethral glands if enlarged, and ductus deferens when displaced or enlarged.

2. Prostatectomy. There are several surgical approaches.

  a. Suprapubic: The incision is through abdominal and vesical walls, avoiding the major portion of the retropubic space.

b. Retropubic: The abdominal wall is incised to gain access to the retropubic space, which is then enlarged by pushing the bladder posteriorly to expose the prostate.

c. Perineal: The incision is through the central tendon and the medial portion of the levator ani muscle. The dissection is anterior to the rectovesical fascia (of Denonvilliers), which "separates the air and water." The rectum is pushed posteriorly, and the urethra pulled forward to visualize the prostate.

d. Transurethral: Removal of hypertrophic tissue in conjunction with cystoscopy

3. Referred pain

a. Because the afferent nerve pathways from the testis accompany the testicular arteries and veins, testicular pain is referred to the lower thoracic and upper lumbar dermatomes.

b. Because the afferent nerve pathways from the epididymis, vas deferens, seminal vesicles and prostate gland accompany the nervi erigentes, the pain associated with these structures is referred to the middle sacral dermatomes.

VI. FEMALE REPRODUCTIVE TRACT

A. Peritoneum and pelvic mesenteries

1. Pelvic peritoneum

a. Parietal peritoneum lines the pelvic cavity.

b. Visceral peritoneum covers those pelvic viscera that protrude into the pelvic cavity. The reflections of serous mesothelium between parietal peritoneum and visceral peritoneum form **mesenteries**.

2. Pelvic mesenteries. The **mesometrium (broad ligament)** is the mesentery of the uterus.

a. Structure. It is formed as the parietal peritoneum is reflected off the walls of the pelvic cavity to course over the uterus and uterine tubes; this mesentery connects lateral margins of the uterus with the side wall of the pelvis (see Fig. 22-9).

b. Relations. It partitions the pelvic peritoneal cavity into two pouches (Fig. 22-8).

(1) The **uterovesical pouch** is an extension of the pelvic peritoneal cavity between the anterior surface of the uterus and the posterior surfaces of the urinary bladder.

(2) The **rectouterine** or **rectovaginal pouch** (of Douglas) is an extension of the pelvic peritoneal cavity between the posterior surfaces of the uterine cervix and the anterior surface of the rectum. The floor of this pouch is apposed to the posterior fornix of the vagina.

c. Divisions. Three peritoneal folds are continuations of the broad ligament.

(1) The **mesosalpinx** supports the uterine tube (see Figs. 22-9 and 22-11).

(a) Beginning at the juncture of the mesovarium, this mesentery represents the superior limit of the broad ligament.

(b) It contains the tubal branches of the uterine vessels as well as the vestigial epoophoron and paroophoron.

(2) The **mesovarium** supports the ovary (see Figs. 22-9 and 22-11).

(a) This mesentery is a posterior extension of the broad ligament.

(b) It becomes continuous with the serosa or "germinal epithelium" of the ovary.

(3) The **infundibulopelvic ligament (suspensory ligament of the ovary)** contains the **ovarian artery** and **ovarian vein** as well as lymphatics and nerves (see Figs. 22-8 and 22-9).

(a) This mesentery is an elevation of peritoneum between the pelvic brim and the mesovarium.

(b) It is formed by the descent of the ovary into the deep pelvis.

d. Contents of the broad ligament (Fig. 22-9)

(1) A **uterine tube** (oviduct) lies in the superior free edge of the broad ligament on each side of the uterus.

(2) The **ovarian ligament (proper ligament of the ovary)** runs between the medial pole of the ovary and the uterine cornu, just below the origin of the uterine tube.

(3) The **round ligament of the uterus** runs from the lateral uterine border near the point of attachment of the ovarian ligament between the anterior and posterior leaflets of the broad ligament and through the inguinal canal (of Nuck) to fuse with the dermis of the labium major.

(4) The **uterine neurovascular bundle** lies at the base of the broad ligament within the transverse cervical (cardinal) ligament and in the mesometrium adjacent to the uterus.

(5) The **ovarian vessels** lie within the suspensory ligament of the ovary and the mesovarium with branches in the mesosalpinx.

(6) Several **vestigial structures** lie within the broad ligament.

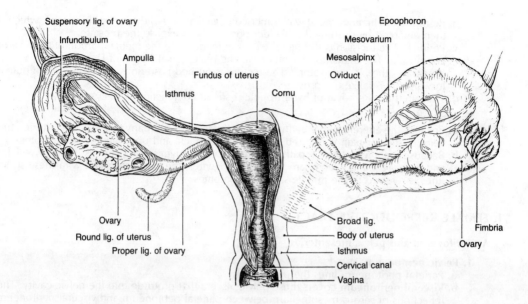

Figure 22-9. *Female reproductive tract*. The posterior aspect of the female reproductive tract is shown. The ligamentous supports of the uterus, right oviduct, and the right ovary are depicted intact. The left side of the uterus, left oviduct, and left ovary are shown dissected.

- **(a)** The **ovarian** and **round ligaments** are homologous to the gubernaculum testis in the male.
- **(b)** Although the mesonephric (wolffian) ducts degenerate in the female, remnants of them may be sites of cyst formation.
  - **(i)** The **epoophoron** constantly lies in the lateral portion of the mesosalpinx and mesovarium, represents residual mesonephric tubules, and is homologous to the epididymis.
  - **(ii)** The **paroophoron** variably lies more medially in the mesosalpinx, also represents residual mesonephric tubules, and is homologous to the paradidymis.
  - **(iii)** **Gartner's ducts** lie in the mesometrium adjacent to the uterus, represent the mesonephric ducts, and are homologous to the vas deferens.
  - **(iv)** A rete ovarii may persist adjacently to the hilus of the ovary and is homologous to the rete testis.
  - **(v)** The hydatid of Morgagni (appendix ovarii) represents the most cranial remnant of the mesonephric duct and is homologous to the appendix of the epididymis.

B. Endopelvic fascia

1. **Location.** The endopelvic fascia is the extraperitoneal tissue of the uterus **(parametrium)**, vagina, urinary bladder, and rectum. It is continuous with the transversalis fascia of the abdomen.

2. **Divisions.** The endopelvic fascia forms thickened bands in the pelvis, termed ligaments, that support the uterus (Fig. 22-10).

   a. Cardinal (transverse cervical) ligaments (of Mackenrodt)
   - **(1)** These fascial condensations arise from the osseomuscular walls of the deep pelvis along the lines of attachment of the levator ani muscle and insert into the lateral walls of the cervix.
   - **(2)** Composed of white fibrous connective tissue with some smooth muscle, they are thickest about the uterine vascular bundles and, thus, support the cervix laterally.

   b. Uterosacral ligaments
   - **(1)** These fascial condensations run between the cervical walls and the sacrum, diverging to either side of the rectum to form the **rectouterine folds** of peritoneum.
   - **(2)** These ligaments contain smooth muscle fibers (rectouterine muscles) and maintain posterior tension on the cervix, drawing the cervix toward the rectum.

   c. Pubocervical (uteropubic, vesicovaginal) ligaments
   - **(1)** This fascial condensation arises from the cervix and cardinal ligament, passes between

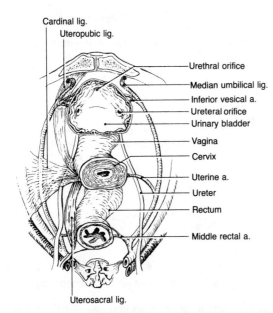

Cardinal lig.
Uteropubic lig.

Urethral orifice
Median umbilical lig.
Inferior vesical a.
Ureteral orifice
Urinary bladder
Vagina
Cervix
Uterine a.
Ureter
Rectum
Middle rectal a.

Uterosacral lig.

Figure 22-10. *Ligamentous support of the cervix.* Anterior, lateral, and posterior condensations of pelvic fascia support the cervix. The close relationship between the ureters and uterine arteries in the cardinal ligament is depicted.

the vagina and bladder, and splits around the urethra to insert onto the pubis.
(2) This layer reinforces the anterior vaginal wall and helps to prevent cystocele.

C. Ovaries

1. External structure. The ovaries are the female gonads. Each almond-shaped ovary is approximately 3 cm x 1.5 cm x 1 cm.
  a. Position. The ovaries are located on the posterior side of the broad ligament and project into the **ovarian fossae**, triangular depressions on each side of the minor pelvic cavity bounded by the external iliac vessels and the internal iliac vessels (see Figs. 22-9 and 22-11).
  b. Support (see Figs. 22-10 and 22-11)
  (1) The **mesovarium** attaches each ovary to the broad ligament.
  (2) The **infundibulopelvic (suspensory) ligament** attaches each ovary to the lateral pelvic wall and contains the ovarian neurovascular bundle.
  (3) The **ovarian ligament (proper)** runs within the broad ligament from the inferior medial pole of each ovary to the uterus in the vicinity of the cornu and is homologous to the cranial portion of the **gubernaculum** in the male.
  (4) Generally, one fimbria of the oviduct infundibulum—the **ovarian fimbria**—attaches to the ovary.

2. Internal structure (see Fig. 22-9)
  a. The cortex is the functional portion of the ovary.
  (1) The ovarian surface is covered by modified peritoneum (''germinal epithelium'') or serosa, which continues from the ovarian hilus as the mesovarium.
  (2) In the mature adult, the cortex contains several notable structures.
    (a) There are a multitude of **immature follicles** and **developing follicles**.
    (b) There may be one or more **mature (graafian) follicles** surrounding maturing oocytes. These structures secrete estrogen.
    (c) Corpora lutea, the remains of follicles immediately after ovulation, secrete estrogen and progesterone. Numerous follicles luteinize each month without reaching ovulation.
    (d) Corpora albicantia are the scars of regressed corpora lutea.
  (3) All the primordial follicles develop in early fetal life.
  (4) Ovulation occurs into the peritoneal cavity. The nearby or overlying oviduct drains the ovum into the infundibulum by a combination of movements of the fimbria and cilia.
  (5) After menopause, the ovary becomes small and atretic, and the follicles disappear.
  b. The medulla contains connective tissue and vasculature.

3. Vasculature of the ovary
  a. Arterial supply (see Figs. 22-1 and 22-12)
  (1) The **ovarian arteries** arise from the abdominal aorta in the vicinity of the renal arteries

and run inferiorly in the suspensory (infundibulopelvic) ligament and mesovarium to reach the ovary.

  (2) The ovarian and tubal branches of the **uterine arteries** anastomose with the ovarian artery in the mesosalpinx and can supply the ovary if the suspensory ligament is transected or the ovarian artery occluded.

b. Venous drainage

  (1) The ovarian veins drain asymmetrically.

    (a) The **right ovarian vein** drains into the inferior vena cava near the renal vein.

    (b) The **left ovarian vein** drains into the left renal vein.

  (2) Twigs from the ovaries also drain along the tubal branches of the uterine veins.

c. Lymphatic drainage of the ovaries is principally toward the para-aortic nodes in the vicinity of the renal arteries.

4. Innervation of the ovaries (see Fig. 22-13)

  a. Although it is known that the autonomic innervation of the gonads is from the aortic plexus, the role of the autonomic nerves with regard to the ovary is unclear.

  b. Visceral afferent fibers from the ovary run along the sympathetic pathways to spinal segments T12–L2. Intractable ovarian pain may be alleviated by transecting the suspensory ligaments, which contain the afferent (GVA) fibers.

5. Clinical considerations

  a. Ovarian torsion

    (1) An abnormally long mesovarium and suspensory ligament predisposes to ovarian torsion (rare).

    (2) Torsion compresses the blood supply, producing ischemia and abdominal symptoms with pain referred to the lower thoracic and upper lumbar dermatomes.

    (3) This condition occurs as a complication in 10%–20% of all ovarian tumors.

  b. Ectopic ovaries

    (1) The normal caudal migration of the ovaries may continue so that the ovaries descend to the anterior abdominal wall in the vicinity of the deep inguinal ring, into the inguinal canal or, very rarely, the labia majora.

    (2) Ovaries that fail to descend are termed retroperitoneal and are located in the posterior abdominal wall superior to the kidneys.

    (3) Ovaries may occasionally prolapse into the pouch of Douglas.

  c. Congenital absence of one or both gonads is usually associated with a unilateral defect of the genital system (very rare).

  d. Ovarian tumors of numerous types are common.

D. Uterus. The uterus and oviducts are considered as a functional unit due to their common development.

1. The oviducts or **uterine (fallopian) tubes** represent the unfused portions of the paramesonephric (müllerian) ducts.

  a. External structure. Approximately 10 cm long, each uterine tube begins at an ostium, which opens freely to the peritoneal cavity.

  b. Support. They are located at the upper free edge of the broad ligament and are supported by **mesosalpinx** (Fig. 22-11).

  c. Divisions. The oviduct is divided into four regions (see Fig. 22-9).

    (1) The **infundibulum**, 2 mm wide, opens into the peritoneal cavity via the **internal ostium**.

      (a) The ostium, is bordered by motile fimbria.

      (b) One fimbria, the **ovarian fimbria**, usually adheres to the ovary.

    (2) The **ampulla**, 2.5 mm wide, is thin-walled, tortuous, and the major portion of the oviduct.

    (3) The **isthmus**, 1 mm wide, is very short, thick-walled, and straight.

    (4) The **intramural** segment, less than 1 mm wide, lies within the uterine wall and communicates with the uterine cavity.

  d. Internal structure

    (1) The serosa is visceral peritoneum.

    (2) The muscularis is comprised of two layers of smooth muscle.

    (3) The mucosa is ciliated columnar epithelium interspersed with mucous secretory cells, the number of which vary with the menstrual cycle.

  e. Functional considerations

    (1) The fimbriae generally sweep over the ovarian surface so that an expelled ovum is directed into the infundibular ostium.

    (2) Fertilization of the ova usually occurs in the ampullary portion of the uterine tube.

    (3) The ova are propelled along the oviduct by peristaltic movement and ciliary action.

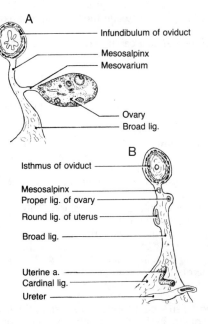

Isthmus of oviduct ——

Mesosalpinx ——
Proper lig. of ovary ——

Round lig. of uterus ——

Broad lig. ——

Uterine a. ——
Cardinal lig. ——
Ureter ——

Figure 22-11. *Ligamentous support of the ovary and oviduct. A,* This lateral parasagittal section through the infundibulum of the oviduct shows the mesosalpinx and mesovarium as extensions of the broad ligament. *B,* This more medial parasagittal section through the isthmus of the oviduct depicts the proper ligament of the ovary within the broad ligament. The relationship of the uterine artery to the ureter in the cardinal ligament at the base of the broad ligament also is indicated.

f. Clinical considerations

    (1) *Pelvic inflammatory disease (PID).* The female genital canal is a direct communicating pathway from the urogenital sinus of the perineum into the peritoneal cavity and is a pathway for infection.

    (2) *Salpingitis*—inflammation of the oviduct—is perhaps the most common cause of female sterility. If fertilization does occur, the zygote may fail to traverse a subsequently scarred or narrowed uterine tube, resulting in ectopic tubal implantation.

    (3) Ectopic implantation

        (a) Because the oviducts open directly into the peritoneal cavity, an ovum may be fertilized in the peritoneal cavity or even in the ovary. If the zygote fails to be taken into one or the other uterine tubes, ectopic implantation will occur; however, ectopic implantations are most common in the ampullary portion of the uterine tube (tubal pregnancy).

        (b) In the case of tubal implantation, rupture of the tube with hemorrrhage and expulsion of the embryo into the peritoneal cavity usually occurs between the fourth week (if in the isthmus) and the tenth week (if in the ampulla). This is a gynecologic emergency because the patient may die of internal hemorrhage.

    (4) Uterosalpingography, in which radiopaque dye is injected into the uterus and oviducts, is used to determine the patency of the uterine tube.

2. The uterus represents the fused portions of the paramesonephric (müllerian) ducts.

    a. External structure. The uterus, although variable in size, is normally 8 cm x 2.5 cm x 5 cm in the nulliparous woman and somewhat larger in the multiparous woman.

    b. Divisions. The uterus is divided into four regions (see Fig. 22-9)

        (1) The **fundus** is the region of the uterus superior to the uterine cornua. This region contributes most of the upper uterine segment during pregnancy.

        (2) The **cornu** on each side defines the point of entrance of a uterine tube.

        (3) The **body (corpus)** is the region inferior to the cornu and superior to the cervix.

            (a) The uterine cavity is shaped like an inverted flattened triangle. At the superolateral angles, the cavity extends into each oviduct.

            (b) The **isthmus** is the dividing line between the cervix and the uterus. It corresponds with the internal os. The isthmus becomes incorporated into the uterus during pregnancy as the lower uterine segment. This is the preferred site for surgical delivery by cesarean section.

        (4) The **cervix** is the narrow inferior portion of the uterus that projects into the vagina.

            (a) It is subdivided into three regions.

                (i) The **internal os** marks the junction between the cervical canal and the uterine body.

      (ii) The **cervical canal**, about 2.5 cm long, lies between the internal and external ostia.

      (iii) The **external os** is the opening of the cervical canal into the vagina. The nulliparous external os is round; the parous os is transverse.

   (b) Changes in the cervix occur during pregnancy.

      (i) The cervix becomes softer, changing in consistency from feeling like a "nose" to feeling like "lips."

      (ii) The glands of the cervical canal secrete a protective mucous plug during pregnancy.

c. Uterine support

   (1) The **levator ani muscle** and **urogenital diaphragm** provide the principal support for the urinary bladder. The **urinary bladder** in conjunction with gravity directly supports the anteverted and anteflexed uterus (see Figs. 22-1 and 22-8).

   (2) The levator ani muscle and urogenital diaphragm also provide the principal support for the vagina, which then indirectly supports the uterus.

   (3) The **round ligament of the uterus** maintains the anteverted position. This ligament stretches considerably during pregnancy (see Fig. 22-8).

   (4) Each **cardinal ligament** with the contained uterine vasculature contributes variable support in this region. In retroversion, in which the uterus does not rest on the bladder, the cardinal ligaments appear to provide the principal support (see Fig. 22-10).

   (5) The **uterosacral** and **pubocervical ligaments** provide variable posterior and anterior support, respectively.

d. Uterine positions

   (1) The normal uterine position is *anteflexed* (uterus bent forward on itself at the level of the internal os) and *anteverted* (the angle of the uterine cervix with the vagina normally is about 90°). Thus, the long axis of the uterus is nearly horizontal and the body of the uterus lies on the bladder (see Fig. 22-8).

   (2) In some individuals the uterus may assume other positions.

      (a) *Retroflexion:* The axis of the body of the uterus passes upward or even backward, but the angle at which the cervix enters the vagina is unchanged. The result is a bending back of the uterus.

      (b) *Retroversion:* The axis of the cervix with the vagina is upward or even backward so that the uterus may rest against the rectum, and either the external os or the posterior lip of the cervix presents first on digital examination.

e. Uterine relations (see Fig. 22-8)

   (1) Anteriorly, the **vesicouterine pouch** separates the uterus from the urinary bladder.

      (a) The supravaginal cervix is separated from the urinary bladder by the pubocervical (vesicovaginal) fascia.

      (b) The vaginal portion of the cervix lies posteriorly to the **anterior fornix** of the vagina.

   (2) Posteriorly, the uterine body is separated from the rectum by the **rectouterine** or **rectovaginal pouch** (of Douglas). The most caudal portion of the rectouterine pouch fuses to form the **rectouterine septum** between the vagina and rectum; this layer corresponds to the rectovesical fascia in the male.

   (3) The broad ligament and its contents lie laterally to the uterus. The ureter is approximately 1 cm lateral to the supravaginal cervix.

f. Internal structure

   (1) The **endometrium** (mucosa) undergoes cyclic changes with the menstrual cycle.

      (a) During the **menstrual phase**, when the corpus luteum is regressing and the progesterone titers are decreasing, there is a sloughing of endometrium.

      (b) During the **proliferative phase**, under the influence of estrogen from the maturing follicle, the endometrium is reformed.

      (c) During the **secretive phase**, under the influence of progesterone and estrogen from the new corpus luteum, there is glandular proliferation and secretion as well as proliferation of the endometrial blood vessels so that the mucosa is ready for an implantation.

   (2) The **myometrium**, forming the bulk of the uterine wall, is smooth muscle.

   (3) The **mesometrium** is **visceral peritoneum**.

g. Vasculature of the uterus (Fig. 22-12)

   (1) **Arterial supply** to the uterus is principally by the **uterine arteries**.

      (a) These vessels arise from each internal iliac artery and run in the transverse cervical ligament at the base of the broad ligament.

      (b) They cross immediately anterior to the ureters in the transverse cervical ligament, which is a most important surgical consideration.

      (c) The branches of the uterine artery include:

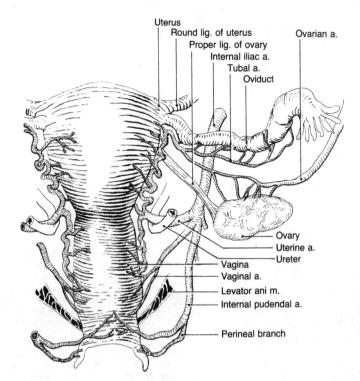

Uterus
Round lig. of uterus
Proper lig. of ovary
Internal iliac a.
Tubal a.
Oviduct
Ovarian a.

Ovary
Uterine a.
Ureter
Vagina
Vaginal a.
Levator ani m.
Internal pudendal a.

Perineal branch

Figure 22-12. *Vasculature of the female reproductive tract.* The ovarian, uterine, and pudendal arteries supply the female tract and genitalia with anastomoses occurring between the ovarian and uterine arteries as well as between the uterine arteries and deep perineal branches of the pudendal arteries.

     (i) The **ascending branch**, which gives off the **tubal branch** and anastomoses with the ovarian artery.
     (ii) The **descending (vaginal) branch**, which anastomoses with other vaginal branches and with the perineal branches of the internal pudendal artery.
   (d) During pregnancy, the uterine artery becomes hypertrophic with an enormous volume of blood flowing to the uterus and placenta.
  (2) Venous return from the uterus is along similarly named veins, which generally parallel the arterial supply.
  (3) Lymphatic drainage of the uterus generally parallels the blood supply.
   (a) The lymph from the uterine fundus drains mainly along the ovarian vessels to the para-aortic nodes.
   (b) The uterine body generally drains along the internal iliac vessels, but some lymphatics follow the round ligament to the superficial inguinal nodes.
   (c) The cervix drains in three directions.
     (i) Principal drainage is along the uterine vessels to the internal iliac nodes.
     (ii) There are also lymph pathways along the internal pudendal vessels to the internal iliac nodes.
     (iii) Other pathways course through the deep pelvis toward the sacral nodes.
   (d) Due to widespread lymphatic drainage, invasive carcinoma of the cervix may require extensive lymphatic dissection.
 h. Innervation of the uterus. The autonomic innervation to the uterus is derived as the **uterovaginal plexus** from the lateral pelvic plexus (Fig. 22-13).
  (1) Sympathetic neurons originate from the lateral horns of the T12–L2 segments of the spinal cord.
   (a) These neurons leave the spinal nerve via the white rami communicantes to gain access to the sympathetic chain, where they may synapse.
   (b) The second-order neurons leave the sympathetic chain via **lumbar splanchnic nerves** and course in sequence through the aortic plexus, superior hypogastric plexus, inferior hypogastric plexus, lateral pelvic plexus, and uterovaginal plexus to reach the uterus.
  (2) Parasympathetic neurons originate from the S2–S4 segments of the spinal cord.
   (a) These neurons leave the spinal nerves in the **pelvic splanchnic nerves (nervi erigentes).**
   (b) These nerves pass through the lateral pelvic plexus and uterovaginal plexus to reach the uterus, where they synapse with a second-order neuron.

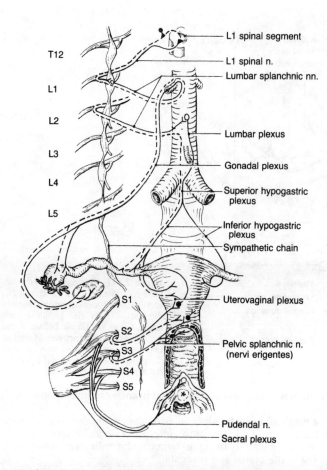

Figure 22-13. *Innervation of the female reproductive tract and genitalia*. The sympathetic pathways arise from the lower thoracic and upper lumbar spinal levels (*black triangle*). There are no white rami below L2. These reach the aortic plexus via thoracic and lumbar splanchnic nerves. Synapse occurs in the aortic plexus (*white circles*). The postsynaptic neurons reach the pelvic viscera via hypogastric plexuses. The parasympathetic pathways arise from the midsacral spinal levels and reach the pelvic viscera via the pelvic splanchnic nerves. Synapse occurs in the walls of the viscera (*white circles*). Visceral afferent fibers (*dashed*) from the pelvic viscera travel specifically along either one or the other autonomic pathways, have their cell bodies in the dorsal root ganglia, and produce specific patterns of referred pain. The pudendal nerve provides somatic innervation to and from the perineum.

(3) Visceral afferents from the uterus generally take two pathways to the spinal cord.
  (a) From the body of the uterus and uterine tubes, the course of the visceral afferents is parallel to the sympathetic pathways. Hence, pain associated with uterine spasm and salpingitis is *referred* to the dermatomes from T12–L2 (i.e., the midback, the inguinal and pubic regions, and the anterior thigh).
  (b) From the cervical region, the course of the visceral afferents is along the nervi erigentes. Thus, pain associated with cervical dilation is *referred* to sacral dermatomes S2–S4 (i.e., the perineum, gluteal region, posterior thigh, and leg).

i. Clinical considerations
  (1) Palpable per vagina are the cervix and ostium of the uterus, the vagina, the body of the uterus if retroverted, the rectouterine fossa, and under certain pathologic conditions, the ovary, uterine tubes, and broad ligament.
  (2) Ureteral risk. Because of the location of the ureters in the transverse cervical ligament, there is the possibility of division of the ureter when clamping and ligating the uterine arteries during hysterectomy. Also, compression of the ureters by a growth in the pelvis may result in hydronephrosis and uremia, a potential cause of death.
  (3) Uterine prolapse
    (a) Retroversion of the uterus predisposes toward prolapse because the weight of the uterus is no longer supported by the bladder.
      (i) First-order prolapse is defined as retroversion with the cervix protruding to an abnormal extent into the vagina (procidentia).
      (ii) Second-order prolapse is defined as protrusion of the cervix from the vagina.
      (iii) Third-order prolapse is complete eversion of the vagina.
    (b) Trauma to the uterine support (particularly to the round ligament of the uterus) as a result of difficult childbirth, as well as general relaxation of connective tissues in old age, may predispose to uterine prolapse.
  (4) Anesthesia. During parturition, pain from dilation of the cervix and stretch of the perineum may be appropriately blocked, although the thoracic and lumbar nerves controlling abdominal musculature and reflex uterine contraction should not be affected.

        (a) *Caudal anesthesia:* The needle is inserted between the sacral cornua, penetrating the sacrococcygeal ligament across the sacral hiatus to gain access to the vertebral canal. The anesthetic is infiltrated into the epidural space external to the dural sac, and the spinal nerves are anesthetized as they leave the dura. This is most effective for the sacral nerves.

        (b) *Epidural anesthesia:* The needle is inserted into the L4–L5 interspace, and the anesthetic is infiltrated into the epidural space about the dural sac. This is most effective for the lower lumbar nerves.

        (c) *Spinal (intrathecal) anesthesia:* The needle is inserted through the dural sac, and the anesthetic is infiltrated directly into the subarachnoid space in which the nerve roots are anesthetized. This is very effective for the lumbar and sacral nerves but is more difficult to control and subject to more complications and side effects.

E. Vagina (see Fig. 22-8)

  1. Location. From its opening at the vestibule of the vulva, the vagina passes superiorly and posteriorly through the pelvic diaphragm and surrounds the cervix of the uterus.

  2. External structure. The size of the vaginal canal is somewhat variable. Normally, the vaginal lumen is *H*-shaped.

    a. The anterior wall, which is approximately 7.5 cm long, terminates anteriorly to the projecting cervix as the **anterior fornix**.

    b. The posterior wall, approximately 10.5 cm long, terminates in the **posterior fornix** and is related to the **rectouterine pouch** (of Douglas).

    c. The lateral walls terminate in **lateral fornices** on either side of the projecting cervix.

  3. Relations of other structures to the vagina.

    a. Anteriorly to the upper portion of the vagina lies the uterus. The midvagina apposes the base of the bladder with the lower half of the urethra embedded in the adventitia of the lower vaginal wall.

    b. Posteriorly, the posterior fornix of the vagina is separated from the rectum by the uterorectal pouch (of Douglas), the upper quarter of the posterior vaginal wall is separated from the rectum by the uterorectal septum, and the lower vagina is separated from the anal canal by the **perineal body**.

    c. Laterally, the vagina is flanked by the levator ani muscle and endopelvic fascia. The ureters and uterine arteries lie close to the lateral fornices.

    d. The upper two-thirds of the vagina are within the pelvic cavity (i.e., superior to the pelvic floor).

    e. The lower one-third of the vagina is within the perineum (i.e., inferior to the pelvic floor). It communicates with the exterior via the **introitus**.

  4. Internal structure

    a. The vaginal epithelium is stratified squamous, lubricated by cervical mucus and desquamated vaginal cells. The walls are rugose in the nulliparous female but become smoother after vaginal childbirth.

    b. The muscular walls are of smooth muscle.

    c. Adventitia blends with the endopelvic fascia.

  5. Vasculature of the vagina (see Fig. 22-12)

    a. Arterial supply is via the **vaginal branch** of the **uterine artery**, occasional **vaginal branches** of the internal iliac arteries, possible twigs from the **middle rectal arteries**, and branches from the **internal pudendal arteries**.

    b. Venous return. The **vaginal venous plexus** of veins drains into the internal iliac vein.

    c. Lymphatic drainage of the vagina is by three routes.

      (1) The upper third of the vagina drains into external and internal iliac nodes.

      (2) The middle third drains into the internal iliac nodes.

      (3) The lower third drains into the internal iliac nodes as well as into the superficial inguinal nodes.

  6. Innervation to the vagina (see Fig. 22-13)

    a. The uterovaginal plexus of nerves, a spray from the lateral pelvic plexus, supplies the upper two-thirds of the vagina with sympathetic, parasympathetic, and afferent fibers. The visceral afferents from this region travel along the nervi erigentes so that pain may be referred to the dermatomal distribution of spinal segments S2–S4.

    b. The pudendal nerve supplies the lower (perineal) part of the vagina and mediates all the cutaneous sensory modalities.

  7. Clinical considerations

    a. Internal examinations. The vagina may be examined by inspection and digitally.

(1) Inspection with a speculum
    (a) The vaginal walls and cervix are observed (the cervical os is usually round in the nulliparous and slit-like in the multiparous woman).
    (b) Biopsy is performed, cytologic smears taken, and first-order prolapse is detected.
    (c) The posterior fornix of the vagina is the site for *culdocentesis*, whereby fluid in the rectouterine pouch may be tapped.

(2) Digital examination
    (a) Through the anterior fornix, the urethra, and bladder are palpable.
    (b) Through the posterior fornix, the rectum, coccyx, and sacral promontory are palpable across the pouch of Douglas.
    (c) Through the lateral fornices, the ovaries, uterine tubes, side wall of the pelvis, and occasionally the ureters, are palpable.
    (d) At the apex of the vagina, the anterior and posterior cervical lips are palpable.
        (i) In **anteversion**, the anterior cervical lip presents first.
        (ii) In **retroversion**, either the **external cervical os** or the posterior lip of the cervix presents first.

(3) Bimanual examination with a finger in the vagina and the other hand palpating or exerting pressure on the lower abdomen, enables the size and position of the uterus to be determined, the ovaries and uterine tubes to be palpated, and pelvic inflammation or neoplasms to be detected.

b. Herniation is often the result of lax pelvic support structures weakened during parturition.
    (1) Cystocele is herniation of the urinary bladder through the anterior vaginal wall into the vaginal lumen.
    (2) Rectocele is herniation of the rectum through the posterior vaginal wall into the vaginal lumen.

c. Anomalies and variations
    (1) Uterine and vaginal duplication occurs to variable extents.
        (a) Didelphis, a complete double uterus and double vagina, is very rare.
        (b) Uterus duplex, or double uterus with a single vagina, is rather uncommon.
        (c) Bicornuate or **septate uterus**, a double or partitioned fundus, is not uncommon.
        (d) Unicornuate uterus with complete absence of the uterine tube on one side is rare.
    (2) Vaginal anomalies are mainly associated with failure of the vaginal plate to canalize.
        (a) Rectovaginal fistulas are commonly coincident with an imperforate anus or perineal trauma.
        (b) Vesicovaginal fistulas (rare) are associated with continuous urinary discharge from the vagina.
        (c) Urethrovaginal fistulas are associated with urinary discharge from the vagina only upon micturition.

VII. FUNCTIONAL CONSIDERATIONS

A. Precoitus. Sexual activity of variable duration and intensity is usually required to establish a receptive psychological attitude before sexual congress occurs.

1. Mechanism of erection
    a. When an individual is sexually aroused, parasympathetic volleys travel via the pelvic nerves to the cavernous plexuses, which supply the penis or clitoris.
    (1) This parasympathetic outflow causes a relaxation of the smooth muscle associated with the arteries supplying the erectile tissue from the terminal branches of the internal pudendal artery.
    (2) The consequent widening of these arteries results in an increased blood flow into the cavernous spaces. The increase in blood flow is especially pronounced because the arteries contain longitudinal ridges that partially occlude the lumina when the circular muscularis is slightly tensed.
    (3) The corpora have a spongy structure with irregular endothelial-lined vascular spaces that are essentially collapsed in the flaccid penis or clitoris and vestibular bulbs. With increased blood flow, the corpora fill, expand, and become turgid.
    b. However, erection is a vascular event that involves not only increased inflow but also reduced efflux, of blood.
    (1) As the erectile tissue expands, the peripheral veins are compressed against the enveloping tunica albuginea, which effectively impedes drainage of blood from the cavernous sinuses.
    (2) Concomitantly, the bulbocavernosus and ischiocavernosus muscles, which envelop the roots of the penis or clitoris, contract periodically under voluntary control, enhancing the establishment of turgor.
    (3) Consequently, the penis or clitoris enlarges, elongates, and becomes hard.

2. In the male, in addition to erection of the penis, a similar stimulation results in emission (movement of the sperm from the testes and epididymis to the terminal portion of the vas deferens), and secretion from the glands associated with the urethra.

   a. Although the turgor of the corpora cavernosa in the male is extreme, the corpus spongiosum (corpus cavernosum urethrae) and its extension, the glans penis, do not become as firm.

   (1) This is due to the thinner and more elastic tunica albuginea surrounding this erectile body, which not only allows for a more flexible vaginal penetration but offers some resiliency when the penis strikes the cervix or wall of the posterior fornix during coitus.

   (2) More importantly the less turgid corpus spongiosum allows the seminal bolus to dilate the urethra so that ejaculation can occur.

   b. Under sexual stimulation, the male genital ducts become increasingly motile, which results in emission (movement of sperm to the ejaculatory duct by peristalsis of the vas deferens).

   (1) Seminal fluid is added to the sperm cells by the seminal vesicles and prostate gland.

   (2) Emission is an involuntary phenomenon, apparently regulated by the parasympathetic nervous system.

   (3) If there is voluntary restraint of ejaculation over a prolonged period of time, in spite of repeated sexual stimuli, the muscular walls of the vas deferens go into spasm with the resulting pain referred to the perineum.

3. In the female, this similar stimulation results in secretion from the cervical glands, erection of the clitoris with accompanying enlargement of the vestibular bulb, swelling of the vaginal walls, and secretion from the vestibular glands. When the vestibular bulbs expand, they compress the lateral walls of the vagina, causing the overlying labia minora to part. The anterior wall of the vagina swells somewhat and protrudes slightly into the vestibule.

   a. The cervical glands secrete profusely into the upper reaches of the vagina.

   b. The vestibular glands also secrete mucinous material at the introitus, facilitating the entry of the penis.

B. Second phase of coitus. Penetration of the vagina by the erect penis can be accomplished in a number of varied postures (Fig. 22-14).

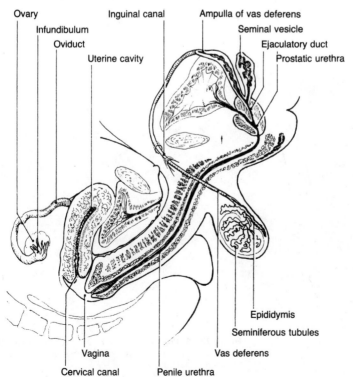

Ovary · Infundibulum · Oviduct · Uterine cavity · Inguinal canal · Ampulla of vas deferens · Seminal vesicle · Ejaculatory duct · Prostatic urethra · Epididymis · Seminiferous tubules · Vagina · Cervical canal · Penile urethra · Vas deferens

Figure 22-14. *Coitus.* The second (penetration) phase of coitus (after Da Vinci).

1. **Penetration.** The postures of coitus are particularly significant if there is repeated failure to effect conception or the need to make adjustments for an especially short vagina, an unusually long penis, retroversion of the cervix, or obesity of one or both partners.
   a. Full penetration of the erect penis elevates the mobile cervix and creates a pocket in the posterior fornix so that the average 9-cm vagina accommodates a substantial portion of the average 15-cm penis.
   b. At the time of penetration and thereafter, male secretions produced by the bulbourethral glands and the glands of the penile urethra are added to the vestibular and cervical secretions.
   c. Coital movements accentuate the stimulation of the penis, clitoris, and vaginal walls while emotional responses may be intensified and glandular secretions may be augmented.

2. **Emission.** The sperm cells move by emission into the prostatic urethra, become activated by seminal fluid, and are motile.

C. **Third, or orgasmic, phase of coitus.** When a particular degree of stimulation is reached, and when the partners so desire, the physical and emotional process of orgasm may occur.

1. **In the male**, orgasm is concomitant with ejaculation, which is brought about by sympathetic activity transmitted along the hypogastric nerve and lateral pelvic plexus and then through the prostatic and cavernous plexuses.
   a. This sympathetic outflow results in ejaculation with the spasmotic and rhythmic contraction of the vas deferens, seminal vesicles, prostate gland, and urethra.
   b. This is accompanied by apparently involuntary contractions of the bulbocavernosus and ischiocavernosus muscles as well as the muscles of the urogenital diaphragm, pelvic floor, and external anal sphincter, all innervated by the pudendal nerve.
   c. Although the onset of ejaculation may be controlled, once started, ejaculation cannot be interrupted.
   d. The ejaculate (2–5 cc) is discharged into the upper part of the vagina where, depending on the coital posture, a seminal pool may form. Regardless of position, the semen have access to the external cervical os.

2. **In the female**, orgasm is concomitant with contractile spasms of the uterus, apparently initiated by sympathetic nerves of the uterovaginal plexus.
   a. The uterine contractions tend to draw sperm cells into the cervical canal.
   b. Additionally, orgasm may involve contraction of the bulbocavernosus and ischiocavernosus muscles as well as the muscles of the urogenital diaphragm, pelvic floor, and external anal sphincters.

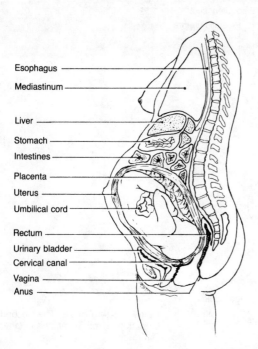

Esophagus
Mediastinum
Liver
Stomach
Intestines
Placenta
Uterus
Umbilical cord
Rectum
Urinary bladder
Cervical canal
Vagina
Anus

Figure 22-15. *Pregnant uterus.* Fetal development at approximately 35 weeks illustrates the relationships within the abdominal cavity.

D. Fourth phase of coitus (postcoital resolution)

1. Resolution

    a. Because the sympathetic outflow to the genital areas is dominant during the orgasmic phase in the male and female, the periarterial muscle increases its tone, thereby reducing the blood flow to the erectile tissues of the penis, clitoris, and vestibular bulbs, which quickly become flaccid.

    b. Usually, but not always, the male (and perhaps the female) enters into a refractory period of variable duration during which a second erection is not possible, probably due to the results of sympathetic innervation during orgasm.

2. Conception. In the absence of physical contraception, once the sperm cells enter the cervical canal it takes approximately 30–60 minutes at a swim-rate of 2 mm a minute to traverse the uterine cavity and enter the ostia of the oviducts.

    a. Once in the oviduct, the sperm are capacitated (attain the ability to penetrate an ovum).

    b. Somewhere in the distal third of the oviduct an ovum, if present, may be fertilized by the penetration of one sperm cell, and procreation of the species continues.

E. Implantation

1. Oviduct transit. Three to four days are required after fertilization for the zygote to reach the uterus.

2. Uterine implantation. At this time, the endometrium is in the secretory phase, and in the absence of an intrauterine device, implantation may occur.

    a. With the development of the chorionic membranes, human chorionic gonadotropin maintains the corpus luteum so that the endometrium remains in the secretory stage.

    b. Development and growth of the fetus requires 38–40 weeks (Fig. 22-15).

STUDY QUESTIONS

Directions: Each question below contains five suggested answers. Choose the **one best** response to each question.

1. In the female, the middle rectal (hemorrhoidal) artery has an anastomotic connection with which of the following arteries?

(A) Internal pudendal
(B) Obturator
(C) Superior gluteal
(D) Umbilical
(E) Uterine

2. In addition to the vas deferens, structures normally found within the internal spermatic fascia of the spermatic cord include all of the following EXCEPT the

(A) testicular artery
(B) pampiniform venous plexus
(C) deferential artery
(D) testicular nerves
(E) genitofemoral nerve

3. The remnant of the abdominal cavity present in the adult scrotum is the

(A) gubernaculum
(B) ductus deferens
(C) spermatic cord
(D) tunica vaginalis
(E) none of the above

4. The female structure occupying the same position in the inguinal region that the vas deferens occupies in the male is the

(A) broad ligament of the uterus
(B) transverse cervical ligament
(C) proper ligament of the ovary
(D) round ligament of the uterus
(E) suspensory ligament of the ovary

Questions 5–8

A 36-year-old man complained to his family physician of occasional dull throbbing pain associated with the right testis and scrotum. Examination indicated a varicocele of the pampiniform plexus. The examining physician remarked to the patient that he had, in all probability, had the condition for most of his adult life and that it should not be a bother to him. The patient emphatically stated that it had arisen within the last few months. The physician was inclined to pass off the complaint, but since the patient was bothered by the pain, surgery was considered.

5. Factors that might be considered by the examining physician include the fact that varicocele of the pampiniform plexus on the right side is associated with

(A) a long, redundant mesorchium
(B) situs inversus abdominis
(C) testicular torsion
(D) hydrocele processus vaginalis
(E) none of the above

6. It was decided that the testicular vein would be ligated just cranial to the deep inguinal ring. It is expected that this procedure would

(A) eliminate pressure by the column of venous blood
(B) reduce countercurrrent heat exchange
(C) reduce blood flow through the testes
(D) reduce sperm viability
(E) abolish the cremasteric reflex

7. During the early stages of the surgical operation, in which an incision was made parallel to the inguinal ligament, a retroperitoneal mass was observed in the vicinity of the lower pole of the right kidney. This growth encroached on the right testicular vein and impeded the flow so that it did not drain normally into the

(A) inferior vena cava
(B) internal iliac vein
(C) hepatic portal vein
(D) right renal vein
(E) right suprarenal vein

8. Since the tumor appeared to be of renal origin, it was decided that a computed tomography scan, an arteriogram, and an intravenous pyelogram should be performed before removal was attempted. Thus, the testicular vein was ligated as planned. After ligation, venous blood from the testes will return via all of the following collateral pathways EXCEPT the

(A) external pudendal veins
(B) cremaster veins
(C) veins of the vas deferens
(D) deep dorsal veins of the penis
(E) posterior scrotal veins

(end of group question)

9. A woman with lower right quadrant pain is prepared for sonography to narrow the differential diagnosis. The sonic probe is placed in the anterior vaginal fornix and aimed anteriorly. Assuming normal anatomy, structures that can be visualized anteriorly, anterosuperiorly, anterolaterally, and anteroinferiorly include all of the following EXCEPT the

(A) body of the uterus
(B) cervix
(C) left and right ovaries
(D) pubic symphysis
(E) urinary bladder

10. Viable fertilization of the ovum normally occurs in the

(A) ampulla of the oviduct
(B) body of the uterus
(C) fundus of the uterus
(D) peritoneal cavity
(E) primary follicle of the ovary

11. Which of the following pelvic structures refers pain to the L3–S1 dermatomes?

(A) Paraurethral glands of Skene
(B) Gonads
(C) Vas deferens
(D) Urinary bladder
(E) None of the above

Directions: Each question below contains four suggested answers of which **one or more** is correct. Choose the answer

A if **1, 2, and 3** are correct
B if **1 and 3** are correct
C if **2 and 4** are correct
D if **4** is correct
E if **1, 2, 3, and 4** are correct

12. Distinguishing features of the female pelvis include

(1) an oval pelvic inlet as compared to heart-shaped in the male
(2) a shallow false pelvis with flaring of the ilia as compared to the deep false pelvis with more vertical ilia in the male
(3) a larger pelvic outlet than the male
(4) a narrow subpubic angle between inferior pubic rami as compared to the wide angle in the male

13. Correct statements about the ischiorectal fossa include which of the following?

(1) It is filled with adipose tissue traversed by irregular connective tissue septa
(2) It is continuous with the superficial (fatty) layer of superficial fascia of the perineum and abdominal wall
(3) It contains inferior hemorrhoidal vessels and nerves
(4) It extends between the pelvic and urogenital diaphragms

14. Which of the following structures plays a significant role in fecal continence?

(1) Ischiorectal fat
(2) External anal sphincter
(3) Puborectalis muscle
(4) Internal anal sphincter

15. External hemorrhoids develop in branches of the hemorrhoidal vein and are characterized by

(1) profuse bleeding
(2) being larger than internal hemorrhoids
(3) a location superior to the pectinate line
(4) pain

16. Correct statements concerning the scrotum include which of the following?

(1) Stroking the inner thigh causes the dartos to contract
(2) The region beneath the dartos is a continuation of the superficial perineal pouch
(3) Scrotal pain is referred to the hypogastric and suprapubic regions
(4) The cremaster reflex tests the integrity of the L1–L2 spinal levels

17. Structures found in the male superficial perineal space include the

(1) ischiocavernosus muscle
(2) corpus spongiosum
(3) bulbospongiosus muscle
(4) bulbourethral glands (of Cowper)

18. Correct statements about the ischiocavernosus muscle include which of the following?

(1) Its contraction is the principal mechanism of erection
(2) It inserts into the central tendon of the perineum
(3) It receives its motor innervation from nervi erigentes
(4) It lies in the superficial perineal space

19. Which of the following statements correctly characterizes the external urethral sphincter?

(1) It receives innervation by the pudendal nerve
(2) It has an identical anatomic structure in both the male and female
(3) It is under voluntary control
(4) It is a portion of the pelvic diaphragm

20. Characteristics of the bulbourethral glands (of Cowper) include which of the following?

(1) They are homologous with the paraurethral glands (of Skene)
(2) They secrete into the penile urethra prior to and during ejaculation
(3) They contribute fructose to seminal fluid
(4) They lie in the deep perineal space

21. In the female perineum, the vestibule has which of the following characteristics?

(1) It receives both the urethra and the vagina
(2) It is bordered by the labia majora
(3) It receives drainage from the greater vestibular and paraurethral glands
(4) It receives sensory innervation from the nervi erigentes

22. Benign prostatic hypertrophy results in obstruction of the prostatic urethra by enlargement of

(1) the posterior lobe
(2) the anterior lobe
(3) either lateral lobe
(4) the median lobe

23. Which of the following may be palpated by digital examination per rectum in the male?

(1) The lateral lobes of the prostate gland
(2) Evidence of urinary retention in the bladder
(3) Disease of a seminal vesicle
(4) The median lobe of the prostate gland

SUMMARY OF DIRECTIONS

A	B	C	D	E
1, 2, 3 only	1, 3 only	2, 4 only	4 only	All are correct

---

24. True statements about the ovary include which of the following?

(1) Its sensory fibers follow the lateral pelvic plexus back to nervi erigentes
(2) Its arteries are found in the suspensory ligament of the ovaries
(3) It is supported by a fold of serosa called the mesosalpinx
(4) Its arteries are direct branches of the aorta

25. Lymph from the body of the uterus drains **directly** to which of the following groups of nodes?

(1) Superficial inguinal nodes
(2) Internal iliac nodes
(3) Para-aortic nodes
(4) Presacral nodes

26. Sensory fibers for pain from the body of the uterus travel to the spinal cord via the

(1) plexus around the uterine arteries
(2) white rami communicantes of L3 and L4
(3) lateral pelvic plexus and inferior hypogastric plexus
(4) nervi erigentes

Directions: The groups of questions below consist of lettered choices followed by several numbered items. For each numbered item, select the **one** lettered choice with which it is **most closely** associated. Each lettered choice may be used once, more than once, or not at all. Choose the answer

    A if the item is associated with **(A) only**
    B if the item is associated with **(B) only**
    C if the item is associated with **both (A) and (B)**
    D if the item is associated with **neither (A) nor (B)**

Questions 27–31

For each function of the reproductive tract, select the appropriate division of the autonomic nervous system that is most likely responsible.

(A) Parasympathetic division
(B) Sympathetic division
(C) Both
(D) Neither

27. Sensation from the glans penis

28. Clitoral erection

29. Male ejaculation

30. Relaxation of the external anal sphincter

31. Detrusor reflex

Questions 32–36

For each question pertaining to the reproductive tract, select the structure that is most likely to be associated with it.

(A) Prostate gland
(B) Seminal vesicles
(C) Both
(D) Neither

32. Derived from paramesonephric (müllerian) duct primordia

33. Produces seminal fluid

34. Site of spermatozoa maturation and storage

35. Homologous to the paraurethral glands (of Skene)

36. Refers pain to the upper lumbar dermatomes

Directions: The group of questions below consists of lettered choices followed by several numbered items. For each numbered item select the **one** lettered choice with which it is **most closely** associated. Each lettered choice may be used once, more than once, or not at all.

Questions 37–40

For each male pelvic structure named, select the most appropriate female homologue from the list below.

(A) Labia majora
(B) Vagina
(C) Vestibular bulbs
(D) Vestibule
(E) None of the above

37. Scrotum

38. Penile (cavernous) urethra

39. Corpus spongiosum

40. Perineal body

ANSWERS AND EXPLANATIONS

1. The answer is A. [*Chapter 20 IV A 3 b (1) (g) (ii)*] The middle rectal artery anastomoses with the superior rectal branch of the inferior mesenteric artery and with the inferior rectal branch of the internal pudendal artery. The middle rectal vein makes similar anastomotic connections.

2. The answer is E. [*Chapter 21 IV A 3 b (2) (b), (3)*] The testicular neurovascular bundle, the ductus deferens, and a remnant of the processus vaginalis lie within the internal spermatic fascia of the spermatic cord. The genital branch of the genitofemoral nerve lies in the cremaster layer beneath the external spermatic fascia.

3. The answer is D. [*Chapter 21 IV A 3 b (4)*] The tunica vaginalis of the testis is a remnant of the processus vaginalis, which extends from the abdominal cavity into the scrotum. The gubernaculum, a condensation of connective tissue between the inferior pole of the testis and the scrotum, guides the descent of the gonad. As the testis descends through into the scrotum, it drags with it components of the abdominal wall, which form the spermatic cord.

4. The answer is D. [*Chapter 22 VI A 2 d (3)*] The gubernaculum fails to lengthen during development in the male, drawing the testis into the scrotum so that the vas deferens comes to lie in the inguinal canal. In the female, the counterpart of the gubernaculum is the combined proper ligament of the ovary and the round ligament of the uterus. The latter passes through the inguinal canal to the deep fascia of the labia majora.

5. The answer is B. [*Chapter 21 IV A 3 b (3) (b), (c)*] Varicocele of the pampiniform plexus on the right side is very uncommon because this condition usually results from compression of the testicular vein by the full sigmoid colon. In cases of situs inversus, where the sigmoid colon is on the right side, the varicocele also would tend to be on the right.

6. The answer is A. [*Chapter 22 V A 3 b (1) (a)*] Testicular vein ligation at the inguinal ligament eliminates the hydrostatic pressure produced by the column of venous blood in the abdomen. Blood return from the testes is through anastomotic pathways. While a varicocele interferes with the countercurrent heat exchange mechanism, thereby raising scrotal temperature and decreasing sperm viability, the ligation reverses this effect.

7. The answer is A. [*Chapter 22 V A 3 b (1) (b)*] The right testicular vein normally drains into the inferior vena cava. The left spermatic vein joins the left renal vein.

8. The answer is D. [*Chapter 22 V A 3 b (2)*] The collateral venous drainage of the testes is primarily along the veins of the vas deferens, but there are anastomoses with the external pudendal veins, the posterior scrotal veins, and the cremaster veins (branches of the veins of the anterior abdominal wall). The deep dorsal vein of the penis, draining the deep structure of the penis, does not offer the possibility for anastomotic connections.

9. The answer is B. [*Chapter 22 VI C 2, D 2 d (1), E 2 a; Figure 22-8*] The cervix lies posterosuperiorly to the anterior fornix. With the probe in the anterior fornix, the urinary bladder, an anteflexed and anteverted uterus, the ovaries and oviducts as well as the pubic symphysis would all lie in the anterior sonic field.

10. The answer is A. [*Chapter 22 VI D 1 e (2); VII D 3 b*] Fertilization of the ovum usually occurs in the ampullary portion of the uterine tube. During the subsequent four days, while moving toward the uterus, the zygote undergoes numerous divisions, which enables implantation of the resultant morula. However, fertilization may occur (rarely) in the ovary or peritoneal cavity, which may result in ectopic implantation. While fertilization may occur in the uterus, the zygote usually is expelled before the implantable morula stage is reached.

11. The answer is E. [*Chapter 22 Figures 22-3 and 22-13*] Because there are no white rami communicantes associated with spinal nerves L3–S1, visceral afferents traveling in association with sympathetic pathways cannot enter the spinal cord at these levels; thus, no visceral pain can be referred to these dermatomes.

12. The answer is A (1, 2, 3). [*Chapter 20 I B 3 b; Table 20-1*] The female pelvis, in comparison with the male pelvis, is larger, has a shallower false pelvis, a more oval inlet, a larger outlet, and a wider inferior pubic angle.

13. The answer is E (all). [*Chapter 21 III A 1, 2*] The fat that fills the ischiorectal fossa is a continuation of the superficial layer of the superficial fascia of the perineum and abdominal wall. Within this layer are the inferior rectal (hemorrhoidal) neurovascular bundles. The anterior horn of the ischiorectal fossa extends between the pelvic and urogenital diaphragms.

14. The answer is A (1, 2, 3). [*Chapter 20 II A 2 a (3); Chapter 21 III A 2 a, B 3 a (1), (2), E 1*] The internal anal sphincter relaxes upon the approach of a peristaltic wave and, as such, is not a major factor in fecal continence. The external anal sphincter can be voluntarily contracted upon the approach of a peristaltic wave, but only for about 15 seconds. Prolonged contraction of the puborectalis portion of the pubococcygeus muscle establishes the carrying angle, which precludes emptying of the rectum. Compression of the anal canal by the ischiorectal fat as a result of gluteus maximus contraction also assists fecal continence.

15. The answer is D (4). [*Chapter 21 III F 2 b*] The portion of the anal canal inferior to the pectinate line is innervated by the inferior rectal branches of the pudendal nerve. Hemorrhoids in this region are painful and demand early attention. Internal hemorrhoids, above the pectinate line, are painless and tend to bleed profusely, frequently causing anemia.

16. The answer is C (2, 4). [*Chapter 21 IV A 2 b, 3 b (2), F*] The cremaster reflex (elevation of the testis within the scrotum when the inner thigh is stroked) tests the integrity of the L1–L2 spinal levels. Cold causes the smooth muscle of the dartos layer of the scrotum to contract so that heat loss is reduced. Because the dartos layer is a continuation of Colles' fascia, the scrotum is considered an extension of the superficial perineal pouch. The scrotum is innervated primarily by the pudendal nerve, which is somatic so that scrotal pain is not referred to another site.

17. The answer is A (1, 2, 3). [*Chapter 21 IV C 2, D 2 c*] The superficial perineal space contains the root of the penis (i.e., the bulb of the corpus spongiosum and crura of the corpora cavernosa) and the associated bulbospongiosus and ischiocavernosus muscles. The bulbourethral glands are located in the deep perineal space.

18. The answer is D (4). [*Chapter 21 IV C 2 b; Chapter 22 VII A 1 b (2)*] The ischiocavernosus muscle, lying in the superficial pouch, contracts periodically under voluntary control. It is innervated by the perineal branch of the pudendal nerve. It inserts just distal to the point at which the crura join to form the body of the penis or into the margin of the pubic arch and into each crus of the clitoris.

19. The answer is B (1, 3). [*Chapter 21 IV D 2 b (2); V D 2 b (1)*] The external urethral sphincter, the portion of the deep transverse perineal muscle adjacent to the urethra, is innervated by the internal pudendal nerve and is, therefore, under voluntary control. In the male, the external urethral sphincter completely surrounds the membranous urethra; in the female, because the urethra is embedded in the anterior vaginal wall, this sphincter is incomplete and less effective. It is part of the urogenital diaphragm.

20. The answer is C (2, 4). [*Chapter 21 IV D 2 c; V A 1 b (3) (d); Chapter 22 V D*] The bulbourethral glands, lying in the deep perineal space, discharge into the penile urethra prior to and during ejaculation. The prostate gland is homologous to the paraurethral glands in the female. The fructose component of seminal fluid is a product of the seminal vesicles.

21. The answer is B (1, 3). [*Chapter 21 V A 1 b (3), F 1 a*] The vestibule is bounded by the labia minora. It receives the urethra and the vagina, as well as the ducts of the paraurethral and vestibular glands. The innervation of the vestibule is by the perineal branches of the pudendal nerve.

22. The answer is D (4). [*Chapter 22 V E 1 c (3) (a)*] The median (middle) lobe of the prostate gland is most commonly involved in proliferative hyperplasia (benign prostatic hypertrophy), which obstructs the prostatic urethra with retention of urine in the bladder. The posterior lobe is most commonly involved in malignant transformation.

23. The answer is A (1, 2, 3). [*Chapter 22 V F 1*] The lateral and (more importantly) the posterior lobes of the prostate gland as well as a distended urinary bladder and diseased seminal vesicles are all palpable per rectum. The median lobe of the prostate, which surrounds the urethra, cannot be palpated, and rectal examination may provide little evidence on the extent of suspected benign prostatic hypertrophy.

24. The answer is C (2, 4). [*Chapter 22 VI C 2, 3 a–c, 4 b, 5 a (1), 6 a*] The blood supply to the ovaries is primarily by the ovarian arteries (direct branches of the aorta), which run in the suspensory ligament. In addition to the suspensory ligament, the mesovarium and the proper ligament of the ovary provide support. Afferent innervation is to the aortic plexus and thence along the least splanchnic and lumbar splanchnic nerves to the spinal levels; thus, pain is referred to the T12, LI, and L2 dermatomes.

25. The answer is A (1, 2, 3). [*Chapter 22 VI D 2 g (3)*] Lymph from the body of the uterus drains along the round ligament to the superficial inguinal nodes, along the ovarian vessels to the para-aortic nodes, and along the uterine vessels to the internal (deep) iliac nodes. The cervix, in addition to draining to these three groups of nodes, albeit by other routes, also drains along the uterosacral ligaments to the presacral nodes.

26. The answer is B (1, 3). [*Chapter 22 VI D 2 h (1) (b), (3)*] The afferent fibers innervating the uterine body travel along the uterine plexus and about the uterine vessels, the lateral pelvic plexus, the inferior and superior hypogastric plexuses, and the aortic plexus to reach the lumbar splanchnic nerves, which arise from spinal segments L1 and L2. (There are no white rami and, thus, no lumbar splanchnic nerves below L2.) Pain from the uterus is referred to the inguinal and pubic regions as well as to the lateral and anterior aspects of the thigh.

27–31. The answers are: 27-D, 28-A, 29-B, 30-D, 31-A. [*Chapter 21 III B 3 a (2); IV F 1; Chapter 22 II B 4 b, 5 b (5); III A–C; VII A 1 a, C 1 a*] The pudendal nerve mediates sensation from the external genitalia and provides motor control for the perineal musculature, including the external anal and urethral sphincters. Erection in both men and women is mediated by parasympathetic nerves, which control the blood flow to the erectile tissues. Ejaculation in men is a sympathetic event as are the uterine contractions that accompany the female orgasm. The detrusor (bladder-emptying) reflex is mediated by visceral afferents and parasympathetic efferents traveling in the nervi erigentes.

32–36. The answers are: 32-D, 33-C, 34-D, 35-A, 36-D. [*Chapter 22 II C 1 a (1) (b) (i); IV B–D; V B 1 b, D, E, F 3 b*] The only significant remnant of the paramesonephric ducts in the male appear to be the prostatic utricle. Both the seminal vesicles and prostate gland contribute significantly to the seminal fluid. However, the epididymis is the site of sperm storage and maturation. The prostate is homologous to the paraurethral glands of Skene in the female; the bulbovestibular glands of Bartholin appear to be homologous to the bulbourethral glands of Cowper in the male. Both seminal vesicles and the prostate gland refer inflammatory pain to the midsacral dermatomes.

37–40. The answers are: 37-A, 38-D, 39-C, 40-E. [*Chapter 21 II B, C; IV A 1 c (2); V A 1 a, C 1 b*] Both the scrotum in men and the labia in women develop from the labial–scrotal folds. Fusion of the urogenital (urethral) groove forms the penile urethra and the corpus spongiosum; in women, the urogenital groove remains as the vestibule flanked by the vestibular bulbs underlying the labia minora. The perineal body (central tendon) is a structure common to both men and women.

Part VII
Lower Extremity

23
Gluteal Region

I. INTRODUCTION

A. Basic principles

1. **Primitive position.** The basic pattern of the human lower extremity may be approximated by simulating the primitive position, that is, by abducting the thigh and leg until horizontal with the sole facing forward so that the lower extremity approximates the pelvic fin of a fish (see Fig. 4-3).

   a. **The primitively ventral surface** generally corresponds to the flexor side; the **primitively dorsal surface** generally corresponds to the extensor side.

   b. **Axial borders.** An imaginary line drawn through the long axis of the extremity will differentiate between a cephalad **preaxial border** and a caudal **postaxial border**.

2. **Early terrestrial adaptation.** From this original fin-like position of the lower extremity, flexing the knee 90° and extending the foot at the ankle 90° approximates the amphibian/reptilian lower extremity.

3. **Late terrestrial adaptation.** In mammals, the amphibian/reptilian thigh has rotated (abducted) 90° cephalad, bringing the lower extremity alongside the lateral body wall so that the primitively dorsal surface of the leg faces cephalad, but the primitively dorsal surface of the thigh continues to face dorsally.

4. **Assumption of an erect posture** effectively extends the hip 90° so that the primitively dorsal surface of the entire lower extremity faces anteriorly and the primitively ventral surface of the limb faces posteriorly.

   a. **Pattern of innervation.** The arrangement of nerves in the limbs reflects its developmental origin as a horizontal bud (see Fig. 4-2).

      (1) Each spinal nerve innervates a discrete sensory area, the **dermatome.** In the pelvic region, the dermatomes are drawn distally and appear to march down the preaxial side and back along the postaxial side (see Fig. 4-3). While there is generally overlap between sequential dermatomes, there is *no* overlap across the axial lines.

      (2) Each spinal nerve innervates a discrete muscle group, **a myotome.** In the lower extremity, there is an orderly preaxial to postaxial progression of muscles of myotomal origin and, therefore, sequential innervation (see Fig. 4-3).

   b. **Functional adaptation.** While the human upper and lower extremities are homologous, structural differences between them reflect different functions. The lower extremity retains its original role of support and propulsion at the expense of dexterity and intrinsic mobility.

      (1) The extensive fusion between the pelvic girdle and the axial skeleton as well as among the bones of the pelvic girdle are adaptations to weight-bearing and maximal stability.

      (2) The bony buttressing and strong ligamentous support at the hip joint permit sufficient movement while providing maximal support.

      (3) The leg portion of the lower extremity becomes fixed in the pronated position and loses its ability to supinate.

      (4) The foot, while retaining some prehensile characteristics, is adapted to support and locomotion.

B. The lower extremity can be divided into five regions.

1. **The pelvic girdle** consists of three bones: the **ilium, ischium,** and **pubis.**

2. **The thigh** contains one bone: the **femur.**

3. **The leg** contains two bones: the **tibia** and **fibula.**

4. **The ankle or tarsus** contains seven bones: a proximal row of three tarsal bones (**talus, calcaneus**, and **navicular**) and a distal row of four tarsal bones (the three **cuneiform** bones and the **cuboid** bone).

5. **The foot** contains nineteen bones: five **metatarsals** and fourteen **phalanges** (two in the great toe and three in each of the other toes).

C. **Basic organization.** The free portion of the lower extremity (thigh, leg, ankle, and foot) is suspended from the axial skeleton by the embedded portion of the lower extremity (the pelvic girdle). The only bony articulation between the pelvic girdle and the axial skeleton is at the strong and immobile sacroiliac joint. The pelvic girdle is very stable at the expense of mobility.

II. GLUTEAL (HIP) REGION

A. **Bony landmarks**

1. **Hip.** Palpable pelvic landmarks include (proceeding clockwise from the right side) the iliac crest, anterior–superior iliac spine, inguinal ligament, pubic tubercle, pubic symphysis, ischiopubic ramus, ischial tuberosity, ischial spine, posterior–superior iliac spine.

2. **Thigh.** Palpable landmarks of the proximal femur include the femoral head, greater trochanter, and femoral shaft.

B. **The pelvic (hip) girdle** consists of a **hip (innominate or coxal) bone** on either side of the sacrum (see Fig. 20-1). Each hip bone consists of three fused pelvic bones: **ilium, ischium**, and **pubis**.

1. **Characteristics**
   a. **The ilium** lies superiorly. The medial **auricular surface** forms the sacroiliac articulation and the expansive **alar plates** provide attachments for muscles of the abdomen and thigh. The **iliac crest** defines the lower limit of the waist and terminates in **anterior** and **posterior–superior iliac spines**. The posterior border defines the upper border of the sciatic notch before fusing with the ischium.
   b. **The ischium** lies posteroinferiorly. The posterior surface defines the lower border of the sciatic notch, which is divided into greater and lesser portions by the **ischial spine**. The lowermost **ischial tuberosity** provides attachment for posterior thigh muscles. The **ischial ramus** fuses with the inferior pubic ramus.
   c. **The pubis** is anterior. The body of the pubis articulates with its contralateral counterpart at the **pubic symphysis**. The **superior ramus** fuses with the ilium, while the inferior ramus fuses with the ischial ramus to form the **obturator foramen**.

2. **Articulations**
   a. **The sacroiliac joint** is the articulation between the pelvic girdle and the axial skeleton. It is formed by the articulation of the alar plates of the sacrum with the ilium.
      (1) Unlike the pectoral girdle, the attachment of the pelvic girdle to the axial skeleton is primarily ligamentous.
      (2) Also, unlike the upper extremity, there normally is no movement between the pelvic girdle and the axial skeleton at the sacroiliac joint so that the pelvic girdle is extremely stable.
   b. **The pubic symphysis**, anteriorly, is the articulation between the left and right hip bones.
   c. **The hip (coxal) joint**, is the distal articulation of each hip bone with the femur.
      (1) The **acetabulum** (*L.* vinegar cup) is the fossa of the hip joint, which is formed by the fusion of the ilium, ischium, and pubis (Fig. 23-1). The nearly hemispherical acetabulum faces laterally but slightly posteriorly and slightly inferiorly.
         (a) The **articular surface** of the acetabulum is horseshoe-shaped and lined with hyaline cartilage. The **acetabular notch** is an inferior discontinuity in the articular rim of the acetabulum. The bone behind the articular surface is thick with bony trabeculae arranged along the lines of force.
         (b) The **acetabular labrum**, a lip of fibrocartilage, increases the depth of the bony acetabulum and contributes to the great stability of the hip joint. The acetabular labrum continues across the acetabular notch as the **transverse acetabular ligament**.
         (c) The **articular fossa**, the rough, nonarticular medial wall of the acetabulum enclosed by the limbus of articular cartilage, is a very thin sheet of bone.
      (2) The **ligamentum teres** of the femur originates from the transverse ligament over the acetabular notch and inserts into the **fovea (foveal notch)** of the femoral head.

C. **The femur** is the long bone of the thigh (see Fig. 23-1).

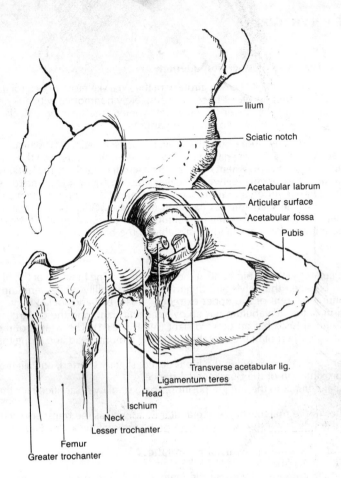

Ilium

Sciatic notch

Acetabular labrum
Articular surface
Acetabular fossa
Pubis

Transverse acetabular lig.
Ligamentum teres
Head
Ischium
Neck
Lesser trochanter
Femur
Greater trochanter

Figure 23-1. *Hip joint.* The femur articulates with the pelvic bone at the acetabulum. The ligamentum teres has been cut and the femur disarticulated.

1. **Characteristics.** The proximal end of the femur consists of a head, neck, greater and lesser trochanters, and shaft.
   a. **The head** is approximately two-thirds of a sphere.
      (1) It fits deeply into the acetabulum and is covered with hyaline cartilage except at the **foveal notch** where it receives the **ligamentum teres**.
      (2) An epiphyseal plate, which usually fuses between the nineteenth and twenty-first years, demarcates the head from the neck.
   b. **The femoral neck** joins the head to the shaft.
      (1) The neck is normally angled upward at about 125° (slightly more in males and less in females).
      (2) It is also angled forward (anteverted) by about 15°.
   c. **The shaft** has a slightly anterior bow.
      (1) The **greater trochanter** on the lateral surface projects above the junction with the neck and provides attachment for the gluteus medius and minimus muscles as well as for the piriformis muscle.
      (2) The **lesser trochanter** is situated on the medial surface distal to the junction of the neck and receives the insertion of the iliopsoas muscle.
      (3) The **intertrochanteric crest** runs obliquely along the posterior surface of the shaft between the trochanters.

2. **The internal structure** of the femur enables the transmission of huge forces between the pelvis and tibia.
   a. The bony trabeculae of the femoral head, neck, and trochanters are arranged along the lines of prevailing force, effectively transmitting these forces to the compact bone of the shaft.
   b. This trabecular arrangement attains the greatest development and strength during the active period of life and is reduced by bone resorption during senescence, thereby predisposing to fractures.

III. ARTICULATIONS OF THE PELVIC GIRDLE

A. Sacroiliac joint

1. **Structure.** The ilium articulates with the **sacrum** posteriorly.

2. **Movement.** Unlike the articulation of the pectoral girdle with the axial skeleton, the sacroiliac joint is only slightly movable. By middle age, it becomes completely nonmobile with ankylosis typically occurring after age 50. However, in pregnant women, the ligaments slacken somewhat under the influence of hormones to facilitate parturition.

3. **Ligamentous support** is provided by numerous strong ligaments (see Fig. 20-1), which include the **dorsal sacroiliac ligaments** (between the posterior–superior iliac spine and the dorsum of the sacrum), **ventral sacroiliac ligaments** (across the sacroiliac joint on the anterior surface of the pelvis), **iliolumbar ligaments** (between the iliac crest on either side to the transverse processes of the fifth lumbar vertebra), **interosseous ligaments** (between the ilium and sacrum across the articulation), the **sacrospinous ligament** (from the ischial spine to the sacrum), and the **sacrotuberous ligament** (from the ischial tuberosity to the sacrum). This support makes the sacroiliac joint the strongest and most stable in the body.

B. Hip (coxal) joint

1. **Structure.** The femur articulates with the hip bone at the acetabulum. The function of the hip joint is to transmit forces between the pelvis and thigh with great stability while permitting mobility. Like the glenohumeral joint or the upper extremity, it is a ball-and-socket joint.
   a. **The bony configuration** of the acetabulum has an overhanging slant, and the congruent configuration of the femoral head has an upward slant (see Fig. 23-1). The weight of the body is transmitted through the roof of the acetabulum to the femoral head and along the femoral neck to the shaft.
      (1) Maximum articular contact occurs when the hip is flexed, slightly everted, and slightly abducted—an approximation of the quadrapedal stance.
      (2) There is less articular contact in the upright position; some stability is sacrificed for the bipedal stance.
   b. **A joint capsule** attaches to the margins of the acetabular lip as well as the transverse acetabular ligament and extends like a sleeve to the base of the femoral neck (see Fig. 23-1; Fig. 23-2).
      (1) The inner surface of the capsule is **synovial membrane**.
      (2) The outer capsular layer is fibrous.
         (a) Some fibers run circumferentially about the femoral neck, the **zona orbicularis**.
         (b) Other groups of fibers are reflected along the neck as the **retinacula**, carrying important blood vessels to the synovium and the femoral head.
         (c) The monolayer of **synovial fluid** between the extensive articular surfaces produces considerable surface tension, providing tremendous adhesion while permitting nearly frictionless gliding.

2. **Movement.** As a ball-and-socket joint, there are three degrees of freedom. Range of motion is sacrificed for stability.
   a. **Abduction/adduction.** Abduction 30°; adduction 20°.
   b. **Flexion/extension.** Flexion 120° (with knee flexed); extension 15°.
   c. **Rotation.** Medial 60°; lateral 30°.
   d. **Circumduction** at the hip joint is not as great as at the shoulder where muscles (not ligaments and bones) stabilize the joint.

3. **Ligamentous support.** Strong ligaments reinforce the hip joint. Also, the dynamic actions of the muscles that cross the joint contribute to its stability.
   a. **Capsular ligaments** are three fibrous condensations of the joint, which lie externally to the fibrous capsule.
      (1) The **iliofemoral ligament (Y ligament of Bigelow)** runs from the anterior–inferior iliac spine to the length of the intertrochanteric line (see Fig. 23-2A) and resists hyperextension and excess lateral rotation of the femur. It is especially thick at the edges, giving the appearance of an inverted **Y**. The lateral limb is termed the **iliotrochanteric band**.
      (2) The **pubofemoral ligament** runs from the iliopubic ramus to the inferior surface of the intertrochanteric line (see Fig. 23-2A) and resists excessive abduction. It is rather thin, and a tear may result in continuity between the joint capsule and the iliopectineal bursa.
      (3) The **ischiofemoral ligament** runs from the posterior surface of the acetabulum to the inner aspect of the greater trochanter and the lateral surface of the intertrochanteric crest (Fig. 23-2B). It is the thinnest of the capsular ligaments.

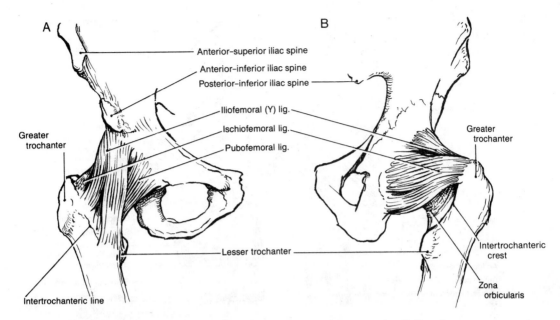

Figure 23-2. *Fibrous joint capsule of the hip joint.* A, Anterior. B, Posterior.

  b. The ligamentum teres of the femur runs from the acetabular notch and transverse acetabu-
    lar ligament to the femoral fovea (see Fig. 23-1).
    (1) This ligament plays little, if any, role in the stability of the hip.
    (2) It carries the **artery of the ligamentum teres**, which is inconsistent but may be an im-
      portant blood supply to the femoral head, especially in children.

4. Clinical considerations
  a. Dislocations. *Posterior dislocation* is more frequent than *anterior dislocation*. While tilting
    of the pelvis with assumption of the bipedal posture tightens the capsular ligaments and
    reinforces the hip joint, flexion tends to unwind the capsular ligaments so that the hip is
    less stable. When the flexed thigh comes into violent contact with an object such as a dash-
    board, posterior dislocation results.
  b. Osteoarthritis of the hip. The hip is a common site for degenerative arthritis. Because the
    abductor group contracts strongly to keep the pelvis level when the contralateral foot is
    raised, a cane used in support will be held in the opposite hand to alleviate pressure within
    the afflicted joint.

IV. MUSCLE FUNCTION AT THE HIP JOINT

  A. Organization. Movement at the hip joint is accomplished by muscles that run from the axial
    skeleton and the pelvic girdle across the hip joint to the femur, tibia, or fibula (Figs. 23-3 and 23-4;
    Tables 23-1, 23-2, 23-3, and 23-4). Many of the movements of the hip joint are accomplished by
    combinations of primitively dorsal and primitively ventral musculature. The musculature of the
    hip joint is arranged into five groups.

  1. The principal flexors are primitively dorsal muscles, which lie in the **anterior compartment**
    of the thigh and comprise two groups:
    a. The iliopsoas group consists of the **iliacus** and **psoas muscles** (see Table 23-1).
      (1) They arise from the iliac fossa or lumbar spine and insert into the lesser trochanter of
        the femur, anteromedially (see Fig. 24-3).
      (2) In addition to flexing the hip, if the leg is planted, they flex the lumbar vertebral col-
        umn.
      (3) The iliopsoas group is innervated by twigs from the second through fourth lumbar
        nerves.
    b. The anterior group, which flexes the hip, includes the **rectus femoris** and **sartorius** (see
      Table 23-1).
      (1) This group originates from the anterior portion of the ilium (see Fig. 24-3). The rectus

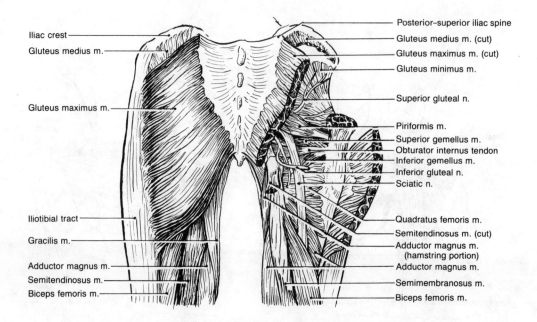

Iliac crest

Gluteus medius m.

Gluteus maximus m.

Iliotibial tract

Gracilis m.

Adductor magnus m.

Semitendinosus m.

Biceps femoris m.

Posterior–superior iliac spine

Gluteus medius m. (cut)

Gluteus maximus m. (cut)

Gluteus minimus m.

Superior gluteal n.

Piriformis m.

Superior gemellus m.

Obturator internus tendon

Inferior gemellus m.

Inferior gluteal n.

Sciatic n.

Quadratus femoris m.

Semitendinosus m. (cut)

Adductor magnus m. (hamstring portion)

Adductor magnus m.

Semimembranosus m.

Biceps femoris m.

Figure 23-3. *Gluteal musculature.* The right sciatic nerve passes through the medial inferior quadrant of the buttock.

femoris inserts into the proximal tibia through the patella; the sartorius inserts into the medial tibial condyle.

(2) The anterior group is innervated by the femoral nerve.

2. The **principal abductors** are mostly primitively dorsal muscles, which lie in the gluteal region (see Table 23-2). Muscles of the **abductor group** include the **gluteus medius**, **gluteus minimus** and, to a lesser extent, the **tensor fasciae latae** (see Fig. 23-3).

a. These muscles originate from the ilium. The gluteus minimus and medius insert onto the greater trochanter; the tensor fasciae latae inserts into the fibula.

b. The abductor group is innervated by the superior and inferior gluteal nerves.

3. The **principal extensors** are mostly primitively ventral muscles, which lie in the **posterior compartment** of the thigh and comprise two groups.

a. Of the **abductor group**, only the **gluteus maximus** extends the thigh (see Table 23-1).

(1) It originates from the iliac crest and inserts into the fibula.

(2) It is innervated by the inferior gluteal nerve.

b. Of the **hamstring group**, the **semitendinosus**, **semimembranosus**, long head of the **biceps femoris**, and posterior portion of the **adductor magnus** all extend the thigh (see Table 23-1).

(1) These originate from the ischial tuberosity and insert in the vicinity of the knee (see Figs. 24-4 and 24-5).

(2) The hamstring muscles included here are innervated by the tibial nerve.

4. The **principal adductors** are mostly primitively ventral muscles, which lie in the **medial compartment** and comprise the **adductor group**, including the **adductors magnus** (anterior portion), **longus**, and **brevis**; the **pectineus**; and the **gracilis** (see Table 23-2).

a. These muscles originate from the ischial rami and pubis and insert into the proximal femur (see Figs. 23-3 and 24-3).

b. The adductor group is innervated primarily by the obturator nerve.

5. The **principal external (lateral) rotators** are largely primitively dorsal muscles, which lie in the deep gluteal region and comprise the **rotator group**. These muscles include the **piriformis**, **superior** and **inferior gemelli**, and **obturators internus and externus** (see Table 23-3).

a. They originate largely from the sacrum or ischium and insert into or adjacent to the greater trochanter (see Fig. 23-3).

b. The external rotators are usually innervated by twigs from the sacral plexus, and the obturator externus by the obturator nerve.

6. While there is no defined group of **internal (medial) rotators**, most muscles that cross the hip joint have secondary actions that assist either medial or lateral rotation (see Tables 23-3 and 23-4).

B. Group actions on the thigh at the hip joint

   1. Flexion/extension occurs about a transverse axis through the femoral head.
   a. Flexion (90° with knee extended, 120° with knee flexed) is accomplished by a muscle group with a line of action passing anteriorly to the transverse axis of the hip. The major flexor muscle of the hip is the iliopsoas, assisted by the sartorius, rectus femoris, pectineus, and tensor fasciae latae, and to a lesser extent by the adductor longus (see Fig. 23-4 and 24-4 and Tables 23-1 and 23-2).
   (1) With the knee extended, flexion is limited primarily by tension in the hamstring muscles.
   (2) With the knee flexed, the hip may be flexed another 30° until it is stopped by apposition of the thigh to the abdominal wall.
   b. Extension (15°) is accomplished by muscles with lines of action that pass posteriorly to the transverse axis of the hip. These major extensors of the hip include two groups (see Fig. 23-3 and 24-5 and Table 23-1).
   (1) The gluteus maximus muscle is the strongest extensor and the one used principally in climbing stairs and rising from the seated position.
   (2) The posterior belly of the adductor magnus muscle and the hamstring group (the long head of the biceps femoris, semitendinosus, and semimembranosus muscles) are also extensors.

   2. Abduction/adduction occurs about an anteroposterior axis through the femoral head.
   a. Abduction (30°) is accomplished by muscles that pass superiorly to the anteroposterior axis, generally from the ilium to the greater trochanter.
   (1) The abductors are flat muscles that overlap in three layers. These include the tensor fasciae latae, gluteus maximus (superficially), and the gluteus medius, which overlies the gluteus minimus (the most anterior and deep) [see Fig. 23-3 and Tables 23-1 to 23-4].
   (2) The abductors are extremely important in pelvic stability during walking because they keep the pelvis level when the opposite foot is raised from the ground.
   (a) This force of abductor contraction is approximately three times the body weight.
   (b) Within the joint, a force develops equal to four times the body weight (abductor action plus body weight).
   (3) The tensor fasciae latae and the gluteus maximus, the superficial muscles of the buttock, insert into the **iliotibial tract**.
   b. Adduction of the abducted limb (20°) is accomplished by muscles that pass inferiorly to the anteroposterior axis of the hip joint.
   (1) The adductors run from the ischiopubic ramus to the medial aspect of the femur.
   (2) The major adductors of the thigh are included in the adductor group, including the adductor magnus, adductor longus, adductor brevis, and gracilis muscles (see Figs. 23-3 and 24-4 and Table 23-2).

   3. Rotation occurs about a vertical axis through the femoral head.
   a. Lateral rotation (eversion) [30°] is accomplished by muscles that act posteriorly to the vertical axis.

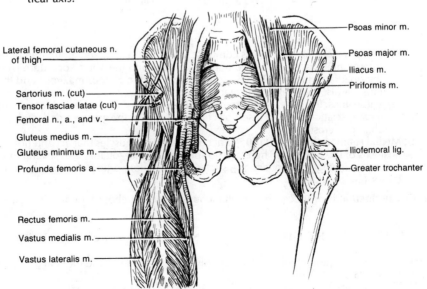

Figure 23-4. *Principal flexor muscles of the thigh.* The right femoral neurovascular bundle lies in the femoral triangle.

Table 23-1. Muscles Acting at the Hip Joint—Anterior and Posterior Compartments

Muscle	Origin	Insertion	Primary Action	Innervation
Anterior Compartment				
Iliopsoas:				
Iliacus	Iliac fossa	Distal to lesser trochanter	Flexes thigh	Femoral n. (L2–L4, posterior)
Psoas	Transverse processes of lumbar vertebrae	Lesser trochanter	Flexes thigh	Lumbar nn. (L2–L5, posterior)
Rectus femoris	Anterior–inferior iliac spine	Tibial tuberosity through patellar tendon	Flexes thigh	Femoral n. (L2–L4, posterior)
Sartorius	Anterior–superior iliac spine	Medial tibial head	Flexes, abducts, and laterally rotates thigh	Femoral n. (L2–L3, posterior)
Pectineus	Iliopectineal line and pubis	Base of lesser trochanter	Flexes and adducts thigh	Femoral n. (L2–L3, posterior) Obturator n. (L2–L3, anterior)
Posterior Compartment				
Gluteus maximus	Posterosuperior ilium and sacrum, associated ligaments and fascia	Gluteal tuberosity of femur and iliotibial tract	Extends, abducts, and laterally rotates thigh	Inferior gluteal n. (L5–S2, posterior)
Adductor magnus:				
Posterior belly	Ischial tuberosity	Adductor tubercle of femur	Adducts, extends, and aids medial rotation of thigh	Tibial n. (L4–L5, anterior)
Semitendinosus	Ischial tuberosity	Medial surface of proximal tibia	Extends and aids medial rotation of thigh	Tibial n. (L5–S1, anterior)
Semimembranosus	Ischial tuberosity	Posterior side of of tibial condyle	Extends and aids medial rotation of thigh	Tibial n. (L5–S1, anterior)
Biceps femoris:				
Long head	Ischial tuberosity	Lateral side of fibular head	Extends and aids lateral rotation of thigh	Tibial n. (L5–S2, anterior)

    (1) The principal lateral rotators comprise the pelvotrochanteric group (the piriformis, obturator internus, superior and inferior gemelli, and the obturator externus) assisted by the quadratus femoris. The posterior portion of the gluteus maximus and the sartorius muscles also have external rotatory function.

    (2) These muscles pass nearly transversely, generally from the ischium or through its greater sciatic notch, to insertions in the vicinity of the greater trochanter and intertrochanteric crest.

  b. Medial rotation (inversion) [60 °] is accomplished by muscles that act anteriorly to the vertical axis of the hip joint. There are no specific internal rotators, but several muscles have secondary medial rotatory actions, such as the anterior portion of the gluteus minimus and tensor fasciae latae.

 4. Circumduction is produced by simultaneous movement about two or more axes.

V. VASCULATURE OF THE GLUTEAL REGION AND HIP (see Chapter 20 IV)

  A. Arteries. The hip region receives blood supply from branches of the internal and external iliac arteries (Fig. 23-5).

Table 23-2. Muscles Acting at the Hip Joint—Abductor and Adductor Groups

Muscle	Origin	Insertion	Primary Action	Innervation
Abductor Group				
Gluteus medius	Superolateral surface of ilium	Greater trochanter	Abducts and aids medial rotation of thigh	Superior gluteal n. (L4–S2, posterior)
Gluteus minimus	Lateral surface of ilium	Greater trochanter	Abducts and aids medial rotation of thigh	Superior gluteal n. (L4–S1, posterior)
Tensor fasciae latae	Anterior–superior iliac spine and iliac crest	Iliotibial tract and lateral fibular head	Abducts, flexes, and medially rotates thigh	Superior gluteal n. (L4–S1, posterior)
Piriformis	Anterior sacrum	Greater trochanter	Laterally rotates and aids abduction of thigh	Nn. to piriformis (S1–S2, posterior)
Adductor Group				
Adductor magnus:				
Anterior belly	Ischial ramus and pubis	Distal portion of linea aspera	Adducts, flexes, and laterally rotates thigh	Obturator n. (L2–L4, anterior)
Posterior belly	Ischial tuberosity	Adductor tubercle	Adducts, extends, and medially rotates thigh	Tibial n. (L4–L5, anterior)
Adductor longus	Pubis between crest and symphysis	Linea aspera	Adducts and aids flexion and medial rotation of thigh	Obturator n. (L2–L3, anterior)
Adductor brevis	Body and inferior pubic ramus	Proximal portion of linea aspera	Adducts and flexes thigh	Obturator n. (L2–L4, anterior)
Pectineus	Iliopectineal line and pubis	Base of lesser trochanter	Adducts and flexes thigh	Femoral n. (L2–L3, anterior) Obturator n. (L2–L3, anterior)
Gracilis	Ischiopubic ramus	Medial side of tibial head	Adducts and flexes thigh	Obturator n. (L3–L4, anterior)

1. **The superior gluteal artery** is a branch of the posterior trunk of the internal iliac artery.
   a. It leaves the pelvic cavity via the suprapiriform portion of the greater sciatic foramen along with the superior gluteal nerve, deep to the posterior edge of the gluteus medius muscle.
   b. The main trunk passes anteriorly deep to the gluteus medius muscle.

2. **The inferior gluteal artery** is a branch of the anterior trunk of the internal iliac artery.
   a. It leaves the pelvic cavity via the infrapiriform region of the greater sciatic foramen along with the inferior gluteal nerve, pudendal nerve, and sciatic nerve.
   b. It passes anteriorly on the deep surface of the gluteus maximus muscle.
   c. It supplies portions of the gluteus maximus muscle, the lateral rotators of the hip, and the most proximal portions of the hamstring group; it also sends twigs to the hip joint.
   d. It anastomoses with branches of the profundus femoris artery, forming part of the **cruciate anastomosis**.

3. **The obturator artery** is most often a branch of the anterior trunk of the internal iliac artery.
   a. Frequently (30%), this artery arises from the inferior epigastric artery and descends to the obturator foramen close to the femoral ring where it may complicate surgical repair of a femoral hernia.
   b. It leaves the pelvic cavity via the obturator foramen.
   c. It supplies the adductor group and sends twigs to the joint, including the artery of the ligamentum teres.

4. **The femoral artery**, the continuation of the external iliac artery, gives off the **profunda femoris artery** from which the **medial** and **lateral femoral circumflex arteries** arise.

Table 23-3. Muscles Acting at the Hip Joint—Lateral Rotators

Muscle	Origin	Insertion	Primary Action	Innervation
Gluteus maximus	Posterosuperior ilium and sacrum	Gluteal tuberosity of femur and iliotibial tract	Extends, abducts, and laterally rotates thigh	Inferior gluteal n. (L5–S2, posterior)
Piriformis	Anterior sacrum	Greater trochanter	Laterally rotates and aids abduction of thigh	Nn. to piriformis (S1–S2, posterior)
Obturator: Externus	Pubic and ischial rami	Greater trochanter	Laterally rotates and aids abduction of thigh	Obturator n. (L3–L4, anterior)
Internus			Laterally rotates thigh	Nn. to obturator internus (L5–S2, anterior)
Gemellus: Superior	Ischial spine	Greater trochanter through obturator internus tendon	Laterally rotate thigh	Nn. to obturator internus (L5–S2, anterior)
Inferior	Ischial tuberosity			Nn. to quadratus femoris (L4–S1, anterior)
Quadratus femoris	Ischial tuberosity	Distal to greater trochanter	Laterally rotates thigh	Nn. to quadratus femoris (L4–S1, anterior)

    a. These supply the proximal portions of the more lateral and anterior muscles of the thigh and are the major supply to the trochanters, neck, and head of the femur.

    b. The two circumflex arteries pass transversely to anastomose with each other, with branches of the first perforating artery (inferiorly), and with the inferior gluteal artery (superiorly). This forms the **cruciate anastomosis** posterolaterally.

  B. Vasculature of the proximal femur is clinically important (see Fig. 23-5).

    1. Arterial supply

      a. The nutrient artery, a branch of the profunda femoris artery, supplies the shaft.

      b. The obturator, medial femoral circumflex, and lateral femoral circumflex arteries supply the trochanters and the neck and head of the femur.

        (1) The **artery of the ligamentum teres**, a branch of the obturator artery, is of principal importance in children, in whom it supplies the femoral head proximal to the epiphyseal plate.

        (2) The **retinacular (extracapsular) arteries** arise from the medial and lateral femoral circumflex arteries and give off branches that supply the greater trochanter and that penetrate the joint capsule at the intertrochanteric line to supply the neck.

          (a) Three groups of retinacular arteries with great variability run along the surface of the neck toward the femoral head.

          (b) These arteries supply the region distal to the epiphyseal plate in young people.

          (c) After fusion of the epiphyseal plate, the retinacular arteries anastomose with those of the ligamentum teres, and the latter may degenerate or become marginally significant in the adult.

          (d) Because of the course of the retinacular arteries, fracture of the neck places the blood supply to the head in jeopardy.

    2. Venous return consists of superficial and deep pathways. The deep veins parallel the arteries (see Chapter 24 V B).

VI. INNERVATION OF THE GLUTEAL REGION (see Chapter 20 V).

  A. Superior gluteal nerve (L4–S1, posterior)

    1. Course. This nerve exits the pelvis via the suprapiriform portion of the greater sciatic foramen along with the superior gluteal vessels (see Fig. 23-3). It lies in the supermedial quadrant of the buttock.

Table 23-4. Muscles Acting at the Hip Joint—Medial Rotators

Muscle	Origin	Insertion	Primary Action	Innervation
Gluteus medius	Superolateral surface of ilium	Greater trochanter	Abducts and aids medial rotation of thigh	Superior gluteal n. (L4–S2, posterior)
Gluteus minimus	Lateral surface of ilium	Greater trochanter	Abducts and aids medial rotation of thigh	Superior gluteal n. (L4–S1, posterior)
Semitendinosus	Ischial tuberosity	Medial surface of proximal tibia	Extends and aids medial rotation of thigh	Tibial n. (L5–S1, anterior)
Semimembranosus	Ischial tuberosity	Posterior side of tibial condyle	Extends and aids medial rotation of thigh	Tibial n. (L5–S1, anterior)
Adductor magnus: Posterior belly	Ischial tuberosity	Adductor tubercle of femur	Adducts, extends, and aids medial rotation of thigh	Tibial n. (L4–L5, anterior)

    2. Distribution. It innervates the gluteus medius, gluteus minimus, and tensor fasciae latae muscles.

    3. Clinical consideration. Palsy results in *"abductor lurch,"* a rolling (Trendelenburg) gait because of an inability to keep the pelvis level when the contralateral foot is raised off the ground. The lurch is a shift of the body weight, directly over the ipsilateral femoral head.

  B. Inferior gluteal nerve (L5–S2, posterior)

    1. Course. This nerve exits the pelvis via the infrapiriform portion of the greater sciatic foramen, along with the inferior gluteal vessels and the sciatic nerve (see Fig. 23-3). It lies in the inferolateral quadrant of the buttock.

    2. Distribution. It innervates the gluteus maximus muscle.

    3. Clinical consideration. Palsy of this nerve results in difficulty in rising from a seated position and in climbing stairs due to weakness of hip extension.

  C. Sciatic nerve (L4–S2, anterior and posterior)

    1. Course. It exits from the pelvis via the infrapiriform recess of the greater sciatic foramen, along with the inferior gluteal neurovascular bundle (see Fig. 23-3). It courses in an arc, starting halfway between the ischial tuberosity and the iliac spine. It passes over the neck of the femur halfway between the ischial tuberosity and the greater trochanter and lies in the inferomedial quadrant of the buttock.

    2. Distribution. This nerve passes through the gluteal region but does not contribute to the gluteal innervation.

    3. Clinical consideration. The intramuscular injection in the gluteal region should be in the upper lateral quadrant to avoid the sciatic nerve.

  D. Summary. The superior gluteal neurovascular bundle, the piriformis muscle, the inferior gluteal neurovascular bundle, the sciatic nerve, the internal pudendal vessels, and the pudendal nerve leave the deep pelvis through the greater sciatic notch (see Figs. 23-3 and 23-4).

VII. CLINICAL CONSIDERATIONS

  A. Congenital dislocation of the hip. A stable hip is essential as a fulcrum for the action of abduction. In congenital dislocation of the hip, the femoral head lies outside the acetabulum, and a failure of abduction results in the waddling (Trendelenburg) gait.

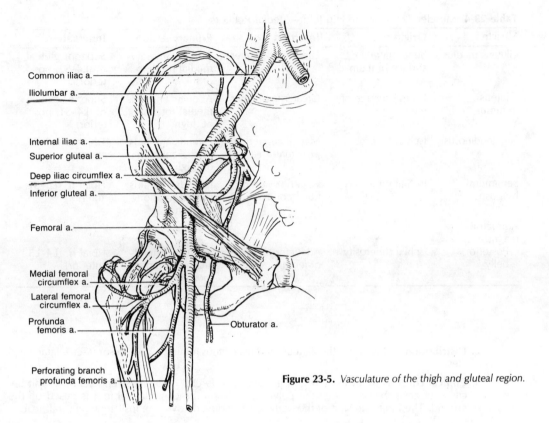

Common iliac a.
Iliolumbar a.
Internal iliac a.
Superior gluteal a.
Deep iliac circumflex a.
Inferior gluteal a.
Femoral a.
Medial femoral circumflex a.
Lateral femoral circumflex a.
Profunda femoris a.
Obturator a.
Perforating branch profunda femoris a.

Figure 23-5. *Vasculature of the thigh and gluteal region.*

B. Septic arthritis. In children, the femoral head gets its arterial supply from the artery of the ligamentum teres. Elevated intracapsular pressure, as a result of septic arthritis, may compress the artery and produce ischemic necrosis of the head.

C. Fractures of the femoral neck occur frequently, particularly in the elderly.

   1. Intracapsular fractures may be subcapital or midcervical.
   a. In the elderly with resorption and weakening of the bony trabeculae, these fractures may be the result of a fall or the result of muscle spasm with the fall subsequent to the fracture. So common is this complication that in an elderly patient the initial treatment of choice is joint replacement.
   b. Intracapsular fractures endanger the blood supply to the proximal fragment, so that nonunion and avascular necrosis may result, depending upon the degree to which the retinacular arteries are torn and the effectiveness of the artery of the ligamentum teres.

   2. Extracapsular fractures (intertrochanteric and subtrochanteric) occur in a region with abundant anastomotic blood supply and present few problems with healing once reduction and internal fixation have been achieved.

   3. Fracture through the epiphyseal plates. The greater trochanter and lesser trochanter have separate ossification centers and may be avulsed along the epiphyseal plates prior to age 18 or 19, especially in heavy teenage males.

   4. Fractures of the proximal femoral shaft are the result of violent trauma as the femoral shaft is especially strong.
   a. In fractures of the proximal third of the femur, the proximal fragment is flexed, abducted, and externally rotated by the iliopsoas, gluteus medius and minimus, and the deep gluteal rotators, respectively.
   b. In fractures of the mid-femur, there is little characteristic pull, but there is marked overriding of the fragments, resulting in a shortened thigh.

D. Dislocations of the hip

    1. Because the hip joint is so stable, dislocation is usually produced by trauma severe enough to fracture the acetabulum.

    2. The hip joint is least stable in the flexed position, which slackens the ligaments of the fibrous capsule.

        a. In posterior dislocation, which is most common, the femoral head comes to lie posteriorly to the iliofemoral ligament, either superiorly or inferiorly to the obturator internus tendon. The sciatic nerve is usually damaged.

        b. In anterior dislocation, the femoral head comes to lie anteriorly to the iliofemoral ligament, either superiorly against the superior pubic ramus or inferiorly in the obturator foramen.

    3. If the articular capsule is torn as a result of dislocation, the blood supply to the femoral head could be jeopardized.

E. Osteoarthritis

    1. Progressive degeneration of the articular cartilage results in pain and limited range of movement.

    2. A cane used as support will be held in the hand *opposite* the affected hip. When the unaffected leg is raised to step, the arm and cane balance the trunk. Abductor action to stabilize the pelvis is, therefore, unnecessary and the weight-bearing joint is relieved of as much as 75% of the compressive loading and the accompanying pain.

I. INTRODUCTION

A. Basic principles. The thigh is that portion of the lower extremity between the pelvic girdle and the knee joint.

1. The primitively dorsal musculature of the thigh lies anteriorly, flexes the hip joint, and extends the knee joint. Conversely, the **primitively ventral musculature** lies dorsally, extends the hip, and flexes the knee.

2. Muscles acting at the knee. Most of the major muscles that act at the knee joint are thigh muscles; however, muscles of the posterior compartment of the leg also act at this joint.

B. Bony landmarks

1. The distal femur has a palpable shaft, medial epicondyle, and lateral epicondyle.

2. The patella is palpable at the knee.

3. The tibia has a palpable medial epicondyle, lateral epicondyle, anterior lip of the tibial plateau, tibial tubercle, medial surface of the shaft, and medial malleolus.

4. The fibula has a palpable head, lateral collateral ligament, and lateral malleolus.

II. BONES AT THE KNEE

A. Distal femur

1. Characteristics

a. The femoral shaft broadens distally as condyles, which provide a stable weight-bearing articulation with the tibia. The **linea aspera**, a ridge, runs along the posterior aspect of the femur and provides attachment for the septa, which define the posterior compartment.

b. The lateral epicondyle, a bony ridge lateral to the lateral condyle, serves as the attachment for the **lateral (fibular) collateral ligament** and the lateral head of the gastrocnemius muscle.

c. The medial epicondyle, a bony ridge medial to the medial condyle, serves as a point of attachment for the **medial (tibial) collateral ligament** and the medial head of the gastrocnemius muscle. Superior to the medial epicondyle is the **adductor tubercle**, where the posterior belly (hamstring portion) of the adductor magnus attaches.

2. Femoral articular surfaces at the knee

a. A deep **intercondylar notch (fossa)** between the lateral and medial condyles divides the femorotibial articular surface inferiorly and posteriorly.

b. The patellar articular surface (trochlear groove) bridges between the condyles and gives the articular surface of the anterior aspect of the distal femur a U-shaped appearance. However, the patellar articular surface is separated from the tibial articular surface by a faint groove in each condyle.

c. The lateral condyle is wider than the medial condyle. It lies along the principal line of force and is probably more important in weight-bearing.

d. The medial condyle has a longer tibial articular surface than the lateral condyle.

(1) This results in lateral rotation of the tibia with reference to the femur during the terminal phase of knee extension.

(2) The medial condyle usually projects slightly further distally than the lateral condyle, producing a physiologic inward angle at the knee, normal valgus, which is usually more pronounced in the female (170°) than in the male (175°).

    (a) *Genu valgum*, or "**knock-knees**," is an exaggerated inward angling at the knee. Valgus occurs as a result of excessive prolongation of the medial femoral condyle or by reduced growth in the lateral portion of the distal epiphyseal plate. If severe, the foot will also be somewhat everted and medially rotated.

    (b) *Genu varum*, or "**bow-legs**," an exaggerated outward angling at the knee, is the reverse situation. If varus is severe, the foot will also be somewhat inverted and laterally rotated.

  e. The axis of the knee is not fixed. The progressive radii of curvature of the condyles forms a spiral-shaped **evolute of rotation**.

    (1) As the knee is extended, the radii of curvature of both condyles increase. The tibia is effectively pushed distally, tightening the ligaments that connect the tibia and femur.

      (a) The radii are shorter posteriorly (17 mm medial, 12 mm lateral).

      (b) The radii are longer anteriorly (38 mm medial, 60 mm lateral).

    (2) In flexion, the transverse axis of rotation passes more posteriorly through the femoral condyles. Ligaments that attach to the femur anteriorly are, therefore, taut in flexion.

    (3) In extension, the transverse axis of rotation passes more anteriorly. Ligaments that attach to the femur posteriorly are, therefore, taut in extension.

3. Fractures of the distal femur are generally classified as supracondylar and intracondylar.

  a. In a *supracondylar fracture*, the distal fragment is displaced posteriorly by gastrocnemius spasm. This jeopardizes the large and important popliteal neurovascular structures, which at this level lie adjacently to the femur.

  b. An *intracondylar fracture* is **T**-shaped, combining a supracondylar fracture with a vertical fracture plane that separates the condyles. This may result in a valgus or varus knee deformity if not reduced properly.

B. Proximal tibia

1. Characteristics

  a. The tibial plateau is formed by the **medial** and **lateral tibial condyles**. On the anterior tibial surface of the lateral condyle, a small elevation serves as insertion for the iliotibial tract.

  b. The tibial tuberosity, the insertion of the patellar tendon, lies inferiorly to the tibial head on the anterior surface of the shaft.

  c. The tibial shaft is triangular in cross section and tapers distally towards the **medial malleolus**.

2. Tibial articular surfaces at the knee

  a. Femorotibial (knee) joint

    (1) The **medial tibial condyle** is concave in both frontal and sagittal planes. Like the medial femoral condyle, it is longer in the anteroposterior diameter.

    (2) The **lateral tibial condyle** is concave in the frontal plane but convex in the sagittal plane and more nearly round.

    (3) An **intercondylar eminence** between the tibial condyles lodges in the intercondylar notch of the femur and acts as a pivot for rotation around a vertical axis.

    (4) Menisci of fibrocartilage lie between the femoral and tibial condyles. These become slightly distorted as flexion/extension occurs, thereby compensating for the lack of congruity between the tibial and femoral articular surfaces and distributing weight evenly over the joint.

  b. Proximal (superior) tibiofibular joint

    (1) The fibula has lost contact with the femur. Instead, the tibial head articulates with the fibula laterally, just inferior to the lateral condyle. This is a gliding joint with extremely limited movement.

    (2) An **interosseous membrane** connects the fibula and tibia along their shafts. The connective tissue fibers run inferolaterally, functioning as shock absorbers and helping to stabilize the ankle joint.

C. Patella

1. Characteristics

  a. A sesamoid bone, the patella (*L.* little plate) sits in the tendon of the quadriceps femoris muscle. It is triangular in shape with the apex directed inferiorly (see Fig. 24-3).

  b. The patellar tendon (often termed ligament) runs between the apex and the tibial tuberosity.

    (1) The **patellar retinaculum** is an encapsulation of dense connective tissue derived from the quadriceps femoris tendon. It extends inferolaterally and inferomedially to insert into the tibial condyles, representing a broad tendinous insertion for the quadriceps femoris muscle. It reinforces the capsule of the knee joint.

(2) Subcutaneous bursae lie anteriorly to the knee.

   (a) The **prepatellar bursa** is superficial to the patella and deep to the epithelium. Inflammation due to repeated abuse results in *prepatellar bursitis* or *"housemaid's knee."*

   (b) The **infrapatellar bursa** lies between the tibial tuberosity and the epithelium. Inflammation results in *infrapatellar bursitis* or *"vicar's knee."*

   (c) The **suprapatellar bursa** extends proximally between the quadriceps femoris tendon and the femur. Effusions of the knee joint distend this bursa (*suprapatellar bursitis* or *"water on the knee"*), and on examination, the patella appears to float over the femur. Postinflammatory adhesions in the suprapatellar bursa reduce knee mobility upon extension (*"stiff knee"*).

2. Femoropatellar joint

   a. Medial and lateral articular surfaces are formed by a vertical ridge on the posterior surface of the patella. As a gliding joint, the lateral and medial articular surfaces slide in the **trochlear groove** of the femur.

   (1) In flexion, the tension from the quadriceps femoris tendon and the patellar retinaculum tends to keep the patella deep in the trochlear groove between the femoral condyles.

   (2) In extension, due to the usual valgus at the knee, the line of action of the rectus femoris, vastus lateralis, and vastus intermedius muscles (through the quadriceps femoris tendon to the patella and thence through the patellar tendon to the tibial tuberosity) is somewhat lateral to the patella. This, together with a shallower trochlear groove in extension, tends to draw the patella laterally.

      (a) The prominent lateral ridge of the trochlear groove and the lateral epicondyle, the patellar retinaculum, and especially the line of action of the vastus medialis muscle all act to prevent lateral dislocation of the patella.

      (b) Underdevelopment of the lateral ridge of the trochlear groove or exaggerated genu valgum may result in recurrent patellar dislocation. This can be corrected surgically by transplanting the tibial tubercle slightly more medially and inferiorly.

   b. Functions of the patella

   (1) As a sesamoid bone, it obviates wear and attrition on the quadriceps tendon as it passes across the trochlear groove.

   (2) It lengthens the lever arm, thereby increasing the mechanical advantage of the quadriceps femoris muscle.

3. Fracture of the patella through the articular surface will result in eventual degenerative *patellofemoral arthritis*. A congenitally *bipartate patella* may be misdiagnosed as a fracture.

III. KNEE JOINT

A. Structure. The knee joint is composed of two articulations, the femoropatellar joint and the femorotibial joint.

   1. The femoropatellar joint functions as a trochlear (*L.* pulley) for the quadriceps femoris muscle group (see II C 2).

   2. The femorotibial joint supports the weight of the body. While adapted for weight-bearing at any degree of flexion, stability is greatest in full extension.

   a. The menisci (semilunar cartilages). Composed of fibrocartilage rings that are triangular in cross section, the menisci lie between the femoral and tibial condyles and incompletely divide the femorotibial joint into a **suprameniscal compartment** and an **inframeniscal compartment** (Fig. 24-1).

   (1) The lateral meniscus is nearly O-shaped (Fig. 24-2).

      (a) The **anterior** and **posterior horns** of the lateral meniscus insert close together into the lateral tubercle of the intercondylar eminence.

      (b) The **coronary ligament** loosely attaches the lateral meniscus to the joint capsule.

      (c) The **meniscofemoral ligament** (of Wrisberg), a portion of the posterior cruciate ligament, attaches the lateral meniscus to the posterior aspect of the medial femoral condyle. There is little or no attachment of the lateral meniscus to the lateral collateral ligament.

      (d) Movement of the lateral meniscus is considerable due to the close insertions of this meniscus, the loose attachment to the joint capsule, and little or no attachment to the lateral collateral ligament. During the terminal phase of extension with the foot planted firmly on the ground, the femur pivots and the lateral femoral condyle slides anteriorly with the meniscus to lock the knee.

A

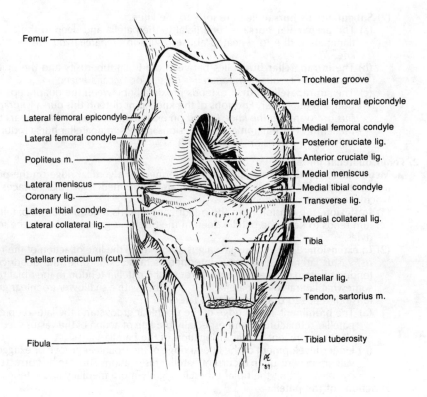

Femur

Trochlear groove

Medial femoral epicondyle

Lateral femoral epicondyle

Medial femoral condyle

Lateral femoral condyle

Posterior cruciate lig.

Popliteus m.

Anterior cruciate lig.

Medial meniscus

Lateral meniscus

Medial tibial condyle

Coronary lig.

Transverse lig.

Lateral tibial condyle

Medial collateral lig.

Lateral collateral lig.

Tibia

Patellar retinaculum (cut)

Patellar lig.

Tendon, sartorius m.

Fibula

Tibial tuberosity

B

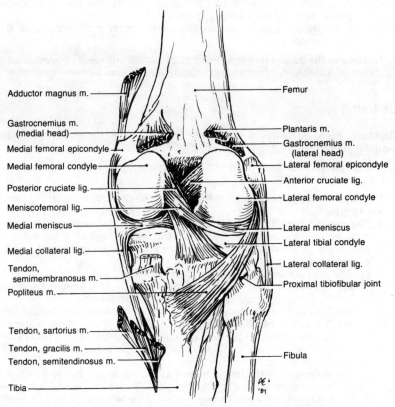

Adductor magnus m.

Femur

Gastrocnemius m.
(medial head)

Plantaris m.

Gastrocnemius m.
(lateral head)

Medial femoral epicondyle

Medial femoral condyle

Lateral femoral epicondyle

Anterior cruciate lig.

Posterior cruciate lig.

Lateral femoral condyle

Meniscofemoral lig.

Medial meniscus

Lateral meniscus

Medial collateral lig.

Lateral tibial condyle

Tendon,
semimembranosus m.

Lateral collateral lig.

Proximal tibiofibular joint

Popliteus m.

Tendon, sartorius m.

Tendon, gracilis m.

Tendon, semitendinosus m.

Fibula

Tibia

Figure 24-1. *Right knee joint. A,* Anterior. *B,* Posterior.

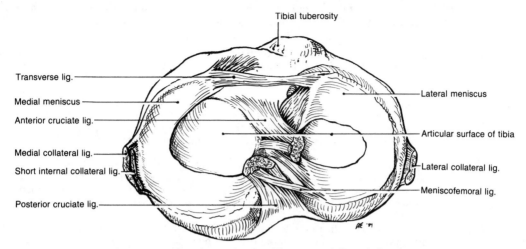

Figure 24-2. *Articular cartilages and ligaments of the right knee joint.*

   (2) The medial meniscus is **C**-shaped (see Fig. 24-2).
      (a) The **anterior** and **posterior horns** insert externally to those of the lateral meniscus (anterior and posterior to the intercondylar eminence, respectively), giving the medial meniscus a somewhat greater diameter than the lateral meniscus.
      (b) The short **internal collateral ligament** runs from the margin of the meniscus to the medial femoral epicondyle and tightly attaches the meniscus to the joint capsule.
      (c) The **medial collateral ligament** is attached to the meniscus posteriorly.
      (d) Movement of the medial meniscus is severely restricted due to the widely separated points of attachment, to the medial attachment to the joint capsule, to the medial epicondyle, and to the medial collateral ligament. This makes the medial meniscus especially prone to *meniscal tears.*
   b. Meniscal tears. During the rolling movements of flexion and extension, the menisci are drawn posteriorly and anteriorly, respectively. The menisci can be torn if they fail to follow the movements of the femoral condyles and can become wedged between condylar surfaces so that the knee becomes locked in partial flexion, making full extension impossible.
      (1) Violent hyperextension leads to transverse meniscal tears or detachment of the anterior horns.
      (2) Violent twisting movements, which combine lateral displacement with lateral rotation (especially in a semiflexed knee), pull the medial meniscus toward the center of the joint where it may be trapped and crushed by the medial femoral condyle. This event may lead to a longitudinal (bucket-handle) tear or partial detachment of the medial meniscus.
   c. Osteoarthritis of the knee. The knee is a common site for degenerative arthritis. A cane used as support should be held in the hand on the same side as the afflicted knee. The cane held in parallel with the leg relieves a portion of the body weight from the joint surface and reduces the accompanying pain.

B. Ligamentous support

   1. The synovial capsule of the knee is extensive.
      a. Circumferentially, the synovial cavity inserts more or less about the margins of the articular surfaces.
      b. Medially, the capsule attaches to the medial meniscus.
      c. Posteriorly, the capsule invaginates into the intercondylar notch so that the cruciate ligaments are technically extracapsular.
      d. Superiorly, the suprapatellar bursa extends from the capsule between the quadriceps femoris tendon and the femur. Numerous other synovial bursae about the knee may or may not communicate with the joint capsule.
      e. Effusion of the knee joint extends the joint capsule, precluding a full range of motion, and the knee assumes approximately 20° of flexion.

   2. The fibrous capsule of the knee is rather weak.
      a. The short internal collateral ligament, a thickening of the fibrous capsule deep to the medial collateral ligament, extends from the medial femoral epicondyle to the medial meniscus.
      b. The coronary ligament is the thickened periphery of the capsule that loosely attaches the

two menisci to the tibia. Frequently (60%), it is thickened anteriorly to form the **transverse ligament**, which connects the anterior margins of the two menisci (see Figs. 24-1*A* and 24-2).

3. **The collateral ligaments and patellar retinaculum** reinforce the fibrous capsule of the knee.
   a. **The medial (tibial) collateral ligament** (see Figs. 24-1 and 24-2) extends from the medial femoral epicondyle and widens to insert into the shaft of the tibia below the level of the tibial tuberosity.
      (1) **Structure.** The medial collateral ligament is broad and fan-shaped so that a portion of this ligament can be effective at any degree of flexion. The posterior portion is fused to the joint capsule and is attached to the medial meniscus. Phylogenetically, this ligament is the degenerated tendon of insertion of the posterior portion of the adductor magnus muscle.
      (2) **Functions**
         (a) The medial collateral ligament prevents abduction at the knee. A torn medial collateral ligament can be recognized by abnormal passive abduction of the extended leg.
         (b) Since it lies posteriorly to the axes of flexion/extension, it becomes taut upon extension and, thus, also limits extension of the leg.
   b. **The lateral (fibular) collateral ligament** (see Figs. 24-1 and 24-2) extends from the lateral femoral epicondyle to insert into the head of the fibula.
      (1) **Structure.** It is a strong narrow cord-like ligament that is not quite parallel to the medial collateral ligament. It is readily palpated when the legs are crossed and the ankle rests on the opposite knee. It is relatively free of the joint capsule and does not attach to the lateral meniscus. Phylogenetically, this ligament is the degenerated tendon of origin of the peroneus longus muscle.
      (2) **Functions**
         (a) The lateral collateral ligament prevents adduction of the leg at the knee. A torn lateral collateral ligament can be recognized by abnormal passive adduction of the extended leg.
         (b) Since the lateral collateral ligament also lies posteriorly to the axes of flexion/extension, it becomes taut upon extension and limits extension of the leg.
   c. **The posterior cruciate ligament** (see Figs. 24-1 and 24-2) extends anteromedially from the tibia posterior to the intercondylar eminence and inserts into the medial femoral condyle anteriorly within the intercondylar notch. It is approximately parallel to the lateral collateral ligament and perpendicular to the anterior cruciate ligament.
      (1) **Structure**
         (a) The posterior cruciate ligament fans out towards its linear insertion along the medial femoral condyle. Because the fibers in different portions of the ligament are of unequal length, some are always under tension at any degree of knee flexion. Fibers inserting anteriorly on the femoral condyle tend to be taut in flexion, whereas fibers inserting posteriorly tend to be taut in extension.
         (b) The **meniscofemoral ligament**, a subdivision of the posterior cruciate ligament (see Figs. 24-1*B* and 24-2), inserts into the posterior horn of the lateral meniscus and draws that meniscus posteriorly as the lateral condyle slides on the tibial plateau.
      (2) **Function.** The posterior cruciate ligament is a key stabilizer of the knee joint, tightening maximally upon extension.
         (a) It checks anterior movement of the femur on the tibial plateau (posterior movement of the tibia with respect to the femur). In the flexed weight-bearing knee (descending stairs or walking down hill), this ligament prevents the femur from sliding forward.
         (b) A torn posterior cruciate ligament produces a perceived instability so that an individual upon descending stairs will lead with the opposite leg for each step. This torn ligament can be recognized by abnormal passive posterior displacement of the tibia (*posterior drawer sign*).
   d. **The anterior cruciate ligament** (see Figs. 24-1 and 24-2) extends posterolaterally from the tibia anterior to the intercondylar eminence and inserts onto the lateral femoral condyle posteriorly within the intercondylar notch. It is approximately perpendicular to the lateral collateral ligament and the posterior cruciate ligament.
      (1) **Structure.** The anterior cruciate ligament, which is slightly shorter than the posterior cruciate, also fans out towards its linear insertion along the lateral femoral condyle. Because the fibers in different portions of the ligament are of unequal length, some are always under tension at any degree of knee flexion. Fibers inserting posteriorly on the femoral condyle tend to be taut in extension, whereas fibers inserting more anteriorly tend to be taut in flexion.
      (2) **Function.** The anterior cruciate ligament is a key stabilizer of the knee joint. Since this ligament inserts posteriorly on the femur, it tightens maximally upon extension and

provides a pivot around which rotation occurs during the terminal phase of extension.
- **(a)** It checks posterior movement of the femur on the tibial plateau (anterior movement of the tibia with reference to the femur).
- **(b)** A torn anterior cruciate ligament will often not be noticed, except that occasionally the knee "suddenly gives way." This is because the leg will not lock upon full extension. A torn anterior cruciate ligament can be recognized by abnormal passive anterior displacement of the tibia (*anterior drawer sign*).

- **e. The oblique popliteal ligament** is a lateral extension of the semimembranosus tendon. It strengthens the joint capsule posteriorly, resisting hyperextension of the leg as well as lateral rotation during the terminal phase of extension.
- **f. The patellar retinaculum** reinforces the capsule of the knee joint medially and laterally while the patellar tendon reinforces the joint capsule anteriorly [see II C 1 b (1)].

4. Stability of the knee is conferred by the collateral and cruciate ligaments.
- **a. Rotational stability.** Because the cruciate ligaments are slightly coiled inward, lateral rotation of the femur tightens both ligaments, preventing further lateral rotation. Similarly, because the collateral ligaments are slightly coiled outward, medial rotation of the femur tightens these coils and prevents further medial rotation.
- **b. Flexion/extension.** Because extension of the knee is checked by the collateral and cruciate ligaments, in the erect posture with the knee extended and locked, the weight of the body at the knees is stabilized primarily by ligaments.
- **c. The dynamic stability** of the knee is largely dependent upon the action of the muscles, which pass over this joint, especially when the knee is partially flexed. The thigh muscles inserting below the knee and the leg muscles originating above the knee contribute in a major way to the dynamic stability of the knee. When the limb is immobilized in a cast, disuse atrophy of the muscles occurs, and physical therapy is necessary to restore full function and stability.

5. Instability of the knee as a result of torn ligaments
- **a. Ligamentous sprains and tears.** Because the knee joint represents a compromise between stability and mobility, there is great potential for sprains (partial tears) and tears. Violent hyperextension, abduction, and adduction may sprain or completely tear one or more collateral or cruciate ligaments.
- **b. Combined knee injury,** such as the "unhappy triad," is the result of fixation of a semiflexed leg with violent abduction and lateral rotation that occurs as the result of "clipping" in football or the "caught-edge" fall in skiing.
  - **(1)** Three elements compose the unhappy triad.
    - **(a)** The medial collateral ligament is torn by excessive abduction, and since this ligament is attached to the joint capsule, the medial capsular ligaments are torn as well.
    - **(b)** The anterior cruciate ligament is torn as a result of forward displacement of the tibia. This is accompanied by a bloody effusion in the joint capsule, which may extend into any connecting bursa.
    - **(c)** The medial meniscus is torn as a result of the medial collateral ligament attachment and the excessive lateral rotation. Often this meniscus is drawn between the medial femoral and tibial condyles and crushed or torn longitudinally.
  - **(2)** Treatment involves surgical reconstruction of the ligaments with every attempt to save as much of the meniscus as possible to prevent or limit future osteoarthritis. An arthroscope inserted into the knee joint through a small incision permits direct visualization of trauma or pathology. Corrective procedures can often be accomplished through the instrument.

C. Dynamic action of the knee
- **1. Movement.** The knee joint has an extensive range of movement. The **axis of rotation** of the knee changes through an evolute so that the knee does not function as a simple hinge joint. Separate and distinct movements occur in the suprameniscal and inframeniscal compartments.
  - **a. Suprameniscal compartment.** The femoral condyles rotate upon the superior meniscal surfaces as a result of rotation about the transverse axis, approximating a simple hinge (Fig. 24-3A).
  - **b. Inframeniscal compartment.** The inferior meniscal surfaces glide on the tibial plateau, the result of an anterior/posterior translation of the transverse axis (see Fig. 24-3A). In addition, the femoral condyles pivot on the tibial plateau about a vertical axis at the end of extension (see Fig. 24-3B).
  - **c. Combined movements** in the suprameniscal and inframeniscal compartments are not independent but occur in conjunction with flexion/extension. The ligaments are arranged so that the femur and tibia are always held in close apposition with stability throughout the

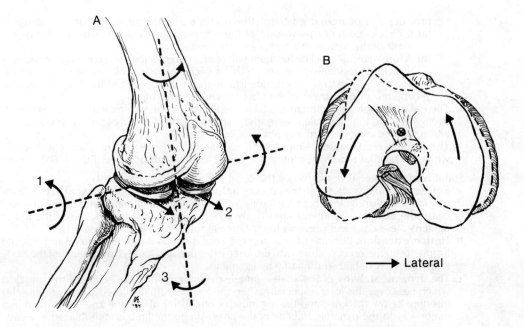

A

B

1

2

3

➙ Lateral

Figure 24-3. *Dynamic action of the knee joint. A,* During extension, rotation occurs in the suprameniscal compartment about the transverse axis (*1*) and forward gliding occurs in the inframeniscal compartment (*2*). The medial rotation of the terminal phase of extension occurs about a vertical axis (*3*). *B,* Looking down on the right tibial plateau in terminal extension, the relative movement of the femur relative to the tibia is indicated by the *arrow* with an axis of rotation about the anterior cruciate ligament.

ranged movement. The proportion of each individual movement is dictated by the ligaments, the shape of the condyles, and the length of the articular surfaces.

  (1) Flexion/extension (140°) approximates a hinge joint.
  (2) Rotation (20°) occurs at the terminal phase of extension or the initial phase of flexion.

2. **Phases of extension**
  a. Initial phase
    (1) In the flexed knee, the anterior and posterior cruciate ligaments tether the femoral condyles on the tibial plateau. As the femorotibial joint begins to extend through the first 20°– 30°, there is a concomitant anterior translation of the condyles on the tibial plateau.
    (2) This complex movement involves rotation in the suprameniscal compartment about a transverse axis through the femoral condyles and a concomitant anterior gliding movement in the inframeniscal compartment as that transverse axis moves anteriorly (see Fig. 24-3*A*).
  b. Intermediate phase
    (1) As extension continues, the increasing radii of curvature of the femoral condyles tightens the anterior cruciate ligament and the posterior fibers of the posterior cruciate ligament. This prevents further forward gliding of the femur in the inframeniscal compartment.
    (2) However, the femoral condyles continue to rotate on the meniscal surfaces in the suprameniscal compartment as full extension is approached (see Fig. 24-3*A*).
  c. Terminal phase
    (1) Toward full extension, the lateral condyle reaches the limit of its articular surface and ceases rotation; however, the longer articular surface of the medial condyle continues to rotate in the suprameniscal compartment. The lateral condyle is, thus, forced to slide forward and medially in the inframeniscal compartment about an axis formed by the taut anterior cruciate ligament (see Fig. 24-3*B*).
    (2) The result is a medial rotation of the femur (or lateral rotation of the tibia) in the inframeniscal compartment about an axis through the intercondylar eminence at the insertion of the anterior cruciate ligament. (Some sources have the vertical axis passing through the lateral femoral condyle, but the fact that in terminal extension this condyle glides about an axis rather than simply pivots negates the notion.)
    (3) This medial rotation of the femur tightens the oblique popliteal ligament, the medial

collateral ligament, and the lateral collateral ligament as well as stretches the popliteal muscle so that the femorotibial joint becomes locked ("screwed home").

(4) At terminal extension, the angle between the femur and the tibia slightly exceeds 180°.

   (a) The weight of the body now passes anteriorly to the transverse axis of the femorotibial joint and tightens the knee ligaments.

   (b) While muscular activity is unnecessary to maintain the extended and locked position when standing, the line of action of the tensor fasciae latae and gluteus maximus muscles passes anteriorly to the transverse axis and ensures the locked position.

IV. FASCIA AND MUSCULATURE OF THE THIGH

A. The fascia lata is the deep fascia of the thigh.

   1. The iliotibial tract (band) is a lateral thickening of the fascia lata. It runs from the iliac crest to the lateral tibial condyle (see Fig. 23-3; Fig 24-4*A*). It functions as the tendon of insertion for the tensor fasciae latae muscle and the gluteus maximus muscle.

   a. At the hip. Because these muscles originate on the pelvis and pass over the hip joint, they function in extension of the thigh.

   b. At the knee. Because the iliotibial tract crosses the knee joint, these muscles function both in extension and flexion of the leg.

   (1) When the knee is extended, the line of action of the iliotibial tract passes anteriorly to the axis of rotation of the knee joint, and the tensor fasciae latae and gluteus maximus muscles, therefore, keep the knee extended.

   (2) When the knee is slightly flexed, the line of action of the iliotibial tract passes posteriorly to the axis of rotation of the knee joint so that these muscles now flex the knee joint. This is evident when a person, standing erect and somewhat relaxed, has the knees knocked from behind. The line of action of the iliotibial tract shifts across the axis of rotation so that the tensor fasciae latae and gluteus maximus muscles, which keep the knee extended in the erect posture, now flex the knee and produce the disconcerting sinking motion.

   2. Fascial septa from the fascia lata divide the thigh musculature into anterior and posterior compartments.

   a. The lateral intermuscular septum arises from the posterior border of the iliotibial tract and inserts into the lateral lip of the linea aspera of the femur. This provides an important insertion onto the femur for the tensor fasciae latae and the gluteus maximus muscles.

   b. The medial intermuscular septum arises from the anteromedial border of the iliotibial tract and attaches to the medial lip of the linea aspera.

B. Musculature

   1. Organization. The thigh is divided into an anterior compartment, containing primitively dorsal musculature, and a posterior compartment, containing primitively ventral musculature.

   a. The anterior compartment is divided into the iliopsoas and anterior muscle groups.

   (1) The iliopsoas group acts at the hip (see Figs. 23-4 and 24-4). This group is discussed in Chapter 23 IV A 1 a.

   (2) The anterior group acts as flexors at the hip and principally as extensors at the knee (see Fig. 24-4; Table 24-1).

   (a) This group consists of the **sartorius** and **quadriceps femoris muscles**, the latter consisting of the rectus femoris and the vasti lateralis, intermedius, and medialis.

   (b) These muscles generally arise from the anterior portion of the ilium (sartorius and rectus femoris) or the femoral shaft (the vasti) and insert into the tibia through the patella, except that the sartorius inserts onto the medial tibial condyle.

   b. The posterior compartment is divided into adductor and hamstring muscle groups.

   (1) The adductor group acts primarily across the hip joint and consists of the **gracilis, adductor longus**, and **adductor brevis muscles** as well as the anterior portion of the **adductor magnus muscle** and a portion of the **pectineus muscle** (see Fig. 24-4*A*; Fig. 24-5*B*). Only the gracilis has an action at the knee (flexion).

   (a) The adductor group lies in three layers.

   (i) The **anterior layer** consists of the gracilis, pectineus, and adductor longus (see Fig. 24-4*A*).

   (ii) The **intermediate layer** consists of the adductor brevis (see Fig. 24-4*B*).

   (iii) The **posterior layer** consists of the adductor magnus (see Fig. 24-5*B*).

   (b) The muscles of the adductor group generally arise from the inferior pubic rami and insert into the femoral shaft medially, except the gracilis, which inserts into the medial tibial condyle.

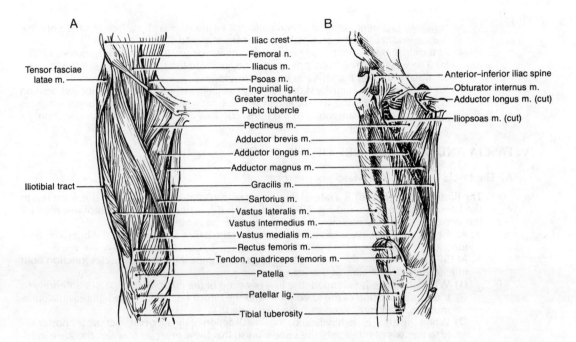

Figure 24-4. *Musculature of the anterior compartment of the thigh. A,* Superficial muscles. *B,* Deep muscles.

    (2) The hamstring group acts at both the hip and knee joints and consists of the **semimembranosus**, **semitendinosus**, the long head of the **biceps femoris**, and the posterior portion of the **adductor magnus muscle**.

        (a) The hamstring group arises from the ischial tuberosity and inserts into the tibial condyles, except that the posterior portion of the adductor magnus inserts onto the adductor tubercle of the medial femoral epicondyle (see Fig. 24-5A).

            (i) Phylogenetically, the sacrotuberous ligament is considered to be a remnant of the hamstring muscles, which had a primitive origin from the sacrum. Similarly, the medial collateral ligament is considered to be a remnant of the primitive insertion of the adductor magnus onto the medial tibial condyle.

            (ii) The semitendinosus tendon joins the **pes anserinus** (*L.* goose foot) with the sartorius and gracilis tendons to insert just distal to the medial tibial condyle (see Fig. 24-5B).

            (iii) The long and short heads of the biceps femoris muscle insert laterally onto the head of the fibula (see Fig. 24-5).

        (b) The hamstrings flex the knee.

2. Group actions

    a. Flexion/extension (140°) occurs about a transverse axis through the femoral condyles (see Fig. 24-3A).

        (1) Extension is accomplished by the anterior group. Because the rectus femoris is a biarticular muscle (flexing the thigh at the hip joint and extending the leg at the knee joint), its efficiency at one joint depends upon the position of the other joint. Thus, when the hip is extended, the rectus femoris is stretched and its force-generating capacity to extend the knee is greatest (see Chapter 3 II B 2).

        (2) Flexion is accomplished by the hamstring group, which crosses the knee posteriorly. Because the hamstrings are biarticular (extending the thigh and flexing the knee), their efficiency at the knee is greatest when the hip is flexed (see Chapter 3 II B 2). The sartorius muscle of the anterior group, and the popliteus and gastrocnemius muscles are also knee flexors.

    b. Rotation (20°) occurs about a vertical axis that passes through the intercondylar eminence of the tibia (see Fig. 24-3B). Rotation occurs concomitantly with the terminal phase of extension. While the knee cannot be actively rotated, the flexed knee may be passively rotated through 70°.

        (1) Lateral tibial rotation occurs during the terminal phase of extension, mainly as a result of the differences in the radii of the lateral and medial femoral condyles and the limits on movement imposed by the cruciate and collateral ligaments.

Table 24-1. Thigh Muscles Acting at the Knee Joint—Anterior and Posterior Compartments

Muscle	Origin	Insertion	Primary Action	Innervation
Anterior Compartment				
Rectus femoris	Anterior–inferior iliac spine	Tibial tuberosity through patellar tendon	Extends knee	Femoral n. (L2–L4, posterior)
Vastus lateralis	Proximal femur lateral to linea aspera	Tibial tuberosity through patellar tendon	Extends knee	Femoral n. (L2–L4, posterior)
Vastus medialis	Proximal femur medial to linea aspera	Tibial tuberosity through patellar tendon	Extends knee	Femoral n. (L2–L4, posterior)
Vastus intermedius	Anterior aspect of proximal femur	Tibial tuberosity through patellar tendon	Extends knee	Femoral n. (L2–L4, posterior)
Sartorius	Anterior–superior iliac spine	Medial tibial side of tibial head	Flexes knee	Femoral n. (L2–L3, posterior)
Posterior Compartment				
Semitendinosus	Ischial tuberosity	Medial surface of proximal tibia	Flexes knee	Tibial n. (L5–S1, anterior)
Semimem-branosus	Ischial tuberosity	Posterior side of tibial condyle	Flexes knee	Tibial n. (L5–S1, anterior)
Biceps femoris: Long head	Ischial tuberosity	Lateral side of fibular head	Flex knee	Tibial n. (L5–S2, anterior)
Short head	Femoral shaft	Lateral side of fibular head	Flex knee	Common peroneal n. (L5–S2, posterior)
Gracilis	Ischiopubic ramus	Medial side of tibial head	Flexes knee	Obturator n. (L3–L4, anterior)
Gluteus maximus	Posterosuperior ilium and sacrum	Gluteal tuberosity and linea aspera of femur and iliotibial band to fibular head	Extends extended knee and flexes flexed knee	Inferior gluteal n. (L5–S2, posterior)
Tensor fasciae latae	Anterior–superior iliac spine and iliac crest	Iliotibial band to fibular head	Extends extended knee and flexes flexed knee	Superior gluteal n. (L4–S1, posterior)

   (2) Medial tibial rotation, unlocking of the knee, requires contraction of the **popliteus muscle**. Once the knee is unlocked, the flexors can initiate flexion.

3. **Group innervations**
   a. **The anterior group** is supplied by the **femoral nerve**.
   b. **The adductor group** is innervated by the **obturator nerve**.
   c. **The hamstrings**, including the posterior portion of the adductor magnus, is innervated by the **tibial nerve**. The short head of the biceps, not technically a part of the hamstring group, is innervated by the **common peroneal nerve**.

V. VASCULATURE OF THE THIGH

   A. **Arterial supply** to the thigh is primarily by the femoral and obturator arteries with some contribution from the gluteal vessels.

   1. **Obturator artery**
      a. **Course.** This artery usually arises from the **internal iliac artery** deep within the pelvis (see Fig. 20-3), but it may arise aberrantly from the inferior epigastric artery (see Fig. 15-6). It exits the pelvic cavity with the obturator nerve through the **obturator foramen** and divides into two branches (see Fig. 23-5).

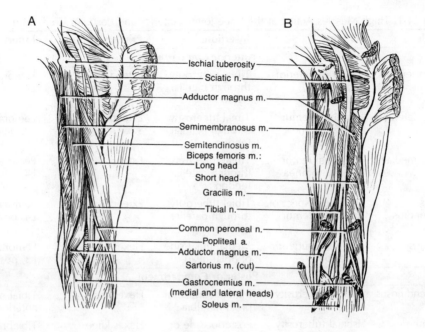

Figure 24-5. *Muscles of the posterior compartment of the thigh. A,* Superficial muscles. *B,* Deep muscles.

b. Distribution
   (1) The **anterior branch** supplies muscles of the adductor group and anastomoses with the medial circumflex branch of the profunda femoris artery.
   (2) The **posterior branch** mainly supplies the hamstring group and gives off the **acetabular branch** from which the **artery of the ligamentum teres** arises.

2. Femoral artery
   a. Course
      (1) The femoral artery is the continuation of the **external iliac artery** distal to the inguinal ligament. It enters the **femoral triangle** by passing between the inguinal ligament and iliopubic ramus (see Fig. 23-5 and Chapter 15 VIII F 1).
         (a) The **femoral triangle** is bounded by the inguinal ligament, the sartorius muscle, and the adductor longus muscle (see Fig. 24-4).
         (b) The **femoral pulse** is palpable high within the femoral triangle at which point, the femoral artery is a convenient source of arterial blood for blood-gas determinations and a preferred site for catheter insertion in angiography.
      (2) At the inferior angle of the femoral triangle, the femoral artery enters the **adductor (of Hunter) canal** along with the **femoral vein**, the **saphenous nerve**, and the **nerve to the vastus medialis**. This canal passes between the adductor (anterior) and hamstring (posterior) portions of the adductor magnus muscle and enters the popliteal fossa through the adductor hiatus (see Fig. 24-5A; Fig 24-6). This is a frequent site for *femoral artery stenosis* and *femoral artery aneurysm*.
      (3) Emerging from the adductor canal into the popliteal fossa, the femoral artery becomes the **popliteal artery** (see Figs. 24-5A and 24-6).
   b. Distribution. The femoral artery gives off several branches in the femoral triangle.
      (1) The **superficial epigastric artery** supplies the hypogastric region of the abdomen.
      (2) The **external pudendal artery** supplies the anterior pudendal region.
      (3) The **deep femoral (profunda femoris) artery**, the largest branch in the femoral artery, initially lies laterally to the femoral artery before passing posteriorly to the femoral artery to lie on the psoas muscle before passing into the posterior compartment (see Figs. 23-5 and 24-6). It gives off several branches.
         (a) The **lateral femoral circumflex artery** arises from the lateral side of the deep femoral artery and gives off three branches (see Fig. 23-5).
            (i) The **ascending branch** passes superiorly and anastomoses with the inferior gluteal artery.
            (ii) The **transverse branch** passes around the femur to anastomose with the medial femoral circumflex artery at the **cruciate anastomosis**.

(iii) The **descending branch** anastomoses with both the descending genicular branch of the femoral artery and with the superior lateral genicular branch of the popliteal artery.

(b) The **medial femoral circumflex artery** arises from the posterior aspect of the deep femoral artery and also gives off three branches (see Fig. 23-5).

(i) The **ascending branch** passes into the gluteal musculature.

(ii) The **transverse branch** passes posteriorly to the femur to anastomose with the transverse branch of the lateral femoral circumflex artery at the **cruciate anastomosis**.

(iii) The **descending branch**, which is rather small, passes inferiorly.

(c) **Perforating arteries**, three or four in number, supply the major portion of the hamstring muscles (see Fig. 24-6).

3. Popliteal artery

a. **Course.** The popliteal artery emerges from the adductor hiatus and descends through the **popliteal fossa** (see Fig. 24-6). The popliteal artery continues into the leg as the **posterior tibial artery**.

(1) The **popliteal fossa** is bounded by the biceps femoris muscle (superolaterally) and the adductor magnus muscle (superomedially) as well as the lateral and medial heads of the gastrocnemius muscle (inferiorly).

(2) The **popliteal pulse** is best palpated by compressing the artery against the popliteus muscle with the leg flexed because the artery lies deeply within the popliteal fossa.

b. **Distribution.** The popliteal artery gives off several branches, which participate in the **geniculate anastomosis** (see Fig. 24-6).

(1) These branches include the **descending genicular artery** with its saphenous and articular branches, the **lateral** and **medial superior genicular arteries** as well as the **lateral** and **medial inferior genicular arteries**.

(2) In addition to the branches of the popliteal artery, the geniculate anastomosis communicates with branches of the deep femoral artery as well as the anterior and posterior tibial arteries to provide abundant collateral circulation around the knee (see Figs. 24-6 and 26-5).

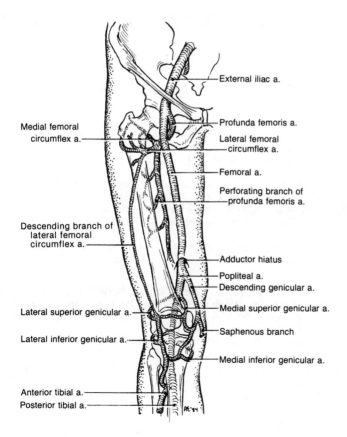

Figure 24-6. *Vasculature of the thigh.*

B. Venous return of the thigh is by way of superficial and deep veins.

    1. The great saphenous vein is the principal superficial vein of the thigh.

        a. Course. It lies on the superficial fascia in the medial aspect of the leg and thigh, passing on the flexor side of the ankle, knee, and hip joints. It contains about eight valves in the thigh. In the femoral triangle, it penetrates the fascia lata through the **saphenous hiatus**, an opening in the fascia lata that is incompletely covered by the **cribriform fascia**.

        b. Branches

            (1) Perforating veins communicate between the saphenous vein and the deep veins of the leg and thigh.

            (2) The **superficial epigastric vein**, **superficial circumflex iliac vein**, and the **external pudendal vein** join the saphenous vein in the femoral triangle.

    2. The femoral vein is the principal deep vein of the thigh.

        a. Course. The femoral vein accompanies the femoral artery, frequently as a subdivided venae comitantes. In the femoral triangle, it is a single vein that lies medially to the femoral artery at which point it is a source of venous blood as well as a site for insertion of venous catheters (see Fig. 15-7 and Chapter 15 VIII F 1).

        b. The major lymphatic pathway from the lower extremity to the deep inguinal lymph nodes lies medially to the femoral vein.

    3. Venous hemodynamics and varicose veins

        a. The course of the deep veins between the deep muscles of the leg and thigh forms a venous pump, which is essential for the return of blood to the heart against gravity.

        b. Valves in the saphenous vein direct the flow upward and protect that vein from the hydrostatic pressure produced by the standing column of blood. Valves in the perforating veins direct the flow inward and also protect the saphenous vein from the pressure that results from the pumping action of the leg and thigh muscles upon the deep veins. Where the saphenous vein joins the femoral vein, there is an important valve.

            (1) Old age and venous stasis associated with long periods of standing predispose the superficial vein to dilation, which reduces the competency of the valves.

            (2) Incompetent valves in the perforating veins allow blood to escape at relatively high venous pressure into the saphenous system, exacerbating the dilation and forming *varicose veins*.

            (3) Treatment for varicose veins varies from identification and ligation of the perforating veins with incompetent valves to removal of the entire saphenous vein.

VI. INNERVATION OF THE THIGH

    A. Anterior compartment innervation is primarily by the femoral and genitofemoral nerves, which arise from the lumbar plexus.

        1. Femoral nerve

            a. Course and composition. The femoral nerve arises from the posterior divisions of lumbar plexus roots L2–L4 (see Fig. 18-6). It exits the greater pelvis through the muscular compartment inferior to the inguinal ligament (Fig. 24-7).

            b. Distribution

                (1) Motor. The femoral nerve innervates the muscles of the anterior compartment of the thigh (see Fig. 24-7). These branches arise in the femoral triangle.

                (2) Sensory. The **anterior femoral cutaneous nerve** innervates the dermatomes of the anterior aspect of the thigh. The cutaneous **saphenous nerve** continues into the leg (see Fig. 24-7).

            c. Femoral nerve injury is usually the result of trauma to the femoral triangle. Such injury produces weakness of hip flexion with an inability to extend the knee as indicated by the loss of the **patellar reflex**.

        2. Genitofemoral nerve. The **femoral branch** of the genitofemoral nerve (L1–L2, anterior) is sensory to the medial aspect of the proximal thigh (see Fig. 18-6). It provides the afferent limb of the **cremaster reflex**, which is used to test the integrity levels L1–L2 of the spinal cord.

    B. Posterior compartment innervation is by the obturator and sciatic nerves.

        1. Obturator nerve

            a. Course and composition. The obturator nerve arises from the posterior division of roots L2–L4 of the lumbar plexus (see Figs. 18-6 and 24-7). It exits the pelvis via the obturator foramen in company with the obturator artery. It divides into two branches.

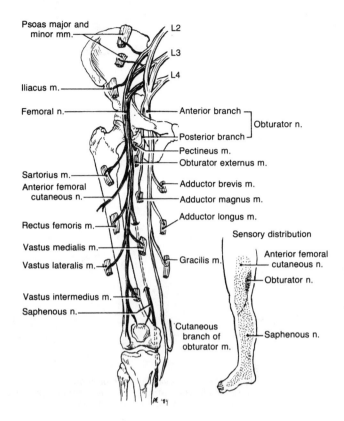

Psoas major and
minor mm.

L2

L3

L4

Iliacus m.

Femoral n.

Anterior branch

Obturator n.

Posterior branch

Pectineus m.

Obturator externus m.

Sartorius m.
Anterior femoral
cutaneous n.

Adductor brevis m.

Adductor magnus m.

Adductor longus m.

Rectus femoris m.

Sensory distribution

Vastus medialis m.

Vastus lateralis m.

Gracilis m.

Anterior femoral
cutaneous n.

Obturator n.

Vastus intermedius m.

Saphenous n.

Cutaneous
branch of
obturator m.

Saphenous n.

Figure 24-7. *Innervation of the anterior compartment of the thigh.* The distribution of the femoral and obturator nerves.

b. Distribution

(1) The **anterior (superficial) branch** is motor to the adductor longus and brevis muscles as well as to the gracilis muscle (see Fig. 24-7). It is sensory to a small area of the distal medial aspect of the thigh.

(2) The **posterior (deep) branch** innervates the obturator externus and the anterior portion of the adductor magnus muscles as well as a portion of the pectineus muscle (see Fig. 24-7). It sends sensory twigs to the hip joint.

2. Sciatic nerve

a. Course and composition.
The sciatic nerve arises from both anterior and posterior divisions of roots L4–S3 of both lumbar and sacral plexuses (see Fig. 20-4).

(1) It exits from the pelvis via the infrapiriform recess of the greater sciatic foramen along with the inferior gluteal neurovascular muscle. It arches through the inferomedial quadrant of the buttock, starting halfway between the ischial tuberosity and the iliac spine, passing over the neck of the femur halfway between the ischial tuberosity and the greater trochanter. It passes through the posterior compartment of the thigh and joins the popliteal artery in the popliteal fossa.

(2) The sciatic nerve is formed by the fusion of the **tibial nerve** and the **common peroneal nerve** (Fig. 24-8). The extent of the fusion is variable.

b. Distribution

(1) The **tibial nerve** (L4–S3, anterior) innervates the semimembranosus and semitendinosus muscles and the long head of the biceps femoris muscle as well as the posterior portion of the adductor magnus muscle (see Fig. 24-8). It has no sensory component in the thigh.

(2) The **common peroneal nerve** (L4–S2, posterior) innervates only one muscle in the thigh, the short head of the biceps femoris muscle (see Fig. 24-8). It has no sensory component in the thigh.

c. Sciatic nerve injury.
The most frequent sciatic nerve injury results from herniation of an intervertebral disk compressing spinal roots that comprise the sciatic nerve [see Chapter 19 I E 4 b (5)]. It may be injured by an intramuscular injection into the inferomedial quadrant of the buttock. Occasionally, the sciatic nerve exits through the piriformis muscle and spasm

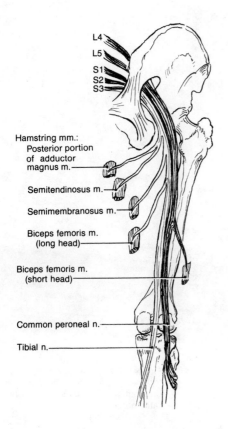

L4
L5
S1
S2
S3

Hamstring mm.:
  Posterior portion
  of adductor
  magnus m.

Semitendinosus m.

Semimembranosus m.

Biceps femoris m.
  (long head)

Biceps femoris m.
  (short head)

Common peroneal n.

Tibial n.

Figure 24-8. *Innervation of the posterior compartment of the thigh.* The distribution of the common peroneal and tibial divisions of the sciatic nerve.

of that muscle may compress the nerve. Injury can also result from traction of the nerve as it passes posteriorly to the neck of the femur. In the thigh, sciatic nerve injury manifests as weakness of knee flexion.

3. **The lateral femoral cutaneous nerve** (L2–L3, posterior) is sensory to the lateral thigh. It may become entrapped beneath the inguinal ligament, especially in an obese person, producing hyperesthesia or pain.

4. **The posterior femoral cutaneous nerve** (S1–S3, anterior and posterior) is sensory to most of the posterior thigh.

25
Anterior Leg
and Dorsal Foot

I. INTRODUCTION

A. Basic principles. The leg is that part of the lower extremity that lies between the knee and ankle.

1. **The primitively dorsal musculature** lies anteriorly. The muscles of the anterior and lateral compartments comprise the dorsiflexors (extensors) of the ankle and the extensors of the pedal digits.

2. **Muscles acting at the ankle and on the foot.** Most of the muscles of the leg cross the ankle joint as well as the joints of the foot. In addition, the gastrocnemius and popliteus muscles cross the knee joint.

B. Bony landmarks

1. **The tibia** has a palpable medial condyle, lateral condyle, anterior lip of the tibial plateau, tibial tubercle, medial surface of the shaft, and medial malleolus.

2. **The fibula** has a palpable head, lateral collateral ligament, and lateral malleolus.

3. **The tarsal bones** (including the calcaneus and navicular bone), metatarsal bones, and phalanges are all palpable.

II. BONES OF THE LEG AND FOOT

A. Leg

1. **Distal tibia**
   a. Characteristics
   (1) The distal tibia corresponds to the radius of the forearm. (The upper and lower extremities have rotated in opposite directions, phylogenetically.)
   (2) The **tibial tubercle** receives the attachment of the patellar tendon. The medial surface or skin is subcutaneous. Distally, the shaft narrows before expanding into the subcutaneous medial malleolus, which is notched posteriorly by the tendon of the tibialis posterior muscle.
   b. Articular surfaces. The distal end articulates with the talus at the **talocrural (ankle) joint**.
   (1) The articular surface is trapezoid in shape (wider anteriorly) and concave. It continues medially as the articular surface of the medial malleolus.
   (2) Laterally, the **fibular notch** accommodates the fibula as the **distal tibiofibular joint**.

2. **Distal fibula**
   a. Characteristics
   (1) The distal fibula corresponds to the ulna of the forearm. In the process of developmental pronation and fixation in the pronated position, it has lost contact with the femur and, thus, takes no part in the knee joint.
   (2) The shaft is narrow. A slightly bulbous head articulates with the lateral tibial condyle at the **proximal tibiofibular joint**. The **styloid process** of the head attaches the lateral collateral ligament. Distally, the shaft expands slightly as the subcutaneous **lateral malleolus**, which extends further distally than the medial malleolus. The posterior surface of the lateral malleolus is grooved by the tendons of the peroneus longus and brevis muscles.
   b. Articular surfaces
   (1) The distal end articulates with the tibia at the **distal tibiofibular joint**.
   (2) On the medial surface is a triangular facet for the talus at the talocrural ankle joint.

3. The interosseous membrane is a broad fibrous ligament connecting the adjacent borders of the tibia and fibula along their entire length. The most distal region is thickened as the **interosseous ligament**. Both of these structures stabilize the tibiofibular joints.

B. Foot. Seven **tarsal bones** and the five **metatarsal bones** comprise the foot.

1. The proximal tarsal group consists of three bones, not in a row (Fig. 25-1).
  a. The talus (astragalus) has a rounded head, a neck, and a cuboid body (Fig. 25-2). Numerous ligaments, but no muscles, attach to the talus.
    (1) The head articulates with the navicular bone (anteriorly) and with the calcaneus (inferiorly).
    (2) The neck has a deep inferior groove, the **sulcus tali**, in which the interosseous ligament of the talocalcanean joint attaches.
    (3) The body has a number of articular surfaces. Superiorly, the **trochlea tali**, which is wider anteriorly, articulates with the tibia. Laterally and medially, the body of the talus articulates with the malleoli. Inferiorly, a flat surface articulates with the calcaneus. Posteriorly, there is a groove for the flexor hallucis longus tendon.
  b. The calcaneus, which forms the heel, is the largest tarsal bone (see Fig. 25-2).
    (1) Posteriorly, the calcaneus receives **calcanean (Achilles) tendon** (see Fig. 25-2). Posteroinferiorly, short plantar-flexor muscles of the foot attach on both sides.
    (2) It has three superior articular surfaces for the talus with the **sulcus tali** separating the middle and posterior surfaces.
    (3) Anteriorly, it articulates with the cuboid bone and has a tubercle for the attachment of the **long plantar ligament** (see Fig. 26-4).
    (4) The **sustentaculum tali** (see Fig. 25-2*B*), a prominent shelf on the anteromedial edge, provides attachment for the **calcaneonavicular (spring) ligament**, which bridges between the calcaneus and navicular bones. Inferiorly, the sustentaculum tali is grooved by the tendon of the flexor hallucis longus muscle.
      (a) The sustentaculum tali and the calcaneonavicular (spring) ligament support the head of the talus (see Fig. 25-2*B*).
      (b) The calcaneonavicular (spring) ligament helps to maintain the **longitudinal plantar arch**. Stretching of this ligament results in *fallen arches* or *flat feet*.

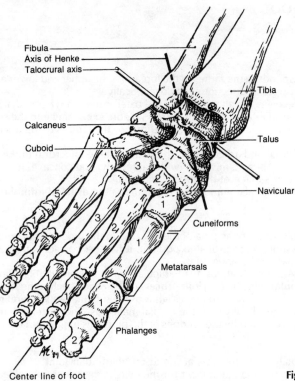

Fibula
Axis of Henke
Talocrural axis
Tibia
Calcaneus
Cuboid
Talus
3
Navicular
5
4
3
Cuneiforms
3 2
1
1
Metatarsals
2
1
Phalanges
3 2
3 2
Center line of foot

Figure 25-1. *Bones of the distal leg and foot.*

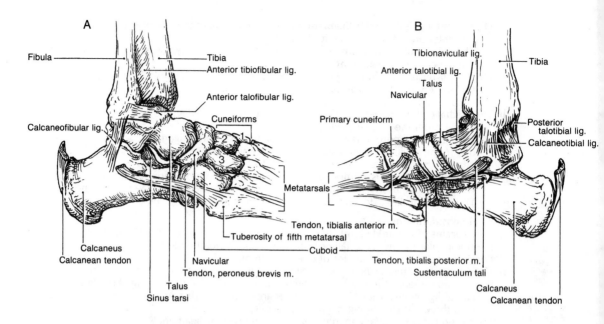

Figure 25-2. *Ligaments of the ankle joint. A,* Lateral. *B,* Medial.

    c. The navicular (scaphoid) bone articulates with all of the tarsal bones, except the calcaneus to which it is strongly connected by the calcaneonavicular (spring) ligament. It is curved posteriorly to receive the head of the talus at the apex of the arch of the foot. It has a prominent tuberosity, which is the principal attachment of the tibialis posterior muscle.

2. The distal tarsal group consists of a row of four bones (see Figs. 25-1 and 25-2).
    a. The three **cuneiform bones (medial, intermediate,** and **lateral)** articulate with the navicular bone, proximally, and with the first three metatarsal bones, distally.
    b. The **cuboid bone** is lateral to the navicular bone and the cuneiform bones. It articulates posteriorly with the calcaneus at the **calcaneocuboid joint** and with the fourth and fifth metatarsals distally.

3. The metatarsals and phalanges (*phalanx, G.* row of soldiers) are rather similar to the metacarpals and phalanges of the hand (see Fig. 25-2).

III. ANKLE AND TARSAL JOINTS

    A. Talocrural (ankle) joint
        1. Structure. The socket for the talocrural joint is formed by the tibia, the two malleoli, and the inferior transverse ligament. These form a mortise for the tendon of the trochlea tali.
            a. The weight of the body is transmitted from the tibia to the talus, which distributes the weight anteriorly and posteriorly within the foot.
            b. The center of gravity in the erect posture passes somewhat anteriorly to the ankle joint.
                (1) The wedge-shaped trochlea tali is forced between the malleoli upon dorsiflexion, thus increasing the inherent stability of the joint.
                (2) The strength of the tibiofibular ligaments prevents the malleoli from separating.
                (3) The fan-shaped collateral ligaments keep the trochlea tali in the joint socket.
        2. Movement. The ankle joint has 1° of freedom about a transverse axis that passes through the lateral malleolus and the trochlea tali (see Fig. 25-1).
            a. It approximates a hinge joint. However, the axis is not in a coronal plane so that the long axis of the foot deviates by about 10° laterally in full dorsiflexion.
            b. Flexion/extension (25°/35°) occurs about this axis.
        3. Ligamentous support. The capsule of the joint is attached to the edges of the articular cartilage and supported by strong collateral ligaments (see Fig. 25-2).
            a. The external collateral ligaments consist of three bands (see Fig. 25-2A).

(1) The **anterior talofibular ligament** passes from the tip of the lateral malleolus to the talus, anteriorly. It tends to limit plantar-flexion.

(2) The **calcaneofibular ligament** passes from the lateral malleolus to the calcaneus with the **talocalcanean ligament** running at its base.

(3) The **posterior talofibular ligament** passes from the tip of the lateral malleolus to the talus, posteriorly. The **posterior talocalcanean ligament** extends this band to the calcaneus. It tends to limit dorsiflexion.

b. The medial collateral (deltoid) ligament has superficial and deep portions (see Fig. 25-2*B*), which run from the medial malleolus.

(1) The superficial portion is composed of the **tibionavicular ligament**, which runs anteriorly to the navicular bone and the calcaneotibial ligament, which runs medially to the edge of the calcaneus. These prevent abduction.

(2) The deep portion is comprised of the **anterior** and **posterior talotibial ligaments**, which run anteriorly and posteriorly between the medial malleolus and the talus. These tend to limit dorsiflexion and plantar-flexion, respectively.

4. Clinical considerations

a. True sprains of the ankle joint. Because the talocrural joint is inherently stable, tearing of the tibiofibular ligaments usually occur as a result of violent adduction and internal rotation in association with fracture of one or both malleoli.

b. Ligamentous sprains usually result from excessive movement at the subtalar joint.

(1) *Inversion sprains*, the most common ankle injury, involve the tearing of the lateral collateral ligaments. There is usually a progression of tearing from anterior to posterior with three corresponding degrees of severity.

(2) *Eversion sprains* usually involve tearing of the deltoid ligament.

c. Midshaft and ''boot-top'' fractures of the tibia and fibula are common. Usually the bones override somewhat due to muscular pull.

d. Distal fractures of the tibia and fibula are the result of severe external rotation, and abduction may fracture the tibia just proximal to the malleoli. This is a typical skiing fracture, particularly with the new plastic boots, because the bones rotate within the boot.

B. The subtalar joint is formed where the talus rests on the calcaneus.

1. Structure. There are three plantar joints at which rotational gliding motions occur.

2. Movement. The movement of this joint can be described as abduction/adduction. However, because the anterior talocalcanean articulation is part of the transverse tarsal joint, the movements are conjoined so that abduction/adduction does not occur in pure form.

3. Ligamentous support. The calcaneofibular and the deltoid ligaments traverse this joint, and the talus and calcaneus are connected by three major ligaments:

a. Lateral talocalcanean ligament

b. Medial talocalcanean ligament

c. Interosseous talocalcanean ligament. This ligament runs from deep within the sulcus tali to the sulcus calcanei and provides an axis of rotation about which movement in the subtalar joint occurs.

4. Sprains. Excessive movements in this joint will produce ligamentous sprains. Severe trauma causes more widespread injury and may produce fractures of the malleoli and ligamentous tears involving the talocrural joint.

a. Abduction stresses the medial aspect of the joint.

b. Adduction stresses the lateral aspect of the joint.

C. Transverse (mid) tarsal joint consists of two conjoined joints.

1. The talocalcaneonavicular joint is formed by the articulation of the ovoid head of the talus with a socket formed by the navicular bone, the calcaneonavicular (spring) ligament, and the anterior articular surface of the calcaneus.

a. Structure. This is a ball-and-socket joint.

b. Movement in this joint can be described as a rotational supination and pronation.

(1) Because the talus and calcaneus also participate in the subtalar joint, the movements are conjoined so that pronation/supination does not occur in pure form.

(2) A slight amount of conjoined dorsiflexion/plantar-flexion also occurs in this joint.

c. Ligamentous support. Strong ligaments support this joint.

(1) The **talonavicular ligament** and **dorsal calcaneonavicular ligament** support the joint dorsally.

(2) The **plantar calcaneonavicular (spring) ligament** supports the head of the talus. Laxity of this ligament results in *fallen arches* or *flat feet*.

(3) Numerous other ligaments provide additional support.

2. The calcaneocuboid joint is part of the transverse tarsal joint.
    a. Structure. This is a gliding joint at which movement is an accommodation to the movements that occur conjointly at the subtalar and talocalcaneonavicular joints.
    b. Ligamentous support is by the **long plantar ligament**.

D. Cuneonavicular, cuneocuboid, intercuneiform, and tarsometatarsal joints

1. Movement. These ligaments permit small gliding movements of accommodation that change the shape of the **transverse plantar arch**.

2. Support. They are all supported by a series of strong dorsal and plantar ligaments.

E. Metatarsophalangeal joints

1. Movement. These joints have two degrees of freedom, permitting flexion/extension and abduction/adduction.
    a. Unlike the hand and as an accommodation to walking, extension of the pedal digits (55°) is greater than flexion (35°).
    b. Also unlike the hand, abduction/adduction take place about a long axis through the second toe.
    c. Unlike the thumb, the great toe is not capable of opposition.

2. Support. They are supported by **deep transverse metatarsal ligaments** and **collateral ligaments**.

F. Interphalangeal articulations

1. Movement. These are basic hinge joints with one degree of freedom, permitting flexion and extension.

2. Support. They are supported by collateral ligaments.

G. Conjoined movements of the foot

1. Movements at the subtalar and transverse tarsal joints (posterior tarsal joints) are mechanically linked by common joints so that they function together with but one degree of freedom.
    a. Inversion of the foot is produced by the conjoined movements of adduction and supination (and to some extent dorsiflexion).
    b. Eversion of the foot is produced by the conjoined movements of abduction and pronation (and to some extent plantar-flexion).
    c. The resultant axis of rotation (of Henke) about which inversion/eversion occurs passes inferomedially through the lateral side of the neck of the talus, through the sinus tali, along the interosseous ligament to the sinus calcanei, and emerges from the medial aspect of the posterior tubercles of the calcaneus (see Fig. 25-1).

2. Movements at the anterior tarsal joints are also conjoined because the articulations are mechanically linked.
    a. The movements are primarily ones of accommodation to movements in the posterior tarsal joints.
    b. Because of the shape of the articular surfaces, these movements change the shape of the **transverse plantar arch**.
        (1) Eversion and dorsiflexion tend to flatten the transverse plantar arch.
        (2) Inversion and plantar-flexion tend to heighten the transverse arch.

IV. ANTERIOR AND LATERAL CRURAL COMPARTMENTS

A. Fascia of the leg

1. The superficial fascia of the leg is continuous with that of the thigh. It supports the cutaneous nerves and superficial veins.

2. The deep (crural) fascia is continuous with the fasciae latae of the thigh posteriorly. It also would be continuous anteriorly if both did not attach to the patella.
    a. Anteromedially, the crural fascia is attached to the medial surface of the tibia.
    b. Laterally, two septa run to the anterior and posterior borders of the fibula, the anterior and posterior intermuscular septa.
        (1) The **posterior intermuscular septum**, the fibula, the interosseous membrane, and the tibia divide the leg into **anterior/lateral** and **posterior compartments**.
        (2) The **anterior intermuscular septa** subdivide the extensor compartment into an **anterior compartment** and a **lateral compartment**.

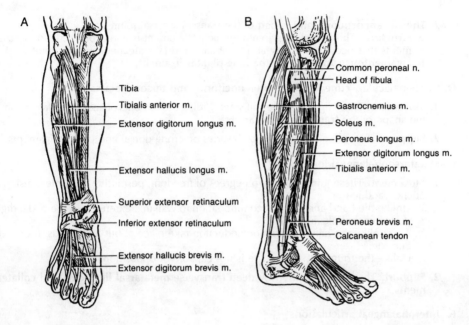

Figure 25-3. *Musculature of the anterior and lateral crural compartments. A,* Anterior compartment muscles. *B,* Lateral compartment muscles.

   c. At the ankle, the crural fascia condenses to form retinacula for the tendons of the extensor, flexor, and peroneal muscles (Fig. 25-3).

     (1) The **superior extensor retinaculum** (transverse crural ligament) passes between the distal shafts of the tibia and fibula and contains the tibialis anterior, the extensor hallucis longus, the extensor digitorum longus tendon, and the peroneus tertius muscle.

     (2) The **inferior extensor retinaculum** is **Y**-shaped (see Fig. 25-5). It diverges from a common lateral attachment on the calcaneus to both the medial malleolus and the deep fascia of the medial aspect of the foot. Unlike the superior extensor retinaculum, the inferior extensor retinaculum is subdivided into compartments for the individual tendons.

     (3) The **superior peroneal retinaculum** runs from the lateral malleolus to the calcaneus. The **inferior peroneal retinaculum** is a continuation of the inferior extensor retinaculum over the lateral surface of the calcaneus. It is attached to the peroneal trochlea between the peroneus longus and the peroneus brevis.

B. Musculature

   1. Organization. The primitively dorsal musculature of the leg is divided by the anterior intermuscular septum into an anterior compartment and a lateral compartment. Because most of the muscles of the anterior and lateral compartments continue into the foot, the leg and foot may be considered as a functional unit.

    a. The anterior crural compartment contains the dorsiflexors of the ankle as well as both invertors and evertors of the foot. This group includes four muscles, all of which arise from the lateral tibial condyle as well as from the proximal two-thirds of the tibial and fibular shafts and the interosseous membrane (see Figs. 25-3*A*, 25-5, and 26-4*B*).

     (1) The **tibialis anterior** inserts on the medial aspect of the medial cuneiform bone (see Figs. 25-2*B* and 25-5).

     (2) The **extensor hallucis longus** inserts into the distal phalanx of the great toe (see Fig. 25-5).

     (3) The **extensor digitorum longus** divides into four slips, which insert into the **extensor expansion (hood)** of the second through fifth digits. The extensor expansion then attaches over the dorsal surfaces of the middle and distal phalanges (see Fig. 25-5).

     (4) The **peroneus tertius** inserts into the bases of the fourth and fifth metatarsals (see Fig. 25-5).

    b. The lateral crural compartment contains the peroneal muscles, which are plantar flexors of the ankle as well as the principal evertors of the foot. The peroneal muscles arise from the fibular shaft (see Figs. 25-3*B* and 25-5).

     (1) The **peroneus longus** tendon grooves the posterior surface of the lateral malleolus. It crosses the lateral and inferior surfaces of the calcaneus to insert into the medial cunei-form bone and the first metatarsal bone (see Figs. 25-3 and 26-4*B*).

(2) The **peroneus brevis** tendon parallels that of the peroneus longus but inserts into the fifth metatarsal bone (see Figs. 25-2*A*, 25-3, and 26-4*B*).

2. **Group actions** are determined by the location of the tendons relative to the axis of rotation of the ankle joint and that of the foot (Table 25-1).

   a. **Flexion/extension** occurs about a transverse axis through the ankle joint. (Fig. 25-4). Muscles that pass anteriorly to the transverse axis dorsiflex the foot at the ankle; those that pass posteriorly to this axis plantar-flex the foot at the ankle.

      (1) **Dorsiflexion.** All of the muscles of the anterior compartment pass anteriorly to the transverse axis and are dorsiflexors of the foot.

      (2) **Plantar-flexion.** Both muscles of the lateral compartment pass behind the lateral malleolus and are, thus, posterior to the transverse axis so that they are plantar-flexors of the foot.

   b. **Inversion/eversion** occurs about the complex axis of Henke (see Fig. 25-4). Muscles that pass medially to this axis are invertors; muscles that pass laterally to it are evertors.

      (1) **Inversion.** The medial muscles of the anterior compartment (tibialis anterior and extensor hallucis longus) are both invertors of the foot.

      (2) **Eversion.** Both muscles of the lateral compartment are strong evertors of the foot. The medial muscles of the anterior compartment (extensor digitorum longus and peroneus tertius) are also evertors.

3. **Group innervations**

   a. **Anterior compartment innervation** is by the **deep (anterior tibial) branch** of the **common peroneal nerve.**

   b. **Lateral compartment innervation** is by the **superficial branch** of the **common peroneal nerve.**

Table 25-1. Anterior and Lateral Compartment Muscles Acting at the Ankle and Foot Joints

Muscle	Origin	Insertion	Primary Action	Innervation
Anterior Compartment				
Tibialis anterior	Proximal half of anterior tibia	Medial cuneiform bone and base of first metatarsal	Dorsiflexion and inversion	Deep peroneal n. (L4–L5, posterior)
Extensor hallucis longus	Front of fibula	Base of second phalanx of great toe	Extends terminal phalanx of great toe	Deep peroneal n. (L4–S1, posterior)
Extensor digitorum longus	Lateral condyle of tibia, interosseous membrane, and fibula	Second and third phalanges of second to fifth toes	Extends metatarso-phalangeal joints	Deep peroneal n. (L4–S1, posterior)
Peroneus tertius	Anterior fibula	Base of fifth metatarsal	Dorsiflexes and everts foot	Deep peroneal n. (L5–S1, posterior)
Lateral Compartment				
Peroneus longus	Proximal half of fibula	Medial cuneiform bone and base of first metatarsal	Plantar-flexion and eversion of foot	Superficial peroneal n. (L5–S1, posterior)
Peroneus brevis	Distal half of fibula	Base of fifth metatarsal	Plantar-flexion and eversion of foot	Superficial peroneal n. (L5–S1, posterior)
Dorsum of Foot				
Extensor hallucis brevis	Calcaneus	Base of proximal phalanx of great toe	Extends proximal phalanx of great toe	Deep peroneal n. (L4–S1, posterior)
Extensor digitorum brevis	Calcaneus	Extensor hood of second to fourth toes	Extends metatarso-phalangeal joints of second, third, and fourth toes	Deep peroneal n. (L4–S1, posterior)

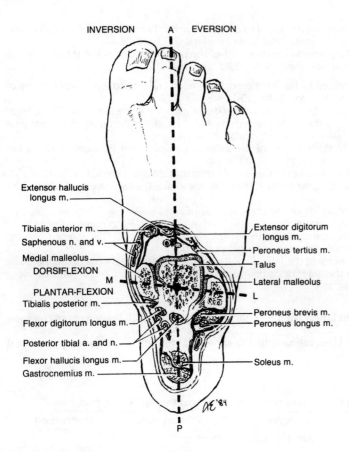

INVERSION A EVERSION

Extensor hallucis longus m.

Tibialis anterior m.
Saphenous n. and v.
Medial malleolus
DORSIFLEXION
                M
PLANTAR-FLEXION
Tibialis posterior m.

Flexor digitorum longus m.

Posterior tibial a. and n.

Flexor hallucis longus m.
Gastrocnemius m.

Extensor digitorum longus m.
Peroneus tertius m.
Talus
Lateral malleolus
                L
Peroneus brevis m.
Peroneus longus m.

Soleus m.

AE '84

P

Figure 25-4. *Action of the crural muscles at the ankle joint.* Those muscles passing anteriorly to the transverse axis of the ankle joint (M–L) dorsiflex the foot; those passing posteriorly to this axis plantarflex the foot. Those muscles passing medially to the resultant axis (A–P) of the subtalar and midtarsal joints invert the foot; those passing laterally to this axis evert the foot.

V. DORSUM OF THE FOOT

A. The crural (deep) fascia of the foot is divided into two layers by the extensor tendons of the foot. The two layers fuse at the margins of the foot.

B. Musculature. The dorsum of the foot, unlike the hand, has two short dorsiflexor muscles.

1. Organization
   a. The extensor digitorum brevis muscle arises from the calcaneus and divides into three slips, which insert into the lateral margins of the extensor hoods of the second, third, and fourth digits (Fig. 25-5).
   b. The extensor hallucis brevis muscle arises, along with the extensor digitorum brevis, from the calcaneus and inserts into the base of the proximal phalanx of the great toe.

2. Group actions. The extensor digitorum brevis assists the action of the extensor digitorum longus in dorsiflexion of the metatarsophalangeal joints. The extensor hallucis brevis is an extensor of the proximal phalanx of the great toe.

3. Group innervations. The intrinsic muscles of the dorsal foot are usually (78%) innervated by the deep peroneal nerve but may be innervated by the superficial peroneal nerve (22%).

VI. VASCULATURE OF THE ANTERIOR AND LATERAL COMPARTMENTS AND DORSUM OF THE FOOT

A. Arteries

1. Anterior tibial artery
   a. Course. The anterior tibial artery arises from the popliteal artery in the popliteal fossa (see Fig. 24-6). It passes inferiorly in the popliteal fossa into the superior portion of the posterior compartment of the leg. Passing through a gap in the interosseous membrane, the anterior

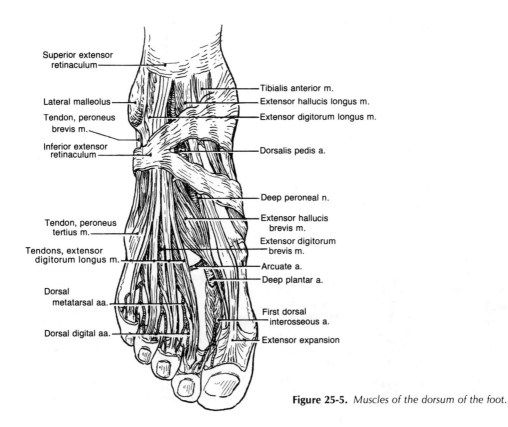

Figure 25-5. *Muscles of the dorsum of the foot.*

tibial artery enters the anterior crural compartment. It descends anteriorly to the interosseous membrane in company with the deep fibular (peroneal) branch of the common fibular (peroneal) nerve (Fig. 25-6).

  b. Distribution

    (1) The anterior tibial artery gives rise to recurrent branches, which communicate with the geniculate anastomosis.

    (2) At the ankle, it gives off medial and lateral malleolar branches, which anastomose with branches of the posterior tibial and fibular (peroneal) arteries (see Fig. 25-6).

  2. The dorsal pedal (dorsalis pedis) artery is the continuation of the anterior tibial artery onto the dorsum of the foot (see Fig. 25-6).

    a. Course. The dorsal pedal artery lies between the tendons of the extensor hallucis longus and extensor digitorum longus muscles where the **dorsal pedal pulse** may be palpated. It runs toward the great toe and terminates as the deep plantar artery.

    b. Distribution (see Fig. 25-6)

      (1) Lateral and **medial tarsal arteries** supply the tarsus.

      (2) The **arcuate artery** courses across the metatarsals, giving off **dorsal metatarsal arteries**, each of which bifurcate into **dorsal digital arteries**.

      (3) The **first dorsal metatarsal artery** supplies adjacent sides of the first and second toes.

      (4) The **deep plantar artery** dives between the heads of the first interosseous muscle to the plantar aspect of the foot and joins with the lateral plantar branch of the posterior tibial artery to form the **plantar arterial arch**.

B. Veins (see Chapter 16 IV C)

VII. INNERVATION OF THE ANTERIOR AND LATERAL CRURAL COMPARTMENTS BY THE COMMON PERONEAL NERVE

  A. Course. The common peroneal nerve arises from the posterior portions of the lumbar and sacral plexuses (L4–S3) and forms the posterior division of the **sciatic nerve** (see Chapter 24 VI B 2 a).

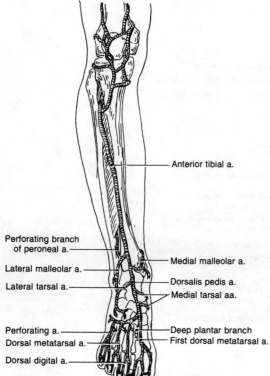

Anterior tibial a.

Perforating branch of peroneal a.

Lateral malleolar a.

Lateral tarsal a.

Medial malleolar a.

Dorsalis pedis a.

Medial tarsal aa.

Perforating a.

Deep plantar branch

First dorsal metatarsal a.

Dorsal metatarsal a.

Dorsal digital a.

Figure 25-6. *Vasculature of the anterior crural compartment and dorsum of the foot.*

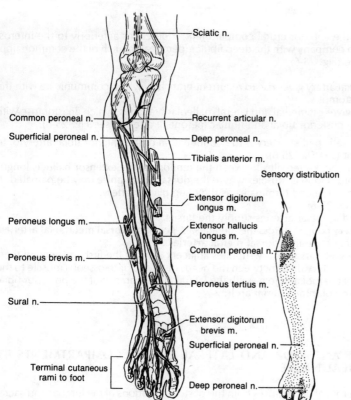

Sciatic n.

Common peroneal n.

Recurrent articular n.

Superficial peroneal n.

Deep peroneal n.

Tibialis anterior m.

Sensory distribution

Extensor digitorum longus m.

Peroneus longus m.

Extensor hallucis longus m.

Common peroneal n.

Peroneus brevis m.

Peroneus tertius m.

Sural n.

Extensor digitorum brevis m.

Superficial peroneal n.

Terminal cutaneous rami to foot

Deep peroneal n.

Figure 25-7. *Distribution of the common peroneal nerve in the leg and foot.*

1. **In the popliteal fossa** the sciatic nerve bifurcates into common peroneal and tibial nerves (Fig. 25-7). The common peroneal nerve diverges laterally in the popliteal fossa and gives rise to one or more articular branches.

2. **Passing posteriorly to the head of the fibula** (from which this nerve gets its name: *peroneus*, *L.* pin) and laterally across the neck of that bone, the common fibular (peroneal) nerve divides into the **sural communicating nerve, superficial fibular (peroneal) nerve**, and **deep fibular (peroneal, anterior tibial) nerve.**

B. Distribution

1. **Motor**
   a. **The superficial fibular nerve** innervates all of the muscles of the lateral compartment.
   b. **The deep fibular (anterior tibial) nerve** innervates all of the muscles of the anterior compartment and usually the intrinsic muscles of the dorsum of the foot.

2. **Sensory**
   a. **The sural communicating nerve** contributes the lateral half of the **sural nerve** (see Fig. 26-6).
   b. **The superficial fibular nerve** sends a terminal cutaneous branch to the dorsum of the foot (see Fig. 25-7).
   c. **The deep fibular nerve** sends a terminal cutaneous branch to a small area between the first and second toes, which is specific for L5 (see Fig. 25-7).

C. Injury to the common fibular nerve results in *foot drop*. There is an inability to dorsiflex the foot or to stand back on the heels as well as a loss of sensation along the lateral aspect of the leg and dorsum of the foot. Fibular nerve palsy produces a characteristic gait with *"foot-slap."*

1. **Proximal injury** to this nerve may result from herniation of an intervertebral disk or intramuscular injection into the inferomedial quadrant of the buttock.

2. **Distal injury** may result from pressure where the common fibular nerve passes across the fibula. This pressure can be as frank as blunt trauma or fracture of the proximal fibula. It may be as subtle as the constant pressure exerted by the edge of a plaster cast, which terminates just below the knee, or by sitting with the knees crossed for a prolonged period.

26
Posterior Leg and Plantar Foot

I. INTRODUCTION

A. Basic principles. The primitively dorsal aspects of the leg and foot may be considered as a functional unit.

1. **The primitively ventral musculature** lies posteriorly. The muscles of the posterior compartment comprise the plantar-flexors of the ankle and the flexors of the pedal digits.

2. **Muscles acting at the ankle and on the foot.** Most of the muscles of the posterior crural compartment cross the ankle joint as well as the joints of the foot. In addition, the gastrocnemius and popliteus muscles cross the knee joint.

B. Bony landmarks are discussed in Chapter 25 I B.

C. Bones of the leg are discussed in Chapter 25 II A.

D. Bones of the foot are discussed in Chapter 25 II B.

II. POSTERIOR CRURAL COMPARTMENT

A. Fascia

1. **The superficial fascia** of the leg is continuous with that of the thigh. It supports the cutaneous nerves and superficial veins.

2. **The deep (crural) fascia** is continuous with the fascia lata of the thigh posteriorly and envelops the muscles of the leg.
   a. **The posterior intermuscular septum** (a septum of the crural fascia), tibia, interosseous membrane, and fibula together form a barrier that separates the **posterior compartment** from the **anterior compartment**.
   b. **The transverse crural septum**, another septum of the crural fascia, subdivides the **posterior compartment** into superficial and deep compartments.
   c. **The flexor retinaculum (tarsal tunnel)** is a condensation of the crural fascia on the medial side of the ankle, extending between the medial malleolus and the calcaneus.
      (1) It is subdivided into four compartments.
      (2) It tethers the tibialis posterior muscle, the flexor digitorum longus muscle, the posterior tibial artery and nerve, and the flexor hallucis longus. (The mnemonic here is **T**om, **D**ick, **AN**d **H**arry.)

3. **The plantar fascia** is a continuation of the deep fascia onto the foot.

B. Musculature

1. **Organization.** The flexor muscles of the posterior compartment (calf) are divided into superficial and deep groups by the transverse crural septum.
   a. **The superficial group** consists of the **triceps surae** (Table 26-1).
      (1) This group is comprised of three muscles (Fig. 26-1).
         (a) The **gastrocnemius muscle** has two heads, which arise from the posterior aspects of the femoral epicondyles and define the lower limits of the popliteal fossa.
         (b) The **plantaris muscle** is very small and arises from the distal femur.
         (c) The **soleus muscle** arises from the proximal posterior surfaces of the fibula and tibia.

Table 26-1. Posterior Compartment Muscles Acting at the Ankle and Foot Joints

Muscle	Origin	Insertion	Primary Action	Innervation
Superficial Group				
Gastrocnemius	Posterior surfaces of medial and lateral femoral condyles	Calcaneus	Plantar-flexion	Tibial n. (L5–S2, anterior)
Soleus	Proximal tibia, interosseous membrane, and fibula	Calcaneus	Plantar-flexion	Tibial n. (L5–S2, anterior)
Plantaris	Lateral supra-condylar ridge	Calcaneus	Plantar-flexion	Tibial n. (L4–S1, anterior)
Deep Group				
Popliteus	Lateral femoral condyle	Medially on proximal posterior tibia	Medially rotates and flexes knee	Tibial n. (L4–S1, anterior)
Flexor hallucis longus	Distal third of fibula	Phalanges of great toe	Flexes great toe	Tibial n. (L5–S2, anterior)
Tibialis posterior	Posterior shafts of tibia, fibula, and interosseous membrane	Navicular and medial cuneiform bones	Plantar-flexion and inversion of foot	Tibial n. (L5–S1, anterior)
Flexor digitorum longus	Posterior surface of tibial shaft	Distal phalanges of second through fifth toes	Flexes distal phalanges of second through fifth toes	Tibial n. (L5–S2, anterior)

    (2) The **calcanean (Achilles) tendon** is formed by the tendons of the triceps surae. It inserts into the calcaneus (see Figs. 25-2 and 26-1). *Achilles tendinitis,* caused by repeated stress upon heel strike, may progress to degeneration and rupture of the tendon.

    (3) These muscles are innervated by the tibial nerve (L4–L5) and plantar-flex the foot (see Fig. 25-4).

  b. The deep group consists of four muscles (Fig. 26-2; see Table 26-1).

    (1) The **popliteus muscle** forms the floor of the popliteal fossa.

      (a) It arises from the lateral femoral condyle and inserts into the posterior medial aspect of the proximal tibia (see Fig. 24-1*B*).

      (b) It acts at the knee either to rotate the tibia medially or to rotate the femur laterally, depending upon which bone is fixed, thereby unlocking and initiating flexion of the knee.

    (2) The **flexor hallucis longus** originates along the posterior aspect of the fibula and inserts into the distal phalanx of the great toe.

      (a) The distal tendon of this muscle is associated with two sesamoid bones where it passes under the distal head of the first metatarsal.

      (b) It is powerful and plays an important role in the propulsive (toe-off) phase of gait.

    (3) The **flexor digitorum longus** arises from the interosseous membrane and posterior surface of the tibia then divides into four slips, which insert into the distal phalanges of the second through fifth toes.

    (4) The **tibialis posterior muscle**, the deepest of the posterior group, arises from the tibia, interosseous membrane, and fibula to insert into the plantar surface of the cuneiform bones with slips to the second through fourth metatarsals (see Figs. 25-2*B* and 26-4*B*). *Shin splints* are most likely tendinitis at the tibialis posterior origin on the tibia and interosseous membrane. The cause is supposedly an imbalance that may be corrected by using different running shoes or orthotics.

  2. Group actions

    a. Flexion/extension occurs about a transverse axis through the talocrural joint.

      (1) Plantar-flexion. All of the muscles of the posterior compartment pass posteriorly to the transverse axis of the talocrural joint and, therefore, are plantar-flexors. In this action, they are synergistic with the muscles of the lateral compartment.

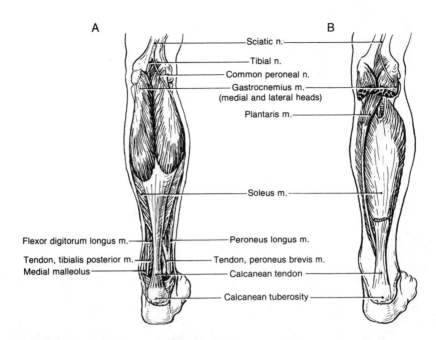

Figure 26-1. *Superficial musculature of the posterior crural compartment. A, Gastroc-nemius muscle. B, Soleus and plantaris muscles.*

    (2) Dorsiflexion is a function of anterior compartment musculature [see Chapter 25 IV B 2 a (1)].
  b. Inversion/eversion occurs about the axis of Henke, which is the resultant axis of the combined subtalar and midtarsal joints.
    (1) Inversion. All of the muscles of the posterior compartment pass medially to the resultant axis (of Henke) and, therefore, are invertors of the foot.
    (2) Eversion is primarily a function of lateral compartment musculature [see Chapter 25 IV B 2 b (2)].
3. Group innervations. The muscles of the posterior compartment are innervated by the **tibial nerve**.

III. PLANTAR FOOT

A. Fascia

  1. The superficial fascia of the leg continues onto the foot and supports the cutaneous nerves and superficial veins.

  2. The plantar (deep) fascia of the foot is continuous with the crural fascia.
    a. Septa run between the skin on the sole of the foot and the plantar fascia, limiting the mobility of the plantar skin.
    b. The **plantar aponeurosis** is a central thickening of the plantar fascia.
      (1) It is attached to the calcaneus and divided into five slips, each of which bifurcates into superficial and deep fascicles. The superficial fascicle of each slip inserts into the skin of the toes; the deep fascicle divides to insert into the bases of the proximal phalanx of the second through fifth digits.
      (2) Strong septa arise from this layer to divide the foot into muscular compartments.
      (3) This layer is instrumental in maintaining the longitudinal plantar arch, along with the calcaneonavicular (spring) ligament and the deep plantar ligament.

B. Intrinsic plantar muscles can be grouped according to two methods.

  1. Organization. The plantar muscles are more commonly divided into four layers (Table 26-2).

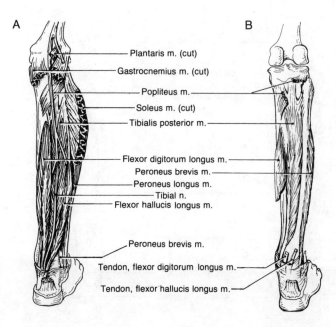

Figure 26-2. *Deep musculature of the posterior crural compartment.* A, Flexor digitorum longus and flexor hallucis longus muscles. B, Tibialis posterior muscle.

Labels in figure:
A / B
Plantaris m. (cut)
Gastrocnemius m. (cut)
Popliteus m.
Soleus m. (cut)
Tibialis posterior m.
Flexor digitorum longus m.
Peroneus brevis m.
Peroneus longus m.
Tibial n.
Flexor hallucis longus m.
Peroneus brevis m.
Tendon, flexor digitorum longus m.
Tendon, flexor hallucis longus m.

a. The first layer includes the **abductor hallucis brevis**, **flexor digitorum brevis**, and **abductor digiti minimi** (Fig. 26-3*A*).

   (1) The abductor hallucis brevis and abductor digiti minimi both arise from the calcanean tuberosity and insert into medial and lateral sides, respectively, of the first phalanx of the great and fifth toes. Their names describe their principal actions.

   (2) The flexor digitorum brevis originates from the calcaneus and divides into four slips, which run to the second through fifth toes, where each splits to insert into the sides of the middle phalanx. In a manner similar to that of the flexor digitorum superficialis of the upper extremity, the split insertions of the flexor digitorum brevis provide passageways for the tendons of the extrinsic flexor digitorum longus. As indicated by its name it flexes the second phalanges.

b. The second layer includes the **quadratus plantae** and four **lumbrical muscles** as well as the tendon sheaths of the flexor hallucis longus and flexor digitorum longus (see Fig. 26-3*B*).

   (1) The quadratus plantae (accessory flexor digitorum) runs from the calcaneus to the distal phalanges of the second through fifth toes, which it flexes.

   (2) The lumbricals are numbered (1–4) from the medial side of the foot. They arise from the tendons of the flexor digitorum longus and insert into the dorsal digital expansions over the proximal phalanges. Thus, they flex the metatarsophalangeal joint and extend the proximal and distal interphalangeal joints.

   (3) The tendon sheaths of the flexor hallucis longus and flexor digitorum longus muscles lie in the second layer. The powerful flexor hallucis longus, passing under the sustentaculum tali and the first metatarsal, bowstrings beneath, thereby supporting the longitudinal plantar arch.

c. The third layer consists of the **flexor hallucis brevis**, **adductor hallucis**, and **flexor digiti minimi** (Fig. 26-4*A*). This layer also contains the **long plantar ligament** and the deeper **short plantar ligament**.

   (1) The principal actions of the flexor hallucis brevis, adductor hallucis, and the flexor digiti minimi are indicated by their names.

   (2) The long plantar ligament spans from the calcanean tuberosity to the lateral three metatarsals; the deeper short plantar ligament spans from the calcaneus to the cuboid bones. These ligaments are important supports of the longitudinal plantar arch.

d. The fourth layer consists of the **dorsal and plantar interossei**, as well as the tendons of the **tibialis posterior** and **peroneus longus** muscles (see Fig. 26-4*B*).

   (1) The dorsal interossei arise from the metatarsals and insert into both sides of the second toe and the lateral sides of the third and fourth toes; they abduct the second through fourth toes away from the center line of the foot (see Fig. 25-1), which passes through the long axis of the second toe.

   (2) The plantar interossei arise from the medial side of the third, fourth, and fifth metatarsals and insert into the medial sides of the same digits; they adduct their respective toes toward the second toe (center line).

Table 26-2. Intrinsic Muscles of the Plantar Foot

Muscle	Origin	Insertion	Primary Action	Innervation
First Layer				
Abductor hallucis	Medial tubercle of calcaneus	Base of first phalanx of great toe	Abducts great toe	Tibial n. (L5–S1, anterior)
Flexor digitorum brevis	Medial tubercle of calcaneus	Splits to sides of middle phalanx of second through fifth toes	Flexes middle phalanx of second through fifth toes	Tibial n. (L5–S1, anterior)
Abductor digiti minimi	Calcaneus	Lateral side of first phalanx of fifth toe	Abducts fifth toe	Tibial n. (S1–S2, anterior)
Second Layer				
Quadratus plantae	Calcaneus	Tendon of flexor digitorum longus	Flexes distal phalanx of second through fifth toes	Tibial n. (S1–S2, anterior)
Lumbricals	Tendons of flexor digitorum longus	Medial sides of phalanx of second through fifth toes	Flex proximal phalanx and extend middle and distal phalanges of second through fifth toes	Tibial n. (L5–S2, anterior)
Third Layer				
Flexor hallucis brevis	Cuneiforms and cuboid	Base of first phalanx of great toe	Flexes proximal phalanx of great toe	Tibial n. (L5–S1, anterior)
Adductor hallucis:				
Oblique head	Calcaneus	Base of first phalanx of great toe	Adduct great toe	Tibial n. (S1–S2, anterior)
Transverse head	Metatarsophalangeal ligaments			
Flexor digiti minimi brevis	Base of fifth metatarsal	Base of first phalanx of fifth toe	Flexes proximal phalanx of fifth toe	Tibial n. (S1–S2, anterior)
Fourth Layer				
Interossei:				
Dorsal (4)	Between adjacent sides of metatarsals	First and second to sides of second toe; third and fourth to sides of third and fourth toes	Abduct second through fourth toes from the second toe	Tibial n. (S1–S2, anterior)
Plantar (3)	Medial sides of third through fifth metatarsals	Medial side of corresponding toe	Adduct third through fifth toes toward the second toe	Tibial n. (S1–S2, anterior)

(3) The tendons of the tibialis posterior and peroneus longus muscles also lie in this layer. These muscles, coming from opposite sides of the foot and inserting onto the same bones, act as a sling to provide dynamic support for the longitudinal plantar arch. Weakness of these muscles may result in stretching of the ligaments and *flat feet*.

2. Alternate organization. In a manner similar to the muscles of the hand, the plantar muscles may be grouped into a **medial plantar group** associated with the great toe, an **intermediate group** associated with the second through fifth digits, and a **lateral group** associated with the fifth digit alone.

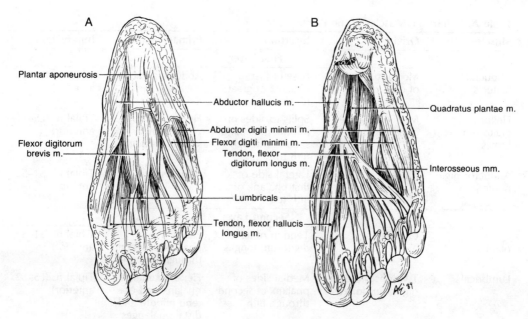

Plantar aponeurosis

Abductor hallucis m.

Quadratus plantae m.

Flexor digitorum
brevis m.

Abductor digiti minimi m.
Flexor digiti minimi m.
Tendon, flexor
digitorum longus m.

Interosseous mm.

Lumbricals

Tendon, flexor hallucis
longus m.

Figure 26-3. *Superficial muscles of the plantar foot. A, First layer. B, Second layer.*

IV. VASCULATURE OF THE POSTERIOR CRURAL COMPARTMENT AND PLANTAR FOOT

A. Arteries

  1. Posterior tibial artery

    a. Course. A direct continuation of the popliteal artery, the posterior tibial artery descends medially in the posterior compartment. It continues posteriorly to the medial malleolus where the **posterior tibial pulse** is normally palpable. The posterior tibial artery then terminates by bifurcating into the plantar arteries (Fig. 26-5).

    b. Distribution. The **peroneal artery** arises from the posterior tibial artery in the upper third of the leg and descends laterally in the posterior compartment between the tibialis posterior and flexor hallucis longus muscles (see Fig. 26-5).

      (1) Proximally, it gives off several branches to the muscles of the posterior and lateral crural compartments.

      (2) Distally in the leg and at the ankle, several communicating branches anastomose with the posterior tibial artery (see Figs. 25-6 and 26-5). In particular, the **posterior lateral malleolar branch** passes laterally about the ankle and anastomoses with the anterior lateral malleolar branch of the anterior tibial artery. The ankle joint has a profuse collateral circulation.

      (3) Occasionally, the peroneal artery is especially large and gives rise to the dorsalis pedis artery via a perforating branch.

  2. Lateral plantar artery

    a. Course. As the terminal branch of the posterior tibial artery, the lateral plantar artery deviates laterally across the long plantar arch and then turns medially across the bases of the metatarsal bones (see Fig. 26-5).

    b. Distribution

      (1) The lateral plantar artery joins with the deep plantar branch of the dorsalis pedis artery in the first interosseous space to form the **plantar arch** (see Fig. 26-5).

      (2) It gives off the **proper digital artery** to the lateral side of the fifth toe and four **plantar metatarsal arteries**. Each plantar metatarsal artery communicates with the corresponding branch of the dorsal metatarsal artery and bifurcates into **proper digital arteries** to supply the sides of adjacent toes (see Figs. 25-6 and 26-5).

  3. Medial plantar artery. The medial plantar artery is smaller than the lateral plantar artery.

    a. Course. The medial plantar artery passes between the abductor hallucis and the flexor digitorum brevis muscles to anastomose with the dorsal metatarsal branch of the dorsalis pedis artery [see Chapter 25 V A 2 b (3)].

    b. Distribution. It supplies the medial side of the plantar portion of the foot. It terminates along the lateral side of the great toe (see Fig. 26-5).

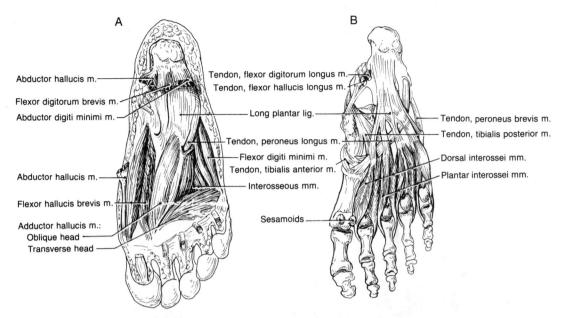

Figure 26-4. *Deep muscles of the plantar foot. A*, Third layer. *B*, Fourth layer with the deep ligaments.

B. Veins

1. **Superficial veins** of the leg are the great saphenous vein and the small saphenous vein.
   a. **The great saphenous vein** arises by the coalescence of a venous network into the **medial marginal vein** of the dorsal foot. The great saphenous vein passes anteriorly to the medial malleolus in company with the saphenous nerve, where it is accessible for venipuncture or insertion of venous lines (see Fig. 25-4). It ascends along the medial aspect of the leg and passes posteriorly to the medial condyles of the knee. It contains about 12 valves in the leg.
   b. **The small saphenous vein** begins as the lateral marginal vein along the lateral side of the dorsum of the foot. The small saphenous vein passes posteriorly to the lateral malleolus and ascends along the lateral aspect of the leg, freely anastomosing with the great saphenous vein. It penetrates the crural fascia to enter the popliteal fossa where it joins the popliteal vein. It contains about 12 valves.
   c. **Varicosities.** The superficial veins of the lower extremity are especially susceptible to varicosities, for which there seems to be a genetic predisposition (see Chapter 24 V B 3).

2. **Deep veins.** The posterior tibial vein accompanies the posterior tibial artery, usually as venae comitantes. The posterior tibial and common peroneal vein join in the popliteal fossa to form the popliteal vein, which continues in the thigh as the femoral vein (see Chapter 24 V B 2 a).

V. INNERVATION OF THE POSTERIOR CRURAL COMPARTMENT AND PLANTAR FOOT

A. **Cutaneous sensory innervation of the leg and foot** is by way of several nerves (Figs. 26-6 and 26-7).

1. **The saphenous branch** (L4) of the femoral nerve innervates the medial aspect of the leg and foot, including the great toe (see Fig. 24-7).

2. **The superficial peroneal branch** (L5) of the common peroneal nerve innervates the anterolateral aspect of the leg and foot (see Fig. 25-7).

3. **The sural nerve** (S2), formed by contributions from the tibial and common peroneal nerves, innervates the posterior aspect of the leg (see Fig. 26-6).

4. **The lateral and medial plantar branches** (S1) of the tibial nerve innervate the central and lateral aspects of the plantar foot, including the small toe (see Fig. 26-6).

B. **Motor innervation of the posterior crural compartment and plantar foot** is by the **tibial nerve** (L4–S3), the anterior division of the **sciatic nerve** (see Figs. 26-6 and 26-7).

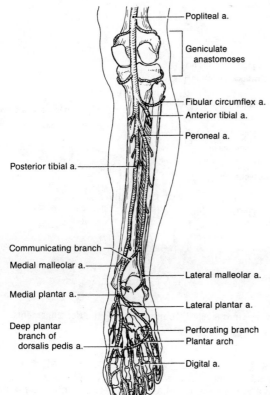

Popliteal a.

Geniculate
anastomoses

Fibular circumflex a.
Anterior tibial a.
Peroneal a.

Posterior tibial a.

Communicating branch
Medial malleolar a.
Lateral malleolar a.

Medial plantar a.
Lateral plantar a.

Deep plantar
branch of
dorsalis pedis a.
Perforating branch
Plantar arch

Digital a.

Figure 26-5. *Vascular supply to the posterior and lateral crural compartments and the plantar foot.*

1. **Course and composition.** The tibial nerve is derived from anterior roots of anterior portions of roots L4–S3 and travels with the common peroneal nerve as the sciatic nerve. It usually separates from the sciatic nerve high in the popliteal fossa and passes between the heads of the gastrocnemius muscle (see Fig. 26-6). Upon entering the foot, it bifurcates into medial and lateral plantar nerves.

2. **Distribution**
   a. Motor
   (1) The **tibial nerve** gives off branches to the musculature of the superficial and deep posterior compartments and accompanies the posterior tibial vessels behind the medial malleolus.
   (2) The **medial plantar nerve** (L4–L5) generally supplies the intrinsic muscles on the medial side of the foot (see Fig. 26-7).
   (3) The **lateral plantar nerve** (S1–S2) generally supplies the intrinsic muscles on the lateral side of the foot (see Fig. 26-7).
   b. Sensory
   (1) The **medial plantar nerve** (L4–L5) generally supplies the skin on the medial side of the foot (see Fig. 26-6).
   (2) The **lateral plantar nerve** (S1–S2) generally supplies the skin on the lateral side of the foot (see Fig. 26-6).

3. **Nerve injury.** Herniated vertebral disks in the lower lumbar region may compress the sacral roots, and intramuscular injection in the inferomedial quadrant of the buttock may injure the sciatic nerve. Injury to the tibial nerve is less frequent than injury to the common peroneal nerve because the course is deeper and generally more protected. However, trauma at the popliteal fossa or at the medial malleolus may injure the tibial nerve.
   a. A high lesion involves wasting of calf musculature and an inability to stand on tiptoe. There is loss of the **ankle-jerk reflex** (L4–L5) when the Achilles tendon is tapped with a rubber mallet. In addition, there is an inability to abduct or adduct the toes and loss of plantar sensitivity.
   b. A low lesion, such as *medial malleolar fracture* or *tarsal tunnel syndrome* (produced by entrapment of the nerve under the flexor retinaculum) results in an inability to abduct or adduct the toes (which many normal people cannot accomplish) and loss of plantar sensitivity.

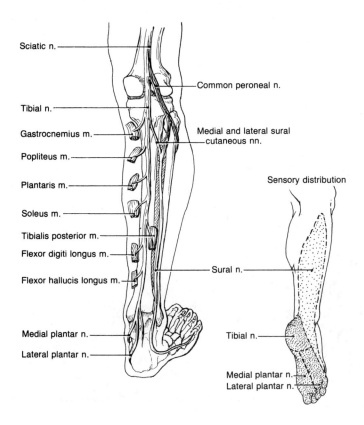

Figure 26-6. *Distribution of the tibial nerve in the leg.*

VI. AMBULATION

A. Normal gait has been described as the translation of the body's center of gravity through space with a minimum expenditure of energy. The center of gravity is repeatedly moved forward, resulting in a momentary imbalance. The individual moves a lower limb in an attempt to regain stability. This results in an alternate shifting of dynamic equilibrium from one foot to the other.

  1. Swing and stance phases. In normal gait, each lower limb passes through a cycle consisting of alternating stance and swing phases.
  a. The stance phase (66%) consists of three parts:
  (1) Heel-strike or the restraining portion
     (a) The hip abductors and adductors stabilize the pelvic girdle over the newly weight-bearing limb as the heel hits the ground.
     (b) The quadriceps femoris stabilizes the slightly flexed knee.
     (c) The gluteus maximus and hamstrings extend the femur, propelling the center of gravity forward.
     (d) Relaxation of the dorsiflexors permits lowering of the foot to midstance, transferring the weight from the heel to the plantar arch.
  (2) Midstance
     (a) The center of gravity is momentarily directly over the limb and distributed evenly by the plantar arch to the heel and metatarsals.
     (b) The dorsiflexors pull the tibia forward, moving the center of gravity anteriorly.
     (c) The triceps surae begin to lift the heel off the ground.
  (3) Toe-off or the propulsive portion
     (a) The triceps surae maximally lift the heel and move the body forward by transferring the weight to the metatarsals and phalanges.
     (b) The quadriceps femoris extends the slightly flexed knee.
     (c) The flexor hallucis longus gives a final strong plantar-flexion, which pushes the body off from the great toe.
     (d) The momentum raises the limb free from the ground into the swing phase.
  b. The swing phase (33%) consists of two parts:
  (1) Early swing portion
     (a) The hip flexors accelerate the limb forward and the knee flexes passively, lifting the foot from the ground.

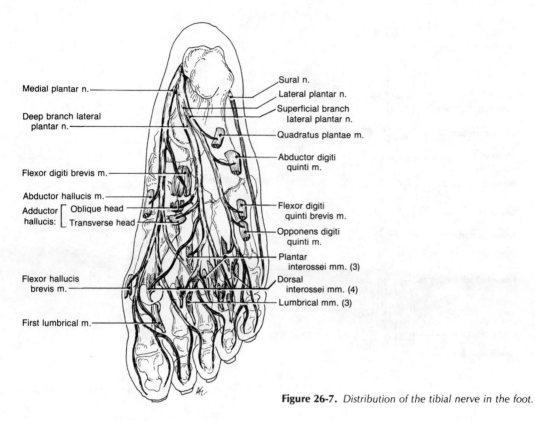

Medial plantar n.

Deep branch lateral plantar n.

Flexor digiti brevis m.

Abductor hallucis m.

Adductor hallucis: Oblique head
Transverse head

Flexor hallucis brevis m.

First lumbrical m.

Sural n.

Lateral plantar n.

Superficial branch lateral plantar n.

Quadratus plantae m.

Abductor digiti quinti m.

Flexor digiti quinti brevis m.

Opponens digiti quinti m.

Plantar interossei mm. (3)

Dorsal interossei mm. (4)

Lumbrical mm. (3)

Figure 26-7. *Distribution of the tibial nerve in the foot.*

(b) The tibialis anterior and synergistic dorsiflexors keep the foot clear of the ground.
(2) **Late swing portion**
(a) The dorsiflexors position the foot for heel-strike.
(b) The hamstrings decelerate the limb just prior to heel-strike.

2. **Ambulation** consists of reciprocally alternating cycles of gait.
a. **Walking** is defined as gait in which there are overlapping stance phases.
b. **Running** has no double-stance phase (i.e., left heel-strike with right toe-off).

B. **Pathologic gait** is an attempt either to minimize pain or to reduce expenditure of energy by compensation.

1. Heel-strike transmits force to the hip joint, so patients with hip disease avoid heel-strike when walking.

2. Disease in the first metatarsophalangeal joint inhibits toe-off, thereby altering normal gait.

STUDY QUESTIONS

Directions: Each question below contains five suggested answers. Choose the **one best** response to each question.

1. All of the following muscles are *primary* adductors at the hip joint EXCEPT

(A) adductor brevis
(B) adductor longus
(C) anterior belly of the adductor magnus
(D) pectineus
(E) posterior belly of the adductor magnus

2. In the adult, avascular necrosis of the femoral head is a likely sequela to

(A) dislocation of the hip with tearing of the ligamentum teres of the femur
(B) intertrochanteric fracture of the femur
(C) intracapsular femoral neck fracture
(D) thrombosis of the obturator artery at its origin
(E) all of the above

3. Meniscal tears in the knee joint usually result from which of the following circumstances?

(A) Compression
(B) Hyperflexion
(C) Rotation in partial flexion
(D) Rotation in full extension
(E) None of the above

4. Which of the following actions takes place during the final phase of knee joint extension?

(A) The femur glides forward on the medial tibial condyle
(B) The medial femoral condyle pivots on the tibial plateau
(C) The leg undergoes medial rotation
(D) The popliteus muscle is stretched
(E) None of the above

5. Walking across a slope requires accommodative movements in

(A) the distal tibiofibular joint
(B) the talocrural joint
(C) the tarsometatarsal joint
(D) the transverse tarsal joint
(E) none of the above

6. Tingling, painful, or itching sensations in the lateral region of the thigh may occur in the older, overweight individual as a result of a bulging abdomen, compressing a nerve beneath the inguinal ligament. Which of the following nerves is most apt to be involved?

(A) Anterior femoral cutaneous
(B) Femoral branch of the genitofemoral
(C) Genital branch of the genitofemoral
(D) Ilioinguinal
(E) Lateral femoral cutaneous

7. The dorsalis pedis artery is most commonly a continuation of which of the following arteries?

(A) Anterior tibial
(B) Lateral plantar
(C) Medial plantar
(D) Peroneal
(E) Posterior tibial

8. Which of the following muscles inserts onto the tuberosity of the fifth metatarsal bone?

(A) Abductor digiti minimi
(B) Peroneus brevis
(C) Peroneus longus
(D) Tibialis anterior
(E) Tibialis posterior

9. While stretching to the left for a backhand return in a game of tennis, a 35-year-old man feels a sharp pain in the calf of the weight-bearing (left) leg. He misses the ball and hobbles off the court. His doubles partner, a physician, insists on examining the leg, whereupon a bulge is felt in the distal part of the popliteal fossa. Severe pain is elicited upon attempted plantar-flexion against resistance. This, in addition to the bulge, indicates that a tendon has been torn. Which of the following muscles could produce the signs and symptoms described?

(A) Peroneus brevis muscle
(B) Peroneus longus muscle
(C) Plantaris muscle
(D) Popliteus muscle
(E) None of the above

10. The contraction of the plantar interossei muscles between the fourth and fifth toes results primarily in

(A) abduction of the fourth toe
(B) adduction of the fourth toe
(C) adduction of the fifth toe
(D) flexion of the interphalangeal joints of the fourth toe
(E) flexion of the interphalangeal joints of the fifth toe

Questions 11–13

The great saphenous vein needs to be cannulated for an intravenous line in a presurgical patient.

11. A convenient site for insertion of an intravenous line in the great saphenous vein is

(A) anterior to the medial malleolus
(B) inferior to the inguinal ligament
(C) posterior to the lateral malleolus
(D) medial to the popliteal fossa
(E) none of the above

12. During the cutdown and preparation of the vein for insertion of the cannula, the patient experiences pain radiating along the medial border of the dorsum of the foot. Which of the following nerves is most likely to be accidently included in a ligature during the cannulation procedure?

(A) Lateral sural
(B) Medial femoral cutaneous
(C) Medial sural
(D) Saphenous
(E) Superficial peroneal

13. Considering the pain distribution along the medial border of the dorsum of the foot, which spinal level is represented in the saphenous nerve?

(A) L2
(B) L3
(C) L4
(D) L5
(E) S1

(end of group question)

Directions: Each question below contains four suggested answers of which **one or more** is correct. Choose the answer

A if **1, 2, and 3** are correct
B if **1 and 3** are correct
C if **2 and 4** are correct
D if **4** is correct
E if **1, 2, 3, and 4** are correct

14. Muscles that insert onto the lesser trochanter include the

(1) obturator externus
(2) piriformis
(3) obturator internus
(4) iliopsoas

Questions 15–17

A person experiences weakness when climbing stairs. As part of the neurologic examination, the muscle action of the thigh is tested by asking the patient to extend the thigh against resistance.

15. Muscles that are active in producing the indicated movement include the

(1) semitendinosus
(2) gluteus maximus
(3) semimembranosus
(4) gluteus medius

16. A patient is placed in the supine position with the hip and knee joints extended. The patient is asked to abduct the limb against resistance supplied by the examiner. This tests which of the following muscles?

(1) Gluteus medius
(2) Semitendinosus
(3) Gluteus minimus
(4) Semimembranosus

17. The neurologic examination reveals that the knee-jerk reflex is normal and that the patient can stand on his heels and toes. From this result, it can be concluded that the innervation involved in this patient's problem includes which of the following nerves?

(1) Common peroneal
(2) Femoral
(3) Tibial
(4) Inferior gluteal

(end of group question)

18. A tumor of the piriformis muscle that compresses the superior gluteal nerve as it passes through the suprapiriform recess of the greater sciatic foramen would produce paralysis of which of the following muscles?

(1) Gluteus minimus
(2) Tensor fasciae latae
(3) Gluteus medius
(4) Gluteus maximus

19. During full active extension of the knee joint the ligaments that are tightened include the

(1) tibial collateral
(2) fibular collateral
(3) patellar
(4) anterior cruciate

20. After lesion of the tibial portion of the sciatic nerve, some flexion still may be possible at the knee joint. The muscles responsible for this remaining flexion include the

(1) long head of the biceps femoris
(2) gracilis
(3) gastrocnemius
(4) short head of the biceps femoris

21. Supracondylar fracture of the femur may injure which of the following structures?

(1) Anterior tibial artery
(2) Sciatic nerve
(3) Deep (profunda) femoral artery
(4) Popliteal artery

22. The adductor canal (of Hunter) between the anterior and posterior portions of the adductor magnus muscle contains the

(1) obturator artery
(2) saphenous nerve
(3) great saphenous vein
(4) femoral artery

23. The longitudinal plantar (tarsal) arch is supported and maintained by the

(1) tibialis anterior muscle
(2) peroneus longus muscle
(3) tibialis posterior muscle
(4) calcaneonavicular ligament

24. Factors that contribute to the stability of the ankle joint include the

(1) deltoid ligament
(2) trapezoidal shape of the trochlea of talus
(3) calcaneotibular ligament
(4) calcaneonavicular ligament

25. Muscles of the lateral compartment of the leg include the

(1) peroneus brevis
(2) peroneus tertius
(3) peroneus longus
(4) tibialis anterior

26. Structures found within the tarsal tunnel (flexor retinaculum) include the

(1) posterior tibial artery
(2) tibialis posterior tendon
(3) tibial nerve
(4) peroneus brevis tendon

27. Muscles attaching to the plantar surface of the distal (third) phalanx of the second through fifth toes include the

(1) group of dorsal and plantar interossei
(2) flexor digitorum brevis
(3) group of lumbricals
(4) flexor digitorum longus

Directions: The groups of questions below consist of lettered choices followed by several numbered items. For each numbered item, select the **one** lettered choice with which it is **most closely** associated. Each lettered choice may be used once, more than once, or not at all. Choose the answer

- **A**   if the item is associated with **(A) only**
- **B**   if the item is associated with **(B) only**
- **C**   if the item is associated with **both (A) and (B)**
- **D**   if the item is associated with **neither (A) nor (B)**

Questions 28–32

For each function or characteristic pertaining to the thigh listed below, select the muscle with which it is most apt to be associated.

(A) Sartorius muscle
(B) Adductor longus muscle
(C) Both
(D) Neither

28. Bounds the femoral triangle

29. Is innervated by the obturator nerve

30. Assists flexion of hip

31. Assists lateral rotation of the thigh

32. Extends knee

Questions 33–35

For each of the muscle compartments listed below, select the action most likely to be associated with it.

(A) Dorsiflex the foot
(B) Evert the foot
(C) Both
(D) Neither

33. Muscles of the anterior crural compartment

34. Muscles of the lateral crural compartment

35. Muscles of the posterior crural compartment

Questions 36–40

For each clinical finding listed below, select the appropriate nerve with which it is most apt to be associated.

(A) Common peroneal nerve
(B) Tibial nerve
(C) Both
(D) Neither

36. Inability to stand on the toes by plantar-flexing the foot

37. Inability to stand on the heels by dorsiflexing the foot

38. Inability to evert the foot

39. Inability to flex the knee

40. Loss of Achilles tendon reflex

ANSWERS AND EXPLANATIONS

1. The answer is E. [*Chapter 23 IV A 3, 4; Table 23-2*] All of the muscles listed in the question (i.e., adductor brevis and longus, anterior belly of the adductor magnus, and pectineus) except the posterior portion of the adductor magnus belong by position to the adductor group of the thigh. The posterior portion of the adductor magnus is a member of the hamstring group of the posterior compartment, which primarily extends the hip.

2. The answer is C. [*Chapter 23 V B 2 a; VII C 1 b*] In the adult, blood flow through the artery of the ligamentum teres (when present) is usually insufficient to supply the head of the femur. The adult blood supply to this region of the femur is primarily via the retinacular branches of the medial and lateral femoral circumflex arteries. Consequently, an intracapsular femoral neck fracture jeopardizes this vascular supply and may well result in necrosis of the femoral head. Proximal thrombosis of the obturator artery, from which the artery of the ligamentum teres arises, could produce a similar problem but only in a child.

3. The answer is C. [*Chapter 24 III A 2 b*] Most meniscal tears are the result of lateral rotation of the partially flexed leg. Because the knee joint is most stable in full extension, meniscal tears are unlikely in this position. Since the full flexion is not compatible with adequate mobility or support, it is a rarely used posture and unlikely to cause problems.

4. The answer is D. [*Chapter 24 III C 2 c*] During the terminal phase of extension, the femur rotates medially about the anterior cruciate ligament, so that the medial femoral condyle glides backward and the popliteus muscle is stretched. The medial rotation of the thigh is equivalent to a lateral rotation of the leg.

5. The answer is D. [*Chapter 25 III C, G 1 a–c*] Walking across a slope necessitates inversion and eversion of the foot. This movement involves the transverse tarsal joint, which consists of the mechanically linked talocalcaneonavicular and calcaneocuboid joints, and takes place about a resultant axis (of Henke).

6. The answer is E. [*Chapter 24 VI B 3*] The lateral femoral cutaneous nerve passes between the inguinal ligament and the iliopubic ramus near the anterior–superior iliac spine. Here the nerve may be entrapped, producing sensory symptoms along the lateral aspect of the thigh.

7. The answer is A. [*Chapter 25 VI A 2; Chapter 26 IV A 1 b (1) (c)*] The dorsalis pedis artery usually is a continuation of the anterior tibial artery. Occasionally, the dorsalis pedis artery arises from the peroneal artery, a branch of the posterior tibial artery.

8. The answer is B. [*Chapter 25 Table 25-1; Figure 25-2A*] The peroneus brevis inserts into the tubercle of the fifth metatarsal. Because it is a powerful evertor, forceful inversion can avulse the tendinous insertion.

9. The answer is C. [*Chapter 26 II B 1 a (1) (b), (2); Table 26-1*] The plantaris, a small but significant part of the triceps surae, acts with the gastrocnemius and soleus to plantar-flex the foot. Originating from the lateral supracondylar ridge of the femur, it passes through the popliteal fossa and becomes tendinous between the soleus and gastrocnemius muscles to insert into the calcanean tendon. Because the plantaris muscle spans both knee and ankle joints, its tendon may be avulsed when sudden plantar-flexion coincides with strong knee extension. The popliteus muscle, a medial rotator of the knee, is not involved in plantar-flexion. The peroneal muscles are in the lateral compartment.

10. The answer is C. [*Chapter 26 III B 1 d (2)*] The plantar interossei adduct the toes toward the axis of the second toe. The dorsal interossei abduct the toes away from the second toe. The plantar and dorsal interossei also extend the proximal and distal interphalangeal joints and flex the metatarsophalangeal joints.

11. The answer is A. [*Chapter 26 IV B 1 a*] The preferred site for great saphenous vein cannulation is where the vein passes immediately anterior to the medial malleolus. The lesser saphenous vein, which is not usually used for venipuncture, passes posteriorly to the lateral malleolus. The saphenous vein passes medially to the popliteal fossa and enters the femoral vein at the level of the inguinal ligament.

12. The answer is D. [*Chapter 26 V A 1*] The saphenous nerve comes into the proximity of the great saphenous vein posteriorly to the medial tibial condyle. It accompanies the vein along the anteromedial aspect of the leg and into the foot anteriorly to the medial malleolus.

13. The answer is C. [*Chapter 4 Figure 4-3; Chapter 24 Figure 24-7*] The saphenous nerve, the terminal cutaneous portion of the femoral nerve, contains the contribution from the L4 spinal root. The medial aspect of the foot is generally innervated by L4, and the lateral aspect is generally innervated by S1.

14. The answer is D (4). [*Chapter 23 II C 1 c (2)*] The iliopsoas muscle alone inserts onto the lesser trochanter of the femur. The piriformis, obturators internus and externus, gemelli, and gluteus medius and minimus all insert into the greater trochanter.

15. The answer is A (1, 2, 3). [*Chapter 23 IV A 3 b; Table 23-1*] The hamstring group, including the biceps femoris, semitendinosus, and semimembranosus, as well as the gluteus maximus, are all extensors of the thigh. The gluteus medius is an abductor.

16. The answer is B (1, 3). [*Chapter 23 IV A 2, 3; Table 23-2*] Abduction of the thigh is primarily by the gluteus medius and minimus muscles, as well as the tensor fasciae latae muscle. The semimembranosus and semitendinosus muscles, the principal extensors of the thigh, are weak adductors.

17. The answer is D (4). [*Chapter 23 VI B 4; Table 23-1*] The inferior gluteal nerve innervates the gluteus maximus muscle, and paralysis of this produces weakness in climbing stairs and in rising from a seated position. The normal knee jerk indicates that the femoral nerve is not involved, and the ability to stand on the heels and toes indicates that the common peroneal and tibial nerves, respectively, are not involved.

18. The answer is A (1, 2, 3). [*Chapter 23 VI A*] The superior gluteal nerve innervates the gluteus medius and minimus muscles, as well as the tensor faciae latae muscle. The gluteus maximus muscle is innervated by the inferior gluteal nerve.

19. The answer is E (all). [*Chapter 24 II C 1 b; III B 3 a (2) (b), b (2) (b), d (2)*] During active extension, the quadriceps femoris pulls on the tibia through the patellar ligament. As the leg comes into full extension, the collateral ligaments become taut, as do the cruciate ligaments, with the anterior cruciate ligament providing the axis about which the femur rotates medially.

20. The answer is C (2, 4). [*Chapter 24 IV B 1 b (2); Table 24-1*] The gracilis muscle, which acts to flex the knee, is innervated by the obturator nerve, and the short head of the biceps femoris muscle, which is also a knee flexor, is innervated by the common peroneal nerve. As such, paralysis of the "true" hamstring group by tibial nerve palsy does not abolish the ability to flex the knee.

21. The answer is D (4). [*Chapter 24 V A 2 b, 3 b (2); Chapter 25 VI A 1; Chapter 26 IV A*] A supracondylar femoral fracture jeopardizes the popliteal neurovascular structures, especially the popliteal artery. The sciatic nerve divides into common peroneal and tibial nerves well above the knee region. The popliteal artery gives rise to the anterior tibial artery behind the tibia at the inferior border of the popliteal fossa before continuing as the posterior tibal artery. Only small anastomotic twigs of the deep femoral artery reach the vicinity of the knee to participate in the geniculate anastomoses.

22. The answer is C (2, 4). [*Chapter 24 V A 2 a (2)*] The adductor canal through the adductor magnus muscle contains the femoral artery, the femoral vein, the saphenous nerve, and the nerve to the vastus medialis; the latter two are continuations of the femoral nerve. The obturator artery does not accompany the obturator nerve very far into the adductor compartment, and the great saphenous vein lies externally to the fascia lata.

23. The answer is E (all). [*Chapter 25 II B 1 b (3), (4); Chapter 26 III B 1 d (3)*] The calcaneonavicular (spring) ligament spans the talocalcaneonavicular joint from the sustentaculum tali on the anteromedial aspect of the calcaneus to the inferoposterior aspect of the navicular bone. It maintains the longitudinal plantar arch and supports the head of the talus. In addition, the dynamic activity of the muscles that insert onto the medial cuneiform bone (the anterior and posterior tibial muscles as well as the peroneus longus muscle) helps to maintain the arch.

24. The answer is A (1, 2, 3). [*Chapter 25 III A 1 b (1), (3), 3 b*] The stability of the ankle joint is provided primarily by ligaments, including the deltoid ligament (between the medial malleolus of the tibia and the naviculum and calcaneus) and the lateral ligament (between the lateral malleolus of the fibula and the talus and calcaneus). The trapezoidal shape of the trochlea of talus increases stability in dorsiflexion. The calcaneonavicular (spring) ligament supports the talus and helps maintain the longitudinal plantar arch.

25. The answer is B (1, 3). [*Chapter 25 IV B 1 b*] The lateral crural compartment of the leg contains the peroneus longus and peroneus brevis muscles. The peroneus tertius is found in the dorsum of the foot, and the tibialis anterior is an anterior crural compartment muscle.

26. The answer is A (1, 2, 3). [*Chapter 26 II A 2 c; V B 3 b*] The tendons of the deep muscles of the posterior compartment of the leg (i.e., tibialis posterior, flexor digitorum, and flexor hallucis longus), the tibial artery, and accompanying tibial nerve pass through the tarsal tunnel inferiorly to the medial malleolus. The peroneal muscles of the lateral compartment pass beneath the peroneal retinaculum inferiorly to the lateral malleolus. The sural nerve passes on the lateral side.

27. The answer is D (4). [*Chapter 26 II B 1 b (3); Table 26-1; Table 26-2*] Only the flexor digitorum longus reaches the plantar surface of the distal phalanx of toes two through five. The flexor digitorum brevis splits at its insertion to the plantar surface of the middle phalanges, so that deeper flexor digitorum longus may reach the distal phalanx. The lumbricals and interossei insert into the dorsal expansions of the pedal digits.

28–32. The answers are: 28-C, 29-B, 30-C, 31-A, 32-D. [*Chapter 23 IV A 1 b, 4 b, B 1 b; Table 23-1; Table 23-2; Chapter 24 V A 2 a (1) (a); Table 24-1*] The femoral triangle is bounded by the inguinal ligament and the sartorius, and adductor longus muscles. The sartorius muscle (a member of the anterior compartment and innervated by the femoral nerve) is a flexor and lateral rotator of the hip as well as a flexor of the knee. The adductor longus (a member of the adductor group and innervated by the obturator nerve) is a flexor and medial rotator of the hip.

33–35. The answers are: 33-A, 34-B, 35-D. [*Chapter 25 IV B 2; Figure 25-4; Chapter 26 II B 2*] The muscles of the anterior crural compartment, passing anteriorly to the transverse axis of the ankle joint and medially to the resultant axis of the subtalar and transverse tarsal joints, dorsiflex and primarily invert the foot. The muscles of the lateral crural compartment, passing posteriorly to the transverse axis of the ankle joint and laterally to the resultant axis of the subtalar and transverse tarsal joints, plantar-flex and evert the foot. The muscles of the posterior crural compartment, passing posteriorly to the transverse axis of the ankle joint and medially to the resultant axis of the subtalar and transverse tarsal joints, plantar-flex and invert the foot.

36–40. The answers are: 36-B, 37-A, 38-A, 39-C, 40-B. [*Chapter 24 VI B 2 a (2), b (2), c; Table 24-1; Chapter 25 VII B 1, C 1; Table 25-1; Chapter 26 V B 3; Table 26-1*] Injury to the tibial nerve will affect the plantar-flexors of the posterior crural compartment and be indicated by loss of the Achilles tendon reflex and the ability to stand on tiptoes. Injury to the common peroneal nerve will affect the lateral crural compartment resulting in *foot-slap* and an inability to stand back on the heels. A high sciatic nerve injury, involving both common peroneal and tibial divisions, which innervate the hamstring muscles, results in an inability to flex the knee in conjunction with paralysis of the leg and foot musculature.

Part VIII
Somatic Neck and Neurocranium

27
Posterior
Cervical Triangle

I. INTRODUCTION

A. Major divisions. The head and neck may be divided, developmentally, functionally, and medically, into an anterior visceral portion and a posterior somatic portion. The course of the sternomastoid muscle approximates the separation of the somatic and visceral portions of the neck and, as such, divides the neck into an **anterior triangle** and a **posterior triangle**.

   1. The somatic portion comprises the neurocranium and the cervical vertebral column with the associated musculature.

   2. The visceral portion comprises derivatives of the upper end of the primitive gut and its associated branchial (gill) structures.

B. Fascia

   1. The superficial fascia in the head and neck regions contains loose connective tissue.
      a. The more superficial portions contain variable amounts of adipose tissue.
      b. The deeper portions enclose the voluntary muscles of facial expression. This group is represented by the platysma muscle in the neck.
      c. In the head and face, the superficial fascia is frequently continuous with and inseparable from the deep fascia overlying bone.

   2. The deep cervical fascia surrounds the neck and gives off septa, which separate the visceral and somatic portions. This fascial layer can be subdivided into four parts.
      a. The superficial (investing) layer of the deep cervical fascia ensheaths the neck beneath the superficial fascia (Fig. 27-1).
         (1) Posteriorly, it attaches to the ligamentum nuchae. Superiorly, it is attached along the mandible, mastoid process, and ligamentum nuchae. Inferiorly, it finds attachment along the acromion, the clavicle, and the manubrium sterni.
         (2) This layer splits several times to ensheath the trapezius, omohyoid, and sternomastoid muscles individually.
            (a) Anteriorly between the sternomastoid muscles, the split layers fail to reunite, producing the small **suprasternal space** (of Burns), which contains the anterior jugular veins and an occasional lymph node.
            (b) Superiorly, it splits to ensheath the parotid gland and the submandibular glands.
      b. The pretracheal layer represents a middle fascial layer, which originates as septa from the superficial layer of the deep fascia (see Fig. 27-1). This layer passes between the infrahyoid (strap or ribbon) muscles and the trachea.
         (1) Superiorly, it attaches to the cricoid cartilage; inferiorly, this layer continues into the middle mediastinum, where it fuses with the fibrous pericardium.
         (2) Anteriorly, it splits to surround and support the thyroid gland.
      c. The carotid sheath is a fascial condensation about the carotid arteries, the internal jugular vein, and the vagus nerve (CN X) [see Fig. 27-1].
         (1) The pretracheal fascia and the superficial layer of the deep cervical fascia both send septa to the carotid sheath.
         (2) The sympathetic chain and superior cervical ganglion lie posteriorly to the carotid sheath, anterior to the prevertebral fascia.
      d. The prevertebral layer encircles the vertebral column and its associated muscles, defining the somatic portion of the neck (see Fig. 27-1).
         (1) After originating from the cervical spinous processes, it passes between the trapezius and the intrinsic muscles of the cervical spine.

    (2) In the posterior triangle, it is apposed to the superficial layer of deep cervical fascia.

        (a) The spinal accessory nerve (CN XI) and lymphatics lie between these two layers.

        (b) Low in the posterior triangle, the prevertebral layer invests the scalene muscles, leaving a space between them and the superficial layer of the deep cervical fascia.

           (i) This space is occupied by the subclavian and external jugular veins, the transverse cervical and suprascapular arteries, and the omohyoid muscle.

           (ii) As the nerves to the upper limb pass between the anterior and middle scalene muscles, a continuation of the prevertebral fascia forms the **axillary sheath**.

    (3) Anteriorly, the prevertebral fascia attaches to the transverse processes of the cervical vertebrae. It then splits into two leaves, the deeper of which inserts onto the anterior aspect of the cervical vertebrae and intervertebral disks to enclose the longus coli muscles. The more superficial (alar) leaf jumps to the contralateral transverse process. The compartment between these two layers of prevertebral fascia is the so-called danger space.

    (4) Inferiorly, the prevertebral fascia extends into the posterior mediastinum.

  3. Cervical spaces. The **fascial planes** form potential spaces in the neck. Superficial or deep extension of infection tends to be limited by the fascial planes but may track up and down the neck along the resultant spaces (see Fig. 27-1).

    a. The **visceral compartment** is that region bounded by the pretracheal and prevertebral fascial planes. It is subdivided into two spaces.

        (1) The **pretracheal space** is deep to the pretracheal fascia, anterior to the esophagus, and descends into the superior mediastinum. (Infection within the pretracheal space may come to the surface in the suprasternal space or may track into the superior mediastinum.)

        (2) The **retropharyngeal (retrovisceral) space** is posterior to the esophagus, anterior to the prevertebral fascia, and descends into the posterior mediastinum. Infection in the retrovisceral space may track deeply into the mediastinum.

    b. The **somatic compartment** is that region enclosed by the prevertebral fascia. The "**danger space**" lies between the two layers of prevertebal fascia anteriorly to the longus coli muscles. (Infection here can extend from the base of the skull into the abdomen.)

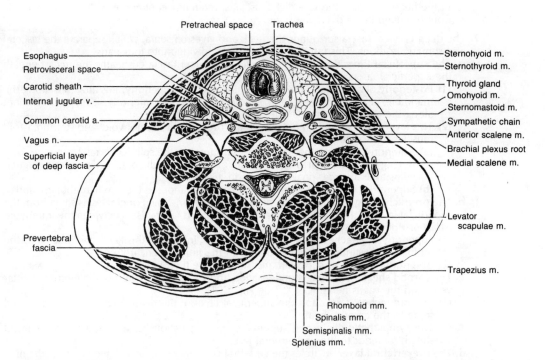

Figure 27-1. *Deep cervical fascia.* The superficial layer of the deep cervical fascia invests the trapezius, sternomastoid, and infrahyoid muscles. The pretracheal fascia invests the visceral structures. The carotid sheath contains the carotid arteries, internal jugular vein, and vagus nerve. The prevertebral fascia invests the musculature of the somatic neck.

II. SOMATIC NECK

A. Boundaries of the posterior triangle are the trapezius muscle, posteriorly, the sternomastoid muscle, anteriorly, and the clavicle, inferiorly (Fig. 27-2). The sternomastoid muscle approximates the division between the somatic neck and the visceral neck.

1. Structures in the posterior cervical triangle include muscles and nerves associated with the cervical vertebral column.

2. Many of the structures of the posterior triangle are en route to the upper limb.

B. Cervical vertebrae (see Chapter 19 I D 1)

C. Cervical musculature

1. The superficial muscles of the posterior cervical triangle are enclosed in the superficial layer of the deep cervical fascia, are derived from one continuous sheet, and are associated with the pectoral girdle.

   a. The sternomastoid (sternocleidomastoid) muscle divides the neck into anterior and posterior triangles (see Fig. 27-2).

      (1) Attachments. It has a narrow tendinous head from the manubrium sterni and a flat muscular head from the medial third of the clavicle. It inserts onto the lateral aspect of the mastoid process of the skull.

      (2) Action. It rotates and laterally flexes the head (i.e., it apposes the side of the head to the ipsilateral shoulder).

         (a) This action is seen in patients with sternomastoid spasm—**spasmodic torticollis**.

         (b) Congenital torticollis is produced by fibrosis of the sternomastoid, usually the result of hyperextension injury of the muscle upon difficult parturition.

      (3) Innervation. The sternomastoid muscle is innervated by the spinal accessory nerve (CN XI) as well as by twigs from the ventral rami of spinal nerves C2 and C3.

   b. The trapezius muscle derives its name from its bilateral appearance (see Fig. 27-2).

      (1) Attachments. It arises from the superior nuchal line of the occiput and the external occipital protuberance as well as from the ligamentum nuchae and the spinous processes of vertebrae C7 to T12. Its insertion into the pectoral girdle is extensive.

      (2) Actions

         (a) The more superior portions insert into the distal third of the clavicle, elevating and rotating the scapula.

         (b) The middle portion inserts into the acromion process and the spine of the scapula, retracting the scapula.

         (c) The more inferior portion inserts onto the medial part of the spine of the scapula, depressing and rotating the scapula.

      (3) Innervation. The trapezius is innervated by the spinal accessory nerve (CN XI) as well as by twigs from the ventral rami of spinal nerves C3 and C4.

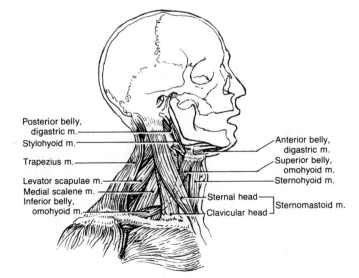

Posterior belly, digastric m.
Stylohyoid m.
Trapezius m.
Levator scapulae m.
Medial scalene m.
Inferior belly, omohyoid m.
Sternal head
Clavicular head
Anterior belly, digastric m.
Superior belly, omohyoid m.
Sternohyoid m.
Sternomastoid m.

Figure 27-2. *Superficial cervical musculature.*

2. The intrinsic muscles lie beneath the **prevertebral fascia** of the posterior triangle.

   a. The lateral group includes the scalenes (anterior, middle, and posterior) and the levator scapulae. These are innervated by lateral branches of ventral primary rami of spinal nerves (Fig. 27-3).

      (1) The **anterior scalene muscle** lies deeply to the sternomastoid muscle. The phrenic nerve courses anteriorly to this muscle to enter the thorax, and the roots of the cervical plexus and of the brachial plexus emerge at the posterolateral edge of this muscle.

         (a) Attachments. It arises from the anterior tubercles of the transverse processes of the third to sixth cervical vertebrae and inserts into the scalene tubercle of the first rib, thus intercalating the subclavian artery and vein.

            (i) The first branches of the subclavian artery (i.e., the vertebral artery and the inferior thyroid branch of the thyrocervical trunk) pass superiorly on the surface of this muscle.

            (ii) The superficial cervical and suprascapular branches of the thyrocervical trunk pass horizontally across this muscle.

         (b) Action. Its action is to elevate the first rib.

         (c) Innervation. It is innervated by twigs from the ventral primary rami of spinal nerves C4–C6.

      (2) The **middle scalene muscle** lies posteriorly to the roots of the brachial plexus and is traversed by the dorsal scapula nerve.

         (a) Attachments. It arises from the posterior tubercles of the transverse process of the cervical vertebrae and inserts along the first rib posteriorly to the subclavian groove.

         (b) Action. It acts to raise the first rib.

         (c) Innervation. It is innervated by twigs from the ventral primary rami of spinal nerves C3–C8.

      (3) The **posterior scalene muscle** is actually that part of the middle scalene that inserts along the outer surface of the second rib.

      (4) The **levator scapulae** crosses the middle of the posterior triangle.

         (a) Attachments. It arises from the posterior tubercles of the transverse processes of the upper four cervical vertebrae and inserts into the superior angle of the scapula.

         (b) Action. It acts to raise the shoulder.

         (c) Innervation. It is innervated by twigs from the ventral primary rami of spinal nerves C3 and C4 and occasionally by a twig from the dorsal scapula nerve (C5), which runs along its deep surface.

   b. The anterior group includes the longus colli, longus capitis, rectus capitis anterior, and rectus capitis lateralis (Fig. 27-4). These flexors are innervated by anterior branches of the ventral primary rami.

      (1) The **longus cervicis (colli) muscle**, which primarily flexes the vertebral column, arises in three parts.

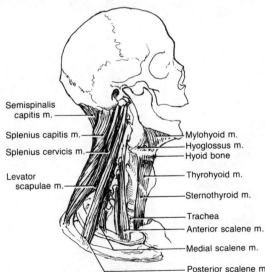

Semispinalis capitis m.

Splenius capitis m.

Splenius cervicis m.

Levator scapulae m.

Mylohyoid m.

Hyoglossus m.

Hyoid bone

Thyrohyoid m.

Sternothyroid m.

Trachea

Anterior scalene m.

Medial scalene m.

Posterior scalene m. **Figure 27-3.** *Lateral cervical musculature.*

(a) The superior oblique portion runs from the anterior tubercles of the transverse processes of the third to sixth cervical vertebrae to the spine of the axis.

(b) The vertical portion runs from the bodies of the upper thoracic vertebrae to insert onto the bodies of the upper cervical vertebrae.

(c) The inferior oblique portion runs from the bodies of the upper thoracic vertebrae to the anterior tubercles of the transverse processes of the fourth and fifth cervical vertebrae.

(2) The **longus capitis muscle** arises from the anterior tubercles of the transverse processes of the third through sixth cervical vertebrae; it inserts onto the base of the occipital bone. It primarily flexes the head.

(3) The **rectus capitis anterior** arises from the lateral side of the atlas and inserts into the basilar part of the occipital bone. It flexes the head at the atlanto-occipital joint.

(4) The **rectus capitis lateralis** arises from the transverse processes of the atlas; it inserts into the jugular process of the occipital bone. It stabilizes the atlanto-occipital joint.

c. **The posterior group** is composed of the splenius group, the semispinalis, and the suboccipital group. These extensors are innervated by dorsal primary rami of spinal nerves.

(1) **Spinotransverse group.** The **splenius muscles** comprise the external layer of the intrinsic muscles of the back (Fig. 27-5). Acting bilaterally, they extend the head and neck; acting singly, they turn the head toward the ipsilateral shoulder. They are, therefore, synergists of the contralateral sternomastoid muscles.

(a) The **splenius capitis muscle** arises from the spines of the upper thoracic vertebrae and the lower portion of the ligamentum nuchae. It emerges from beneath the trapezius muscle in the posterior triangle and inserts into the mastoid process of the skull.

(b) The **splenius cervicis muscle** arises from the spines of the third through sixth thoracic vertebrae. It lies deeply to the splenius capitis and inserts onto the transverse processes of the upper four cervical vertebrae.

(2) **Sacrospinalis group.** The **longissimus capitis** and **longissimus cervicis** muscles, comprising the intermediate layer of intrinsic musculature, run between transverse processes (see Fig. 27-5).

(a) The **longissimus capitis muscle** arises from the transverse processes of the lower four cervical vertebrae and inserts into the mastoid process.

(b) The **longissimus cervicis muscle** runs between the transverse processes of the neck and thorax.

(3) **Transversospinalis group.** The **semispinalis muscles**, comprising the deep layer of the intrinsic musculature, run between the transverse processes of the thoracic and cervical vertebrae (see Fig. 27-5). They act as extensors of the head and neck.

(a) The **semispinalis cervicis** inserts onto the transverse processes of the atlas.

(b) The **semispinalis capitis** inserts onto the occipital bone between the superior and inferior nuchal lines.

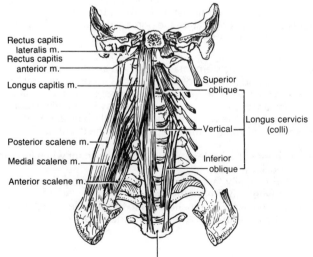

Rectus capitis
lateralis m.
Rectus capitis
anterior m.
Longus capitis m.
Superior
oblique
Longus cervicis
(colli)
Vertical
Posterior scalene m.
Medial scalene m.
Inferior
oblique
Anterior scalene m.
Third thoracic vertebra

Figure 27-4. *Anterior cervical musculature.*

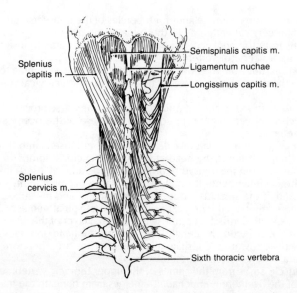

Splenius capitis m.

Semispinalis capitis m.
Ligamentum nuchae
Longissimus capitis m.

Splenius cervicis m.

Sixth thoracic vertebra

Figure 27-5. *Posterior cervical musculature.*

    (4) The suboccipital group comprises the deepest layer of intrinsic musculature and is innervated by branches from spinal nerve C1 (Fig. 27-6).

       (a) The **rectus capitis posterior major** arises from the spine of the axis and then diverges from its counterpart, leaving a small triangular space. It inserts onto the occipital bone immediately caudal to the lateral part of the inferior nuchal line and acts to extend the head at the atlanto-occipital joint.

       (b) The **rectus capitis posterior minor** arises from the posterior tubercle of the atlas and inserts onto the occipital bone immediately caudal to the medial part of the inferior nuchal line. It acts to extend the head at the atlanto-occipital joint.

       (c) The **obliquus capitis inferior** arises from the spine of the axis and inserts into the transverse process of the atlas. It rotates the head to the ipsilateral side at the atlantoaxial joint.

       (d) The **obliquus capitis superior** arises from the transverse process of the atlas and inserts between the superior and inferior nuchal lines of the occiput. It rotates the head to the ipsilateral side at the atlantoaxial joint and extends the head at the atlanto-occipital joint.

       (e) The **suboccipital triangle** on each side is bounded by the rectus capitis posterior major and the two obliquus muscles.

          (i) The floor of this triangle is formed by the atlanto-occipital membrane, which represents the missing posterior articulation between the axis and atlas.

          (ii) It contains the posterior arch of the atlas, the primary dorsal rami of the first cervical nerve, and the vertebral artery.

  3. Clinical considerations

    a. Respiration. The scalene muscles and the sternomastoid muscle assist extreme inspiratory effort by elevating the first and second ribs. Use of these muscles accompanied by labored breathing is a sign that a patient is in extreme respiratory distress, as in an asthma attack.

    b. Scalene syndrome. The subclavian artery and brachial plexus may be subject to compression where they pass over the first rib. The cause may be a cervical rib or spasm of the scalene muscles. The results are ischemia of the limb and pain along the distribution of the affected nerves.

    c. Central venous lines. The relations of the scalenus anterior are important when inserting a central venous line into the subclavian vein.

      (1) The needle must be guided medially, approximating the long axis of the clavicle, to reach the posterior surface where the vein runs over the first rib.

      (2) If the large bore needle is passed directly perpendicular to the clavicle, then the subclavian artery and the cervical pleura are at risk of being punctured.

D. Nerves of the posterior triangle

  1. The dorsal primary rami of the cervical nerves innervate the intrinsic musculature of the back and the dorsal portions of the cervical dermatomes.

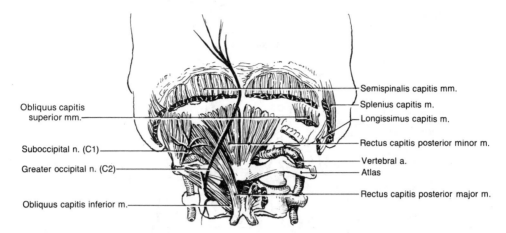

Obliquus capitis
superior mm.

Suboccipital n. (C1)

Greater occipital n. (C2)

Obliquus capitis inferior m.

Semispinalis capitis mm.

Splenius capitis m.

Longissimus capitis m.

Rectus capitis posterior minor m.

Vertebral a.

Atlas

Rectus capitis posterior major m.

Figure 27-6. *Suboccipital muscles and triangle.*

> **a. The suboccipital nerve** (dorsal branch of C1) supplies the suboccipital muscles and, unlike the other spinal nerves, has no sensory root (see Fig. 27-6).
> **b. The greater occipital nerve** arises from the medial branch of C2 to pierce the semispinalis and trapezius. It supplies sensory innervation to the posterior regions of the scalp (see Fig. 27-6).
> **c. The third (least) occipital nerve** arises from the medial branch of C3 to innervate the posteroinferior portion of the scalp.

2. The ventral primary rami of the cervical nerves pass along a shallow groove on the superior surface of the cervical transverse processes and come to lie in the space between the anterior and middle scalene muscles.

> **a. The ventral primary rami of nerves C1–C4** intermingle in a predictable pattern to form the **cervical plexus**, which gives rise to cutaneous nerves, muscular branches, and communicating branches (Fig. 27-7).
>> **(1)** The **cutaneous nerves** emerge into the posterior triangle at the posterior border of the sternomastoid muscle (see Fig. 27-7).
>>> **(a)** The **lesser occipital nerve** (C2) ascends to the scalp posterior to the auricle.
>>> **(b)** The **great auricular nerve** (C2–C3) passes vertically by the angle of the jaw before bifurcating about the auricle.
>>> **(c)** The **transverse cervical (colli) nerve** (C2–C3) crosses the sternomastoid horizontally to innervate the anterior triangle of the neck.
>>> **(d)** The **supraclavicular nerves** (C3–C4) supply the dermatomes of the shoulder and upper chest.
>> **(2) Muscular branches**
>>> **(a)** Small branches pass from the cervical plexus to all muscles arising from the ventral and lateral aspects of the cervical vertebral column as well as to the trapezius, sternomastoid, and diaphragm.
>>> **(b)** The **ansa cervicalis** innervates the strap muscles of the visceral neck (see Fig. 27-7).
>>>> **(i)** A communicating branch from C2 joins the hypoglossal nerve (CN XII). Some of the C2 fibers leave the hypoglossal nerve as the **superior root** of the **ansa cervicalis** (whence the former name **descendens hypoglossi**) to innervate most of the infrahyoid muscles. Other C2 fibers continue with the hypoglossal nerve to innervate the geniohyoid, thyrohyoid, and sternohyoid muscles.
>>>> **(ii)** The **inferior root** of the **ansa cervicalis** is formed by the union of branches from C2 and C3 (whence the former name **descendens cervicalis**) and innervates the omohyoid and sternothyroid muscles.
>> **(3)** The sympathetic chain sends **gray rami communicantes** to each cervical root (with perhaps the exception of C1), carrying secretomotor, vasomotor, and pilomotor fibers.
> **b. The ventral primary rami of the spinal nerves C4–C8** contribute to the formation of the **brachial plexus**, which is primarily involved with innervation of the upper extremity (see Fig. 6-6). Several branches of the brachial plexus course through the posterior triangle.
>> **(1)** The **dorsal scapular nerve** (C5, posterior) pierces the scalenus medius muscle, runs posteriorly on the deep surface of the levator scapulae muscle, and descends along the medial border of the scapula to supply the rhomboid muscles.

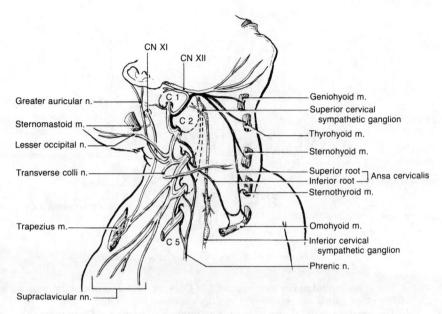

Figure 27-7. *Cervical plexus.*

(2) The **suprascapular nerve** (C5–C6, posterior) runs along the base of the posterior triangle just superior to the clavicle. It innervates the supraspinatus muscle before passing through the great scapular notch to innervate the infraspinatus muscle.

3. Spinal accessory nerve (CN XI) [see Fig. 27-7; Fig. 27-8]
   a. Composition and course
      (1) Rootlets from C1–C5 unite within the vertebral canal and pass superiorly through the foramen magnum into the cranial cavity and exit through the jugular foramen.
      (2) As the spinal accessory nerve transits the jugular foramen, it is joined for a short distance by the **cranial accessory nerve** before the latter splits off to join the vagus nerve to be distributed to the recurrent laryngeal nerve.
      (3) From the jugular foramen, the spinal accessory nerve courses posteriorly and passes laterally (70%–80%) to the jugular vein.
   b. Distribution
      (1) It descends across the medial surface of the stylohyoid and digastric muscles to enter the deep surface of the sternomastoid muscle, to which it sends branches.
      (2) It emerges at the posterior border of the sternomastoid muscle in the posterior triangle and passes superficially to the levator scapulae. Its position approximates the perpendicular bisector of a line joining the mastoid process to the angle of the mandible. It passes deeply to the trapezius muscle, to which it sends branches, about 5 cm superior to the clavicle.
      (3) At the superior angle of the scapula, it accompanies the dorsal scapular artery, and they run together along the medial border of the scapula superficially to the rhomboid muscles.
   c. Clinical considerations. The spinal accessory nerve must be avoided when exploring the posterior triangle. Paralysis of the spinal accessory nerve results in an inability to shrug the shoulder. Swollen lymph nodes in the posterior triangle may irritate this nerve, producing **spasmodic torticollis**.

E. Vasculature of the posterior triangle

   1. Arterial supply
      a. The occipital artery contributes a profuse blood supply to the scalp (Fig. 27-9).
         (1) Course. This large vessel is a branch of the external carotid artery. It is crossed near its origin by the hypoglossal nerve. It then passes along the inferior border of the posterior belly of the digastric muscle, deeply to the sternomastoid muscle. It grooves the medial surface of the mastoid bone, where it may be torn in a fracture of the base of the skull, producing a pathognomonic hematoma—Battle's sign.

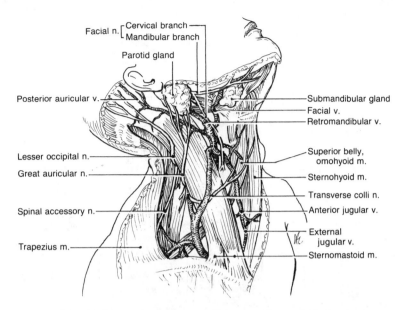

Figure 27-8. *Superficial structures of the posterior cervical triangle.*

(2) **Distribution.** It emerges from behind the sternomastoid muscle at the apex of the posterior triangle.
b. **The subclavian artery** arises from the brachiocephalic artery on the right and from the aorta on the left. It can be divided into three sections by the overlying anterior scalene muscle. It gives off three major branches.
  (1) The **vertebral artery** courses through the transverse foramina of the seventh through first cervical vertebrae and the foramen magnum to supply the brain.
  (2) The **internal thoracic artery** supplies the anterior walls of the chest and abdomen.
  (3) The **thyrocervical artery (trunk)** supplies portions of the neck and shoulder by three or four major branches.
    (a) The **inferior thyroid artery** ascends toward the inferior pole of the thyroid gland and gives off the **ascending cervical artery**, supplying deep muscles of the neck and the **inferior laryngeal artery**, which accompanies the recurrent laryngeal nerve.
    (b) The **transverse cervical artery** passes anteriorly and laterally to the anterior scalene muscle and across the base of the posterior triangle to reach the edge of the levator scapulae, which divides it into a **superficial branch** and a **deep (dorsal, descending) scapular artery**. About half of the time, there is no transverse cervical artery and a **superficial cervical artery** as well as the **dorsal (descending) scapular artery** arise separately from either the thyrocervical trunk or subclavian artery.
      (i) The superficial branch of the transverse cervical artery or the **superficial cervical artery**, whichever is present, accompanies the spinal accessory nerve on the deep surface of the trapezius muscle and anastomoses with the occipital artery.
      (ii) The deep branch (dorsal or descending scapular artery) accompanies the dorsal scapular nerve deep to the rhomboids and anastomoses with the circumflex scapular and subscapular arteries.
    (c) The **suprascapular artery** runs laterally with the suprascapular nerve, passing superiorly to the transverse ligament bridging the scapular notch (while the nerve passes inferiorly to the ligament) to supply the supraspinatus and infraspinatus muscles.
2. **Venous return.** The **external jugular vein** is formed at the angle of the jaw by the union of the posterior division of the **retromandibular vein** and the **posterior auricular vein** (see Fig. 27-8).
  a. **Course.** Superficial and variable, the external jugular vein passes in the superficial fascia and posteriorly to the middle of the clavicle, where it joins the subclavian vein.
  b. **Distribution.** It receives four tributaries.
    (1) The suprascapular vein, which accompanies the like-named artery
    (2) The superficial cervical vein, which accompanies the like-named artery
    (3) The anterior jugular vein from the anterior triangle
    (4) The posterior jugular vein from the apex of the posterior triangle

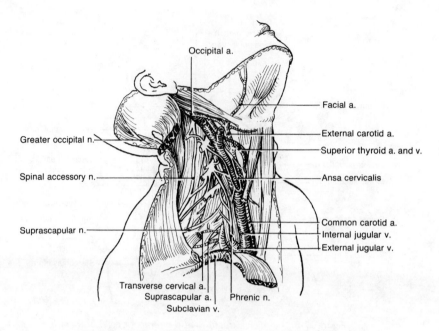

Figure 27-9. *Deep structures of the posterior cervical triangle.*

c. **Clinical considerations.** The external jugular vein may be demonstrated by obstructing its drainage with light finger pressure or by having the patient perform Valsalva's maneuver. Careful observation of the external jugular will usually reveal venous pressure waves. The level of the column of blood in the external jugular vein is a useful indication of right atrial pressure.

3. **Lymphatic drainage**
   a. The **superficial cervical lymph nodes** lie along the external jugular vein.
   b. The **occipital nodes**, at the apex of the posterior triangle, and the **retroauricular nodes**, over the mastoid process, drain the posterior scalp either through the superficial cervical nodes or through the deep cervical nodes (related to the internal jugular vein).
   c. The **nodes of the neck** are numerous and accessible. A lump in the posterior triangle is a common presentation of malignancy within the lymphatic drainage. No physical examination is complete without palpation of the lymph nodes of the neck.
   d. The **thoracic duct** passes anteriorly to the insertion of the anterior scalene to drain into one of the great veins at the root of the neck on the left side.
   e. The **bronchomediastinal trunks** also pass into the base of the neck and enter the great veins on their respective sides. Tumors of the abdomen and thorax may, therefore, involve the nodes about the subclavian vein.
   f. The breast drains via the axillary nodes into the **subclavian trunk** in the base of the neck.

28
Neurocranium

I. INTRODUCTION

A. Skull. By definition, the skull is the skeleton of the head.

B. Cranium. By definition, the cranium is the portion of the skull without the mandible. The cranium may be divided into an anterior (visceral) portion and a posterior (somatic) portion.

1. **The neurocranium** is the somatic portion of the cranium. It consists of eight bones.
   a. **The calvaria** (L. skullcap) is the vault of the neurocranium and is composed of portions of several bones. It serves to cover and protect the cerebral hemispheres of the brain.
      (1) The brain is protected by meninges, which attach to the calvaria and to the brain.
      (2) The brain is suspended in a pool of cerebrospinal fluid (CSF) confined to the subarachnoid space between arachnoid and pial meningeal layers.
   b. **Cranial nerves** pierce the base of the neurocranium to reach the face and the visceral neck. Conversely, many blood vessels in the face and the visceral neck traverse the bones of the neurocranium to reach the inner aspect of the cranial vault.
   c. **The scalp** covers the neurocranium.

2. **The facial cranium** is made up of derivatives of the upper end of the primitive gut and its associated branchial (gill) structures. It consists of 14 bones.

II. SCALP

A. Composition. The scalp comprises five layers, which correspond to the letters of the word itself: **S**kin, **C**onnective tissue, **A**poneurosis, **L**oose connective tissue, and **P**eriosteum of the calvaria (Fig. 28-1).

1. **The skin** of the scalp has the greatest concentration of hair and sebaceous glands on the body.
   a. The distribution of the scalp hair is sexually determined. Androgens produce a variable loss of hair in a wide sagittal band from the forehead to the lambdoid suture (pattern baldness).
   b. The scalp is a particularly common site for sebaceous cysts, formed by obstructed sebaceous glands.

2. **The subcutaneous tissue** of the scalp is dense and binds the skin strongly to the underlying epicranial aponeurosis.
   a. The **vasculature** of the scalp runs primarily in the subcutaneous layer. It is rich and widely anastomotic.
   b. Wounds of the scalp bleed profusely but heal well.
   c. Swelling as a result of bleeding or edema in this layer is limited by the denseness of the connective tissue and, therefore, appears as a firm tender lump.

3. **The occipitofrontalis muscle and the epicranial aponeurosis** (galea aponeurotica) comprise the musculotendinous layer of the scalp.
   a. Several flat muscles insert into the epicranial aponeurosis.
      (1) The **occipitofrontalis muscle** is formed from the frontalis and occipitalis muscles, together with the common tendon.
         (a) The paired **frontalis muscles** have no direct bony attachment. They arise above the eyebrows within the dense superficial fascia, which is in turn bound by fascial septa to the supraorbital ridges of the frontal bones. Contraction of the frontalis muscles raises the eyebrows, as in the expression of surprise.

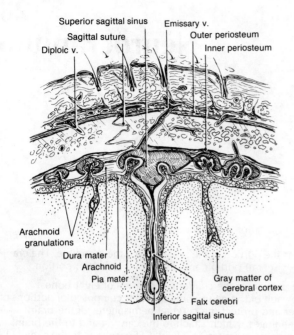

Superior sagittal sinus
Sagittal suture
Diploic v.
Emissary v.
Outer periosteum
Inner periosteum

Arachnoid granulations
Dura mater
Arachnoid
Pia mater
Falx cerebri
Inferior sagittal sinus
Gray matter of cerebral cortex

Figure 28-1. *Scalp, cranium, and meninges in coronal section.*

    (b) The paired **occipitalis muscles** arise from the highest nuchal line of the occipital bone and the mastoid process of the temporal bone. Contraction of the occipitalis muscles draws the scalp posteriorly.

    (2) The **anterior** and **superior auricularis muscles** (but not the **posterior auricularis muscle**) converge from a broad origin on the epicranial aponeurosis to insert at the base of the auricle. Some individuals have remarkable control over these muscles.

    (3) All of these muscles are derivatives of primitive second branchiomeric arch musculature and are innervated by the facial nerve (CN VII).

    b. The epicranial aponeurosis is under tension from the various muscles and, therefore, deep lacerations of the scalp gape widely.

4. The loose connective tissue between the epicranial aponeurosis and the periosteum permits considerable movement of the scalp. This forms the **subaponeurotic space**.

    a. Bleeding in the subaponeurotic space (**extracranial hematoma**) can extend over the cranium. It can extend posteriorly to the highest nuchal line; anteriorly into the eyelids to produce the "black eye," and laterally to the temporal line.

        (1) Extracranial hematoma frequently occurs in conjunction with normal childbirth, the lumpy clot being resorbed within a few weeks.

        (2) In conjunction with a depressed cranial fracture, the pressure from arterial bleeding into the closed subaponeurotic space can exacerbate the depressed fracture, compressing the brain or even driving bone fragments into the brain.

    b. The loose connective tissue is the plane of separation in any injury that tears the scalp from the calvaria.

5. The periosteum is fused firmly with the bone at the sutures. It fuses with the periosteum of the adjacent bone, thus limiting the subperiosteal space.

B. Vasculature of the scalp

    1. Arterial supply

        a. The external carotid artery provides major blood supply to the scalp through its **occipital**, **retroauricular**, and **superficial temporal branches**.

        b. The internal carotid artery supplies some of the scalp by way of the **supraorbital** and **supratrochlear branches** of the **ophthalmic artery**.

    2. Venous return

        a. The superficial veins in the subcutaneous layer generally parallel the arteries.

        b. The deep veins in the subaponeurotic space communicate with the diploic veins of the cranium as well as with the dural sinuses within the cranial vault via **emissary veins**. Due to this potential route for hematogenous spread of infection, the term "**danger space**" is applied to the subaponeurotic space.

C. Innervation to the scalp

1. **Somatic innervation.** The posterior half of the scalp is innervated by somatic nerves: the **greater occipital nerve** (C2, posterior), the **lesser occipital nerve** (C2–C3, anterior), the **least (third) occipital nerve** (C3, posterior), and the **great auricular nerve** (C2–C3, posterior).

2. **Branchiomeric innervation**
   a. **Sensory.** The anterior half of the scalp (forehead) is innervated by branches of the **trigeminal nerve** (CN V): the **supraorbital** and **supratrochlear nerves** (from CN V_1), the **zygomaticotemporal nerve** (from CN V_2), and the **auriculotemporal nerve** (from CN V_3).
   b. **Motor.** Both heads of the occipitofrontalis muscle are innervated by the facial nerve (CN VII).

III. NEUROCRANIUM

A. Basic structure

1. **Composition.** The neurocranium consists of eight bones formed from three midline endochondral bones and three pairs of lateral dermal (intramembranous) bones. However, the most anterior of the lateral pairs (the frontal bones) fuse during development.
   a. The unpaired midline bones include the **ethmoid bone**, anteriorly; the **sphenoid bone**, centrally; and the **occipital bone**, posteriorly. They form the base of the neurocranium.
   b. The paired lateral bones include the fused **frontal bone**, anterolaterally; the **temporal bones**, laterally; and the **parietal bones**, superiorly and lateroposteriorly. They form the calvarium.

2. **Articulations.** While some of the bones of the cranium (such as the frontal) fuse early in life (synchondroses), most interdigitate at unyielding sutures. The vault of the neurocranium is for the most part provided by the articulation of the flat (squamous) bones.
   a. In children, the sutures permit centrifugal growth of flat bones.
   b. In adults, the boundaries of the various bones have little significance but provide useful landmarks for the identification of related structures.

3. **Development and growth**
   a. The squamous bones are formed by intramembranous ossification early in fetal life. These plates "float" within the periosteal membranes on the surface of the developing brain and do not articulate with each other. During passage through the birth canal, the bony plates ride over one another to some extent, facilitating parturition.
   b. In infancy, the bony plates resolve into compact inner and outer **tables** separated by cancellous **diploë**, and the bones either fuse with each other at synchondroses or form sutures. In an infant, there are gaps (fontanelles) in the calvaria at the rounded corners of the parietal bones (Fig. 28-2).

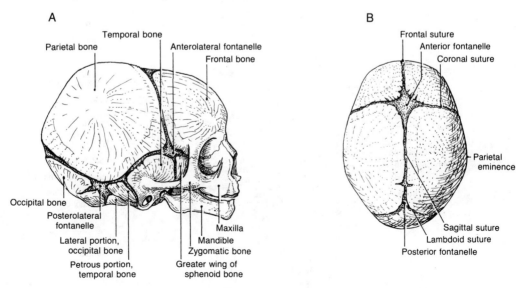

Figure 28-2. *Skull of the newborn. A,* Lateral aspect. Note the relative sizes of the neurocranium and facial skeleton, the fontanelles, and the absence of a mastoid process. *B,* Superior aspect.

(1) The anterior gap is the **anterior fontanelle**.
    (a) The anterior fontanelle is diamond-shaped. This enables the obstetrician to determine the orientation of the fetal head by vaginal examination during labor.
    (b) The anterior fontanelle permits access to intracranial veins and also enables the perinatologist to assess intracranial pressure.
(2) The posterior gap is the **posterior fontanelle**.
(3) There are also anterolateral and posterolateral fontanelles.

4. **The marrow cavity** of the diploë contains veins.
  a. These **diploic veins** drain into four main trunks: the occipital, posterior temporal, anterior temporal, and frontal diploic veins.
  b. Some of the diploic veins form communicating channels (**emissary veins**) between the superficial veins and the dural sinuses.

B. **External surface**

1. **The base** of the cranium slopes upward from posterior to anterior. This oblique surface forms the juncture of the neurocranium and the facial cranium (see Fig. 28-3).

2. **The lateral aspect** displays features of the frontal bone, zygomatic bone, temporal bone, and parietal bone (Fig. 28-3).
  a. **Frontal bone landmarks** (see Fig. 28-3)
    (1) The **glabella** (L. hairless) is the smooth prominence immediately superior to the bridge of the nose (**nasion**) [see Fig. 31-2].
    (2) The **superciliary crest** is a prominent ridge on either side of the midline just superolateral to the glabella.
    (3) A **supraorbital notch** (occasionally **foramen**) marks the point where the supraorbital artery and nerve leave the orbit and run onto the forehead.
    (4) More laterally, the supraorbital margin continues as the **supraorbital ridge**.
  b. **Zygomatic bone landmarks** (see Figs. 28-3 and 28-4)
    (1) The orbital margin turns sharply inferior along the frontal process of the zygomatic bone.
    (2) The **temporal process** of the zygomatic bone projects posteriorly to fuse with the zygomatic frontal process of the temporal bone, forming the **zygomatic arch**.
    (3) The zygomatic arch bridges the **temporal fossa** and conceals the base of the cranium from lateral view.
  c. **Temporal bone landmarks** (see Fig. 28-3)
    (1) At the temporal root of the **zygomatic frontal process**, the **articular tubercle** lies just anterior to the **condylar notch**.
    (a) These form a sigmoidal articular surface for the condylar process of the mandible.
    (b) An articular disk is interposed between the bones of the **temporomandibular joint**.

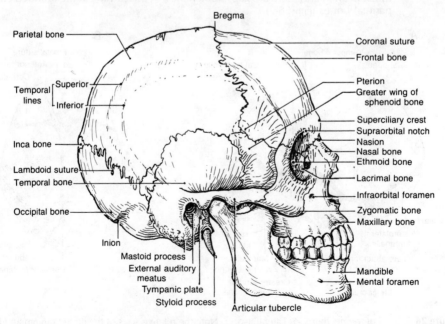

Figure 28-3. *Lateral aspect of the skull.*

(2) A flat, nonarticular **tympanic plate** intervenes between the condylar notch and the **external auditory meatus**.

(3) Superior to the external auditory meatus and mastoid process, the zygomatic arch continues as a bony ridge, the **supramastoid crest**.

   (a) A small **suprameatal spine** lies at the posterosuperior edge of the external auditory meatus. This spine and the supramastoid crest bind the small **suprameatal triangle**, which is a surgical landmark for the **mastoid antrum**.

   (b) The supramastoid crest passes along the lateral aspect of the neurocranium as the **temporal line** from which the temporalis muscle originates (see Fig. 28-3).

(4) Inferior to the external auditory meatus, the long **styloid process** projects at a right angle to the base of the cranium (see Fig. 28-3).

   (a) This process gives rise to several muscles of the visceral neck, as well as to the stylohyoid and stylomandibular ligaments.

   (b) It is a remnant of the second branchial arch.

(5) The stout **mastoid process** lies posterolaterally to the styloid process (see Fig. 28-3).

   (a) The mastoid process is not present at birth but develops as a result of traction from the sternomastoid muscles. The splenius capitis and longissimus capitis muscles insert onto its posterolateral aspect.

   (b) The mastoid process is filled with air cells, which communicate, via the **mastoid antrum**, with the cavity of the middle ear. The spread of a middle ear infection into the mastoid air cells is difficult to treat because of poor drainage. Radical surgery involves scraping and extirpation of the infected air cells (mastoidectomy).

   (c) On the internal aspect, the mastoid air cells are related to the sigmoid sinus.

   (d) There are foramina for a few emissary veins posterior to the mastoid process (see Fig. 28-5).

  d. **The squamous temporal and frontal bones** do not meet; the small gap is filled by the greater wing of the **sphenoid bone** (see Fig. 28-3).

   (1) The **pterion** is the point at which the sphenoid bone joins the parietal bone. It is located two finger-breadths above the midpoint of the zygomatic arch, and in infants, it is the location of the anterolateral fontanelle.

   (2) The intracranial surface of the temporal bone is grooved and occasionally tunneled by the middle meningeal vessels, so that a blow to the side of the head, which either fractures the skull or tears the periosteum at this point, produces rapid intracranial hemorrhage (*epidural hematoma*).

3. Posterior aspect (see Figs. 28-3 and 28-4)

  a. The attachment of the ligamentum nuchae to the occipital bone is marked by the **external occipital protuberance (inion)** and a thin **external occipital crest**, extending in the sagittal plane to the foramen magnum.

  b. The **superior nuchal line** connects the inion with the mastoid process and gives origin to the trapezius muscle.

  c. The **inferior nuchal line** divides the insertions of the splenius capitis muscle from the insertion of the posterior rectus capitis muscle.

  d. The **highest nuchal line** provides the origin for the occipital belly of the occipitofrontalis muscle.

4. Superior aspect (see Fig. 28-3)

  a. Sutures

   (1) The **coronal (frontoparietal) suture** is between the frontal bone and the parietal bones.

   (2) The parietal bones are separated by the **sagittal (parietal) suture**.

   (3) The **bregma** is the point where the sagittal and coronal sutures meet; it is the location of the **anterior fontanelle**.

   (4) The **lambdoid (occipitoparietal) suture** is between the parietal bones and the occipital bone.

  b. The lambda is the point where the sagittal and lambdoid sutures meet; it is the location of the **posterior fontanelle**.

  c. Sutural (wormian) bones are small squamous bones, which occasionally form within sutures. They are most common in the lambdoid suture, where they tend to be symmetrically placed and may be called "Inca bones."

5. Inferior aspect (Fig. 28-4)

  a. Anteriorly, the midline bones of the base of the neurocranium roof the nasal cavity, and the corrugated orbital plates of the frontal bones roof the orbital cavities.

  b. Posteriorly, the **occipital bone** is composed of two parts.

   (1) The more posterior **squamous part** of the occipital bone has an inferior triangular midline notch. The anterior basilar part of the occipital bone is Y-shaped. Fused together they bound the diamond-shaped **foramen magnum**.

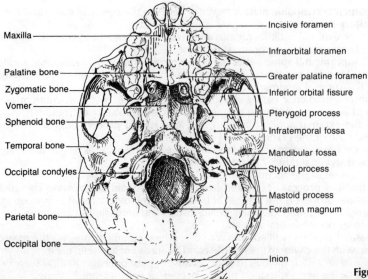

Maxilla

Palatine bone

Zygomatic bone

Vomer

Sphenoid bone

Temporal bone

Occipital condyles

Parietal bone

Occipital bone

Incisive foramen

Infraorbital foramen

Greater palatine foramen

Inferior orbital fissure

Pterygoid process

Infratemporal fossa

Mandibular fossa

Styloid process

Mastoid process

Foramen magnum

Inion

Figure 28-4. *Inferior aspect of the cranium.*

(2) The **basilar part** of the occipital bone has (along the anterior edges of the foramen magnum) paired **occipital condyles** for articulation with the atlas.
  (a) The **hypoglossal (anterior condylar) canal**, emerging beneath the anterolateral edge of each condyle, transmits the hypoglossal nerve (CN XII).
  (b) A **posterior condylar canal** at the posterolateral end of each condyle transmits an emissary vein.
c. Laterally, the petrous portions of the **temporal bones** consist of two columns of dense bone that abut the anterolateral edges of the limbs of the basioccipital bone. Each petrous portion of the temporal bone and the adjacent edges of the squamous portions of the temporal and occipital bones contain numerous foramina.
  (1) Important bony landmarks lie in three nearly parallel rows that run obliquely along the petrous portion of the temporal bone.
    (a) **Anterolateral row** (Fig. 28-5)
      (i) The **petrosquamous fissure** lies between the petrous and squamous portions of the temporal bone. Its medial portion, the **petrotympanic fissure**, transmits the chorda tympani nerve.
      (ii) The **foramen spinosum** transmits the middle meningeal artery.
      (iii) The **foramen ovale** transmits the motor and sensory roots of the mandibular division of the trigeminal nerve (CN V), together with the accessory meningeal artery, the lesser superficial petrosal nerve, a recurrent meningeal branch of the trigeminal nerve, and an emissary vein.
      (iv) The **pterygoid (vidian) canal** lies at the base of the **medial pterygoid plate**. It transmits the vidian nerve (formed by the greater superficial petrosal nerve and the deep petrosal nerve) to the pterygopalatine fossa.
    (b) The **middle row** lies in the anterior edge of the petrous bone (see Fig. 28-5).
      (i) The **mastoid process** is grooved on its medial side for the digastric muscle and for the occipital artery.
      (ii) The **tympanic plate** forms the sheer vertical anterior wall of the external auditory meatus.
      (iii) The middle ear cleft continues anteriorly as the **tympanic (auditory, eustachian) tube**. The bony portion emerges from the anteromedial end of the tympanic plate. It is continued by a cartilaginous portion, which runs in a groove along the anterolateral margin of the petrous portion of the temporal bone.
    (c) **Posteromedial row** (see Fig. 28-5)
      (i) The **jugular foramen** runs posteriorly into the cranial cavity. It transmits the internal jugular vein, glossopharyngeus nerve (CN IX), vagus nerve (CN X), and spinal accessory nerve (CN XI).
      (ii) The **carotid canal** curves sharply anterior along the petrous temporal bone.

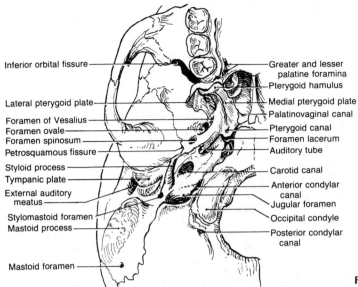

Inferior orbital fissure

Lateral pterygoid plate

Foramen of Vesalius
Foramen ovale
Foramen spinosum
Petrosquamous fissure

Styloid process
Tympanic plate
External auditory
  meatus

Stylomastoid foramen
Mastoid process

Mastoid foramen

Greater and lesser
  palatine foramina
Pterygoid hamulus

Medial pterygoid plate
Palatinovaginal canal
Pterygoid canal
Foramen lacerum
Auditory tube

Carotid canal
Anterior condylar
  canal
Jugular foramen
Occipital condyle
Posterior condylar
  canal

Figure 28-5. *Foramina and landmarks of the infratemporal fossa.*

    (iii) The **foramen lacerum** results from a discontinuity at the ragged apex of the temporal bone. The **internal carotid** artery enters the cranial cavity alongside the sphenoid bone, which forms the anterior margin of the foramen lacerum.

  (2) The **pterygoid processes (plates)** arise from the inferior surface of the greater wing of the sphenoid bone (see Fig. 28-5), and these form the lateral boundaries of the **choanae**, which demarcate the nasal cavity from the nasopharynx.

    (a) The **medial pterygoid plate** gives rise to the superior constrictor of the pharynx and terminates in the **pterygoid hamulus**, which functions as a trochlea for the tendon of the tensor veli palatini muscle.

    (b) The **lateral pterygoid plate** gives origin to the pterygoid muscles (lateral and medial).

  (3) The **pterygopalatine fossa** is a small, inverted pyramidal space deep to the infratemporal fossa, through which arteries and nerves gain access to the nasal cavity, the palate, the orbit, and the face (see Fig. 31-4).

    (a) The **pterygomaxillary fissure** opens into the pterygopalatine fossa.

    (b) It is bounded by the maxilla anteriorly and by the pterygoid process of the sphenoid bone posteriorly.

    (c) Medially, the thin perpendicular plate of the palatine bone separates it from the nasal cavity.

  (4) The **inferior orbital fissure** lies between the greater wing of the sphenoid and the maxilla (see Fig. 28-4).

C. Interior surface

  1. Calvaria (Fig. 28-6)

    a. The **cranial sutures** become obliterated with old age. The inner calvarial surface is deeply grooved and occasionally tunneled by the meningeal vessels.

    b. A **frontal crest** provides attachment for the falx cerebri, a midsagittal fold of dura mater. It diminishes and becomes the **sagittal groove**, which deepens posteriorly to accommodate the **sagittal sinus**.

    (1) On either side of the midline are **granular lacunae**.

      (a) These accommodate the venous lacunae, into which CSF drains from arachnoid granulations.

      (b) They become larger and deeper with age.

    (2) The sagittal groove becomes a raised gutter over the occipital bone.

      (a) At the **internal occipital protuberance**, it meets similar gutters (sulci for the transverse sinuses), converging from both sides in the transverse plane.

      (b) This marks the **confluence of sinuses** (torculus of Herophilus) [see Fig. 28-7].

    c. The **internal occipital crest**, below the confluence of sinuses, provides attachment for the falx cerebelli.

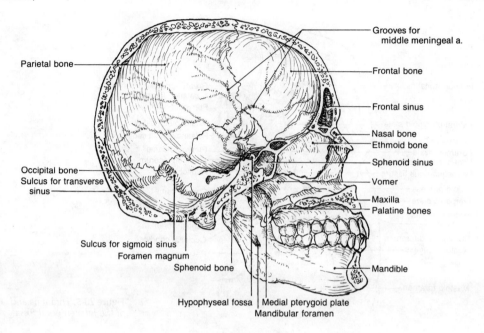

Parietal bone

Grooves for
middle meningeal a.

Frontal bone

Frontal sinus

Nasal bone
Ethmoid bone

Sphenoid sinus

Vomer

Maxilla
Palatine bones

Occipital bone
Sulcus for transverse
sinus

Mandible

Sulcus for sigmoid sinus
Foramen magnum
Sphenoid bone

Hypophyseal fossa │ Medial pterygoid plate
Mandibular foramen

Figure 28-6. *Lateral aspect of the interior skull.*

2. The floor of the cranial cavity (see Fig. 28-6; Fig. 28-7). In the midline, the floor of the cranial cavity declines in a continuous convex curve, interrupted only by the **tuberculum sellae** and the **dorsum sellae** of the sphenoid bone. Laterally, the **cranial fossae** (anterior, middle, and posterior) decline in steps from anterior to posterior.

  a. The anterior cranial fossa, formed by the ethmoid, frontal, and sphenoid bones, contains the frontal lobes of the brain.
  (1) The anterior fossa is bounded anteriorly and laterally by the frontal bone.
  (2) The **crista galli** of the ethmoid bone provides attachment for the falx cerebri.
    (a) Between the crista galli and the crest of the frontal bone is a small **foramen cecum**, transmitting an emissary vein that passes from the veins of the frontal sinus and nasal cavity to the superior sagittal sinus.
    (b) On either side of the crista galli lie the olfactory bulbs of the brain.
    (c) The thin **cribriform plate** of the ethmoid bone is perforated on each side by 15–20 foramina through which the olfactory nerves pass from the nasal mucosa to each olfactory bulb.
  (3) The orbital plates of the frontal bone form the floor of the anterior fossa and the roof of the orbit.
  (4) The smooth arc of the lesser wing of the sphenoid bone demarcates the sharp posterior margin of the anterior cranial fossa.
  b. The middle cranial fossa, formed by the sphenoid and temporal bones, is deeper than the anterior fossa and is related to the pituitary gland and the temporal lobes of the brain (see Fig. 28-7).
  (1) Most of the anterior wall of the middle cranial fossa is formed by the greater wing of the sphenoid bone.
    (a) Anteriorly, the lesser wing meets the orbital plate. Medially, the lesser wing forms the **anterior clinoid process**.
    (b) Laterally, the greater wing of the sphenoid bone fuses with frontal, parietal, and temporal bones to close the pterion.
    (c) The space between the greater and lesser wings is the **superior orbital fissure**.
  (2) The **optic groove (sulcus chiasmatis)** grooves the sphenoid bone between the **optic foramina** and the tuberculum sellae. It is so named because it contains the optic chiasm, rather than the optic nerve.
  (3) The **sella turcica** is named for its fanciful resemblance to a premedieval Turkish saddle with its anterior tuberculum sellae, central fossa, and posterior dorsum sellae.
    (a) The **tuberculum sellae** rises from the body of the sphenoid bone, dividing the hypophyseal fossa from the optic groove.

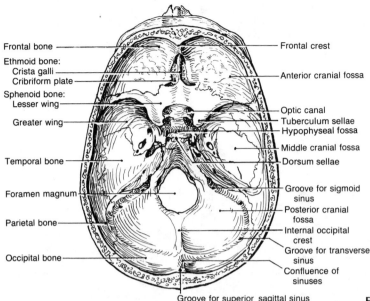

Frontal bone

Ethmoid bone:
Crista galli
Cribriform plate

Sphenoid bone:
Lesser wing

Greater wing

Temporal bone

Foramen magnum

Parietal bone

Occipital bone

Frontal crest

Anterior cranial fossa

Optic canal
Tuberculum sellae
Hypophyseal fossa

Middle cranial fossa

Dorsum sellae

Groove for sigmoid sinus

Posterior cranial fossa

Internal occipital crest

Groove for transverse sinus

Confluence of sinuses

Groove for superior sagittal sinus

Figure 28-7. *Floor of the cranium.*

(b) The **hypophyseal (pituitary) fossa**, a midline depression, contains the **pituitary gland (hypophysis cerebri)**.

(c) The **dorsum sellae** is a vertical plate between the the posterior clinoid processes, forming the posterior wall of the sella.

(d) The terms in (a), (b), and (c) above tend to be used somewhat interchangeably (e.g., ''. . .an expanding pituitary adenoma eroding the hypophyseal fossa with supracellar extension compressing the optic chiasm. . .'').

(4) The **posterior clinoid processes** are continuations of the dorsolateral corners of the flat dorsum sellae.

(5) The **optic canal** contains the optic nerve, the ophthalmic artery, and the central vein of the retina.

(6) A group of foramina form an arc around the medial edge of the greater wing of the sphenoid bone (Fig. 28-8).

(a) The **superior orbital fissure** passes between the wings of the sphenoid into the orbital cavity. The oculomotor nerve (CN III), the trochlear nerve (CN IV), the ophthalmic division of the trigeminal nerve (CN V_1), and the abducens nerve (CN VI) all pass through the medial end of the superior orbital fissure to gain access to the orbit. The ophthalmic veins leave the orbit through this fissure.

(b) The **foramen rotundum** conveys the maxillary division of the trigeminal nerve (CN V_2) into the pterygopalatine fossa.

(c) The **foramen ovale** transmits the motor and sensory roots of the **mandibular division** of the **trigeminal nerve** (CN V_3) into the infratemporal fossa, together with an accessory meningeal artery, the lesser superficial petrosal nerve, a recurrent meningeal branch of the trigeminal nerve, and an emissary vein.

(d) The **foramen spinosum** transmits the **middle meningeal artery**.

(e) There are other small, variable foramina that lie along this arc (e.g., the emissary foramen, the foramen of Vesalius, and the innominate canal).

(7) The **carotid canal** passes through the body of the petrous portion of the temporal bone to emerge laterally to the dorsum sellae.

(a) The **foramen lacerum** is formed by superior and inferior defects in the carotid canal (i.e., the petrous temporal bone fails to meet the sphenoid bone). In the floor of the middle cranial fossa, the foramen lacerum is covered by a layer of fibrocartilage and periosteum.

(b) No structure passes all the way through the foramen lacerum.

(8) The posterior margin of the middle cranial fossa is formed by the superior margin of the petrous temporal bone laterally to the posterior clinoid process.

(a) The roof of the middle ear cavity is the thin **tegmen tympani**. It is easy to understand how a severe middle ear infection may erode into the cranial cavity.

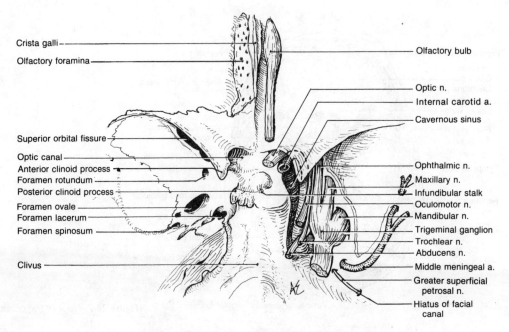

Crista galli
Olfactory foramina
Olfactory bulb
Optic n.
Internal carotid a.
Cavernous sinus
Superior orbital fissure
Optic canal
Anterior clinoid process
Foramen rotundum
Posterior clinoid process
Foramen ovale
Foramen lacerum
Foramen spinosum
Clivus
Ophthalmic n.
Maxillary n.
Infundibular stalk
Oculomotor n.
Mandibular n.
Trigeminal ganglion
Trochlear n.
Abducens n.
Middle meningeal a.
Greater superficial petrosal n.
Hiatus of facial canal

Figure 28-8. *Foramina of the middle cranial fossa.*

    (b) The anterior semicircular canal is accommodated by the **arcuate eminence**.
    (c) The superior edge of the petrous portion of the temporal bone is sharp and grooved for the superior petrosal sinus.
    (d) At the apex of the petrous portion of the temporal bone is a depression for the trigeminal (semilunar) ganglion and, medially to that, a groove for the abducens nerve (CN VI).

  c. The posterior cranial fossa, the deepest of the fossae, is formed by the temporal and occipital bones and contains the cerebellum and brain stem (see Figs. 28-6 and 28-7).

    (1) The **clivus** of the basal portion of the occipital bone climbs from the foramen magnum to the dorsum sellae (see Fig. 28-8).
    (2) The **transverse sinus** and **sigmoid sinus** deeply groove the walls of the posterior cranial fossa. The sigmoid sinus joins the **inferior petrosal sinus** to become the **superior bulb** of the **internal jugular vein** as it exits the cranial cavity through the **jugular foramen**.
    (3) On the posterior surface of the petrous portion of the temporal bone are numerous minor fossae: a fossa for the endolymphatic sac, one for the canaliculus (aqueduct) of the cochlea, and the subarcuate fossa.
    (4) Several foramina pass through the basal walls of the posterior cranial fossa.
      (a) The **internal auditory meatus** emerges through the posteromedial surface of the petrous portion of the temporal bone, transmitting the facial nerve (CN VII) and the vestibulocochlear nerve (CN VIII).
      (b) The **jugular foramen** transmits the internal jugular vein through its posterior part. The glossopharyngeus nerve (CN IX), the vagus nerve (CN X), and the spinal accessory nerve (CN XI) all pass through the jugular foramen anteriorly to the internal jugular vein.
      (c) The **anterior condylar (hypoglossal) canal** transmits the hypoglossal nerve (CN XII).
      (d) At the **foramen magnum**, the brain stem is continuous with the spinal cord. The spinal roots of the spinal accessory nerve (CN XI) and the vertebral arteries pass into the cranial cavity through this foramen.
      (e) The **posterior condylar canal** transmits an emissary vein from the sigmoid sinus to the veins of the pericranium.

D. Summary of the foramina of the neurocranium

  1. The anterior cranial fossa is pierced by numerous **olfactory foramina** in the cribriform plate of the ethmoid bone.

2. The major foramina of the middle cranial fossa all pass through the sphenoid bone.
    a. Optic canal: CN II, the ophthalmic artery, and the central vein of the retina
    b. Superior orbital fissure: CN III, CN IV, CN V$_1$, CN VI, and the ophthalmic vein
    c. Foramen rotundum: CN V$_2$
    d. Foramen ovale: CN V$_3$, accessory meningeal artery, recurrent meningeal nerve, and lesser superficial petrosal nerve
    e. Foramen spinosum: the middle meningeal artery
    f. Foramen lacerum: Nothing traverses it completely.
    g. Hiatus of the facial canal: greater superficial petrosal nerve

3. The major foramina of the posterior cranial fossa pass through the temporal or occipital bones.
    a. Internal auditory meatus: CN VII and CN VIII
    b. Jugular foramen: CN IX, CN X, CN XI, and the internal jugular vein
    c. Hypoglossal (anterior condylar) canal: CN XII
    d. Foramen magnum: roots of CN XI (retrograde), the brain stem, and vertebral arteries

E. Clinical considerations

  1. Fracture of the cranium may involve the scalp, dura, cranial nerves, and cerebral tissue.
    a. The prominence of the parietal bones makes them most susceptible to fracture. The occipital bones are less often fractured.
    b. Bone fragments may tear the venous sinuses, dura, or a meningeal vessel or even lacerate the brain.
    c. Fractures of the base of the skull frequently cross foramina (weaker portions of the bone) and jeopardize the contained structures.

  2. Space-occupying lesions (depressed skull fractures, tumors, hematomas, and even swellings of cerebral tissue) encroach upon the space occupied by the incompressible brain because the neurocranium is a rigid box.
    a. Regions of the temporal lobes may be forced through the tentorial notch, causing oculomotor nerve dysfunction.
    b. The brain stem may be forced through the foramen magnum, compressing nerves and arteries; the resultant medullary ischemia rapidly results in death.
    c. The bony confines of a foramen or canal cannot accommodate swelling of a nerve. Inflammatory swelling produces nerve compression and dysfunction (e.g., Bell's palsy).

IV. MENINGES AND VENOUS SINUSES

  A. Structure. The central nervous system is enclosed in three layers of meninges: the dura mater, the arachnoid mater, and the pia mater (see Fig. 28-1; Fig. 28-9).

    1. The dura mater is a tough and fibrous protective layer divided into two portions.
      a. The outer dural layer is derived from the internal layer of cranial periosteum. It adheres closely to the bone.
        (1) The **middle meningeal artery** is primarily associated with this layer. Even though this artery and its major branches run between the periosteal and meningeal dural layers, the small branches are within the periosteal layer.
        (2) The corresponding veins are completely periosteal.
      b. The inner dural layer is the true meningeal dura mater. It adheres in most places to the outer layer, explaining the confusion, as this structure was named 2300 years ago.
        (1) Folds of the inner dural layer (back-to-back layers) extend deeply between the major divisions of the brain.
          (a) The **falx cerebri** (*falx, L.* sickle) lies in the midsagittal plane between the cerebral hemispheres.
            (i) It attaches to the crista galli and the frontal crest anteriorly and joins the tentorium cerebelli posteriorly.
            (ii) Superiorly, the two layers (one from either side of the cranial vault) split and attach to the lateral edges of the sagittal groove to enclose the **superior sagittal sinus**.
          (b) The small **falx cerebelli** lies in the midsagittal plane, attaching to the internal occipital crest.
          (c) The **tentorium cerebelli** (*tendo, L.* to stretch) lies in a more or less transverse plane between the cerebellum and the cerebrum. It attaches around the occipital bone and along the ridge of the petrous portion of the temporal bone.

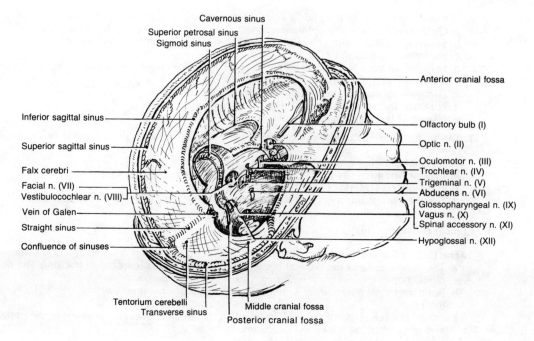

Figure 28-9. *Dural folds, venous sinuses, and cranial nerves.* Note that the cranial nerves pass through the meninges in numerical order.

(i) The midline **tentorial notch (incisura tentorii)** fits snugly about the midbrain and divides the posterior cranial fossa from the rest of the cranial cavity.

(ii) Laterally, the two layers split and attach to the calvarial walls to enclose the **transverse sinus** and attach along the ridge of the petrous portion of the temporal bone to enclose the **superior petrosal sinus**.

(iii) At the apex of the petrous portion of the temporal bone, the inferior layer evaginates into the middle cranial fossa as the **trigeminal cave** (of Meckel) for the trigeminal ganglion.

(iv) Medially to this, the corner of the tentorium cerebelli twists so that the free edge continues anteriorly toward its attachment on the anterior clinoid process, and the attached border continues medially to the posterior clinoid process.

(d) The **diaphragma sellae** covers the hypophyseal fossa except for a small hiatus through which the infundibular stalk and the hypophyseal portal veins pass.

(2) Along the attached borders of the dural folds, the two layers diverge before joining with the periosteal dura.

(a) These spaces, bounded by two leaves of meningeal dura and with a base of periosteal dura, are lined by endothelium to form **dural venous sinuses** (see Figs. 28-1 and 28-9).

(b) The triangular shape of the sinus prevents collapse, so that if a sinus is torn by a bone fragment from a compound depressed cranial fracture or accidently incised at surgery, air can be sucked into the venous system sufficient to fill the right ventricle with froth and cause death.

(c) Major sinuses at dural folds include the following:

(i) The **superior sagittal sinus** lies in the root of the falx cerebri (see Figs. 28-1 and 28-9).

(ii) The **inferior sagittal sinus** lies in the free edge of the falx cerebri (see Figs. 28-1 and 28-9).

(iii) A small vein (misnamed the **great cerebral vein of Galen**) drains the posterior aspect of the midbrain and unites with the inferior sagittal sinus to form the **straight sinus**, which runs along the intersection of the falx cerebri and the tentorium cerebelli.

(iv) The small **occipital sinus** lies in the root of the falx cerebelli (see Fig. 28-10).

(v) The superior sagittal, straight, and occipital sinuses meet at the **internal occipital protuberance** to form the **confluence of sinuses (torculus of Herophilus)**.

(vi) The **left** and **right transverse sinuses** drain from the confluence. The superior sagittal sinus drains predominantly into the right transverse sinus; the straight and occipital sinuses drain predominantly into the left transverse sinus.

   (vii) The **transverse sinuses** lie along the edge of the tentorium cerebelli. On reaching the petrous portion of the temporal bone, they turn sharply inferior as the **sigmoid sinus**, which becomes the internal jugular vein.

   (viii) The **superior petrosal sinus** enters the transverse sinus at the point where it becomes the sigmoid sinus.

(3) Numerous small sinuses (e.g., the inferior petrosal, sphenoparietal, and cavernous sinuses) are unrelated to dural folds (Fig. 28-10). Of these, only the **cavernous sinus** is of major importance.

   (a) The cavernous sinus is like a plexus of sinuses lying laterally to the body of the sphenoid bone, extending from the superior orbital fissure to the petrous portion of the temporal bone.

   (b) Venous flow through the cavernous sinus is a slow percolation.

   (i) It receives the ophthalmic vein, the central vein of the retina and the middle and inferior cerebral veins, as well as emissary veins from the deep face and visceral neck.

   (ii) It drains into the superior and inferior petrosal sinuses.

   (c) The carotid artery and abducens nerve (CN VI) run through the cavernous sinus medially within sheaths of endothelium. Cranial nerves III, IV, V_1, and V_2 lie more laterally along the wall of the cavernous sinus (Fig. 28-11).

   (i) Because the ophthalmic veins drain a small area of the face via connections with the angular facial vein and because emissary veins communicate with the visceral neck, infections in these areas may spread into the cavernous sinus.

   (ii) Infection may result in *cavernous sinus thrombosis* with pain behind the eye, cranial nerve palsies affecting eye movement (ophthalmoplegia), thrombosis of the internal carotid artery with the possibility of stroke, and edema of the optic disk (papilledema) from venous engorgement of the retina.

(4) The meningeal layer of dura continues through the foramen magnum. The dura also continues over the optic and olfactory tracts. About the other cranial nerves, it is represented by perineurium.

c. From anterior to posterior, the cranial nerves pass through the dura in numerical order to leave the cranial cavity. However, this numerical relationship does not hold for the foramina of the bony skull.

d. The potential space between the dura and arachnoid layers, the **subdural space**, normally contains no more than a thin film of serous secretion.

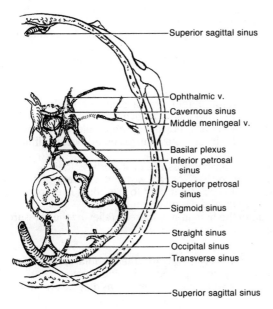

Figure 28-10. *Venous sinuses of the cranium.*

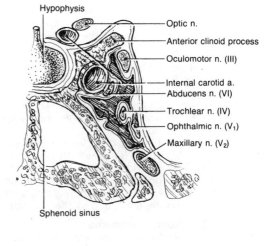

Figure 28-11. *Contents of the cavernous sinus.*

2. The arachnoid layer lies deeply to the dura to which it adheres somewhat (see Fig. 28-1).

a. **Arachnoid trabeculae** (fine strands of subarachnoid tissue) cross to the underlying pia mater.

b. **The subarachnoid space**, the actual space between the arachnoid layer and the pia mater, is filled with cerebrospinal fluid (CSF).

(1) **CSF** is produced by the choroid plexuses, which project into the ventricles of the brain.

(a) It flows through the ventricular system, entering the subarachnoid space through a midline **foramen of Magendie** and two lateral **foramina of Luschka** in the roof of the fourth ventricle.

(b) CSF flows through the subarachnoid space of the spinal cord and brain, suspending the brain and functioning as a hydraulic shock absorber.

(c) It reenters the circulatory system through the **arachnoid villi** into the venous sinuses, primarily the superior sagittal sinus.

(i) These small, nipple-like tufts of arachnoid project through the inner layer of dura into the venous sinuses.

(ii) CSF pressure is usually somewhat greater than venous pressure. When it is greater, there is drainage of CSF into the venous system. When venous pressure is greater than CSF pressure, the minute tubules are compressed, preventing seepage of blood into the CSF, where it is an extreme irritant.

(iii) The flow of CSF may be interrupted by congenital malformations or by the scarring associated with the healing of meningitis. The fluid accumulates as *hydrocephalus*. In children with hydrocephalus, the neurocranium enlarges and the fontanelles fail to close. This condition can be treated by early placement of a shunt to drain the CSF into the venous system or into the peritoneal cavity.

(2) The major distributing arteries to the brain course in the subarachnoid space.

(a) A vascular bleed from a major artery—for example, a ruptured cerebral aneurysm—will disseminate widely in the subarachnoid space.

(b) The presence of blood in the CSF from a lumbar tap is evidence of arterial bleeding.

3. The pia mater is a delicate layer that intimately follows the surface of the brain through all of its convolutions. It descends into sulci and ensheaths the blood vessels as they enter the brain parenchyma (see Fig. 28-1).

B. Clinical considerations

1. Intracranial bleeding creates a space-occupying hematoma.

a. The raised intracranial pressure causes headache, nausea, depressed consciousness, and edema of the optic disk.

b. Tentorial herniation (squeezing the temporal lobe through the tentorial notch) compresses the oculomotor nerve to extraocular and intraocular muscles. One of the first signs of temporal lobe herniation is dilation of the pupil on the side of the lesion, then bilateral dilation as continuing herniation involves the contralateral nerve. This is the reason for monitoring pupillary responses in cases of head injury.

c. Compression of the medulla causes anoxia to the vital centers of the brain stem and subsequent death.

2. Epidural hematoma

a. Epidural hematoma is most commonly the result of temporal or parietal bone fracture.

(1) The middle meningeal vessels, lying in the meningeal groove, may be torn (see Fig. 28-6). Bleeding may be either arterial or venous; however, arterial bleeding is rapid and at arterial pressure. Symptoms of brain compression generally occur within 3 hours.

(2) Usually, head trauma is followed by a transient loss of consciousness (concussion), which is followed by a lucid interval, which lasts from a few minutes to several hours. Individuals often recover from the initial concussion only to die hours later from an undiagnosed epidural hematoma—the *talk-and-die syndrome*.

b. Emergency craniotomy is indicated to drain the hematoma, thereby relieving the pressure on the brain, and to effect hemostasis.

3. Subdural hematoma

a. As the larger cerebral veins pierce the arachnoid layer and cross the subdural space to gain access to the dural sinuses, they are relatively unsupported and vulnerable.

(1) As a result of a blow to the head or even a sudden jarring, the veins may be torn in the subdural space.

(2) Bleeding extends beneath the dura and superficially to the arachnoid; it seldom appears in the CSF if the arachnoid is intact.

b. The venous ooze is often very slow, and, because CSF production is a function of intra-cranial pressure, the effects of the subdural hematoma are in part compensated for by a reduction in CSF production. Hence, the characteristic chronic fluctuation of symptoms.

4. Subarachnoid hemorrhage

a. Rupture of a cerebral artery, which has been weakened by atheroma or idiopathic (berry) aneurysm, may be rapidly fatal. There is usually a history of sudden severe headache with onset correlated to strenuous activity.

b. Blood is present in a lumbar tap.

5. Pial hemorrhage

a. Cerebral trauma that causes the brain to strike the cranial surfaces may bruise the small vessels of the pia and the brain tissue. This occurs not only directly under the blow (coup), but often also at the opposite side of the brain (contracoup) as a result of rebound.

b. The aftermath of hemorrhage is atrophy, degeneration, and corresponding loss of function, illustrated by the ''punch-drunk'' boxer.

I. INTRODUCTION

A. Brain

1. Composition. The brain consists of a central stem partially surrounded by two expansions, the cerebrum and the cerebellum.

   a. The brain stem is continuous with the **spinal cord** within the foramen magnum (see Fig. 29-5).

   (1) In the posterior cranial fossa, the brain stem lies on the basioccipital bone.

   (2) It passes through the **tentorial notch (incisura tentorii)** of the tentorium cerebelli to reach the middle cranial fossa where it lies on the body of the sphenoid bone.

   b. The cerebellum extends dorsally from the brain stem and fills the posterior cranial fossa (see Fig. 29-4).

   c. The cerebrum occupies the greater portion of the middle and anterior cranial fossae. It is divided into a number of lobes named by the overlying bone (see Fig. 29-1).

2. Substructure. The **cerebral cortex** consists of 10^{15} individual neurons that are connected in specific patterns.

   a. Some neurons make synaptic contact with several thousand other neurons; others are more specific and synapse with very few neurons.

   b. In the adult, neurons may die if subjected to damage or ischemia. Neuronal death is accompanied by alterations in function. In many instances, function may be lost or altered permanently because brain tissue does not regenerate.

B. Cranial nerves.
Generally arising from the ventral and lateral surfaces of the brain stem, the twelve cranial nerves exit the cranium through foramina in the floor.

II. CEREBRUM (FOREBRAIN)

A. Telencephalon

1. Overview. The left and right cerebral hemispheres consist of the cerebral cortex, white matter, and the basal ganglia (see Fig. 29-3). Originally related to the olfactory bulbs, the cerebral hemispheres expanded superiorly and posteriorly during development. Their numerous connecting axon pathways (the internal capsule) flank the thalamus and hypothalamus (diencephalic structures). The connections between the cerebral hemispheres and the thalamus provide a very close functional association.

2. Structure. The **cerebral hemispheres** are separated by a deep cleft, the **longitudinal cerebral fissure**, into which the falx cerebri projects.

   a. Divisions. Each cerebral hemisphere may be divided, like any other part of the central nervous system (CNS), into white and gray matter.

   (1) The **gray matter**, covering the surface, consists of nerve cells (neurons) arranged in vertical functional columns with a complex circuitry.

   (a) Input into the columns is principally (99%) from cortical association areas and the thalamus with only a small (1%) input from subcortical areas.

   (b) The output of each functional column is primarily to other cortical areas, to the thalamus, and to the basal ganglia as well as to the cerebellum. The neurons of columns in specific areas influence the lower motor neurons of the cranial nerve nuclei and the spinal cord.

   (2) The deeper **white matter** consists of tracts or fascicles of axons from the input and output neurons.

  b. The surface is folded into **gyri** (ridges) separated by **sulci** (grooves). These cortical folds increase the surface area, thereby increasing the number of functional columns of neurons that can be accommodated (Fig. 29-1). While the pattern of gyri and sulci is quite variable, some are remarkably consistent in location and provide convenient landmarks. Several of the sulci are deeper than the others and may be used to demarcate various lobes.

   (1) The **lateral (sylvian) fissure** is a deep groove between the frontal and temporal lobes.

    (a) The **parietal** and **frontal lobes** lie superiorly to this fissure.

    (b) The **temporal lobe** lies inferiorly to this fissure.

   (2) A **central sulcus** (fissure of Rolando) extends in a coronal plane from the posterior end of the sylvian fissure over the lateral surface and for a short distance on the medial surface in the longitudinal fissure.

    (a) The **frontal lobe** lies anteriorly to this sulcus.

    (b) The **parietal lobe** lies posteriorly to this sulcus.

   (3) The **parieto-occipital fissure** more or less defines the boundary between the parietal and occipital lobes, especially on the medial surface.

   (4) The **calcarine fissure**, discernible only in the medial aspect, bisects the occipital lobe in a transverse plane.

 3. Functions of the hemispheres. The subdivision of the cerebral lobes is useful because it closely approximates the division of function (see Fig. 29-1).

  a. Dominance. In general, each cerebral hemisphere deals with the contralateral side of the body. Usually one cerebral hemisphere (the left in 98% of all individuals) is "dominant." However, some higher functions are lateralized to one or the other hemisphere in a predictable way.

   (1) The **left hemisphere** in most individuals is concerned with verbal, calculating, and analytical thinking as well as interpretation of speech, stereognosis, and motor function to the right hand.

   (2) The **right hemisphere** in most individuals is the seat of nonverbal, spatial, temporal, and synthetic function, appreciation of art and music, and motor function to the left hand. Lesions affecting the right hemisphere may produce loss of visuospatial awareness such that paralysis of the left side is compounded by a tendency to ignore the paralyzed part.

  b. The **primary cortical areas** are regions most directly related to specific functions. These usually are adjacent to major sulci. In some cortical areas (especially the precentral and postcentral gyri), the neurons are **somatotopically** arranged. Areas of the body may be mapped onto the corresponding points on the surface of the cortex (see Fig. 29-2).

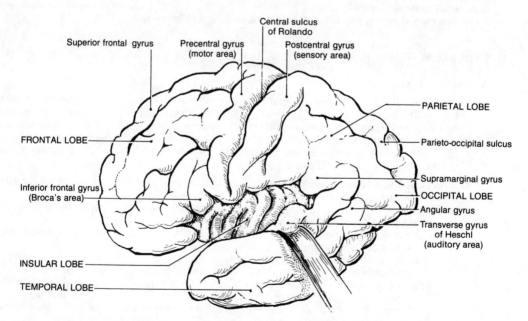

Figure 29-1. *Cerebrum.* A lateral view of the left (dominant) cerebral hemisphere. The lateral fissure has been spread open by traction on the temporal lobe to show the underlying insular lobe.

  c. The **secondary cortical (association) areas** are adjacent to the primary areas and are concerned with a higher level of organization and integration. The functional areas are really much more diffuse than the regions defined in the classical description and overlap one another.

4. **Functions of the lobes**
    a. **The frontal lobe**, lying anteriorly to the central sulcus and lateral fissure, serves motor function, speech, cognition, and the highest levels of affective behavior (see Fig. 29-1).
        (1) The **primary motor area (precentral gyrus)** of the frontal lobe contains the cell bodies of primary motor pathways.
            (a) These project down the pyramidal pathways to the contralateral lower motor neurons in the brain stem and spinal cord. They also influence the lower motor neurons less directly through relays in the brain stem.
            (b) Connections from the ventral lateral and ventral anterior nuclei of the thalamus convey modulating influences from the basal ganglia and from the cerebellum to this motor area.
            (c) This area has a high degree of somatotopic organization (the motor homunculus) [Fig. 29-2]. Damage to specific locations correlates with loss of specific function (*spastic upper motor neuron paralysis*) on the opposite side of the body.
        (2) The **supplementary motor area** of the medial surface of the frontal lobe (**prefrontal gyrus**) is somehow involved in the integration of voluntary movements. At the inferior end of the left supplementary motor area is the **speech area** (of Broca). Damage in this area results in the inability to say what is thought (*motor aphasia*).
    b. **The parietal lobe**, between the central sulcus and the parieto-occipital fissure, is involved with somatosensory processing (see Fig. 29-1).
        (1) The **primary sensory area (postcentral gyrus)** of each parietal lobe receives general sensation from the contralateral side of the body via relays in the ventral posterolateral and posteromedial nuclei of the thalamus.
            (a) This area has a high degree of somatotopic organization (the sensory homunculus) [see Fig. 29-2].
            (b) Damage to specific locations correlates with **paresthesias** on the opposite side of the body.
        (2) The **sensory association area** posterior to the postcentral gyrus of the parietal lobe integrates tactile and visual stimuli to create an understanding of shape and form or evoke memory.

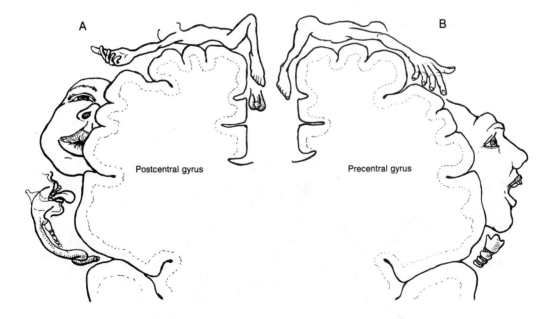

Figure 29-2. *Somatotopic organization of the cerebral cortex. A*, Coronal section through the postcentral gyrus. The relative sizes of the cortical areas associated with the various regions of the body are depicted by the sensory homunculus. *B*, Coronal section through the precentral gyrus. The relative sizes of the cortical areas that innervate the muscles of the various regions of the body are depicted by the motor homunculus.

(3) The **supramarginal gyrus** of the parietal lobe is involved in higher perceptual mechanisms. Lesions to this area produce *tactile agnosia* (inability to identify objects by feel).

c. **The occipital lobe** lies posteriorly to the parieto-occipital sulcus and contains the visual cortex (see Fig. 29-1).

(1) The **primary visual areas (striate cortex)** lie on either side of the calcarine sulcus in the occipital lobe.

(a) The retinal pathways that convey the visual fields of each eye project to the lateral geniculate bodies of the thalamus in such a manner that the left visual field for each eye is projected to the right lateral geniculate body (see V B 2 c).

(b) From the lateral geniculate body of the thalamus, the pathways pass posteriorly to the occipital cortex in the optic radiation (see Fig. 29-6).

(c) Lesions of the optic cortex produce contralateral homonymous visual field defects.

(2) The **visual association (peristriate) areas** of each occipital lobe surround the primary visual cortex in concentric bands.

(a) Here dots are resolved into lines and recognized as shapes.

(b) Injury to these areas produces defective spatial orientation and visual disorganization in contralateral homonymous visual fields.

(c) Lesions between the angular gyrus and the occipital cortex interrupt the interpretation of written language (*alexia* or *visual agnosia*).

d. **The temporal lobe** lies inferiorly to the lateral sulcus and is involved in memory and audition (see Fig. 29-1).

(1) The **primary auditory area (transverse gyrus of Heschl)** lies within the temporal surface of the lateral fissure at the caudal end of the **superior temporal gyrus**.

(a) Sensory neurons from the cochlear nerve project to the cortex via relays in the cochlear nuclei, the inferior colliculus, and the medial geniculate body of the thalamus.

(b) The main element of auditory processing deals with appreciation and interpretation of language, which is subserved by the dominant hemisphere.

(2) The **auditory association areas** (the angular gyrus) are in some manner involved in the comprehension of language. A patient with a lesion in this area cannot understand what he hears (*sensory aphasia*).

e. **The central (insular) lobe** lies at the base of the lateral fissure and seems to be involved in the autonomic nervous system (see Fig. 29-1).

f. **The limbic lobe** lies deeply in the sagittal fissure (cingulate gyrus and angular gyrus) and is concerned with olfaction, regulation of the viscera, primitive emotions, and behavioral activity (see Fig. 29-4).

5. Projections of the cerebral cortex

a. Commissural fibers pass between hemispheres above the thalamus as the **corpus callosum** (Fig. 29-3; see Fig. 29-4).

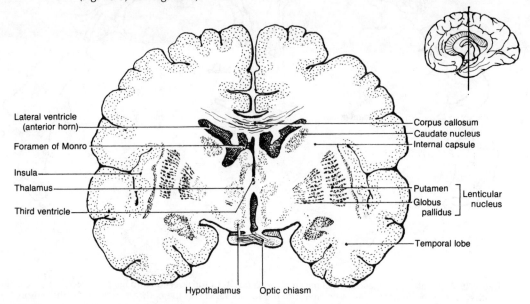

Figure 29-3. *Coronal section through the cerebrum.* The section is through the basal ganglia, just anterior to the thalamus.

(1) The corpus callosum contains over 10^6 crossing axons. These connections are important because the hemispheres have different capabilities.

(2) Cutting the corpus callosum results in a *split-brain* individual in whom information cannot be relayed to the contralateral side for specialized association (e.g., a common object placed in one hand cannot be recognized or matched with a similar object in the other hand).

b. The primary motor areas project via the **corticobulbar pathways** to the lower motor neurons in the brain stem nuclei and via the **pyramidal pathways** to lower motor neurons in the gray matter of the spinal cord. They also influence the lower motor neurons less directly through relays in the brain stem, the **extrapyramidal pathways**.

B. Basal ganglia

1. Structure. These large nuclei lie deeply within the base of the cerebral hemispheres (see Fig. 29-3). The basal ganglia comprise the **caudate nucleus**, the **lenticular nucleus** (consisting of the **putamen** and **globus pallidus**), the **subthalamic nucleus**, and the **substantia nigra**.

2. Projections. Many areas of the cerebral cortex project to the basal ganglia. The outflow from the basal ganglia is primarily to the thalamus, whence they project via a feedback circuit to the motor areas of the cerebral cortex.

3. Functions of the basal ganglia seem to involve the modulation of the motor outflow of the cortex, smoothing voluntary actions.

4. Dysfunction of the basal ganglia produces *release dyskinesias*, such as tremors, choreas, ballism, and athetosis.

a. Cerebral primary motor areas project through the **internal capsule** (between the caudate nucleus and thalamus, medially, and the lenticular nucleus, laterally) before passing into the **cerebral peduncle**.

b. Small lesions here, caused by thrombosis or hemorrhage of the striate arteries, have devastating effects (*stroke*).

C. Diencephalon.
The most rostral portion of the brain stem consists predominantly of the thalamus with small epithalamic, subthalamic, and hypothalamic regions (see Fig. 29-3).

1. The thalamus with over 25 separate nuclei serves as a major synaptic relay station.

a. All afferent modalities (except olfaction) synapse in the thalamus on the way to the sensory areas of the cerebral cortex.

b. The output from the basal ganglia also projects to the thalamus before being relayed to the motor cortex.

c. This massive and complex nucleus is also concerned with crude appreciation of subconscious sensations and vague levels of awareness, including sleep and the level of consciousness. The rhythms established by certain thalamic nuclei form the major contributions to the electroencephalogram (EEG).

d. Thalamic dysfunction (*thalamic syndrome*), usually the result of tumor or thrombosis of the posterior choroidal artery, lowers the threshold for pain, temperature, and tactile sensation to a point where non-noxious stimuli produce marked unpleasant effects.

2. The subthalamus contains the **subthalamic nucleus** and the **substantia nigra**, which are included as basal ganglia.

a. Damage to the subthalamic nuclei results in *ballism*.

b. Reduced dopamine production in the substantia nigra results in *paralysis agitans (Parkinson's disease)*, which is characterized by rigidity, tremors at rest, and difficulty in initiating or ceasing movements.

3. The hypothalamus, which lies below the thalamus, has two primary outflow pathways by which it exerts influence.

a. It modulates visceral activities, such as thermoregulation, appetite, thirst, sexual impulses, and emotion through neural mechanisms mediated by the autonomic nervous system.

b. It also modulates and regulates the release of hormones from the **pituitary gland (hypophysis)**, which is suspended by the thin **infundibulum**.

(1) The **neurohypophysis**, the posterior section of the pituitary, is an extension of the hypothalamus and releases antidiuretic hormone (vasopressin) and milk let-down factor (oxytocin) into the bloodstream.

(2) The **adenohypophysis**, the anterior section, is a pharyngeal derivative (Rathke's pouch).

(a) It releases various trophic hormones, under hypothalamic control, which regulate the major endocrine axes.

(b) It is prone to the formation of benign tumors, which erode the walls of the pituitary fossa and may impinge on the optic chiasm, causing visual field loss.

III. MIDBRAIN (MESENCEPHALON)

A. **Location.** The mesencephalon lies inferiorly to the tentorium cerebelli (Fig. 29-4).

B. **Function.** The midbrain is associated through the substantia nigra and the basal ganglia with modulation of movement. Its **tectum** (roof) also contains centers for auditory and visual reflexes. It contains the nuclei of the oculomotor nerve and the trochlear nerve (see Fig. 29-7).

   1. **The oculomotor nerve** (CN III) emerges from the anterior aspect of the midbrain (Fig. 29-5).

   2. **The trochlear nerve** (CN IV) emerges from the posterior aspect of the midbrain (see Fig. 29-5).

IV. HINDBRAIN

A. **The metencephalon** consists of the cerebellum and pons (see Fig. 29-4).

   1. **The cerebellum** is a dorsal expansion of the brain stem.
      a. The expression of cerebellar function is ipsilateral.
      b. The cerebellum is concerned with the coordinated contraction and relaxation of agonistic and antagonistic muscle groups to produce smooth precise movements.
      c. Cerebellar dysfunction produces jerky uncoordinated movements (**ataxia**).

   2. **The pons** has three pairs of **cerebellar peduncles**.
      a. The middle peduncles project fibers to the cerebellum from the pontine nuclei, which receive input from the cerebral cortex.
      b. The superior peduncles primarily relay to the thalamus.
      c. The inferior peduncles project relay fibers from the peripheral proprioceptors and brain stem to the cerebellum and from the cerebellum to the brain stem.

   3. **The trigeminal nerve** emerges from the pons laterally, and the **abducens nerve** emerges from the anterior aspect of the pontomedullary junction (see Fig. 29-5).

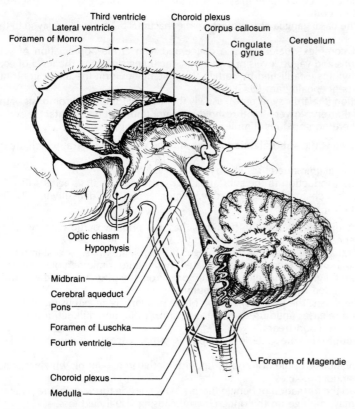

Figure 29-4. *Sagittal section through the brain, brain stem, and cerebellum.*

B. Myelencephalon

1. **Location.** The **medulla oblongata** lies caudally to the pons.

2. **Function.** It contains the nuclei of cranial nerves VII to XII.
   a. The abducens (CN VI) and hypoglossal (CN XII) nerves (somatic motor) emerge from the ventral surface of the medulla. The facial (CN VII), glossopharyngeal (IX), and vagus (X) nerves emerge from the lateral side of the medulla.
   b. In addition to numerous nuclei, the brain stem contains important pathways.
      (1) Ascending sensory pathways to the thalamus are relayed to the cerebral hemispheres.
      (2) Descending motor pathways from the cortex project to the brain stem nuclei and to the gray matter of the spinal cord.

V. CRANIAL NERVE NUCLEI

A. Overview

1. **Classifications**
   a. Nerves are classified as **somatic** (associated with the body wall) or **visceral** (associated with splanchnic structures).
   b. Nerves are also classified as **afferent** (sensory) or **efferent** (motor).
      (1) Cranial afferent nerves are further subdivided as **general** (e.g., touch, pressure, and temperature) or **special** (e.g., smell, sight, vision, and taste).
      (2) Cranial efferent nerves are also subdivided as **general**, which supply musculature derived from myotomes, and **special**, which are motor to structures derived from the branchial or gill arches.

2. **Organization**
   a. **Within the spinal cord**, the motor and sensory areas have specific regions.
      (1) The somatic sensory neurons relay to the posterior (dorsal) horn.
      (2) The motor nuclei are located ventrally (somatic motor) and laterally (visceral motor).
   b. **In the brain stem**, the basic arrangement is modified by the large size of the fourth ventricle, which essentially separates the dorsal horns. The dorsal sensory areas are connected by the thin roof (tegmentum) as though the walls never came together dorsally to fuse in the midline.

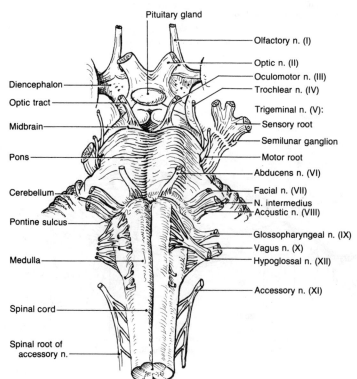

Figure 29-5. *Brain stem.* The anterior aspect of the brain stem and the cranial nerves.

    (1) The somatic sensory nuclei are dorsolateral (see Fig. 29-8).

    (2) The motor nuclei are arranged in three longitudinal columns that extend through the brain stem (see Fig. 29-7). The somatic motor nuclei are most ventral, the parasympathetic nuclei are intermediate, and the branchiomotor nuclei are more lateral.

  c. Composition. Cranial nerves might carry only one modality or innervate one type of structure. They may also be variably mixed, containing motor, sensory, and parasympathetic fibers as well as branchiomotor and special sensory fibers.

B. Special somatic afferent (SSA) nerves. Afferent information related to the special senses, olfaction, vision, audition, and balance is conveyed by special somatic nerves.

  1. The olfactory nerves (CN I) arise in close association with the cerebrum.

    a. Approximately 20 **olfactory nerves** pass from the nasal mucosa through the cribriform plate to the olfactory bulbs.

      (1) The delicate olfactory nerves are frequently damaged by inflammatory processes or severed in head injury (even without fracture of the ethmoid bone), causing *anosmia*.

      (2) The olfactory nerves may provide entry for infections to the brain or meninges.

    b. The **olfactory bulbs** lie on the cribriform plate of the ethmoid bone.

    c. The **olfactory tracts** extend from the olfactory bulbs to the brain.

  2. The optic nerves (CN II) arise in close association with the cerebrum (Fig. 29-6).

    a. They are really tracts of the brain and carry meningeal sheaths (complete with dura, arachnoid, subarachnoid space, and pia), which are continuous with those of the brain.

      (1) The pressure of cerebrospinal fluid (CSF) in the subarachnoid space surrounding the optic nerve reflects the intracranial pressure.

      (2) High intracranial pressure compresses the central vein of the retina, thereby obstructing the venous drainage of the retina and producing observable edema at the optic disk (*papilledema*).

    b. The portion of the optic tract anterior to the optic chiasm is often termed the optic nerve.

      (1) The true equivalent of optic nerves are represented in the retina by the **bipolar cells**, which contact the rod and cone receptor cells.

      (2) The ganglion cells of the retina, the optic "nerve," the optic tracts, and the lateral geniculate bodies are diencephalic structures.

    c. Visual pathways (see Fig. 29-6)

      (1) The **lens** of each eye projects an inverted and reversed visual field onto each retina. Thus, lateral visual fields project onto the nasal halves of each retina; medial visual fields project onto the temporal halves of each retina.

      (2) The optic nerves leave the retina at the **optic disk (blind spot)** and course to the optic chiasm. Lesions of the optic nerve anterior to the chiasm will cause *total blindness* in one eye.

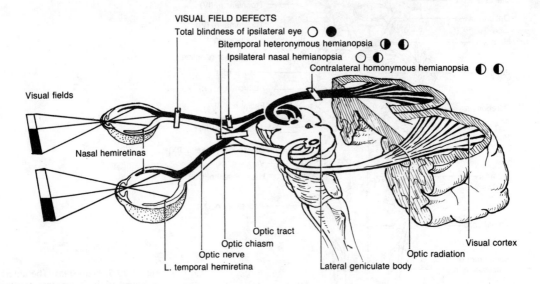

Figure 29-6. *Optic pathways.* The locations of typical lesions and the resultant blindness are shown.

(3) At the **optic chiasm**, pathways from both retinas are sorted such that similar visual fields from each eye are collected on the opposite side of the brain.

    (a) The pathways from the nasal half of each retina cross at the optic chiasm; the pathways from the temporal half of each retina remain ipsilateral.

    (b) The optic chiasm is in close relation to the pituitary fossa, so that suprasellar extension of a pituitary tumor may press on the optic chiasm and interrupt input from the nasal retina of both eyes to produce loss of both temporal visual fields (*bitemporal heteronymous hemianopsia* or *tunnel vision*).

(4) From the optic chiasm, the optic tracts containing common visual fields pass posteriorly to enter the **lateral geniculate body** of the thalamus.

(5) From each lateral geniculate body of the thalamus, the pathways project through the temporal lobe along the **optic radiations** to the primary visual area of each occipital lobe. Lesions of the optic tracts, of the optic radiations, or of the occipital cortex result in *cortical blindness* with *contralateral homonymous hemianopsia* (i.e., loss of the contralateral half of the visual fields of both eyes).

(6) There are also connections with midbrain nuclei that are involved with reflex eye movements.

  d. The optic nerve is one of the regions affected by the widespread demyelination of the central nervous system, which characterizes *disseminated sclerosis*.

  3. The vestibulocochlear nerve (CN VIII) arises in association with the hindbrain.

    a. The vestibular nuclei are involved with balance and equilibrium.

      (1) Through connections to the nuclei of the external ocular muscles, they enable the eye to fix on an object when the head is moving.

      (2) These nuclei also project to the cervical spinal cord and enable the head to remain stable when the body moves.

      (3) Abnormal ocular movements (*nystagmus*) are associated with vestibular dysfunction.

    b. The cochlear nuclei relay acoustic information to the medial geniculate body of the thalamus for subsequent relay to the primary auditory cortex.

C. General somatic efferent (GSE) nerves. Arising from the column of somatic motor nuclei, these nerves are equivalent to motor roots and innervate muscle derived from embryologic somites.

  1. The oculomotor, trochlear, and abducens nerves (CN III, IV, and VI) innervate the extraocular muscles.

    a. The oculomotor nucleus gives rise to the somatic component of the **oculomotor nerve (CN III)** [Fig. 29-7].

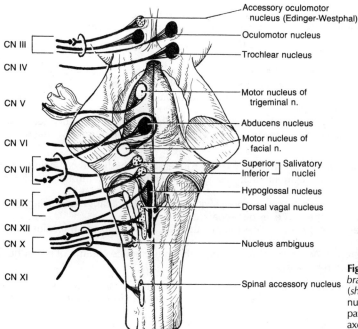

Figure 29-7. *Motor nuclei of the brain stem.* The somatic motor nuclei (*shaded*), the branchiomeric motor nuclei (*unshaded*), and the parasympathetic nuclei (*stippled*) contribute axons to the motor components of the cranial nerves.

(1) Axons from this nucleus innervate five of the extraocular muscles (levator palpebrae superioris, superior rectus, medial rectus, inferior rectus, and inferior oblique).

(2) The axons of each nerve are both ipsilateral (medial rectus) and contralateral.

(3) Lesions will produce *lateral strabismus (wall eyes)* with nearly complete *ophthalmoplegia* of the eye as well as *ptosis (drooping)* of the eyelid.

b. The trochlear nucleus gives rise to the **trochlear nerve (CN IV)** [see Fig. 29-7].

(1) Axons from this nucleus innervate the superior oblique muscle.

(2) All axons are contralateral.

(3) Lesions will produce a *diplopia* upon looking down and out.

c. The abducens nucleus gives rise to the **abducens nerve (CN VI)** [see Fig. 29-7].

(1) Axons from this nucleus innervate the lateral rectus muscle.

(2) All axons are ipsilateral.

(3) Lesions will produce a *medial strabismus (crossed eyes)*.

d. A medial longitudinal fasciculus connects the extraocular nuclei, permitting conjugate eye movements.

e. The oculomotor and trochlear nerves pass through the tentorial notch.

(1) An intracranial mass (tumor, hematoma, or edema) may squeeze portions of the temporal lobes through the tentorial notch (temporal lobe herniation).

(2) Compression of cranial nerves within the tentorial notch causes *ophthalmoplegia with mydriasis* (a dilated pupil that does not respond to light).

2. The hypoglossal nucleus gives rise to the **hypoglossal nerve (CN XII)** [see Fig. 29-7].

a. This nerve is formed by the union of a series of rootlets, which emerge from the anterolateral sulcus of the medulla.

b. It innervates the ipsilateral muscles of the tongue.

c. Lesions result in deviation of the tongue toward the side of the lesion upon protrusion.

D. Special visceral efferent (SVE) nerves. Arising from the most lateral column of motor nuclei, these efferent nerves innervate muscles derived from branchial arch derivatives (i.e., the musculature associated with the embryonic gills) [see Fig. 29-7].

1. The trigeminal motor nucleus, located in the mid pons, gives rise to the branchiomotor component of the **trigeminal nerve (CN V)** [see Fig. 29-7].

a. It is the most rostral special visceral nucleus.

b. The axons pass through the trigeminal ganglion and exit the cranium via the foramen ovale to join the mandibular nerve (CN V_3).

c. It innervates the ipsilateral muscles of mastication and a few other muscles.

2. The facial motor nucleus gives rise to the branchiomotor component of the **facial nerve (CN VII)** [see Fig. 29-7].

a. This nucleus lies just caudal to the trigeminal motor nucleus.

b. The facial nerve innervates the ipsilateral muscles of facial expression.

3. The nucleus ambiguus gives rise to the branchiomotor component of the **glossopharyngeal**, **vagus**, and **cranial accessory nerves** (see Fig. 29-7).

a. The glossopharyngeal nerve (CN IX) innervates the ipsilateral stylopharyngeus muscle.

b. The vagus nerve (CN X) innervates the ipsilateral pharyngeal muscles via the pharyngeal plexus, as well as the ipsilateral cricothyroid muscle via the superficial branch of the superior laryngeal nerve.

c. The cranial accessory nerve (CN X), which briefly joins the spinal accessory nerve, innervates the ipsilateral muscles of the larynx (except the cricothyroid muscle) via the recurrent laryngeal nerve.

4. The spinal accessory nerve (CN XI) is formed by rootlets from the lateral cell column of the upper five or six cervical segments that pass into the cranial cavity through the foramen magnum.

a. This nerve joins the cranial accessory nerve within the jugular canal. After passing through the jugular foramen, the two divisions separate.

b. The spinal accessory nerve innervates the ipsilateral sternomastoid and trapezius muscles.

E. General visceral efferent (GVE) nerves. Arising from the intermediate column of motor nuclei, are the **parasympathetic nerves** (see Fig. 29-7).

1. The accessory oculomotor nucleus (of Edinger-Westphal), the most rostral parasympathetic nucleus, gives rise to the parasympathetic component of the **oculomotor nerve (CN III)** [see Fig. 29-7].

a. The preganglionic parasympathetic neurons from this nucleus travel via the oculomotor nerve (CN III) to the ciliary ganglion.

b. The postganglionic axons of the ciliary ganglion course to the sphincter and ciliary muscles of the pupil, controlling pupillary constriction and accommodation to near vision.

2. The superior salivatory nucleus gives rise to the parasympathetic secretomotor fibers of the **facial nerve (CN VII)** [see Fig. 29-7].
   a. The preganglionic axons travel with the **nervus intermedius** of the facial nerve to the pterygopalatine ganglion (via the greater superficial petrosal and vidian nerves) and the submandibular ganglion (via the chorda tympani and lingual nerve).
   b. The postganglionic axons of the pterygopalatine ganglion supply the lacrimal, nasal, and palatine glands.
   c. The postganglionic axons of the submandibular ganglion supply the submandibular gland, the sublingual gland, and the glands of the anterior tongue.

3. The inferior salivatory nucleus gives rise to the parasympathetic secretomotor fibers of the **glossopharyngeal nerve (CN IX)** [see Fig. 29-7].
   a. The preganglionic axons travel with the glossopharyngeal nerve, then with the tympanic nerve (of Jacobson) and the lesser superficial petrosal nerve to reach the otic ganglion.
   b. The postganglionic axons of the otic ganglion then supply the parotid gland.

4. The dorsal vagal nucleus, the most caudal of the cranial parasympathetic nuclei, gives rise to the parasympathetic component of the **vagus nerve (CN X)** [see Fig. 29-7].
   a. It provides the long preganglionic parasympathetic fibers, which synapse in small ganglia associated with the viscera of the thorax and much of the abdomen.
   b. The postganglionic neurons from the small parasympathetic ganglia are secretomotor and stimulate visceral muscular activity.

F. General visceral afferent (GVA) nerves. Afferent sensation from the thoracic and abdominal viscera is conveyed to the brain stem by general visceral nerves.

1. The solitary nucleus receives input from the **glossopharyngeal nerve** with cell bodies in the **inferior glossopharyngeal (petrosal) ganglion** to mediate the gag and carotid reflexes (see Fig. 29-8).

2. The dorsal nucleus of the vagus receives input from the **vagus nerve** with cell bodies in the **inferior vagal (nodosal) ganglion** to mediate the visceral reflexes involving the vagus nerve (e.g., cough, cardiac, and respiratory reflexes) [see Fig. 29-7].

G. Special visceral afferent (SVA) nerves. These nerves convey taste sensation to the brain stem.

1. The nucleus solitarius receives taste input from the lingual, glossopharyngeal, and vagus nerves.
   a. Taste from the anterior two-thirds of the tongue finds its way via the **lingual nerve**, to the **chorda tympani** to the **facial nerve (CN VII)**. The cell bodies of these afferent neurons are located in the **geniculate ganglion** and synapse in the solitary nucleus (Fig. 29-8).
   b. Taste from the posterior surface of the tongue is carried via the **glossopharyngeal nerve (CN IX)**. The cell bodies of these afferent neurons are located in the **inferior glossopharyngeal (petrosal) ganglion** and synapse in the nucleus solitarius (see Fig. 29-8).
   c. Taste from the epiglottis is carried via the **vagus nerve (CN X)**. The cell bodies of these afferent neurons are located in the **inferior vagal (nodosal) ganglion** and synapse in the nucleus solitarius (see Fig. 29-8).

2. From the solitary nucleus, the second neurons convey taste sensation to the thalamus.

H. General somatic afferent (GSA) nerves. These cranial nerves are equivalent to dorsal roots. The **trigeminal nuclei** receive somatic afferent sensation from the trigeminal, glossopharyngeal, and vagus nerves.

1. The trigeminal nerve (CN V) is responsible for the general sensation from the face, the anterior scalp, the mouth, the pharynx, the nose, and most of the ear.
   a. The cell bodies of most of these afferent neurons are located in the **trigeminal (gasserian, semilunar) ganglion**.
   b. Each of the modalities of pain, touch, and temperature is associated with a different brain stem nucleus where the second-order neurons are located.
      (1) The **principal nucleus of V** in the pons (see Fig. 29-8) receives fibers carrying light touch and two-point discrimination.
      (2) The **mesencephalic nucleus of V** (see Fig. 29-8) mediates proprioception (e.g., the **jaw-jerk reflex**). This nucleus contains first-order sensory neurons, the only exception to the rule that the first-order neurons are located in dorsal root ganglia or cranial equivalents.

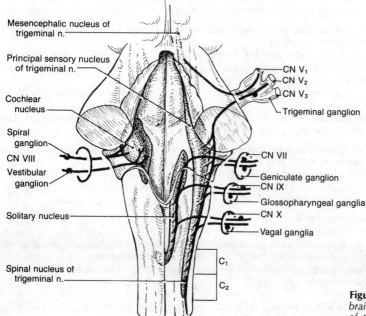

Mesencephalic nucleus of trigeminal n.

Principal sensory nucleus of trigeminal n.

Cochlear nucleus

Spiral ganglion

CN VIII

Vestibular ganglion

Solitary nucleus

Spinal nucleus of trigeminal n.

CN V₁
CN V₂
CN V₃
Trigeminal ganglion

CN VII
Geniculate ganglion
CN IX
Glossopharyngeal ganglia
CN X
Vagal ganglia

C₁
C₂

Figure 29-8. *Sensory nuclei of the brain stem.* The sensory components of the cranial nerves terminate in sensory nuclei.

(3) The **spinal nucleus of V**, which extends into the cervical spinal cord (see Fig. 29-8), receives fibers conveying pain, coarse touch, and temperature sensation. This nucleus is somatotopically organized such that neurons from the ophthalmic nerve enter most caudally while those from the mandibular nerve enter most rostrally.

2. **The glossopharyngeal nerve** is responsible for a small area of somatic sensation within the external auditory meatus.
   a. The cell bodies of these afferent neurons are located in the **superior glossopharyngeal (jugular) ganglion**.
   b. The fibers synapse in the spinal nucleus of V (see Fig. 29-8).

3. **The vagus nerve (CN X)** is responsible for a small area of somatic sensation on the posterior surface of the ear and a small area within the external auditory meatus.
   a. The cell bodies are located in the **superior vagal (jugular) ganglion**.
   b. The fibers synapse in the spinal nucleus of V (see Fig. 29-8).

VI. VENTRICULAR SYSTEM

A. **Structure.** The ventricles represent the rostral end of the central canal of the neural tube. The terminal expansion of the neural tube is divided into two lateral ventricles, the midline third ventricle, the cerebral aqueduct, and the midline fourth ventricle (see Fig. 29-4).

1. **The lateral ventricles** are cavities within each cerebral hemisphere (see Fig. 29-3).
   a. Each has three horns that extend into the frontal, temporal, and occipital lobes.
   b. Each opens into the third ventricle through the **interventricular foramen** (of Monro) [see Fig. 29-4].

2. **The third ventricle**, a cleft-like cavity that lies between the thalami (see Figs. 29-3 and 29-9), is continuous with the **cerebral aqueduct** posteriorly.

3. **The iter or cerebral aqueduct (of Sylvius)** lies in the midbrain and connects the third and fourth ventricles (Fig. 29-9).

4. **The fourth ventricle** lies in the pons beneath the cerebellum.
   a. Posteriorly, it extends into the **central canal** of the spinal cord.
   b. It communicates with the **cisterna magna** of the subarachnoid space (located between the medulla and the cerebellum) through three openings in the roof: the median **foramen of Magendie** and two lateral **foramina of Luschka** (see Fig. 29-9).

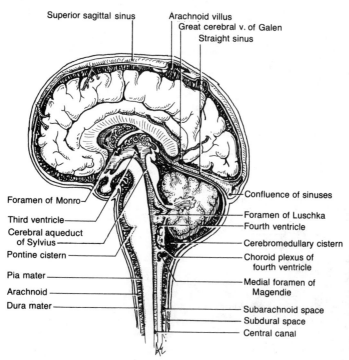

Superior sagittal sinus

Arachnoid villus
Great cerebral v. of Galen
Straight sinus

Foramen of Monro

Third ventricle

Cerebral aqueduct of Sylvius

Pontine cistern

Pia mater

Arachnoid

Dura mater

Confluence of sinuses

Foramen of Luschka
Fourth ventricle

Cerebromedullary cistern

Choroid plexus of fourth ventricle

Medial foramen of Magendie

Subarachnoid space
Subdural space
Central canal

Figure 29-9. *Ventricular system of the brain.* Cerebrospinal fluid (CSF), secreted by the choroid plexuses of the ventricles, enters the subarachnoid space through three foramina in the roof of the fourth ventricle. The brain is suspended in this layer of CSF, which is resorbed into the venous system at the arachnoid villi.

B. Cerebrospinal fluid (CSF)

1. **Within the ventricles**, CSF is formed by a combination of ultrafiltration and active transport across the vascular tufts of the **choroid plexuses**. The nervous tissue around the ventricles is deficient at several sites where the pia mater and ependyma fuse. Overlying vascular mesenchyme invades this membrane to form choroid plexuses, which occur in several locations (see Fig. 29-4):
   a. Along the medial wall of the body and inferior horns of the lateral ventricles
   b. In the roof of the third ventricle
   c. Along the roof of the fourth ventricle

2. **In the subarachnoid space**, CSF escapes from the ventricles into the **cisterna magna** of the subarachnoid space through three foramina in the roof of the fourth ventricle (see Fig. 29-9).
   a. If flow is obstructed, hydrocephalus develops.
   b. Fracture of the ethmoid bone may allow CSF to drain through the nose (*rhinorrhea*); fracture of the base of the skull may allow CSF to drain from the ear (*otorrhea*), a serious sign.
   c. CSF is readily available for clinical analysis by lumbar tap.
   d. CSF returns to the venous system through the **arachnoid villi**, which project into the superior sagittal sinus (see Figs. 28-1 and 29-9).

3. **Clinical considerations.** CSF is normally a clear fluid low in protein and with few cells.
   a. An increased number of white blood cells in the CSF indicates bacterial meningitis.
   b. Blood in the CSF indicates subarachnoid (arterial) hemorrhage.

VII. CEREBRAL VASCULATURE

A. Overview. The brain is a highly metabolic organ. While it comprises but 2% of the body weight, it uses about 20% of the blood supply. It requires a sufficient and constant circulation of blood. Ischemia rapidly leads to loss of consciousness, followed by convulsions and (since the neurons begin to die after 4 minutes) brain death.

B. Arterial supply. Blood to the brain is supplied by paired internal carotid arteries and paired vertebral arteries (Fig. 29-10).

1. **Carotid arteries**
   a. The **right common carotid artery** originates from the brachiocephalic artery, while the **left common carotid artery** arises directly from the arch of the aorta.

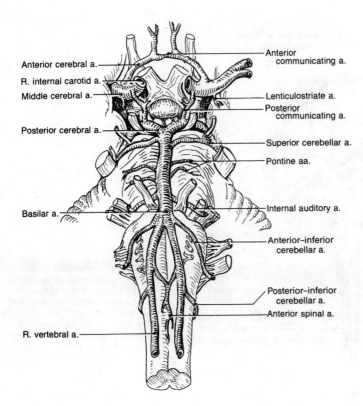

Anterior cerebral a.

R. internal carotid a.

Middle cerebral a.

Posterior cerebral a.

Basilar a.

R. vertebral a.

Anterior communicating a.

Lenticulostriate a.

Posterior communicating a.

Superior cerebellar a.

Pontine aa.

Internal auditory a.

Anterior–inferior cerebellar a.

Posterior–inferior cerebellar a.

Anterior spinal a.

Figure 29-10. *Arterial supply to the brain.*

   b. In the neck, each common carotid bifurcates into an **external carotid artery**, which supplies the face and neurocranium, and an **internal carotid artery**, which supplies the brain.
      (1) At the carotid bifurcation, the **carotid sinus** monitors the blood pressure, and the **carotid body** monitors the oxygen saturation.
      (2) These arteries are instrumental in reflex control of cardiac rate and output.
         (a) The carotid branch of CN IX is the afferent limb.
         (b) The vagus nerve is the efferent limb.
   c. The **internal carotid artery** ascends in a stepwise manner from the mouth of the carotid canal just medial to the styloid to the anterior clinoid process of the sphenoid.
      (1) Within the carotid canal, it is surrounded by a plexiform sheath of **postganglionic sympathetic fibers**.
      (2) At the apex of the petrous bone, it turns up and medially along the body of the sphenoid, where it turns anteriorly again within the cavernous sinus.
      (3) On leaving the cavernous sinus, it arches backwards beneath the anterior clinoid process and the optic nerve.
      (4) An **ophthalmic branch** accompanies the optic nerve and gives off the central retinal artery, which enters the optic nerve.
      (5) The **posterior communicating branch** passes posteriorly above the oculomotor nerve.
      (6) The internal carotid artery divides into anterior and middle cerebral arteries.
         (a) The **anterior cerebral artery** passes dorsally around the corpus callosum at the base of the sagittal sulcus in company with its counterpart of the opposite side. The two are connected by a small **anterior communicating branch**. They supply the medial aspect of the frontal and parietal cerebral cortex.
         (b) The **middle cerebral artery** passes superiorly in the lateral fissure.
            (i) It supplies most of the lateral aspect of the cerebral cortex.
            (ii) **Lenticulostriate arterial branches** supply the basal ganglia, the thalamus, and the internal capsule. They are commonly the site of hemorrhage or thrombosis that causes the familiar pattern of contralateral paralysis, termed a **stroke** after its sudden and devastating nature.
   2. The vertebral arteries arise from the subclavian arteries.
      a. They pass through the transverse foramina of the sixth through first cervical vertebrae.
      b. They pierce the dura mater and arachnoid at the atlanto-occipital membrane.

    c. They pass through the foramen magnum and ascend on the clivus of the basioccipital bone, giving off several branches.

        (1) The **posterior–inferior cerebellar arteries** pass laterally over the inferior surface of the cerebellum. These, in turn, give off **posterior spinal arteries**.

        (2) The **anterior spinal artery** is formed by the union of a small branch from each vertebral artery. It runs down the midline in the anterior median fissure of the medulla.

    d. At the pontomedullary junction, the vertebral arteries unite to form the **basilar artery**, which gives off numerous branches.

        (1) The **anterior–inferior cerebellar artery** passes close to the abducens nerve, the vestibulocochlear nerve, and the glossopharyngeal nerve. Each anterior branch passes laterally superior to the cerebellum.

        (2) There are numerous small **pontine branches**.

        (3) At the anterior border of the pons, the basilar artery gives off the **superior cerebellar arteries**.

        (4) At this point, the basilar artery bifurcates into the **posterior cerebral arteries**. Each of these arteries runs laterally along the pons to supply the tentorial surface of the occipital lobe.

        (5) The oculomotor nerve (CN III) emerges between the superior cerebellar and posterior cerebral arteries. An aneurysm of either of these vessels is often readily apparent as the nerve is compressed.

3. Anastomotic circle (of Willis). The carotid and vertebral arteries anastomose with each other on the ventral surface of the brain stem.

    a. Structure. The anastomotic circle (of Willis) is formed by seven vessels.

        (1) The **anterior cerebral arteries** of the internal carotids

        (2) The **anterior communicating branch** between the left and right anterior cerebral arteries

        (3) **Posterior communicating branches** between the internal carotid arteries and the posterior cerebral arteries

        (4) The **posterior cerebral arteries**

    b. Function. The circle of Willis is complete in 90% of cases, although in a large number of these the communicating arteries are so small as to be of little functional significance.

        (1) An occlusion within one of the vessels forming the circle of Willis may be partially overcome by reversing the flow of the blood.

        (2) The vessels comprising the circle of Willis are prone to the formation of *berry aneurysms*, which may rupture with disastrous subarachnoid hemorrhage.

4. Major branches. The cerebral arteries pass dorsally around the sides of the brain within deep fissures. All of the major arteries lie in the subarachnoid space.

    a. Subarachnoid hemorrhage occurs upon rupture of an atheromatous artery or of a berry aneurysm. In many victims, there is a stepwise progression of disease prior to a major crippling attack. This may involve transient episodes, such as dizziness, weakness, and loss of memory.

        (1) Angiograms can be used to localize the site of blockage.

        (2) Blockages or aneurysms of the major cerebral vessels may be treated surgically.

    b. Cerebral hemorrhage may be caused by a rupture of cerebral vessels within the brain parenchyma (stroke) or by penetrating trauma. There is no surgical cure.

C. Venous return

1. The cerebral veins generally parallel the terminal portion of the arterial tree in the pial layer. The large collecting veins run in the subarachnoid layer.

2. They pierce the arachnoid layer and traverse the subdural space to gain access to the dural venous sinuses.

3. The veins receive little support when traversing the subdural space and may be torn by moderate jarring trauma. Hemorrhage into the subdural space (*subdural hematoma*) has no pathway of escape and may not pass across the subarachnoid layer to present in the CSF.

30
Orbit,
Eye, and Ear

I. ORBIT (ORBITA, ORBITAL CAVITY)

A. Bony orbit. A shell of bone nearly surrounds and protects the eyeball (Fig. 30-1).

    1. Structure. It is a pyramidal cavity with the axis directed about 23° from the sagittal plane.

        a. The superior wall, the floor of the anterior cranial fossa, is formed largely by the frontal bone with a minor contribution from the lesser wing of the sphenoid bone.

            (1) Laterally, the roof contains a depression for the lacrimal gland, the **lacrimal fossa**.

            (2) The **supraorbital notch** or **foramen** at the supraorbital margin transmits the supraorbital branch of the frontal nerve.

            (3) Anteriorly, the frontal bone contains the **frontal sinuses**.

        b. The medial wall (in the sagittal plane) is largely composed of the ethmoid bone and the small lacrimal bone.

            (1) That portion of the medial wall that separates the orbit from the underlying ethmoid air cells—the **lamina papyracea**—is very thin.

            (2) The anterior and posterior ethmoid foramina transmit neurovascular structures from the orbit to the nasal cavity and paranasal sinuses.

            (3) Anteriorly in the lacrimal bone, the fossa of the lacrimal sac (**lacrimal groove**) continues as the **nasolacrimal canal**, which contains the nasolacrimal duct and terminates in the inferior nasal meatus.

        c. The lateral wall (about 45° from the sagittal plane) is formed largely by the zygomatic bone and the greater wing of the sphenoid bone.

        d. The orbital floor is formed by the maxilla with a minute contribution from the palatine bone.

            (1) The floor is traversed by the **infraorbital groove**, which transmits the infraorbital branch of the maxillary nerve. This groove becomes the **infraorbital canal** just before reaching the infraorbital margin and exits via the **infraorbital foramen**.

            (2) The floor of the orbit is also the roof of the maxillary sinus.

    2. Foramina and fissures. At the apex of the bony orbit are three apertures (see Fig. 30-1).

        a. The optic canal (foramen) connects the orbit with the middle cranial fossa and contains the optic nerve and the ophthalmic artery.

        b. The superior orbital fissure connects the orbit with the middle cranial fossa. It transmits several neurovascular structures (see Fig. 30-4).

            (1) Oculomotor nerve (CN III) with its superior and inferior divisions

            (2) Trochlear nerve (CN IV)

            (3) Three branches of the ophthalmic division of the trigeminal nerve (CN V_1), that is, the lacrimal nerve, the frontal nerve, and the nasociliary nerve

            (4) Abducens nerve (CN VI)

            (5) Ophthalmic vein

        c. The inferior orbital fissure connects the orbit with the pterygopalatine fossa. It transmits the infraorbital branch of the maxillary division of the trigeminal nerve. It is spanned in large part by the involuntary **orbitalis muscle** (Müller's first).

        d. The superior and inferior orbital fissures are continuous at their medial end around the orbital surface of the greater wing of the sphenoid bone.

B. Eyelids (palpebrae)

    1. Overview. The eyelids protect the anterior aspect of the eye (see Fig. 30-2).

        a. They enclose a potential space, the **conjunctival sac**, which opens into the skin of the face at the palpebral fissure.

        b. The **palpebral fissure** is bounded by the upper and lower palpebral margins.

        c. The **palpebral margins** meet at the **medial** and **lateral canthi**.

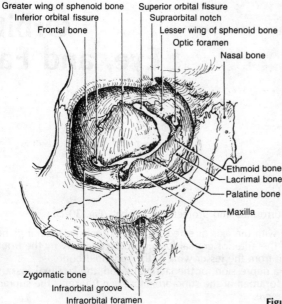

Greater wing of sphenoid bone
Inferior orbital fissure
Frontal bone

Superior orbital fissure
Supraorbital notch
Lesser wing of sphenoid bone
Optic foramen
Nasal bone

Ethmoid bone
Lacrimal bone
Palatine bone
Maxilla

Zygomatic bone
Infraorbital groove
Infraorbital foramen

Figure 30-1. *Bones of the right orbit.*

2. Structure. The upper eyelid is the larger and more mobile. Each lid contains a **tarsal plate** of dense connective tissue, which maintains the shape of the lid and to some extent protects the underlying eyeball (see Fig. 30-2).

    a. The superior and inferior tarsal plates merge to form the **medial palpebral ligament** and the **lateral palpebral ligament**, which insert into their respective orbital margins.

    b. The **orbital septum** is a fascial sheet that attaches the tarsal plates to the superior and inferior orbital margins.

    c. The levator palpebrae superioris muscle attaches to the tarsal plate of the superior eyelid.

    d. The subcutaneous tissue of the eyelids is very loose and may swell significantly with collection of fluid (puffy eyelids) or extravasated blood (black eye).

3. Muscles of the eyelid

    a. The palpebral portion of the orbicularis oculi muscle runs beneath the skin of the eyelid.

        (1) The palpebral portion is a muscle of facial expression.

        (2) Contraction of this portion produces the blink.

        (3) The fibers of the palpebral portion are so arranged that when they shorten they move toward geodesic lines—the shortest distance between two points over the surface of a sphere—and thus close the palpebral fissure.

        (4) Along with the other muscles of facial expression, the palpebral portion of the orbicularis oculi is innervated by the facial nerve (CN VII).

            (a) This nerve forms the efferent limb of the blink reflex.

            (b) Paralysis of the palpebral portion of the orbicularis oculi results in an inability to close the eyelids. This is a major problem in *Bell's* (facial nerve) *palsy* because the conjunctiva dries and the cornea may ulcerate when the eyes cannot blink.

    b. The levator palpebrae superioris muscle is an extraocular muscle (see Figs. 30-2 and 30-6).

        (1) This muscle arises from the apex of the orbit. It widens into an aponeurosis, which inserts into the superior tarsal plate.

        (2) It draws the lid upward when the eyeball is elevated.

        (3) It is innervated by the superior division of the oculomotor nerve. Paralysis produces *ptosis*, an inability to lift the lid when the gaze is directed upward.

    c. The superior tarsal muscle (Müller's third) is composed of smooth (involuntary) muscle.

        (1) It runs from the superior border of the tarsal plate of the upper lid to the tendon of the levator palpebrae superioris muscle.

        (2) The superior tarsal muscle is innervated by sympathetic nerves that have their preganglionic cell bodies in the upper thoracic levels of the spinal cord and their postsynaptic cell bodies in the superior cervical ganglion.

        (3) It accentuates the opening of the palpebral fissure under sympathetic stimulation ("wide-eyed fear").

        (4) A lesion involving the cranial sympathetics (*Horner's syndrome*) with paralysis of this muscle produces a slight drooping of the lid, *pseudoptosis*.

4. The margins of the eyelids contain a double or triple row of hairs, as well as numerous sebaceous tarsal (meibomian) glands that lubricate the lids, sebaceous glands (of Zeis) associated with the hair follicles, and sweat glands (of Moll). However, the medial end of the palpebral margin is hairless (see Fig. 30-2).

  a. The tarsal glands occasionally become infected (*acute meibomianitis*).

  b. Infection of the glands of Zeis produces the common **sty**.

5. The inner surfaces of the eyelid and the eyeball are lined with the **conjunctivae**.

  a. The palpebral conjunctiva of the inner surface of the eyelids is reflected off the bases of the lids onto the eyeball (see Fig. 30-6).

  (1) Superior and inferior **conjunctival fornices** are formed by these reflections.

  (2) The lacrimal ducts drain into the superior fornix.

  (3) The palpebral conjunctiva is richly supplied with blood. It is normally pink but becomes pallid in anemia.

  b. The bulbar conjunctiva protects the cornea.

  (1) It is transparent and contains a great number of small blood vessels, which produce the *bloodshot eye* or *pinkeye* when dilated due to allergy, irritation, or inflammation.

  (2) The conjunctiva is innervated by the ophthalmic and maxillary divisions of the trigeminal nerve (CN V). This is the afferent limb of the blink reflex.

  (a) The conjunctiva is very sensitive and any foreign contact with the conjunctiva causes blinking. Interconnections in the brain stem produce reflex blinking in both eyes simultaneously—the consensual blink reflex.

  (b) The afferent and efferent portions of the facial and trigeminal nerves are routinely tested in this manner.

  (i) If both eyes blink when the right eye is touched but not when the left eye is touched, the problem must lie with the left trigeminal nerve.

  (ii) Conversely, if only the right eye blinks when either eye is touched, the problem must lie with the left facial nerve.

6. The lacrimal glands are located in the superior lateral region of the orbit (Fig. 30-2).

  a. The lacrimal gland has an intimate relationship with the tendon of the levator palpebrae superioris muscle. The major portion of the lacrimal gland lies above it, while a smaller portion lies beneath it. Thus, movement of the eyelids tends to milk the gland, so that continuous lubrication is provided and the conjunctiva is kept moist.

  b. About a dozen ductules drain the lacrimal gland into the lateral region of the superior conjunctival fornix.

  c. The lacrimal gland is controlled by parasympathetic innervation.

  (1) Presynaptic parasympathetic neurons arise from the **superior salivatory nucleus** (see Fig. 29-7).

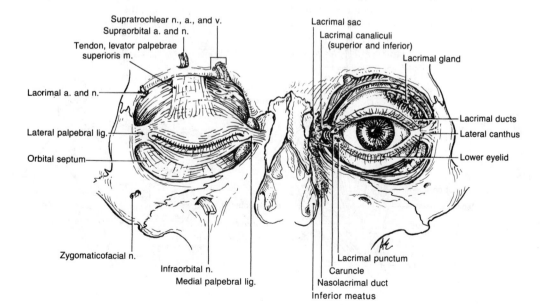

Figure 30-2. *Palpebrae and lacrimal apparatus.* The structure of the eyelids and the insertion of the levator palpebrae superioris muscle of the right eye and the lacrimal apparatus of the left eye.

(2) The preganglionic axons travel with the **nervus intermedius** of the facial nerve, the greater superficial petrosal nerve, and the nerve of the vidian canal to the pterygopalatine ganglion in the pterygopalatine fossa.

(3) The postganglionic axons of the **pterygopalatine ganglion** travel along the maxillary nerve, its zygomatic branch, and a communicating branch to the lacrimal branch of the ophthalmic nerve, finally reaching the lacrimal gland.

d. With cholinergic stimulation, tears flush across the eyeball and are drawn by capillary action along the palpebral fissure.

7. Lacrimal apparatus

a. At the **medial canthus**, a small fold of conjunctiva, the plica semilunaris, encloses a triangular area, the **lacus lacrimalis**.

b. Bordering the lacus lacrimalis, each lid has a small **lacrimal papilla** with an apical punctum at the orifice of the **lacrimal canaliculus**.

c. The upper and lower canaliculi drain behind the medial palpebral ligament and then join to drain into the lacrimal sac.

d. The **lacrimal sac** is the blind upper end of the nasolacrimal duct.

(1) It rests in the **lacrimal groove** of the lacrimal and maxillary bones.

(2) The lacrimal fascia passes around the sac from the crest on the frontal process of the maxilla to the lacrimal crest.

(3) Blinking deforms the lacrimal sac, which, acting like a bulb syringe, aspirates the collecting fluid into the lacrimal canaliculi.

(4) The **nasolacrimal duct** drains into the inferior meatus of the nasal cavity.

(a) A small flap of mucosa at its nasal orifice usually prevents the percolation of air into the conjunctival sac.

(b) If the duct is blocked, tears cannot drain into the nose and, therefore, run out onto the cheek (*epiphora*).

(5) The early stages of lacrimation involve ''sniffling,'' as the lacrimal fluid drains into the nose. As the lacrimal canaliculi become overwhelmed, the fluid overflows the palpebrae and tears flow down the cheeks.

C. Extraocular musculature. The orbital cavity contains seven voluntary muscles of somatic origin, six of which move the eyeball (Fig. 30-3).

1. Ocular axes. The eyeballs, suspended within the orbit by a fibrous capsule, are very mobile, with movement occurring about three axes.

a. Elevation and depression occur about the transverse axis through the equator of the eyeball.

b. Abduction and adduction occur about a vertical axis through the poles.

c. Intorsion and extorsion occur about an anteroposterior axis through the pupil. Intorsion is thus a medial rotation of the upper portion of the eye toward the midline; extorsion is the lateral rotation of the upper portion of the eye away from the midline.

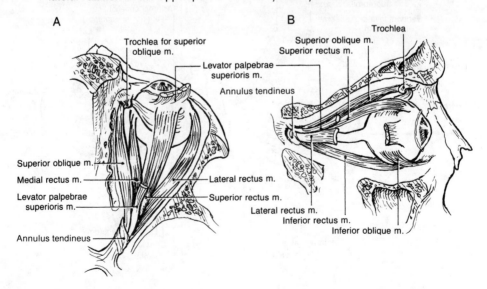

Figure 30-3. *Extrinsic muscles of the right eye. A, Viewed from above. B, Viewed from the right side.*

2. The annulus tendineus (of Zinn) lies at the apex of the periorbita (see Fig. 30-3).
    a. It gives rise to the four rectus muscles.
    b. The optic nerve, ophthalmic artery, superior and inferior divisions of the oculomotor nerve, abducens nerve, and nasociliary branch of the ophthalmic nerve enter the orbital cavity through the annulus tendineus as they leave the optic canal and the superior orbital fissure.
    c. The ophthalmic vein and trochlear nerve, as well as the frontal and lacrimal branches of the ophthalmic nerve, pass superiorly to the annulus tendineus as they pass through the superior orbital fissure.

3. Four rectus muscles diverge from the annulus tendineus to insert into the sclera of the eyeball (see Fig. 30-3).
    a. The superior rectus muscle inserts onto the anterosuperior aspect of the eyeball and primarily elevates the eyeball.
    b. The inferior rectus muscle inserts onto the anteroinferior aspect of the eyeball and primarily depresses the eyeball.
    c. The lateral rectus muscle inserts onto the anterolateral aspect of the eyeball and abducts the eye.
    d. The medial rectus muscle inserts onto the anteromedial aspect of the eyeball and adducts the eye.

4. The two oblique muscles have different origins and courses (see Fig. 30-3).
    a. Superior oblique muscle
      (1) This muscle originates from the sphenoid bone superomedially to the annulus tendineus.
      (2) It has a unique course: Its tendon passes through a fibrous trochlea (*L.* pulley) attached to the frontal bone at the anterosuperior aspect of the medial orbital margin. The tendon then passes backward to insert onto the posterolateral quadrant of the superior surface of the eyeball.
      (3) This muscle draws the posterior portion of the eyeball upward, depressing the eye.
    b. Inferior oblique muscle
      (1) This muscle originates from the anteroinferior aspect of the medial orbital margin.
      (2) It seems to represent the distal portion of a primitive muscle that paralleled the superior oblique muscle. It passes backward to insert onto the posterolateral quadrant of the inferior surface of the eyeball.
      (3) This muscle draws the posterior portion of the eyeball downward, elevating the eye.

5. The levator palpebrae superioris muscle originates superiorly to the annulus tendineus, inserts into the tarsal plate of the superior eyelid, and acts to draw the eyelid upward when the eyeball is elevated.

6. Group actions of the extraocular muscles (Table 30-1)
    a. Because the bony orbit is directed somewhat laterally and because the pupil normally is directed anteriorly, the lines of action of the extraocular muscles above and below the eye

Table 30-1. Extraocular Musculature

Muscle	Origin	Insertion	Primary Action	Secondary Actions	Innervation
Superior rectus	Annulus tendineus	Anterosuperior aspect	Elevation	Adduction and intorsion	Oculomotor (CN III)
Inferior oblique	Lacrimal bone	Posterolateral inferior quadrant	Elevation	Abduction and extorsion	Oculomotor (CN III)
Inferior rectus	Annulus tendineus	Anteroinferior aspect	Depression	Adduction and extorsion	Oculomotor (CN III)
Superior oblique	Sphenoid bone	Posterolateral superior quadrant	Depression	Abduction and intorsion	Trochlear (CN IV)
Medial rectus	Annulus tendineus	Anteromedial aspect	Adduction	. . .	Oculomotor (CN III)
Lateral rectus	Annulus tendineus	Anterolateral aspect	Abduction	. . .	Abducens (CN VI)
Levator palpebrae superioris	Sphenoid bone	Tarsal plate of upper eyelid	Raises upper eyelid	. . .	Oculomotor (CN III)

pass medially to the vertical axis. Therefore, they tend to draw the respective points of attachment medially.

(1) The superior rectus and superior oblique muscles secondarily rotate the superior portion of the eyeball medially (intorsion).

(2) The superior and inferior oblique muscles draw the posterior portion of the eyeball medially, thereby secondarily abducting the eye.

b. Similarly, the lines of action of the extraocular muscles above and below the eye pass medially to the anteroposterior axis. Therefore, they tend to draw the respective points of attachment medially about this axis.

(1) The superior rectus and superior oblique muscles secondarily rotate the superior portion of the eyeball medially (intorsion).

(2) The inferior rectus and the inferior oblique secondarily rotate the inferior portion of the eyeball medially (extorsion, because the top of the eye rotates laterally).

c. It is interesting to note that the secondary actions of the extraocular muscles normally tend to cancel. For example, the superior rectus and inferior oblique muscles are used to elevate the eyeball to direct the gaze upward. The secondary actions of the superior rectus cancel the secondary actions of the inferior oblique, so that pure elevation is obtained (see Table 30-1).

d. Paralysis of one or more extraocular muscles produces imbalance of the secondary actions. Complex forms of double vision occur when gaze is directed in certain directions.

7. The fascia bulbi (Tenon's capsule) forms a connective tissue socket in which the eyeball is suspended.

a. The extraocular muscles pass through Tenon's capsule to insert into the sclera.

b. The sheaths of the medial and lateral rectus tendons are thickened, forming the medial and lateral **check ligaments**, which check the action of opposite muscles.

c. These fascial condensations continue around the inferior surface of the eyeball to form a fascial hammock that suspends the eyeball between the medial and lateral margins of the bony orbit.

D. Innervation of the orbit

1. Nerves of the extraocular musculature

a. Oculomotor nerve (CN III) [Fig. 30-4; see Fig. 29-7]

(1) Axons from the oculomotor nucleus innervate five of the extraocular muscles.

(a) The **superior division** innervates the levator palpebrae superioris and the superior rectus muscles.

(b) The **inferior division** innervates the medial rectus, the inferior rectus, and the inferior oblique muscles.

(2) Axons from the accessory oculomotor nucleus (Edinger-Westphal) contribute the parasympathetic component of the oculomotor nerve (see Fig. 29-7).

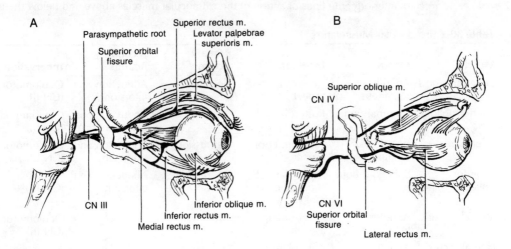

Figure 30-4. *Innervation of the extrinsic muscles of the eye. A, Oculomotor nerve. B, Trochlear and abducens nerves.*

(a) The preganglionic parasympathetic neurons leave the oculomotor nerve via the **parasympathetic (motor) root** to reach the ciliary ganglion.

(b) The postganglionic axons of the **ciliary ganglion** course through the **short ciliary nerves** to the iridial sphincter for pupillary constriction and to the ciliary muscle for accommodation to near vision.

(3) Lesions of the oculomotor nerve produce lateral strabismus (*cocked eye* or *wall eye*) and nearly complete *ophthalmoplegia* of the eye as well as *ptosis* (drooping) of the eyelid. In addition, the pupil is dilated, and there is an inability to accommodate for near vision.

(4) The oculomotor nerve passes through the incisura tentoria. An intracranial mass (tumor, hematoma, or edema) may squeeze portions of the temporal lobes through the tentorial notch (*temporal lobe herniation*), compressing the oculomotor nerve within the tentorial notch and causing *ophthalmoplegia* and *mydriasis* (dilated pupil).

b. Trochlear nerve (CN IV) [see Fig. 29-7]

(1) Axons arise from the trochlear nucleus and innervate the superior oblique muscle.

(2) Lesions produce diplopia upon looking down. Individuals with diplopia usually experience difficulty and apprehension upon descending stairs.

c. Abducens nerve (CN VI) [see Fig. 29-7]

(1) Axons arise from the abducens nucleus and innervate the lateral rectus muscle.

(2) Lesions produce *medial strabismus (crossed eyes)*. The diplopia is minimal when looking toward the side opposite the lesion.

2. The ophthalmic division of the trigeminal nerve (CN V) provides the sensory branches within and about the orbit (Fig. 30-5).

a. The lacrimal nerve runs superolaterally within the orbit.

(1) It supplies the lateral portion of the eyelids and conjunctiva.

(2) Proximally, it contains no secretomotor fibers to the lacrimal gland.

(3) Distally, it receives postsynaptic parasympathetic secretomotor fibers from the pterygopalatine ganglion via the zygomatic branch of the maxillary nerve.

b. The frontal nerve, the largest branch, bifurcates.

(1) The **supraorbital nerve** runs superiorly to the axis of the orbit.

(a) It passes through the **supraorbital notch** or **foramen**.

(b) It supplies the skin of the eyelid, forehead, and scalp.

(2) The **supratrochlear nerve** runs more medially and supplies the skin of the medial portions of the eyelids and the central portion of the forehead.

c. The nasociliary nerve runs superomedially within the orbit and becomes the anterior ethmoidal nerve.

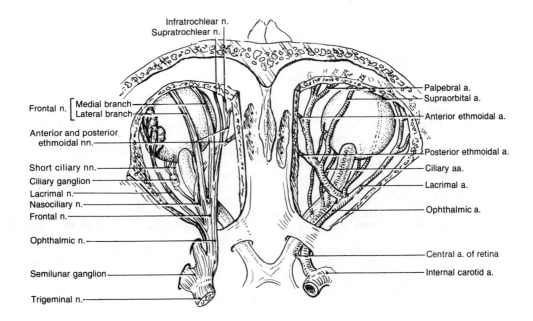

Figure 30-5. *Nerves and vessels of the orbit.* The ophthalmic division of the left trigeminal nerve and the right ophthalmic artery.

(1) A **sensory root (ramus communicans)** connects to the ciliary ganglion. Sensation from the cornea passing along the **short ciliary nerves** gains access to the nasociliary nerve via this ramus.

(2) Long ciliary nerves also convey corneal sensation directly from the eyeball.

(3) The **posterior ethmoidal nerve** passes through the posterior ethmoidal foramen to supply the mucosa of the sphenoidal and ethmoidal air sinuses as well as a portion of the nasal mucosa.

(4) The **infratrochlear nerve** supplies the medial conjunctiva and skin of the medial portion of the eyelids.

(5) The **anterior ethmoidal nerve**, the continuation of the nasociliary nerve, passes through the anterior ethmoidal foramen and descends through the ethmoid bone.

    (a) The **internal nasal branch** supplies the mucosa of the ethmoidal and frontal air sinuses as well as portions of the nasal mucosa.

    (b) The **external nasal branch** supplies the skin at the tip of the nose and about the external nares.

3. The optic nerves are really tracts of the brain and carry meningeal sheaths complete with dura, arachnoid, subarachnoid space, and pia, which are continuous with those of the brain.

  a. The optic nerves pass through the **optic foramen** in company with the ophthalmic artery.

  b. The **central artery of the retina**, a branch of the ophthalmic artery, runs within the optic nerve.

4. Autonomic innervation of the orbit

  a. Sympathetic pathways

    (1) The preganglionic neurons lie in the upper thoracic segments of the spinal cord.

      (a) These axons leave the spinal nerves and gain access to the sympathetic chain via **white rami communicantes**.

      (b) The axons then course superiorly along the **sympathetic chain** to reach the superior cervical ganglion.

    (2) From the **superior cervical ganglion**, the postsynaptic neurons course in the carotid plexus within the adventitia of the internal carotid artery. From the **carotid plexus**, the sympathetic fibers may reach the eye by a number of routes.

      (a) The sympathetic fibers continue along the internal carotid artery and the ophthalmic artery and its posterior ciliary branches to gain access to the orbit and eyeball.

      (b) In the cavernous sinus, sympathetic fibers may pass to the ophthalmic nerve and its nasociliary branch and thence along the long ciliary nerves to the eyeball or, alternatively, along the sensory root (ramus communicans) of the ciliary ganglion and then along the short ciliary nerves to the eyeball.

    (3) Injury to the cervical sympathetic chain results in *Horner's syndrome*. Signs include:

      (a) *Miosis* (pupillary constriction) due to unopposed action of the parasympathetic nerves

      (b) *Pseudoptosis* (a slight drooping of the eyelid) due to paralysis of the superior tarsal muscle (Müller's third)

      (c) Slight *enophthalmos* (retraction of the eyeball into the orbit) due to paralysis of the orbitalis muscle (Müller's first), which spans the inferior orbital fissure

      (d) *Anhidrosis* (lack of sweating)

      (e) *Flushing* due to release from the vasoconstrictive effects of the sympathetic nerves

  b. Parasympathetic pathways

    (1) Presynaptic neurons from the accessory oculomotor nucleus (of Edinger-Westphal) course in the oculomotor nerve (CN III).

      (a) The preganglionic parasympathetic neurons leave the oculomotor nerve via the **parasympathetic (motor) root**.

      (b) They then reach the **ciliary ganglion**, which is located laterally to the optic nerve.

    (2) The postganglionic axons of the ciliary ganglion follow the **short ciliary nerves**.

      (a) They course to the iridial sphincter for pupillary constriction and to the ciliary muscle of the pupil for accommodation to near vision.

      (b) Some nonparasympathetic pathways may pass through the ciliary ganglion to gain access to or from the short ciliary nerves.

        (i) Sympathetic neurons from the ophthalmic artery may pass through a **sympathetic root** and continue through the ciliary ganglion to reach the eyeball via the short ciliary nerves.

        (ii) Sensory fibers from the cornea that pass along the short ciliary nerves continue through the ciliary ganglion to reach the nasociliary nerve via the **sensory root** (ramus communicans).

E. Orbital vasculature

1. **The ophthalmic artery** is a branch of the internal carotid artery (see Fig. 30-5).
    a. It enters the orbit through the **optic canal** beneath the optic nerve.
    b. It supplies the eyeball through the **central artery of the retina** as well as by anterior ciliary and posterior ciliary branches.
    c. Within the orbit, it winds around the medial surface of the optic nerve in company with the nasociliary nerve.
    d. The branches correspond to the branches of the ophthalmic division of the trigeminal nerve: ethmoidal, supratrochlear, supraorbital, and lacrimal as well as anterior and posterior ciliary.

2. **Ophthalmic vein**
    a. The **superior ophthalmic vein** drains the superior regions of the orbit, eyelids, and forehead. It anastomoses with the angular vein, a branch of the facial vein.
    b. The **inferior ophthalmic vein** drains the inferior regions of the orbit and eyelids as well as the **vorticose veins** from the eyeball. It also anastomoses with the angular vein and the pterygoid plexus.
    c. The superior and inferior ophthalmic veins leave the orbit via the superior orbital fissure to drain, either separately or by one trunk, into the cavernous sinus.
    d. The anastomoses between the angular and ophthalmic veins may result in spread of infection from the periorbital and perinasal regions to the cavernous sinus.
        (1) Inflammatory thrombosis of the cavernous sinus interferes with venous drainage of the retina, thereby resulting in engorgement of the retinal arteries, followed by retinal ischemia and ultimate blindness.
        (2) Because the ophthalmic, oculomotor, trochlear, and abducens nerves also pass through the cavernous sinus, infection here may result in *ophthalmoplegia*, *mydriasis*, and sensory deficits in the periorbital region.

II. EYEBALL (BULBUS OCULI)

A. Overview. The eyeball is formed as an outgrowth of the brain.

1. The two layers of the retina of the adult eye represent an extension of the brain. Other parts of the eyeball, such as the lens, are derived from ectoderm.

2. The eye inverts and focuses the visual fields onto the retina; the retina transduces electromagnetic radiation into electrical impulses.

B. Structure. The eyeball is a very durable structure because it has a tough fibrous coat and a fluid-filled cavity that maintains the shape and distributes hydraulic pressures uniformly.

1. **The fibrous tunic** comprises the sclera and cornea.
    a. **The sclera** (*skleros*, G. hard), the white of the eye (Fig. 30-6), comprises approximately five-sixths of the eyeball and provides insertion for the extraocular muscles.
        (1) It is composed of dense connective tissue.
        (2) Anteriorly, the sclera is covered by the **bulbar conjunctiva**, which is transparent and contains a great number of small blood vessels and nerve endings.
            (a) Inflammation of the conjunctiva causes vascular engorgement (*bloodshot eye* or *pinkeye*). Although the usual cause of conjunctivitis is infection or allergy, occasionally the pathology is within the eyeball (*iritis, glaucoma*), in which case there is often pain or impaired vision.
            (b) The conjunctiva is innervated by the ophthalmic and maxillary divisions of the trigeminal nerve (CN V), which form the afferent limb of the blink reflex.
        (3) The sclera joins the cornea at the **limbus**.
    b. **The cornea**, which is transparent, comprises the anterior one-sixth of the eyeball (see Fig. 30-6).
        (1) It is composed of dense, regularly arranged collagen fibers; the extreme regularity of the tissue results in a liquid crystalline structure that is transparent to light.
            (a) The cornea is the principal refractor of the eye, much more so than the lens. The small radius of curvature of the cornea and the fact that the cornea separates media of two different refractive indices (air and aqueous humor) result in strong bending of light waves.
            (b) The shape of the cornea and the anteroposterior diameter of the eye determines the focal point.

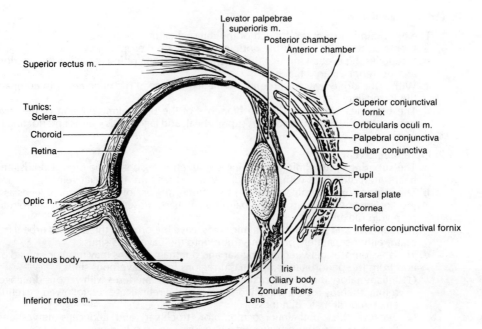

Figure 30-6. *Eyeball and conjunctival sac.*

- **(i)** Insufficient corneal refraction that cannot be corrected by the lens results in a focal point behind the retina (*hyperopia* or *hypermetropia*). The result is far-sightedness because distant objects can be focused on the retina by accommodation of the lens.
- **(ii)** Too much corneal refraction results in a focal point in front of the retina (*myopia*). The result is nearsightedness because close objects focus on the retina without accommodation by the lens.
- **(iii)** Irregularities in the shape of the cornea produce variation in the focal points (*astigmatism*).
- **(2)** The cornea is covered by **corneal epithelium**, a continuation of the conjunctiva.
- **(3)** Injury or inflammation of the cornea (*keratitis*) resolves with healing but with a loss of the high degree of organization and, consequently, loss of transparency. Corneal scars can be treated by corneal transplant (keratoplasty).

2. **Vascular tunic**
   a. **The choroid layer** consists primarily of blood vessels supplied by the short ciliary arteries and drained by the vorticose veins (see Fig. 30-6).
   b. **The ciliary body** is the anterior continuation of the choroid layer (see Fig. 30-6; Fig. 30-7).
      (1) The ciliary body suspends the lens by a multitude of **zonular fibers** (of Zinn), which insert into the capsule of the lens.
         (a) The normal elastic tension exerted on the capsule of the lens through the zonular fibers tends to flatten the lens, so that there is minimal refraction of light rays—far accommodation.
         (b) The **ciliary muscle** runs from the base of the ciliary body to the scleral spur at the limbus. Contraction of this muscle draws the ciliary body anteriorly, thereby releasing tension on the zonular fibers. This movement allows the elastic lens to assume a more spherical shape, so that there is increased refraction and convergence of light rays—near accommodation.
         (c) Accommodation is controlled by complex central pathways and effected by the parasympathetic nerves of the accessory oculomotor nucleus (of Edinger-Westphal).
            (i) These nerves travel to the orbit along the oculomotor nerve (CN III), leave via the parasympathetic root, and synapse in the ciliary ganglion.
            (ii) The postsynaptic fibers travel along the short ciliary nerves to reach the eyeball and ciliary body.
      (2) The **iris** arises from the ciliary body, anteriorly.
         (a) It divides the space between the cornea and lens into **anterior** and **posterior chambers**, which contain variable amounts of pigment.

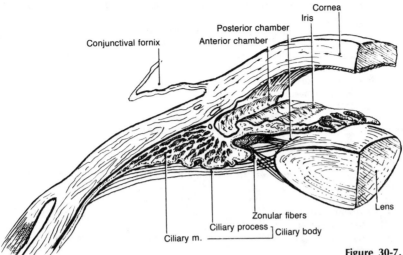

Figure 30-7. *Limbus of the eyeball and the ciliary body.*

(i) If it is heavily pigmented in the anterior layer, the result is brown eyes.

(ii) If the pigment is only in the posterior layer, the iris appears blue or gray because light refracts within the unpigmented anterior layer.

(b) The iris contains primitive myoepithelial cells surrounding a central aperture, the **pupil**.

(i) The **sphincter pupillae** consist of myoepithelial cells arranged in a circle under parasympathetic control.

(ii) The **dilator pupillae** consist of radially arranged myoepithelial cells under sympathetic control.

(c) The size of the pupil is normally a reflex response to the intensity of light reaching the retina.

(i) Injury to the oculomotor nerve releases parasympathetic influence and produces a dilated pupil (*mydriasis*).

(ii) Injury to the upper thoracic spinal cord or to the cervical sympathetic chain (*Horner's syndrome*) releases sympathetic influence and produces a constricted pupil (*miosis*).

(3) The epithelium of the ciliary body secretes **aqueous humor** into the posterior chamber.

3. Retina

a. The optic vesicle, an evagination of the brain, forms the retina.

(1) The anterior half of the optic vesicle involutes against the posterior half, forming the **optic cup**.

(2) Two primitive layers, thus, are formed by occlusion of the optic vesicle.

b. The neural retina (the anterior primitive layer) consists of four major layers of nerve and supporting cells through which light rays must pass to reach the photosensitive cells.

(1) The first layer consists of nerve axons that collect at the **optic disk (blind spot)** and pass through the **cribriform plate** of the sclera to form the optic nerve.

(2) The second layer is formed by the so-called ganglion cells and is equivalent to a brain stem nucleus.

(3) The third layer is composed of bipolar cells, equivalent to dorsal root ganglia.

(4) The fourth layer contains the light-sensitive rods and cones. At the **fovea centralis**, a great concentration of cones maximizes visual acuity in the center of the visual field.

c. The pigmented retina (the posterior primitive layer) is heavily pigmented to absorb any light that passes completely through the anterior layer and, thus, prevents confusing backscatter.

d. Clinical considerations

(1) Retinal detachment is the result of separation between the anterior and posterior layers, re-establishing the primitive optic vesicle.

(a) The separation usually starts anteriorly and, therefore, is easily missed until well advanced.

(b) The predisposition to detachment is bilateral, so care must be taken to protect the other eye.

(c) The layers may be reattached by coagulation procedures.

(2) Papilledema. Because the optic nerve is a portion of the central nervous system (CNS) and is, as such, surrounded by layers of meninges, elevated cerebrospinal fluid (CSF) pressure produces edema of the optic disk (*papilledema*), which can be detected by ophthalmoscopic examination.

(3) Funduscopic examination

(a) The **central artery of the retina** enters the eyeball with the optic nerve and branches over the retina. The retinal vessels can be examined with an ophthalmoscope.

(b) These are the only arteries in the body that can be examined directly for signs of systemic disease, such as *hypertension* and *diabetes*.

4. Chambers of the eye

a. The **anterior chamber** lies between the cornea and the iris; the **posterior chamber** lies between the iris and the lens (see Fig. 30-7).

(1) These chambers contain the thin, watery aqueous humor.

(a) Aqueous humor is secreted by the ciliary process into the posterior chamber.

(b) It passes through the pupil into the anterior chamber.

(c) It drains into the venous system through **Schlemm's canal** at the angle of the anterior chamber.

(2) If drainage is impaired, intraocular pressure increases and the retinal blood flow is impaired, producing *retinal ischemia (glaucoma)* and blindness.

b. The **vitreous body**, a transparent and semigelatinous material, fills the **vitreous chamber** behind the lens (see Fig. 30-6).

5. The lens separates the aqueous humor from the vitreous body (see Figs. 30-6 and 30-7).

a. It is composed of highly ordered connective tissue cells; the high degree of order confers transparency.

b. It is enclosed in an elastic capsule into which the zonular fibers insert.

c. The lens itself is deformable and elastic.

d. Its refractive index is slightly different from the aqueous and vitreous humors, providing some degree of refraction.

(1) The tension exerted on the lens by the zonular fibers adjusts the shape of the lens, thereby altering the refracting power.

(a) When the ciliary muscle contracts and releases tension on the zonular fibers, the lens assumes a more spherical shape and strongly refracts light rays, accommodating the eye for near vision.

(b) When the ciliary muscle relaxes, the tension on the zonular fibers is restored, and the lens is flattened so that it weakly refracts light rays, accommodating the eye for far vision.

(2) With aging, the lens tends to harden and lose intrinsic elasticity. The aged lens cannot be deformed sufficiently to accommodate to the near and far extremes (*presbyopia*), both of which may require correction with bifocal lenses.

(3) The most common cause of blindness is *cataracts*, which are progressive opacities that result from degenerative changes. When of sufficient size to impair vision, the lens can be removed surgically.

III. EAR

A. External ear The auricle (pinna), the external acoustic (auditory) meatus, and the tympanic membrane comprise the external ear.

1. The auricles, lying on either side of the head and directed slightly forward, concentrate sound waves and enable stereophonic localization of the source (Fig. 30-8).

a. The skeleton of the ear is a single convoluted plate of elastic cartilage. The skin of the auricle is firmly attached to the underlying perichondrium.

b. Anteriorly, the auricle is attached to the calvaria; posteriorly, there is a free rolled edge that forms the **helix**.

(1) The helix begins as the **crus** in the concha superiorly to the external auditory meatus.

(2) A small **superior (darwinian) tubercle** is a vestige of the point of the ear.

(3) The helix ends as a flabby **lobule** or **ear lobe**.

c. A second ridge, the **antihelix**, runs approximately parallel to the helix, dividing the auricle into an outer **scaphoid fossa** and the deeper **concha** (*konche*, G. shell).

(1) The antihelix begins anterosuperiorly as two crura that define the **triangular fossa**.

(2) Inferiorly, the antihelix terminates as the **antitragus**.

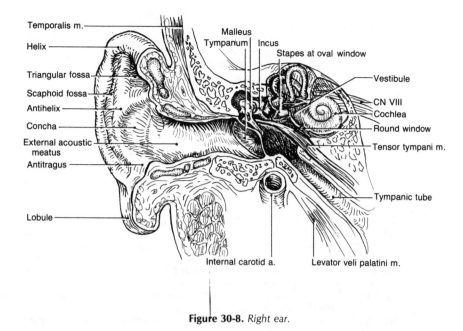

Temporalis m.
Helix
Triangular fossa
Scaphoid fossa
Antihelix
Concha
External acoustic meatus
Antitragus
Lobule

Malleus
Tympanum | Incus
Stapes at oval window

Vestibule
CN VIII
Cochlea
Round window
Tensor tympani m.

Tympanic tube

Internal carotid a.   Levator veli palatini m.

Figure 30-8. *Right ear.*

d. The **tragus** is a projection from the anterior portion of the external ear.
  (1) Directed posteriorly, it partially covers and protects the external auditory meatus.
  (2) The tragus is separated from the antitragus by the **intertragic incisure**.

2. The external auditory (acoustic) meatus extends from the concha to the tympanic membrane (see Fig. 30-8).
  a. It is somewhat S-shaped and has two distinct portions.
    (1) The external one-third is formed by a continuation of the elastic cartilage of the concha. It is directed upward and backward.
      (a) The skin lining the external auditory meatus contains sebaceous glands, associated with hair follicles, and ceruminous glands.
      (b) Ear wax (cerumen) consists of the secretions of both glands in addition to desquamated cells and dust. Excessive accumulations of ear wax can clog the external auditory meatus, especially if one tends to remove it with the fifth digit, formerly named the auricularis (*L.* ear).
    (2) The internal two-thirds of the external auditory meatus runs within a bony canal (see Fig. 30-8). It is directed slightly downward. The epithelium contains fewer glands and is devoid of hairs.
  b. The shape of the concha and external auditory meatus amplifies sound waves by a factor of five to ten times (5–10 dB).
  c. The external auditory meatus is a remnant of the first branchial (gill) groove.

3. The tympanum (*G.* drum) [**tympanic membrane** or **ear drum**] separates the external ear from the internal ear (see Fig. 30-9).
  a. It consists of a sheet of fibrous tissue covered on both sides by epithelium. The fibers are both circularly and radially arranged to keep the membrane moderately tense and, thus, receptive to sonic vibrations.
  b. The tympanum is tilted across the external auditory meatus so that the anteroinferior quadrant is deeper than the posterosuperior quadrant. It is slightly concave.
    (1) It is attached to the **malleus** along the length of the **manubrium**.
    (2) The central portion of the concavity, the **umbo**, marks this attachment.
    (3) Superior to the attachment of the malleus, the tympanic membrane appears less tense and is termed the **pars flaccida**.
  c. The pressure changes associated with sonic waves are transduced into mechanical vibrations at the tympanum.
    (1) This mechanism is so sensitive that the movement of air molecules against the tympanum by brownian motion is just below the threshold of hearing in a young individual with acute hearing.
    (2) The distance through which the tympanum moves in response to normal (60 dB) conversation is measured in nanometers.

4. Innervation of the external ear
   a. The anterior aspect of the auricle and a variable part of the external auditory meatus is innervated by the **auriculotemporal branch** of the **mandibular nerve** (CN V$_3$) and by the **posterior auricular branch** of the **facial nerve** (CN VII); the posterior aspect of the auricle is innervated by the **great auricular** and **lesser occipital nerves**, which arise from spinal levels C2 and C3.
   b. The external auditory meatus is also innervated by twigs from the glossopharyngeal nerve (CN IX); and the vagus nerve (CN X). This explains why a patient may gag or cough when an insect enters the external auditory meatus or when cerumen is removed by curettage.

B. Middle ear. The tympanic cavity with its extensions and the auditory ossicles comprises the middle ear.

   1. The tympanic cavity (antrum) is a space between the squamous and petrous portions of the temporal bone (Fig. 30-9).
      a. Lateral wall
         (1) The **tympanic membrane** and the lateral wall of the **epitympanic recess** are major features.
         (2) The **chorda tympani**, a branch of the facial nerve, passes between the fibrous and mucous layers of the tympanum and crosses the manubrium of the malleus.
            (a) It enters the tympanic cavity at the posterior canaliculus of the chorda tympani (iter chordae posterius) in the posterior wall.
            (b) It leaves the tympanic cavity through the anterior canaliculus (iter chordae anterius) in the petrotympanic fissure.
            (c) This nerve conveys taste sensation from the anterior two-thirds of the tongue as well as parasympathetic presynaptic secretomotor fibers to the submandibular ganglion.
      b. The roof (tegmen tympani) separates the epitympanic recess from the middle cranial fossa. The **epitympanic recess**, a superior extension of the antrum, contains the head of the malleus and the body of the incus.
      c. The floor of the tympanic cavity is a plate of bone separating the middle ear cavity from the jugular canal.
      d. The posterior wall of the tympanic cavity has numerous communications with mastoid air cells.
         (1) Infection of the middle ear may spread into these spaces and be difficult to treat.
         (2) Projecting from the posterior wall is a pyramidal eminence that contains the stapedius muscle.

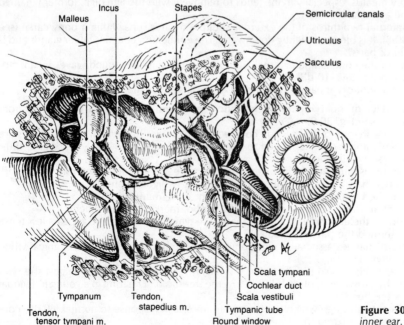

Figure 30-9. *Right middle ear and inner ear.*

e. The anterior wall separates the tympanic cavity from the carotid canal.
  (1) The **auditory (pharyngotympanic, eustachian) tube** opens into the anterior wall at the **tympanic orifice**.
    (a) It connects the middle ear with the nasopharynx and serves as a means of equalizing the pressure across the tympanum. A pressure differential of 100–150 mm Hg will rupture the tympanum.
    (b) It is a pathway for the spread of infection from the nasopharynx to the middle ear.
    (c) It is a remnant of the first branchial pouch.
  (2) The canal of the **tensor tympani muscle** opens into the anterior wall just superior to the tympanic orifice.
f. The medial wall is the most complex.
  (1) A **superior prominence** marks the position of the lateral semicircular canal of the inner ear.
  (2) The **prominence of the facial (fallopian) canal** marks the course of the horizontal portion of the facial nerve. Occasionally, there is only a layer of periosteum between the middle ear and the nerve. *Otitis media* may involve the facial nerve.
  (3) The **oval window** (fenestra vestibuli) receives the foot-plate of the stapes and transmits the sonic vibrations of the ossicles to the perilymph of the scala vestibuli.
  (4) The **tympanic bulla (promontory)** is formed by the basal turn of the cochlea.
    (a) The **tympanic plexus** (of Jacobson) passes across this promontory and contains sensory contributions from the glossopharyngeal and vagus nerves, which distribute to the tympanum and external auditory meatus.
    (b) Anteriorly from this plexus, the **lesser superficial petrosal nerve** conveys the presynaptic parasympathetic secretomotor fibers of the glossopharyngeal nerve (CN IX) origin to the otic ganglion.
  (5) The **round window** (fenestra tympani) is covered by an elastic membrane (the secondary tympanic membrane). It accommodates the pressure waves transmitted to the perilymph of the scala tympani.

2. The three auditory ossicles bridge the tympanic cavity and transmit sonic vibrations from the external ear to the inner ear (see Fig. 30-9).
  a. Malleus (*L.* hammer)
    (1) The manubrium (*L.* handle) of the malleus is attached along its length to the tympanum.
    (2) The head projects into the epitympanic recess and articulates with the incus. It is stabilized by a superior ligament to the tegmentum tympani.
    (3) An anterior process provides attachment for the anterior ligament, which passes through the petrotympanic fissure and seems to be developmentally continuous with the sphenomandibular ligament. Both are remnants of Meckel's cartilage and are, thus, first branchial arch derivatives.
    (4) The **tensor tympani muscle** (see Fig. 30-8) originates within the like-named canal in the anterior wall and inserts into the neck of the malleus.
      (a) As a first arch derivative, it is innervated by the motor division of the mandibular nerve (CN V_3).
      (b) A reflex, mediated by the mandibular nerve, damps the vibration of the malleus in response to loud noises and upon swallowing.
    (5) The chorda tympani nerve crosses the manubrium of the malleus.
  b. Incus (*L.* anvil)
    (1) This ossicle lies primarily in the epitympanic recess.
    (2) The body articulates with the head of the malleus at the incudomalleal joint.
    (3) A short posterior crus provides the attachment for the posterior ligament, which runs to the posterior wall.
    (4) A longer descending crus articulates with the stapes at the incudostapedial joint.
    (5) Like the malleus, the incus is a first branchial arch derivative.
  c. Stapes (*L.* stirrup)
    (1) The body of the stapes bifurcates into two limbs, which end in a single foot-plate.
    (2) The foot-plate inserts into the oval window, and the articulation is maintained by an annular ligament. *Otosclerosis* at the edge of the oval window impedes movement and is the most common cause of adult deafness.
    (3) The **stapedius muscle** originates within the pyramidal eminence on the posterior wall of the antrum and inserts into the head of the stapes.
      (a) As a second branchial arch derivative, it is innervated by the facial nerve (CN VII).
      (b) Reflex contraction of the stapedius muscle damps the vibrations of the stapes. Paralysis of this muscle as a result of facial nerve palsy produces *hyperacusis*, whereby normal sounds are perceived as annoyingly loud.
    (4) The stapes is derived from the second branchial arch.

3. Middle ear function. The auditory ossicles not only transmit sonic vibrations from the outer ear to the inner ear but also amplify the force.

    a. The area ratio of the tympanum to the oval window is about 18 to 1. However, the inferior crus of the incus is not as long as the handle of the malleus, so the excursion of the stapes is only about half that of the tympanum. The result is a net mechanical advantage of about 9 times.

    b. This amplification of force compensates for the differences in impedance between the air on one side of the tympanum and the perilymph on the other side of the oval window.

C. Inner ear. This division is contained within the petrous portion of the temporal bone and consists of a vestibular portion concerned with balance and a cochlear portion concerned with audition (see Fig. 30-9).

1. Structure. The inner ear is composed of a series of bony canals, the **osseous labyrinth**, within which is a system of continuous membranous canals, the **membranous labyrinth**.

    a. The osseous labyrinth is filled with perilymph, in which is suspended the membranous labyrinth.

    b. The membranous labyrinth is filled with endolymph and contains the sensory organs (Fig. 30-10).

2. Vestibular apparatus

    a. Labyrinth of the vestibular portion

        (1) The **vestibule** is a chamber in the osseous labyrinth situated behind the oval window.

            (a) From this chamber radiate three bony semicircular canals and one of the chambers (scala vestibuli) of the spiral cochlea.

            (b) Within these chambers and canals lie the comparable portions of membranous labyrinth.

        (2) The **utricle** and **saccule** are dilations of the membranous labyrinth within the vestibule.

            (a) Within each of these dilations is a sensory **macula** that projects into the endolymph.

            (b) The cytoarchitecture of the **macula** is such that sensory nerve endings are deformed in response to static gravity and inertia as well as in response to vibrations.

    b. Three semicircular canals of the labyrinth

        (1) The bony canals are arranged in mutually perpendicular planes.

            (a) The **anterior (superior) semicircular canal** projects vertically with the long axis directed anteromedially at about 45°.

            (b) The **lateral semicircular canal** is nearly horizontal and projects slightly into the middle ear cavity.

            (c) The **posterior semicircular canal** projects vertically with the long axis directed posterolaterally at about 45°.

            (d) The **anterior semicircular canal** is parallel to the contralateral posterior semicircular canal.

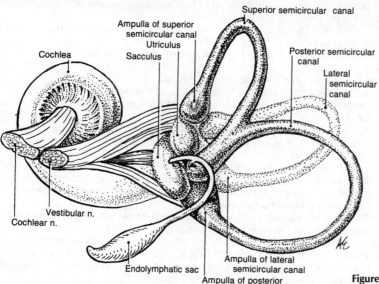

Figure 30-10. *Right membranous labyrinth.*

(2) The membranous semicircular canals are suspended in perilymph.
  (a) A dilation of the membranous labyrinth at one end of each semicircular canal contains a **crista ampullaris**.
  (b) The cytoarchitecture of the **crista ampullaris** is such that sensory nerve endings are stimulated on structural deformation caused by rotational inertia on the enclosed endolymph.
  c. The nerves from the maculae and cristae ampullaris travel in the vestibular portion of the **vestibulocochlear nerve (CN VIII)** [see Figs. 29-8 and 30-10].

3. Cochlear apparatus (Fig. 30-11)
  a. The bony cochlea consists of two adjacent ducts, each somewhat less than semicircular in cross section, that spiral two and three-quarters turns about a central **modiolus**.
    (1) The upper **scala vestibuli** (scala, *L.* stairway) begins in the vestibule and receives the vibrations transmitted to the perilymph at the oval window.
    (2) The lower **scala tympani** connects with the scala vestibuli through the **helicotrema** at the apex of the cochlea and terminates at the round window, at which the sound pressure waves are dissipated.
  b. The membranous cochlear duct (scala media) is wedged distally between the scala vestibuli and scala tympani as far as the the helicotrema.
    (1) This duct contains the **spiral organ** (of Corti), which is suspended in endolymph.
    (2) The cytoarchitecture of the spiral organ is such that a specific portion of this structure resonates harmonically with each audible frequency.
      (a) The width of the spiral organ is greater toward the apex of the cochlea than at the base; thus, the lower frequencies resonate near the helicotrema and the higher frequencies near the oval window.
      (b) Sensory cells detect the resonant vibration.
  c. The nerves travel through the hollow modiolus to form the acoustic division of the **vestibulocochlear nerve** (CN VIII) [see Fig. 30-11].
    (1) Sensory neurons from specific portions of the spiral organ project to the medial geniculate body and thence to the auditory cortex where tone is perceived.
    (2) Slight differences in the quality, amplitude, and phasing of the sound result in stereophonic localization of the source.

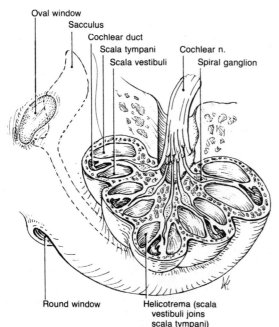

Figure 30-11. *Right cochlea.*

Part VIII Somatic Neck and Neurocranium

STUDY QUESTIONS

Directions: Each question below contains five suggested answers. Choose the **one best** response to each question.

1. Which of the following groups of structures is enclosed by the prevertebral fascia?

(A) Infrahyoid (ribbon) muscles
(B) Muscles of the cervical vertebral column
(C) Sternomastoid and trapezius muscles
(D) Thyroid gland
(E) Trachea and esophagus

2. Which of the following nerves of the cervical plexus has motor function?

(A) Greater occipital
(B) Lesser occipital
(C) Posterior auricular
(D) Suboccipital
(E) Transverse cervical

3. Infection may spread from the nasal cavity to the meninges along the olfactory nerves. Olfactory fibers pass from the mucosa of the nasal cavity to the olfactory bulb via the

(A) anterior and posterior ethmoidal foramina
(B) cribriform plate of the ethmoid bone
(C) hiatus semilunaris
(D) nasociliary nerve
(E) sphenopalatine foramen

4. The crista galli serves as an attachment for

(A) the diaphragma sella
(B) the falx cerebelli
(C) the falx cerebri
(D) the tentorium cerebelli
(E) none of the above

5. Cerebrospinal fluid enters the subarachnoid space at the

(A) arachnoid villi
(B) choroid plexuses
(C) foramina of Luschka and Magendie
(D) foramen of Monro
(E) iter

6. Cerebrospinal fluid enters the venous system

(A) at arachnoid villi
(B) at the cisterna magna
(C) through subarachnoid veins
(D) via capillaries in the ependyma
(E) by none of the above routes

7. A subdural hematoma is usually caused by

(A) fracture of the diploic space
(B) laceration of the superficial temporal artery
(C) leakage from a cerebral vein
(D) rupture of a cerebral artery
(E) tearing of a meningeal vessel

8. A cerebral vascular accident in the primary visual area of the left cerebral hemisphere results in

(A) binasal heteronymous hemianopsia
(B) bitemporal heteronymous hemianopsia
(C) contralateral homonymous hemianopsia
(D) ipsilateral homonymous hemianopsia
(E) total blindness of the right eye

9. All of the following arteries are branches of the internal carotid artery EXCEPT the

(A) anterior cerebral
(B) middle cerebral
(C) ophthalmic
(D) posterior cerebral
(E) posterior communicating

10. All of the following statements concerning lacrimation are correct EXCEPT

(A) blinking of the eyelids expresses small amounts of lacrimal fluid from the gland
(B) lacrimal secretion is controlled by parasympathetic nerves
(C) the lacrimal canaliculi drain through an ampulla into the lacrimal sac
(D) blinking causes the lacrimal sac to aspirate lacrimal fluid from the lacus lacrimalis
(E) the nasolacrimal duct ends in the hiatus semilunaris of the middle nasal meatus

11. All of the following extrinsic extraocular muscles have a posterior site of attachment in the orbit EXCEPT the

(A) inferior oblique
(B) inferior rectus
(C) lateral rectus
(D) levator palpebrae superioris
(E) superior oblique

12. The cell bodies of the neurons responsible for pupillary dilation are located in the

(A) accessory oculomotor nucleus
(B) ciliary body
(C) ciliary ganglion
(D) pterygopalatine ganglion
(E) superior cervical ganglion

Directions: Each question below contains four suggested answers of which **one or more** is correct. Choose the answer

    A if **1, 2, and 3** are correct
    B if **1 and 3** are correct
    C if **2 and 4** are correct
    D if **4** is correct
    E if **1, 2, 3, and 4** are correct

13. The infrahyoid (ribbon) muscles are innervated by the

(1) superior ramus of the ansa cervicalis
(2) hypoglossal nerve
(3) inferior ramus of the ansa cervicalis
(4) spinal accessory nerve

14. Which of the following structures might be involved by a cranial fracture that passes through the jugular foramen?

(1) Spinal accessory nerve
(2) Vagus nerve
(3) Cranial accessory nerve
(4) Hypoglossal nerve

15. Loud, low-pitched sounds cause

(1) reflex activity along a branch of CN V
(2) reflex activity along a branch of CN VII
(3) a bulging of the round window membrane into the middle ear cavity
(4) maximal vibration in the apical portion of Corti's spiral organ

16. Nerves passing through the cavernous sinus include the

(1) oculomotor
(2) abducens
(3) trochlear
(4) ophthalmic

17. True statements concerning the precentral gyrus of the brain include which of the following?

(1) It is the primary motor area
(2) It receives input from the basal ganglia and cerebellum
(3) It projects to brain stem nuclei and the gray matter of the spinal cord
(4) It receives direct sensory input

18. Which of the following nerves provide sensory innervation to the external ear?

(1) Mandibular division of the trigeminal nerve (CN V₃)
(2) Glossopharyngeal nerve (CN IX)
(3) Facial nerve (CN VII)
(4) Vagus nerve (CN X)

19. Palsy of the right abducens nerve results in diplopia when the gaze is directed

(1) up and to the right
(2) laterally to the right
(3) down and to the right
(4) laterally to the left

20. In moving the eye outward, the lateral rectus as well as the superior and inferior oblique muscles are used. These muscles are innervated by which of the following nerves?

(1) Abducens nerve
(2) Inferior division of the oculomotor nerve
(3) Trochlear nerve
(4) Superior division of the oculomotor nerve

Directions: The groups of questions below consist of lettered choices followed by several numbered items. For each numbered item, select the **one** lettered choice with which it is **most closely** associated. Each lettered choice may be used once, more than once, or not at all. Choose the answer

A if the item is associated with **(A) only**
B if the item is associated with **(B) only**
C if the item is associated with **both (A) and (B)**
D if the item is associated with **neither (A) nor (B)**

Questions 21–23

For each muscle action listed below, select the effect with which it is most commonly associated.

(A) Rotation of the head to the right
(B) Lateral flexion of the head to the right
(C) Both
(D) Neither

21. Spasmotic contracture (torticollis) of the right sternomastoid muscle

22. Normal action of the right scalene group of muscles

23. Normal action of the right splenius group of muscles

Questions 24 and 25

For each case history listed below, select the nerve damage with which it is most likely to be associated.

(A) Right CN V damage
(B) Right CN VII damage
(C) Both
(D) Neither

24. After an injury to the face, stimulation of the right cornea resulted in blinking of the left eye but not the right eye

25. After sustaining a fracture that separated the facial skull from the somatic skull (Le Fort type III), stimulation of the right cornea failed to produce blinking in either eye. Stimulation of the left cornea produced blinking in the left eye only.

Directions: The groups of questions below consist of lettered choices followed by several numbered items. For each numbered item select the **one** lettered choice with which it is **most closely** associated. Each lettered choice may be used once, more than once, or not at all.

Questions 26–30

For each structure named below, select a foramen through which it passes at some point along its course.

(A) Foramen magnum
(B) Foramen ovale
(C) Foramen rotundum
(D) Foramen spinosum
(E) Stylomastoid foramen

26. Facial nerve

27. Middle meningeal artery

28. Mandibular division of the trigeminal nerve

29. Maxillary division of the trigeminal nerve

30. Spinal accessory nerve

Questions 31–35

For each structure named below, select the fissure or canal through which it passes at some point along its course.

(A) Anterior condylar canal
(B) Hiatus of the facial canal
(C) Inferior orbital fissure
(D) Optic canal
(E) Superior orbital fissure

31. Abducens nerve

32. Greater superficial petrosal nerve

33. Hypoglossal nerve

34. Ophthalmic artery

35. Ophthalmic vein

Questions 36–40

For each question concerning the blood supply to the brain, select the most appropriate artery from the list below.

(A) Anterior cerebral artery
(B) Basilar artery
(C) Internal carotid artery
(D) Middle cerebral artery
(E) Vertebral artery

36. Branches of this artery serve the region above the corpus callosum

37. A branch of this artery supplies the eyeball

38. Branches of this artery supply the basal ganglia and thalamus

39. Branches of this artery supply the inferior cerebellum

40. Branches of this artery serve the speech area

ANSWERS AND EXPLANATIONS

1. The answer is B. [*Chapter 27 I B 2 d, 3 b; Figure 27-1*] The prevertebral fascia encloses and invests the somatic musculature of the neck—that is, muscles of the cervical vertebral column. The sternomastoid and trapezius muscles are associated with the superficial layer of the deep cervical fascia, while the strap muscles lie between that layer and the pretracheal fascia. The visceral structures of the neck are bounded by the pretracheal fascia.

2. The answer is D. [*Chapter 27 II D 1 a, b, 2 a (1), (2)*] The suboccipital nerve (C 1, posterior) supplies the muscles of the occipital triangle and has no sensory function. All of the other named branches of the cervical plexus (C2–C4, anterior), with the exception of the ansa cervicalis, are sensory nerves.

3. The answer is B. [*Chapter 28 III C 2 a (2) (c); Chapter 29 V B 1 a (1); VI B 2 b*] The cribriform plate of the ethmoid bone provides the passageways for the olfactory nerves from the olfactory mucosa of the superior nasal meatus to the olfactory bulb of the brain. Infection may track along these nerves, thereby spreading to the meninges. Fracture of the ethmoid bone may result in leaking of cerebrospinal fluid through the nose.

4. The answer is C. [*Chapter 28 III C 2 a (2); IV A 1 b (1) (a)–(d)*] In the anterior cerebral fossa, the falx cerebri attaches to the crista galli, the continuation of the vertical plate of the ethmoid bone. The diaphragma sella covers the pituitary fossa; the tentorium cerebelli separates the middle from the posterior cranial fossae; and the falx cerebelli lies in the posterior cranial fossa.

5. The answer is C. [*Chapter 28 IV A 2 b (1) (a); Chapter 29 VI B*] Cerebrospinal fluid (CSF), secreted into the ventricles by the choroid plexuses, enters the subarachnoid space through the foramina of Luschka and Magendie in the roof of the fourth ventricle. CSF passes from the lateral ventricles through the foramen of Monro into the third ventricle and thence through the iter into the fourth ventricle.

6. The answer is A. [*Chapter 28 IV A 2 b (1) (c); Chapter 29 VI B 2 d; Figure 29-9*] Cerebrospinal fluid is secreted by the choroid plexuses of the lateral ventricles, the third ventricle, and the fourth ventricle. It circulates through the ventricles and enters the subarachnoid space through the foramina of Luschka and Magendie in the roof of the fourth ventricle. The cerebrospinal fluid enters the superior sagittal sinus via arachnoid villi.

7. The answer is C. [*Chapter 28 IV B 3*] Cerebral veins are most vulnerable to tearing as they pass between the arachnoid and dura mater to drain into the venous sinuses. Injury here results in extravasation of blood in the subdural space (subdural hematoma). A meningeal artery tear produces an epidural hematoma, while a ruptured cerebral artery bleeds into the subarachnoid space.

8. The answer is C. [*Chapter 29 II A 4 c (1) (c)*] Because the optic projections to the visual cortex contain fibers that convey information from the contralateral fields, loss of the primary visual area on one side produces loss of the same contralateral visual field in each eye—*contralateral homonymous hemianopsia*.

9. The answer is D. [*Chapter 29 VII B 1 c, 2 d (4)*] The posterior cerebral artery is a branch of the basilar artery, the continuation of the joined vertebral arteries. The basilar artery not only supplies the brain stem but also supplies the occipital lobe of the cerebrum.

10. The answer is E. [*Chapter 30 I B 6, 7*] The nasolacrimal duct enters the inferior meatus of the nose. The hiatus semilunaris in the middle meatus receives drainage from the frontal, ethmoidal, and maxillary sinuses.

11. The answer is A. [*Chapter 30 I C 4 b; Table 30-1*] The inferior oblique muscle of the eye originates from the maxilla at the anteromedial border of the orbit. All the other extraocular muscles, including the levator palpebrae superioris, originate from the apex of the orbit, which surrounds the optic foramen.

12. The answer is E. [*Chapter 30 I D 4 a*] The cell bodies of the neurons that innervate the dilator pupillae muscle are located in the superior cervical ganglion. These receive presynaptic stimulation from .neurons located in the uppermost thoracic levels of the spinal cord. Generally, the sympathetic neurons follow perivascular pathways to the site of innervation.

13. The answer is B (1, 3). [*Chapter 27 II D 2 a (2) (b)*] The superior ramus of the ansa cervicalis from spinal nerve C1 runs with (but is not a part of) the hypoglossal nerve to innervate the geniohyoid and thyrohyoid, and sternohyoid muscles. The inferior ramus of the ansa cervicalis from spinal nerves C2 and C3 innervates the omohyoid and sternothyroid muscles. The two limbs of the ansa communicate and provide cross innervation to some of these muscles, such as the sternothyroid.

14. The answer is A (1, 2, 3). [*Chapter 28 III B 5 c (1) (c) (i)*] In addition to the jugular vein, the jugular foramen transmits the glossopharyngeal nerve (CN IX), the vagus nerve (CN X) with its cranial accessory component, and the spinal accessory nerve (CN XI). The hypoglossal nerve leaves the cranium through the anterior condylar (hypoglossal) canal.

15. The answer is E (all). [*Chapter 30 III B 1 f (5), 2 a (4), c (3), C 3 b (2) (a)*] Low-pitched sounds cause harmonic resonance in the apical portion of the spiral organ. Loud sounds result in reflex contraction of the tensor tympani muscle (innervated by CN V) and the stapedius muscle (innervated by CN VII). Every action of the stapes at the oval window produces an opposite action at the round window.

16. The answer is E (all). [*Chapter 28 IV A 1 b (3) (c); Figure 28-11*] The oculomotor, trochlear, abducens, and ophthalmic nerves run through the cavernous sinus en route to the orbit. The carotid artery also lies in the medial wall of this sinus.

17. The answer is A (1, 2, 3). [*Chapter 29 II A 4 a (1), C 1 a*] The precentral gyrus of the frontal lobe is the primary motor area. Its somatotopically organized pyramidal cells project to the lower motor neurons of the brain stem and spinal cord. It receives input from the cortical sensory areas as well as modulating influences from the basal ganglia and cerebellum. The motor area receives no direct sensory projections.

18. The answer is E (all). [*Chapter 30 III A 4*] The anterior aspect of the pinna is innervated by the auriculotemporal branch of the mandibular division of the trigeminal nerve, and the posterior aspect is innervated by the auricular branch of the facial nerve. The external auditory meatus in innervated by branches from both the glossopharyngeal and vagus nerves.

19. The answer is A (1, 2, 3). [*Chapter 29 V C 1 c (3); Chapter 30 I D 1 c*] Because the abducens nerve innervates the lateral rectus muscle, abducens palsy results in a medial strabismus. The diplopia is minimized when the gaze is directed toward the opposite side.

20. The answer is A (1, 2, 3). [*Chapter 30 Table 30-1*] The lateral rectus muscle, the principal abductor of the eye, is innervated by the abducens nerve (CN VI). The superior oblique muscle, an elevator and abductor of the eye, is innervated by the trochlear nerve (CN IV). The inferior oblique muscle, a depressor and abductor of the eye, and the inferior rectus muscles are innervated by the inferior division of the oculomotor nerve (CN III). The superior rectus and medial rectus muscles, as well as the levator palpebrae superioris, are all innervated by the superior division of the oculomotor nerve.

21–23. The answers are: 21-B, 22-D, 23-A. [*Chapter 27 II C 1 a (2), 2 a (1) (b), (2) (b), (4) (b), c (1)*] The sternomastoid muscles turn the head toward the contralateral side and flex the neck toward ipsilateral side. The scalene muscles, acting as accessory respiratory muscles, elevate the first and second ribs and have no action on the head. The splenius group, acting bilaterally, extends the head; acting unilaterally, it turns the head toward the ipsilateral side.

24 and 25. The answers are: 24-B, 25-C. [*Chapter 30 I B 3 a (4), 5 b (2)*] The trigeminal nerve (CN V) is the afferent limb of the blink reflex. Complex neural pathways in the brain stimulate both ipsilateral and contralateral facial nerve nuclei so that a bilateral (consensual) blink response is mediated by the facial nerves, forming the efferent limb of the blink reflex. The results of the tests may be deduced from this knowledge.

26–30. The answers are: 26-E, 27-D, 28-B, 29-C, 30-A. [*Chapter 28 III D*] The facial nerve enters the temporal bone through the internal acoustic meatus to gain access to the facial canal, which terminates at the stylomastoid foramen. The middle meningeal artery transits the foramen spinosum. The mandibular and maxillary nerves lie in the foramina ovale and rotundum, respectively. The roots of the spinal accessory nerve enter the cranial cavity by passing upward through the foramen magnum; this nerve then leaves via the jugular foramen.

31–35. The answers are: 31-E, 32-B, 33-A, 34-D, 35-E. [*Chapter 28 III D*] The nerves to the extraocular muscles, including the abducens (CN VI) enter the bony orbit through the superior orbital fissure along with the ophthalmic branch of the trigeminal nerve (CN V$_1$) and the ophthalmic vein. However, the ophthalmic artery enters the orbit through the optic canal along with the optic nerve. The greater superficial petrosal nerve leaves the facial nerve by way of the hiatus of the facial canal before entering the pterygoid (vidian) canal. The hypoglossal nerve passes through the anterior condylar (hypoglossal) canal in the basal occiput.

36–40. The answers are: 36-A, 37-C, 38-D, 39-E, 40-D. [*Chapter 29 VII B 1 c, 2, 3*] The ophthalmic branch of the internal carotid artery supplies the eyeball and a portion of the forehead. The anterior cerebral artery passes dorsally to the corpus callosum to supply the frontal and parietal cortexes. The middle cerebral artery, the third major branch of the internal carotid, lies in the lateral fissure and supplies the basal ganglia and thalamus through lenticulostriate arteries as well as the more lateral areas of the cerebral cortex, including the speech area. The vertebral–basal artery complex supplies the brain stem, cerebellum, and occipital lobe.

Part IX
Facial Cranium and Visceral Neck

31
Facial Skeleton

I. INTRODUCTION

A. Developmental considerations. The facial cranium develops in conjunction with the sensory organs for vision and smell, the nasal passages for respiration, and the oral stoma for taste and ingestion.

1. **Visceral derivatives.** Portions of the facial skeleton and the anterior skeletal structures of the neck are the visceral derivatives.
   a. The face and anterior neck superior to the hyoid bone have lost the usual somatic covering, and the visceral (branchiomeric) musculature has come to lie on the surface.
   b. Because these structures derive from the primitive gill–arch system, they exhibit **branchiomeric segmentation** (*branchios*, G. gill).

2. **Basic innervation** of the visceral portion of the head and neck
   a. One series of cranial nerves is related to the organs of special senses: the olfactory nerve (CN I), the optic nerve (CN II), and the vestibulocochlear nerve (CN VIII).
   b. Another series of cranial nerves is related to muscles with somatic segmentation in the head.
      (1) The extraocular muscles are derived from cephalic somites and innervated by the oculomotor nerve (CN III), the trochlear nerve (CN IV), and the abducens nerve (CN VI).
      (2) In addition, many of the muscles of the tongue are derived from cephalic somites and are innervated by the hypoglossal nerve (CN XII).
   c. A third series of cranial nerves is associated with muscles of visceral (branchiomeric) segmentation and is related to the gill (branchial) arches.

B. Organization

1. **Branchial arches.** In the pharynx of lower vertebrates (e.g., fish), a series of gill clefts passes completely through the pharyngeal wall. Respiration occurs by passing swallowed water through the gill slits, where gaseous exchange occurs between the water and the very superficial branchial blood vessels. Both cephalad and caudal to the branchial cleft (trema) are branchial arches that are composed of bone (or cartilage), muscles, blood vessels, and nerves. In man, remnants of these gill arches remain, many being incorporated into other structures and taking on other functions (Fig. 31-1).
   a. **First arch derivatives**
      (1) Mandible, sphenomandibular ligament, malleus, and incus
      (2) Muscles of mastication as well as the mylohyoid, anterior belly of the digastric, tensor tympani, and tensor veli palatini muscles
      (3) Trigeminal nerve (CN V)
   b. **Second arch derivatives**
      (1) Lesser horn of hyoid bone, stylohyoid ligament, styloid process, and stapes
      (2) Muscles of facial expression as well as the stapedius, stylohyoid, and posterior belly of the digastric
      (3) Facial nerve (CN VII)
   c. **Third arch derivatives**
      (1) Body and greater horn of hyoid bone
      (2) Stylopharyngeus muscle
      (3) Glossopharyngeal nerve (CN IX)
   d. **Fourth arch derivatives**
      (1) Laryngeal cartilages
      (2) Most of the pharyngeal musculature and cricothyroid muscle
      (3) Superior laryngeal branch of the vagus nerve (CN X)

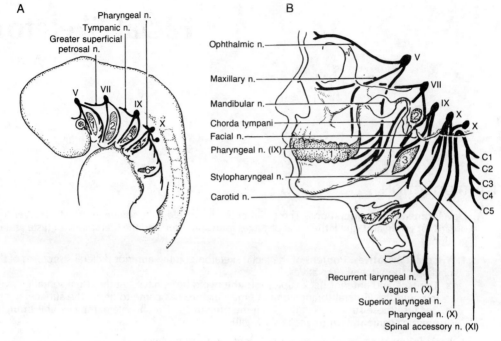

Figure 31-1. *Developmental organization of the cranial nerves. A,* Each branchiomeric nerve has a sensory pretrematic division and a mixed post-trematic division, which innervate the adjacent sides of each gill cleft. *B,* The same pattern can be discerned in the adult.

    e. Fifth and sixth arch derivatives
        (1) Intrinsic laryngeal cartilages
        (2) Most of the laryngeal musculature
        (3) Recurrent laryngeal branch of the vagus nerve (CN X) [the cranial accessory portion of CN XI]

2. Many branchiomeric muscles maintain relationships with the bones associated with the arch, but some muscle migration also occurs.
    a. The muscles of facial expression migrate cranially from the second arch to cover the facial skeleton and to overlie the muscles of the first arch.
    b. The diaphragm, while not branchiomeric, migrates caudally into the thoracic region.

3. The branchiomeric nerves are all mixed nerves, containing afferent and efferent fibers. These mixed nerves run from the brain stem to the region of the cleft and divide in a characteristic manner forming **pretrematic** (*trema, G.* hole) and **post-trematic** branches (see Fig. 31-1).
    a. The post-trematic branches of the branchiomeric nerves contain both sensory and motor nerves.
    b. The pretrematic branches of the branchiomeric nerves are sensory only and tend to join the post-trematic branch of the preceding branchial arch. Thus, each arch has, to some extent, a dual innervation.
        (1) This dual innervation accounts for the maxillary nerve (pretrematic branch of the trigeminal nerve for the first branchial arch) being totally sensory and the mandibular nerve (the post-trematic branch of the trigeminal nerve) being mixed.
        (2) It explains why the chorda tympani (sensory pretrematic branch of the facial nerve for the second branchial arch) joins the lingual branch of the mandibular nerve, why the tympanic nerve (sensory pretrematic branch of the glossopharyngeal nerve for the third branchial arch) joins the acoustic branch of the facial nerve to innervate the external auditory meatus, and why the pharyngeal branch of the vagus nerve (sensory pretrematic) joins the pharyngeal branch of the glossopharyngeal nerve (post-trematic).

II. FACIAL CRANIUM

    A. Basic structure. The facial cranium fills a space bounded posteriorly by the sphenoid bone and superiorly by the floor of the anterior cranial fossa. The facial cranium is the external counterpart of the step between the anterior and middle cranial fossae.

1. **Composition.** The facial cranium consists of 14 bones.
   a. The unpaired midline bones include the **ethmoid bone** and **vomer**.
   b. The lateral bones that fuse in the midline include the **maxilla** and **mandible**. The **hyoid bone** sometimes is included.
   c. The five pairs of separate lateral bones include the **nasal bones, lacrimal bones, inferior nasal conchae, palatine bones**, and **zygomatic bones**.
   d. In addition, the facial skeleton shares a number of the bones of the neurocranium, including the frontal, ethmoid, sphenoid, temporal, and basioccipital bones.

2. **In the upper face**, the orbital margin is bounded by the frontal, zygomatic, and maxillary bones. The sphenoid, palatine, ethmoid, lacrimal, and nasal bones also contribute to the orbital walls (Fig. 31-2).

3. **In the midface**, on either side of the sagittal plane, a pyramidal stack of hollow bones surrounds the nasal cavity (see Fig. 31-2).
   a. Its apex is the ethmoid bone; its floor (the hard palate) is formed by the maxilla and palatine bones; and in between, the vomer contributes to the nasal septum.
   b. On either side are the maxilla, palatine, inferior nasal concha, lacrimal, and nasal bones.

4. **In the lower face**, the mandible surrounds the floor of the mouth and pharynx. It articulates with the temporal bone of the neurocranium to complete the skull (see Fig. 31-2).

5. **Severe fracture** may separate the visceral face from the neurocranium (Le Fort type III), allowing the face to move posteriorly and inferiorly to obstruct the airway.

B. **Bones of the facial cranium**

   1. **The frontal bone** is shared with the neurocranium (Figs. 28-3 and 31-2).
      a. The squamous portion of the frontal bone turns sharply toward the posterior at the orbital margins to form the **orbital plates**.
      b. Between the orbital plates is the **ethmoid notch**, which receives the cribriform plate of the ethmoid bone. On each side in the frontal bone are two tiny canals, which transmit the anterior and posterior ethmoid nerves and arteries to the superior surface of the cribriform plate and to the nasal cavity.
      c. Anteriorly, the **frontal sinuses (air cells)** lie between the bony tables of the frontal bone. The frontal sinuses drain via the **frontonasal duct (infundibulum)**, which empties into the **hiatus semilunaris** of the **middle nasal meatus**.
      d. Anteriorly in the midline, the **nasal notch** articulates with the nasal bone and the frontal process of the maxilla. This is the location of the **nasion**.
      e. In the center of the supraorbital margin, the **supraorbital notch** or **foramen** transmits the supraorbital neurovascular bundle to the forehead.

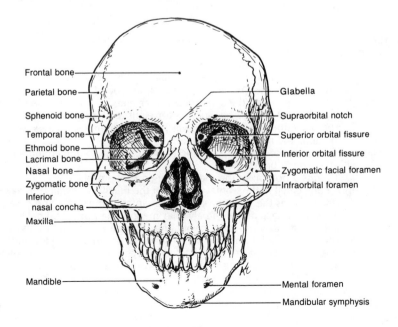

Figure 31-2. *Facial skeleton.*

2. **The zygomatic bones** are laterally placed (Figs. 28-3 and 31-2).
   a. The zygomatic processes of the maxillary and frontal bones transmit forces from the facial skeleton to the zygomatic bone.
   b. These forces are transmitted along the temporal process to the neurocranium.
   c. Zygomatic fractures result from trauma to the cheek bone with loss of stability in the zygomatic arch. Because the zygomaticotemporal and zygomaticofacial branches of the maxillary nerve pass through foramina in the zygomatic bone, there may be associated paresthesias.

3. **The ethmoid bone** is also shared with the neurocranium (Figs. 28-3 and 31-2).
   a. The ethmoid sits at the apex of a pyramidal stack of bones that define the nasal opening. It contributes to the roof, walls, septum, sinuses, and conchae of the nasal cavity (see Fig. 31-3*A*).
   b. The **cribriform plate** forms the roof of the nasal cavity (see Fig. 31-3*B*).
   c. The **ethmoid labyrinths** on either side form the posterosuperior walls of the nasal cavity, medially, and a portion of the orbital wall, laterally.
      (1) Thin, scroll-shaped **superior** and **middle conchae** divide the lateral wall into a **sphenoethmoidal recess**, a **superior meatus**, and a **middle meatus**.
      (2) The ethmoid bone contains about a dozen **ethmoid sinuses (air cells)**, which open medially into the nasal cavity.
         (a) The posterior ethmoid sinuses open into the superior meatus.
         (b) The middle and anterior ethmoid sinuses open into the **hiatus semilunaris** of the middle meatus (see Fig. 31-3*A*).
      (3) The lateral plates of the ethmoid labyrinth form the **lamina papyracea** of the medial orbital wall.
   d. In the midline, the **vertical plate** extends above the cribriform plate as the **crista galli**, and articulates with the vomer below to form a portion of the bony **nasal septum**.
   e. **Ethmoid fracture**
      (1) Fracture of the cribriform plate may produce rhinorrhea [discharge of cerebrospinal fluid (CSF) through the nose].
      (2) Fractures of the lamina papyracea breach separation of the nasal and orbital cavities so that blowing the nose may produce momentary exophthalmos.

4. **The nasal bones** form the bridge of the nose and articulate with the frontal processes of the maxilla as well as with the ethmoid and frontal bones of the neurocranium (see Figs. 30-1 and 31-2).

5. **The lacrimal bones** not only are the smallest bones of the facial cranium but also are very thin (see Fig. 30-1).
   a. Each lies between the frontal process of the maxilla and the ethmoid labyrinth. They contribute to the medial wall of the orbit.
   b. Each lacrimal bone bears a groove (lacrimal fossa) for the lacrimal sac.

6. **The maxillae** are fused to form the largest bone of the facial cranium (see Fig. 31-2).
   a. Each maxilla lies below the frontal bone and ethmoid labyrinth. It is more or less pyramidal in shape and from it extend several processes.
      (1) The lateral **zygomatic process** articulates with the zygomatic bone and marks the boundary of the anterior face and the infratemporal fossa.
      (2) The **alveolar process** (the superior alveolar margin) supports the upper teeth.
         (a) The alveolar process has sockets for eight teeth on each side.
         (b) The posterior end of the alveolar process continues a little way beyond the third molar as the **maxillary tubercle**.
      (3) The **frontal process** extends around the anterior end of the ethmoid bone to articulate with the frontal bone.
         (a) It forms a portion of the medial margin of the orbit.
         (b) A bony canal in the frontal process of the maxilla transmits the **nasolacrimal duct**, which drains the lacrimal sac into the inferior meatus of the nasal cavity.
      (4) On its inner surface above the alveolar margin, the horizontal **palatine process** forms the anterior floor of the nasal cavity and the roof of the oral cavity.
         (a) These are fused at the midline to form the anterior four-fifths of the **hard palate**.
         (b) The horizontal palatine process is pierced anteriorly by the **incisive foramen**, which transmits the nasopalatine neurovascular bundle.
   b. The **maxillary tuberosity** marks the posterior end of the maxilla. The narrow space behind the maxilla (the pterygopalatine fossa) is closed medially by the perpendicular plate of the palatine bone and posteriorly by the pterygoid process of the sphenoid bone (see Fig. 31-4).

c. The superior (orbital) surface of the maxilla contains the **infraorbital canal**, which contains the infraorbital nerve, and terminates in the face as the **infraorbital foramen**.

d. The large **maxillary sinus** (Highmore's antrum) lies within the maxillary bone.

    (1) The opening is high on the medial surface so that it does not drain by gravity when the head is held erect. It drains with the frontal sinus and several ethmoid sinuses into the **hiatus semilunaris**.

        (a) This explains why infection so readily spreads among these nasal sinuses.

        (b) Once infected, it is difficult to clear because the cilia, which would normally sweep the mucus up to the opening, are disabled. A collection of stagnant and putrid mucus may have to be drained surgically.

    (2) The floor of the maxillary sinus is formed by the alveolar process; only a thin layer of bone separates the sinus cavity from the roots of the teeth. *Sinusitis* may produce a toothache.

    (3) The infraorbital canal frequently projects from the root of the sinus.

e. Maxillary fractures

    (1) Fracture of the orbital floor may result in herniation of the periorbital fat into the maxillary sinus with enophthalmus. Blowing the nose may produce momentary exophthalmos.

    (2) Collapse of the thin anterior wall of the maxillary sinus may denervate the anterior maxillary teeth because the anterior–superior alveolar nerves run in tunnels through this bone.

    (3) The maxilla may be fractured transversely at the level of the nasal floor (Le Fort type I).

7. The inferior nasal concha, a separate bone, lies in the lateral wall of the nasal cavity and curls above the inferior meatus (Fig. 31-3*A*).

    a. It articulates anteriorly with the perpendicular plate of the ethmoid bone.

    b. It forms the inferior boundary of the **hiatus semilunaris**, into which the frontal, anterior, and middle ethmoidal sinuses as well as the maxillary sinus drain.

8. The vomer is a thin, midline plate of bone that completes the posterior portion of the bony nasal septum (Fig. 31-3*B*).

    a. Anteriorly, it is grooved for articulation with the septal cartilage.

    b. Posteriorly, it splits into two alae to accommodate the sphenoidal crest.

    c. It is usually deviated to one side. Simple preliminary observation of the nasal septum can save much time when a nasogastric or nasotracheal tube must be inserted.

9. The palatine bones have both vertical and horizontal processes.

    a. The perpendicular plate of the palatine bone is flush with the posteromedial edge of the maxilla and forms the lateral walls of the posterior portion of the nasal cavity (see Fig. 31-3*A*). Posteriorly, nerves and vessels within the laterally situated perpendicular plate of the palatine bone pierce the adjacent hard palate through the **greater** and **lesser palatine foramina**.

    b. The horizontal palatine process parallels the palatine process of the maxilla and forms the posterior one-fifth of the **hard palate** (see Fig. 31-6*A*).

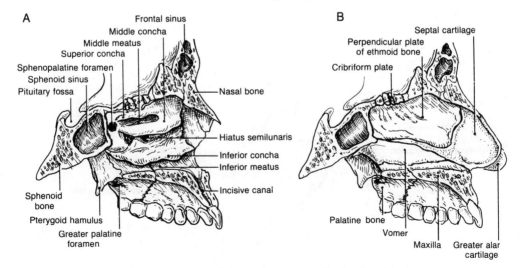

Figure 31-3. *Bones of the nasal cavity. A, Lateral wall. B, Nasal septum.*

(1) The horizontal processes of the maxilla and palatine bones articulate to form the hard palate. Medially, the horizontal plate of the palatine bone articulates with the vomer, completing the nasal septum.

(2) The posterior edge of the hard palate is sharp and concave on either side of the nasal spine.

  c. Superiorly, the sphenopalatine notch articulates with the sphenoid bone to form the **sphenopalatine foramen**, which transmits the sphenopalatine neurovascular bundle.

10. The sphenoid bone in the midline also is shared with the neurocranium. It lies between the left and right middle cranial fossae and separates the cranial cavity from the nasal cavity, the orbit, and the infratemporal fossa (see Fig. 31-3A).

  a. The body of the sphenoid bone abuts the posterior surface of the ethmoid bone.

  b. Its inferior surface has a median crest that contributes to the bony nasal septum.

  c. The **sphenoid sinus**, within the body of the sphenoid bone on each side, opens into the **sphenoethmoidal recess** above the superior concha. Because its opening is high on the anterior wall, it does not drain by gravity with the head held in the erect position.

  d. The greater and lesser sphenoidal wings reach laterally beyond the ethmoid bone and articulate with the orbital plates of the frontal bone.

    (1) The vertical surface of the greater wing comprises the anterior wall of the middle cranial fossa on either side as well as the posterior walls of the orbital cavities.

    (2) The greater wing of the sphenoid bone also contributes to the floor of the middle cranial fossa.

    (3) The pterygoid processes project perpendicularly from the infratemporal skull surface.

      (a) These processes broaden into the **medial** and **lateral pterygoid plates** for the attachment of pterygoid muscles.

      (b) Between the two plates, a triangular defect (the scaphoid fossa) is roofed by the pyramidal process of the palatine bone.

  e. The **pterygopalatine fossa** (Fig. 31-4). The maxilla does not extend as far posteriorly as does the ethmoid bone, leaving a gap between the pterygoid process of the greater wing of the sphenoid bone and the maxilla.

    (1) The pterygopalatine fossa is open laterally, communicating with the infratemporal fossa through the **pterygomaxillary fissure**.

    (2) Inferiorly, the pterygomaxillary fissure is closed by the approximation of the maxillary process, the palatine bone, and the pterygoid process of the sphenoid bone.

      (a) As the pterygoid process passes inferiorly, it approaches the posterior wall of the maxilla, and the pterygopalatine fossa tapers.

      (b) The maxillary tubercle closes the most dependent portion of the pterygopalatine fissure laterally, forming the **greater palatine canal** between the alveolar process of the maxilla, the palatine bone, and the pterygoid process.

      (c) The **greater palatine canal** transmits the palatine neurovascular bundle to the oral cavity.

    (3) The posterior wall is formed by the pterygoid process of the greater wing of the sphenoid bone. It contains three foramina.

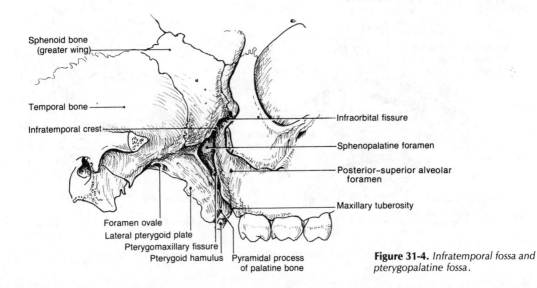

Figure 31-4. *Infratemporal fossa and pterygopalatine fossa.*

(a) The **pterygoid (vidian) canal** traverses the root of the pterygoid process and contains the presynaptic parasympathetic neurons of the greater superficial petrosal branch of the facial nerve, which synapse in the pterygopalatine ganglion. It also contains postsynaptic sympathetic fibers from the carotid perivascular plexus via the deep petrosal nerve.

(b) The **foramen rotundum** pierces the greater wing of the sphenoid bone just anterior to the root of the pterygoid process and transmits the maxillary division of the trigeminal nerve (CN V_2).

(c) The **pterygovaginal canal** contains vessels that communicate with the pharynx.

(4) The medial wall is formed by the perpendicular plate of the palatine bone. The superior portion of this bone is deficient (the sphenopalatine notch) and forms the **sphenopalatine foramen** at the articulation with the sphenoid bone, through which these sphenopalatine nerves and vessels reach the nasal cavity.

(5) The anterior wall is formed by the posterior surface of the maxilla and the greater wing of the sphenoid bone. However, these bones fail to fuse, forming the **inferior orbital fissure**, through which the maxillary nerve reaches the orbit and face.

(6) The greater wing of the sphenoid bone is separated from the lesser wing, forming the **superior orbital fissure**, which also communicates with the orbit and transmits the oculomotor nerve (CN III), the trochlear nerve (CN IV), the ophthalmic division of the trigeminal nerve (CN V_1), and the abducens nerve (CN VI), as well as the ophthalmic vein.

III. MANDIBLE

A. Basic structure. The U-shaped mandible forms the lower jaw and bounds the floor of the mouth.

1. The rami constitute the two posterior vertical portions of the mandible.

 a. External surface of the ramus (Fig. 31-5A)

  (1) At the **angle** of the jaw, the base turns sharply upward as the thick rounded posterior border of the ramus, which continues through a **neck** to the **condyloid process**. The anterior border of the ramus continues into the base as the **oblique line**.

   (a) The condyloid process has a rounded **head (condyle)** for articulation with the temporal bone.

   (b) The lateral pterygoid muscle inserts into the **neck**.

   (c) A sharp **coronoid process** arises from the anterior border of the ramus and provides insertion for the temporalis muscle.

   (d) The **mandibular incisure** lies between the coronoid and condyloid processes.

  (2) A series of oblique ridges at the angle of the mandible provides insertion for the masseter muscle.

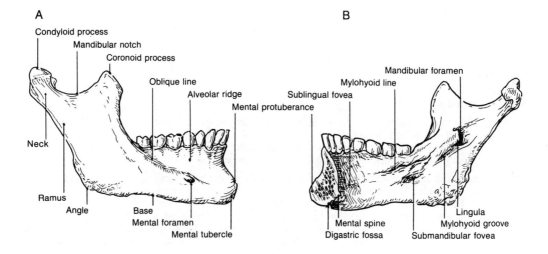

Figure 31-5. *Mandible. A,* External aspect. *B,* Internal aspect.

    b. Internal surface of the ramus (Fig. 31-5*B*)

      (1) The **mandibular foramen** opens into the **mandibular canal**, which carries the inferior alveolar branch of the trigeminal nerve.

        (a) The small triangular **lingula** guards the anterior margin of the mandibular foramen and provides attachments for the **sphenomandibular ligament** from which the mandible swings.

        (b) The **mylohyoid groove**, immediately inferior to the lingula and running just inferior to the mylohyoid line, carries the mylohyoid branch of the inferior alveolar nerve.

      (2) A series of oblique ridges near the angle of the mandible provides insertion for the medial pterygoid muscle. At the angle, the periosteum is not attached to the bone because there are few Sharpey's fibers. The medial pterygoid and masseter muscles together form the **mandibular sling** at the angle of the ramus.

2. The body constitutes the horizontal portion of the mandible. Its thick, rounded inferior margin is the **base**.

    a. External surface of the body (see Fig. 31-5*A*)

      (1) At birth, the halves of the body are joined at the **mandibular symphysis (menti)** by dense, fibrous tissue. Bony fusion, occurring after the age of 2 years, is marked by a faint midline ridge, which widens inferiorly as the **mental tubercles**.

      (2) The **mental foramen**, located on each side inferiorly to the second premolar tooth, transmits the **mental branch of the inferior alveolar nerve**, which, in turn, is a branch of the mandibular division of the trigeminal nerve.

    b. Internal surface of the body (see Fig. 31-5*B*)

      (1) The **mylohyoid line** (into which the mylohyoid muscle inserts) separates the sublingual fossa (for the sublingual gland) from the submandibular fossa (for the submandibular gland) and digastric fossa (for the anterior belly of the digastric muscle).

      (2) The **genial tubercles** for the genioglossus and geniohyoid muscles are just superior to the anterior end of the mylohyoid line.

    c. The alveolar ridge runs along the superior margin of the body (see Fig. 31-5*A*).

      (1) On each side, the alveolar ridges have sockets for eight teeth.

      (2) The alveolar ridges form in response to the presence of teeth and transmit masticatory forces to the jaw. In the edentulous individual, the alveolar ridges atrophy.

B. Mandibular fracture. The shape of the mandible makes it particularly susceptible to fracture, second in frequency only to the nasal bone.

1. Subcondylar fracture interferes with protrusion of the jaw by the lateral pterygoid muscle on the injured side.

2. Fractures of the body may result in malocclusion of the teeth anterior to the fracture line as well as damage to the inferior alveolar nerve within the mandibular canal. Fractures of the body that are parallel to the ramus produce less displacement than fractures that are perpendicular to the ramus because the anterior digastric and the masseter muscles pull the fragments in opposite directions.

3. Parasymphyseal fractures produce malocclusion between left and right sides as well as loss of jaw stability.

IV. TEETH

A. Introduction

1. In children, there are 20 **deciduous** (milk) **teeth**. In each jaw on each side, there are:

    a. Two incisors, which come in at approximately 6 and 8 months

    b. One canine, which comes in at approximately 10 months

    c. Two premolars, which come in during the second year

2. In the adult, there are 32 **permanent teeth**. In each jaw on each side, there are:

    a. Two incisors, which erupt in the seventh and eighth years

    b. One canine, which erupts in the tenth year

    c. Two premolars (bicuspid teeth), which erupt in the ninth and eleventh years

    d. Three molars (tricuspid teeth), the first of which comes in during the sixth year, followed by the second molar in the early teen years and the "wisdom teeth" in the late teen years

B. Structure of the tooth. The basic structural element is **dentine**—a yellowish substance that is nurtured through the fine dentinal tubules of odontoblasts lining the central pulp space.

1. **The root** is embedded in the alveolar part of the mandible or maxilla. It is covered with **cement**, which is connected to the bone of the socket by a layer of modified periosteum, the **periodontal ligament**.

2. **The crown** projects into the oral cavity, composed of one or more cusps. It is covered with **enamel**—a hard, white crystalline substance formed before the tooth erupts.
   a. After eruption, this inert material changes little. Its only change is to adsorb fluoride ions, which reduce its solubility in the acid metabolites of oral bacteria.
   b. Like bone, enamel is stained by tetracycline antibiotics but only while it is being formed. Therefore, tetracycline should not be given to children or pregnant women.

3. **An intermediate cervical part** is related to the gingiva (gum).

C. **Organization of the teeth**

   1. **Types.** Each alveolar margin bears 16 teeth—2 incisors, 1 canine, 2 premolars, and 3 molars on either side, each classified according to the shape of the crown (Fig. 31-6).
      a. Incisors have a thin cutting edge.
      b. Canines have a single prominent cone.
      c. The premolar crown is divided by a sagittal groove into two cusps.
      d. The molars of adults have three or more cusps.
         (1) Many years of grinding an unrefined diet may flatten the molar surfaces completely.
         (2) The upper molars have three roots, and the lower molars have two.

   2. **Sockets.** The bone of the socket has a thin cortex, the **lamina dura**, separated from the adjacent labial and lingual cortices by a variable amount of trabeculated bone.
      a. The labial wall of the socket is particularly thin over the incisor teeth.
      b. This is the surface to break in order to remove an incisor tooth.
      c. The reverse is true of the molars, where the lingual route is the easier.

D. **Innervation of the teeth** (see Fig. 32-4)

   1. **The maxillary teeth** are innervated by the anterior, middle, and posterosuperior alveolar branches of the maxillary nerve (CN V_2).

   2. **The mandibular teeth** are innervated by the inferior alveolar branch of the mandibular nerve (CN V_3).

E. **Clinical considerations**

   1. **Advanced infection of the root** involves the periodontal ligament and erodes through the lamina dura (a sign of chronic infection that is apparent on an x-ray plate).

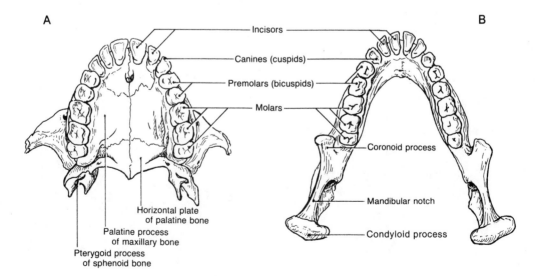

Figure 31-6. *Teeth. A,* Maxilla. *B,* Mandible.

2. Abscess involving the mandibular teeth may spread through the lower jaw to emerge on the face or in the floor of the mouth. The mandibular nerve also innervates a portion of the ear, and the pain of an infected lower tooth may be referred to the ear.

3. Abscess involving the maxillary teeth may spread through the upper jaw to emerge on the face or in the roof of the mouth. The relations of the upper teeth may influence the mode of presentation.

   a. Infection of a second incisor may spread along the palate.

   b. An infected canine may point in the face, thrombose the angular facial vein, and spread along the superior ophthalmic vein to the cavernous sinus [see Chapter 28 IV A 1 b (3)].

   c. Because the roots of the maxillary molars project into the maxillary sinus, infection of an upper tooth may produce symptoms of sinusitis with pain referred to the distribution of the maxillary nerve.

   d. Similarly, sinus infection may irritate the nerves to these teeth, causing toothache.

I. INTRODUCTION

A. Developmental considerations. The face comprises derivatives of the primitive gut (the nose and mouth) and special sensory receptors (the eye and ear).

B. Facial features. The orifices and their associated superficial structures are the features of the face. Variations in the proportions of each feature determines an individual's facial identity.

   1. Facial openings. At the oral fissure, the nostrils, and the palpebral fissures, the skin of the face is continuous with the mucous membrane lining these structures.

   a. The palpebral fissures and eyes have been discussed (see Chapter 30 I, II).

   b. The external nose is a pyramidal anterior extension of the nasal cavity.

      (1) Its skeleton is in part bony and in part cartilaginous.

      (2) The frontal process of the maxilla and the nasal bones articulate with the nasal notch of the frontal bone to form the arch of the nose.

      (3) The septal cartilage is a quadrangular plate of fibrocartilage thickened around its margin.

         (a) It lies approximately in the midline sagittal plane.

         (b) Its posterior edge meets the septal plate of the ethmoid bone and the vomer.

         (c) The anterior edge emerges from behind the nasal bones as the ridge of the nose.

         (d) It articulates with the lateral and alar plates to complete the cartilaginous portion of the nose.

   c. The oral fissure is bounded by the superior and inferior labial folds.

      (1) These meet laterally at the commissures.

      (2) The labial margin has prominent superficial vascular papillae (thelia), which give it a vermilion hue.

      (3) A thin median frenulum extends from the inner surface to the gum.

      (4) Between the epithelial surfaces are the pea-sized labial salivary glands and the orbicularis oris muscle.

   d. The external ear has been discussed (see Chapter 30 III).

   2. Facial musculature. Around each orifice, a system of subcutaneous muscles is arranged as sphincters and dilators, the muscles of facial expression. These muscles are innervated by the facial nerve (CN VII).

II. MUSCLES OF FACIAL EXPRESSION

A. Basic principles

   1. Development. All of the **muscles of facial expression** are derived from a single plate of muscle, derived from the second branchial arch, which migrates over the skull.

   a. Most of these muscles comprise sphincters and dilators that guard the facial openings.

   b. The muscles of facial expression are important for the display of emotions in humans.

   2. Innervation. The **facial nerve** (CN VII) is the nerve of the second branchial arch and, therefore, innervates all of the muscles of facial expression.

B. Oral musculature (Fig. 32-1 and Table 32-1)

   1. Sphincter. The **orbicularis oris muscle** is the sphincter of the mouth.

   a. It lies circumferentially to the margin of the lips with muscle fascicles passing in a series of loops around the mouth. Peripherally, some fascicles decussate at the commissure and

Table 32-1. Major Muscles of Facial Expression

Muscle	Origin	Insertion	Action	Innervation
Oral musculature				
Orbicularis oris	Modiolus	Skin of lips	Closes lips	Facial n. (CN VII)
Zygomaticus major	Zygomatic bone	Modiolus	Draws angle of mouth upward	Facial n. (CN VII)
Zygomaticus minor	Zygomatic bone	Skin of upper lip	Raises upper lip	Facial n. (CN VII)
Levator anguli oris	Maxilla	Modiolus	Draws angle of mouth upward	Facial n. (CN VII)
Levator labii superioris	Maxilla	Skin of upper lip	Raises upper lip	Facial n. (CN VII)
Depressor anguli oris	Mandible	Modiolus	Draws angle of mouth down	Facial n. (CN VII)
Depressor labii inferioris	Mandible	Skin of lower lip	Depresses lower lip	Facial n. (CN VII)
Mentalis	Mandible	Skin of lower lip	Depresses and protrudes lower lip	Facial n. (CN VII)
Buccinator	Maxilla, mandible, and pterygomandibular raphe	Skin of lips	Compresses cheeks	Facial n. (CN VII)
Platysma	Skin of upper thorax and shoulder	Skin of lower lips and modiolus	Tenses skin of anterior neck and depresses lower lip and angle of mouth	Facial n. (CN VII)
Orbital musculature				
Orbicularis oculi:				
Orbital portion	Orbital margin	Palpebral skin	Eye wink	Facial n. (CN VII)
Palpebral portion	Lateral palpebral ligament	Medial palpebral ligament	Eye blink and compresses lacrimal sac	Facial n. (CN VII)
Epicranial musculature				
Occipitofrontalis:				
Occipitalis	Nucal line	Epicranial aponeurosis	Draws scalp posteriorly	Facial n. (CN VII)
Frontalis	Epicranial aponeurosis	Skin at the supraciliary crest	Draws scalp anteriorly	Facial n. (CN VII)
Corrugator supercilii	Frontal bone	Skin of medial end of upper eyelid	Draws eyebrows inferomedially	Facial n. (CN VII)

pass into the face along the various dilator muscles, and other fascicles pass into the **modiolus** (of Lightoller) on either side of the mouth.

    b. It purses or puckers the lips and closes the oral cavity.

  2. Dilators. The dilator muscles of the mouth have either facial or bony origins, and their insertions blend with the muscular substance of the lip.

    a. They are numerous and small (often the name in print is longer than the muscle itself). Many of these muscles insert into the **modiolus**, an arrangement that provides an almost infinite range of facial expression.

    b. The dilator muscles may be resolved into four layers.

      (1) Subcutaneous layer. The **risorius muscle** is small and variable. It arises from the parotid fascia and runs horizontally to insert at the angle of the mouth. It retracts the modiolus.

      (2) Superficial layer. Three muscles form two inverted V's as they diverge from their origins at the orbital margins to their insertions.

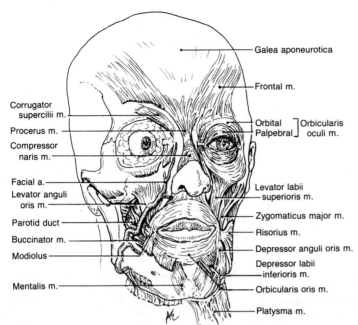

Galea aponeurotica

Frontal m.

Corrugator supercilii m.

Procerus m.

Compressor naris m.

Facial a.

Levator anguli oris m.

Parotid duct

Buccinator m.

Modiolus

Mentalis m.

Orbital ⎤ Orbicularis
Palpebral ⎦ oculi m.

Levator labii superioris m.

Zygomaticus major m.

Risorius m.

Depressor anguli oris m.

Depressor labii inferioris m.

Orbicularis oris m.

Platysma m.

Figure 32-1. *Muscles of facial expression.* The left superficial muscles and the right deeper muscles are depicted.

   (a) The **zygomaticus major** inserts on the angle of the mouth; the **zygomaticus minor** inserts on the lip. They act to raise the lip.
   (b) The **levator labii superioris alaeque nasi** inserts on the lip laterally and on the major alar cartilage medially. It also acts to raise the lip.
 (3) Middle layer. Four muscles form this layer and are named according to function.
   (a) The **triangular depressor anguli oris**, arising from the oblique line on the body of the mandible, and the **levator anguli oris**, arising from the incisive fossa of the maxilla, converge at the modiolus. The names are descriptive of their actions.
   (b) The quadrangular **depressor labii inferioris**, arising from the oblique line on the body of the mandible, and the **levator labii superioris**, arising from the orbital margin, insert on lower and upper lips, respectively. The names describe their actions.
 (4) Deep layer. Two muscles form this layer.
   (a) The **mentalis muscle** arises from the incisive fossa of the mandible, inserts on the lower lip near the midline, and depresses the lower lip.
   (b) The **buccinator muscle** (*L.* trumpeter) arises from the maxilla and the mandible, posterior to the third molar teeth.
     (i) Between these two points, it interdigitates with fibers of the superior pharyngeal constrictor in the **pterygomandibular raphe**.
     (ii) The middle fibers decussate at the angle of the mouth; the peripheral fibers pass directly into the lips.
     (iii) Most importantly, the buccinator serves to draw the cheeks against the teeth in mastication.

 C. Nasal musculature (see Fig. 32-1 and Table 32-1)

   1. Sphincter. The **compressor naris muscle**, the transverse part of the **nasalis muscle**, comprises the nasal sphincter. Its function is rudimentary in humans.
     a. It is connected with its counterpart of the opposite side by an aponeurosis.
     b. It pulls down on the bridge of the nose, slightly constricting the nostrils.

   2. Dilators. The **dilator naris**, the alar part of the **nasalis muscle**, comprises the nasal dilator.
     a. It inserts along with the nasal part of the levator labii superioris alaeque nasi on the alar cartilage.
     b. Both act to flare the nostrils.

 D. Orbital musculature (see Fig. 32-1 and Table 32-1)

   1. Sphincters
     a. The orbicularis oculi, the orbital sphincter, has two parts.

(1) The thick **orbital portion** encircles the eyes in continuous loops.

    (a) It arises from the medial orbital margin and the medial palpebral ligament.

    (b) When it contracts, the eye is tightly closed (the wink).

(2) The **palpebral portion** crosses the eyelid beneath the skin.

    (a) Contraction of the palpebral portion of this muscle produces the blink.

    (b) Its fibers are so arranged that when they shorten they move toward geodesic lines—the shortest distance between two points over the surface of a sphere—and, thus, close the palpebral fissure.

    (c) As a muscle of facial expression, it is innervated by the facial nerve (CN VII).

        (i) This nerve forms the efferent limb of the **blink reflex**.

        (ii) Paralysis of the palpebral portion of the orbicularis oculi results in an inability to close the eyelids. This is a major problem in Bell's (facial nerve) palsy because, if the eye cannot blink, the conjunctiva dries and the cornea may ulcerate.

    (d) The origin encircles the lacrimal sac, and those fibers posterior to the sac may be referred to as the lacrimal portion. Thus, blinking compresses the lacrimal sac, thereby aspirating fluid from the medial corner of the eye.

    (e) At the lateral canthus, the fibers decussate in a fiberous raphe.

  b. The procerus and corrugator supercilii, two small muscles, pull the eyebrows downward and inward in an expression of concentration or to shield the eyes from bright light.

2. Dilators

  a. The levator palpebrae superioris muscle acts directly on the upper eyelid to open the palpebral fissure.

  b. The occipitofrontalis muscle widens the eyes very slightly by raising the eyebrows as in an expression of fear or surprise.

E. Auricular musculature

  1. The anterior and superior auricular muscles converge on the auricle from the epicranial aponeurosis.

  2. The posterior auricular muscle arises from the mastoid process.

F. Platysma. This muscle migrates over the anterior surface of the neck and may extend onto the upper thoracic wall (see Fig. 32-1).

III. INNERVATION OF THE SUPERFICIAL FACE

A. Facial nerve (CN VII)

  1. Origin. The facial nerve is formed by the union of two roots, which represent pretrematic and post-trematic divisions (see Fig. 31-1).

    a. The motor root innervates muscles derived from the second branchial arch.

    b. The sensory root (nervus intermedius) contains sensory and autonomic fibers.

  2. Course and distribution (Figs. 32-2 and 32-3)

    a. The facial nerve reaches the middle ear cavity via the **internal auditory meatus** in the company of the vestibulocochlear nerve (CN VIII). Here it turns sharply backward and runs in a canal in the medial wall of the tympanic cavity.

    (1) At the angle (genu) is the sensory **geniculate ganglion** (*genu, L.* knee).

    (2) The **greater superficial petrosal nerve** leaves the facial nerve at this point and passes anteriorly in the canal of the same name.

        (a) It conveys parasympathetic secretomotor fibers.

        (b) It joins with the deep petrosal nerve (carrying sympathetic fibers from the carotid perivascular plexus) to form the **nerve of the pterygoid canal** before entering that structure.

        (c) The parasympathetic nerves synapse in the **pterygopalatine ganglion** before continuing along various nerves to the lacrimal gland as well as to the glands of the palate and nasal mucosa.

    b. The facial nerve descends in the **facial canal** behind the posterior wall of the middle ear cavity.

    (1) The **nerve to the stapedius muscle** leaves in the posterior portion of the canal.

    (2) The **chorda tympani nerve** leaves just as the facial nerve exits the skull.

        (a) The chorda tympani (the pretrematic branch of the nerve of the second branchial arch) runs forward between the malleus and incus to exit the skull at the **petrotympanic fissure** just medial to the mandibular condyle.

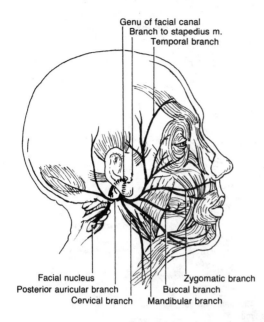

Genu of facial canal
Branch to stapedius m.
Temporal branch

Facial nucleus
Posterior auricular branch
Cervical branch   Mandibular branch
                Buccal branch
          Zygomatic branch

Figure 32-2. *Branchiomotor component of the facial nerve (CN VII).* This portion of the facial nerve innervates the muscles of facial expression—the posterior belly of the digastric, the stylohyoid, and the stapedius muscle.

    (b) It runs forward, medial to the mandibular condyle, and continues along the medial surface of the lateral pterygoid muscle. It joins the lingual nerve (the post-trematic branch of the nerve of the first branchial arch) at the lower border of this muscle.

    (c) It conveys parasympathetic secretomotor fibers, which synapse in the **submandibular ganglion** before continuing along the lingual nerve to the submandibular and sublingual salivary glands.

    (d) Sensory fibers convey **taste** from the anterior two-thirds of the tongue.

  c. The facial nerve exits the skull via the **stylomastoid foramen**.

    (1) As soon as it exits the skull, the facial nerve gives off digastric, stylohyoid, and posterior auricular branches.

    (2) The posterior auricular nerve divides into a small auricular branch (to the posterior auricular muscle) and a larger occipital branch (to the occipital belly of the occipitofrontalis muscle).

  d. The facial nerve passes anterolaterally through the parotid gland deep to the lobule of the auricle.

    (1) A plexus of branches divides the parotid gland into superficial and deep lobes. Each branch is separated from the parotid parenchyma by a fascial plane.

    (2) Five major branches emerge onto the face where they are very superficial and easily damaged by facial trauma.

      (a) The **temporal branch** supplies the anterior and superior auricular muscles, the frontal belly of occipitofrontalis, the corrugator supercilii, and the orbicularis oculi (all of the muscles above the eyes).

        (i) This branch conveys the efferent limb of the **corneal blink reflex**.

        (ii) The motor nuclei whose fibers run in the temporal nerves receive a bilateral corticomedullary innervation. Therefore, unilateral supranuclear lesions (such as the common stroke) do not affect the forehead to the degree that they affect the rest of the face.

      (b) The **zygomatic branch** passes to the orbicularis oculi. This branch also mediates the efferent limb of the blink reflex.

      (c) The **buccal branch** supplies the numerous muscles of the cheek, including the buccinator muscle, the muscles of the upper lip, and the muscles of the nose.

      (d) The **mandibular branch** loops down over the mandible. It supplies the risorius muscle and the depressors of the lower lip.

      (e) The **cervical branch** innervates the platysma muscle.

3. Signs and symptoms produced by injury to the facial nerve are related to the site of the lesion.

  a. Lesions involving the terminal branches of the facial nerve produce an imbalance of the muscles of facial expression with unilateral expressionless drooping of the face.

    (1) Paralysis of the orbicularis oculi can result in epiphora with conjunctival and corneal ulceration. Obviously, the corneal blink reflex is absent.

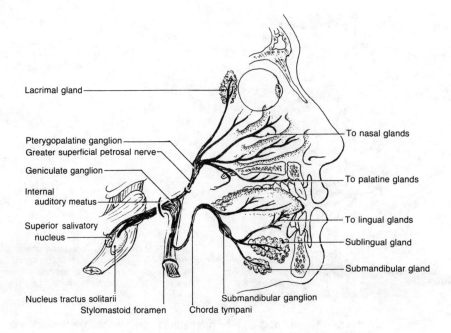

Figure 32-3. *Parasympathetic and sensory components of the facial nerve (CN VII)*. The autonomic portions of the facial nerve run to the pterygopalatine and submandibular ganglia. The sensory portion conveys taste from the anterior portion of the tongue with cell bodies of the sensory neurons located in the geniculate ganglion.

    (2) Paralysis of the orbicularis oris results in drooling and difficulty in masticating food.
  b. Lesions at the stylomastoid foramen produce both of the above symptoms.
  c. Inflammation of the nerve within the facial canal is the most common cause of Bell's palsy. There is no effective treatment. Most cases recover spontaneously, often with little or no nerve damage.
    (1) All of the above symptoms occur.
    (2) Involvement of the chorda tympani results in loss of taste from the anterior two-thirds of the tongue and reduced salivary secretion on the ipsilateral side.
  d. Lesions in the facial canal
    (1) All of the above symptoms occur.
    (2) Hyperacusis results from involvement of the nerve to the stapedius muscle.
  e. Lesions in the internal auditory meatus
    (1) All of the above symptoms (except epiphora) occur.
    (2) Involvement of the greater superficial petrosal nerve results in loss of lacrimation and in loss of secretion by the nasal and palatine glands.
    (3) Upon regeneration, if the secretomotor fibers regrow along each other's pathways and innervate the wrong gland, the anticipation of food then produces lacrimation instead of salivation and vice versa (*syndrome of crocodile tears*).

B. Trigeminal nerve (CN V) [Fig. 32-4]

  1. Origin. A coronal line drawn over the scalp and connecting the tragi of the ears marks the boundary between branchiomeric and somatic innervation.

  2. Course and distribution. The trigeminal nerve (as the name implies) has three divisions that correspond to the three parts of the first branchial arch: the frontal, the maxillary, and the mandibular. The cutaneous supply of the face conforms to the boundaries of these regions (see Fig. 32-4).
  a. The ophthalmic division supplies the forehead and nose.
    (1) It exits the cranium via the **supraorbital fissure** and exits the orbit via the **supraorbital notch** or foramen.
    (2) The major branches include the **lacrimal, supraorbital, supratrochlear, infratrochlear**, and **external nasal nerves**.
    (3) These nerves supply the afferent limb of the **corneal blink reflex**.

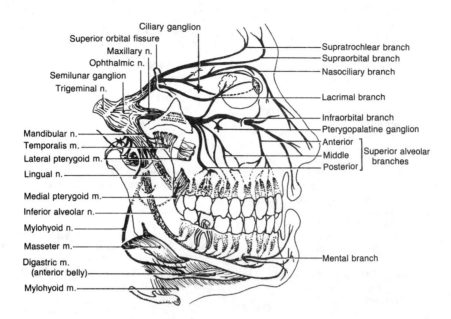

Figure 32-4. *Trigeminal nerve (CN V).* The three divisions of the trigeminal nerve provide sensory innervation to the face. The motor root of the mandibular division innervates the muscles of mastication, the tensor tympani muscle, and the tensor veli palatini muscle.

    b. The maxillary division supplies an area of skin between the lower eyelid, the upper lip, and the maxillary teeth.
        (1) It exits the cranium via the **foramen rotundum** and exits the facial skeleton via the **infraorbital foramen**.
        (2) The major branches include the **infraorbital, zygomaticotemporal**, and **zygomaticofacial nerves**.
        (3) This nerve supplies the afferent limb of the **sneeze reflex**.
    c. The mandibular division supplies a lateral area that curves around the side of the face (but the angle of the jaw is supplied by the great auricular nerve, C2–C3), as well as general sensation to the supratentorial meninges, all of the mucous membranes of the face, and the mandibular teeth.
        (1) It exits the cranium via the **foramen ovale** and exits the skull via the **mental foramen**.
        (2) It has several major branches.
            (a) The **auriculotemporal nerve** innervates the lateral aspect of the scalp.
            (b) The **buccal nerve** innervates the cheek (but not the buccinator muscle).
            (c) The **mental nerve** innervates the chin (but not the mentalis muscle).
            (d) The **mandibular nerve** is the only division to contain motor fibers.
                (i) This nerve innervates the **muscles of mastication** (masseter, temporalis, medial, and lateral pterygoids), which are deep muscles of the face. This nerve also innervates the mylohyoid muscle, the anterior belly of the digastric muscle, the tensor tympani muscle, and the tensor veli palatini muscle.
                (ii) This nerve supplies both afferent and efferent limbs for the **jaw-jerk reflex**.
  3. Signs and symptoms produced by lesions of the trigeminal nerve.
    a. Trigeminal neuralgia (tic douloureux) produces excruciating facial pain that may be initiated by touching a trigger area. So severe and unremitting is the pain that patients may be driven to suicide.
        (1) The pain is often in the distribution of the mandibular nerve.
        (2) The trigger area is often maxillary.
    b. Inflammation of the cavernous sinus or a lesion at the superior orbital fissure results in paresthesias of the forehead and nose with loss of the **corneal blink reflex**.
    c. A lesion at the foramen rotundum or within the infraorbital canal results in paresthesias of the midface and maxillary teeth with loss of the **sneeze reflex**.
    d. A lesion at the foramen ovale results in paresthesias along the mandible, the mandibular teeth, and the side of the face, as well as paralysis of the muscles of mastication, possible hearing aberrations, and loss of the **jaw-jerk reflex**.

IV. VASCULATURE OF THE SUPERFICIAL FACE

A. Arterial supply. Arteries of the face are branches of the external carotid artery with a small contribution from the internal carotid artery (Fig. 32-5).

1. **The ophthalmic artery** is a branch of the **internal carotid artery** and generally follows the distribution of the ophthalmic division of the trigeminal nerve.

2. **The external carotid artery** gives rise to several major branches.
   a. **Ascending pharyngeal artery**
   b. **Superior thyroid artery**
   c. **Lingual artery**
   d. **Facial (external maxillary) artery.** This branch winds around the mandible (where a pulse may be palpated) at the anterior edge of the masseter muscle.
      (1) It runs obliquely to the medial canthus of the eye, giving off numerous branches.
      (2) Profuse anastomotic connections between the branches of the left and right facial arteries occur across the midline.
   e. **Occipital artery**
   f. **Posterior auricular artery**
   g. **Superficial temporal artery.** This artery gives off a transverse facial branch before it crosses the zygomatic arch just anterior to the ear. Where the superficial temporal artery courses along the bone, a pulse may be palpated.
   h. **Maxillary (internal) artery.** This artery passes through the pterygomaxillary fissure, the pterygopalatine fossa, the infraorbital groove, and the infraorbital canal to emerge below the eye.

B. Venous return. Veins of the face follow the arteries (see Fig. 32-5).

1. **The facial vein** drains to the internal jugular.

2. **The superficial temporal vein** drains to the retromandibular, which, in turn, drains into the internal jugular vein.

3. **The maxillary vein** drains into the pterygoid plexus.

4. **The superior ophthalmic vein** drains to the cavernous sinus.

C. Lymphatic drainage. The lymph vessels of the superficial face generally parallel the superficial veins.

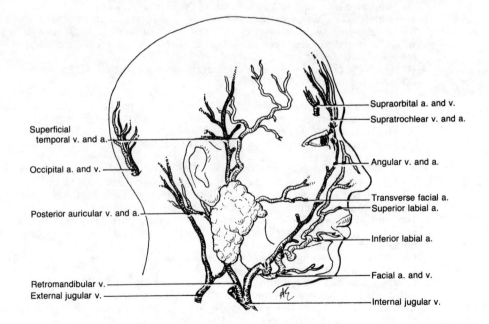

Figure 32-5. *Vasculature of the superficial face.*

V. PAROTID GLAND

A. Relations. The parotid salivary gland is wrapped around the ramus of the mandible, below the ear. The superficial part of the gland is divided into superficial and deep lobes by branches of the facial nerve.

1. **Medially**, it extends between the mandible and the medial pterygoid muscle just superficial to the styloid process and the external carotid artery.

2. **Laterally**, it lies superficially to the masseter muscle.
   a. An anterior extension (or facial process) surrounds the parotid duct.
   b. It may be separate from the remainder as the **accessory parotid gland**.

3. **Posteriorly** are the external auditory meatus and the mastoid process.

B. Drainage. The parotid gland drains a thin serous saliva into the mouth via the **parotid duct** (of Stensen). The firm cord-like parotid duct dives medially at the anterior edge of the masseter to enter the mouth opposite the second upper molar. The oblique course of the duct between the mucous membrane and the buccinator acts as a valve.

C. Innervation. Parasympathetic secretomotor fibers reach the parotid gland by a very circuitous route (Fig. 32-6).

1. **The presynaptic neurons** lie in the **inferior salivatory nucleus** of the glossopharyngeal nerve (CN IX).
   a. These fibers leave the **glossopharyngeal nerve** in the jugular canal as the **tympanic branch** (of Jacobson), which enters the middle ear cavity from the posterior wall and forms a plexus on the medial wall.
   b. The **lesser superficial petrosal nerve** leaves this plexus through a canal of the same name in the anterior wall and gains access to the middle cranial fossa. The lesser superficial petrosal nerve courses briefly across the petrous temporal bone and exits via either the foramen ovale or the innominate foramen.
   c. The fibers synapse in the **otic ganglion**.

2. **The postsynaptic secretomotor fibers** leave the otic ganglion and are distributed by the **auriculotemporal nerve** to the parotid gland. Because the auriculotemporal nerve also contains sudomotor fibers to the sweat glands of the scalp, if the nerve is severed, the fibers can regenerate along each other's pathways and innervate the wrong gland. The anticipation or taste of food then produces sweating instead of salivation (*Frey's syndrome*).

D. Tumors of the parotid. Usually these arise in the superficial lobe without involvement of the facial nerve. The surgeon is presented with a challenge since iatrogenic facial nerve injury has major effects.

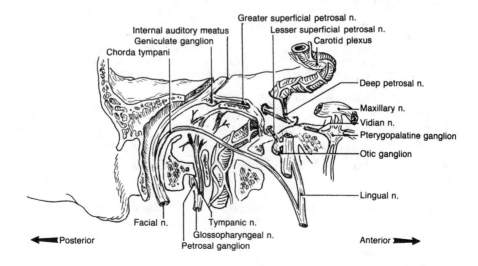

Figure 32-6. *Course of the facial and glossopharyngeal nerves in the middle ear.*

VI. DEEP FACE

A. **Infratemporal fossa.** This region lies between the ramus of the mandible and the pharynx.

1. **Boundaries**
   a. **The roof** forms the infratemporal surface of the sphenoid bone (see Fig. 31-4).
      (1) It is pierced by the foramen ovale and foramen spinosum.
      (2) It bears a pterygoid process, which expands inferiorly into medial and lateral pterygoid plates.
         (a) The medial pterygoid plate provides attachment for the superior constrictor muscle.
         (b) The lateral pterygoid plate provides attachments for the medial and lateral pterygoid muscles.
   b. **The anterior wall** is formed by the maxilla.
   c. **The medial wall** is formed by the temporal bone and the lateral pterygoid plate of the sphenoid bone.

2. **Contents.** The parotid gland curves around the posterior border of the mandible into the infratemporal fossa. The carotid sheath passes between the parotid gland and the intrinsic pharyngeal musculature. In addition, this fossa contains the muscles of mastication, pterygoid venous plexus, and branches of the trigeminal nerve.

B. **Muscles of mastication.** All four of these muscles arise within the infratemporal fossa, insert onto the ramus of the mandible, and are innervated by the mandibular division of the trigeminal nerve (Table 32-2).

1. **The temporalis muscle** occupies the temporal fossa (Fig. 32-7A).
   a. Fibers arise from beneath the temporal fascia from a broad origin on the lateral aspect of the skull inferiorly to the temporal line. They pass deeply to the zygomatic arch and converge to insert onto the coronoid process of the mandible.
   b. The anterior and medial fascicles elevate the jaw; the posterior fascicles retract the jaw.

2. **The masseter** lies deeply to the superficial lobe of the parotid gland and completely covers the outer surface of the mandible. A **buccal fat pad** separates the masseter from the buccinator muscle.
   a. It arises from medial and lateral surfaces of the zygomatic arch and inserts into the periosteum at the angle of the mandible.
   b. It is a strong elevator of the jaw. The masseter may hypertrophy as a result of *bruxism*, the nocturnal habit of clenching the jaw at times of nervous tension.
   c. The masseteric neurovascular bundle passes through the mandibular notch to reach the deep surface of this muscle.

3. **The lateral pterygoid muscle** lies in a horizontal plane and has two heads that converge to their insertions (Fig. 32-7B). The medial surface is adjacent to the mandibular nerve, the maxillary artery, and the pterygoid venous plexus.
   a. **The superior head** arises from the infratemporal surface of the sphenoid bone and inserts mainly into the articular capsule and disk of the temporomandibular joint. The **inferior head** arises from the lateral surface of the lateral pterygoid plate and inserts onto the condylar process of the mandible.
   b. This muscle pulls the condyle and its articular disk anteriorly (protrusion) and downward (depression) so that the mandible effectively rotates not about the condyle but about a resultant axis through the lingula at the mandibular foramen.

Table 32-2. Muscles of Mastication

Muscle	Origin	Insertion	Action	Innervation
Temporalis	Temporal and parietal bones	Coronoid process of mandible	Elevates and retracts jaw	Mandibular n. (CN V_3)
Masseter	Zygomatic arch	Angle of mandible	Elevates jaw	Mandibular n. (CN V_3)
Lateral pterygoid: Superior head	Sphenoid bone	Temporomandibular disk	Draws articular disk forward	Mandibular n. (CN V_3)
Inferior head	Lateral pterygoid plate	Neck of mandible	Protracts jaw	Mandibular n. (CN V_3)
Medial pterygoid	Medial and lateral pterygoid plates	Angle of the mandible	Elevates jaw	Mandibular n. (CN V_3)

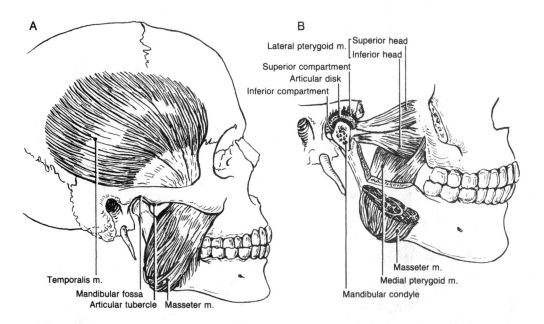

Figure 32-7. *Muscles of mastication. A*, Temporalis and masseter muscles. *B*, Medial pterygoid and lateral pterygoid muscles.

      (1) Together, the lateral pterygoid muscles protrude the jaw.
      (2) Singly, they swing the jaw from side-to-side to produce grinding movements.

  4. The medial pterygoid muscle is separated from the mandible by the deep lobe of the parotid gland, the lingual nerve, the chorda tympani nerve, the inferior alveolar nerve, and the inferior alveolar artery. The medial surface is adjacent to the tensor veli palatini muscle, the styloglossus muscle and the stylopharyngeus muscle.
    a. The deep part arises from the medial surface of the lateral pterygoid plate and the pyramidal process of the palatine bone; the smaller superficial part arises from the tubercle of the maxilla. It inserts into the periosteum of the ramus at the angle of the mandible.
    b. The medial pterygoid is a strong elevator of the mandible. This thick quadrilateral muscle resembles its counterpart—the masseter—with which it forms the **masseteric (mandibular) sling**. This sling supports and elevates the angle of the mandible but, lacking Sharpy's fibers, is not attached to the bone.

  5. Accessory muscles of mastication. While not grouped with the muscles of mastication, the digastric, mylohyoid, and geniohyoid muscles insert into the mandible and act to produce motion at the temporomandibular joint.

C. Temporomandibular joint (Fig. 32-8)

  1. Structure
    a. The temporomandibular joint is formed by the **condyle** of the mandible together with the **articular tubercle** and the **mandibular fossa** of the temporal bone.
    b. A fibrocartilage biconcave **articular disk** is interposed between the rounded condyle of the mandible and the corresponding sigmoid articular surface of the temporal bone, dividing the joint into suprameniscal and inframeniscal compartments.
      (1) In the **inframeniscal compartment**, the condyle simply rotates on the articular disk.
      (2) In the **suprameniscal compartment**, the condyle and disk slide anteriorly and posteriorly over the articular tubercle of the temporal bone.

  2. Movement
    a. The masseter and medial pterygoid muscles act together to close the jaw. The temporalis muscle both closes and retracts the jaw. The lateral pterygoid muscles can act together to protract the jaw or, by acting on one side alone, rotate the mandible about a vertical axis.
    b. During simple opening and closing of the mouth, the compound axis passes through the lingula of the ramus.
      (1) This action might be expected because of the neurovascular bundle and sphenomandibular ligament, which join the mandible at this point.

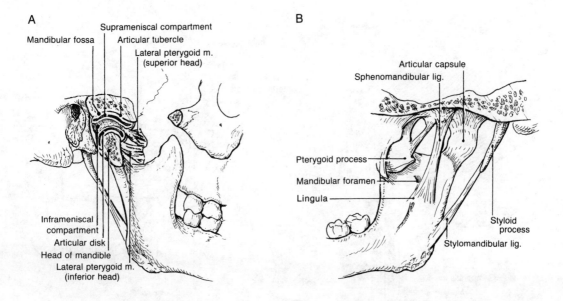

Figure 32-8. *Temporomandibular joint. A,* Lateral aspect with the joint capsule opened to show the articular disk. *B,* Medial aspect.

(2) As the mouth is opened, the condyle slides anteriorly with the disk due to the traction on both by the two heads of the lateral pterygoid muscle. Concomitantly, the condyle rotates.

(3) If the condyle passes beyond the apex of the articular tubercle, spasm of the temporalis muscle may jam it under the zygoma, locking the jaw.

    (a) The physician or oral surgeon faced with this problem must first pull downward on the mandible to overcome the pull of temporalis, masseter, and medial pterygoid muscles.

    (b) As the jaw snaps back into place, he or she should then beware of the forceful apposition of the molars upon sudden release.

D. Pterygopalatine fossa

1. Structure. This fossa is a narrow cleft between the maxilla and the pterygoid process (*pteryx,* G. wing) of the sphenoid bone, separated from the nasal cavity by the perpendicular plate of the palatine bone (see Fig. 31-4).

2. Communications. The pterygopalatine fossa communicates with several other cavities or spaces, including the:

    a. Middle cranial fossa via the foramen rotundum and the pterygoid canal

    b. Infratemporal fossa via the pterygomaxillary fissure

    c. Orbit via the inferior orbital fissure

    d. Nasal cavity via the sphenopalatine foramen

    e. Oral cavity via the palatine canal

    f. Pharynx via the pterygovaginal canal

3. Contents

    a. It contains branches of the pterygopalatine ganglion, the vidian nerve, the maxillary nerve, and the maxillary artery. It has been called the neurovascular junction of the deep face.

    b. Vascular structures enter this fossa laterally; nerves enter through the posterior wall (Fig. 32-9).

VII. INNERVATION OF THE DEEP FACE

A. Maxillary division of the trigeminal nerve (CN V₂)

1. Course and composition. This nerve represents the pretrematic division of the trigeminal nerve and is entirely sensory (see Figs. 31-1 and 32-4). It enters the pterygopalatine fossa through the **foramen rotundum**. While it is supported by fatty tissue, it appears to hang free in the pterygopalatine fossa until it exits through the inferior orbital fissure.

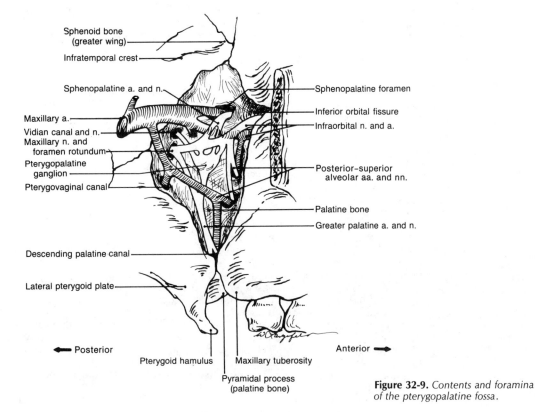

Sphenoid bone (greater wing)

Infratemporal crest

Sphenopalatine a. and n.

Maxillary a.

Vidian canal and n.

Maxillary n. and foramen rotundum

Pterygopalatine ganglion

Pterygovaginal canal

Descending palatine canal

Lateral pterygoid plate

Sphenopalatine foramen

Inferior orbital fissure

Infraorbital n. and a.

Posterior–superior alveolar aa. and nn.

Palatine bone

Greater palatine a. and n.

◀ Posterior

Anterior ▶

Pterygoid hamulus

Maxillary tuberosity

Pyramidal process (palatine bone)

Figure 32-9. *Contents and foramina of the pterygopalatine fossa.*

2. **Distribution.** It gives off numerous branches (see Fig. 32-9).
   a. **Communicating branches** through the pterygopalatine ganglion carry sensation from the nose and palate. These branches appear to suspend the pterygopalatine ganglion.
   b. **The zygomatic nerve** passes into the orbit via the **inferior orbital fissure**. It divides into a **zygomaticotemporal branch** and a **zygomaticofacial branch**, both of which traverse the lateral wall of the orbit to supply a small area of skin lateral to the eye.
   c. **The posterior–superior alveolar nerves** run along the posterior surface of the maxillary tubercle and enter that bone to supply the maxillary molars.
   d. **The posterior–superior nasal nerves** pass through the **sphenopalatine foramen** into the nasal cavity where it divides into medial and lateral groups.
      (1) The medial group arches over the roof to the nasal septum. The longest of the group, the **nasopalatine nerve**, descends through the **incisive foramen** to the oral palate.
      (2) The lateral group innervates the posterior ends of the nasal conchae.
   e. **The palatine nerves** descend in the pterygopalatine fossa to the **pterygopalatine canal**.
      (1) The **greater palatine nerve** passes through the **greater palatine foramen**. It then runs anteriorly along the inner aspect of the alveolar margin in a groove on the surface of the hard palate.
      (2) The **lesser palatine nerves** emerge through the **lesser palatine foramen**. They innervate the soft palate, the tonsils, and the palatoglossal folds.
   f. **The main trunk of the maxillary nerve** continues through the **inferior orbital fissure** to become the **infraorbital nerve** in the floor of the orbital cavity where it lies in the infraorbital canal.
      (1) It gives off **middle–superior alveolar** and **anterior–superior alveolar nerves**, which run within the walls of the maxilla to the maxillary premolars and incisors. The anterior–superior alveolar nerve also gives off branches to the lateral wall of the nasal cavity.
      (2) It traverses the orbital margin in a bony canal, the **infraorbital foramen**, and terminates in a spray of small branches to the nose, lip, and lower eyelid.

3. **Clinical considerations**
   a. The afferent limb of the **sneeze reflex** is provided by sensory branches of the maxillary nerve, which convey general sensation from the nasal cavity and palate.
   b. The maxillary nerve may be anesthetized at the foramen rotundum by inserting a needle

through the mandibular notch and guiding it anteriorly along the pterygoid plate until it slips into the pterygopalatine fossa. Alternatively, a curved needle may be passed into the greater palatine foramen and upward along the pterygopalatine canal.

B. Mandibular division of the trigeminal nerve (CN V₃)

1. **Course and composition.** This nerve represents the post-trematic branch of the trigeminal nerve (see Figs. 31-1 and 32-4) and is a mixed nerve. It leaves the cranium through the **foramen ovale**, where the large sensory root is joined by a small motor root. The otic ganglion lies just medial to the mandibular nerve at this point.

2. **Distribution**
   a. **The main trunk** gives off a recurrent meningeal branch and motor branches to the muscles of mastication before dividing into anterior and posterior trunks. All of these structures lie on the medial surface of the lateral pterygoid muscle.
   b. **The anterior trunk** is primarily motor and passes inferiorly to the lateral pterygoid muscle.
      (1) The branch to the medial pterygoid muscle emerges from the medial aspect of the mandibular nerve.
         (a) It gives off slender filaments to the tensor veli palatini muscle.
         (b) It also supplies the tensor tympani muscle.
      (2) The branches to the masseter, lateral pterygoid, and temporalis muscles emerge from the lateral aspect of the mandibular nerve.
      (3) The meningeal branch enters the cranium through the foramen spinosum.
      (4) The terminal sensory branch passes inferiorly to the lateral pterygoid muscle and courses toward the anterior aspect of the condylar process of the mandible.
   c. **The posterior trunk** gives off several branches.
      (1) The sensory **auriculotemporal nerve** is formed by the union of two branches of the posterior trunk. These branches pass on either side of the middle meningeal artery.
         (a) This nerve runs between the neck of the mandible and the external auditory meatus to reach the temple where it has a cutaneous distribution.
         (b) Parasympathetic secretomotor fibers from the otic ganglion pass along this nerve to innervate the parotid gland.
      (2) The sensory **lingual nerve** is joined by the **chorda tympani nerve** (the pretrematic division of the facial nerve) at the inferior border of the lateral pterygoid muscle.
         (a) It descends beyond the lateral pterygoid muscle and onto the lateral surface of the medial pterygoid muscle. It enters the floor of the oral cavity on the lateral surface of the styloglossus muscle below the origin of the superior constrictor muscle. At the posterior end of the inferior alveolar margin, it lies adjacently to the third molar. On the surface of the hyoglossus muscle, it passes superficially to the submandibular duct.
         (b) It conveys general sensation from the anterior two-thirds of the tongue with cell bodies located in the **semilunar (gasserian) ganglion.**
         (c) Taste sensation from the anterior two-thirds of the tongue is carried centrally in the chorda tympani nerve; the cell bodies are located in the **geniculate ganglion.**
         (d) Parasympathetic secretomotor fibers from the chorda tympani leave the lingual nerve by a few short communicating branches to synapse in the **submandibular ganglion.**
      (3) The mixed **inferior alveolar nerve** runs inferiorly to the lingual nerve to enter the mandibular foramen on the inner surface of the mandibular ramus.
         (a) A **mylohyoid branch** passes along the inner surface of the mandible below the mylohyoid line. It contains the only motor branches of the posterior trunk and innervates the mylohyoid muscle and the anterior belly of the digastric muscle.
         (b) Within the mandibular canal, the inferior alveolar nerve provides sensory innervation to the lower teeth and gums.
         (c) At the mental foramen, it bifurcates into an **incisive nerve** and the cutaneous **mental nerve.** The mental nerve exits through the **mental foramen** and provides cutaneous innervation to the anterior aspect of the chin.
   d. **Clinical considerations**
      (1) The mandibular nerve may be anesthetized where it passes through the foramen ovale by inserting a needle through the mandibular notch and marching the needle posteriorly along the pterygoid plate.
      (2) The inferior alveolar nerve may be anesthetized intraorally by inserting a needle laterally to the pterygomandibular raphe and then walking the needle posteriorly along the medial aspect of the ramus and injecting the anesthetic in the vicinity of the mandibular foramen. Fracture of the body of the mandible may damage the nerve.
      (3) The mental and incisive nerves may be anesthetized by injection of anesthetic directly into the mental foramen.

C. Maxillary autonomic components of the facial nerve (CN VII)

1. **Presynaptic course.** Secretomotor nerves originate from the superior salivatory nucleus.
   a. The **greater superficial petrosal nerve** carries presynaptic secretomotor fibers derived from the nervus intermedius of the facial nerve to the maxillary nerve. It arises from the facial nerve at the genu of the facial canal (see Fig. 32-6).
   b. It is joined at the foramen lacerum by sympathetic fibers from the **deep petrosal nerve** to form the **(vidian) nerve of the pterygoid canal**. This canal runs anteriorly in the base of the pterygoid process to enter the pterygopalantine fossa (see Fig. 32-9).
   c. Entering the pterygopalatine fossa from the vidian canal, the presynaptic parasympathetic fibers synapse in the pterygopalatine (sphenopalatine) ganglion.

2. **Postsynaptic course.** From the **pterygopalatine ganglion**, postsynaptic fibers travel along the branches of the maxillary nerve.
   a. Secretomotor fibers are distributed medially along the nasopalatine branch to the nasal mucosa.
   b. Secretomotor fibers are distributed inferiorly along the palatine branches to the hard and soft palates.
   c. Secretomotor fibers are distributed anteriorly along the infraorbital nerve to the maxillary sinus.
   d. Secretomotor fibers for lacrimation are distributed superiorly along the infraorbital nerve and its zygomaticotemporal branch to the lacrimal branch of the ophthalmic branch of the trigeminal nerve, which carries them to the lacrimal gland.

D. Mandibular autonomic components of the facial nerve (CN VII)

1. **Presynaptic course.** Secretomotor nerves originate from the superior salivatory nucleus.
   a. The **chorda tympani nerve** carries presynaptic secretomotor fibers derived from the nervus intermedius of the facial nerve to the mandibular nerve and represents the pretrematic division of the second branchial arch nerve (see Fig. 31-1).
   b. It arises from the facial nerve in the vicinity of the stylomastoid foramen (see Fig. 32-6) and re-enters the temporal bone through the posterior canaliculus of the chorda tympani, which opens into the middle ear. The chorda passes across the handle of the malleus to re-enter the temporal bone in the anterior canaliculus of the chorda tympani, which emerges in the infratemporal fossa through the petrosquamous fissure.
   c. The chorda tympani quickly joins the mandibular division of the trigeminal nerve and continues along the lingual branch to the submandibular ganglion.

2. **Postsynaptic course.** From the **submandibular ganglion**, postsynaptic secretomotor fibers travel directly to the submandibular and sublingual glands as well as along branches of the lingual nerve to the tongue.

VIII. VASCULATURE OF THE DEEP FACE

A. Arterial supply. The terminal branches of the **external carotid artery** are the superficial temporal artery and the maxillary artery. They pass on either side of the ramus of the mandible in the parotid gland.

1. **The superficial temporal artery** arises from the external carotid artery in the parotid gland and runs superiorly to the scalp.

2. **The maxillary artery** runs anteriorly out of the parotid gland on the deep surface of the mandibular ramus, where it crosses superficially to the inferior alveolar and lingual nerves.
   a. The maxillary artery gives rise to several branches within the infratemporal fossa (Fig. 32-10).
      (1) Small tympanic and deep auricular branches supply the ear.
      (2) The **middle meningeal artery** enters the cranium through the foramen spinosum surrounded by a plexus of postganglionic sympathetic neurons.
      (3) An **accessory meningeal artery** accompanies the mandibular nerve through the foramen ovale.
      (4) The **inferior alveolar artery** runs with the inferior alveolar nerve and its branches to the lower jaw.
   b. The maxillary artery usually passes deeply to the inferior head of the lateral pterygoid muscle. Branches of the pterygoid part of the maxillary artery correspond to the branches of the anterior division of the mandibular nerve. They are the masseteric, buccal, deep temporal, and pterygoid arteries.
   c. The terminal portion of the maxillary artery passes through the pterygomaxillary fissure to

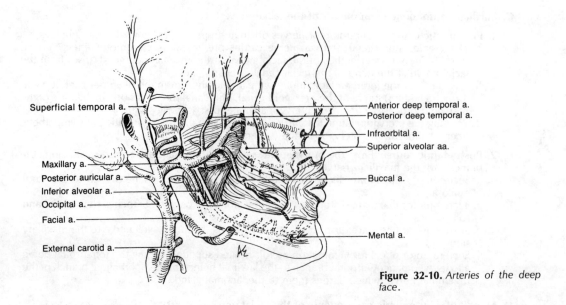

Figure 32-10. *Arteries of the deep face.*

enter the pterygopalatine fossa. Branches in the pterygopalatine fossa accompany the branches of the maxillary nerve into the nose, palate, and upper jaw (see Fig. 32-9).

   (1) The **pharyngeal branch** courses posteriorly through the palatovaginal canal.

   (2) The **posterior–superior alveolar arteries** supply the maxillary molars.

   (3) The **sphenopalatine artery** has a similar distribution to the posterior nasal nerves, supplying the medial and lateral walls of the nasal cavity.

   (4) The **descending palatine artery** supplies the hard and soft palates.

   (5) The **infraorbital artery** accompanies the infraorbital nerve and supplies the remainder of the maxillary teeth.

B. Venous return

   1. The pterygoid venous plexus lies in close association with the lateral pterygoid muscle and the middle portion of the maxillary artery. It receives drainage from the veins of the nasal cavities, the paranasal sinuses, and the oral cavity as well as from the structures of the infratemporal fossa.

      a. The vessels that drain into the pterygoid plexus have anastomotic connections with the superficial veins of the face by numerous pathways, including the **deep facial (buccal) vein**, the **inferior ophthalmic vein**, and the **pharyngeal plexus**.

      b. The pterygoid plexus anastomoses with the cavernous sinus via the **sphenoid emissary vein** (of the foramen of Vesalius).

      c. Thus, infection in the danger zone about the nose may spread to the pterygoid plexus and from there into the cranial, orbital, or pharyngeal regions.

   2. The maxillary vein, which drains into the external jugular vein via the **retromandibular vein**, receives a major contribution from the pterygoid plexus.

C. Lymphatic drainage is primarily to the deep cervical (internal jugular) nodes.

33
Anterior
Cervical Triangle

I. INTRODUCTION

A. Boundaries of the anterior triangle. The anterior triangle is bounded by the anterior border of the sternomastoid, the base of the mandible, and the ventral midline. Posteriorly, the flat anterior wall of the vertebral column lies in a coronal plane between the somatic and visceral necks.

B. Skin and fascia of the anterior triangle

1. **The cleavage lines (Langer's lines)** run horizontally around the neck. A transverse incisional scar is almost invisible (see Fig. 2-1).

2. **The subcutaneous tissue** is sparce and contains the platysma muscle.
   a. This cutaneous muscle acts to flatten the skin of the neck, which can be useful in shaving.
   b. It is innervated by the cervical branch of the facial nerve.

3. **The deep cervical fascia** was discussed in Chapter 27.

4. **The carotid sheath** encloses the neurovascular bundle that runs cranially anterior to the cervical transverse processes on either side of the visceral structures (see Fig. 27-1).

C. Muscles of the anterior triangle

1. A wall of musculature surrounds the floor of the mouth, the pharynx, the larynx, the esophagus, the trachea, the salivary glands, and the thyroid gland. These branchiomeric muscles are derived from primitive gill structures. They surround and support the cranial and cervical portions of the gut and are best understood by reference to the basic structure of the oral cavity, the pharynx, and the larynx (see Fig. 33-1).

2. The U-shaped hyoid bone, the thyroid cartilage, and the cricoid cartilage support the anterior aspect of the pharynx and larynx. They also provide attachment for many branchiomeric muscles.

II. HYOID BONE

A. Characteristics

1. This U-shaped bone consists of a median **body**, paired **lesser cornua (horns)** laterally, and paired **greater cornua (horns)** posteriorly.

2. The **stylohyoid ligament** runs between the styloid process and the lesser horn.

B. Muscle attachments

1. **Along the greater cornu**, the middle constrictor attaches posteriorly; the hyoglossus, digastric and stylohyoid muscles attach superiorly; the thyrohyoid muscle attaches inferiorly.

2. **Along the body**, the geniohyoid and mylohyoid muscles attach superiorly; the omohyoid and sternohyoid muscles attach inferiorly.

III. MUSCULATURE OF THE ANTERIOR TRIANGLE

A. Sternomastoid muscle (Table 33-1)

1. **Orientation.** This muscle passes anteriorly and medially from the mastoid process of the cranium to its sternal and clavicular attachments (Fig. 33-1). It defines the anterior triangle.

Table 33-1. Muscles of the Anterior Triangle

Muscle	Origin	Insertion	Action	Innervation
Superficial musculature				
Sternomastoid	Mastoid process of temporal bone	Sternum and clavical	Rotates head to opposite side and flexes neck	Spinal accessory n. (CN XI)
Infrahyoid musculature				
Geniohyoid	Anterior mandible	Hyoid bone	Protracts hyoid	Superior ramus of ansa cervicalis (C1–C2)
Sternohyoid	Sternum	Hyoid bone	Lowers hyoid	Superior ramus of ansa cervicalis (C1–C2)
Thyrohyoid	Hyoid bone	Thyroid cartilage	Raises larynx	Superior ramus of ansa cervicalis (C1–C2)
Sternothyroid	Sternum	Thyroid cartilage	Lowers larynx	Inferior ramus of ansa cervicalis (C2–C3)
Omohyoid	Scapular notch	Hyoid bone	Lowers hyoid	Inferior ramus of ansa cervicalis (C2–C3)
Suprahyoid musculature				
Digastric:				
Posterior belly	Mastoid process of temporal bone	Anterior belly through hyoid trochlea	Elevates hyoid bone	Facial n. (CN VII)
Anterior belly	Posterior belly through hyoid trochlea	Mandible	. . .	Mylohyoid branch of inferior alveolar n. (CN V_3)
Mylohyoid	Body of mandible	Midline raphe	Elevates floor of mouth and tongue	Mylohyoid branch of inferior alveolar n. (CN V_3)
Stylohyoid	Styloid process	Lesser horn of hyoid	Elevates and retracts hyoid	Facial n. (CN VII)

    2. Actions. It acts to rotate the head to the opposite side and flex the head to the same side.

    3. Innervation. It is innervated by the spinal accessory nerve and fibers from the cervical plexus (C2 and C3).

  B. Digastric and omohyoid muscles insert onto the hyoid bone (see Table 33-1). They divide the anterior triangle into digastric, mylohyoid, carotid, and muscular triangles.

    1. The anterior and posterior bellies of the digastric muscle radiate toward the superior corners of the anterior triangle from a central attachment on the hyoid bone and, thus, define the **digastric triangle** superiorly and the **mylohyoid triangle** anteriorly (see Fig. 33-2).

    2. The omohyoid muscle divides the rest of the anterior triangle into a **carotid triangle** posteriorly and a **muscular triangle** anteriorly (see Fig. 27-8).

  C. Infrahyoid (strap or ribbon) muscles (see Fig. 33-1 and Table 33-1)

    1. Organization. These muscles are named for their attachments.
      a. The sternohyoid muscle, running between the hyoid bone and the sternum, is the longest and most superficial strap muscle.
      b. The short thyrohyoid muscle runs between the hyoid bone and the oblique line of the thyroid cartilage.
      c. The sternothyroid muscle continues the path of the thyrohyoid muscle between the oblique line on the lamina of the thyroid cartilage and the sternum.
      d. The omohyoid muscle (*omos*, G. shoulder) runs between the hyoid bone and the superior transverse ligament of the scapular notch.

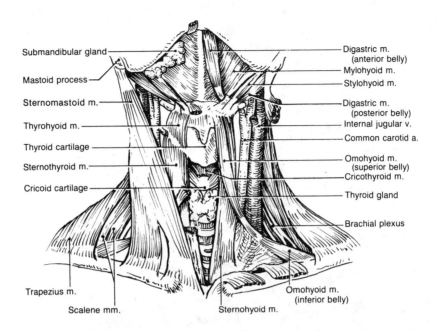

Figure 33-1. *Musculature of the anterior cervical triangle.*

(1) The anterior belly runs upwards across the anterior triangle to the hyoid bone.
(2) The posterior belly runs along the base of the posterior triangle, where it is bound to the clavicle by the investing layer of deep cervical fascia.
(3) A tendinous intersection between the two bellies is attached to the investing fascia deep to the sternomastoid muscle and allows a directional change from vertical to nearly horizontal.

2. Group actions
  a. These external laryngeal muscles raise, depress, and stabilize the hyoid bone and larynx during swallowing and phonation.
  b. The sternohyoid and sternothyroid muscles bind the thyroid gland to the anterior surface of the neck.

3. Innervation. The infrahyoid muscles are innervated by the **ansa cervicalis** (C1–C3).

D. Suprahyoid muscles (see Table 33-1)
  1. Organization. These muscles surround the floor of the mouth (Fig. 33-2).
    a. The digastric muscle, as the name implies, consists of two bellies with an intervening tendon.
      (1) The **posterior belly** originates from the mastoid process and passes through a trochlea on the superior surface of the hyoid bone near the lesser horn.
      (2) The **anterior belly** continues from the trochlea to the digastric fossa of the mandible.
    b. The stylohyoid muscle runs from the styloid process to the lesser horn of the hyoid bone, where its tendon of insertion splits for the passage of the digastric muscle.
    c. The mylohyoid muscle arises from the mylohyoid line on the medial surface of the body of the mandible.
      (1) Fibers run anteroinferiorly to meet the corresponding fibers from the opposite side in a midline fibrous raphe, which is tethered to the hyoid bone.
      (2) This muscle forms a sling beneath the floor of the mouth.
    d. Three muscles pass through the interval between the superior and middle constrictors—the **styloglossus muscle**, the **stylopharyngeus muscle**, and the **hyoglossus muscle** (see Fig. 34-7). These muscles are discussed in Chapter 34 III D and IV E.
    e. The three **pharyngeal constrictors** that comprise the innermost layer are discussed in Chapter 34 IV E.

  2. Innervation
    a. Digastric muscle
      (1) The **posterior belly** is innervated by the digastric branch of the facial nerve.
      (2) The **anterior belly** is innervated by the mylohyoid branch of the inferior alveolar nerve.

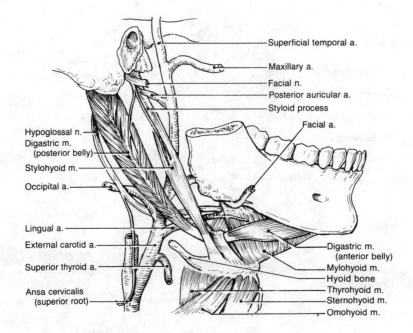

Figure 33-2. *Nerves and vessels of the submandibular triangle.*

b. The stylohyoid muscle is innervated by the facial nerve.

c. The mylohyoid muscle is innervated by the mylohyoid branch of the mandibular division of the trigeminal nerve.

IV. THYROID AND PARATHYROID GLANDS

A. Thyroid gland

1. Structure

a. Each pear-shaped **lobe** extends around the trachea and esophagus to the carotid sheath (see Fig. 33-1). Superiorly, it is limited by the insertion of the sternothyroid muscle onto the thyroid cartilage. Its base reaches the level of the fifth tracheal ring.

b. An **isthmus** connects the left and right lobes and lies anteriorly to the upper tracheal rings.

2. Development

a. The gland develops from a diverticulum in the floor of the mouth, which is marked by the **foramen caecum**.

(1) The thyroid primordium migrates caudally in the neck anterior to the hyoid bone, trailing the **thyroglossal duct**, to reach its final position on the anterior surface of the trachea.

(a) The **median pyramidal lobe**, when present, is a remnant of the thyroglossal duct.

(b) The apex of the pyramidal lobe may be joined to the hyoid bone by a fibrous band, which occasionally contains some muscle fibers.

(2) Rests of thyroid tissue may remain along the track of the thyroglossal duct. These may develop into midline **thyroglossal cysts**.

b. The **parafollicular cells**, which secrete calcitonin, arise from the fourth branchial pouches.

3. Support. The pretracheal layer of deep cervical fascia binds the gland strongly to the trachea and larynx (see Fig. 27-1).

a. The fascia splits at the posterior border of each lobe.

(1) The anterior layer is tightly bound to the larynx.

(2) The posterior layer extends around the esophagus.

(3) Branches of the inferior thyroid artery and the recurrent laryngeal nerve lie between the anterior and posterior layers.

b. The gland may be felt to move up and down during swallowing.

4. Function. The thyroid gland produces thyroxine and thyrocalcitonin, which regulate metabolism and calcium balance, respectively.

B. Parathyroid glands

1. **Structure.** There are four parathyroid glands. These glands are small and brownish-pink due to their vascularity. They can be difficult to find at operation.

2. **Development**
   a. **The superior parathyroid glands** are fourth branchial pouch derivatives. They lie posteriorly to the apex of each lobe of the thyroid gland.
   b. **The inferior parathyroid glands** are third branchial pouch derivatives. They are related to the base of each lobe. However, they occasionally may follow the thymus (also a third pouch derivative) into the thorax, which is frustrating if the gland develops a tumor requiring excision.

3. **Function.** The parathyroid glands secrete parathyroid hormone, which controls calcium metabolism.

C. Thyroid vasculature

1. **Arterial supply.** The superior thyroid artery (a branch of the external carotid artery), the inferior thyroid artery (a branch of the thyrocervical trunk of the subclavian artery), and occasionally a median thyroidea ima artery from the brachiocephalic artery supply the thyroid and parathyroid glands.

2. **Venous return.** A venous plexus on the surface of the gland and over the trachea drains into superior, middle, and inferior thyroid veins.

D. Clinical considerations

1. **Goiter.** An enlarged thyroid gland may extend superiorly in the carotid triangle and down into the thoracic inlet. Pressure of the trachea at the thoracic inlet may cause shortness of breath and a wheeze, especially when the patient is asked to raise his arms above his head.

2. **Tracheostomy.** The surgeon may identify and ligate bleeding vessels or even divide the thyroid isthmus, placing the tracheal stoma just above the jugular notch. In an emergency, tracheostomy is frequently accomplished through the cricothyroid membrane, a relatively avascular region.

3. **Total parathyroidectomy** (usually inadvertent) is followed by a gradual fall in blood calcium levels, which can cause fatal tetany. These glands are extremely hardy and will continue to function if located and transplanted from an excised thyroid into the sternomastoid muscle.

V. VASCULATURE OF THE ANTERIOR NECK

A. Carotid sheath

1. **Composition.** Formed by septa from the superficial layer of deep cervical fascia, it contains the neurovascular bundle of the neck (see Fig. 27-1).

2. **Location.** For the most part, the carotid sheath is under cover of the sternomastoid muscle and is anterior to the transverse processes of the cervical vertebrae.
   a. Arterial pulsations can be felt by gently pressing the vessels against the unyielding bone. Further pressure occludes the artery. It is, therefore, unwise to feel for the carotid pulses bilaterally, as one would do anywhere else in the body when evaluating blood flow.
   b. In the elderly, atheromatous plaques may be dislodged upon carotid palpation. It is wise, therefore, to take the carotid pulse on the right side since a stroke induced in the nondominant cerebral hemisphere would be less devastating.

B. Arterial supply (Fig. 33-3)

1. **The common carotid arteries** supply most of the head and neck.
   a. **The left common carotid artery** arises from the aortic arch.
   b. **The right common carotid artery** arises with the right subclavian artery from the brachiocephalic artery, a remnant of the artery of the right fourth branchial arch.
   c. **The carotid bifuration** occurs at the level of the superior edge of the thyroid cartilage (C4). Two receptors at the bifurcation provide feedback to the vasomotor and respiratory centers concerning the pressure and oxygenation of the blood en route to the brain.
      (1) The **carotid sinus** is a fusiform dilatation that functions as a baroreceptor.
         (a) The afferent limb of the inhibitory **carotid reflex** is mediated by the carotid branch of the glossopharyngeal nerve (CN IX). This branch is part of the post-trematic division of the nerve of the third branchial arch (see Fig. 31-1).

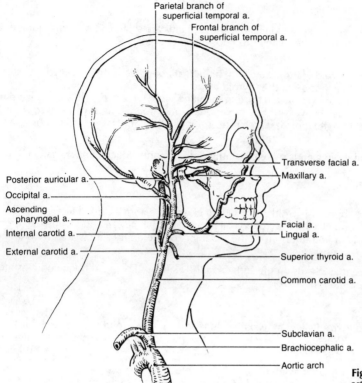

Parietal branch of
superficial temporal a.

Frontal branch of
superficial temporal a.

Transverse facial a.

Maxillary a.

Posterior auricular a.

Occipital a.

Ascending
pharyngeal a.

Internal carotid a.

External carotid a.

Facial a.

Lingual a.

Superior thyroid a.

Common carotid a.

Subclavian a.

Brachiocephalic a.

Aortic arch

Figure 33-3. *Branches of the carotid arteries in the neck.*

(b) The efferent limb of the inhibitory **carotid reflex** is mediated by the vagus nerve (CN X).

(c) *Supraventricular tachycardia*, is susceptible to vagal effects and may be converted to normal sinus rhythm by carotid sinus massage.

(2) The **carotid body** is very vascular and lies between the bifurcation.

(a) It is a chemoreceptor sensitive to blood oxygenation and to a lesser extent to pH and carbon dioxide partial pressure (P_{CO_2}).

(b) It occasionally gives rise to tumors that are difficult to excise because of their location.

2. The **external carotid artery** supplies most of the face and visceral neck with eight major branches (see Fig. 33-3).

a. The **ascending pharyngeal artery** is a long slender branch, which arises close to the bifurcation of the common carotid artery.

(1) It runs cranially between the internal carotid artery and the pharyngeal constrictors.

(2) It anastomoses freely with the ascending palatine branch of the facial artery.

b. The **superior thyroid artery** arises at the tip of the greater horn of the hyoid bone. Occasionally, it arises with the lingual artery from a common trunk.

(1) It runs along the posterior border of the thyrohyoid muscle. Here it gives off an internal laryngeal branch, which passes beneath the muscle and pierces the thyrohyoid membrane in company with the internal laryngeal nerve.

(2) The main trunk runs down the medial edge of the upper pole of the thyroid gland.

(a) At the thyroid isthmus, it anastomoses with the contralateral superior thyroid artery.

(b) It supplies the anterior surface of the gland.

c. The **lingual artery** also arises opposite the greater horn of the hyoid bone.

(1) It gives off a **tonsillar branch**, which also supplies the lateral pharyngeal walls.

(2) After a preliminary loop, it dives deeply to the hyoglossus muscle.

(a) This course contrasts with the paths of the hypoglossal nerve, the lingual nerve, and the duct of the submandibular salivary gland, all of which pass superficially to the hyoglossus muscle.

(b) Deep to the hyoglossus muscle, the lingual artery gives off the **dorsal lingual branch**, which supplies the posterior half of the oral cavity.

(3) The **deep lingual branch** supplies the anterior tongue.

(4) The **sublingual branch** supplies the anterior half of the oral cavity.

d. The facial artery arises opposite the angle of the mandible. However, it has to continue cranially to the level of the superior edge of the middle constrictor to pass laterally over the submandibular salivary gland.

    (1) At the apex of this loop, it is separated from the tonsil by the styloglossus muscle. Here it gives off a **tonsillar branch**, which is usually the largest branch to the tonsil.

    (2) A small **ascending palatine branch** parallels the ascending pharyngeal artery with which it anastomoses.

    (3) The **submental branch** arises as the facial artery emerges between the mandible and the submandibular gland. This branch runs forward to the chin along the insertion of the mylohyoid muscle.

    (4) In a sinuous course, the facial artery continues over the ramus of the mandible to enter the face, where the facial pulse is palpable. It then gives off **inferior** and **superior labial branches**.

    (5) It continues laterally to the nose as the **angular artery**.

e. The occipital artery arises from the posterior aspect of the external carotid artery.

    (1) Deep to the origin of the digastric muscle, the occipital artery continues posteriorly, grooving the medial surface of the mastoid process, to emerge at the apex of the posterior cervical triangle.

    (2) It gives off two branches to the sternomastoid. The upper branch is associated with the spinal accessory nerve; the lower with the hypoglossal nerve.

f. The posterior auricular artery also arises posteriorly from the external carotid artery.

    (1) It ascends between the superior surface of the posterior belly of the digastric muscle and the parotid gland.

    (2) It grooves the base of the skull between the mastoid process and the auricle.

g. The superficial temporal artery ascends anteriorly to the external auditory meatus and supplies the scalp.

h. The (internal) maxillary artery supplies the deep face as well as portions of the nasal and oral cavities. It was discussed in Chapter 32.

3. The internal carotid artery ascends to the **carotid canal** in the base of the cranium without giving off any branches in the neck.

  a. The styloid process and its associated muscles intervene between the internal and external carotid arteries.

  b. Its **ophthalmic branch** supplies the upper portion of the face.

C. Venous return (Fig. 33-4)

  1. The internal jugular vein exits the skull through the **jugular foramen** in company with the glossopharyngeal, vagus, and spinal accessory nerves.

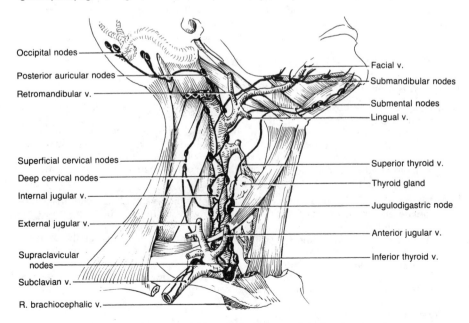

Occipital nodes
Posterior auricular nodes
Retromandibular v.
Facial v.
Submandibular nodes
Submental nodes
Lingual v.
Superficial cervical nodes
Deep cervical nodes
Internal jugular v.
External jugular v.
Superior thyroid v.
Thyroid gland
Jugulodigastric node
Anterior jugular v.
Supraclavicular nodes
Subclavian v.
Inferior thyroid v.
R. brachiocephalic v.

Figure 33-4. *Venous and lymphatic drainage of the head and neck.*

    a. As it descends through the neck in the carotid sheath, the internal jugular vein winds laterally.

    b. It joins the **subclavian vein** by an acute angle behind the sternoclavicular joint to form the **brachiocephalic vein**.

    c. In the root of the neck, the subclavian and jugular veins lie anteriorly to the corresponding arteries, facilitating emergency venous access.

2. The external jugular vein is superficial and variable. It may be demonstrated by obstructing its drainage with light finger pressure or by having the patient perform Valsalva's maneuver. Careful observation of the external jugular vein will usually reveal venous pressure waves. The level of the column of blood in the external jugular veins is a useful indication of right atrial pressure.

    a. It is formed at the angle of the jaw by the union of the posterior division of the **retromandibular vein** and the **posterior auricular vein** (see Fig. 27-8).

    b. It usually receives four tributaries.

        (1) The **suprascapular vein** accompanies the like-named artery.

        (2) The **superficial cervical vein** accompanies the like-named artery.

        (3) The **anterior jugular vein** comes from the anterior triangle.

        (4) The **posterior jugular** vein comes from the apex of the posterior triangle.

    c. It passes in the superficial fascia of the posterior triangle to the middle of the clavicle, where it joins the subclavian vein.

D. Lymphatic drainage of the anterior neck (see Fig. 33-4)

1. Cervical nodes. The lymph from the face and anterior neck drains by the superficial cervical nodes that parallel the external jugular vein and by deep cervical nodes that parallel the internal jugular vein.

    a. The superficial (external jugular) nodes receive drainage from the **occipital nodes**, the **retroauricular nodes**, the **parotid nodes**, and some of the **anterior cervical nodes**.

    b. The deep (internal jugular) nodes generally receive drainage from the lymphatics of the superficial face and deep structures via the **submandibular**, **submental**, and **jugulodigastric nodes**.

        (1) Just below the posterior belly of the digastric muscle, the jugulodigastric node drains the tonsillar region and is valuable in the diagnosis of pharyngeal inflammation.

        (2) Some anterior cervical nodes also drain into the deep lymphatic chain.

    c. The jugular trunk forms from the deep lymphatics in the root of the neck.

        (1) On the right side, the jugular trunk ends at the junction of the internal jugular vein and the subclavian vein.

        (2) On the left side, it joins the thoracic duct.

2. The thoracic duct enters the neck behind the left subclavian artery.

    a. From the thorax, it ascends into the neck to the level of C6, where it crosses anteriorly to the vertebral artery and the thyrocervical trunk.

    b. It drains into either the jugular or the subclavian vein.

VI. INNERVATION OF THE ANTERIOR TRIANGLE

A. Hypoglossal nerve (CN XII) [Fig. 33-5]

1. Course and composition. The hypoglossal nerve exits the cranium through the **anterior condylar (hypoglossal) canal**.

    a. It passes laterally between the internal carotid artery and the internal jugular vein.

    b. It winds over the surface of the inferior ganglion of the vagus to the angle of the mandible, where it turns anteriorly to pass superficially to the occipital artery.

    c. It crosses the external carotid artery, the loop of the lingual artery, and the hyoglossus muscle to reach the root of the tongue.

2. Distribution. The purely motor hypoglossal nerve innervates all of the intrinsic muscles of the tongue and three extrinsic muscles—the styloglossus, hyoglossus, and genioglossus. These are somatic muscles.

    a. As a rule of thumb, the hypoglossal nerve innervates all muscles that end in the suffix "glossus" (with the exception of the palatoglossus, which is innervated by the pharyngeal branch of the vagus nerve).

    b. In hypoglossal nerve injury, the protruded tongue deviates toward the affected side due to the unopposed action of the genioglossus muscle.

B. Cervical plexus

1. The transverse cervical (colli) nerve (C2 and C3) innervates the skin.

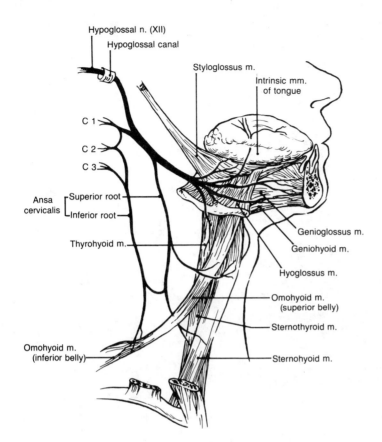

Figure 33-5. *Hypoglossal nerve.*

2. **The ansa cervicalis** innervates the strap muscles.
   a. **Course and composition.** Fibers from cervical nerves C1–C3 form the ansa cervicalis (see Fig. 33-5).
   b. **Distribution**
      (1) Some of these fibers (C1) accompany the hypoglossal nerve to be distributed to the thyrohyoid muscle and the geniohyoid muscle.
      (2) Other fibers (C1, C2) leave the company of the hypoglossal nerve as the **superior ramus** of the **ansa cervicalis** (whence the former name **descendens hypoglossi**) to innervate the sternohyoid and sternothyroid muscles.
      (3) The **inferior ramus** of the **ansa cervicalis** is formed by the union of branches from C2–C3 (whence the former name **descendens cervicalis**) and innervates the omohyoid muscle and a portion of the sternothyroid muscle.
      (4) The superior and inferior roots anastomose to form a dependent loop on the surface of the carotid sheath.

C. **Cervical sympathetic nerves**

1. **Presynaptic pathway.** The cervical sympathetic chain runs cranially in the neck posterior to the carotid sheath (see Fig. 27-1).
   a. It is thinner and more medially placed than the vagus nerve.
   b. Preganglionic fibers run up the chain from the upper thoracic levels of the spinal cord to synapse in the superior, middle, and inferior cervical ganglia. The large **superior cervical ganglion** sits immediately below the skull.

2. **Postsynaptic pathway**
   a. **Gray rami communicantes** pass postganglionic fibers to the upper four cervical nerves carrying secretomotor, vasomotor, and pilomotor fibers.
   b. Fibers to the head form a perivascular plexus around the internal carotid artery.
      (1) They innervate, among other things, the pupillodilators of the eyes, the palpebral muscle (of Müller), and the orbitalis muscle (of Müller), as well as the sweat glands and the vasoconstrictors.
      (2) Lesions of the cervical sympathetic chain, therefore, produce meiosis, ptosis, enophthalamos, anhydrosis, and flushing (Horner's syndrome).

34
Cranial and Cervical Viscera

I. INTRODUCTION

A. Developmental considerations. The cranial end of the primitive gut develops into two systems: the **respiratory tract** and the **alimentary canal (gastrointestinal tract)**. Despite the functional dichotomy of these two systems, the anatomic division is incomplete between the palate and the larynx (i.e., the pharynx is shared between the two systems).

B. Anatomic divisions

1. The **nasal cavity** serves primarily in respiration and olfaction.

2. The **oral cavity** functions primarily in ingestion and taste but may also function in respiration and articulation of speech.

3. The **pharynx** serves both respiration and ingestive functions with a minor taste function.

4. The **larynx** functions in respiration and phonation.

II. NASAL CAVITY

A. Basic structure. The nasal cavity is a pyramidal bony chamber that lies beneath the anterior cranial fossa (see Fig. 31-2). It is capped by the cartilaginous external nose.

1. The **nares (nostrils)** open into the nasal cavity anteriorly.

2. The **choanae** open into the nasopharynx posteriorly (see Fig. 34-3).

3. The **nasal septum** extends medially from the roof to the floor and has both bony and cartilaginous portions (see Fig. 31-3*B*). It usually deviates to one side of the midline.

4. **Conchae (turbinate bones)** project scroll-like into the nasal cavity from the lateral walls (see Fig. 31-2). The superior and middle conchae project from the ethmoid bone; the inferior concha is a separate bone (see Fig. 31-3A). The conchae divide the nasal cavity into recesses (meatuses).

   a. The **sphenoethmoidal recess** is the narrow space above the small superior concha. The **olfactory epithelium** lies superiorly in this recess, and the **ostium of the sphenoid sinus** lies posteriorly.

   b. The **superior meatus** below the superior concha receives several ducts from the posterior ethmoid sinuses.

   c. The **middle meatus** has a prominent bulge, the **ethmoid bulla**, which is the medial wall of the middle ethmoid sinuses. The slit-like **hiatus semilunaris** curves anteriorly and beneath this rounded prominence (see Fig. 31-3A).

      (1) The **infundibulum of the frontal sinus** as well as numerous ducts from the anterior and middle ethmoid sinuses enter the anterosuperior end of the hiatus semilunaris.

      (2) The **ostium of the maxillary sinus** opens into the posteroinferior end of the hiatus semilunaris.

   d. The **inferior meatus** receives the **nasolacrimal duct** anteriorly. A small fold of mucous membrane at the nasolacrimal orifice acts as a valve.

   e. The **antrum** is a shallow depression just anterior to the middle meatus, which is limited superiorly by a small rudimentary concha (agger nasi). Inferiorly, the antrum is continuous with the **vestibule** of the nose, just inside the nares.

B. Mucous membranes

1. **Vestibular mucosa.** The vestibule is lined by skin that bears short thick hairs, called **vibrissae**.

2. **Olfactory mucosa.** The mucous membrane lining the apex of the nasal cavity contains olfactory cells.
   a. Approximately 20 olfactory nerves convey the sensation of smell from the olfactory cells through the cribriform plate of the ethmoid bone to synapse in the olfactory bulbs.
   b. These olfactory nerves have small meningeal sheaths.
   c. They are delicate and prone to injury.

3. **Nasal mucosa.** Elsewhere, the mucosa is of the pseudostratified columnar type, containing numerous mucus-secreting glands.
   a. The layer of mucosa, covering the middle and inferior concha, traps particulate matter and moistens the inspired air.
   b. The polluted mucus is swept backwards by cilia into the nasopharynx and is swallowed.
      (1) The cilia are affected by low temperatures and cease to beat at about 10°C, explaining, in part, why the nose "runs" in cold weather.
      (2) Upper respiratory tract infections and cigarette smoke also inhibit ciliary action.

C. Vasculature of the nasal cavity

1. **Arterial supply.** A profuse arterial anastomosis is fed by branches of the internal and external carotid arteries (Fig. 34-1).
   a. **Distribution**
      (1) The **ethmoidal arteries**, branches of the ophthalmic artery, supply the anterior and superior regions.
      (2) The **sphenopalatine artery**, a branch of the maxillary artery, supplies the posterior region.
      (3) The **greater palatine artery**, also a branch of the maxillary artery, supplies an anteroinferior region.
      (4) The **facial artery**, a branch of the external carotid, supplies the inferior and interior portions through its superior labial and ascending palatine branches.
   b. **Clinical considerations.** Trauma, hypertension, infection, and clotting disorders are associated with epistaxis (nosebleeds).
      (1) Treatment depends upon the cause, but topical pressure, vasoconstrictors, and cautery are usually effective.
      (2) Ligation is impractical because the branches are deep, although limited success has been reported with external carotid ligation.

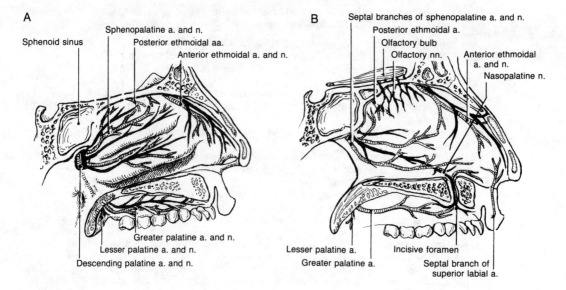

Figure 34-1. *Nasal cavity. A,* Neurovascular structures of the lateral wall. *B,* Neurovascular structures of the nasal septum.

2. Venous return. There is an extensive **submucous venous plexus**.
   a. Course. The veins tend to drain into the pterygoid plexus.
   b. Function. A submucous plexus of veins functions as a heat exchanger. The vasomotor control of this plexus is a delicately balanced system capable of accommodating changes in humidity, temperature, and air flow.
   c. Clinical considerations. Vasodilation and venous engorgement are accompanied by a certain amount of mucosal edema, which may occlude the nasal airway.
      (1) Disturbances in control are sometimes the result of allergic responses (such as those to pollen and house dust) but often are of uncertain etiology.
      (2) Vasoconstrictor agents produce temporary relief because rebound vasodilation occurs and tolerance quickly develops.

D. Innervation of the nose. Like the arterial supply, the nerves are divided into an anterior ethmoidal (ophthalmic) area and a posterior sphenopalatine (maxillary) area.

   1. Ophthalmic branches. The **anterior ethmoid nerve** and the **posterior ethmoid nerve** (see Fig. 34-1), which supply the anterior half, are branches of the nasociliary nerve, a division of CN V$_1$.

   2. Maxillary branches
      a. Posterior nasal branches of the maxillary nerve, which pass through the sphenopalatine foramen in company with the artery of that name, supply the posterior half of the nasal cavity (see Fig. 34-1).
      b. The anterior–superior alveolar branch of the maxillary nerve supplies a small area around the anterior end of the inferior concha, while the external nares are largely supplied by **branches of the intraorbital branches** of the maxillary nerve (see Fig. 34-1).

E. Paranasal sinuses

   1. The frontal sinus (see Fig. 31-3A) develops as the superior extension of an anterior ethmoid sinus.
      a. The left and right frontal sinuses are usually asymmetrical.
      b. They usually drain into the hiatus semilunaris of the middle meatus via the infundibulum.

   2. The ethmoid labyrinth is composed of 6–18 thin-walled air cells (sinuses).
      a. These open medially above and below the middle concha with considerable variation.
      b. Infection of these sinuses may erode through the lamina papyracea into the orbit.

   3. The sphenoid sinuses fill the body of the sphenoid bone (see Fig. 34-1A). They are divided into the right and left by a thin septum.
      a. The septum of the sphenoid process rarely lies in the midline.
      b. The sphenoid sinus opens by a small ostium high on the anterior wall, so that drainage cannot occur by gravity with the head held erect.
      c. Just lateral to the sphenoid sinus are the optic nerve, internal carotid artery, maxillary nerve, and pyterygoid (vidian) nerve. Infection of the sinus may erode the thin walls and involve these structures.
      d. The pituitary gland may be reached through the nose and sphenoid sinus.
         (1) This transsphenoidal approach follows the septum of the nose through the body of the sphenoid.
         (2) Care must be taken to avoid the cavernous sinus and the internal carotid artery in order to avoid potentially catastrophic hemorrhage.

   4. The maxillary sinus (antrum of Highmore) is located in the body of the maxilla.
      a. Much of the medial wall of the maxillary sinus is composed of cartilage.
      b. The slit-like osteum is high on the medial surface, so that it does not drain by gravity with the head held erect.
         (1) It is closely related to the orifices of the frontal and ethmoid sinuses at the hiatus semilunaris. Infection in any one of these quickly spreads to the maxillary sinus.
         (2) The maxillary sinus is particularly prone to infection, which tends to be complicated by the collection of a stagnant pool of mucus. This pool may be drained surgically by piercing the bony nasal wall of the sinus beneath the inferior concha.
      c. The sinuses are lined by mucoperiosteum, which is thinner and less richly supplied with blood vessels and glands than the mucosa of the nasal cavity. Cilia sweep mucus towards the ostia. In mucosal infections, the cilial action is inhibited, and mucus collects.
      d. *Maxillary sinusitis* mimics the clinical sign of maxillary tooth abscess.
         (1) Approximately 89% of all cases of maxillary sinusitis are related to an infected tooth.
         (2) Infection may also spread from the maxillary sinus to the upper teeth.

III. ORAL CAVITY

A. Basic structure. The oral cavity extends from the oral fissure to the oropharyngeal isthmus between the palatoglossal arches (Fig. 34-2).

1. **The roof of the mouth** is formed by the **palate**.

2. **The oral cavity** has two walls: an inner bony wall (anterior), which supports the teeth, and an outer fleshy wall (lateral). Between the walls is the **vestibule**, and inside the alveolar ridges is the **oral cavity proper**.

3. **The floor of the oral cavity** is supported inferiorly by the mylohyoid muscle.

B. Palate

1. **The hard palate** forms the anterior four-fifths of the palate (see Fig. 34-5).
   a. The hard palate is formed by the maxillary and palatine bones.
   b. It separates the respiratory and digestive tracts. Because it is found only in mammals, it can be surmised that its primary function is to isolate the mouth in suckling.

2. **The soft palate (velum)** contains muscle fibers, glands, lymphoid tissue, and an aponeurosis.
   a. The **uvula** hangs from the posterior border of the soft palate in the midline (see Figs. 34-2 and 34-3).
   b. The mucous membrane of the palate is stratified squamous epithelium.
   c. The aponeurosis, which is the tendinous expansion of the tensor veli palatini muscle, continues the plane of the hard palate.

3. **Musculature of the palate**
   a. **Intrinsic musculature.** The levator veli palatini muscle and the tensor veli palatini muscle run nearly parallel but on opposite sides of the superior constrictor muscle of the pharynx and auditory tube (Fig. 34-3 and Table 34-1).
      (1) The **tensor veli palatini muscle** arises from the lateral (membranous) surface of the auditory tube and from the scaphoid fossa at the base of the pterygoid process.
         (a) It descends along the lateral surface of the superior constrictor muscle. The tendon of tensor veli palatini passes around the hamulus of the medial pterygoid plate, pierces the buccinator, which also arises from the pterygoid plate, and spreads out within the soft palate to become the palatine aponeurosis.
         (b) Contraction of the tensor veli palatini muscle tightens the aponeurosis, thereby tensing the palate as well as opening the auditory tube.
         (c) Developmentally, it is a first arch muscle that has been appropriated by the developing pharynx. Therefore, it is innervated by the mandibular nerve.
      (2) The **levator veli palatini muscle** arises from the petrous portion of the temporal bone and the medial (cartilaginous) surface of the auditory tube.

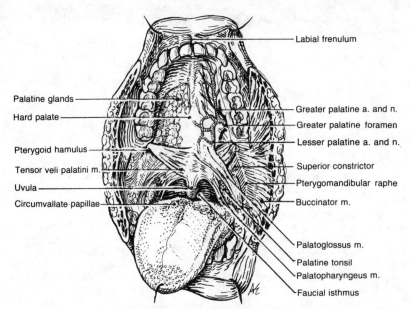

Figure 34-2. *Oral cavity.*

(a) It follows the auditory tube over the superior constrictor muscle and inferiorly along its inner surface. It inserts into the tensor aponeurosis of the soft palate.

(b) It elevates the palate to establish the nasopharyngeal seal and opens the auditory tube.

(c) Since it lies within the pharynx, it is innervated by the vagus nerve (via the pharyngeal plexus).

(3) The paired **uvulae muscles** descend from the spine of the hard palate into the uvula on each side (see Fig. 34-3).

(a) They are innervated by the vagus nerve (CN X).

(b) These muscles contract when the palate is raised.

(i) When one muscle contracts alone, it draws the uvula to the same side.

(ii) The integrity of the vagus nerve is observed when the examining physician has the patient say ''ahhhhhh.''

Table 34-1. Muscles of the Oral Cavity and Pharynx

Muscle	Origin	Insertion	Action	Innervation
Palatine musculature				
Tensor veli palatini	Lateral surface of auditory tube	Velum via hamulus	Tenses velum	Mandibular n. (CN V₃)
Levator veli palatini	Temporal bone and medial surface of auditory tube	Velum	Raises velum	Pharyngeal branch of vagus n. (CN X)
Palatoglossus	Velum	Tongue	Raises tongue	Pharyngeal branch of vagus n. (CN X)
Palato-pharyngeus	Velum	Pharynx	Raises larynx	Pharyngeal branch of vagus n. (CN X)
Uvulae	Spine of hard palate	Velum	Tenses velum	Pharyngeal branch of vagus n. (CN X)
Tongue musculature				
Hyoglossus	Hyoid bone	Lateral tongue	Lowers sides of tongue	Hypoglossal n. (CN XII)
Genioglossus	Anterior mandible	Tongue	Protracts tongue	Hypoglossal n. (CN XII)
Styloglossus	Styloid process of temporal bone	Tongue	Retracts tongue	Hypoglossal n. (CN XII)
Palatoglossus	Velum	Tongue	Raises tongue	Pharyngeal branch of vagus n. (CN X)
Intrinsics	Tongue	Tongue	Changes shape of tongue	Hypoglossal n. (CN XII)
Pharyngeal musculature				
Superior constrictor	Pterygoid plate, pterygomandibular raphe, and mandible	Occipital bone and pharyngeal raphe	Constricts upper part of pharynx in deglutition	Pharyngeal branch of vagus n. (CN X)
Middle constrictor	Hyoid bone	Pharyngeal raphe	Constricts mid-pharynx in deglutition	Pharyngeal branch of vagus n. (CN X)
Inferior constrictor	Thyroid cartilage	Pharyngeal raphe	Constricts lower part of pharynx	Pharyngeal branch of vagus n. (CN X)
Palato-pharyngeus	Velum	Pharynx	Raises pharynx and larynx	Pharyngeal branch of vagus n. (CN X)
Salpingo-pharyngeus	Auditory tube	Pharynx	Raises pharynx and larynx	Pharyngeal branch of vagus n. (CN X)
Stylo-pharyngeus	Styloid process	Pharynx	Raises pharynx	Glossopharyngeal n. (CN IX)

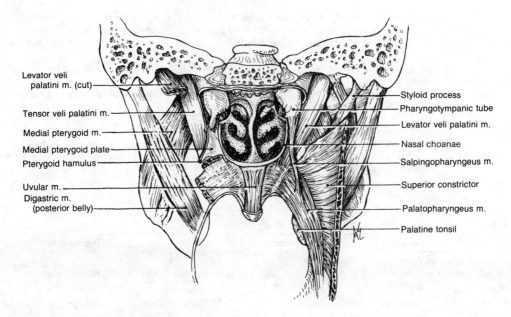

Levator veli palatini m. (cut)
Tensor veli palatini m.
Medial pterygoid m.
Medial pterygoid plate
Pterygoid hamulus
Uvular m.
Digastric m. (posterior belly)

Styloid process
Pharyngotympanic tube
Levator veli palatini m.
Nasal choanae
Salpingopharyngeus m.
Superior constrictor
Palatopharyngeus m.
Palatine tonsil

Figure 34-3. *Musculature of the soft palate.* The viewpoint is posterior.

  b. Extrinsic musculature (see Table 34-1)
    (1) The **palatoglossus muscle** arches upward from the side of the tongue to the inferior surface of the soft palate (see Fig. 34-2).
      (a) It underlies the **anterior faucial pillar** (palatoglossal fold).
      (b) It acts as a sphincter (in conjunction with the base of the tongue) between the oral cavity and the pharynx.
      (c) It is innervated by the vagus nerve (CN X) through the pharyngeal plexus.
    (2) The **palatopharyngeus muscle** arises laterally from the palate and inserts into the pharyngeal musculature.
      (a) It underlies the **posterior faucial pillar** (palatopharyngeal fold).
      (b) It also acts as a sphincter (in conjunction with the base of the tongue) between the oral cavity and the pharynx.
      (c) It helps to raise the larynx in swallowing (deglutition).
      (d) It is innervated by the vagus nerve (CN X) through the pharyngeal plexus.
  c. Muscle function. Closure of the nasopharyngeal isthmus by elevation of the palate (establishment of the velopharyngeal seal) is important in swallowing, speaking, and blowing.
    (1) During quiet nasal respiration, the soft palate hangs vertically.
    (2) In swallowing, the soft palate is raised to prevent food from entering the nose.
    (3) In blowing or in the production of the explosive consonants, the escape of air through the nose is prevented so that the pressure can build up within the mouth.
    (4) *Cleft palate,* a developmental anomaly, prevents the establishment of the velopharyngeal seal. If surgically uncorrected, the individual experiences difficulty with speech and deglutition.

4. Vasculature of the palate
  a. Arterial supply is from several sources (see Figs. 34-1 and 34-2).
    (1) The **greater palatine artery**, which gives off the lesser palatine artery, which supplies the hard and soft portions, respectively
    (2) The **ascending palatine branch** of the facial artery
    (3) The **tonsillar branch** of the facial artery
  b. Venous return is to the **pterygoid plexus** and a peritonsillar plexus.

5. Innervation of the palate
  a. General sensation for the hard palate is carried by the **greater palatine nerve** and **nasopalatine nerve** (branches of the maxillary nerve); the soft palate is innervated by the **lesser palatine nerve** (a branch of the maxillary nerve) and the **glossopharyngeal nerve** (see Fig. 34-1).
  b. Motor innervation is provided by the **mandibular nerve** (CN V$_3$) and the **vagus nerve** (CN X).

C. Walls of the oral cavity

1. **The cheeks** or fleshy walls of the oral cavity comprise a layer of muscle between the skin of the face and the mucous membrane of the **vestibule**.
   a. **The labial folds (lips)** are separated by the oral fissure anteriorly.
      (1) The labial folds contain the **orbicularis oris muscle** and some pea-sized **labial salivary glands**.
      (2) The junction between lip and cheek is marked by the nasolabial folds.
   b. **The cheek** contains the **buccinator muscle**, which arises from the posterior ends of the alveolar margins and the anterior aspect of the **pterygomandibular raphe**. The superior constrictor muscle arises from the posterior aspect of this raphe (see Fig. 34-2).
      (1) At its origin, it is pierced by the tendon of the tensor veli palatini muscle. At the anterior border of the masseter muscle, it is pierced by the parotid duct, and small **buccal salivary glands** surround the duct on the outer surface of the muscle. Fibers pass around the cheeks and into the lips.
      (2) The buccinator muscle's main function is to compress the vestibule and, together with the tongue, to keep the food on the cusps of the molars for mastication.
      (3) The buccinator muscle is innervated by the facial nerve (CN VII).

2. **The alveolar margins** and associated dentition form the inner wall of the **vestibule**.
   a. The alveolar margins are covered with the dense fibrous tissue and thinly keratinized stratified squamous mucosa of the **gums (gingivae)**.
   b. In a young person, the gums surround the cervical portion of the teeth.
   c. The teeth and their sockets were discussed in Chapter 31.
   d. In the midline, there are small folds of mucosa related to the inner and outer surfaces of the gums: **frenulum labia** of the upper and lower lips and the **frenulum linguae** of the tongue.

D. The tongue is a muscular organ concerned with ingestion, swallowing, speech, taste, and general sensation (Fig. 34-4).

1. **The oral and pharyngeal portions of the tongue** have different origins, different mucosa, and different innervation. A V-shaped sulcus lies between the second and third arch derivatives, which comprise the oral and lingual parts. The apex of the V points posteriorly and marks the **foramen caecum** from which the **thyroglossal duct** originated.
   a. **The oral portion** of the tongue has a tip, a dorsal surface, a ventral surface, and a root.
      (1) Small projections of the lamina propria over the dorsum of the oral (presulcal) tongue create papillae, which increase the surface area of mucosa available for taste receptors.

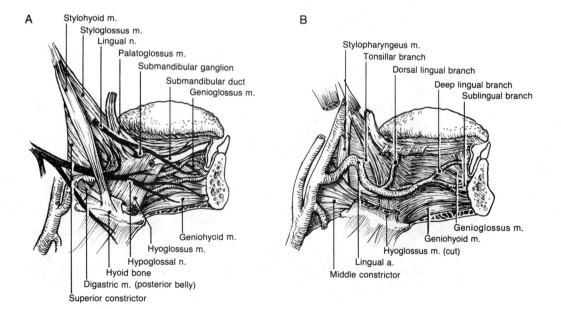

Figure 34-4. *Tongue. A,* Lingual musculature. *B,* The hyoglossus muscle has been partially removed to show the deep neurovascular structures.

      (a) They vary in size from the large flat-topped **vallate (circumvallate) papillae** through the **fungiform** to the conical **filiform**.

      (b) The vallate papillae are often confined to the area immediately adjacent to the sulcus.

      (c) Small transverse mucosal folds at the lateral edges of the tongue create **foliate papillae**.

    (2) The mucosa of the ventral surface of the tongue is smooth.

      (a) Two fringed folds of mucosa, the **plica fimbriata**, extend from the floor of the mouth almost to the tip of the tongue.

      (b) A median frenulum also extends the length of the ventral surface.

      (c) At the root of the tongue, the frenulum is related to the paired **sublingual papillae** for the ducts (of Wharton) of the submandibular salivary glands.

      (d) The associated sublingual glands create sublingual folds around the root of the tongue.

      (e) The ventral surface is highly vascular, and the vessels are close to the surface. This is a route for sublingual absorption of substances, such as nitroglycerin and nicotine.

  b. The pharyngeal portion of the tongue has no papillae.

    (1) The boundaries between the oral and pharyngeal tongue are blurred a little by the extension of third arch primordia superficially across the sulcus, so that the area marked by the vallate papillae (in front of the sulcus) is also innervated by the glossopharyngeal nerve.

    (2) The lumpy surface of the mucosa covers the lymphoid tissue of the **lingual tonsil** (see Fig. 34-5).

2. The root of the tongue lies deeply to the mandible and above the hyoid bone.

  a. Extrinsic muscles course through the root of the tongue to the hyoid bone, the soft palate, the styloid process, and the genial tubercle of the mandible (Fig. 34-4*A*; see Table 34-1)).

    (1) The **hyoglossus muscle** originates along the ramus of the hyoid bone and interdigitates with the styloglossus muscle and the lateral intrinsic musculature of the tongue.

      (a) Superficial to the hyoglossus muscle are:

        (i) The hypoglossal nerve (CN XII), which passes near the inferior border of the muscle

        (ii) The submandibular duct, which passes superiorly to the muscle to empty into the oral vestibule

        (iii) The lingual nerve, which winds around the submandibular duct, crossing it first on its lateral aspect and then on its medial aspect

        (iv) A portion of the submandibular gland

      (b) Deep to the hyoglossus muscle are:

        (i) The lingual artery, which runs its tortuous course deeply to the hyoglossus muscle and superficially to the genioglossus muscle, giving off dorsal and deep branches and the sublingual artery

        (ii) The terminal portion of the deep branch of the glossopharyngeal nerve, which passes posteriorly to this muscle to enter the posterior portion of the tongue

      (c) The hyoglossus draws the side of the tongue inferiorly.

      (d) It is innervated by the hypoglossal nerve.

    (2) The **genioglossus muscle** arises from the superior genial tubercle, just superior to the geniohyoid (see Fig. 34-4).

      (a) It passes posteriorly deep to the hyoglossus muscle.

      (b) Some fibers continue beyond the tongue into the middle constrictor muscle.

      (c) The genioglossus muscle pulls the base of the tongue anteriorly in protrusion.

        (i) Acting singly, they push the protruded tongue to the opposite side.

        (ii) It is innervated by the hypoglossal nerve, the integrity of which is checked by the examining physician when he or she has the patient stick out the tongue.

    (3) The **styloglossus muscle** arises from the styloid process and interdigitates with the hyoglossus muscle and the lateral intrinsic musculature of the tongue (see Fig. 34-4).

      (a) It draws the posterior portion of the tongue superiorly and posteriorly and is the principal retractor of the tongue.

      (b) It is innervated by the hypoglossal nerve.

    (4) The **palatoglossus muscle** lies within the palatoglossal arch at the oropharyngeal isthmus (see Figs. 34-2 and 34-4*A*).

      (a) It acts more on the palate than on the tongue, establishing the palatoglossal sphincter.

      (b) Unlike the other muscles of the tongue, it is innervated by the vagus nerve (CN X).

  b. Intrinsic muscles of the tongue are arranged in longitudinal, transverse, and vertical groups (see Fig. 34-4*A*).

(1) They arise and insert within the substance of the tongue.

(2) The two halves of the tongue are separated by a fibrous septum into which some of these muscles insert.

(3) They act to change the shape of the tongue (Fig. 34-4*B*).

(4) All are striated muscles of somatic origin and are innervated by the hypoglossal nerve (CN XII).

3. Vasculature of the tongue

a. Arterial supply. The **lingual artery** provides the principal blood supply to the tongue (see Fig. 34-4*B*).

(1) It usually arises from the external carotid artery but may on occasion arise from a common trunk with the facial artery.

(2) It courses deeply to the posterior belly of the digastric muscle and continues deeply to the hyoglossus muscle to enter the tongue through the root.

(3) Branches

(a) The **tonsillar branch** supplies the palatine tonsil and the pharyngeal portions of the tongue.

(b) The **dorsal lingual branch** supplies the posterior portions of the tongue.

(c) The **deep lingual branch**, as the name implies, supplies the deep portions of the central part of the tongue.

(d) The **sublingual branch** passes along the genioglossus muscle to supply the anterior portion of the tongue.

b. Venous return. The **lingual vein** drains most of the tongue.

(1) It is formed by like-named veins, which drain along the arterial branches.

(2) The lingual vein usually drains into the retromandibular vein but may drain directly into the external jugular vein.

c. Lymphatic drainage is to the deep cervical nodes but takes three routes from different regions of the tongue. The lymphatic drainage of the lips and tongue is an important consideration in the spread of tobacco-associated carcinoma of these structures.

(1) The **posterior two-thirds** of the tongue drains unilaterally into the deep cervical nodes via the **jugulodigastric nodes**, which lie at the point where the internal jugular vein is crossed by the posterior belly of the digastric muscle.

(2) The **marginal portions of the anterior two-thirds** of the tongue drain unilaterally through the mylohyoid muscle to the **submandibular nodes**, which in turn drain through the **juguloomohyoid nodes** into the deep cervical nodes.

(3) The **central portions of the anterior two-thirds** of the tongue drain **bilaterally** through the mylohyoid muscle to the **submental nodes** and the **juguloomohyoid nodes**, which drain into the deep cervical nodes.

4. Innervation of the tongue is by branches of the trigeminal, facial, glossopharyngeal, and hypoglossal nerves (see Figs. 34-4*A* and 34-6).

a. The lingual nerve (CN V$_3$), a portion of the post-trematic nerve of the first arch, supplies branchiomeric sensation for the anterior two-thirds of the tongue.

b. The chorda tympani, the pretrematic division of the **facial nerve** (CN VII), the nerve of the second branchial arch, carries taste and secretomotor fibers for the submandibular gland, the sublingual gland, and the intrinsic glands (of Nuhn) of the anterior two-thirds of the tongue.

c. The glossopharyngeal nerve (CN IX) supplies branchiomeric sensation and taste for the posterior third of the tongue.

d. The hypoglossal nerve (CN XII) supplies general somatic motor innervation to the tongue. This nerve innervates the hyoglossus, genioglossus, and styloglossus muscles, as well as the intrinsic muscles of the tongue.

E. Submandibular and sublingual salivary glands

1. Boundaries

a. The submandibular gland is folded around the posterior edge of the mylohyoid muscle, so that a portion lies within the floor of the oral cavity and a portion is external to the oral cavity.

(1) The submandibular gland is ensheathed by the investing layer of deep cervical fascia.

(2) The **submandibular duct** (of Wharton) runs from the anterior end of the deep portion superficially to the hyoglossus muscle (see Fig. 34-4*A*). It empties into the oral cavity at a small papilla at the side of the frenulum of the tongue.

(3) The lingual nerve winds around the duct from lateral to medial.

b. The sublingual gland surrounds the terminal portion of the submandibular duct.

(1) It empties directly into the floor of the mouth by 10–20 short ductules.

(2) A few ductules also drain via the submandibular duct.

2. Glandular innervation. These glands are innervated by parasympathetic secretomotor fibers from the facial nerve (CN VII).

   a. Parasympathetic fibers from the facial nerve join the lingual nerve via the chorda tympani.

   b. The submandibular ganglion is attached to the lingual nerve by a few short connecting branches.

   c. The presynaptic fibers synapse in the submandibular ganglion and return to the lingual nerve to be distributed to the submandibular and sublingual glands as well as the intrinsic glands of the anterior tongue.

IV. PHARYNX

A. Basic structure. The pharynx extends between the rami of the mandible and the base of the cranium.

   1. Divisions. The upper end of the pharynx may be divided in the plane of the palate into the **nasopharynx** and the **oropharynx**, which open anteriorly into the nasal and oral cavities, respectively. It continues inferiorly as the **hypopharynx** into the larynx and esophagus. The walls consist of mucosa and striated muscle.

   2. Relations. On either side of the pharynx, the muscles of mastication pass in the infratemporal fossa from the cranium to the mandible. Posteriorly, the pharynx is apposed to the prevertebral fascia of the somatic neck.

B. Nasopharynx

   1. The nasal choanae mark the beginning of the nasopharynx (Fig. 34-5). The floor of the nasopharynx is formed by the soft palate. The posterior and superior aspects of the nasopharynx are related to the basilar occipital bone and the arch of the atlas.

   a. The pharynx narrows at the level at which the soft palate establishes the velopharyngeal seal. This section of the pharynx is called the **nasopharyngeal isthmus**.

   b. When the pharyngeal musculature is drawn superiorly in deglutition, a slightly posterior ridge (of Passavant) helps close this seal.

   2. The auditory (pharyngotympanic) tube opens on the lateral wall of the nasopharynx at the level of the inferior meatus of the nasal cavity (see Fig. 34-5).

   a. The cartilaginous wall of the tube raises a **tubal elevation** (torus tubarius).

   b. A small aggregation of lymphoid tissue forms the **tubal tonsil** in this region. Hypertrophy or edema of the tubal tonsil may occlude the auditory tube with accumulation of secretions in the middle ear.

   c. Infection in the nasopharynx may track along the auditory tube to produce *otitis media*.

   d. The auditory tube is a remnant of the first branchial pouch.

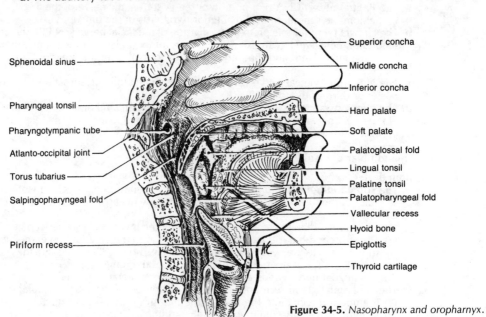

Figure 34-5. *Nasopharynx and oropharnyx.*

3. **The salpingopharyngeus muscle** (*salpynx*, G. horn) originates from the end of the auditory tube and inserts into the musculature of the pharynx (see Fig. 34-6).
   a. It acts to raise the pharynx during swallowing.
   b. The **salpingopharyngeal fold** overlies this muscle (see Fig. 34-5).
   c. It is innervated by the pharyngeal branch of the vagus nerve (CN X).

4. **The pharyngeal tonsil (adenoid)** is formed by lymphoid tissue embedded in the posterior wall of the nasopharynx (see Fig. 34-5). Hypertrophy of the pharyngeal tonsil (swollen adenoids) may interfere with nasal respiration and phonation.

C. Oropharynx

1. **The faucial pillars** (arches or folds) bound the oropharyngeal isthmus.
   a. The palatoglossus muscles raise palatoglossal folds anteriorly on the lateral wall of the oropharynx (see Figs. 34-2 and 34-5).
   b. The palatopharyngeus muscles raise palatopharyngeal folds posteriorly on the lateral wall of the oropharynx (see Figs. 34-2 and 34-5).

2. **The tonsillar fossae** lie between the diverging fauces on each side.
   a. This triangular fossa contains a mass of lymphoid tissue, the **palatine tonsil** (see Fig. 34-5).
      (1) The palatine tonsil extends from the base of the tongue to the edge of the soft palate.
      (2) The medial surface has a superior **intratonsillar cleft** and 12–15 **tonsillar crypts** that extend deeply into the lymphoid tissue.
      (3) The lateral surface has a fibrous **tonsillar capsule**.
         (a) This capsule limits the peritonsillar space.
         (b) The capsule is easily separated from the pharyngeal wall in all places except at the root of the tongue. The tonsillar arteries enter the gland at this point.
   b. Not only does the tonsillar fossa lie between the palatine muscles, but it is also related to the space between the superior and middle pharyngeal constrictors (Fig. 34-6).
      (1) Posteriorly, the gap is limited by the origin of the middle constrictor and the stylohyoid ligaments from which it arises.
      (2) Anteriorly, the hyoglossus muscle ascends to the lateral surface of the tongue.
      (3) The stylopharyngeus muscle carries the glossopharyngeal nerve between the superior and middle constrictors.
         (a) Therefore, both structures are laterally related to the tonsil. The nerve is at risk during tonsillectomy.
         (b) If the styloid process is unusually long (4% of individuals), a swollen tonsil may compress the glossopharyngeal nerve against the bone with pain referred to the pharynx and ear.
   c. The tonsillar fossa represents the second branchial pouch.

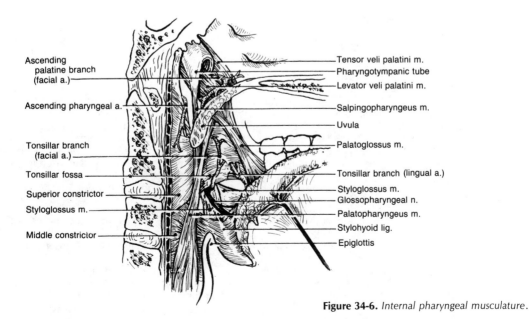

Figure 34-6. *Internal pharyngeal musculature.*

3. Tonsillar vasculature

a. Arterial supply. The tonsil is supplied by tonsillar branches from several sources (see Fig. 34-6), including the:

(1) Lesser palatine branch of the maxillary artery

(2) Ascending palatine branch of the facial artery

(3) Tonsillar branch of the lingual artery

(4) Ascending pharyngeal artery

(5) Tonsillar branch of the facial artery is the primary source

b. Venous return. A **peritonsillar venous plexus** is drained by veins that parallel the arterial branches.

(1) The principal drainage is by the tonsillar branch of the lingual vein.

(2) However, any of the veins that drain the tonsil may be quite large and be a source of a profuse venous hemorrhage after tonsillectomy.

4. Waldeyer's ring, a ring of lymphoid tissue extending around the pharynx, includes the pharyngeal, tubal, palatine, and lingual tonsils.

a. The adenoids hypertrophy and regress by 8 years.

b. The palatine tonsils hypertrophy somewhat later and regress by puberty.

c. The lingual tonsils enlarge at the time of puberty and regress very little during adult life.

5. Tonsillectomy. Because the palatine tonsils usually regress, the current treatment of choice for **tonsillitis** is conservative (antibiotic) therapy. Tonsillectomy is indicated in persistent infection, particularly if complicated by sinusitis, otitis media, or direct spread into the loose tissue of the pharynx (quinsy).

D. Hypopharynx

1. The base of the tongue forms the anterior wall.

2. The epiglottis marks the upper surface of the larynx and guards the opening into the larynx (see Figs. 34-6 and 34-9A).

a. A median glossoepiglottic fold extends from the base of the tongue to the epiglottis with a **vallecular recess** to each side.

b. Lateral glossoepiglottic (pharyngeoepiglottic) folds run from the anterolateral walls of the hypopharynx to the base of the epiglottis and define the inferolateral boundaries of the vallecular recesses.

c. The vallecular recesses represent the remnants of the third branchial pouches.

3. The piriform recesses of the hypopharynx extend inferiorly to the lateral glossoepiglottic folds on either side of the larynx (see Fig. 34-5).

a. These dilate considerably. Food and liquid are diverted to either side of the larynx into these recesses upon deglutition.

b. Swallowed foreign bodies may lodge in these recesses.

E. Pharyngeal musculature consists of three overlapping constrictors and three diagonal muscles (see Fig. 34-6; Fig. 34-7; see Table 34-1).

1. The superior constrictor lies deeply to the ramus of the mandible in the infratemporal fossa.

a. Anteriorly, it is attached to the pterygoid plate, the pterygomandibular raphe, which it shares with the buccinator muscle, and the posterior portions of the maxillary and mandibular alveolar processes.

b. Posteriorly, it attaches to the pharyngeal tubercle of the basioccipital bone as well as to the posterior pharyngeal raphe.

c. The lateral superior edge is free and does not meet the cranium.

(1) The auditory tube passes through this hiatus to open into the nasopharynx.

(2) The tensor veli palatini muscle descends vertically from the auditory tube, lateral to this hiatus, to reach the hamulus; it then turns medially through the hiatus to insert into the soft palate.

(3) The levator veli palatini muscle remains deep to the superior constrictor.

2. The middle constrictor is fan-shaped.

a. Anteriorly, it arises from the stylohyoid ligament and the greater and lesser horns of the hyoid bone.

b. Its fibers pass posteriorly, external to those of the superior constrictor, to insert into the pharyngeal raphe.

c. A gap exists between the lowest origin of the superior constrictor and the uppermost origin of the middle constrictor. The stylopharyngeus muscle, the pharyngeal branch of the glossopharyngeal nerve, and the tonsillar branch of the facial artery pass through this gap.

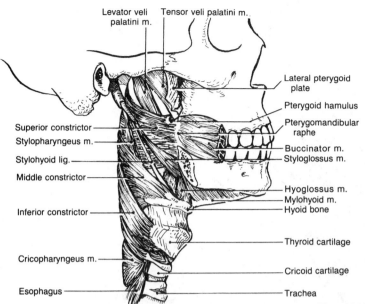

Figure 34-7. *External pharyngeal musculature.*

3. **The inferior constrictor** is fan-shaped superiorly but tubular inferiorly.
   a. It arises from the oblique line on the lamina of the thyroid cartilage and from the cricoid cartilage.
   b. The fibers pass posteriorly, external to the middle constrictor, to insert into the pharyngeal raphe.
   c. The cricoid portion of this muscle (the **cricopharyngeus muscle**) functions as a sphincter at the superior end of the esophagus. Failure of the cricopharyngeus to relax during swallowing occasionally causes the mucosa to herniate through the inferior constrictors, forming a pharyngeal diverticulum.

4. **The palatopharyngeus muscle** underlies the posterior faucial pillar (palatopharyngeal fold) (see Fig. 34-6).
   a. It arises laterally from the palate and inserts into the pharyngeal musculature with some fascicles inserting onto the posterolateral border of the thyroid cartilage.
   b. It acts as a sphincter (in conjunction with the base of the tongue) between the oral cavity and the pharynx. It helps to raise the larynx in swallowing (deglutition).
   c. It is innervated by the vagus nerve (CN X) through the pharyngeal plexus.

5. **The salpingopharyngeus muscle** underlies the **salpingopharyngeal fold** (see Fig. 34-6).
   a. It originates from the end of the auditory tube at the torus tubarius and inserts into the musculature of the midpharynx.
   b. It acts to raise the pharynx and larynx during swallowing.
   c. It is innervated by the vagus nerve (CN X) through the pharyngeal plexus.

6. **Stylopharyngeus muscle** (see Fig. 34-4*B*)
   a. It originates from the styloid process. It passes through the hiatus between the superior and middle constrictors to interdigitate with the pharyngeal musculature with some fascicles inserting onto the posterolateral border of the thyroid cartilage.
   b. It raises the pharynx during deglutition.
   c. It is the only muscle innervated by the glossopharyngeal nerve (CN IX).

F. **Innervation of the pharynx** is by the glossopharyngeal and vagus nerves.

1. **The glossopharyngeal nerve** (CN IX) comprises the pretrematic and post-trematic portions of the nerve of the third branchial arch. It has fibers belonging to all of the functional categories except somatic efferent (Fig. 34-8).
   a. The glossopharyngeal nerve passes out of the cranium in a deep groove on the medial margin of the **jugular foramen.**

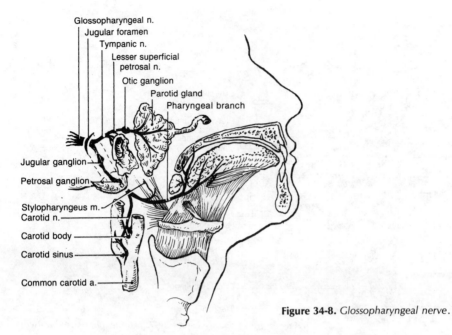

Figure 34-8. *Glossopharyngeal nerve*.

(1) The pretrematic **tympanic branch** conveys sensory neurons to the ear and parasympathetic neurons continue along the **lesser superficial petrosal nerve** to the **otic ganglion** for parotid gland secretion.

(2) It has a small **superior (jugular) ganglion** and **inferior (petrosal) ganglion** just below the jugular foramen. These contain the cell bodies of the sensory neurons.

   b. Distally it passes laterally between the internal jugular vein and internal carotid artery; it then winds around the stylopharyngeus muscle to enter the pharynx between the superior and middle constrictors.

(1) The **pharyngeal branch** (also pretrematic) supplies branchiomeric sensation from the posterior third of the tongue and the walls of the pharynx.

   (a) This is the afferent limb of the **gag reflex**.

   (i) The integrity of the glossopharyngeal nerve is tested by the examining physician with a wooden tongue depressor.

   (ii) Stimulation in the external auditory meatus may initiate a referred gag reflex.

   (b) The pharyngeal branch also supplies taste sensation from the posterior third of the tongue.

(2) The post-trematic motor branch supplies the stylopharyngeus muscle.

   (a) This seems to be the only branchiomeric muscle supplied by the glossopharyngeal nerve, although some sources also claim the middle constrictor.

   (b) The post-trematic sensory branch is associated with the carotid body and sinus and mediates the **carotid reflex**.

2. The vagus nerve (CN X) comprises pretrematic and post-trematic portions of the nerves of the fourth and sixth branchial arches. It supplies all of the branchiomeric musculature of the pharynx (except the stylopharyngeus muscle) through its pharyngeal branch (see Fig. 34-13).

G. Deglutition (swallowing) may be divided into three phases: oral, pharyngeal, and esophageal.

1. The oral (first) phase is voluntary.

   a. Elevation of hyoid bone by the digastric and mylohyoid muscles raises the tongue to the roof of the mouth.

   b. The intrinsic muscles press the tip of the tongue against the maxillary incisors and the hard palate to squeeze the food toward the pharynx.

   c. The palatoglossus muscle elevates the base of the tongue to squeeze the food through the fauces into the pharynx.

   d. The styloglossus muscle draws the base of the tongue posteriorly, propelling the bolus into the oropharynx.

   e. Together with the elevated and retracted tongue, the palatoglossus and palatopharyngeus muscles close the oropharyngeal isthmus behind the bolus.

2. The pharyngeal phase is also voluntary, but once started, it cannot be comfortably interrupted.

 a. The soft palate is elevated by the action of the levator veli palatini and tensor veli palatini muscles to seal off the nasopharyngeal isthmus.

 b. The superior constrictor, palatopharyngeus, and salpingopharyngeus muscles draw the upper portion of the pharynx upward over the bolus, accentuating Passavant's ridge and thereby reinforcing the velopharyngeal seal.

 c. The concomitant contraction of the stylohyoid and digastric muscles draws the hyoid bone cranially. This action also draws the attached larynx cranially under the tongue so that the epiglottis assumes a more horizontal posture—that is, the voice box (larynx) rises to close itself against the lid (epiglottis).

 d. The arytenoid cartilages are tilted forward and approximated to assist in closing off the larynx, a mechanism, which is itself sufficient to prevent the inhalation of food if the epiglottis is excised.

 e. Sequential contraction of the superior, middle, and inferior pharyngeal constrictors propels the bolus through the piriform recesses to either side of the larynx.

3. The esophageal (final) phase is involuntary.

 a. The inferior constrictor and the cricopharyngeus muscles squeeze the bolus into the esophagus; peristalsis moves the bolus toward the stomach.

 b. The lingual and pharyngeal musculature relax, breaking the oropharyngeal and velopharyngeal seals.

 c. The infrahyoid muscles contract to draw the larynx inferiorly.

  (1) Contraction of the hyoglossus and genioglossus muscles returns the tongue to the floor of the oral cavity.

  (2) With depression of the hyoid bone, the epiglottis assumes a more vertical position, opening the larynx for respiration.

V. LARYNX

A. Overview

1. Basic structure. The larynx is formed by a rigid framework of bones, cartilages, and ligaments.

 a. The hyoid, thyroid, cricoid, and epiglottis are single symmetrical structures.

 b. The arytenoid, corniculate, and cuneiform cartilages are paired.

2. Function. The larynx functions as a compound sphincter (Fig. 34-9). It closes the airway during swallowing and during Valsalva's maneuver (as in coughing, urination, and defecation). With fine motor control, it constricts the airway for phonation.

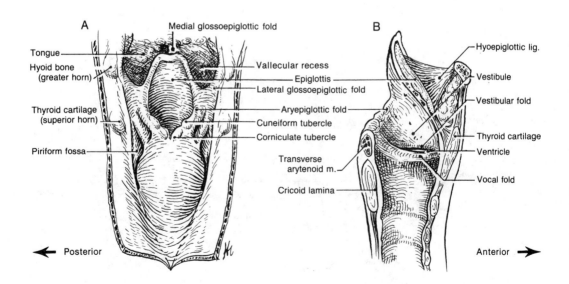

Figure 34-9. *Hypopharynx and larynx. A,* Posterior aspect. *B,* Lateral aspect.

B. Laryngeal ossicles (Fig. 34-10)

1. **The hyoid bone** is U-shaped with a median **body**, paired **lesser cornua (horns)** laterally, and paired **greater cornua (horns)** posteriorly.
   a. The **stylohyoid ligament** runs between the styloid process and the lesser horn.
   b. Numerous muscles are attached along the outer surface of the greater horn and body.
   c. The **thyrohyoid membrane** runs from the smooth medial surface of the hyoid bone to attach along the upper border of the thyroid cartilage (see Fig. 34-10).
      (1) The thyrohyoid membrane is thickened anteriorly to form the **medial thyrohyoid ligament**.
      (2) It is thickened laterally to form the **lateral thyrohyoid ligament**, which frequently contains a small cartilaginous nodule.
      (3) It is pierced superolaterally by the internal branch of the superior laryngeal neurovascular bundle.

2. **The epiglottis** is a leaf-shaped stalk that projects at an angle posterosuperiorly into the hypopharynx (see Fig. 34-9).
   a. It attaches to the anterior inner surface of the thyroid cartilage.
   b. The anterior surface of the epiglottis is joined to the hyoid by the **hyoepiglottic ligament**.
   c. The superior surface of the epiglottis has a **medial glossoepiglottic fold** between two **vallecular recesses**, which are limited inferolaterally by the **lateral glossoepiglottic folds**.
   d. On either side of the laryngeal opening inferior to the lateral glossoepiglottic folds are the **piriform fossae** of the hypopharynx.
   e. The innervation of the epiglottis is divided.
      (1) Branchiomeric sensation from the upper epiglottic surface is carried by the glossopharyngeal nerve.
      (2) The taste buds of the epiglottis are innervated by the internal branch of the superior laryngeal nerve, which arises from the vagus nerve and represents the pretrematic division of the nerve of the fourth branchial arch.
      (3) Branchiomeric sensation from the lower surface of the epiglottis and supraglottic larynx also is carried by the internal branch of the superior laryngeal branch of the vagus nerve. This provides the afferent limb of the **cough reflex**.

3. **The thyroid cartilage** comprises two laminae fused in the anterior midline (see Fig. 34-10A).
   a. **The upper border** has a median **thyroid notch**.
      (1) Below the notch is a **laryngeal prominence**.
      (2) In men, the angle between the laminae is more acute; the thyroid notch and laryngeal prominence (Adam's apple) are more apparent than in women. This results in longer vocal cords that vibrate at a lower frequency.
   b. **The outer surface** is smooth and ends posteriorly on either side at an **oblique line** between the superior and inferior tubercles. The linear origins of the thyrohyoid muscle and inferior

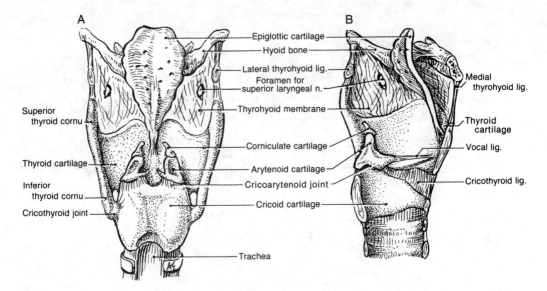

Figure 34-10. *Laryngeal ossicles. A,* Posterior aspect. *B,* Lateral aspect.

constrictor are separated by the insertion of the sternothyroid muscle along the crest of the **oblique line**.

    c. The posterolateral border is thickened for the insertion of the stylopharyngeus and palatopharyngeus muscles.

      (1) Its ends are drawn out into the **superior cornu** and the **inferior cornu**.

      (2) The short, inferior cornu has a synovial articulation with the lamina of the cricoid cartilage. The joint allows a hinge-like rotation about a single axis between the thyroid and cricoid cartilages.

  4. The cricoid cartilage is likened to a signet ring with a broad lamina posteriorly and narrow arch anteriorly (see Fig. 34-10*A*).

    a. The inferior cornua of the thyroid articulate with the lateral surface.

    b. The **lamina** has a median ridge for the tendons of the longitudinal fibers of the esophagus.

    c. The arytenoid cartilages articulate with the upper posterolateral borders of the cricoid cartilage.

    d. A posterior cricoarytenoid ligament attaches to the base of each arytenoid cartilage.

    e. A medial cricothyroid ligament extends between the cricoid and thyroid cartilages anteriorly.

    f. A thick **cricotracheal ligament** binds the cricoid cartilage to the first tracheal ring.

  5. The arytenoid cartilages have a **base**, an **apex**, an **anterior vocal process**, and a **lateral muscular process** (see Fig 34-10*B*).

    a. The base forms a shallow ball-and-socket articulation with the upper border of the cricoid cartilage. There are three axes and, therefore, three degrees of freedom at this joint.

    b. A small detached portion of the apex is the **corniculate cartilage**.

  6. Fibroelastic plates extend between the arytenoid, cricoid, and thyroid cartilages (see Fig. 34-10*B*).

    a. The fibroelastic **quadrangular membrane** runs between the arytenoid cartilages and the epiglottis.

      (1) The free upper border of this membrane forms the **aryepiglottic fold**.

        (a) This fold forms the wall separating the larynx from the piriform recesses of the hypopharynx.

        (b) A small, rod-like **cuneiform cartilage** lies within the free edge of each fold.

      (2) The free lower border of this membrane lies within and forms the horizontal **vestibular fold** (false vocal cord).

      (3) The laryngeal **vestibule** lies between the aryepiglottic folds and the vestibular folds.

    b. Inferior to the **vestibular folds** is a fusiform **laryngeal sinus (ventricle)** [see Figs. 34-9*B* and 34-12*A*].

      (1) The pouch of mucous membrane, the **saccule**, extends superiorly from the anterior end of the sinus.

        (a) It lies between the quadrangular membrane and the small thyroepiglottic muscle.

        (b) Numerous mucous glands open into the saccule, which empty their secretion when the thyroepiglotticus contracts.

      (2) The laryngeal sinus seems to represent the fourth branchial pouch.

C. Intrinsic musculature of the larynx (Fig. 34-11)

  1. Articulations. The movements of the larynx occur at the cricothyroid and cricoarytenoid joints.

    a. Cricothyroid joint. This joint occurs between the inferior horns of the thyroid cartilage and the posterior lateral walls of the cricoid cartilage. Movement at this joint adjusts tension on the vocal cords.

    b. Cricoarytenoid joint. The arytenoid cartilage can slide along the cricoid in abduction and adduction, rotate about a vertical axis, and tilt forward and backward slightly. These movements are conjoined so that the abduction always occurs with external rotation, and adduction with internal rotation.

  2. Musculature. Muscles radiate from the muscular process and posterior surface of the arytenoid (see Fig. 34-11; Table 34-2).

    a. Cricothyroid muscle

      (1) This muscle, crossing the cricothyroid joint, runs between the external surfaces of the thyroid and cricoid arches.

      (2) In contraction, it approximates the anterior edges of the cricoid and thyroid cartilages, thereby stretching and tensing the vocal folds.

      (3) This is the only muscle innervated by the external branch of the superior laryngeal branch of the vagus nerve.

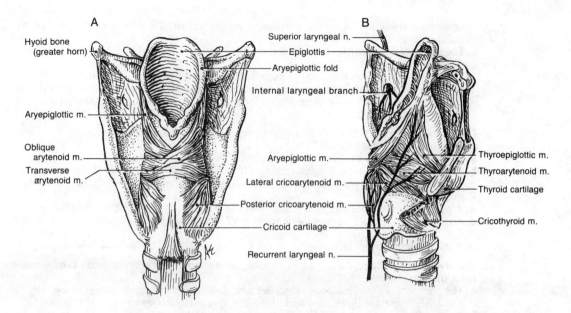

Figure 34-11. *Laryngeal musculature. A,* Posterior aspect, *B,* Lateral aspect.

b. Posterior cricoarytenoid muscle
  (1) Fibers converge to the muscular process of the arytenoid cartilage from a broad origin on the posterior lamina of the cricoid cartilage.
  (2) The horizontal fibers rotate the arytenoid laterally, thereby swinging the vocal process outward to abduct the vocal folds.
  (3) The vertical fibers tilt the arytenoid backwards, thereby tensing the vocal cords.
  (4) This is the only dilator muscle of the rima glottidis.
c. Transverse arytenoid and oblique arytenoid muscles
  (1) These fibers run between the medial surfaces of the arytenoid cartilages and fibers from the oblique arytenoid muscles, which continue to the epiglottis as the **aryepiglotticus muscle**.
  (2) These adduct the arytenoid cartilages, thereby adducting the vocal cords.
d. Lateral cricoarytenoid muscle
  (1) It arises from the outer surface of the cricoid arch and inserts onto the muscular process of the arytenoid cartilage.
  (2) It rotates the arytenoid cartilage medially and tilts it forward, thereby adducting the vocal cords and reducing the tension on the cords.
e. Superior thyroarytenoid muscle
  (1) This muscle passes obliquely from the thyroid cartilage to the aryepiglottic folds and arytenoid cartilage superficial to the quadrangular membrane.
  (2) Superiorly, it is the **thyroepiglotticus muscle** that acts to widen the laryngeal inlet.
  (3) Inferiorly, it is the **thyroarytenoid muscle** (see Fig. 34-11*B*; Fig. 34-12).
    (a) The thyroarytenoid is another internal rotator of the arytenoid cartilage and thereby adducts the vocal folds.
    (b) At the same time, it draws the arytenoids forward, slackening the vocal cords.
    (c) The most medial portion of the thyroarytenoid is termed the **vocalis muscle**.
      (i) Although the vocalis slackens the posterior portion of the cord, it tenses the anterior portion and, thus, raises the pitch of the voice.
      (ii) It also contracts during deglutition to close the rima glottidis, the space between the vocal cords.

D. Laryngeal vasculature

  1. Arterial supply is through laryngeal branches of the superior thyroid arteries (branches of the external carotid) and the inferior thyroid arteries (branches of the thyrocervical trunk of the subclavian artery).

  2. Venous return. The veins follow these arteries.

Table 34-2. Intrinsic Muscles of the Larynx

Muscle	Origin	Insertion	Action	Innervation
Cricothyroid	Cricoid cartilage anteriorly	Thyroid cartilage laterally	Stretches and tenses vocal cords	Superior laryngeal branch of vagus n. (CN X)
Posterior cricoarytenoid	Cricoid cartilage posteriorly	Arytenoid cartilage	Abducts vocal cords	Inferior laryngeal branch of vagus n. (CN X)
Transverse arytenoid	Arytenoid cartilage	Arytenoid cartilage	Adducts vocal cords	Inferior laryngeal branch of vagus n. (CN X)
Lateral cricoarytenoid	Cricoid cartilage laterally	Arytenoid cartilage	Adducts vocal cords	Inferior laryngeal branch of vagus n. (CN X)
Thyroaryte-noid (vocalis)	Thyroid cartilage anteriorly	Arytenoid cartilage	Adducts and tenses vocal cords	Inferior laryngeal branch of vagus n. (CN X)

E. Laryngeal innervation is by vagus nerve (CN X) [Fig. 34-13].

1. **Composition and course.** The **vagus nerve** also has fibers belonging to all of the functional categories except somatic efferent. The vagal trunk passes down the neck between the internal jugular vein and the carotid arteries.

2. **Distribution.** There are numerous vagal branches in the neck (see Fig. 34-13).
   a. **The auricular branch** from the external auditory meatus and a **meningeal branch** provide the majority of the cell bodies in the **superior (jugular) ganglion.**
   b. **The pharyngeal nerve** passes between the carotids and onto the pharynx in the interval between the superior and middle constrictors to contribute motor fibers to the pharyngeal plexus.
   c. **The superior laryngeal nerve** passes deeply to the carotids onto the middle constrictor, where it divides into motor and sensory branches.
      (1) The **internal branch** pierces the thyroid membrane.
         (a) It is sensory between the inferior surface of the epiglottis and the vocal folds and also conveys taste fibers from the epiglottis.

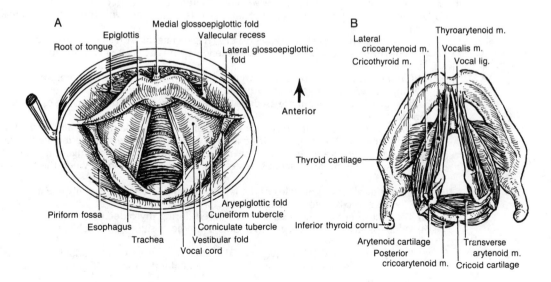

Figure 34-12. *Vocal cords. A,* As seen through a laryngoscope. *B,* Vocalis muscle.

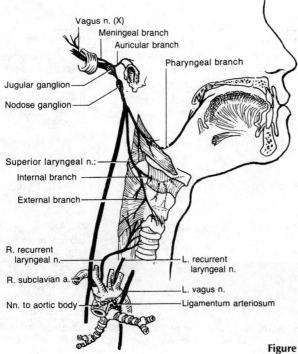

Figure 34-13. *Vagus nerve.*

        (b) It provides the afferent **limb** of the **cough reflex**.
        (c) Its cell bodies lie in the **inferior (nodose) ganglion**.
        (d) It seems to represent pretrematic and post-trematic divisions of the nerve of the fourth branchiomeric arch.
    (2) The **external branch** descends on the surface of the inferior constrictor to the cricothyroid muscle.
        (a) It has a posterior relation to the upper pole of the thyroid gland and must be avoided when clamping the upper vascular peduncle in thyroidectomy.
        (b) This seems to represent the post-trematic division of the nerve of the fourth branchial arch.
  d. The inferior (recurrent) laryngeal nerve takes a different course on each side of the neck.
    (1) During development, the recurrent laryngeal is associated with the sixth branchial arch.
        (a) On the right side, the fifth and sixth arches degenerate, allowing the recurrent laryngeal nerve to pass around the subclavian artery, which is associated with the fourth branchial arch.
        (b) On the left side, the recurrent laryngeal is trapped below the ligamentum arteriosum, the remnant of the artery of the sixth branchial arch.
    (2) Each recurrent laryngeal nerve ascends the neck in the groove between the esophagus and the trachea.
    (3) It is intimately related to the posterior surface of the thyroid gland, lying among the terminal branches of the inferior thyroid artery between two layers of pretracheal fascia.
    (4) At the level of the inferior border of the inferior constrictor (cricopharyngeus muscle), it dives into the larynx and ascends posteriorly to the articulation between the thyroid and cricoid cartilages.
    (5) It is motor to all of the muscles of the larynx except the cricothyroid muscle and sensory to the mucosa below the vocal folds.
        (a) The recurrent nerve is occasionally injured during thyroidectomy.
        (b) When paralyzed, the vocal fold adopts a middle position. This lack of tension produces hoarseness.
        (c) Because the vocal folds have several adductors and only one abductor, laryngospasm apposes the vocal folds. This can be fatal if the patient is not intubated immediately.
  e. The majority of fibers in the vagus are general visceral afferents and efferents associated with the thoracic and abdominal viscera.

F. Phonation and speech

   1. Phonation is accomplished by forcing air between the closely apposed vocal cords (lateral cricoarytenoid and transverse arytenoic muscles), thereby causing the column of air in the larynx to vibrate.

   a. By tensing the cords (cricothyroid muscle), the vibrations become strong and the pitch rises.

   b. By effectively shortening the cords (vocalis muscle), the pitch also rises.

   c. The vibrations are transmitted to the pharyngeal, oral, and nasal passages, as well as to the paranasal, thereby producing resonance.

   2. Speech is accomplished by varying the resonance characteristics of the nasopharynx, oropharnyx, and oral cavity to produce the vowel sounds; interrupting the resonance generally produces the consonants.

   a. The posterior tongue and the palatine musculature change the shape of the oropharynx and intermittently limit the access to the nasopharynx.

   b. The mandibular musculature, the tongue, and the facial musculature about the lips change the shape of the oral cavity, thereby varying the resonance characteristics.

   c. Finally, the tongue, in combination with the palate, teeth, and lips interrupt the resonance.

35
Cranial Nerve Summary

I. SEQUENTIAL ORGANIZATION. The twelve cranial nerves are numbered in the order that they penetrate the dura (Table 35-1).

  A. CN I, olfactory nerve (see Chapter 29 V B 1)

  B. CN II, optic nerve (see Chapter 29 V B 2)

  C. CN III, oculomotor nerve (see Chapter 30 I D 1 a)

  D. CN IV, trochlear nerve (see Chapter 30 I D 1 b)

  E. CN V, trigeminal nerve (see Chapter 32 III B, VII)

  F. CN VI, abducens nerve (see Chapter 30 I D 1 c)

  G. CN VII, facial nerve (see Chapter 32 III A)

  H. CN VIII, vestibulocochlear nerve (see Chapter 29 V B 3)

  I. CN IX, glossopharyngeal nerve (see Chapter 34 IV F 1)

  J. CN X, vagus nerve (see Chapter 34 V E)

  K. CN XI, spinal accessory nerve (see Chapter 29 V D 4)

  L. CN XII, hypoglossal nerve (see Chapter 33 VI A; Chapter 34 III D 4 d)

II. DEVELOPMENTAL ORGANIZATION. The cranial nerves can be grouped by three criteria as follows: first, whether they are special (i.e., special senses, branchiomotor) or general; second, whether they innervate somatic or visceral structures; and third, whether they are afferent or efferent.

  A. Special somatic afferent (SSA)

    1. The olfactory nerve (CN I) mediates smell.
      a. It reaches the nasal olfactory mucosa through the cribriform plate of the ethmoid bone.
      b. It terminates in the olfactory bulb.

    2. The optic nerve (CN II) mediates vision.
      a. It reaches the eyeball through the optic foramen.
      b. It terminates primarily in the lateral geniculate body of the thalamus.

    3. The vestibulocochlear nerve (CN VIII) mediates balance and audition.
      a. It reaches the inner ear through the internal auditory meatus.
      b. It terminates primarily in the medial geniculate body of the thalamus.

  B. General somatic efferent (GSE)

    1. The oculomotor nerve (CN III) innervates several extraocular muscles as well as the iris and ciliary body.
      a. This nerve reaches the orbit through the superior orbital fissure.

    b. The somatic motor component from the oculomotor nucleus innervates the:
- **(1)** Levator palpebrae superioris muscle
- **(2)** Superior rectus muscle
- **(3)** Medial rectus muscle
- **(4)** Inferior rectus muscle
- **(5)** Inferior oblique muscle

    c. The parasympathetic component (GVE) originates in the accessory oculomotor nucleus.
- **(1)** Presynaptic fibers leave the ciliary root of the inferior division.
- **(2)** Synapse occurs in the ciliary ganglion.
- **(3)** Postsynaptic fibers course in the short ciliary nerves to the eyeball.
- **(4)** The parasympathetic component mediates pupillary constriction and accommodation.

2. The trochlear nerve (CN IV) innervates the superior oblique muscle only.
- **a.** It originates in the trochlear nucleus.
- **b.** It enters the orbit through the superior orbital fissure.

3. The abducens nerve (CN VI) innervates the lateral rectus muscle only.
- **a.** It originates in the abducens nucleus.
- **b.** It enters the orbit through the superior orbital fissure.

4. The hypoglossal nerve (CN XII) mediates movement of the tongue.
- **a.** It originates in the hypoglossal nucleus.
- **b.** It leaves the cranium through the anterior condylar (hypoglossal) canal.
- **c.** It innervates the:
  - **(1)** Intrinsic muscles of the tongue
  - **(2)** Styloglossus muscle
  - **(3)** Hyoglossus muscle
  - **(4)** Genioglossus muscle
- **d.** It causes protrusion of the tongue toward the injured side in the presence of a lesion.

C. Mixed branchiomeric nerves

1. The trigeminal nerve (CN V) has sensory and branchiomeric function.
- **a. The ophthalmic division** (CN V$_1$) is sensory to the forehead and orbital area with cell bodies in the trigeminal (semilunar, gasserian) ganglion.
  - **(1)** It enters the orbit through the superior orbital fissure.
  - **(2)** It has the following major branches.
    - **(a)** Lacrimal
    - **(b)** Frontal (which bifurcates into supraorbital and supratrochlear)
    - **(c)** Nasociliary
  - **(3)** It comprises the afferent limb of the corneal blink reflex.
- **b. The maxillary division** (CN V$_2$) is sensory to the mid-face, nose, palate, and maxillary teeth with cell bodies in the trigeminal (semilunar, gasserian) ganglion.
  - **(1)** It exits the cranium through the foramen rotundum and passes through the pterygopalatine fossa.
  - **(2)** It has the following major branches:
    - **(a)** Infraorbital
    - **(b)** Greater palatine
    - **(c)** Nasal
    - **(d)** Anterior, middle, and posterior–superior alveolar
  - **(3)** It mediates the afferent limb of the sneeze reflex.
- **c. The mandibular division** (CN V$_3$) is both sensory and branchiomotor.
  - **(1)** It leaves the cranium through the foramen ovale.
  - **(2)** It is branchiomotor (SVE) to the following muscles.
    - **(a)** Masseter
    - **(b)** Temporalis
    - **(c)** Lateral pterygoid
    - **(d)** Medial pterygoid
    - **(e)** Tensor tympani
    - **(f)** Tensor veli palatini
    - **(g)** Mylohyoid
    - **(h)** Anterior belly of the digastric
  - **(3)** It is sensory to the jaw, mandibular teeth, and anterior two-thirds of the tongue with cell bodies in the trigeminal ganglion.
  - **(4)** It mediates the jaw-jerk reflex.

2. **The facial nerve** (CN VII) has both general and special sensation as well as branchiomeric and parasympathetic function.
   a. It leaves the cranium through the stylomastoid foramen.
   b. It is branchiomotor (SVE) to the following muscles:
      (1) Muscles of facial expression, mediating the efferent limb of the corneal blink reflex
      (2) Stapedius
      (3) Stylohyoid
      (4) Posterior belly of the digastric
   c. It has a parasympathetic component (GVE) that arises from the superior salivatory nucleus and contributes to the nervus intermedius.
      (1) Lacrimation and nasopalatine secretion
         (a) Presynaptic fibers leave the facial nerve at the genu to course in the greater superficial petrosal nerve and then the nerve of the pterygoid canal.
         (b) Synapse occurs in the pterygopalatine ganglion.
         (c) Postsynaptic fibers course along the maxillary nerve and the zygomaticofacial branch to reach the lacrimal branch of the ophthalmic nerve.
         (d) Other postsynaptic fibers join the nasopalatine branch of the maxillary nerve and enter the sphenopalatine foramen to innervate the glands of the nose.
         (e) Still other postsynaptic fibers join the descending palatine branch of the maxillary nerve and pass through the palatine canal to innervate the glands of the palate.
      (2) Salivation
         (a) Presynaptic fibers leave the facial nerve in the vicinity of the stylomastoid foramen as the chorda tympani, which passes through the middle ear and joins the lingual nerve.
         (b) Small rami communicantes convey these fibers to the submandibular ganglion, where they synapse.
         (c) Postsynaptic fibers rejoin the lingual nerve to reach the submandibular, sublingual, and anterior lingual glands.
   d. It has sensory (SVA) fibers that mediate taste from the anterior two-thirds of the tongue via the chorda tympani. The cell bodies of these fibers are located in the geniculate ganglion.

3. **The glossopharyngeal nerve** (CN IX) has somatic, visceral, and special sensation as well as branchiomeric and parasympathetic motor functions.
   a. It leaves the cranium through the jugular foramen.
   b. It is branchiomotor (SVE) to the stylopharyngeus muscle only.
   c. It conveys parasympathetic secretomotor fibers (GVE) to the parotid gland.
      (1) Fibers from the inferior salivatory nucleus, as well as general sensory fibers, leave the nerve in the jugular canal to form the tympanic nerve.
      (2) From the tympanic nerve, fibers run in the lesser superficial petrosal nerve to the otic ganglion.
      (3) The postsynaptic fibers pass along the auriculotemporal branch of the mandibular nerve to the parotid gland.
   d. It has sensory cell bodies in the superior and inferior glossopharyngeal ganglia.
      (1) General sensation (GSA) from the external auditory meatus
      (2) Visceral sensation (GVA) from the posterior third of the tongue and pharynx, mediating the afferent limb of the gag reflex
      (3) Taste (SVA) from the posterior third of the tongue
      (4) Reflex afferents (GVA) from the carotid body and sinus

4. **The vagus nerve** (CN X) has somatic, visceral, and special sensation as well as branchiomeric and parasympathetic motor function.
   a. It leaves the cranium via the jugular foramen.
   b. It is branchiomotor (SVE) to pharyngeal and laryngeal muscles.
      (1) The pharyngeal branch is motor to the following muscles:
         (a) Levator veli palatini
         (b) Uvulae
         (c) Palatopharyngeus
         (d) Palatoglossus
         (e) Salpingopharyngeus
         (f) Superior, middle, and inferior pharyngeal constrictors
      (2) The superior laryngeal branch is motor to the cricothyroid muscle.
      (3) The inferior laryngeal branch is motor to all of the remaining laryngeal muscles.
   c. It provides parasympathetic innervation (GVE) to the thoracic and much of the abdominal viscera.

d. It has sensory cell bodies in the superior and inferior vagal ganglia.
  (1) General sensation (GSA) to the external auditory meatus
  (2) Visceral sensation (GVA) to the larynx and trachea. It mediates the afferent limb of the cough reflex.
  (3) Taste fibers (SVA) from the epiglottis

  5. The spinal accessory nerve (CN XI) does not fall into a clear category.
  a. This nerve leaves the cranium through the jugular foramen.
  b. It innervates the sternomastoid and trapezius.
  c. A portion of the vagus nerve runs with the spinal accessory nerve as the so-called cranial accessory nerve, which innervates laryngeal musculature via the inferior laryngeal nerve.

III. FUNCTIONAL ORGANIZATION. The cranial nerves can be grouped according to their functional components. Conversely, each cranial nerve may be subdivided into its functional components (see Table 35-1).

Table 35-1. Summary of the Cranial Nerves

Functional Component	Branches	Cell Bodies	Course and Cranial Foramen	Distribution	Associated Nucleus
Olfactory Nerve (CN I)					
Afferent SSA	. . .	Olfactory epithelium	Cribriform plate	Olfactory epithelium	Olfactory bulb and tract
Optic Nerve (CN II)					
Afferent SSA	. . .	Bipolar cells of retina	Optic foramen	Rods and cones of nasal and temporal retina	Ganglion cells of retina to lateral geniculate body to occipital lobe
Oculomotor Nerve (CN III)					
Efferent GSE	Superior	Oculomotor nucleus	Superior orbital fissure	Levator palpebrae, superioris muscle, and superior rectus muscle	. . .
	Inferior	Oculomotor nucleus	Superior orbital fissure	Medial rectus muscle, inferior rectus muscle, and inferior oblique muscle	. . .
GVE	Ciliary	Accessory oculomotor nucleus and ciliary ganglion	Superior orbital fissure, motor root, ciliary ganglion, and short ciliary nerves	Ciliary muscle (accommodation) and iris (constriction)	. . .
Trochlear Nerve (CN IV)					
Efferent GSE	. . .	Trochlear nucleus	Superior orbital fissure	Superior oblique muscle	. . .

Continued on next page

Table 35-1. Continued

Functional Component	Branches	Cell Bodies	Course and Cranial Foramen	Distribution	Associated Nucleus
Trigeminal Nerve (CN V)					
Afferent GSA	Ophthalmic: Lacrimal Nasociliary Supratrochlear Supraorbital	Semilunar ganglion	Superior orbital fissure	Forehead, mucous membranes of nasal cavity and sinuses, and conjunctiva (corneal blink reflex)	Trigeminal sensory nucleus
	Maxillary: Infraorbital Nasopalatine Descending palatine Superior alveolar Zygomatico-facial Zygomatico-temporal	Semilunar ganglion	Foramen rotundum	Midface, upper teeth, mucous membranes of nasal cavity, maxillary sinus, and hard palate (sneeze reflex)	Trigeminal sensory nucleus
	Mandibular: Inferior alveolar Lingual Mental Mylohyoid Buccal Auriculotemporal	Semilunar ganglion	Foramen ovale	Mandible, lower teeth, cheeks, and anterior two-thirds of tongue (jaw-jerk reflex)	Trigeminal sensory nucleus
Efferent SVE	Motor root of mandibular n.: Muscular Mylohyoid	Motor nucleus of CN V	Foramen ovale	Muscles of mastication, tensor tympani muscle, tensor veli palatini muscle, mylohyoid muscle, and anterior belly of the digastric muscle (jaw-jerk reflex)	. . .
Abducens Nerve (CN VI)					
Efferent GSE	. . .	Abducens nucleus	Superior orbital fissure	Lateral rectus muscle	. . .
Facial Nerve (CN VII)					
Afferent GSA	Auricular	Geniculate ganglion	Facial canal	External ear	Spinal nucleus of CN V
GVA	Greater superficial petrosal	Geniculate ganglion	Hiatus of the facial canal	Nasal mucosa	Nucleus solitarius

Continued on next page

Table 35-1. Continued

Functional Component	Branches	Cell Bodies	Course and Cranial Foramen	Distribution	Associated Nucleus
SVA	Chorda tympani via lingual (V)	Geniculate ganglion	Petrotympanic fissure, middle ear, facial canal, and internal auditory meatus	Taste, anterior two-thirds of tongue	Nucleus solitarius
Efferent					
GVE	Greater superficial petrosal to nerve of pterygoid canal	Superior salivatory nucleus and pterygopalatine ganglion	Hiatus of the facial canal and pterygoid canal	Secretomotor, lacrimal, nasal, and palatine glands	. . .
	Chorda tympani	Superior salivatory nucleus and submandibular ganglion	Facial canal, middle ear, petrotympanic fissure, and lingual nerve (V)	Secretomotor, submandibular, and intrinsic lingual glands	. . .
SVE	Temporal Zygomatic Buccal Mandibular Cervical	Facial nucleus	Stylomastoid foramen	Muscles of facial expression, stapedius muscle, posterior belly of digastric muscle, and stylohyoid muscle (corneal blink reflex)	. . .
Vestibulocochlear Nerve (CN VIII)					
Afferent					
SSA	Vestibular	Vestibular ganglion thelium	Internal auditory meatus	Utricle, saccule, and semicircular canals	Vestibular nuclei
	Cochlear	Spiral ganglion	Internal auditory meatus	Spiral organ	Cochlea nuclei to medial geniculate body to temporal lobe
Glossopharyngeal Nerve (CN IX)					
Afferent					
GSA	Tympanic	Jugular ganglion (IX)	Tympanic canal	External auditory meatus and middle ear	Spinal nucleus of CN V
GVA	Lingual	Petrosal ganglion	Jugular foramen	Sensation, posterior third of tongue	Nucleus solitarius
	Pharyngeal	Petrosal ganglion	Jugular foramen	Nasopharynx and oropharynx (gag reflex)	Nucleus solitarius
	Carotid	Petrosal ganglion	Jugular foramen	Carotid sinus and body (baroreflex and chemoreflex)	Nucleus solitarius

Continued on next page

Table 35-1. Continued

Functional Component	Branches	Cell Bodies	Course and Cranial Foramen	Distribution	Associated Nucleus
SVA	Lingual	Petrosal ganglion	Jugular foramen	Taste, posterior third of tongue	Nucleus solitarius
Efferent GVE	Tympanic and lesser superficial petrosal	Inferior salivatory nucleus and otic ganglion	Tympanic canal, middle ear, lesser superficial petrosal, and foramen ovale	Secretomotor, parotid gland	. . .
SVE	Stylopharyngeal	Nucleus ambiguus	Jugular foramen	Stylopharyngeus muscle	. . .

Vagus Nerve (CN X)					
Afferent GSA	Auricular	Jugular ganglion (X)	Tympanic canal	External auditory meatus	Spinal nucleus of CN V
GVA	Aortic plexus	Nodose ganglion	Jugular foramen	Aortic body (chemoreceptor reflex)	Nucleus solitarius
	Superior laryngeal: Internal	Nodose ganglion	Jugular foramen	Superior larynx (cough reflex)	Nucleus solitarius
	Inferior laryngeal	Nodose ganglion	Jugular foramen	Inferior larynx	Nucleus solitarius
	Vagal	Nodose ganglion	Jugular foramen	Thoracic and upper abdominal viscera	Nucleus solitarius
SVA	Superior laryngeal: Internal	Nodose ganglion	Jugular foramen	Taste from epiglottis	Nucleus solitarius
Efferent GVE	Vagal	Dorsal motor nucleus of CN X and distal ganglion	Jugular foramen	Visceral smooth muscle and gland control	. . .
SVE	Pharyngeal	Nucleus ambiguus	Jugular foramen	Palate muscles (except tensor veli palatini) and pharyngeal muscles (except stylopharyngeus)	. . .
	Superior laryngeal: External	Nucleus ambiguus	Jugular foramen	Cricothyroid muscle	. . .
	Contributes to inferior laryngeal	Nucleus ambiguus	Jugular foramen	Laryngeal muscles (except cricothyroid muscle)	. . .

Continued on next page

Table 35-1. Continued

Functional Component	Branches	Cell Bodies	Course and Cranial Foramen	Distribution	Associated Nucleus
		Spinal Accessory Nerve (CN XI)			
Efferent					
SVE	Contributes to inferior laryngeal	Nucleus ambiguus	Jugular foramen	Laryngeal muscles (except cricothyroid muscle)	. . .
SSE	Sternomastoid and trapezius	Lateral column of upper cervical cord	Vertebral canal, foramen magnum, and jugular foramen	Sternomastoid muscle and trapezius muscle	. . .
		Hypoglossal Nerve (CN XII)			
Efferent					
GSE	. . .	Hypoglossal nucleus	Hypoglossal (anterior condylar) canal	Intrinsic tongue muscles, genioglossus muscle, hyoglossus muscle, and styloglossus muscle	. . .

SSA = special somatic afferent; GSE = general somatic efferent; GSA = general somatic afferent; SVE = special visceral efferent; GVA = general visceral afferent; SVA = special visceral afferent; and GVE = general visceral efferent.

Part IX Facial Cranium and Visceral Neck

STUDY QUESTIONS

Directions: Each question below contains five suggested answers. Choose the **one best** response to each question.

1. The buccal area has a dual nerve supply. Which of the following statements concerning the two buccal nerves is true?

(A) The fibers of one nerve are associated with the geniculate ganglion, while those of the other are not associated with any ganglion
(B) The fibers of one nerve are associated with the submandibular ganglion, while those of the other are associated with the pterygopalatine ganglion
(C) The fibers of one nerve are associated with the trigeminal ganglion, while those of the other are associated with the geniculate ganglion
(D) The fibers of one nerve are associated with the trigeminal ganglion, while those of the other are associated with the pterygopalatine ganglion
(E) The fibers of one nerve are associated with the trigeminal ganglion, while those of the other are not associated with a ganglion

2. Anesthesia of the maxillary premolar teeth can be effected by infiltrating the nerve as it leaves the

(A) foramen rotundum
(B) greater palatine foramen
(C) incisive canal
(D) infraorbital foramen
(E) lesser palatine foramen

3. The tissues of the hard and soft palates receive an autonomic neural innervation that is described by all of the following statements EXCEPT

(A) preganglionic parasympathetic fibers travel along the greater superficial petrosal nerve, a branch of CN VII
(B) postganglionic sympathetic fibers arrive via the deep petrosal nerve
(C) the lesser superficial petrosal nerve contributes to the nerve of the pterygoid canal
(D) the greater and lesser palatine nerves pass through the pterygopalatine canal
(E) the anterior portion of the hard palate is supplied by the nasopalatine nerve, which enters the nose through the sphenopalatine foramen

4. The epiglottis is described by all of the following statements EXCEPT that

(A) during swallowing, the epiglottis becomes horizontal to close the laryngeal aditus
(B) it contains taste buds innervated by the vagus nerve
(C) it is connected to the root of the tongue
(D) the piriform recesses lie on either side of it in the hypopharynx
(E) two lateral aryepiglottic folds connect it to the laryngeal cartilages

Directions: Each question below contains four suggested answers of which **one or more** is correct. Choose the answer

A if **1, 2, and 3** are correct
B if **1 and 3** are correct
C if **2 and 4** are correct
D if **4** is correct
E if **1, 2, 3, and 4** are correct

5. Signs and symptoms produced by a fracture passing through the left stylomastoid foramen and injuring the contained nerve include which of the following?

(1) Hyperacusis in the left ear
(2) Loss of lacrimation on the left side
(3) Loss of left parotid gland secretion
(4) Facial palsy

6. Which of the following statements about the carotid sinus and carotid body are correct?

(1) The body functions to regulate cardiac and respiratory rates
(2) They are found in the bifurcation of the common carotid artery
(3) The sinus functions to regulate cardiac and respiratory rates
(4) The carotid sinus is a pressure receptor and the carotid body a chemoreceptor

7. When cell bodies in the left semilunar ganglion are damaged by a viral infection, the signs and symptoms include

(1) loss of sensation of pain from the anterior two-thirds of the tongue
(2) loss of taste from the anterior two-thirds of the tongue
(3) inability to elicit a corneal blink reflex from the left side
(4) left facial paralysis

8. The articular disk, or meniscus, of the temporomandibular joint is characterized by

(1) a fibrocartilage composition
(2) an attachment from the medial pterygoid muscle
(3) a downward and forward movement during protrusion of the mandible
(4) separation of the medial and lateral joint compartments

9. An infection that spreads into the pterygopalatine fossa from the pterygoid venous plexus may subsequently track into which of the following cavities?

(1) Nasal cavity
(2) Middle cranial fossa
(3) Oral cavity
(4) Orbital cavity

10. Symptoms that result from the administration of local anesthetic into the greater palatine canal far enough to reach the ganglion situated superiorly to the canal include

(1) dry nasal mucosa from loss of secretion of the nasal glands
(2) dry mouth from loss of secretion of the parotid gland
(3) dry eyes from loss of secretion of the lacrimal glands
(4) dry mouth from loss of secretion of the submandibular and sublingual glands

11. Structures that drain into the deep cervical lymph nodes include the

(1) occipital region
(2) nasal sinuses
(3) parotid gland
(4) teeth and gingivae

12. Muscles that are innervated by the ansa cervicalis include the

(1) omohyoid
(2) thyrohyoid
(3) sternohyoid
(4) geniohyoid

13. Muscles that remain functional when the hypoglossal nerve deteriorates from a tumor in the brain stem include the

(1) genioglossus
(2) palatoglossus
(3) hyoglossus
(4) geniohyoid

14. Structures that drain into the middle nasal meatus include the

(1) frontal sinus
(2) posterior ethmoid sinuses
(3) maxillary sinus
(4) nasolacrimal duct

15. Sinuses that drain by gravity when the head is erect include the

(1) frontal sinus
(2) sphenoid sinus
(3) ethmoid sinuses
(4) maxillary sinus

16. The palatine tonsil receives blood supply from the

(1) facial artery
(2) ascending pharyngeal artery
(3) maxillary artery
(4) lingual artery

17. Correct statements concerning the inferior laryngeal nerve include which of the following?

(1) It has an internal branch that conveys sensation from the larynx superior to the vocal cords
(2) It produces muscle contraction that lengthens the (true) vocal folds
(3) It conveys taste from the epiglottis
(4) It innervates all of the laryngeal musculature by an external branch except the cricothyroid muscle

Directions: The groups of questions below consist of lettered choices followed by several numbered items. For each numbered item select the **one** lettered choice with which it is **most closely** associated. Each lettered choice may be used once, more than once, or not at all.

Questions 18–23

For each reflex below, select the nerve that mediates the afferent limb of that reflex.

(A) CN V_1
(B) CN V_2
(C) CN V_3
(D) CN IX
(E) CN X

18. Carotid reflex

19. Corneal blink reflex

20. Cough reflex

21. Gag reflex

22. Jaw-jerk reflex

23. Sneeze reflex

Questions 24–28

For each muscle listed below, select the branchial arch from which it is derived.

(A) First
(B) Second
(C) Third
(D) Fourth
(E) Sixth

24. Orbicularis oris muscle

25. Stylopharyngeus muscle

26. Stapedius muscle

27. Vocalis (thryoarytenoid) muscle

28. Lateral pterygoid muscle

Questions 29–32

Deglutition (swallowing) is a complex and highly coordinated activity involving several cranial nerves and their muscles. Match the function described in each question with the innervation from the list below.

(A) CN V
(B) CN VII
(C) CN IX
(D) CN X
(E) CN XII

29. A muscle that keeps food in the oral cavity

30. A muscle that retracts the tongue to form the oropharyngeal seal

31. A muscle that elevates the soft palate to form the nasopharyngeal seal

32. Muscles that enable the licking of the lips

Questions 33–36

To exit the cranial cavity, each cranial nerve passes through a foramen or fissure which may be involved in a cranial fracture. For each cranial nerve injury described below, select the foramen or fissure that is most likely to be involved in a cranial fracture.

(A) Foramen ovale
(B) Jugular foramen
(C) Posterior condylar canal
(D) Stylomastoid foramen
(E) None of the above

33. The tongue deviates to the left upon protrusion

34. The uvula deviates to the right when the palate is raised

35. The left eye cannot blink

36. The jaw deviates to the left upon protrusion

Questions 37–40

For each of the following cell bodies listed below, select the ganglion that is most appropriately associated with it.

(A) Ciliary ganglion
(B) Geniculate ganglion
(C) Otic ganglion
(D) Trigeminal ganglion
(E) None of the above

37. Contains cell bodies for taste afferent

38. Contains cell bodies for the carotid reflex afferent

39. Contains cell bodies for the jaw-jerk reflex

40. Contains cell bodies for conjunctival sensation

ANSWERS AND EXPLANATIONS

1. The answer is E. [*Chapter 32 III A 2 d (2) (c), B 2 c (2) (b); Chapter 35 Table 35-1*] The buccal branch of the facial nerve is motor to the muscles of facial expression, including the buccinator muscle. These fibers originate in the facial nucleus of the brain stem and are not associated with a ganglion. The buccal branch of the trigeminal nerve is sensory to the cheek and has its cell bodies located in the trigeminal (semilunar) ganglion.

2. The answer is A. [*Chapter 32 VII A 2 f, 3 b*] The maxillary premolar teeth are innervated by the middle–superior alveolar nerves. These teeth can be anesthetized effectively by infiltrating the maxillary nerve as it exits through the foramen rotundum into the pterygopalatine fossa.

3. The answer is C. [*Chapter 32 V C 1 b; VII C 1 a–c*] The lesser superficial petrosal nerve, a continuation of the tympanic branch of the glossopharyngeal nerve, conveys parasympathetic preganglionic fibers to the otic ganglion for parotid secretion. The greater superficial petrosal nerve, a branch of the facial nerve, and the deep petrosal nerve convey the parasympathetic parasynaptic and sympathetic postsynaptic fibers, respectively, to the pterygopalatine fossa.

4. The answer is D. [*Chapter 34 IV D 2, 3; V B 2 a–e*] The vallecular recesses lie to either side of the epiglottis. The piriform recesses lie inferiorly to the epiglottis and caudally to the lateral glossoepiglottic folds, which connect the epiglottis to the tongue.

5. The answer is D (4). [*Chapter 32 III A 2 c, 3 a, b*] That portion of the facial nerve that leaves the stylomastoid foramen is motor to the muscles of facial expression, so that injury here will cause facial paralysis only. The branch of the facial nerve to the stapedius muscle comes off higher in the facial canal, and the nerve to the lacrimal gland arises at the genu. The parotid gland is innervated by the tympanic branch of the glossopharyngeal nerve.

6. The answer is E (all). [*Chapter 34 IV F 1 b (2) (b); Chapter 35 II C 3 d (4)*] The carotid body and sinus, located at the bifurcation of the common carotid artery, monitor the partial pressure of the dissolved oxygen in the blood and the blood pressure, respectively. They are innervated by the carotid branch of the glossopharyngeal nerve, which functions as the afferent limb of a reflex that controls the heart and respiratory rates.

7. The answer is B (1, 3). [*Chapter 29 V H 1 a; Chapter 32 III A 2 b (2) (d), 3 a, B 2*] The semilunar or trigeminal ganglion, the equivalent of a dorsal root ganglion, contains the cell bodies of the sensory neurons of the trigeminal nerve (CN V). Selective damage to these neurons results in loss of sensation from the face and anterior two-thirds of the tongue. The motor division of the facial nerve supplies the muscles of facial expression and taste from the anterior two-thirds of the tongue.

8. The answer is B (1, 3). [*Chapter 32 VI C 1 b, 2 a, b*] The articular disk of the temporomandibular joint separates the superior and inferior compartments and receives the superior head of the lateral pterygoid muscle—the jaw protruder muscle. The inferior head inserts primarily onto the mandibular condyle and neck. The medial pterygoid muscle inserts onto the mandibular at the angle.

9. The answer is E (all). [*Chapter 32 VI D 2*] The branches of the nerves and vessels of the pterygopalatine fossa reach the nose, eye, and mouth through several foramina. The pterygopalatine fossa communicates with the middle cranial fossa via the foramen rotundum and by the vidian canal, with the nasal cavity via the sphenopalatine foramen, with the orbital cavity via the inferior orbital fissure, and with the oral cavity via the palatine canal.

10. The answer is B (1, 3). [*Chapter 32 VII A 3 b, C 2*] The postsynaptic parasympathetic neurons of the pterygopalatine ganglion control secretion of the nasal and oral mucosa as well as lacrimation. Injection of anesthetic into the pterygopalatine fossa, either by the inferior approach through the palatine canal or by the lateral approach through the pterygomaxillary fissure, anesthetizes the ganglion as well.

11. The answer is C (2, 4). [*Chapter 33 V D 1 a, b*] The lymphatic drainage of the anterior–inferior portion of the face, the nasal cavities, and the anterior portion of the oral cavity, including the anterior margin of the tongue, gingivae, and teeth, is through the submandibular lymph nodes to the deep cervical nodes. The lymphatic drainage of the occipital region, the external ear, the parotid gland, and the anterior–superior portion of the face is toward the superficial cervical lymph nodes.

12. The answer is E (all). [*Chapter 33 VI B, 2 a, b*] The superior ramus of the ansa cervicalis, arising from C1 and accompanying the hyoglossal nerve, innervates the geniohyoid and thyrohyoid muscles. The inferior ramus, arising from C1 and C3, innervates the omohyoid and sternothyroid muscles.

13. The answer is C (2, 4). [*Chapter 33 VI B 2; Chapter 34 III D 4 d*] The hypoglossal nerve (CN XII) innervates the somatic muscles of the tongue. The geniohyoid muscle is innervated by nerves that arise from C1 of the cervical plexus and along the distal portion of the hyoglossal nerve. Thus, a brain stem lesion in the vicinity of the hypoglossal nucleus would not affect the geniohyoid innervation. The palatoglossus muscle is innervated by the branchiomeric motor division of the vagus nerve (CN X).

14. The answer is B (1, 3). [*Chapter 34 II A 4 c*] The maxillary, frontal, and anterior ethmoid sinuses, as well as the middle ethmoid sinus, drain into the hiatus semilunaris within the middle nasal meatus. The sphenoid sinus and posterior ethmoid sinuses drain into the sphenoethmoid recess. The nasolacrimal duct drains into the inferior meatus.

15. The answer is B (1, 3). [*Chapter 34 II E 3 b, 4 b*] The frontal and ethmoid sinuses drain by gravity with the head erect. The maxillary and sphenoid sinuses drain when the head is flexed forward or when the body is prone.

16. The answer is E (all). [*Chapter 34 IV C 3 a*] The blood supply to the palatine tonsil is derived from several sources, including two branches of the facial artery (tonsillar and ascending palatine), the tonsillar branch of the lingual artery, the ascending pharyngeal artery, and the lesser palatine branch of the maxillary artery.

17. The answer is D (4). [*Chapter 34 V E 2 d (5)*] The recurrent (inferior) laryngeal nerve, a branch of the vagus nerve (CN X), innervates all of the muscles of the larynx except the cricothyroid muscle, which is innervated by the external branch of the superior laryngeal nerve and functions to lengthen the vocal folds. Taste from the epiglottic taste buds is conveyed by the superior laryngeal branch.

18–23. The answers are: 18-D, 19-A, 20-E, 21-D, 22-C, 23-B. [*Chapter 32 III B 2 a (3), b (3), c (2) (d) (ii); Chapter 33 V B 1 c (1) (a); Chapter 34 IV F 1 b (1) (a), (2) (b); V E 2 c (1) (b)*] The trigeminal nerve mediates three reflexes: The ophthalmic division provides the afferent limb of the corneal blink reflex; the maxillary division, the afferent limb of the sneeze reflex; and the mandibular division, both limbs of the jaw-jerk reflex. The glossopharyngeal nerve provides the afferent limbs of the gag and carotid reflexes. The vagus nerve provides the afferent limb of the cough reflex.

24–27. The answers are: 24-B, 25-C, 26-B, 27-D, 28-A. [*Chapter 31 I B 1 a–e*] The muscles of facial expression, including the orbicularis oris, as well as the stapedius, stylohyoid, and posterior belly of the digastric are all innervated by the facial nerve (CN VII) and are, thus, second branchial arch derivatives. The stylopharyngeus is the only muscle innervated by the glossopharyngeal nerve (CN IX), the nerve of the third branchial arch. The vocalis (thryoarytenoid) muscle and the intrinsic laryngeal musculature (except the cricothryoid muscle) are sixth branchial arch derivatives and are all innervated by the recurrent laryngeal nerve (a branch of the vagus). The muscles of mastication, including the lateral pterygoid muscle, as well as the tensors tympani and veli palatine, mylohyoid, and posterior belly of the digastric, are all derived from the first branchial arch and are innervated by the mandibular division of the trigeminal nerve (CN V). The cricothryoid muscle, innervated by the superior laryngeal nerve (a branch of the vagus), is a fourth arch derivative. The fifth branchial arch never appears in humans.

29–32. The answers are: 29-B, 30-E, 31-D, 32-E. [*Chapter 32 II A 2, B 1 b; Chapter 34 III B 3 a (2), D 4 a, b*] The orbicularis oris muscle, the principal sphincter of the mouth, is innervated by the facial nerve (CN VII). The styloglossus muscle, innervated by the hypoglossal nerve (CN XII), retracts the tongue to help form the oropharyngeal seal. The hypoglossal nerve also innervates the genioglossus muscle, which protrudes the tongue, and the intrinsic musculature, which provides lingual motility and dexterity. The levator veli palatini muscle, innervated by the pharyngeal branch of the vagus nerve (CN X), raises the palate to form the nasopharyngeal seal during deglutition and phonation.

33–36. The answers are: 33-E, 34-B, 35-D, 36-A. [*Chapter 32 III A 3 b, B 3 d; Chapter 34 III B 3 a (3), D 2 a; Chapter 35 Table 35-1*] The hypoglossal nerve (CN XII) passes through the anterior condylar (hypoglossal) canal en route to the tongue. If the genioglossus muscle is paralyzed, the tongue will deviate toward the injured side upon protrusion. The glossopharyngeal (CN IX), vagus (CN X), and spinal accessory (CN XI) nerves exit the cranial cavity via the jugular foramen. Injury to the vagus nerve at this point results in deviation of the uvula away from the injured side, due to the unopposed action of the contralateral levator veli palatini muscle as well as hoarseness due to paralysis of the intrinsic laryngeal musculature. A fracture involving the stylomastoid foramen can injure the facial nerve with paralysis of the muscles of facial expression on the injured side. Injury to the mandibular division of the trigeminal nerve as it passes through the foramen ovale would paralyze the muscles of mastication with deviation of the jaw toward the injured side due to the unopposed action of the lateral pterygoid muscle.

37–40. The answers are: 37-B, 38-E, 39-D, 40-B. [*Chapter 35 Table 35-1*] Taste afferents, which run from the anterior two-thirds of the chorda tympani, have their cell bodies in the geniculate ganglion. The neurons that mediate the afferent limbs of the carotid and gag reflexes are in the petrosal ganglion of the glossopharyngeal nerve. The trigeminal ganglion contains afferent cell bodies that innervate the face and, therefore, provide the afferent limbs of the blink, sneeze, and jaw-jerk reflexes. The ciliary pterygopalatine, submandibular, and otic ganglia contain only postsynaptic neurons of parasympathetic motor pathways.

Challenge
Exam

Introduction

One of the least attractive aspects of pursuing an education is the necessity of being examined on what has been learned. Instructors do not like to prepare tests and students do not like to take them.

However, students are required to take many examinations during their learning careers, and little if any time is spent acquainting them with the positive aspects of tests and with systematic and successful methods for approaching them. Students perceive tests as punitive and sometimes feel that they are merely opportunities for the instructor to discover what the student has forgotten or has never learned. Students need to view tests as opportunities to display their knowledge and to use them as tools for developing prescriptions for further study and learning.

A brief history and discussion of the National Board of Medical Examiners (NBME) examinations (i.e., Parts I, II, and III and FLEX) are presented in this preface, along with ideas concerning psychological preparation for the examinations. Also presented are general considerations and test-taking tips as well as ways to use practice exams as educational tools. (The literature provided by the various examination boards contains detailed information concerning the construction and scoring of specific exams.)

National Board of Medical Examiners Examinations

Before the various NBME exams were developed, each state attempted to license physicians through its own procedures. Differences between the quality and testing procedures of the various state examinations resulted in the refusal of some states to recognize the licensure of physicians licensed in other states. This made it difficult for physicians to move freely from one state to another and produced an uneven quality of medical care in the United States.

To remedy this situation, the various state medical boards decided they would be better served if an outside agency prepared standard exams to be given in all states, allowing each state to meet its own needs and have a common standard by which to judge the educational preparation of individuals applying for license.

One misconception concerning these outside agencies is that they are licensing authorities. This is not the case; they are examination boards only. The individual states retain the power to grant and revoke licenses. The examination boards are charged with designing and scoring valid and reliable tests. They are primarily concerned with providing the states with feedback on how examinees have performed and with making suggestions about the interpretation and usefulness of scores. The states use this information as partial fulfillment of qualifications upon which they grant licenses.

Students should remember that these exams are administered nationwide and, although the general medical information is similar, educational methodologies and faculty areas of expertise differ from institution to institution. It is unrealistic to expect that students will know all the material presented in the exams; they may face questions on the exams in areas that were covered only superficially in their classes. The testing authorities recognize this situation, and their scoring procedures take it into account.

Scoring the Exams

The diversity of curriculum necessitates that these tests be scored using a criteria-based normal curve. An individual score is based not only on how many questions are answered correctly by a specific student but also on how this one performance relates to the distribution of all scores of the criteria group. In the case of NBME, Part I, the criteria group consists of those students who have completed 2 years of medical training in the United States and are taking the test for the first time and those students who took the test during the previous four June sittings.

Since this test has been constructed to measure a wide range of educational situations, the mean, or average, score generally can be achieved by answering 64% to 68% of the questions correctly. Passing the exam requires answering correctly 55% to 60% of the questions. The competition for acceptance into medical school and the performance levels necessary to stay in school are so high that many students who have always achieved these high levels naturally assume they must perform in a similar fashion and attain equivalent scores on the NBME exams. This is not the case. In fact, among students who are accustomed to performing at levels exceeding 80% to 90%, fewer than 4% taking these tests perform at that high level. Unrealistically high personal expectations leave students psychologically unprepared for these tests, and the anxiety of the moment renders them incapable of doing their best work.

Actually, **most students have learned quite well**, but they fail to display this learning when they are tested because they do not understand the construction, purpose, or scoring procedures of board exams. It is imperative that they understand that they are **not** expected to score as well as they have in the past and that the measurement criteria is group performance, not only individual performance.

While preparing for an exam, it is important that students learn as much as they can about the subject they will be tested on as well as prepare to discover just how much they may not know. Students should study to acquire knowledge, not just to prepare for tests. **For the well-prepared candidate, the chances of passing far exceed the chances of failing**.

Materials Needed for Test Preparation

In preparation for a test, many students collect far too much study material only to find that they simply do not have the time to go through all of it. They are defeated before they begin because either they cannot get through all the material leaving areas unstudied, or they race through the material so quickly that they cannot benefit from the activity.

It is generally more efficient for the student to use materials already at hand; that is, class notes, one good outline to cover or strengthen areas not locally stressed and to quickly review the whole topic, and one good text as a reference for looking up complex material needing further explanation.

Also, many students attempt to memorize far too much information, rather than learning and understanding less material and then relying on that learned information to determine the answers to questions at the time of the examination. Relying too heavily on memorized material causes anxiety, and the more anxious students become during a test, the less learned knowledge they are likely to use.

Positive Attitude

A positive attitude and a realistic approach are essential to successful test taking. If concentration is placed on the negative aspects of tests or on the potential for failure, anxiety increases and performance decreases. A negative attitude generally develops if the student concentrates on "I must pass" rather than on "I can pass." "What if I fail?" becomes the major factor motivating the student to **run from failure rather than toward success**. This results from placing too much emphasis on scores rather than understanding that scores have only slight relevance to future professional performance.

The score received is only one aspect of test performance. Test performance also indicates the student's ability to use information during evaluation procedures and reveals how this ability might be used in the future. For example, when a patient enters the physician's office with a problem, the physician begins by asking questions, searching for clues, and seeking diagnostic information. Hypotheses are then developed, which will include several potential causes for the problem. Weighing the probabilities, the physician will begin to discard those hypotheses with the least likelihood of being correct. Good differential diagnosis involves the ability to deal with uncertainty, to reduce potential causes to the smallest number, and to use all learned information in arriving at a conclusion.

This same thought process can and should be used in testing situations. It might be termed **paper-and-pencil differential diagnosis**. In each question with five alternatives, of which one is correct, there are four alternatives that are incorrect. If deductive reasoning is used, as in solving a clinical problem, the choices can be viewed as having possibilities of being correct. The elimination of wrong choices increases the odds that a student will be able to recognize the correct choice. Even if the correct choice does not become evident, the probability of guessing correctly increases. Just as differential diagnosis in a clinical setting can result in a correct diagnosis, eliminating incorrect choices on a test can result in choosing the correct answer.

Answering questions based on what is incorrect is difficult for many students since they have had nearly 20 years experience taking tests with the implied assertion that knowledge can be displayed only by knowing what is correct. It must be remembered, however, that students can display knowledge by knowing something is wrong, just as they can display it by knowing something is right. **Students should begin to think in the present as they expect themselves to think in the future**.

Paper-and-Pencil Differential Diagnosis

The technique used to arrive at the answer to the following question is an example of the paper-and-pencil differential diagnosis approach.

> A recently diagnosed case of hypothyroidism in a 45-year-old man may result in which of the following conditions?
>
> (A) Thyrotoxicosis
> (B) Cretinism
> (C) Myxedema
> (D) Graves' disease
> (E) Hashimoto's thyroiditis

It is presumed that all of the choices presented in the question are plausible and partially correct. If the student begins by breaking the question into parts and trying to discover what the question is attempting to measure, it will be possible to answer the question correctly by using more than memorized charts concerning thyroid problems.

- The question may be testing if the student knows the difference between "hypo" and "hyper" conditions.
- The answer choices may include thyroid problems that are not "hypothyroid" problems.
- It is possible that one or more of the choices are "hypo" but are not "thyroid" problems, that they are some other endocrine problems.
- "Recently diagnosed in a 45-year-old man" indicates that the correct answer is not a congenital childhood problem.
- "May result in" as opposed to "resulting from" suggests that the choices might include a problem that **causes** hypothyroidism rather than **results from** hypothyroidism, as stated.

By applying this kind of reasoning, the student can see that choice **A**, thyroid toxicosis, which is a disorder resulting from an overactive thyroid gland ("hyper") must be eliminated. Another piece of knowledge, that is, Graves' disease is thyroid toxicosis, eliminates choice **D**. Choice **B**, cretinism, is indeed hypothyroidism, but it is a childhood disorder. Therefore, **B** is eliminated. Choice **E** is an inflammation of the thyroid gland—here the clue is the suffix "itis." The reasoning is that thyroiditis, being an inflammation, may **cause** a thyroid problem, perhaps even a hypothyroid problem, but there is no reason for the reverse to be true. Myxedema, choice **C**, is the only choice left and the obvious correct answer.

Preparing for Board Examinations

1. **Study for yourself.** Although some of the material may seem irrelevant, the more you learn now, the less you will have to learn later. Also, do not let the fear of the test rob you of an important part of your education. If you study to learn, the task is less distasteful than studying solely to pass a test.

2. **Review all areas.** You should not be selective by studying perceived weak areas and ignoring perceived strong areas. This is probably the last time you will have the time and the motivation to review **all** of the basic sciences.

3. **Attempt to understand, not just to memorize, the material.** Ask yourself: To whom does the material apply? When does it apply? Where does it apply? How does it apply? Understanding the connections among these points allows for longer retention and aids in those situations when guessing strategies may be needed.

4. Try to **anticipate questions that might appear on the test.** Ask yourself how you might construct a question on a specific topic.

5. **Give yourself a couple days of rest before the test.** Studying up to the last moment will increase your anxiety and cause potential confusion.

Taking Board Examinations

1. In the case of NBME exams, be sure to **pace yourself** to use the time optimally. As soon as you get your test booklet, go through and circle the questions numbered 40, 80, 120, and 160. The test is constructed so that you will have approximately 45 seconds for each question. If you are at a circled number every 30 minutes, you will be right on schedule. A 2-hour test will have 150–170 questions and a 2½-hour test will have approximately 200 questions. You should use all of your allotted time; if you finish too early, you probably did so by moving too quickly through the test.

2. **Read each question and all the alternatives carefully** before you begin to make decisions. Remember, the questions contain clues, as do the answer choices. As a physician, you would not make a clinical decision without a complete examination of all the data; the same holds true for answering test questions.

3. **Read the directions for each question set carefully.** You would be amazed at how many students make mistakes in tests simply because they have not paid close attention to the directions.

4. It is not advisable to leave blanks with the intention of coming back to answer the questions later. Because of the way board examinations are constructed, you probably will not pick up any new information that will help you when you come back, and the chances of getting numerically off on your answer sheet are greater than your chances of benefiting by skipping around. If you feel that you must come back to a question, mark the best choice and place a note in the margin. Generally speaking, it is best not to change answers once

you have made a decision, unless you have learned new information. Your intuitive reaction and first response are correct more often than changes made out of frustration or anxiety. **Never turn in an answer sheet with blanks**. Scores are based on the number that you get correct; you are not penalized for incorrect choices.

5. **Do not try to answer the questions on a stimulus–response basis.** It generally will not work. Use all of your learned knowledge.

6. **Do not let anxiety destroy your confidence.** If you have prepared conscientiously, you know enough to pass. Use all that you have learned.

7. **Do not try to determine how well you are doing as you proceed.** You will not be able to make an objective assessment, and your anxiety will increase.

8. **Do not expect a feeling of mastery** or anything close to what you are accustomed to. Remember, this is a nationally administered exam, not a mastery test.

9. **Do not become frustrated or angry** about what appear to be bad or difficult questions. You simply do not know the answers; you cannot know everything.

Specific Test-Taking Strategies

Read the entire question carefully, regardless of format. Test questions have multiple parts. Concentrate on picking out the pertinent key words that might help you begin to problem solve. Words such as "always," "all," "never," "mostly," "primarily," and so forth play significant roles. In all types of questions, distractors with terms such as "always" or "never" most often are incorrect. Adjectives and adverbs can completely change the meaning of questions—pay close attention to them. Also, medical prefixes and suffixes (e.g., "hypo-," "hyper-," "-ectomy," "-itis") are sometimes at the root of the question. The knowledge and application of everyday English grammar often is the key to dissecting questions.

Multiple-Choice Questions

Read the question and the choices carefully to become familiar with the data as given. Remember, in multiple-choice questions there is one correct answer and there are four distractors, or incorrect answers. (Distractors are plausible and possibly correct or they would not be called distractors.) They are generally correct for part of the question but not for the entire question. Dissecting the question into parts aids in discerning these distractors.

If the correct answer is not immediately evident, begin eliminating the distractors. (Many students feel that they must always start at option A and make a decision before they move to B, thus forcing decisions they are not ready to make.) Your first decisions should be made on those choices you feel the most confident about.

Compare the choices to each part of the question. **To be wrong**, a choice needs to be incorrect for only part of the question. **To be correct**, it must be **totally** correct. If you believe a choice is partially incorrect, tentatively eliminate that choice. Make notes next to the choices regarding tentative decisions. One method is to place a minus sign next to the choices you are certain are incorrect and a plus sign next to those that potentially are correct. Finally, place a zero next to any choice you do not understand or need to come back to for further inspection. Do not feel that you must make final decisions until you have examined all choices carefully.

When you have eliminated as many choices as you can, decide which of those that are left has the highest probability of being correct. Remember to use paper-and-pencil differential diagnosis. Above all, be honest with yourself. If you do not know the answer, eliminate as many choices as possible and choose reasonably.

Multiple True–False Questions

Multiple true–false questions are not as difficult as some students make them. These are the questions in which you must mark:

A if **1, 2, and 3** are correct,

B if **1 and 3** are correct,

C if **2 and 4** are correct,

D if only **4** is correct, or

E if **all** are correct.

Remember that the name for this type of question is multiple true–false and then use this concept. Become familiar with each choice and make notes. Then concentrate on the one choice you feel is definitely incorrect. If you can find one incorrect alternative, you can eliminate three choices immediately and be down to a fifty–fifty probability of guessing the correct answer. In this format, if choice 1 is incorrect, so is choice 3; they go together. Alternatively, if 1 is correct, so is 3. The combinations of alternatives are constant; they will not be mixed. You will not find a situation where choice 1 is correct, but 3 is incorrect.

After eliminating the choices you are sure are incorrect, concentrate on the choice that will make your final decision. For instance, if you discard choice 1, you have eliminated alternatives A, B, and E. This leaves C (2 and 4) and D (4 only). Concentrate on choice 2, and decide if it is true or false. Rereading and concentrating on choice 4 only wastes time; choice 2 will be the decision maker. (Take the path of least resistance and concentrate on the smallest possible number of items while making a decision.) Obviously, if none of the choices is found to be incorrect, the answer is E (all).

Comparison-Matching Questions

Comparison-matching questions are also easier to address if you concentrate on one alternative at a time. Choose option:

A if the question is associated with **(A) only**,

B if the question is associated with **(B) only**,

C if the question is associated with **both (A) and (B)**, or

D if the question is associated with **neither (A) nor (B)**.

Here again, the elimination of obvious wrong alternatives helps clear away needless information and can help you make a clearer decision.

Single Best Answer–Matching Sets

Single best answer–matching sets consist of a list of words or statements followed by several numbered items or statements. Be sure to pay attention to whether the choices can be used more than once, only once, or not at all. Consider each choice individually and carefully. Begin with those with which you are the most familiar. It is important always to break the statements and words into parts, as with all other question formats. **If a choice is only partially correct, then it is incorrect.**

Guessing

Nothing takes the place of a firm knowledge base, but with little information to work with, even after playing paper-and-pencil differential diagnosis, you may find it necessary to guess at the correct answer. A few simple rules can help increase your guessing accuracy. Always guess consistently if you have no idea what is correct; that is, after eliminating all that you can, make the choice that agrees with your intuition or choose the option closest to the top of the list that has not been eliminated as a potential answer.

When guessing at questions that present with choices in numerical form, you will often find the choices listed in an ascending or descending order. It is generally not wise to guess the first or last alternative, since these are usually extreme values and are most likely incorrect.

Using the Challenge Exam to Learn

All too often, students do not take full advantage of practice exams. There is a tendency to complete the exam, score it, look up the correct answers to those questions missed, and then forget the entire thing.

In fact, great educational benefits could be derived if students would spend more time using practice tests as learning tools. As mentioned earlier, incorrect choices in test questions are plausible and partially correct or they would not fulfill their purpose as distractors. This means that it is just as beneficial to look up the incorrect choices as the correct choices to discover specifically why they are incorrect. In this way, it is possible to learn better test-taking skills as the subtlety of question construction is uncovered.

Additionally, it is advisable to go back and attempt to restructure each question to see if all the choices can be made correct by modifying the question. By doing this, four times as much will be learned. By all means, look up the right answer and explanation. Then, focus on each of the other choices and ask yourself under what conditions they might be correct. For example, the entire thrust of the sample question concerning hypothyroidism could be altered by changing the first few words to read:

> "Hyperthyroidism recently discovered in. . . ."
>
> "Hypothyroidism prenatally occurring in. . . ."
>
> "Hypothyroidism resulting from. . . ."

This question can be used to learn and understand thyroid problems in general, not only to memorize answers to specific questions.

The Challenge Exam that follows contains 180 questions and explanations. Every effort has been made to simulate the types of questions and the degree of question difficulty in the various licensure and qualifying exams (i.e., NBME Parts I, II, and III and FLEX). While taking this exam, the student should attempt to create the testing conditions that might be experienced during actual testing situations. Approximately 1 minute should be allowed for each question, and the entire test should be finished before it is scored.

Summary

Ideally, examinations are designed to determine how much information students have learned and how that information is used in the successful completion of the examination. Students will be successful if these suggestions are followed:

- Develop a positive attitude and maintain that attitude.
- Be realistic in determining the amount of material you attempt to master and in the score you hope to attain.
- Read the directions for each type of question and the questions themselves closely and follow the directions carefully.
- Guess intelligently and consistently when guessing strategies must be used.
- Bring the paper-and-pencil differential diagnosis approach to each question in the examination.
- Use the test as an opportunity to display your knowledge and as a tool for developing prescriptions for further study and learning.

National Board examinations are not easy. They may be almost impossible for those who have unrealistic expectations or for those who allow misinformation concerning the exams to produce anxiety out of proportion to the task at hand. They are manageable if they are approached with a positive attitude and with consistent use of all the information the student has learned.

Michael P. O'Donnell

QUESTIONS

Directions: Each question below contains five suggested answers. Choose the **one best** response to each question.

1. Paralysis of the superior gluteal nerve results in an inability to

(A) fully extend the hip joint on the affected side
(B) fully extend the knee joint on the affected side
(C) keep the pelvis level so that it tilts toward the affected side with each step
(D) keep the pelvis level so that it tilts toward the unaffected side with each step
(E) perceive cutaneous sensation over the posterior aspect of the thigh

2. All of the following muscles are flexors of the thigh EXCEPT the

(A) iliopsoas
(B) pectineus
(C) rectus femoris
(D) sartorius
(E) vastus intermedius

3. Occlusion of the inferior mesenteric artery usually does not cause necrosis of the rectal mucosa because the

(A) arterial supply from the left colic artery compensates via anastomoses across Sudeck's point
(B) inferior rectal artery, a major branch of the external iliac artery, also supplies the rectum
(C) major arterial supply to the rectum is from anastomotic connections with the superior mesenteric artery
(D) middle rectal artery, a major branch of the internal iliac artery, also supplies the rectum
(E) supply from the inferior mesenteric artery to the rectum is insignificant

4. Reposition of the thumb (antagonistic to opponens action) is due to all of the following muscles EXCEPT the

(A) abductor pollicis brevis
(B) abductor pollicis longus
(C) adductor pollicis
(D) extensor pollicis brevis
(E) extensor pollicis longus

5. Which of the following muscles may be involved in flexion of the hip joint?

(A) Gluteus maximus
(B) Gluteus medius
(C) Gluteus minimus
(D) Piriformis
(E) Tensor fasciae latae

6. An expanding pituitary adenoma eroding the pituitary fossa and compressing the optic chiasm produces which of the following symptoms?

(A) Binasal heteronymous hemianopsia
(B) Bitemporal heteronymous hemianopsia
(C) Contralateral homonymous hemianopsia
(D) Ipsilateral homonymous hemianopsia
(E) Total blindness of the right eye

7. A young man suffers a knee injury in a rugby scrum. The physician diagnoses tears of the medial collateral ligament, the medial meniscus, and the anterior cruciate ligament. The signs that lead to this diagnosis include all of the following EXCEPT

(A) abnormal abduction of the knee
(B) abnormal adduction of the knee
(C) abnormal anterior movement of the tibia on the femur
(D) bloody effusion in the joint capsule
(E) locking of the knee in partial flexion

8. The most important landmark for the preferred location for a spinal tap or induction of spinal anesthesia is the

(A) iliac crests
(B) posterior–superior iliac spines
(C) prominent spinous process of T12
(D) sacral cornua
(E) sacral hiatus

9. A gray ramus communicans, which extends between a sympathetic trunk ganglion and an anterior primary ramus of a spinal nerve, contains

(A) axons of a preganglionic neuron
(B) fibers that activate smooth muscles (arrector pili) of the skin
(C) fibers that conduct general somatic afferent impulses
(D) fibers that conduct visceral afferent impulses
(E) myelinated fibers

10. The movement of the ribs during relaxed thoracic expiration involves all of the following EXCEPT

(A) contraction of the intercostal portion of the internal intercostal muscles
(B) decrease of the transverse thoracic diameter
(C) inward rotation of the ribs
(D) movement at the costovertebral joints
(E) release of stored energy within the costal cartilages

11. The arterial supply to the maxillary and mandibular teeth comes from

(A) a single branch of the maxillary artery
(B) branches of the internal carotid artery
(C) branches of the maxillary and sublingual artery, respectively
(D) the maxillary and facial arteries
(E) separate branches of the maxillary artery

12. All of the following structures pass through the lesser sciatic foramen EXCEPT the

(A) internal pudendal artery
(B) internal pudendal vein
(C) piriformis muscle
(D) pudendal nerve
(E) tendon of the obturator internus muscle

13. The cell bodies for the sensory fibers that convey pain, touch, and temperature from the anterior two-thirds of the tongue are located in the

(A) geniculate ganglion
(B) pterygopalatine ganglion
(C) semilunar ganglion
(D) submandibular ganglion
(E) petrosal ganglion

14. Characteristics of external hemorrhoids, which develop in branches of the inferior hemorrhoidal vein, include all of the following EXCEPT

(A) they appear in the rectum
(B) they are inferior (distal) to the pectinate line
(C) they are painful
(D) they lie underneath the mucosa
(E) they occur in anastomoses between branches of the internal iliac and internal pudendal veins

15. The secretory function of the parotid gland is likely to be effected by

(A) a severe or prolonged middle ear infection
(B) anesthesia of structures exiting the stylomastoid foramen
(C) facial nerve palsy
(D) severing the glossopharyngeal nerve as it passes between the superior and middle constrictors
(E) severance of the cervical sympathetic chain

16. A fracture of the sphenoid bone that passes through the foramen rotundum may result in paralysis to which of the following muscles?

(A) Buccinator
(B) Lateral pterygoid
(C) Occipitalis portion of occipitofrontalis
(D) Orbicularis occuli (orbital portion)
(E) None of the above

17. Intramuscular injections in the superomedial gluteal quadrant may injure which of the following nerves?

(A) Inferior gluteal
(B) Posterior femoral cutaneous
(C) Pudendal
(D) Sciatic
(E) Superior gluteal

18. An *epidural hematoma* is usually caused by

(A) fracture of the diploic space
(B) laceration of the superficial temporal artery
(C) leakage from a cerebral vein
(D) rupture of a cerebral artery
(E) tearing of a meningeal vessel

19. Correct statements describing the pelvic diaphragm include all of the following EXCEPT

(A) it contains a potentially weak area in the midline
(B) it does not share the same plane with the urogenital diaphragm
(C) it functions to suspend and support pelvic organs
(D) it is comprised of the levator ani and coccygeus muscles and their fasciae
(E) it is innervated by the inferior hemorrhoidal nerve

20. When the tibial nerve is injured in the popliteal fossa, all of the following functions remain EXCEPT

(A) ability to stand on heels
(B) Achilles reflex
(C) flexion of the knee
(D) foot inversion
(E) knee-jerk reflex

21. The gallbladder, which stores and concentrates bile, is located

(A) between the right and caudate lobes of the liver
(B) between the right and quadrate lobes of the liver
(C) in the coronary ligament
(D) in the falciform ligament
(E) in the lesser omentum

22. If cell bodies in the geniculate ganglion are damaged, you would expect the symptoms to be

(A) loss of sensation of pain from the face
(B) loss of sensation of touch from the anterior two-thirds of the tongue
(C) loss of taste from the anterior two-thirds of the tongue
(D) partial facial paralysis
(E) none of the above

23. Which of the following portions of the gastrointestinal tract become predominantly secondarily retroperitoneal during development?

(A) Appendix
(B) First part of the duodenum
(C) Liver
(D) Sigmoid colon
(E) Uncinate process of the pancreas

24. The femoral artery becomes the popliteal artery as it passes

(A) between the medial and lateral heads of the gastrocnemius muscle
(B) deep to the popliteus muscle
(C) deep to the soleus muscle
(D) into the posterior compartment of the leg
(E) through the adductor magnus muscle

25. All of the following muscles are involved in dorsiflexion of the ankle joint EXCEPT the

(A) extensor digitorum longus
(B) extensor hallucis longus
(C) peroneus longus
(D) peroneus tertius
(E) tibialis anterior

26. Cell bodies that are located in the superior cervical ganglion have all of the following functions EXCEPT

(A) contraction of the ciliary muscle
(B) dilation of the pupil
(C) absence of facial perspiration
(D) prevention of pseudoptosis
(E) vasodilation of the face

27. *Epiphora* (constant tearing down the face) may be caused by

(A) blockage of the nasolacrimal duct
(B) excessive activity of the parasympathetic components of CN III
(C) excessive activity of the parasympathetic components of CN IX
(D) injury to the cervical sympathetic chain
(E) lesion of the greater superficial petrosal nerve

28. In addition to the popliteal artery, the popliteal fossa contains the

(A) anterior tibial artery
(B) great saphenous vein
(C) peroneal artery
(D) posterior tibial artery
(E) venae comitantes

29. Femoral hernias pass deeply (inferior) to the inguinal ligament. A segment of small intestine in this instance is in *direct* contact with the

(A) falx inguinalis
(B) lacunar ligament
(C) parietal peritoneum
(D) transversus abdominis muscle
(E) transversalis fascia

30. During a spinal tap, all of the following structures would be penetrated if the needle were inserted precisely in the midline EXCEPT the

(A) arachnoid layer
(B) dura mater
(C) interspinous ligament
(D) posterior longitudinal ligament
(E) supraspinous ligament

31. Which of the following nerves is jeopardized by a fracture of the proximal neck of the fibula?

(A) Common peroneal
(B) Saphenous
(C) Sural
(D) Tibial
(E) None of the above nerves

32. The neural supply to the maxillary incisor teeth is described by all of the following statements EXCEPT

(A) these fibers are accompanied by motor fibers that exit via the infraorbital foramen
(B) these fibers are at risk if the anterior maxillary wall is fractured
(C) these fibers exit from the skull via the foramen rotundum
(D) these fibers cross the pterygopalatine fossa
(E) these fibers run in the inferior orbital fissure

33. The annular ligament prevents dislocation of the

(A) distal end of the clavicle from the acromion process
(B) head of the humerus from the glenoid cavity
(C) head of the ulna from the triquetrum
(D) head of the radius from the radial notch of the ulna
(E) head of the third metacarpal from the base of the proximal phalanx

34. A typical thoracic vertebra includes all of the following components EXCEPT

(A) a heart-shaped vertebral body
(B) inferior articular facets
(C) a neural canal
(D) superior costal facets
(E) transverse foramina

35. A fall on the outstretched hand can result in a broken clavicle rather than a clavicular dislocation. Resistance to dislocation of the clavicle is provided proximally by the articular disk and its attachments at the sternoclavicular joint. Stability is provided distally by the

(A) actions of the rotator cuff muscles
(B) coracoacromial ligament
(C) coracoclavicular ligament
(D) coracohumeral ligament
(E) tendon of the long head of the biceps muscle

36. Tapping the patellar ligament activates afferent impulses from muscle spindles and elicits a knee-jerk reflex (patellar reflex), testing which of the following spinal nerves or spinal cord levels?

(A) T12–L2
(B) L2–L4
(C) L4–S1
(D) S1–S3
(E) S3–S5

37. A 24-year-old woman is taken to the emergency room following 12 hours of nausea, vomiting, and right lower quadrant pain. She has a temperature of 101°F, white blood count (WBC) of 13,700/mm³ (normal 5,000–10,000), and evidence of peritonitis in the lower right quadrant. She is brought to surgery for an emergency appendectomy. At operation, the appendix is normal. Alternative causes of this woman's illness include all of the following EXCEPT

(A) ovarian torsion
(B) Meckel's diverticulitis
(C) salpingitis
(D) sigmoid diverticulitis
(E) urinary tract infection

38. Which of the following statements best characterizes the tunica dartos of the scrotum?

(A) It contains striated muscle fibers
(B) It is a continuation of the two layers of the superficial fascia
(C) It is involved in the cremaster reflex
(D) It is invested with adipose tissue
(E) It responds to cold temperature by lowering the scrotum away from the body

39. A surgeon identifies a segment of gut as jejunum. He forms his decision based upon all of the following external characteristics of the jejunum EXCEPT

(A) an arterial supply with few arcades
(B) a greater diameter
(C) a mesentery that is thicker due to fat deposits
(D) a relatively thick wall
(E) long arteriae rectae

40. An occlusion of the axillary artery does not seriously interfere with the blood supply to the upper limb because of circumscapular anastomoses. Branches of the axillary and subclavian arteries that contribute to this anastomosis include all of the following EXCEPT the

(A) deep brachial artery
(B) posterior humeral circumflex artery
(C) suprascapular artery
(D) subscapular artery
(E) transverse cervical artery

41. A fracture that passes through the right occipital condyle from the anterior condylar canal to the posterior condylar canal may involve

(A) inability to direct the tongue to the opposite side
(B) inability to turn the head to the opposite side and shrug the ipsilateral shoulder
(C) loss of the gag reflex on the ipsilateral side
(D) paralysis of the laryngeal musculature (except the cricothyroid muscle) on the ipsilateral side
(E) none of the above

42. True statements about the prostate gland include all of the following EXCEPT

(A) it contributes fructose and choline to seminal fluid
(B) it empties into the prostatic urethra on either side of the urethral crest
(C) if its median lobe is enlarged, it will obstruct the urethra
(D) the dorsal vein of the penis empties into a plexus of veins around its base
(E) the posterior lobes are most prone to malignant transformation

43. One annoying symptom of Bell's palsy that affects the facial nerve (CN VII) is hyperacusis, which is caused by involvement of the

(A) cochlear nerve
(B) chorda tympani nerve
(C) stapedius muscle
(D) tensor tympani muscle
(E) tympanic membrane

44. In portal hypertension where flow of blood to or through the liver is impeded, blood will return to the systemic circulation by all of the following veins EXCEPT the

(A) esophageal veins
(B) gonadal veins
(C) hemorrhoidal veins
(D) renal veins
(E) veins of Retzius associated with the ascending colon

45. Which of the following statements best describes the teres major muscle?

(A) It contributes to the stability of the posterior part of the shoulder joint
(B) It divides the axillary artery into three parts
(C) It inserts on the humerus just distal to the infraspinatus muscle
(D) It is active in adduction of the shoulder joint
(E) It is innervated by the same nerve that supplies the deltoid muscle

46. Characteristics of an indirect inguinal hernia include all of the following EXCEPT

(A) it is located adjacently to the spermatic cord
(B) it is located laterally to the inferior epigastric artery
(C) it originates laterally to the inguinal triangle
(D) it passes through the deep inguinal ring
(E) it passes through the superficial inguinal ring

47. During full active extension of the knee joint, all of the following are tightened EXCEPT the

(A) anterior cruciate ligament
(B) fibular collateral ligament
(C) tendon of popliteus muscle
(D) patellar tendon
(E) tibial collateral ligament

48. Several days after a myocardial infarct limited to the interventricular septum, the attending physician picks up a new low-pitched systolic murmur. Knowing the location of the infarct, he attributes the new murmur to

(A) an atrial septal defect
(B) aortic valvular insufficiency
(C) mitral valve regurgitation
(D) pulmonary valve regurgitation
(E) tricuspid valve prolapse

49. Which of the following statements describing the lunate bone is correct?

(A) It articulates maximally with a fibrocartilaginous disk upon ulnar deviation (adduction)
(B) It is a component of the carpometacarpal joint
(C) It may produce carpal tunnel syndrome if displayed anteriorly
(D) It provides an attachment for the flexor retinaculum
(E) None of the above

50. The tendon of which of the following muscles is involved when the tuberosity of the fifth metatarsal bone is avulsed (pulled off) in a sprain?

(A) Abductor digiti minimi
(B) Peroneus brevis
(C) Peroneus longus
(D) Tibialis anterior
(E) Tibialis posterior

51. A surgeon making a deep lumbar incision along the 12th rib to perform a right adrenalectomy notes the bright red blood in the field grow suddenly dark. He knows that he has probably

(A) divided a suprarenal artery
(B) entered the costodiaphragmatic recess
(C) incised the spleen
(D) opened the parietal peritoneum
(E) punctured the renal capsule

52. Correct statements concerning the superior laryngeal nerve include all of the following EXCEPT

(A) It contains taste afferents
(B) It has an internal branch that pierces the thyrohyoid membrane to innervate the laryngeal mucosa
(C) It innervates all of the laryngeal musculature by an external branch, except the cricothyroid muscle
(D) It produces muscle contraction that lengthens the (true) vocal folds
(E) It provides the afferent limb of the cough reflex

53. Structures found in the deep perineal pouch include all of the following EXCEPT the

(A) bulbourethral glands of the male
(B) deep transverse perineal muscle
(C) external urethral sphincter
(D) vestibular glands of the female

54. Which of the following statements concerning the transverse vertebral foramen is correct?

(A) It is associated with the transverse processes of the thoracic vertebrae
(B) It is not patent in the adult
(C) It is not present in the atlas
(D) It provides a pathway for the vertebal artery except for C7
(E) It transmits the spinal nerve

55. A 64-year-old man with severe angina pectoris is found by coronary angiography to have a 90% left coronary artery occlusion. A venous homograph is not possible because of severe venous varicosities; thus, the patient is treated surgically by diverting his left internal thoracic artery into the left coronary artery distal to the occlusion. Postoperatively, the chest wall region originally supplied by the left internal thoracic artery receives blood flow from all of the following arteries EXCEPT the left

(A) inferior epigastric artery
(B) pericardiophrenic artery
(C) posterior intercostal arteries
(D) musculophrenic artery
(E) superior epigastric artery

56. The cell bodies that contribute to the deep petrosal nerve and dilate the pupil are located in the

(A) ciliary ganglion
(B) geniculate ganglion
(C) otic ganglion
(D) superior cervical ganglion
(E) superior glossopharyngeal ganglion

57. In the adult, the blood supply to the head of the femur is by the

(A) artery of the ligamentum teres
(B) deep iliac circumflex artery
(C) lateral and medial circumflex femoral arteries
(D) nutrient artery of the femur
(E) obturator artery

58. Which of the following statements concerning internal (medial) rotation of the arm is true?

(A) It cannot occur if the pectoralis major muscle is paralyzed
(B) It is accomplished in part by a muscle that inserts onto the lesser tubercle
(C) It is accomplished in part by muscles innervated by the long thoracic nerve
(D) It is weakened by interruption of the suprascapular nerve
(E) It occurs around the anteroposterior axis of the glenohumeral joint

59. A diastolic murmur heard most distinctly in the 2nd intercostal space to the right of the sternum is most attributable to

(A) aortic insufficiency
(B) mitral stenosis
(C) patent ductus arteriosus
(D) pulmonary stenosis
(E) tricuspid insufficiency

60. Correct statements concerning the kidney include all of the following EXCEPT

(A) destruction of the sympathetic ganglia associated with levels T10–L2 produces diuresis secondary to renal vasoconstriction
(B) major collateral circulation is provided to the kidney from supernumerary renal arteries
(C) nephroptosis produces pain caused by traction on the renal vessels
(D) the perirenal fascia permits spread of infection along the ureters and between the kidneys and pelvic structures
(E) the position of the kidneys is maintained primarily by perinephric and paranephric fat

61. The carotid sheath contains all of the following structures EXCEPT the

(A) common carotid artery
(B) external carotid artery
(C) internal carotid artery
(D) sympathetic chain
(E) vagus nerve

62. The anterior longitudinal ligament has which of the following characteristics?

(A) It anchors the emerging spinal nerves in place
(B) It limits the direction of nucleus pulposus extrusion during disk herniation
(C) It narrows anteriorly to the intervertebral disk
(D) It resists kyphosis
(E) None of the above

63. The tendon that can be seen and felt just posterior to the medial malleolus of the tibia during inversion of the foot is the tendon of which of the following muscles?

(A) Flexor hallucis longus
(B) Peroneus brevis
(C) Peroneus longus
(D) Tibialis anterior
(E) Tibialis posterior

64. Structures that drain into the deep cervical lymph nodes include all of the following EXCEPT the

(A) nasal mucosa
(B) palatine tonsils
(C) parotid gland
(D) teeth and gingivae
(E) tongue

65. Which of the following joints has only one degree of freedom (a one-arc joint)?

(A) Humeroulnar
(B) Metacarpophalangeal
(C) Radiocarpal
(D) Radiohumeral
(E) Sternoclavicular

66. A young man with no prior medical problems is confined in bed for 6 weeks after breaking his leg. The day after sitting in a chair for the first time since the accident, the patient develops mild weakness in the right arm and facial muscles and slurring of speech. A cerebral embolus from the legs is suspected, but a pulmonary embolus is expected. The anatomic explanation for this "paradoxical embolus" might be

(A) a large atrial septal defect
(B) a patent ductus arteriosus
(C) a portacaval anastomosis
(D) a thrombus in the right atrial appendage
(E) dextrocardia (situs inversus thoracis)

67. The vestibule in the female perineum is characterized by which of the following statements?

(A) It is bordered by the labia majora
(B) It is completely covered by the hymen until first coitus
(C) It receives drainage from the greater vestibular and paraurethral glands
(D) It receives sensory innervation from the nervi erigentes
(E) It receives just the vagina

68. A posterolateral herniation of the L4–L5 intervertebral disk may affect

(A) the fourth lumbar nerve
(B) the fifth lumbar nerve
(C) the fourth and fifth lumbar nerves only
(D) the fourth and fifth lumbar and first and second sacral nerves
(E) none of the above

69. The perineal sensation is supplied by all of the following nerves EXCEPT the

(A) genitofemoral
(B) ilioinguinal
(C) inferior rectal
(D) nervi erigentes
(E) pudendal

70. If the motor root of the trigeminal nerve is injured, paralysis occurs in all of the following muscles EXCEPT the

(A) anterior belly of the digastric
(B) buccinator
(C) masseter
(D) mylohyoid
(E) tensor tympani

71. All of the following problems usually accompany the carpal tunnel syndrome EXCEPT

(A) paralysis of the opponens pollicis muscle
(B) paralysis of the flexor pollicis brevis muscle
(C) paralysis of lumbricals 1 and 2
(D) paralysis of the oblique head of the adductor pollicis muscle
(E) sensory disturbances on the radial half of the distal palm and first three digits

72. A longitudinal incision just lateral to the medial border of the rectus sheath to gain access to the abdominal cavity will

(A) be parallel to Langer's lines of cleavage
(B) paralyze a portion of the rectus muscle
(C) require division of the transversus abdominis muscle
(D) spare the blood supply to the rectus muscle
(E) result in all of the above

73. All of the following areas in the male are drained by inguinal lymph nodes EXCEPT the

(A) leg and foot
(B) lower abdominal wall
(C) lower part of the anal canal
(D) penis
(E) testes

74. Which of the following arteries frequently arises as a branch of the external iliac or inferior epigastric artery, instead of as a branch of the internal iliac artery?

(A) Internal pudendal
(B) Obturator
(C) Superior vesical
(D) Umbilical
(E) Uterine

75. True statements about the postcentral gyrus of the parietal lobe include all of the following EXCEPT

(A) it is concerned with the opposite side of the body
(B) it is the site of interpretation and recognition of sensory information
(C) it is somatotopically organized
(D) it lies immediately posterior to the central sulcus
(E) it receives sensory input relayed from the thalamus

76. Which of the following structures is normally both medial to the biceps brachii tendon and deep to the bicipital aponeurosis?

(A) Brachial artery
(B) Deep brachial artery
(C) Median cubital vein
(D) Radial artery
(E) Ulnar artery

77. Which of the following structures passes through the foramen lacerum?

(A) Internal jugular vein
(B) Maxillary division of the trigeminal nerve
(C) Middle meningeal artery
(D) Ophthalmic vein
(E) None of the above structures

78. A drug addict ("mainliner") is diagnosed as having tricuspid insufficiency secondary to bacterial endocarditis (infection of the endocardium and valve leaflets). All of the following findings might be noted EXCEPT

(A) abdominal ascites
(B) distended neck veins
(C) pulmonary edema
(D) swelling of the ankles
(E) systolic murmur in the 5th right intercostal space

79. One means of alleviating portal hypertension is by a side-to-side anastomosis between the hepatic portal vein and the inferior vena cava. The location of the inferior vena cava is

(A) across the epiploic foramen from the portal vein
(B) anterior and to the left of the portal vein
(C) immediately to the right of the portal vein
(D) posterior and to the right of the portal vein
(E) in none of the above places

80. Which of the following structures drains into the inferior meatus of the nose?

(A) Ethmoidal sinuses
(B) Frontal sinus
(C) Maxillary sinus
(D) Nasolacrimal duct
(E) Sphenoidal sinus

81. Cranial nerves with intrinsic parasympathetic components include all of the following EXCEPT the

(A) CN III
(B) CN V
(C) CN VII
(D) CN IX
(E) CN X

82. The *median* umbilical fold is created by the

(A) falx inguinalis
(B) inferior epigastric arteries
(C) lateral borders of the rectus sheath
(D) obliterated umbilical arteries
(E) urachus

83. To equalize air pressure on both sides of the tympanic membrane, which of the following pairs of muscles might contract?

(A) Levator veli palatini and tensor veli palatini
(B) Palatoglossus and palatopharyngeus
(C) Salpingopharyngeus and palatopharyngeus
(D) Tensor tympani and stapedius
(E) Tensor veli palatini and tensor tympani

84. An elderly man is taken to the emergency room because he has an object lodged in his throat. A small steak bone is removed by esophagoscopy and a small posterior tear of the esophagus is noted. The patient is admitted and placed on broad-spectrum antibiotic therapy to prevent infection from developing in the

(A) anterior mediastinum
(B) carotid sheath
(C) pretracheal space
(D) retrovisceral space
(E) superior mediastinum

85. The external urethral sphincter functions to maintain urinary continence and

(A) has an identical anatomic structure in both the male and female
(B) is a portion of the pelvic diaphragm
(C) is under involuntary control
(D) receives innervation by the pudendal nerve
(E) surrounds the prostatic portion of the urethra in the male

86. Infection within the pterygopalatine fossa may track directly into all of the following cavities EXCEPT the

(A) middle cranial fossa
(B) middle ear
(C) nasal cavity
(D) oral cavity
(E) orbital cavity

87. Bilateral lumbar sympathectomy does not affect autonomic control of the descending colon because

(A) the descending colon receives its parasympathetic innervation from the pelvic splanchnics
(B) the descending colon receives its parasympathetic innervation from the vagus nerve
(C) the descending colon receives its sympathetic innervation from thoracic splanchnic nerves
(D) lumbar splanchnics innervate the pelvic viscera via the hypogastric nerve
(E) only presynaptic sympathetic fibers are severed

88. The posterior muscles of the back receive motor innervation from

(A) dorsal primary rami
(B) dorsal roots
(C) posterior branches of the lateral perforating nerves
(D) ventral primary rami
(E) none of the above

89. The ethmoid bone, an endochondral bone, contributes to all of the following structures EXCEPT the

(A) cribriform plate
(B) inferior concha
(C) lamina papyracea of the orbit
(D) nasal septum
(E) superior concha

90. A vertical incision through the anterior wall of the rectus sheath between the umbilicus and pubis will cut through all of the following layers EXCEPT the

(A) aponeurosis of the external oblique muscle
(B) aponeurosis of the internal oblique muscle
(C) aponeurosis of the transverse abdominis muscle
(D) transversalis fascia

91. Hysterectomy (surgical removal of the uterus and ovaries) may result in injury to adjacent anatomic structures. One structure commonly injured is the

(A) external iliac artery
(B) rectum
(C) triangular ligament
(D) ureter
(E) urethra

92. Which of the following muscles is most efficient at extending the knee when the hip joint is extended?

(A) Biceps femoris
(B) Rectus femoris
(C) Vastus intermedius
(D) Vastus lateralis
(E) Vastus medialis

93. The right adrenal gland receives its blood supply from all of the following EXCEPT the

(A) aorta directly
(B) right inferior phrenic artery
(C) right renal artery
(D) superior mesenteric artery

94. All of the following structures are drained by the superficial inguinal lymph nodes EXCEPT the

(A) lateral compartment of the leg
(B) lower abdominal wall
(C) lower part of the anal canal
(D) superficial thigh
(E) testes

95. Pronounced dilation of a smooth muscular tube in the body generally produces pain. Sensory fibers for pain from the cervix travel to the spinal cord via the

(A) deep perineal branches of the pudendal nerve
(B) gray rami communicantes of S2–S4
(C) inferior hypogastric plexus
(D) nervi erigentes
(E) white rami of L1 and L2

96. A woman, aged 57, is diagnosed as having metastatic tumor blocking Cloquet's lymph node, which is located in the right lymph duct before it joins the venous system. In addition to the large and palpable node, diagnostic symptoms include

(A) swelling of the right breast
(B) swelling of the right side of the face
(C) swelling of the right side of the neck
(D) swelling of the right upper extremity
(E) all of the above

Directions: The groups of questions below consist of lettered choices followed by several numbered items. For each numbered item, select the **one** lettered choice with which it is **most closely** associated. Each lettered choice may be used once, more than once, or not at all. Choose the answer

  A   if the item is associated with **(A) only**
  B   if the item is associated with **(B) only**
  C   if the item is associated with **both (A) and (B)**
  D   if the item is associated with **neither (A) nor (B)**

Questions 97–100

For each organ listed below, select the region to which pain in that organ is usually referred.

(A) Inguinal and pubic regions
(B) Perineum, posterior thigh, and leg
(C) Both
(D) Neither

97. Ovary

98. Uterus

99. Epididymis

100. Testis

Questions 101–106

For each situation listed below, select the artery that is most likely to be associated with it.

(A) Right coronary artery
(B) Circumflex branch of left coronary artery
(C) Both
(D) Neither

101. Is found in the coronary sulcus

102. Arises from the left anterior aortic sinus

103. Supplies the anterior portion of the interventricular septum

104. Supplies the atrioventricular node

105. Supplies the sinoatrial node

106. Terminates as the posterior interventricular (descending) artery in most people

Questions 107–110

For each functional event listed below, select the stage of the cardiac cycle in which it is most likely to occur.

(A) Atrial systole
(B) Ventricular diastole
(C) Both
(D) Neither

107. The atrioventricular valves close

108. The semilunar valves close

109. Blood flow through the myocardium is maximal

110. Blood flows into the ventricles

Directions: The groups of questions below consist of lettered choices followed by several numbered items. For each numbered item select the **one** lettered choice with which it is **most** closely associated. Each lettered choice may be used once, more than once, or not at all.

Questions 111–115

For each vein or artery involved in coronary circulation, select the appropriate lettered structure shown in the figures below

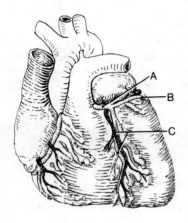

111. A branch of the left coronary artery

112. Supplies about 65% of the interventricular septum

113. Usually supplies the atrioventricular node

114. Can anastomose with the anterior interventricular artery when pathology develops slowly

115. Accompanies the coronary sinus

Questions 116–120

Match each description of muscles listed below with the appropriate rotator cuff component.

(A) Infraspinatus
(B) Subscapularis
(C) Supraspinatus
(D) Teres minor
(E) None of the above

116. A primary medial (internal) rotator of the arm

117. A muscle that forms the posterior wall of the axilla

118. A muscle that initiates humeral abduction (first 15°)

119. A muscle innervated by the axillary nerve

120. A muscle innervated by a branch of the musculocutaneous nerve

Questions 121–125

For each description of the innervation of the face and visceral neck, select the nerve that is most likely to be responsible.

(A) Facial nerve
(B) Glossopharyngeal nerve
(C) Hypoglossal nerve
(D) Mandibular nerve
(E) Vagus nerve

121. Supplies sensory innervation to the anterior two-thirds of the tongue

122. Supplies taste fibers to the anterior two-thirds of the tongue

123. Supplies motor innervation to the anterior two-thirds of the tongue

124. Supplies taste fibers to the epiglottis

125. Supplies motor innervation to most of the pharyngeal muscles

Questions 126–130

For each description of the blood supply to the brain, select the artery that is most apt to be responsible for it.

(A) Anterior cerebral artery
(B) Basilar artery
(C) Internal carotid artery
(D) Middle cerebral artery
(E) Vertebral artery

126. Gives off a branch that supplies the eyeball

127. Supplies the area above the corpus callosum

128. Supplies most of the lateral surfaces of the cerebral cortex

129. Supplies the basal ganglia and thalamus

130. Supplies the inferior cerebellum

Questions 131–135

For each artery or vein comprising the vasculature of the cranium, select the lettered foramen or fissure through which it courses, shown on the illustration of the inferior aspect of the cranium below.

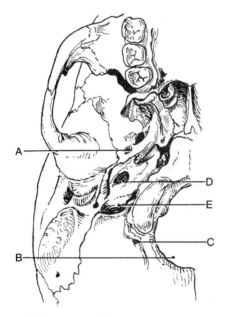

131. Middle meningeal artery

132. Internal carotid artery

133. Internal jugular vein

134. Emissary vein

135. Vertebral artery

Questions 136–140

Match the following.

(A) External oblique muscle
(B) Interchondral portion of internal intercostal muscles
(C) Internal intercostal muscles proper
(D) Levator scapulae muscle
(E) None of the above

136. An inspiratory muscle

137. Decreases the transverse diameter of the thoracic cage

138. Innervated by the phrenic nerve

139. Antagonistic to the diaphragm

140. Split by the intercostal neurovascular bundle

Questions 141–145

For each autonomic pathway listed below, select the correct locations for the presynaptic and postsynaptic cell bodies.

(A) Inferior salivatory nucleus and otic ganglion
(B) Superior salivatory nucleus and ciliary ganglion
(C) Superior salivatory nucleus and pterygopalatine ganglion
(D) Superior salivatory nucleus and submandibular ganglion
(E) None of the above

141. Pathway for accommodation

142. Pathway for lacrimation

143. Pathway for sublingual secretion

144. Pathway for nasal and palatine gland secretion

145. Pathway for parotid secretion

Questions 146–150

Match the following.

(A) Adductor magnus muscle
(B) Iliopsoas muscle
(C) Piriformis muscle
(D) Rectus femoris muscle
(E) None of the above

146. Acts across two joints

147. Extends the thigh

148. Innervated by the obturator nerve

149. Inserts into the lesser trochanter

150. Primarily rotates the thigh laterally

Questions 151–155

Match the following.

(A) Genitofemoral nerve
(B) Hypogastric nerve (plexus)
(C) Nervi erigentes
(D) Pudendal nerve
(E) None of the above

151. Provides the efferent limb of the cremaster reflex

152. Mediates voluntary urinary retention

153. Mediates male ejaculation

154. Mediates perineal sensation

155. Mediates the detrusor (micturition) reflex

Questions 156–160

Match the following.

(A) Arachnoid layer
(B) Denticulate ligament
(C) Dura mater
(D) Pia mater
(E) None of the above

156. Fuses with the periosteum of the cranial vault

157. Lines the dural sac

158. Contains the small blood vessels that supply the spinal cord

159. Supports the spinal cord

160. Forms the filum terminale

Questions 161–165

For each of the muscles listed below, select the nerve that is responsible for its innervation.

(A) Ansa cervicalis
(B) Facial nerve
(C) Greater auricular nerve
(D) Lesser occipital nerve
(E) None of the above

161. Sternomastoid muscle

162. Platysma muscle

163. Occipitofrontalis muscle

164. Omohyoid muscle

165. Geniohyoid muscle

Questions 166–170

Match the following.

(A) Gastroduodenal artery
(B) Left gastric artery
(C) Right gastric artery
(D) Splenic artery
(E) None of the above

166. Supplies the distal esophagus

167. May give rise to an aberrant left hepatic artery

168. Provides the major supply to the greater curvature of the stomach corpus and fundus

169. Supplies the portion of the pancreas where most of the pancreatic islets (of Langerhans) are located

170. Provides branches that anastomose directly with branches of the superior mesenteric artery

ANSWERS AND EXPLANATIONS

1. The answer is D. [*Chapter 23 VI A 3*] The superior gluteal nerve innervates the gluteus medius and gluteus minimus muscles, which, as strong abductors of the thigh, act to keep the pelvis level when the opposite leg is lifted from the ground. A superior gluteal nerve palsy results in a downward tilt of the pelvis when the leg of the unaffected side is lifted off the ground (abductor lurch).

2. The answer is E. [*Chapter 23 IV A 2 a, b; Table 23-1*] The iliopsoas, rectus femoris sartorius, and pectineus muscles are all flexors of the thigh and, as such, oppose the actions of the hamstring group and gluteus maximus muscle. The vastus intermedius does not act across the hip joint.

3. The answer is D. [*Chapter 17 IV D 3*] The rectum receives its blood supply from the superior rectal branch of the inferior mesenteric artery and the middle rectal branches of the internal iliac arteries. Because there usually are anastomoses between these two systems, occlusion of one or the other would not necessarily cause ischemic necrosis of the rectal mucosa.

4. The answer is C. [*Chapter 10 IV C 2 a–c; Table 10-2*] Reposition of the thumb is the result of contraction of the abductors and extensors. The adductor pollicis draws the thumb alongside the first digit.

5. The answer is E. [*Chapter 23 IV B 1 a*] The tensor fasciae latae muscle, originating at the iliac crest and inserting into the iliotibial tract, is a flexor of the thigh at the hip joint. Because the iliotibial band passes anteriorly to the axis of rotation of the knee joint when the knee is extended, the tensor fasciae latae muscle also keeps the extended knee in the extended position.

6. The answer is B. [*Chapter 29 V B 2 c (3) (b)*] Compression of the optic chiasm involves the decussating nerve fibers from the nasal portions of each retina. Because the lateral visual fields are projected onto the nasal portion of the retina, the result is *bitemporal heteronymous hemianopsia (tunnel vision)*.

7. The answer is B. [*Chapter 24 III B 3 a (2), d (2) (b), 5 b*] The "unhappy triad" described in the case is the result of rotation injury of the partially flexed knee. A torn medial collateral ligament permits abnormal abduction and a torn anterior cruciate permits anterior displacement of the tibia on the femur (*anterior drawer sign*). A torn medial meniscus may get caught between the articular surfaces, locking the knee in partial flexion. There will also be bloody effusion into the joint.

8. The answer is A. [*Chapter 19 III D 1 c*] The preferred site for a spinal tap is between vertebrae L4 and L5. The intertubercular plane through the iliac crests passes through the level of the fourth lumbar vertebra or the fourth intervertebral disk. The sacral cornua define the sacral hiatus, which is used for caudal anesthesia. The spinous process of T12 is most difficult to locate and is, therefore, not a landmark. The posterior–superior iliac spines lie at the L5–S1 level.

9. The answer is B. [*Chapter 4 V D 1 b (1) (a)*] Each gray ramus communicans of the sympathetic chain is composed of unmyelinated postsynaptic sympathetic neurons, which reenter the spinal nerve and travel to the integument. These neurons innervate the sweat glands as well as smooth muscle associated with the hair follicles and the arterioles. Only white rami convey visceral afferent sensation.

10. The answer is A. [*Chapter 11 VII B 3, 4; Figure 11-7A*] The expiratory action of the internal intercostal muscles and the release of stored energy within the costal cartilages (extrinsic elastic recoil) result in downward movement of the ribs about an axis described by the costovertebral joints as well as an inward rotation of the ribs. The transverse diameter of the thoracic cage is thereby diminished. Due to the elastic energy stored in the costal cartilages and the effects of gravity, no muscular effort is needed for relaxed expiration. The internal intercostal musculature is used in forced expiration.

11. The answer is E. [*Chapter 32 VIII A 2 a (4), c (2), (5)*] The blood supply to the maxillary teeth is from the posterior–superior alveolar and anterior–superior alveolar branches of the infraorbital branch of the (internal) maxillary artery. The blood supply to the mandibular teeth is from the inferior alveolar branch of the (internal) maxillary artery.

12. The answer is C. [*Chapter 20 I A 1 b (4); IV A 3 b (2), B 1; V B 5 a*] The piriformis muscle, superior and inferior gluteal neurovascular bundles, sciatic nerve as well as the pudendal neurovascular bundle leave the deep pelvis through the greater sciatic foramen. The pudendal nerve and internal pudendal vessels pass through the lesser sciatic foramen to reach the perineum. In addition, the tendon of the obturator internus muscle passes out of the deep pelvis via the lesser sciatic foramen on its way to the greater trochanter of the femur.

13. The answer is C. [*Chapter 29 V H 1 a; Chapter 34 III D 4 a, b, E 2*] The semilunar (trigeminal) ganglion, the equivalent of a dorsal root ganglion, contains the cell bodies of the sensory neurons of the trigeminal nerve (CN V). The lingual branch of the mandibular division of the trigeminal nerve conveys pain, touch, and temperature sensation from the anterior two-thirds of the tongue.

14. The answer is A. [*Chapter 17 IV D 4 c; VI D 1 c; Chapter 21 III F 2 b*] Hemorrhoids are varicosities of the mucosal venous plexus. External hemorrhoids form as a result of enlargement of venous anastomotic connections between the middle and inferior rectal (hemorrhoidal) veins. External hemorrhoids are painful because the anal canal in which they form is innervated by rectal branches of the pudendal nerve.

15. The answer is A. [*Chapter 30 III B 1 f (4) (a); Chapter 32 V C 1 a, b; Chapter 34 IV F 1 a (1)*] The parotid gland is innervated by the tympanic branch of the glossopharyngeal nerve. This nerve passes through the temporal bone and enters the middle ear cavity where it forms the tympanic plexus on the tympanic bulla. It is at risk during *otitis media*.

16. The answer is E. [*Chapter 32 II B 2 (4) (b), D 1 a (1), 2 b; III B 2 b, c (2) (d)*] The maxillary portion of the trigeminal nerve (CN V_2) passes through the foramen rotundum. This pretrematic division of the first branchial arch nerve is only sensory. The buccinator, occipitofrontalis, and orbicularis occuli are all muscles of facial expression and are innervated by the facial nerve (CN VII). The lateral pterygoid, a muscle of mastication, is innervated by the mandibular division of the trigeminal nerve (CN V_2).

17. The answer is E. [*Chapter 23 VI A–C*] The superior gluteal neurovascular bundle lies in the superomedial quadrant of the gluteal region. The sciatic nerve and the inferior gluteal neurovascular bundle lie in the inferolateral and inferomedial gluteal quadrants, respectively. The safest location for intramuscular injection is in the superolateral gluteal quadrant.

18. The answer is E. [*Chapter 28 IV B 2 (1)*] A meningeal artery tear produces an epidural hematoma. Leakage from cerebral veins results in extravasation of blood in the subdural space (subdural hematoma). A ruptured cerebral artery bleeds into the subarachnoid space. Rupture of a superficial vessel, such as the superficial temporal artery, produces an epicranial hematoma.

19. The answer is E. [*Chapter 20 II A 1–4*] The pelvic diaphragm, consisting of the levator ani and coccygeus muscles, is innervated by twigs from the sacral plexus. It functions to support the pelvic viscera. Its weak anterior area is reinforced by the urogenital diaphragm.

20. The answer is B. [*Chapter 24 VI B 2 b; Table 24-1*] Loss of the Achilles reflex and inability to stand on the toes are indicative of tibial nerve injury. Knee flexion is a function of the tibial nerve, but proximal to the level of injury in this case. The knee-jerk reflex is a function of the femoral nerve. The peroneal nerve mediates the ability to stand on the heels and invert the foot.

21. The answer is B. [*Chapter 17 II E 2 a*] The gallbladder develops in the ventral mesogastrium and lies between the right and quadrate lobes of the liver. Because the quadrate lobe is functionally a part of the left lobe, it is reasonable that the gallbladder be located between the two functional divisions of the liver.

22. The answer is C. [*Chapter 29 V G 1*] The geniculate ganglion, the equivalent of a dorsal root ganglion, contains the cell bodies of the sensory neurons of the facial nerve (CN VII). Selective damage to these neurons results in loss of taste sensation from the anterior two-thirds of the tongue. The motor division of the facial nerve supplies the muscles of facial expression.

23. The answer is E. [*Chapter 16 VI A; Table 16-1*] The pancreas, including its uncinate process, become secondarily retroperitoneal. The liver is a peritoneal structure. It is suspended in the peritoneal cavity by the gastrohepatic, hepatoduodenal, falciform, and coronary ligaments. The appendix and sigmoid colon are peritoneal structures supported by the mesoappendix and sigmoid mesocolon, respectively. The first part of the duodenum is peritoneal, supported by the hepatoduodenal ligament.

24. The answer is E. [*Chapter 24 V A 2 a (2), (3)*] Upon passing through the adductor canal (Hunter's), which is between the hamstring (posterior) and adductor (anterior) portions of the adductor magnus muscle, the femoral artery becomes the popliteal artery.

25. The answer is C. [*Chapter 25 IV B 1 b (1), 2 a (2); Figure 25-4*] The peroneus longus muscle, passing posteriorly to the lateral malleolus, is posterior to the transverse axis of the ankle joint and is, therefore, a plantar-flexor of the foot.

26. The answer is A. [*Chapter 33 VI C 1, 2*] Postsynaptic sympathetic neurons that innervate the head are located in the superior cervical ganglion. These neurons innervate the smooth muscle (of Müller) in the eyelid (which prevents ptosis), dilator pupillae muscle of the eye, sweat glands, and vasodilators of the skin. Pupillary constriction and ciliary muscle contraction are functions of the parasympathetic division of the oculomotor nerve (CN III).

27. The answer is A. [*Chapter 30 I B 7 d (4) (b)*] Blockage of the nasolacrimal duct results in *epiphora* as does excessive stimulation of the parasympathetic component of the facial nerve (CN VII). Conversely, a lesion of the greater superficial petrosal nerve results in a loss of lacrimation. Sympathetic stimulation, or lack thereof, appears to have little effect on lacrimation.

28. The answer is E. [*Chapter 24 V A 3 a, B 2*] Venae comitantes accompany the popliteal vein in the popliteal fossa. The anterior tibial peroneal and posterior tibial arteries all originate distally to the popliteal fossa. The great saphenous vein ascends along the medial aspect of the leg, passing posteriorly to the medial condyles of the bone.

29. The answer is C. [*Chapter 15 VIII F 3 a, b*] A femoral hernia passes through the femoral ring, where it is surrounded by the inguinal ligament (anteriorly and superiorly), the lacunar ligament (medially), the pectineal ligament (posteriorly and inferiorly), and the femoral vein (laterally). The hernial sac is in contact with peritoneum, which is covered by attenuated transversalis fascia.

30. The answer is D. [*Chapter 19 I F; III D 1*] During a midline spinal tap, the supraspinous ligament, interspinous ligament, dura mater, and the arachnoid layer are penetrated. The flaval ligaments lie to each side of the midline, extending between laminae of adjacent vertebrae. The posterior longitudinal ligament lies along the posterior surface of the vertebral bodies.

31. The answer is A. [*Chapter 25 VII C 2*] The common peroneal nerve passes laterally to the neck of the fibula, where it is subject to blunt trauma or involvement in a fibular fracture. Common peroneal nerve palsy results in "foot drop."

32. The answer is A. [*Chapter 31 IV D 1; Chapter 32 VII A 2 c (2), (5)*] The innervation of the maxillary incisor teeth is by the anterior, middle, and posterosuperior alveolar branches of the maxillary division of the trigeminal nerve (CN V). The maxillary division of the trigeminal is purely sensory.

33. The answer is D. [*Chapter 7 III A 3 b (1)*] The proximal radioulnar joint, which is formed by the side of the head of the radius and the radial notch of the ulna, is stabilized by two ligaments, the annular ligament and the interosseous membrane. The annular ligament provides the major reinforcement, allowing the radius to rotate relative to the ulna in pronation and supination.

34. The answer is E. [*Chapter 19 I C 1–4, D 1 c (1)*] Transverse foramina are characteristic of cervical vertebrae. These foramina are formed by a partial fusion of the transverse and costal processes of the cervical vertebrae. Except for the seventh cervical vertebra, the transverse foramina contain the vertebral artery as it passes toward the foramen magnum.

35. The answer is C. [*Chapter 6 III A 3; IV A 3, B 3*] The articular disk and strong sternoclavicular ligaments stabilize the clavicle proximally; the coracoclavicular ligament stabilizes the clavicle distally. This stability, along with the pronounced sigmoid shape of the clavicle, predisposes the clavicle to fracture rather than dislocation. The other muscles and ligaments listed in the question contribute to the stability of the glenohumeral joint.

36. The answer is B. [*Chapter 24 VI A 1 a, c*] The afferent and efferent limbs of the knee-jerk reflex are carried by the femoral nerve, which arises mainly from spinal segments L2–L4.

37. The answer is D. [*Chapter 17 III B 6 b (3), C 3 c (1) (c); Table 17-2; Chapter 22 II B 4; III; VI C 4 b, 5 a, D 2 h (3)*] The urinary tract, the ovary, and the oviduct refer pain to the right lower quadrant, while the appendix and Meckl's diverticulum initially refer to the umbilical region. Peritonitis secondary to inflammation of any of these structures will produce peritoneal pain in the right lower quadrant. Sigmoid diverticulitis initially produces diffuse pubic and inguinal pain, which subsequently localizes in the lower left quadrant.

38. The answer is B. [*Chapter 21 IV A 2 b (1) (b)*] The tunica dartos of the scrotum is formed by fusion of the superficial and deep layers of the superficial fascia. This layer contains smooth muscle that functions in temperature regulation. When the scrotal temperature falls below 33.9°C, the scrotum contracts to bring the scrotum toward the body to conserve heat; when it is too warm, the scrotum relaxes and becomes pendulous to dissipate heat.

39. The answer is C. [*Chapter 17 Table 17-1; Figure 17-12*] The mesentery of the jejunum is thin and transparent because there are few fat deposits in this portion. In addition, the jejunum has a greater diameter and thicker wall than the ileum as well as fewer arterial arcades with longer arteriae rectae.

40. The answer is A. [*Chapter 6 VI B 1 b (3) (b), (c), 2 b (3) (a), (b)*] The axillary artery anastomoses in the circumscapular area with all of the arteries listed (i.e., transverse cervical artery, posterior humeral circumflex artery, suprascapular artery, and subscapular artery) except the deep brachial artery. Because of the profuse anastomotic vascular circulation, blockage or ligation of the axillary artery proximal to the posterior humeral circumflex artery has little effect on the circulation to the upper extremity.

41. The answer is A. [*Chapter 28 III B 5 b (2) (a), (b); Chapter 29 V C 2*] The anterior condylar (hypoglossal) canal transmits the hypoglossal nerve (CN XII), which innervates the muscle of the tongue. The posterior condylar canal transmits an emissary vein, and hemorrhage may produce a hematoma in the vicinity of the mastoid process. The glossopharyngeal nerve (the afferent limb of the gag reflex), the inferior laryngeal branch of the vagus nerve (motor to all the laryngeal musculature except the cricothyroid muscle), and the spinal accessory (motor to the sternomastoid and trapezius muscles) all transit the jugular foramen.

42. The answer is A. [*Chapter 22 II C 1 a; V E 1–3*] The prostate gland discharges on either side of the urethral crest of the prostatic urethra. It contributes phosphatase, citric acid, and fibrinogen to the seminal fluid; the seminal vesicles contribute fructose. The prostatic venous plexus receives the dorsal vein of the penis. An enlarged median lobe (benign prostatic hypertrophy) projects into the urinary bladder and prevents complete emptying. Its posterior lobe is most prone to malignant transformation.

43. The answer is C. [*Chapter 32 III A 2 b (1), 3 d; B 2 c (2) (d) (i)*] The stapedius muscle, innervated by the facial nerve (CN VII), functions to dampen loud sounds by limiting the movement of the stapes at the oval window. Nerve dysfunction with paralysis of the stapedius muscle produces hyperacusis. The tensor tympani muscle, which dampens the movement of the tympanic membrane, is innervated by the motor division of the trigeminal nerve (CN V). The chorda tympani conveys parasympathetic secretomotor and taste fibers, both of which are affected by Bell's palsy.

44. The answer is B. [*Chapter 17 VI D 1, 2*] The hepatic portal anastomoses with the systemic system about the esophagus, the rectum, the umbilicus, and between secondarily retroperitoneal sections of the gut and dorsal body wall. There are no anastomoses between the gondal veins and those of the gut.

45. The answer is D. [*Chapter 6 Table 6-2*] The prime action of the teres major muscle is adduction of the humerus at the shoulder joint. Because this muscle inserts on the lesser tubercle of the humerus, it also acts as a medial rotator and, to a lesser degree, an extensor.

46. The answer is A. [*Chapter 15 VIII E 3 a (1)*] An indirect inguinal hernia is located *within* not adjacent to the spermatic cord. The direct inguinal hernia is located adjacent to the spermatic cord, medially. The indirect inguinal hernia passes through the deep and the superficial inguinal rings. It must also pass laterally to the inferior epigastric artery to enter the inguinal canal, making its origin lateral to the inguinal triangle.

47. The answer is C. [*Chapter 24 II C 1 b; III B 3 a (2) (b), b (2) (b), d (2)*] The lateral and medial collateral ligaments as well as the anterior cruciate ligament are tightened on full extension of the knee. Because extension is primarily a function of the quadriceps femoris muscle, the patellar tendon is also under considerable tension. The popliteus muscle contracts during the initial phase of flexion.

48. The answer is E. [*Chapter 13 VI B 1 b; VII B 2*] An infarct within the interventricular septum can produce ventricular arrhythmias caused by involvement of the common atrioventricular conduction bundle (of His) or either the left or right bundle branch. In addition, because only the tricuspid (right atrioventricular) valve has septal papillary muscles, infarct of the septal region can lead to ischemic necrosis of the septal papillary muscles and avulsion of the chordae tendineae with resultant tricuspid incompetence. Prolapse of an atrioventricular valve produces a low-pitched systolic murmur.

49. The answer is C. [*Chapter 8 II A 1 d, B 1 b*] The lunate bone, located in the proximal row of carpal bones between the scaphoid and triangular bones, participates in the radiocarpal (wrist) joint, where it articulates with the radius and the articular disk of the wrist. Articulation between the lunate bone and the triangular articular disk is maximal during radial deviation (abduction). Anterior displacement of the lunate bone produces the *carpal tunnel syndrome* by compressing the superficial and deep digital flexors as well as the median nerve.

50. The answer is B. [*Chapter 25 IV B 1 b (2); Table 25-1*] The peroneus brevis muscle, a plantar-flexor and pronator of the foot, inserts into the tuberosity of the fifth metatarsal bone. The peroneus longus inserts onto the first metatarsal; the tibialis posterior, onto the navicular bone; the tibialis anterior, onto the first cuneiform bone and first metatarsal; and the abductor digiti minimi, onto the first phalanx of the fifth toe.

51. The answer is B. [*Chapter 12 II A 3, D 2, 3; Chapter 18 IV A 4 c (2)*] The pleural cavity extends inferiorly to the 12th rib posteriorly. Extension of an abdominal incision into the costodiaphragmatic recess of the pleural cavity will result in pneumothorax. Incomplete oxygenation produces cyanosis.

52. The answer is C. [*Chapter 34 V C 2 a (3), E 2 c, d*] The large internal branch of the superior laryngeal nerve, a branch of the vagus nerve (CN X), innervates the laryngeal mucosa and provides the afferent limb of the cough reflex. The superior laryngeal nerve also contains taste afferents for the epiglottic taste buds. The smaller external branch of this nerve innervates the cricothyroid muscle, which functions to lengthen the vocal folds. The other muscles of the larynx are innervated by the inferior (recurrent) laryngeal nerve.

53. The answer is D. [*Chapter 21 IV D 2; V C 2, D 2*] The deep perineal pouch contains the deep transverse perineal muscle and its specialized central portion, the external urethral sphincter, and the bulbourethral glands in the male. The vestibular glands of the female are in the superficial pouch.

54. The answer is D. [*Chapter 19 I D 1 c (1) (b)*] The costal and transverse processes of the developing cervical vertebrae fuse about the vertebral artery to form the transverse foramina. While each cervical vertebrae has a pair of transverse foramina, the vertebral artery does not pass through those of C7.

55. The answer is B. [*Chapter 11 V B 2 a–c*] The anterior intercostal branches of the internal thoracic artery and its musculophrenic branch anastomose with the posterior intercostal arteries. The superior epigastric branch anastomoses with the inferior epigastric artery. The pericardiophrenic artery supplies the pericardium and central diaphragm.

56. The answer is D. [*Chapter 30 I D 4 a (2); Chapter 33 VI C 1, 2*] Postsynaptic sympathetic neurons that innervate the head are located in the superior cervical ganglion. Some of these form the deep petrosal nerve, which bring sympathetic fibers into the pterygopalatine fossa. Other neurons innervate the smooth muscle (of Müller) in the eyelid, which prevents ptosis, and also innervate the iridial dilator muscle of the eye.

57. The answer is C. [*Chapter 23 V A 4, B 1*] The retinacular arteries, which arise from anterior and posterior femoral circumflex arteries, supply the neck and head of the *adult* femur. The artery of the ligamentum teres, a branch of the obturator artery, supplies the femoral head in children but degenerates in the adult and becomes of marginal significance at most. This circumstance is responsible for the high incidence of femoral head necrosis accompanying fracture of the femoral neck.

58. The answer is B. [*Chapter 6 V B 2 c, 3; Table 6-2*] The prime internal rotator of the glenohumeral joint is the subscapularis muscle, which inserts on the lesser tubercle of the humerus. It is innervated by the upper subscapular nerve as well as in small part by the lower subscapular nerve. Rotation of the arm occurs about a vertical axis through the glenohumeral joint.

59. The answer is A. [*Chapter 13 IX C 3 c*] The closure of the aortic valve is heard most distinctly in the 2nd intercostal space to the right of the sternum. An insufficient aortic valve produces a diastolic murmur. The pulmonic valve is best heard in the 2nd intercostal space to the left of the sternum, and a patent ductus arteriosus produces a constant rumbling murmur to the left of the pulmonic area.

60. The answer is B. [*Chapter 18 I B 1 b (3), 2 c, F 1 d (1) (b), G 1 b (2)*] The arteries to the kidney are end-arteries, each supplying a segment of renal parenchyma. Ligation of a supernumerary or polar renal artery results in ischemic necrosis of a portion of the kidney.

61. The answer is D. [*Chapter 27 I B 2 c*] The carotid sheath encloses the common carotid artery and, after its bifurcation, the internal and external carotid arteries. In addition, it contains the internal jugular vein, the vagus nerve, and the carotid branch of the glossopharyngeal nerve. The sympathetic chain lies posteriorly to the carotid sheath.

62. The answer is B. [*Chapter 19 I E b (4), (5), F 4*] The anterior longitudinal ligament provides support for the vertebral column and reinforces the anterior and lateral aspects of the intervertebral disks. Consequently, herniation of an intervertebral disk tends to be in a posterolateral direction; spinal nerves may become involved as they pass through the intervertebral foramina.

63. The answer is E. [*Chapter 25 II A 1 a (2); Chapter 26 II A 2 c*] The tibialis posterior tendon is palpable just posterior to the medial malleolus. Adjacent to this tendon is the tendon of the flexor digitorum longus and the posterior tibial artery. The posterior tibial pulse may be palpated here. The peroneus longus and brevis course posteriorly to the lateral malleolus, whereas the tibialis anterior is on the dorsum of the foot.

64. The answer is C. [*Chapter 33 V D 1 a, b*] The lymphatic drainage of the anterior–inferior portion of the face, the nasal cavities, and the anterior portions of the oral cavity, including the anterior margin of the tongue, gingivae, and teeth, is through the submandibular lymph nodes to the deep cervical nodes. The palatine tonsils drain via the jugulodigastric nodes to the deep cervical nodes. The lymphatic drainage of the external ear as well as the parotid gland and anterior–superior portion of the face is toward the superficial cervical lymph nodes.

65. The answer is A. [*Chapter 7 III A 1 a*] The humeroulnar joint has only one degree of freedom, which allows flexion/extension at the elbow joint. The humeroradial joint has two degrees of freedom, permitting rotation as well as flexion/extension at the elbow. The metacarpophalangeal and radiocarpal joints have two degrees of freedom, permitting abduction/adduction and flexion/extension. The sternoclavicular joint has two degrees of freedom, permitting circumduction.

66. The answer is A. [*Chapter 13 X D 2 a; Chapter 5 Figure 5-1*] An atrial septal defect can be asymptomatic because there is normally a left-to-right shunt in affected individuals. Problems only arise if a right-to-left shunt should occur due to a pathophysiologic disturbance or, as in this case, an embolus traverses the atrial septal defect. The ductus arteriosus is not only on the systemic side but enters the aorta after the carotids are given off. A thrombus originating from the right atrial appendage should result in a pulmonary embolus. Dextrocardia is not a consideration as function is normal.

67. The answer is C. [*Chapter 21 V A 1 b (3)*] The vestibule, bordered by the labia minora, is innervated by the perineal branches of the pudendal nerve. It receives the vagina and urethra as well as drainage from the paraurethral and vestibular glands. The hymen incompletely covers the vaginal introitus.

68. The answer is B. [*Chapter 19 I E 4 b (5) (b)*] Because the inferior intervertebral notch of the lumbar vertebrae is so deep, a posterolateral herniation of the L4–L5 intervertebral disk would not affect the fourth lumbar nerve. However, the herniation will place pressure on the subsequent nerve (L5) as it passes toward its intervertebral foramen. If the herniation is large, it may also involve subsequent roots and even involve roots across the midline.

69. The answer is D. [*Chapter 21 IV F; V F; Figures 21-8 and 21-13*] The principal innervation of the perineum is by the pudendal nerve. The perineal branches supply the urogenital triangle, and the inferior rectal branches supply the anal triangle. However, the ilioinguinal nerve and genital branch of the genitofemoral nerve supply the anterior portion of the labia majora or the scrotum. In addition, the posterior femoral cutaneous nerve innervates a posterolateral portion of the perineum. The nervi erigentes, while conveying sensation from pelvic viscera, does not contain somatic afferent nerves.

70. The answer is B. [*Chapter 32 III B 2 c (2) (d) (i)*] The trigeminal nerve provides motor innervation to those muscles derived from the first branchial arch. These include the muscles of mastication, of which the masseter is one, the tensor tympani, the tensor veli palatini, the mylohyoid, and the anterior belly of the digastric. The posterior belly of the digastric muscle and the buccinator muscle are both innervated by the facial nerve (CN VII).

71. The answer is D. [*Chapter 10 VI B 2 a, b, 3 b*] Both the oblique and transverse heads of the adductor pollicis muscle, as well as the deep head of the flexor pollicis and lumbricals 3 and 4, are innervated by the ulnar nerve. Because the ulnar nerve does not pass beneath the transverse carpal ligament, the muscles it innervates are unaffected by carpal tunnel syndrome.

72. The answer is D. [*Chapter 15 VII A 1 a–c*] The blood supply to the rectus abdominis muscle is primarily from the superior epigastric artery, a branch of the internal thoracic artery, and the inferior epigastric artery, a branch of the external iliac artery. The innervation of the rectus abdominis muscle is provided by terminal branches of the spinal nerves, which enter the rectus sheath laterally. A longitudinal incision just lateral to the medial border of the rectus sheath avoids compromising the vascular and nerve supply to the muscle.

73. The answer is E. [*Chapter 22 V A 3 c*] The lymphatic drainage of the testes is primarily along the testicular neurovascular bundle to the para-aortic lymph nodes. Portions of the penis, lower abdominal wall, leg, and foot drain to the inguinal nodes. Portions of the penis and anal canal drain along the internal pudendal vessels to the pelvic nodes.

74. The answer is B. [*Chapter 15 VIII F 4 c; Chapter 20 IV C 3 b (c) (iii)*] The obturator artery, usually a branch of the internal iliac artery, may arise instead from the inferior epigastric artery or even from the external iliac artery. An aberrant obturator artery crosses the femoral ring, where it may become involved in a femoral hernia or complicate the repair of such a hernia.

75. The answer is B. [*Chapter 29 II A 4 b (1)*] The postcentral gyrus of the parietal lobe lies posteriorly to the central sulcus and is the primary sensory area. It receives information from the opposite side of the body relayed through the thalamus. This region of the cortex is organized somatotopically so that specific regions correspond to various regions of the body. The adjacent sensory association areas interpret and integrate sensory stimuli and enable recognition.

76. The answer is A. [*Chapter 7 V A 3, B 2 c (2); VI B 1, 2*] In the cubital fossa, the median nerve and the brachial artery lie beneath the bicipital aponeurosis and just medial to the biceps tendon. The median cubital vein is superficial to the bicipital aponeurosis. While the brachial artery may bifurcate anywhere in the brachium, the radial and ulnar arteries usually arise just distal to the cubital fossa.

77. The answer is E. [*Chapter 28 III B 5 c (1) (c) (iii); C 2 b (7)*] The foramen lacerum is a defect in the floor of the middle cranial fossa and the carotid canal. It is covered by fibrocartilage and periosteum. Normally, it does not transmit any structure into or out of the cranial cavity.

78. The answer is C. [*Chapter 13 IX D 1 a, b*] Mitral insufficiency produces pulmonary edema and pleural effusion. Tricuspid insufficiency, evident as a blowing systolic murmur in the right 5th intercostal space, produces systemic venous congestion, which in turn results in pitting edema and ascites.

79. The answer is A. [*Chapter 16 V E 3; Chapter 17 II E 1 b (4) (a) (i)–(iii); VI D 1, 2 a*] The inferior vena cava lies posteriorly to the epiploic foramen. The hepatic portal vein lies anteriorly to the epiploic foramen in the hepatoduodenal ligament, where it is posterior to the hepatic artery and common bile duct.

80. The answer is D. [*Chapter 34 II A 4 d*] The nasolacrimal duct drains into the inferior meatus of the nose. The maxillary, frontal, and anterior ethmoidal sinuses drain into the middle meatus. The sphenoidal sinus drains into the superior meatus.

81. The answer is D. [*Chapter 29 V E 1–4; H 1–3*] The parasympathetic neurons for pupillary reflexes and accommodation run in the oculomotor nerve (CN III). Those for lacrimation and those for nasal, submandibular, and sublingual secretion initially run with the facial nerve (CN VII). Those for parotic secretion run initially with the glossopharyngeal nerve (CN IX). The vagus nerve contains many parasympathetic functions associated with the thoracic and abdominal viscera. The trigeminal nerve (CN V) has no intrinsic parasympathetic component, although some of its distal branches (lacrimal, sphenopalatine, descending palatine, and lingual) carry parasympathetic nerves, which originate in association with the facial and glossopharyngeal nerves (CN VII).

82. The answer is E. [*Chapter 16 II D 1 d; VI B 2 h*] The median umbilical fold is formed by the peritoneum draping over the underlying urachus. If the urachus is patent to any extent, an incision carried across this structure may result in leakage of urine into the peritoneal cavity. The paired lateral umbilical folds are formed by the inferior epigastric arteries, whereas the obliterated umbilical arteries form the paired medial umbilical folds.

83. The answer is A. [*Chapter 34 III B 3 a (1) (b), (2) (b)*] The tensor veli palatini and the levator veli palatini arise in part from the lateral and medial sides of the cartilaginous portion of the tympanic tube. Tension in these muscles, such as raising the palate during swallowing, opens the auditory tube so that air pressure equalizes between the middle ear and the nasopharynx.

84. The answer is D. [*Chapter 27 I B 3 a (1), (2)*] The retrovisceral space is posterior to the esophagus, and infection within this space may spread inferiorly to produce posterior mediastinitis. The pretracheal space is anterior to the trachea and is continuous with the superior and anterior mediastina. The carotid sheath is lateral to the visceral compartment.

85. The answer is D. [*Chapter 21 IV D 2 b; V D 2 b; Chapter 22 II C 1 b, 2 a*] The external urethral sphincter, which surrounds the membranous urethra, is a portion of the deep transverse perineal muscle and a component of the urogenital diaphragm. This sphincter is innervated by the perineal branch of the pudendal nerve and is under voluntary control.

86. The answer is B. [*Chapter 31 II B 10 e; Chapter 32 VI D 2*] The branches of the nerves and vessels of the pterygopalatine fossa reach the nose, eye, and mouth through several foramina. The pterygopalatine fossa communicates with the middle cranial fossa via the foramen rotundum and the vidian canal, with the nasal cavity via the sphenopalatine foramen, with the orbital cavity via the inferior orbital fissure, and with the oral cavity via the palatine canal.

87. The answer is A. [Chapter 17 IV B 3; VIII A 2 b (4) (b)] The control of peristalsis is primarily a function of parasympathetic innervation. The descending colon receives parasympathetic innervation from the pelvic splanchnic nerves, which arise from levels S2–S4. Bilateral lumbar sympathectomy for the relief of intractable visceral pain, therefore, has little effect on the motility of the descending colon.

88. The answer is A. [Chapter 19 II A 2] The intrinsic musculature of the back receives innervation from the dorsal primary rami of the spinal nerves. The ventral primary rami innervates the superficial muscles of the back as well as those of the anterior and lateral body wall and extremities.

89. The answer is B. [Chapter 28 III C 2 a (2) (c); Chapter 30 I A 1 b (1)] The ethmoid bone is shared between the neurocranium and the facial skeleton. Medially, it forms the cribriform plate and a portion of the nasal septum. Laterally, it forms the lamina papyracea portion of the medial orbital walls as well as the superior and middle conchae. The inferior concha is a separate bone.

90. The answer is D. [Chapter 15 IV C 1 a; VII A 1] The rectus sheath is formed by fusion of the aponeurosis of the internal oblique, external oblique, and transverse abdominis muscles. The anterior leaf of the rectus sheath superior to the umbilicus is formed by the fused aponeuroses of the external oblique and internal oblique muscles. In the pubic region, the anterior leaf also contains the fused aponeuroses of the internal oblique muscle; the posterior leaf is absent, and only transversalis fascia separates the rectus abdominis muscle from the peritoneum.

91. The answer is D. [Chapter 22 VI D 2 i (2)] The ureters, which course through the deep pelvis, lie in the transverse cardinal ligament beneath the uterine arteries. Ligation of the uterine arteries during hysterectomy must be accomplished with care so that the ureters are not injured.

92. The answer is B. [Chapter 3 II B 2; Chapter 24 IV B 2 a (1)] The rectus femoris, spanning both hip and knee joints, is stretched when the hip is extended and, therefore, has greater intrinsic force-generating capacity. The three vasti muscles do not span the hip joint, and the biceps femoris, also a two-joint muscle, is a knee flexor.

93. The answer is D. [Chapter 18 III B 1] Each adrenal gland receives arterial branches from the inferior phrenic arteries, the aorta, and the renal arteries. There normally are no arterial anastomoses between systemic arteries and the superior mesenteric artery.

94. The answer is E. [Chapter 22 V A 3 c] The lymphatic drainage from the testes is along the testicular neurovascular bundle to the para-aortic lymph nodes in the vicinity of the renal arteries. The entire lower extremity as well as portions of both the abdominal wall and perineum drain either primarily or secondarily to the inguinal lymph nodes.

95. The answer is D. [Chapter 22 VI D 2 g (3) (b)] Visceral afferent nerves from the cervix course along the uterine arteries to the lateral pelvic plexus and thence along the nervi erigentes to sacral segments S2–S4 of the spinal cord. Consequently, pain of cervical dilation is referred to the midsacral dermatomes, which include the perineal region, posterior leg, and lateral aspect of the foot. Conversely, afferents from the body of the uterus course to lumbar levels L1–L2 via the inferior hypogastric plexus and refer pain to the back, groin, and thigh. The gray rami of S2–S4 convey sympathetic motor fibers to the skin.

96. The answer is E. [Chapter 5 VII C 6] The right lymph duct drains lymph from the right side of the head and neck, the right upper extremity, and the right side of the thorax. A large lymph node (of Cloquet) lies in this pathway just before the duct drains into the venous system at the junction of the lateral jugular vein and the subclavian vein. Lymph from the rest of the body returns via the thoracic duct.

97–100. The answers are: 97-A, 98-C, 99-B, 100-A. [Chapter 22 II A, V A 4 b, B 4 b; VI C 4 b, D 2 h (3)] Afferent nerves from the pelvic viscera travel along autonomic pathways. Afferents from the ovary, testis, upper to middle ureter, uterine tubes, urinary bladder, and uterine body travel along the least splanchnic nerve to the lower thoracic segment and along the lumbar splanchnic nerves to the upper lumbar segments of the spinal cord; thus, pain is referred to the inguinal and pubic regions as well as the lateral and anterior aspects of the thigh. Afferents from the epididymis, uterine cervix, and distal ureter travel along the pelvic splanchnic nerves to the midsacral spinal segments; thus, pain is referred to the perineum, posterior thigh, and leg.

101–106. The answers are: 101-C, 102-D, 103-D, 104-A, 105-A, 106-A. [*Chapter 13 IV A 1, 2*] Both the right coronary artery and the circumflex branch of the left coronary artery lie in the coronary sulcus. The left coronary artery arises from the left posterior aortic sinus, whereas the right coronary artery arises from the right aortic sinus. The right coronary artery usually supplies both the sinoatrial node and the atrioventricular node and usually terminates as the posterior interventricular (descending) artery. The anterior portion of the interventricular septum is normally supplied by the anterior interventricular (descending) branch of the left coronary artery.

107–110. The answers are: 107-D, 108-B, 109-B, 110-C. [*Chapter 13 IV A 4 b; IX A 1–4*] The ventricles fill during ventricular diastole and atrial systole. The atrioventricular valves close in early ventricular systole. The aortic and pulmonary valves close in early ventricular diastole. The blood flow through the coronary circulation is maximal during ventricular diastole when the pressure differential between the aorta and myocardium is maximal.

111–115. The answers are: 111-B, 112-C, 113-E, 114-D, 115-B. [*Chapter 13 IV A 1, 2, 4, B 1; Figure 13-2*] The circumflex artery (**B**) is a branch of the left coronary artery (**A**) that runs in the coronary sulcus in company with the coronary sinus. The right coronary artery (**E**) usually supplies both sinoatrial and atrioventricular nodes. The anterior descending artery (**C**) supplies the anterior two-thirds of the interventricular septum. The posterior descending (interventricular) artery (**D**), a terminal branch of the right coronary artery (**E**), has the potential for developing anastomoses with the anterior interventricular artery (**C**) with slow-onset occlusion.

116–120. The answers are: 116-B, 117-B, 118-C, 119-D, 120-E. [*Chapter 6 V B 2 a (1), c (1); VI A 3; Figure 6-4; Table 6-2*] The supraspinatus, infraspinatus, teres minor, and subscapularis comprise the rotator cuff that acts across the glenohumeral joint and dynamically stabilizes the shoulder. The subscapularis, forming the posterior wall of the axilla, is a strong medial rotator of the arm. It is innervated by the upper and lower subscapular branches of the posterior cord of the branchial plexus. The supraspinatus, innervated by the suprascapular nerve, initiates abduction of the arm through the first 15°, at which point the deltoid muscle assumes this function. The infraspinatus and teres minor are both lateral rotators, the former of which is innervated by the suprascapular nerve and the latter, by the axillary nerve. The musculocutaneous nerve innervates the coracobrachialis, biceps, and brachialis muscles.

121–125. The answers are: 121-D, 122-A, 123-C, 124-E, 125-E. [*Chapter 35 Table 35-1*] The lingual branch of the submandibular division of the trigeminal nerve mediates general sensation from the anterior two-thirds of the tongue. The taste fibers for this region arise in association with the facial nerve. The hypoglossal nerve supplies all of the intrinsic and most of the extrinsic muscles of the entire tongue. The glossopharyngeal nerve mediates taste from the posterior third of the tongue as well as general sensation from both the posterior third of the tongue and the pharynx. The vagus nerve mediates sensation from the inferior surface of the epiglottis through the bronchial tree as well as mediating taste from the epiglottic taste buds. In addition, the vagus nerve innervates most of the pharyngeal muscles and all of the laryngeal muscles.

126–130. The answers are: 126-C, 127-A, 128-D, 129-D, 130-E. [*Chapter 29 VII B 1 c (4)–(6), 2 c (1)*] The internal carotid artery gives off the ophthalmic branch, which supplies the orbit, eyeball, and supraciliary region of the face. The anterior cerebral and middle cerebral arteries are terminal branches of the internal carotid artery. The former runs above the corpus callosum deep in the superior saggital sulcus; the latter runs in the lateral fissure, supplying most of the lateral surfaces of the cerebral cortex, and gives off lenticulostriate branches to the basal ganglia and thalamus. The inferior cerebellum is supplied by branches of the vertebral artery; the superior cerebellum and posterior surface of the cerebral cortex are supplied by branches of the basilar artery.

131–135. The answers are: 131-A, 132-D, 133-E, 134-C, 135-B. [*Chapter 28 III C 2 b (6) (d), c (4) (d), (e); Figure 28-5*] The foramen spinosum transmits the middle meningeal artery (**A**), a branch of the maxillary artery. The carotid canal (**D**) transmits the internal carotid artery, while the jugular foramen (**E**) contains the internal jugular vein in addition to the glossopharyngeal, vagus, and spinal accessory nerves. Each posterior condylar canal transmits a large emissary vein (**C**). The vertebral arteries enter the cranial cavity through the foramen magnum (**B**) along with the spinal accessory nerve; the spinal cord also transits the foramen magnum.

136–140. The answers are: 136-B, 137-C, 138-E, 139-A, 140-C. [*Chapter 11 IV B 1, 2, C 4, 5; VII B 1–4, C 2 a*] The intrinsic muscles of inspiration include the external intercostals and the interchondral portion of the internal intercostal layer as well as the diaphragm. The external intercostal muscles and the interchondral portion of the internal intercostals increase the transverse and anterior–posterior thoracic diameters; the internal intercostals proper decrease them. The intercostal musculature is innervated by the intercostal nerves, which divide the internal intercostal muscle into two layers. The diaphragm (innervated by the phrenic nerve) increases the vertical dimension of the thoracic cavity. The abdominal musculature, including the external oblique muscle, is antagonistic to the diaphragm.

141–145. The answers are: 141-E, 142-C, 143-D, 144-C, 145-A. [*Chapter 35 Table 35-1*] The presynaptic neurons for accommodation arise in the accessory oculomotor nucleus (of Edinger-Westphal), travel along the oculomotor nerve (CN III) to the ciliary ganglion. The postsynaptic neurons travel to the pupillary constrictors and the ciliary muscle via the short ciliary nerves.

The presynaptic neurons for lacrimation as well as sublingual, submandibular, and nasopalatine secretion arise from the superior salivatory nucleus and travel along the facial nerve (CN VII). Those axons mediating lacrimation and nasopalatine secretion travel along the greater superficial petrosal nerve to the pterygopalatine ganglion; the postsynaptic neurons for lacrimation travel along the maxillary, zygomaticofacial, and lacrimal nerves to reach the lacrimal gland, while the postsynaptic neurons for nasopalatine secretion travel along the sphenopalatine and descending palatine nerves. Those axons mediating submandibular and sublingual salivation leave the facial nerve via the chorda tympani, which joins the lingual branch of the mandibular nerve to reach the submandibular ganglion; the postsynaptic neurons travel directly to the glands.

The presynaptic neurons for parotid salivation arise from the inferior salivatory nucleus, travel with the glossopharyngeal nerve (CN IX), and leave via the tympanic branch, which courses through the middle ear before becoming the lesser superficial petrosal nerve, which terminates in the otic ganglion. The postsynaptic neurons travel along the auriculotemporal branch of the mandibular nerve to reach the parotid gland.

140–150. The answers are: 146-D, 147-A, 148-A, 149-B, 150-C. [*Chapter 23 Table 23-1; Table 23-2; Chapter 24 IV B 2*] The rectus femoris muscle, originating on the anterior–inferior iliac spine and inserting onto the tibial tuberosity, flexes the thigh at the hip joint and extends the leg at the knee joint. The deep (posterior) portion of the adductor magnus extends the thigh at the hip joint and is innervated by the tibial nerve, making it functionally a part of the hamstring group. The anterior portion of the adductor magnus is an adductor innervated by the obturator nerve. The iliopsoas muscle, inserting onto the lesser trochanter, is a flexor of the thigh. The piriformis muscle, inserting onto the greater trochanter, is a lateral rotator of the thigh.

151–155. The answers are: 151-A, 152-D, 153-B, 154-D, 155-C. [*Chapter 21 IV A 3 b (2) (b), F 1, G 1, 2; V F 1, 2; Chapter 22 II B 4 b (2); III A, B*] The genitofemoral nerve provides both afferent and efferent limbs of the cremaster reflex, which tests the integrity of spinal levels L1–L2. The pudendal nerve is sensory to most of the perineum and motor to the perineal musculature, including the external urethral sphincter for voluntary control of bladder retention. The hypogastric nerves convey to the pelvis the sympathetic outflow from spinal levels L1–L2, which generally mediates ejaculation. The nervi erigentes (pelvic splanchnic nerves) convey parasympathetic outflow from spinal levels S2–S4 to pelvic and perineal structures and is responsible for erection, emission, and glandular secretion. The nervi erigentes also convey sensory and motor fibers for the detrusor (bladder-emptying) reflex.

156–160. The answers are: 156-C, 157-A, 158-D, 159-B, 160-D. [*Chapter 19 III A 3 a–c*] The tough dura, the outer meningeal layer, fuses with the periosteum of the cranial vault at the foramen magnum and terminates as the coccygeal ligament, inferiorly. The dura is lined by the arachnoid layer, which delimits the dural sac about the cauda equina. The pia is intimately attached to the spinal cord and contains a plexus of small blood vessels, which supply the neural tissue. The pial layer extends past the conus medullaris to the caudal end of the dural sac as the filum terminale. The denticulate ligament, formed by lateral reflections of pia, attach to the dura and support the spinal cord within the subarachnoid space.

161–165. The answers are: 161-E, 162-B, 163-B, 164-A, 165-A. [*Chapter 32 III A 2 c (2), d (2) (a); Chapter 33 VI B 2 b*] The spinal accessory nerve innervates the sternomastoid and trapezius muscles. The facial nerve innervates the muscles of facial expression, including the occipitofrontalis and platysma. The ansa cervicalis from C2–C4 innervates the strap muscles, including the omohyoid and geniohyoid muscles. The greater auricular and lesser occipital nerves as well as the greater occipital, transverse cervical and supraclavicular nerves are sensory branches of the cervical plexus.

166–170. The answers are: 166-B, 167-B, 168-D, 169-D, 170-A. [*Chapter 17 II A 2 a, B 4 a, D 5 a (3) (a), F 2 b (3) (a), (b), F 4 a (1)–(3)*] The left and right gastric arteries anastomose along the lesser gastric curvature. The left gastric artery supplies the distal esophagus and frequently gives rise to an aberrant left hepatic artery. The splenic artery, through its short gastric and left gastric branches, supplies the greater curvature in the regions of the fundus and corpus. The splenic artery also gives off dorsal pancreatic, great pancreatic, and numerous short pancreatic arteries to the tail of the pancreas, the region containing most of the endocrine islets. The gastroduodenal artery, through its superior pancreaticoduodenal branch, anastomoses with the inferior pancreaticoduodenal branches of the superior mesenteric artery.

Index

Note: Page numbers in italics denote illustrations; those followed by (t) denote tables; those followed by Q denote questions; and those followed by E denote explanations.

A